The frontiers are not east or west,
north or south,
but wherever a man confronts
a fact.

HENRY DAVID THOREAU

Preface

The rationale for this book is that a text about biology should reflect the diversity of living things and that it should be about the plants and animals that comprise this diversity. The whole world of life is of human relevance, some aspects of it more obviously so than others. Rather than emphasize any one of the past or present pre-occupations of biologists, or a single currently fashionable point of view, we have chosen to present a full spectrum of biological knowledge.

Such an approach (if the authors get carried away) can lead either to a dis-connected smorgasbord or an endless encyclopedic recitation of facts, leaving students both bewildered and exhausted. But there are threads that weave together the apparently limitless variations in the phenomena of life: the pattern of cellular structure, the unity of common biochemical pathways, the levels of organization which channel the flow of matter and energy in complex organisms, the mathematical predictability of inheritance, the universality of adaptation, the geological record of evolution on this planet, and the interdependencies of the biosphere.

Underlying these themes should be an awareness of the impact of biological science on the life and thought of mankind as well as an understanding of science as an ongoing human enterprise, two aspects of biology we have been at pains to include throughout this book. As the story unfolds, it should become clear that the route of discovery is sometimes more important than its immediate goals because of the un-foreseen ultimate and occasionally serendipitous discoveries that result. Biology, like the other sciences, is both cumulative and open-ended.

In chapter 1 (where hopefully it will be read by students) we have outlined the plan of this book and our reasons for presenting various aspects of biological knowledge in the sequence we have chosen. No single book can present all the facts. Obviously the topics selected and the length at which they have been presented re-flect the interests and philosophies, indeed the prejudices, of the authors.

We have drawn conclusions and injected our personal points of view where they seemed appropriate. We feel that a book should be readable and even entertaining. The science of biology has to do with the wonders of life, the excitement of dis-covery, and the challenge of the unknown. We hope that this text will be a vehicle for sharing our enthusiasms for the subject with our students.

Lastly we wish to express our gratitude to the Houghton Mifflin Company for permission to use pertinent materials from the senior author's *General Zoology*. We are also indebted to our colleagues, artists, and publishers for their expert assistance and to the students who have always been our sharpest critics and most loyal supporters. Their responses have been one of our most valued guideposts.

Gairdner B. Moment
Helen M. Habermann

Baltimore, Maryland

Contents

1

BASIC CONCEPTS

Biology: The Science of Life

The DNA double helix, the Pill, and the environmental crisis have forced themselves on the attention of people everywhere regardless of race, religion, or economic condition. Yet these things are only three of the more conspicuous points where biological knowledge impinges on our lives. So deeply interconnected are the diverse areas of biological knowledge that no satisfactory understanding of any one is possible without some understanding of the others. Without such understanding, appropriate action is seldom possible. Ecologists concerned with finding effective substitutes for DDT need to know about the biochemistry of the social hormones, or pheromones, which are sex attractants for insects. They also need to know about the physiology, the life history, and the behavior of insects. The poet-naturalist sees downy plumed milkweed seeds drifting on the sunlit autumn air as objects of great beauty and symbols of life's perpetuation; the biochemist sees little traveling packets of DNA. The ecologist realizes that both views are correct and appropriate for a complete understanding of life on this planet.

In the text which follows, the authors hope that the student will find the means to become truly literate in modern biology and to gain a valid insight into the nature of this science and its implications for human life.

Thus it is our conviction that nothing less than a full spectrum biology, from the molecular to the ecological, will suffice to equip a man or a woman for intelligent living and participation in the contemporary world. In a general college course both the student who plans to devote most of his time to literature and the arts and the student who intends to enter a career in the biological sciences have very much the same basic needs. Both need to experience something of the rigor, drive, and intellectual power of modern biology and to feel something of the excitement of discovery. At the same time, both need to gain a lasting sense of biology as an on-going human enterprise, as something which is part of the cultural heritage of all mankind.

It is also important to remember that the biological sciences possess great human interest in themselves because of the amazing things they reveal about the vast panorama of life on this planet. It is no less important to remember that without biological knowledge the loftiest dreams for a better life for individuals or for mankind cannot be converted into reality. It requires more than love to save a dying lake or a dying child. As Slobodkin, a well known American ecologist, put it, saving our environment will take "more than a good heart."

Like all the sciences, biology is both cumulative and open-ended. No one knows what the future may hold. We do know that classical 19th-century physics held nothing which suggested that X-rays or radioactivity existed. In fact, the world's leading physicists said that all major discoveries in physics had been made. We also know that the new discoveries which will be made will develop old knowledge rather than invalidate it. An understanding of this cumulative aspect of scientific knowledge is essential to an appreciation of science as a type of human activity.

Fig. 1-1. Milkweed seeds, traveling packets of DNA, symbolize the continuity and evolution of life.

SCIENCE: ITS LIMITATIONS AND POTENTIAL FOR MANKIND

There have always been sharp differences of opinion about the true purposes, methods, limitations, and possibilities of the natural sciences. Perhaps this is why most biologists continue to work and occasionally to make a discovery without giving much thought to the question of what science *is*. This attitude can be justified. In a real sense, a scientist is like an athlete or a musician who must have the most exquisite muscular coordination, yet needs to know nothing whatever about the latest theories of how a nervous impulse passes from nerve to muscle in order to perform superbly either on the field or in the concert hall. Also, there are very real logical difficulties in justifying the inductive methods of science, *i.e.*, generalizing from a series of observations, on purely logical grounds. For example, the so-called "black swan problem" continues to haunt us. No matter how many white swans an investigator may record, he will never logically be able to claim that all swans are white.

Nevertheless, it is important to take a look at the problem of the nature of the sciences because they are now fundamental in the life and thought of contemporary man whether in a megalopolis or a remote hamlet. It is also important for both the scientist and the non-scientist to gain a clear idea of the kinds of things science can do for mankind and those that it cannot do.

When the Royal Society of London, one of the oldest scientific societies in the world and comparable to the National Academy of Sciences in the United States, was founded in the mid-1600's, a basic rule forbade the discussion of religion and politics. This rule gives the clue: scientific enterprise lies outside the realm of value judgments which play a central role in religion and politics. A clear expression of this point of view can be found in the recent writings of Garrett Hardin, a prominent American biologist. He maintains that there is a class of problems, such as the population explosion, for which there is no technical solution. By this he does not mean that methods of birth control are not technologically possible but rather that the extent to which such techniques should be used, if they are to be used at all, is a question lying outside the field of the natural sciences since it depends on moral judgments.

Most biologists, including the authors of this book, take this orthodox position. Put succinctly, scientific study can tell you whether or not, statistically speaking, Swedes have big ears (they do); however, the natural sciences cannot

Fig. 1-2. Fish kill caused by pollution. It takes more than love to save a dying lake. (Photo by Weyman Swagger—courtesy the Baltimore *Sunpapers*.)

tell you whether big ears are beautiful. Science is a matter of B follows A, not that B is better or worse than A. Such a position certainly does not mean that scientists are freed from the obligation of making value judgments. On the contrary, their responsibility to exercise their role as citizens is, if anything, greater rather than less because they may be better informed. For example, a biologist can foresee environmental effects of pollutants from an internal combustion engine that the automobile manufacturer may not envision.

The distinction between the kinds of questions that science can answer and those that it cannot has been brought into sharp focus by many recent biological advances. For example, it is now possible to obtain living cells from a human fetus by **amniocentesis**, *i.e.*, tapping the amnion, growing the cells thus obtained *in vitro*, examining their chromosomes or testing their metabolic capabilities, and so determining very early in pregnancy whether the resulting child will be a victim of such serious hereditary afflictions as Down's syndrome or Tay-Sachs disease, both of which result in gross mental retardation and early death. If such an afflic-

tion is found, the non-scientific question that arises is obviously whether or not an abortion should be performed. Applied ecology, like medicine, presents tough questions and hard choices which involve value judgments as essential components.

Consider the case of a new nuclear power plant on the Chesapeake Bay. The governor of the state involved set up a Task Force to study the environmental effects to be expected. Opponents of the plant told the Task Force that the additional radiation from the plant would result in a small but statistically significant increase in the cases of leukemia among populations living around the bay. Others pointed out that if the electric power, which the people of Washington and Baltimore are convinced they require, is to be supplied from energy derived from coal, there will surely be at least an equal number of additional deaths among miners from accidents and **pneumoconiosis** (black lung disease), and that freedom from additional cases of leukemia would be paid for by the lives of at least an equal number of miners. If you were a member of the Task Force, how could you reach a decision? In all

Fig. 1-3. Blessing or blight? Calvert Cliffs nuclear power plant shown under construction in 1972. (Courtesy Baltimore Gas and Electric Co.)

Fig. 1-4. Nuclear power plant locations in the United States. ■ in operation; ▲ under construction; ● planned.

such problems value judgments are heavily involved.

Clearly every biologist has fully as much responsibility as any other citizen to make sure that biological knowledge is used for what he considers constructive social purposes. If any-thing, anyone who has studied the biological sciences has more responsibility because he has more knowledge. To those like writer Lewis Mumford who claim that science is at the root of all our troubles, a biologist can point out that murder can be committed with a paleo-lithic stone axe and that evil did not enter the world with the atom bomb. The men, women, and children of Carthage could not have been destroyed more completely by such a bomb than they were by Roman broadswords nor their soil more severely poisoned by modern chemicals than by Roman salt.

Alfred North Whitehead, one of the most outstanding philosophers and students of sci-ence in modern times, claimed that the scien-tists of today are like the Benedictine monks of the late Middle Ages in three important respects: their great respect for learning, their serious purpose, and their insistence on manual labor, *i.e.,* laboratory work. Perhaps this is as balanced a view of the nature of scientists as we can attain.

There are a number of specific and useful ideas about what science can do and how it operates. One important view is that science is the human activity which asks questions that can be answered in terms of grams, centimeters, and seconds. This is a good working definition that reveals a very important aspect of all the sciences. T. H. Huxley, Darwin's great defender, used to say that science is organized common sense. Such an opinion is not much different from that of Percy Bridgman, a Nobel Prize winner and Harvard physicist, that "there is no special scientific method, rather science is merely an all-out attempt to discover the truth, and no holds barred."

It is often said that the scientific method begins with a problem—a question. It then proceeds with the formulation of a tentative explanation, guess, or hypothesis which is followed by testing, which, in turn, usually leads to a modification of the original hypothesis, further testing, and eventually the formulation of a theory and ultimately a law. However, this picture is simplistic. When young Galileo stood in the cathedral watching the swaying of those great lamps on their long chains, did he already have questions in his mind about the laws of motion or did his observations stimulate his questions? Perhaps he could not have answered such a question. There is no clear-cut point at which hypothesis becomes theory or theory becomes law.

The deep-seated social nature of scientific work can hardly be exaggerated. In no other area of human activity do individuals watch more closely or take more seriously the work of others both at home and abroad. This aspect of science brings important rewards not only by making scientific progress possible, but, as Bertrand Russell once said: "If knowledge be true and deep it brings with it a sense almost of comradeship with other seekers after truth in other lands and distant times."

ADVANCES IN SCIENCE

In studying any science it is of great importance to remember that new discoveries do not usually negate old ones. It simply is not true that science is advancing so fast that by the time a student graduates a large part of what he has learned will have been shown to be false. When T. H. Morgan and his students demonstrated that the genes, the Mendelian factors of heredity, are located on the chromosomes, they did not prove that Mendel was wrong, even though Mendel knew nothing about chromosomes. In the same way, the recent discoveries about DNA in no way indicate that Morgan was in error in thinking that the chromosomes are the physical basis of inheritance and that the genes are arranged in a linear order along the chromosomes. The truth about advances in scientific knowledge is quite the contrary. A more complex and interesting example of how biology grows and how the discoveries of one age come to be reinvestigated and reinterpreted is seen in the development of theories about the origin of life. The chapter on evolution will show how the ancient belief that living things frequently arise spontaneously from non-living

Fig. 1-5. T. H. Huxley. Engraved by Lacour. (From *Life and Letters of T. H. Huxley*, D. Appleton and Company.)

matter changed into the Pasteur doctrine that "all life comes from previous life" (which remains the basis of surgical asepsis and the canning and brewing industries), and how more recently we have arrived at the modern view about the origin of life from non-living materials on a developing planet such as ours.

New horizons open in many directions. Some of the men who helped discover how the information in DNA is translated into the characteristic proteins, which in turn determine the character of any plant or animal, feel they have exhausted the possibilities of advance here and have turned to other fields, notably, animal behavior. Others are now exploring the possibilities of gene therapy and even genetic engineering with the hope of not only curing the many diseases of hereditary origin but of producing a general enhancement of human life. The biochemical basis of certain mental diseases, and even of memory and learning, appears to be yielding to modern research. Recent advances in embryology have already made it possible, should we so choose, to produce large numbers of people with identical heredities by asexual methods. Today there is an urgent need to apply long-known ecological principles to environmental management, for manage it we must or die; but there is an equally acute need to fill the wide gaps in our ecological knowledge, even to identify what we do and do not know about our environment. Yet, however far and diverse the explorations of biology's horizons will prove to be, we can be certain that they will be built on the knowledge already won.

MOTIVES FOR BIOLOGICAL INVESTIGATIONS

One of the oldest motives behind the search for new biological knowledge is the search for cures for illness and pain, which are no respecters of age or sex, wealth or poverty, race or ideology. From remote antiquity plants and plant products of many kinds have been used for their therapeutic properties, both real and imagined. New plants have been sought which might provide cures or even some slight relief. It was no accident that Hippocrates, the father of medicine, was a botanist.

In modern times it has become clear that so deep-set is the unity of life that discoveries made on one organism can often, although not always, be applied to others far removed in the plant or animal realms. For example, the results from studying the effectiveness of drugs in experiments with warm-blooded mammals, such as dogs or monkeys, can usually be relied on to furnish valid indications of their probable action on human beings. Sometimes the basic investigations can best be carried out on cold-blooded animals like frogs or mollusks. Many of the recently developed antibiotics used in the treatment of bacterial infections are obtained from the fungi, a group of relatively simple plants.

Most of the modern understanding about how the human body and brain function has come from studies of other animals. Modern theories of the way heartbeat is controlled were greatly aided by work on the heart of *Limulus,* the horseshoe crab. The ideal place to study nerve conduction is not in the nerve of a man or a frog but in certain giant nerve fibers of *Loligo,* the squid. At first sight what animals could be more different from man than one-celled protozoans such as *Amoeba, Paramecium,* and *Tetrahymena*? Yet the requirements for vitamins and other dietary constituents of these microorganisms are very similar, and in some cases virtually identical, to those of man. This is not to say that one-celled animals living in a test tube will soon replace white rats as the standard test animals for nutritional studies; yet

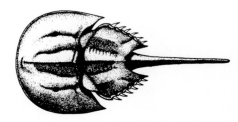

Fig. 1-6. *Limulus,* the horseshoe crab. Modern theories about control of heartbeat are based on studies with this animal.

Fig. 1-7. *Loligo,* the squid. The giant nerve fibers of this animal are ideal for studies of nerve conduction

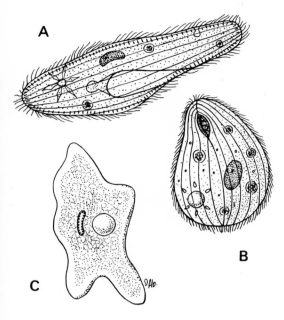

Fig. 1-8. One-celled protozoans that have vitamin requirements similar to those of man. A, *Paramecium;* B, *Tetrahymena;* C, *Amoeba.*

studies with these microscopic animals are uncovering important new facts of wide and perhaps universal importance among living things.

It should not be forgotten (for the penalties of forgetting are potentially very great) that many of the most destructive diseases of man and of his domestic plants and animals on which his life depends are due to fungi, bacteria, viruses, and animal parasites. Malaria, caused by a blood parasite and transmitted by a mosquito, still brings debilitating illness and death to more millions of people than any other disease. Bilharziasis (schistosomiasis), caused by a blood fluke harbored by a snail, is still a scourge in Southeast Asia and in Egypt as it was in the days of the pharaohs. Southern corn leaf blight, caused by a highly virulent fungus, *Helminthosporium maydis,* is a real and present danger that resulted in over a hundred million dollars of crop damage in 1970 and if not controlled could produce devastating effects on the North American economy due to the suceptibility of widely used strains of hybrid corn. Thus by searching for controls of plant

and animal disease, biological knowledge provides a basis for healthy and abundant living.

A second and very ancient universal motive for biological investigation is economic. Every modern nation maintains agricultural research stations to develop new kinds of plants and animals or to develop better ways of obtaining higher yields from existing varieties. "What kind of wheat will produce the highest yield under the soil and climatic conditions of a particular region?" or, perhaps, "What is the best way to feed a pig?" are typical of the practical questions that arise. Humped zebu cattle from India are being crossed with Jersey and Holstein breeds from Europe in attempts to combine general hardiness against heat and poor grazing with superior milking and beef qualities. Vastly improved varieties of rice have recently been developed by cross and selective breeding experiments in the Philippines. Most nations with seacoasts have long supported marine fisheries research and now some enthusiasts are talking of the day when new kinds of marine plants and possibly animals will enable us to farm the oceans as we now farm the land.

A newer and obviously important objective of biological study is to gain insight into the great unsolved problems of human psychology and of the social sciences. The personalities and behavior of men and women cannot be adequately understood without a reasonably complete knowledge of the structure and function of the vertebrate nervous system and its

Fig. 1-9. Santa Gertrudis bull. This relatively new breed of cattle was obtained by crossing the older humped Zebu, Jersey, and Holstein breeds. (Courtesy John A. Cypher, King Ranch.)

interrelation with the hormonal systems. However, the relationship of human heredity to human intelligence and behavior is a highly controversial topic. In recent years the explosion of knowledge about animal behavior has had a marked impact on thinking about human behavior. Popular writers like to jump lightly from the behavior of rats to the behavior of man and seem to forget the dangers of such simple extrapolation.

The study of animal behavior provides insights and suggests new approaches to human problems, but they are suggestive ideas, not definitive answers. Consider, for example, the recent studies on three species of North American bears: the black bear, the grizzly bear, and the polar bear. The Canadian investigator, Jonkel, has shown that when these three species are cornered or caught in a trap or snare each behaves very differently. The black bear will try to hide when approached and, if there is nothing to hide behind, may even cover its face with its paws. The grizzly bear goes into a vicious tantrum. In fact, before the trapper approaches, everything within the grizzly's reach will have been chewed, clawed, or dug up. Polar bears behave in neither of these ways. They calmly sit and watch as you approach. They may even sneak up on you but more probably they will turn their heads far to one side, exposing their long necks. It seems likely that this is one of the "I give up" signs that certain animals make when defeated. Which species of bear exhibits the behavior that reveals the most about the behavior of a cornered man? Obviously, conclusions drawn from any one of these behavior patterns can be extrapolated to human behavior only with great caution.

Within the past few years much biological study has been motivated by the need to gain the knowledge essential for understanding **ecology**, the science of the relationship between living things and their entire environment. This is essential if informed and intelligent action is to be possible in the face of impending environmental crises. It is necessary to know not only what different species of plants and animals actually inhabit our planet, but also something about their structure, function, and interrelationships in the vast multidimensional web of life. Much more knowledge will be required to give us satisfactory answers to all the tough questions that must be answered if we are to continue to live here. Fortunately, however, some of these answers can be had without waiting for great new scientific breakthroughs. No very special knowledge is required to know that mercury is poisonous, or if unrestricted deforestation is allowed to occur, disastrous results are certain to follow. In many other cases so little is known about the life of a given area whether lake, mountain, or valley, that little can be done until more is known.

DESIRE FOR UNDERSTANDING

Together the diverse motives just discussed are not sufficient to explain the sustained drive that has pushed forward the boundaries of biological knowledge generation after generation. A final factor must be understood. As Aristotle, the ancient Greek philosopher, said: "All men by nature desire to know." This thirst to know and to understand is as natural to man as the need for food and air. It is one of the characteristics that distinguishes us from what our ancestors called "the beasts" and is a true part of our humanity.

It is essential to remember that the general objectives and the personal motivations of the investigator are no test whatsoever of the truth or falsity, or the importance or triviality, of a discovery. The validity of Mendel's laws of heredity is completely independent of the nationality, religion, skin color, or table manners of Gregor Mendel. We do not know whether Mendel spent those long hours in his garden crossing peas primarily as refreshment from the routines of monastery life, because he was fascinated with the problem of inheritance, as some sublimation of an interest in sex, or through a determination to show that an Augustinian could make scientific discoveries of the first rank. All, some, or none of these factors may have been involved. The only relevant questions are whether his experiments can be repeated and whether his conclusions hold true for organisms other than peas.

The importance of objectives lies in their ability to lead an investigator toward certain discoveries and to blind him to others. There is an old saying that lucky accidental discoveries come to prepared minds. Many a bacteriologist

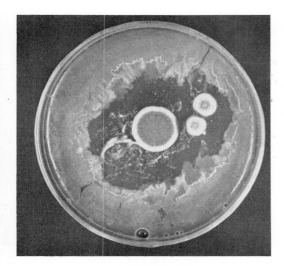

Fig. 1-10. Bacterial culture contaminated with *Penicillium notatum*. Note clear zones of bacterial inhibition. Observations of such cultures led Fleming to the discovery of penicillin. (From *The Story of Penicillin*, Merck & Co., Inc., 1942.)

had seen that troublesome mold, *Penicillium*, ruin his cultures of bacteria before Alexander Fleming had the insight to see that in this obstacle to routine bacteriological research lay a spectacular method of curing bacterial diseases, an objective he had worked toward for many years.

LEVELS OF ORGANIZATION

During the latter part of the 19th century and the earlier years of our own, the theory of organic evolution formed the theoretical basis for most, if not all, biological thought and investigation. The theory of evolution furnished the frame of reference within which a multitude of diverse facts could be meaningfully organized. It provided the biologist with points of orientation in his research and thinking.

In our own times a broad new concept has come into prominence as the most inclusive framework for biological thought. This new frame of reference does not negate the theory of organic evolution but goes beyond it to include the non-living world from which life arose. The new concept is simple on the surface; it postulates that the known universe is composed of a series of levels of organization of such a nature that the units of one level of complexity form the building blocks for the units of the next higher level. Simple as it appears, the concept of a series of levels of organization, each linked to successively higher levels, involves profound philosophical problems. It is by no means a new concept to the philosopher, but has recently come into greatly increased use by biologists as a framework that enables us to know where we are and to find our way among vast mountains of data. It suggests new areas of research and illuminates the whole landscape of biology.

The first, but not necessarily the simplest, level of organization is the realm of the hundred or so subatomic particles or **wave particles** such as electrons, protons, and neutrons. The next level comprises the **atoms** built up from the subatomic particles. The units of this level are the well-known chemical elements hydrogen, oxygen, nitrogen, carbon, phosphorus, iron, etc.

The third or **molecular** level can be roughly equated with the biochemical level. The mole-

Fig. 1-11. Alexander Fleming. (Courtesy The Bettman Archives.)

cules of this level are composed of the atoms of the previous level organized in hundreds of thousands—in fact, in millions—of different patterns. Some are as simple as water, H_2O, and some as intricate as a molecule of hemoglobin as large as DNA, the deoxyribonucleic acid that carries the chemical basis of the genetic code of heredity in the chromosomes.

Above the molecules is the level of living **cells**. Except for the problematic viruses, cells are the smallest and primary independent living units. Each cell is bounded by a complex and physiologically active membrane constructed of large molecules. Within the cell is an assortment of molecules ranging from simple water and salts to complexities like deoxyribonucleic acid and conjugated proteins.

Above the level of single cells lies the level of **multicellular organisms**, the familiar plants and animals which may be composed of millions of

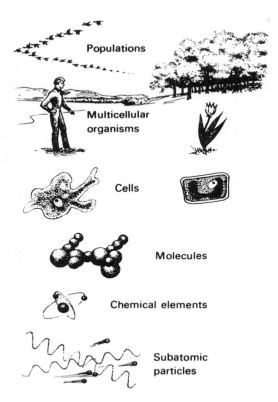

Populations

Multicellular organisms

Cells

Molecules

Chemical elements

Subatomic particles

Fig. 1-12. Levels of organization important to life. Units of one level of complexity form the building blocks for the next higher level.

cells. It will be noted that organs, *e.g.*, liver, brain, or oak leaf, do not constitute a primary level of organization. A primary level is composed of units that can have an independent existence. Electrons or protons, for example, can exist independently. So also can atoms of nitrogen or iron, cells, and whole multicellular animals. A liver or oak leaf can exist for any length of time only as part of an organism. Such structures plainly belong to a secondary level of organization not comparable to the primary levels discussed here.

Above the level of many-celled organisms is the level of **populations**. A school of fish, a swarm of bees, or a stand of redwoods is regarded as a new sixth level for two important reasons. Clearly each can do things which no single individual can do. It takes more than one goose to migrate in a V formation. The "language" of the bees presupposes a community of bees. Single individuals do not evolve into new species. Only populations existing through long periods of time can evolve. In a larger, and perhaps more important, sense, all plants and animals are members of that great population of living things which inhabit the earth. The lonely albatross flying far out over the trackless ocean and the very local field mouse are each not only members of a population of albatrosses or field mice but active participants in the great web of life which is the subject of ecology.

Each level has its own characteristic properties, laws, and independent validity. A fact established on one level remains true regardless of what is discovered about it on another level, even though the meaning we give to the discovery may change. Consider again the case of the phenomena of heredity. Gregor Mendel worked on the level of whole multicellular organisms. His discoveries of the mathematical laws of heredity remain entirely valid regardless of any discovery about cell structures such as chromosomes. And, of course, biochemical discoveries have their own validity in terms of DNA and will remain so no matter what may be discovered in the future.

In the shift from one level of organization to another, new properties emerge. The properties of water molecules are neither the sum nor the average of the properties of the gaseous hydrogen and oxygen atoms of which water is composed. Freezing point, boiling point, chemical attributes—all these are new. This is what biologists mean when they say that the whole is more than the mere sum of its parts.

New properties also emerge when you break up complex patterns and descend to "lower" levels. Any freshman student of chemistry knows that the pure metallic element sodium has many exciting properties not found in sodium chloride (table salt). An oak tree which has grown all its life as a single individual in the center of a field spreads out and becomes a far different thing from the tall, straight-trunked tree with only a small crown of branches at the top found in a forest.

WHAT IS LIFE?

Any science should define its basic concepts. Yet life is almost as difficult to define as the "redness" of red. The essential difference between the living and the non-living has, however, become increasingly clear. It is not a difference in the kind of matter, as was once supposed, for the same atoms which make up the one also compose the other. Nor has the difference anything to do with an escape from the laws of conservation of matter and energy. The famous laws of thermodynamics state that matter and energy can be neither created nor destroyed, only transformed. The conservation of matter and energy holds firmly for respiration and all other activities of living things.

Modern physicists have sometimes said that life is something that feeds on "negative entropy." Positive entropy is characterized by the absence of form or pattern, where everything remains at a "dead" level. Negative entropy means diversity, or greater concentrations of matter and energy in some regions than in others. Such differences are necessary to furnish the energy required for any living process and any machine as well.

Life, in contrast to the random disorder or relatively simple organization found in inanimate things, represents a high degree of patterning and organization of matter and energy. Only when atoms are organized into the largest and most complex molecules, proteins and nucleic acids, does that constellation of properties emerge which we can define as life. These include: respiration (the "breath of life"), assimilation, excretion, growth, reproduction, and sensitivity. Taken separately, each can be duplicated in the non-living world, with the possible exception of reproduction. As one wit put it, you need have no fear that you will wake up some morning to find your desk crawling with little typewriters.

THE UNITY OF LIFE

The basic unity of all living things extends into the most remote corners of the living world and reaches down into the most intimate biochemical details. The greatest achievement of Charles Darwin is held by many to have been his convincing demonstration that man himself is a part of the world of animal life.

It is the fact of this unity that justifies much of the biological research being done today. William Harvey discovered the circulation of blood not only by studying the human body but also by observing worms and shrimp from the Thames River. Stephen Hales discovered blood pressure after he discovered that sap is forced up tubes attached to the cut ends of grape vines. Indeed, cells and nuclei were first found in plants and only later shown to be common to animals as well. The first virus to be isolated was one which causes a disease in the tobacco plant. The relationship of nerves to the beating of the human heart was first elucidated by investigating the problem in the heart of the horseshoe crab.

As familiar as all this is to the biologist, for it forms the basis of much of his day-to-day work, it is not at all clear to many otherwise intelligent and well-informed people. This is shown by the fact that in a recent session of the Congress of the United States the National Institutes of Health were severely criticized for supporting the careful and detailed three-dimensional charting of the interior of the brains of several mammals. It should have been obvious that no study could be more relevant not only to the problems of brain injuries and brain tumors in man but also to the problems of the treatment of mental diseases.

The actual areas and representative details of this unity deserve a closer look. The laws of heredity which Mendel discovered by studying garden peas are now known to be universally applicable. The inheritance of blood groups in man follows no different laws than does the inheritance of shape in pumpkins or of the nutritional capabilities of molds. It is now clear that all living organisms are essentially the expression, or "readout" if you will, of the information coded in their genes. The way a blade of grass or a developing lobster responds to its environment is based on the capabilities built into it by its DNA. The chromosomes which carry this genetic information behave in basically the same ways wherever they are found.

Throughout the living world, if we disregard the problematic viruses for the moment, the stuff of life is packaged in units called cells, all of which share the same basic biochemical pathways. All cells utilize energy in the same ways: to synthesize new molecules from pre-existing ones, to transport molecules or sometimes ions across living membranes, and to produce various kinds of motion whether ciliary beating, cytoplasmic streaming, or muscular contraction. The processes of development are universally under the control of DNA and subject to the laws of evolutionary change through mutation and natural selection.

SPACE-AGE PERSPECTIVES ON LIFE

The exploration of outer space has dramatized both the limits imposed on life and the possibilities open to it that are governed by the laws of physics, chemistry, and mathematics. Because these laws are valid throughout the known universe, the same basic constraints and possibilities will exist for living organisms on any planet in any solar system, even those among the most distant stars. These factors play a fundamental role in determining the properties of the molecules in living things. They have an equally fundamental role in natural selection, independent of the special conditions which may happen to be present on any particular planet. To examine living things in the light of this enlarged space-age perspective is to gain a clearer understanding of why living organisms are what they are and behave as they do. Some of these inherent cosmic factors that govern life are well understood; others are but dimly perceived or completely unknown.

Fig. 1-13. Earth photographed from the Apollo 11 spacecraft 98,000 nautical miles from earth. Most of Africa and portions of Europe and Asia can be seen. (Courtesy National Aeronautics and Space Administration, Washington, D.C.)

Size and Gravity

Consider a very simple case where prediction has been possible based on considerations of gravity and the possibilities of molecular structure. Female moths of various species such as the gypsy moth and the large North American silk moths like *cecropia* or *luna* give off a sex attractant into the air which attracts males from a distance of well over a mile away. Scientists who investigated these remarkable molecules predicted that they would consist of a core of carbon atoms. They further predicted that the number of carbon atoms in these moth sex attractants would have no fewer than five or six carbon atoms and not more than about 20. Why? Because from the biological point of view each moth must produce its own unique kind of messenger molecule. It would be biologically useless for a male gypsy moth to fly to a female *cecropia*. The differences between one molecule and another depend on the carbon core and the kinds and arrangement of atoms attached to it. With fewer than five or six carbons, there are too few possibilities to provide each species with its own identifying sex attractant. What sets the upper limit of molecular size? Gravity. If the molecules become too large they become so heavy that they soon sink to the ground and are hence ineffective as long distance attractants.

The sex **pheromones** of about two dozen insects are now known. All fall within the predicted size limits. It seems evident that this prediction would hold for such chemical messengers on any planet possessing an atmosphere. The upper size limit would be controlled by the size of the planet. On very large planets the upper limit of molecular size would be lower, while on small planets it would be higher, for the very simple reason that large planets have a stronger gravitational pull than small ones. A molecule with a given number and kinds of atoms would be "heavier" on a large planet than it would on a small one.

In contrast, inherent limitations on the size of living things have been extensively analyzed and much is known. Paradoxical as it may seem, much more is understood about size limitations due to geometrical and physical factors than about how the code in the genes produces, for example, domestic cats of one size and Siberian tigers of a very different size. Perhaps this is due to the fact that the biologists have had before them for many decades a very remarkable book which has often been reprinted and translated into many languages, D'Arcy Thompson's *On Growth and Form.* Here are laid out the geometrical principles that determine the shapes and sizes of living things from pollen grains to human skulls.

Erwin Schrödinger, a Nobel Prize-winning modern physicist, introduced his discussion of the place of life in the universe by asking why atoms are so very small, so small that even a bacterium is inconceivably gigantic by comparison. He answered his own question by turning the question right side up and asking why living organisms must be so large compared with the atom: "We cannot see or feel or hear single atoms because every one of our sense

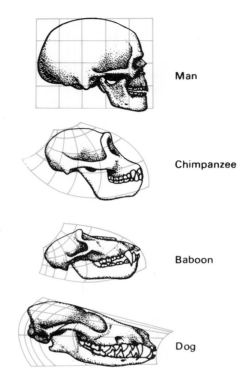

Man

Chimpanzee

Baboon

Dog

Fig. 1-15. Cartesian transformations showing the relationships between skull shapes of man, chimpanzee, baboon, and dog. (Redrawn after D'Arcy Thompson's *On Growth and Form.*)

organs. . .being itself composed of innumerable atoms, is much too coarse to be affected by the impact of a single atom." Schrödinger then asks: "Must this be so? Is there an intrinsic reason for it? Can we trace back this state of affairs to some kind of first principle in order to ascertain and to understand why nothing else is compatible with the laws of Nature? Now this, for once [he continues], is a problem which the physicist is able to clear up completely. The answer to all the queries is in the affirmative."

The explanation is that if sense organs were so small as to respond to single atoms, perception would be utterly chaotic due to the randomness of atomic motions. An ultra-micro-micro nervous system would not be able to function because its components would be too unstable.

Most biologists are more familiar with the factors which govern the lower limit of cell size.

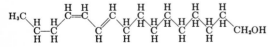

Fig. 1-14. Structure of pheromone secreted by the female silkworm moth, *Bombyx mori.*

Cells are composed of proteins, lipids, water, and nucleic acids. To construct a cell with at least one chromosome, a nuclear membrane, some surrounding cytoplasm, and a cell membrane obviously requires some minimum number of protein, lipid, nucleic acid, and other molecules. You cannot make a cell any smaller, once you have reduced the size to this minimum number of molecules. You cannot get smaller molecules of any given kind unless you can find smaller atoms from which to make them. But atoms of hydrogen, oxygen, nitrogen, and the other needed elements come only in the standard sizes. To find toy-sized atoms you would have to go not just to another world but to another universe.

So, the old verse, usually attributed to Jonathan Swift, the author of *Gulliver's Travels*,

Great fleas have little fleas
Upon their backs to bite 'em
And little fleas have lesser fleas,
And so *ad infinitum.*

is false when it gets to the *ad infinitum.* There is a lower limit set by the structure of matter itself.

Of more immediate interest are the factors which set the lower limit of size to any **homoiothermic**, *i.e.,* warm-blooded, organism. This limit would exist on any planet because size is related to the metabolic rate and surface area. According to the laws of solid geometry, as any body decreases in size without radically changing shape, the volume decreases much faster than the surface area. Therefore, the smaller the animal, the greater is its surface in proportion to its volume. Since any solid object loses heat to its environment through its surface, it follows that the smaller any warm-blooded animal is, the higher its metabolic rate will have to be to provide the heat that is lost. This means also that the smaller any animal that maintains a constant body temperature above its environment is, the more it will have to eat in proportion to its size. Shrews, which are one of the very smallest of mammals, weighing only 3 to 4 grams, have to eat almost continuously. Every 24 hours they eat approximately their own body weight (as if a 150-pound man were to eat three 50-pound meals a day).

What about an upper limit to size? When one thinks of the giant sequoias and the whales it might seem that the upper limits are very large indeed. Yet any tree has many of the same geometrical problems that beset tall office buildings. As the building becomes higher and higher, more and more elevator shafts and service conduits are required to transport people and utilities to the upper stories until virtually all of the lower floors must be devoted to such nonproductive structures and supportive foundations, and any further increase in height is self-defeating. For trees there are also limits imposed by the feasible length of the water transport system of the stem and the size of the roots which provide the feeder system and support for the trunk. The upper limit of size in terrestrial animals is restricted by geometrical factors similar to those operative in setting the lower limit of size in warm-blooded animals. As linear dimensions increase, body volume (which means body weight) increases much faster. Thus any animal that is twice as long as another requires legs which are more than twice as thick to support a more than twice as heavy body. So, because the legs have to become not just absolutely larger but larger in proportion to body size, very large animals would be practically all legs, an obvious impossibility. It is interesting to note here a principle already seen with airborne sex attractants. The larger the planet, the greater the pull of gravity. Hence, if a planet the size of Jupiter supported organisms which could walk, they would have to be very small.

Thus living organisms are restricted within maximum and minimum size limits. These may seem very widely separated limits, but that is only because we ourselves live within them. Compared with the sizes of an atom and of intergalactic space, life is restricted to a very narrow size band indeed.

Life as a Troika

One of the most conspicuous features of life on our planet is the presence of two major forms of life, plants and animals. A closer look reveals actually three basic forms: green plants, animals, and a group of plants called fungi. What is the explanation for this three-part division? Is it due to something so basic that it would be very likely to appear on other planets? Or is it due to some idiosyncrasy of Earth? We have already seen that on the biochemical and cellular levels all plants and animals are basically alike. Furthermore, they are subject to the same laws of inheritance and the same processes of natural selection. However, it is impossible to overlook the differences between the higher green plants like trees or the grasses and the higher animals like the octopus and man. One group is stationary, while the other is motile and has a neuromuscular system from which emerges intelligence. How did such a profound difference arise?

According to current theory, for which there is very convincing evidence, this distinction between green plants on one side and animals and fungi on the other arose very early in the evolution of life. It was based on nutrition, that is to say, on differences in the way energy and the macromolecules involved in cell structures are obtained. Present evidence indicates that living organisms arose in the ocean after a long period when organic molecules of greater and greater complexity were formed under the influence of various energy sources, notably ultraviolet radiation from the sun. If such molecules should be formed today they would quickly be taken up by living organisms. However, before life appeared there were no living organisms to devour such organic molecules and they therefore accumulated eon after eon.

The first living things fed on this rich "soup." However, as living things increased and these food molecules became more and more scarce, a crisis in the history of life must have occurred. There were two possible responses to this challenge. One depended on mutations that conferred greater and greater synthetic abilities on the organism. For example, if the original organisms required substances A, B, C, and D to make their own cytoplasm, and one underwent a mutation that enabled it to make D out of A, B, and C, it would clearly be favored by natural selection over the rest of the population. If one of its descendants had a mutation which enabled it to make B or C out of A, it would have a still greater advantage since it would require only two instead of four kinds of molecules to survive. This line of evolution, suggested here in the simplest terms, obviously

led to green plants which have developed the ability to utilize the sun's energy to manufacture their own food from carbon dioxide (CO_2) and water (H_2O). The other viable series of mutations did not confer the ability to synthesize complex molecules from simpler and fewer compounds but instead enabled their possessors to move around in their environment to reach places where useful prefabricated molecules existed. Ultimately they became able to engulf not only other molecules but other organisms. This line of mutations obviously led to animals.

On our planet there is a kind of "third world" of organisms, the fungi, consisting of the bacteria, some strange forms called slime molds, and the more familiar molds and mushrooms. They are probably diverse in evolutionary origin but they are characterized by dependence on other organisms, ultimately green plants, in the same way animals are for nutrition. They have met the nutritional crisis by evolving in the same nutritional direction as the animals but without evolving a neuromuscular system to move about.

What is the probability that a similar divergence in the basic life styles of living things would occur on other planets? There is every reason to believe that if life arises anywhere in the universe, it must obey the same laws of physics and chemistry and utilize the same kinds of atoms that exist throughout the known cosmos. Therefore, the same nutritional crisis would probably arise and the same two basic ways of meeting it might evolve.

The probability that a group of organisms similar to our fungi would exist is very high. Without them the essential cycling of carbon compounds and nitrogen and phosphorus compounds would cease. The three basic ways of life on our planet may not be the result of some odd quirk. Rather, they appear to be due to a fundamental aspect of evolution which would reappear where ever life arose.

Sexual Reproduction

In looking at life on this planet, there are several additional features which appear to be due to what Schrödinger called "intrinsic reasons" rather than to chance circumstances peculiar to our world. One of these features is sexual reproduction, a virtually universal phenomenon among all the hundreds of thousands of kinds of plants and animals. Even in bacteria there is a mechanism for the passage of genetic information between individuals. There are of course many organisms which regularly reproduce by budding or some other asexual process—the coelenterates, the group to which the jellyfish belong, for example, and most groups of plants. However, in the life cycles of these organisms there is also a sexual phase. In sexual reproduction there is a great evolutionary advantage. Mutations, or changes in the genes, of course form the raw material of evolution. Sexual exchange makes it possible to bring together advantageous mutations which have occurred in different individuals and thus produce superior offspring. On top of this, in the random assorting of chromosomes (genes) in the formation of eggs and sperms and in the randomness of fertilization, a vast amount of variation, the raw material of evolution, is produced. Given the kind of thing life is, sexual reproduction is to be expected wherever life appears.

Intelligence

Everyone has at some time wondered if there are intelligent creatures native to planets in outer space. Is intelligence due to some special peculiarities of earth or is it one of those things which would have a very high probability of emerging wherever there is life? We have already seen that the probability of both plants and animals existing is very great. It is characteristic of animals to move around and "seek" food. This means that to have a head with sense organs at the front end is a selective advantage. You only have to look at animals to see that this is so. Only in animals which have become **sessile** (sedentary or fixed) is there no head. But a head with sense organs lacking a nervous system to control and coordinate the input of the sense organs and regulate the responses would be useless. From the biological point of view, intelligence, the ability to solve problems, is a tool for living just as much as is the antenna cleaner of a honey bee or the hand of a man.

TAXONOMY AND THE DIVERSITY OF LIFE ON THIS PLANET

The world of living things on this planet presents a bewildering jungle of thousands, tens of thousands, and in some groups, hundreds of thousands of different kinds; over 14,000 species of bony fish, nearly 2,000 species of tapeworms, a quarter of a million flowering plants, and well over half a million species of insects have been described. From whatever point of view biology is approached, whether that of agriculture, medicine, economics, ecology, or pure research, classification is necessary to bring manageable order out of this confusion.

Intelligible study of the intricate food chains of the world ecosystem would be impossible without accurate identification of the plants and animals which constitute those interweaving pathways. Control of diseases transmitted by insects depends in large part on knowing what species of mosquito, fly, or louse is involved. Attempts to increase the yield of seaweeds, so that farming the oceans may someday be possible, depends on identifying the kind of alga grown. Only by knowing with certainty the species of plant or animal under observation can any discoveries about it be integrated into the great body of scientific knowledge. Thus **taxonomy**, as the science of classification is called, forms the indispensable framework of biological science.

A further aspect of taxonomy must be recognized. The naming of animals and plants appears to fill some deep and all but universal human need. In the biblical story of creation, one of the first acts of Adam was to name the animals. Even today when a strange creature is washed up on the beach, the most insistent cry is "What is it?"

Since the latter part of the 18th century, taxonomy has become increasingly concerned with relationships of living things. At first the similarities among different kinds of organisms were not thought of as due to descent from common ancestors. Under the leadership of the great Swedish botanist Carl Linnaeus (1707 - 1778) and the anatomists Georges Cuvier (1769 - 1832) in France and Richard Owen (1804 - 1892) in England, the similarities among living things were regarded as the expression of a logical relationship, just as different kinds of triangles or other geometrical figures are related.

With the publication in 1859 of the *Origin of Species* by Charles Darwin, the study of classification became part of the study of evolution. Ever since, taxonomic arrangements of animals and plants have attempted to reflect

Fig. 1-17. Georges Cuvier, founder of comparative zoology. (From G. R. Taylor. *The Science of Life.* McGraw-Hill Book Co., New York, 1963.)

Fig. 1-16. Carl Linnaeus, founder of the binomial system of classification, in Lapland apparel worn during an Arctic collecting trip. (From oil painting by M. Hoffman, courtesy Peabody Library, Baltimore.)

Fig. 1-18. Richard Owen. (From *Life and Letters of Richard Owen,* D. Appleton and Company.)

evolutionary relationships. The closer two kinds of plants or animals are thought to be to a common ancestral population, the closer they are placed on the taxonomic scale.

Many biologists regard the present state of taxonomy as a scientific scandal. This feeling of frustration arises from the fact that in more than 200 years since Linnaeus, it has not been possible to develop an agreed upon and reasonably objective system such as the chemists have for organic compounds. Vigorous attempts are now under way to discover more objective criteria for determining the degree of relatedness of species. Computers are being programmed to compare species in a large number of traits. The structures of large molecules are being analyzed and compared, *e.g.*, the amino acid sequences in a given enzymatic protein from a variety of plants and animals. The base sequences in DNA from many different species are being determined and compared. The success or failure of these methods is still very uncertain. Meanwhile, we have to make do with our present system of classification, which, for all its faults, is still immensely useful.

The Binomial System of Classification

The system of classification in use today is the binomial system, established by Linnaeus, a contemporary of George Washington. According to this scheme each distinct kind of organism receives two names which together constitute the scientific name of the species. Thus all human beings belong to the species *Homo sapiens*, all wood frogs to the species *Rana sylvatica*, and all white oaks to the species *Quercus alba*. The name written first is the name of the **genus**, which is defined as a group of related **species**. Most frogs belong to the genus *Rana*, and oaks to the genus *Quercus*. Following the generic name is the specific name. The wood frog bears the specific name *sylvatica*, and the white oak bears the specific name *alba*. Note that the generic name is capitalized while the specific name is not. Both names are italicized. The word species is the same in both singular and plural. Specific names are commonly adjectives referring to some characteristic of the species; thus *sylvatica* (of the woods) is the specific name for the wood

frog, and *pipiens* (chirping) for the common laboratory frog. However, the specific name may be taken from a person, as in the case of *Rana catesbeiana*, the bullfrog (whether male or female) named after Mark Catesby, an early American naturalist.

The system is an international one. The rules governing this system are formulated and interpreted by international commissions which meet approximately every five years in conjunction with the International Botanical and Zoological Congresses. According to these rules, the form of both generic and specific names must be in Latin. The first person to publish the description of a new species has the right to name it. As fair and inevitable as this rule seems, it has frequently been the cause of much confusion and annoyance when it was found that an organism long familiar under one name had been given a different name earlier by someone who published in a little-known journal.

The American Association for the Advancement of Science, in an official report in 1949, strongly advocated the use of uniform endings for all taxonomic categories above the genus. Although many biologists continue to use the

Table 1-1
Classification of some familiar animals

Taxonomic Categories	Amoeba	Man
Phylum	Protozoa	Chordata
Subphylum	Sarcomastigophora	Vertebrata
Class	Rhizopoda	Mammalia
Order	Amoebiformes	Primatiformes
Family	Amoebidae	Hominidae
Genus	*Amoeba*	*Homo*
Species	*proteus*	*sapiens*

Table 1-2
Classification of some familiar plants

Taxonomic Categories	Edible Mushroom	White Oak
Division	Eumycophyta	Spermatophyta
Subdivision	Basidiomycophyta	Angiospermae
Class	Holobasidiomycetes	Dicotyledoneae
Order	Agaricales	Fagales
Family	Agaricaceae	Fagaceae
Genus	*Agaricus*	*Quercus*
Species	*campestris*	*alba*

old endings, which can be described as a deplorable hodgepodge, standard endings are coming into increasingly wider acceptance. The advantages of a logical system are so great that such endings will be used throughout this text for orders and families. The standard ending for orders of animals is -*iformes* and for orders of plants is -*ales*. Endings of family names are -*idae* for animals and -*aceae* for plants. The classification of several familiar organisms is given in Tables 1-1 and 1-2.

SPECIALIZED BIOSCIENCES

For both theoretical and practical reasons biology is divided into a great many subsidiary specialties. Some, such as medicine and agriculture, deal with very practical aspects of human existence and have evolved from ancient and non-scientific beginnings. Others have rather recent origins in the explosion of knowledge during the 20th century. The specialized areas of biology are useful in a variety of ways from indicating areas of biological work centering around common problems, techniques, or organisms to defining the limits of university departments or the work of professional biologists, and sometimes all three. The human desire to substitute names for long verbal descriptions of things appears to be universal and is just as evident in the vocabulary describing the science of biology as in that used to refer to the kinds of plants and animals that are its subjects. Some of these specialties, like anatomy and physiology, the study of form and function, extend far back into history while still remaining basic. Others, like oceanography, the science of the seas, and ecology, the science of the interrelations of all living things and their total environment, are of recent origin.

All the biological sciences have their anatomical and physiological aspects. **Anatomy** is the science of structure, whether macroscopic gross structure, microscopic cellular structure, or the structures revealed by the electron microscope. **Physiology** is the science of function. Animal physiologists investigate the action of the heart, how a muscle contracts, the processes of digestion, the control and the nature of a nervous impulse, the way the kidney works, and how the eye sees. Plant physiologists study how plants function: the mechanisms by which

water and inorganic salts are absorbed from the soil and transported to all parts of a plant, the ways in which plants respond to light (whether this is in the conversion of solar energy into chemical energy in photosynthesis or the control of flowering by day length), what mineral nutrients are required for healthy plant growth, and the role of each in plant metabolism.

THE PLAN OF THIS BOOK

Obviously some of the biological specialties are more fundamental and generalized than others. Those that contribute the theoretical framework on which present-day biological concepts are built are the ones that ought to be emphasized in a general biology text. It is our intention to begin with considerations of the cell, structural unit of most plants and animals; to proceed to the metabolic, energy-utilizing processes that are virtually universal in living systems; and then to the ways in which the genetic information needed to make new organisms is translated in embryonic development and passed on from one generation to the next. Thus we will begin with some basic cytology, biochemistry, and genetics; however, because we believe that biology is the science of life, we are convinced that theories and generalizations are largely meaningless without frequent references to the kinds of plants and animals best illustrating them. Therefore, chapters about the structure and chemistry of cells are followed by chapters about relatively simple organisms; the chapter on genetics is followed by chapters on development which, after all, is what much of an individual's genetic information is all about. Additional chapters about plants and animals precede more detailed chapters on function. Accurate and extensive knowledge of the plant and animal kingdoms is an essential foundation for any sound understanding of ecology or wise choice of methods for maintaining our environment in a healthy condition. We do not intend to give a full length presentation of all the groups of plants and animals. Rather, we will discuss in detail only certain strategic groups and give only the salient facts about the less important groups. This plan necessitates some hard but necessary choices in any book of reasonable size. Considerations of life at the level of populations where evolution occurs and of the interrelatedness and interactions of living things ought to come last to be truly meaningful and to serve as unifying themes to the diversity that precedes.

There is no single chapter devoted to the human importance of biological science because we believe that to do so is to accept a false separation of biology into two parts, *i.e.* what is important to the human race and what is not. Our conviction is that the whole spectrum of biology is relevant to mankind. Different aspects of it are of greater or lesser importance to different people and for different reasons. Consequently, we point out the human relevance with each topic, whether it is an understanding of the role of plants in the world ecosystem or the astonishing possibilities presented by the newer knowledge of the centers of emotion in the brain. Likewise, we have no chapter on the scientific method. We believe it is far better to include this as part of the deeply humanizing historical perspective which is emphasized at many points. Out of this inclusive focus we hope the student will gain a lasting feeling for biology as an on-going human enterprise with great accomplishments already achieved and with a vast unknown waiting to be explored.

Citizens of the future will be faced with the responsibility of deciding how a whole constellation of discoveries from genetic engineering to the farming of the seven seas can best be used for the welfare of all mankind. It is a staggering responsibility. It is our conviction that all these problems have important social, economic, political, and moral dimensions, but that without understanding the biological bases, sound value judgments are impossible. Consequently, as biologists, we will concentrate on the biological dimensions.

USEFUL REFERENCES

Dubos, R. *So Human an Animal.* Charles Scribner's Sons, New York, 1968.

Ehrlich, P. R., and Ehrlich, A. H. *Population Resources Environment.* W. H. Freeman and Co., San Francisco, 1970.

Ghiselin, B. (ed.) *The Creative Process.* Mentor Books, New York, 1963.

Glass, B. *Science and Ethical Values* University of North Carolina Press. Chapel Hill, 1965.

Hanson, E. D. *Animal Diversity,* 2nd ed. Prentice-Hall Inc., Englewood Cliffs, N. J., 1964.

Hardin, G. The tragedy of the commons. Science, *162:* 1243-1248, 1968.

Harris, R. J. *Plant Diversity.* Wm. C. Brown Co., Dubuque, Iowa, 1969.

Taylor, G. R. *The Science of Life.* McGraw-Hill Book Co., New York, 1963.

Cells: The Units of Life

The realization that all living things are built up of semi-independent units called cells has come gradually over a period of more than 400 years. It is the work of many investigators in many countries. The idea itself is simple, but it is difficult to assimilate into everyday thinking. After all, what one sees is an oak tree or a rose, not a beautifully organized collection of millions of cells. No one feels subjectively that he is composed of vast numbers of minute, more or less independent, units. Nor is it obvious to common sense that the actual bridge between parents and children is a single cell, the fertilized egg.

THE CELL THEORY

The realization that life is not a property of the whole intact organism but rather is based on independent, or potentially independent, units, began in the 17th century with a discovery by the versatile English mathematician and microscopist, Robert Hooke. What Hooke saw was the honeycomb-like structure in the substance of cork, which is the bark of a species of oak. Not surprisingly, Hooke named these units "cells." The box-like structure of cells is relatively easy to see in plant material (the skin of a red Italian onion, for instance); but the outlines of the cells which Hooke saw are the non-living cellulose walls.

No one at the time foresaw the central importance of cells for understanding living things, but almost everyone who had one of the newly invented microscopes observed cells. In Holland Anton van Leeuwenhoek, the gifted Dutch lensmaker, described many kinds of cells, from blood corpuscles to human sperms.

In the south of Europe Marcello Malpighi, a professor on the Faculty of Medicine in Bologna, turned his microscope on parts of large and small animals, and thus became the founder of microscopic anatomy. The innermost cellular layer of the skin is named, after its discoverer, the **Malpighian layer**.

The cell theory as we know it today is mainly the achievement of the 19th century. Microscopists all over Europe began to notice some kind of "mucus" within the boxlike "cells" of Hooke. Then in 1830 Robert Brown, a young Scottish army surgeon and botanist, discovered a nucleus in the cells of orchids and subsequently in many other kinds of cells. By 1839 a Czechoslovakian, Johannes Purkinje, usually known for his studies on vision, showed in a classic paper, "On the Structural Elements of Plants and Animals," that both plants and animals are constructed of cells. It was Purkinje who introduced the term **protoplasm** to denote the living material of the cell.

At about the same time M. J. Schleiden and M. Schwann in Germany popularized similar ideas about the cellular structures of all living organisms, although their belief that cells originate from a noncellular matrix turned out to be very wrong. Finally, in the 1860's, Max Schultze drew together the diverse facts that adult plants and animals are made of cells, that eggs and sperms are cells, and that microscopic animals like *Amoeba* and *Paramecium* are essentially single cells. At the same time he formulated the famous definition of protoplasm as "the physical basis of life."

It remained for a group of investigators in the later part of the 19th century, mostly in Germany and Belgium, to show that all cells

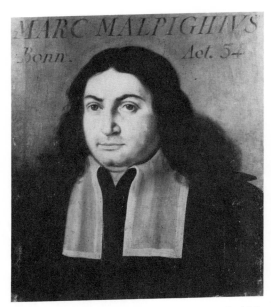

Fig. 2-1. Marcello Malpighi, the long-haired 17th-century Italian Professor of Medicine who discovered blood capillaries in animals and conducting vessels in plants. His extensive discoveries established microscopic anatomy as a science. (Courtesy Johns Hopkins Institute of the History of Medicine.)

problems from the diagnosis of hereditary disease before the birth of a child to the way hormones produce their results or a muscle contracts. Before considering modern cell research, it is essential to have clearly in mind the 19th-century or "classical" knowledge of cells because it is the indispensable foundation of present-day biology.

It is not too difficult to see that most cells are much alike. All consist of a **nucleus** and its surrounding living material, the **cytoplasm**. Both nucleus and cytoplasm are surrounded by a membrane so thin that its structure cannot be discerned with the highest powers of a light microscope. Cells fitting this description can easily be scraped off the inside of your cheek.

A conspicuous difference between the cells of most adult plants and adult animals is that plant cells typically are surrounded by a more or less rigid envelope of **cellulose**, called the **cell wall**, and contain within the cytoplasm a large **vacuole**. In many adult plant cells the cytoplasm is only a thin layer pressed against the

come from previous cells and, in the higher organisms, by a process usually called **mitosis**. Strictly speaking, mitosis is a synonym for the term karyokinesis which refers only to the duplication of the nucleus, including, of course, the **chromosomes**. **Cytokinesis** is a word introduced over a decade later to refer to the division of the cytoplasm. These two terms are coming into use again and are thus worth at least a nodding acquaintance. A survey of recent publications of outstanding cytologists (investigators of cells) shows that mitosis is commonly used by them in the broad sense of the sum total of events connected with cell division. All of this preoccupation with terminology may be very unfortunate but it is a commonplace situation in the biological sciences.

THE BASIC ANATOMY OF LIFE— THE CELL

Modern cell research uses powerful new methods which are yielding important new insights over a very wide range of biological

Fig. 2-2. Robert Brown, the early 19th-century Scottish friend of Darwin and von Humboldt and discoverer of the cell nucleus. (Courtesy Johns Hopkins Institute of the History of Medicine.)

Fig. 2-3. Jan Purkinje, the 19th-century Czech who showed that both plants and animals are composed of cells and coined the word protoplasm. (Courtesy Johns Hopkins Institute of the History of Medicine.)

cellulose wall by the large central vacuole. An examination of most animal cells reveals only nuclei and cytoplasm. The cell membranes separating the cytoplasm of the different cells of the liver, thyroid gland, or a muscle cannot be readily seen without special staining techniques, and the large central vacuole is missing. Some animal cells, such as cartilage and bone cells, secrete extracellular material which surrounds them in the manner of the cellulose wall. The chromosomes within the nucleus are essentially the same in both plant and animal cells in their structure and in their behavior during cell division and sexual reproduction. This is why the laws of inheritance are the same in plants and animals.

MODERN CELL RESEARCH:
THE METHODS

One of the big differences between modern science and the science of the ancient Greeks, perhaps the crucial difference, is the modern realization of the central importance of experimentation, of laboratory work, of technique. Not only is no theory or supposed fact any better than the evidence on which it is based, but how it can be known is a non-disposable part of that fact.

The indispensable workhorse for the study of cells remains the familiar light microscope. Progress in both optics and methods of preparing material for observation has continued ever since Robert Hooke cut that piece of cork with a penknife "sharpened as keen as a razor" and looked at it under a 1665-model microscope. The latest improvements on the light microscope are the phase contrast and the Nomarski lens assemblies. Both show objects within cells with greater clarity and brilliance than ordinary light microscopes. So far their chief contribution has been to confirm in the living cell what had been discovered by the so-called "classical" or standard methods.

The great problem in investigating cellular structure is that cells and their components are almost transparent. Because cells are fragile and disintegrate very easily, they have to be preserved, *i.e.*, fixed and stained. Thick tissue must be sliced thin for the same reason Hooke sliced his piece of cork: microscopes can only "see" by light transmitted through paper-thin slices.

The **fixatives** used are mixtures of chemicals such as formaldehyde, alcohol, and acids in such proportions that the tendencies to shrink protein and other protoplasmic constituents are counterbalanced by the tendencies to cause swelling. The result is that the protoplasm is "fixed" approximately as it was in life. The next step is to infiltrate the tissue with paraffin to make it firm enough to cut. It is then sliced into thin sections with a **microtome,** a device that works much like a bacon-slicing machine. The slices are then pasted on a glass slide and stained with aniline or other dyes.

If there is a pressing need for speed, as when a piece of suspected cancer is to be examined while the patient lies on the operating table, or if the object is to be tested for substances like enzymes that may be removed or changed by the procedures of fixation and infiltration with paraffin, another method is used. The tissue is quickly frozen by evaporating liquid carbon

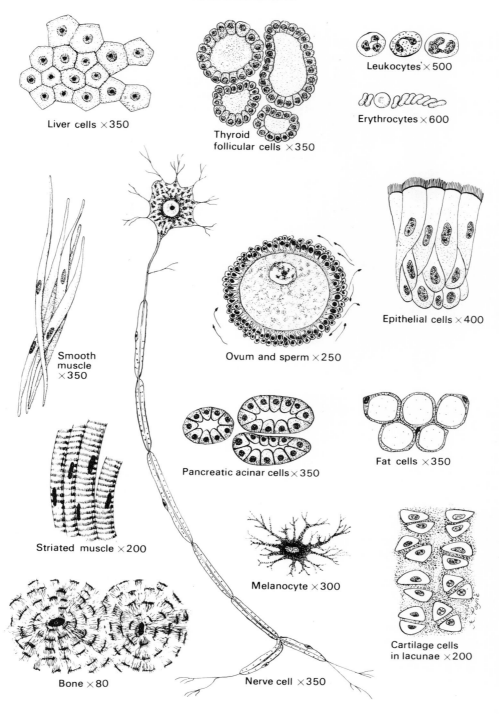

Liver cells ×350

Thyroid
follicular cells ×350

Leukocytes ×500

Erythrocytes ×600

Smooth
muscle
×350

Ovum and sperm ×250

Epithelial cells ×400

Pancreatic acinar cells ×350

Fat cells ×350

Striated muscle ×200

Melanocyte ×300

Cartilage cells
in lacunae ×200

Bone ×80

Nerve cell ×350

Fig. 2-4. A variety of specialized animal cells as seen under a light microscope.

dioxide. It can then be immediately sliced and examined. This method of frozen sections is important in the rapidly developing field of cell chemistry (cytochemistry).

Two of the **stains** mostly commonly used in research and hospital laboratories for plant, animal, and human cells are hematoxylin (a dark blue stain from a South American tree), which dyes the nucleus black or blue, and eosin (a synthetic aniline dye), which stains the cytoplasm orange.

Many biologically interesting molecules such as the flavins and chlorophylls are naturally fluorescent and others, such as proteins, can be made so by using fluorescent antibodies. For example, to identify lens protein from the eye or muscle fiber protein, the specific protein is injected into a rabbit which will produce a specific antibody against that protein. The antibody is then made fluorescent and applied to the section containing the cells to be tested. Wherever the corresponding protein is located in the cells, the antibody will adhere and fluoresce.

New Methods for Studying Cells

Four of the most significant new developments in cell study are, as is commonly the case, not completely new but radical improvements or new combinations of familiar techniques. New ways of growing human and other cells *in vitro, i.e.,* in glass dishes, permit accurate study of the chromosomes. Cells and their components can be labeled with radioactive atoms. The electron microscope is opening up a whole new world within the cell to exploration, and special methods of centrifugation enable us to separate and collect for investigation the ultramicroscopic constituents of cells.

Radioactive Labels

When they are built into the molecules of amino acids, hormones, or whatever substance is to be followed, the labeled molecules undergo reactions that are the same as those of "normal" molecules, except that from time to time a labeled atom will disintegrate, producing radiation which can be detected on a photographic plate or by an instrument, such as the Geiger counter, that can measure the amount of

radioactivity. Tritium, *i.e.,* radioactive hydrogen (^{3}H), is frequently used. It has a half-life of 12 years. This means that of a group of tritium atoms, one-half will have "decayed" to ordinary hydrogen within 12 years. Phosphorus-32 and iodine-131, also frequently used as tracers, have half-lives of 14.3 and 8.1 days, respectively.

After exposure to the radioactively labeled material, a tissue is fixed and sectioned, and the sections are placed in contact with a photographic emulsion. Radioactive disintegrations will occur where the labeled amino acid, hormone, or other material is located. Each disintegration produces a black spot on the developed photographic film. The way chromosomes duplicate themselves has been determined by this method (a chromosome organizes a duplicate of itself along one side for its entire length; the new and the old then separate). Much of the behavior and life history of cells can be investigated by such labeling.

Electron Microscopy

In addition to new methods of cell culture to study chromosomes and new ways of labeling

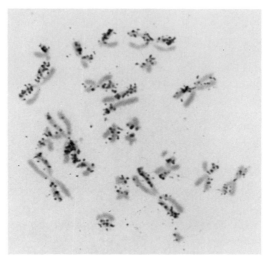

Fig. 2-5. Radioactively labeled chromosomes. Sister chromosomes (chromatids) of the Chinese hamster at the second division after feeding the cells with radioactive thymidine which is incorporated into DNA. Note that only one chromatid of each pair is labeled. (Courtesy T. C. Hsu.)

cells, the electron microscope has added greatly to modern knowledge in many fields. The use of an electron microscope requires basically the same kind of preparative methods that the light microscope does. The material to be studied is usually fixed in a vacuum at freezing temperatures. The material is then embedded in a special plastic and sectioned ultra-thin with a special microtome using a sharp glass or diamond edge. The section is placed on a supporting membrane and then put into the electron microscope.

The source of the electrons, which take the place of the light waves of an ordinary microscope, is a cathode filament, much like a television tube. The electron beam is focused by magnets rather than by glass or quartz lenses. Since electrons travel only very short distances in air, the entire path of the beam is enclosed in a vacuum. The magnified image is projected either onto a fluorescent viewing screen or a photographic plate. Magnifications of over 10,000 diameters are common and up to 100,000 diameters are possible. The ordinary light microscope can magnify only up to about 1,500 times. Table 2-1 indicates the size of some biologically important objects.

The light microscope can **resolve**, *i.e.*, distinguish as separate, two lines separated by 0.2 μ; an electron microscope can resolve lines separated by 3 to 5 Å, *i.e.*, 0.0003 to 0.0005 μ. Resolving power depends both on the type of lens and the wave length of the source of illumination. In general, an object cannot be resolved in the light microscope if its diameter is less than half the wave length of green light—about 5,500 Å—or in the electron microscope if its diameter is less than half the wave length of the electrons that are used— about 0.05 Å.

Analysis of Cell Constituents

The fortunate coincidence which has made so many of the discoveries with the electron microscope meaningful has been the simultaneous development of ultrasensitive methods to separate cell constituents. If cells are homogenized and centrifuged at low temperatures enzyme activity is not destroyed. With relatively light centrifugation the nuclei, being the largest and heaviest constituents, move to the bottom of the tube first and can be

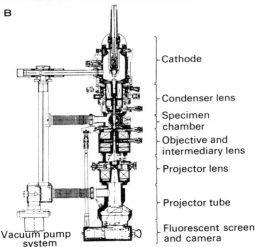

Fig. 2-6. An electron microscope. A, exterior (kindly supplied by Philips Electronic Instruments, Inc.); B, interior construction. (From F. S. Sjöstrand: *Electron Microscopy of Cells and Tissues.* Academic Press, New York, 1967.)

separated from the cytoplasmic components. Somewhat higher speeds throw down mitochondria and plastids. Still higher speeds precipitate minute cytoplasmic particles called **ribosomes.** By filling the centrifuge tube with

Table 2-1
Sizes of some biological objects

Object	Diameter	
	μ	Å
Human egg	100	1,000,000
Red blood cell	10	100,000
Bacterium	1	10,000
Virus	0.1	1,000
Protein molecule	0.01	100
Amino acid	0.001	10

One micron, 1 μ (also called micrometer), is equal to 1/1,000 of a millimeter, *i.e.,* 1 μ = 0.001 mm. One Ångstrom is equal to 1/10,000 of a micron, *i.e.,* 1 Å = 0.0001 μ.

fluids of different densities, a **density gradient** can be produced and very accurate separations achieved. Particles come to rest when they reach a layer with a density closely approximating their own. Each cell component can then be investigated by biochemical methods to study its function and with the electron microscope to study its structure.

Modern Cell Research—The Results

One generalization that has emerged from modern studies of cells is that there are two fundamentally different types. Bacteria and blue-green algae consist of cells which are **prokaryotic** (*pro*, before or early, + *karyon,* kernel). All animals and all green plants including the algae other than the blue-greens, and all the plants commonly called fungi are **eukaryotic** (*eu,* true, + *karyon,* kernel). The problematical viruses, which some biologists do not regard as living things, are called **akaryotic** since they are not cells but consist merely of a bit of genetic material, either DNA or RNA, covered with a coating of protein. They will be considered in Chapter 4.

A typical prokaryotic cell, whether a bacterium or a blue-green alga, lacks a nuclear membrane surrounding definite chromosomes.

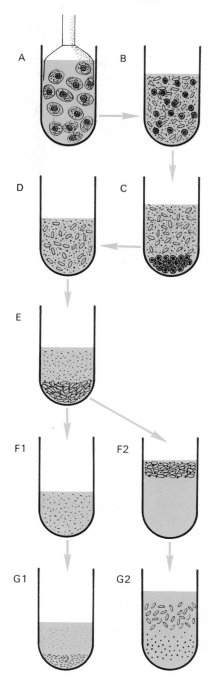

Fig. 2-7. Separation of cell components by centrifugation. A, whole cells in sucrose solution; B, homogenate of cellular components; C, after centrifugation at 600 × *g* (nuclei thrown to bottom); D, homogenate with nuclei removed; E, mitochondria thrown to bottom after centrifugation at 8,500 × *g;* F1, ribosomal portion of E; F2, mitochondrial portion of E placed on a sucrose gradient with the more concentrated and therefore heaviest sucrose solution at the bottom; G1, ribosomes thrown to bottom by centrifugation; G2, lighter mitochondria in a layer above the lysosomes after further centrifugation. (After de Duve.)

Instead there is a vaguely defined region called the nucloid where DNA is located. Cell division is direct by binary fission without the complicated processes of mitosis. Sexual reproduction is rudimentary and consists of occasional transfer of genetic material. Cytoplasmic bridges between bacteria of different morphological types (so they could be identified) have been revealed by the electron microscope. These bridges are believed to be the physical basis of the genetic recombinations that are now well established in bacteria.

An extremely wide degree of metabolic traits characterize the prokaryotes. Some cannot live in the presence of oxygen. Others are facultative anerobes and can live either using oxygen or without it; others are aerobes. A number of species metabolize sulfur, iron, and even atmospheric nitrogen. Some are photosynthetic but their photosynthetic pigment is not contained in discrete plastids: instead it is found on membranes located within the cytoplasm. Other cell organelles found in higher organisms are conspicuously absent, most notably the mitochondria, thread-like or sausage-shaped structures that in the cells of higher plants and animals are sites of oxidative respiration. Little wonder that C. B. van Niel, whose work with bacteria provided one of the keys to our modern understanding of photosynthesis, decided that the primary division of life was not into plants and animals, but rather between what he terms prokaryotes and eukaryotes.

The most distinguishing feature of eukaryotic cells is their discrete chromosomes enclosed in a nuclear membrane. Within the nucleus there is usually a nucleolus, which has an abundance of RNA and certain proteins. Cell division occurs by that complex of chromosomal maneuvers called mitosis. Probably of equal importance, the cytoplasm is endowed with organelles, notably mitochondria, which carry on oxidative respiration and provide the cell with a supply of available energy packaged in molecules of adenosine triphosphate (ATP). In green plants the cells are further provided with photosynthetic organelles, the chloroplasts.

How eukaryotic cells arose in the history of life, no one knows. It has been suggested for nearly a century that mitochondria and chloro-

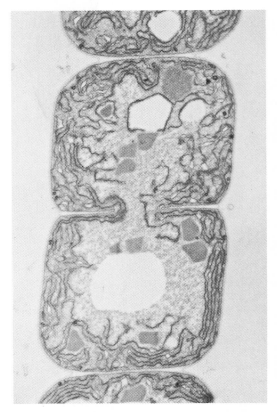

Fig. 2-8. A prokaryotic cell in division—the blue-green alga *Anabaena*. Note the absence of a nucleus, although vacuoles are present. × 13,000. (Courtesy N. J. Lang.)

plasts may be symbiotic bacteria and algae living within the cytoplasm. Although this theory does not explain the origin of mitosis, it explains a lot, and strong evidence has been accumulating for it in recent years. Both mitochondria and chloroplasts are self-duplicating and contain their own DNA. They do not undergo mitosis any more than bacteria do. When the "host" eukaryotic cell divides, they are merely distributed to the two daughter cells in a more or less random fashion. At present the outstanding exponent of the theory that eukaryotic cells arose by multiple symbiosis is Lynn Margulis, an investigator of cilia in Protozoa. Intracellular symbiotic algae with nuclei are well known in some of the lower animals like the green species of *Paramecium* and the green hydras and corals.

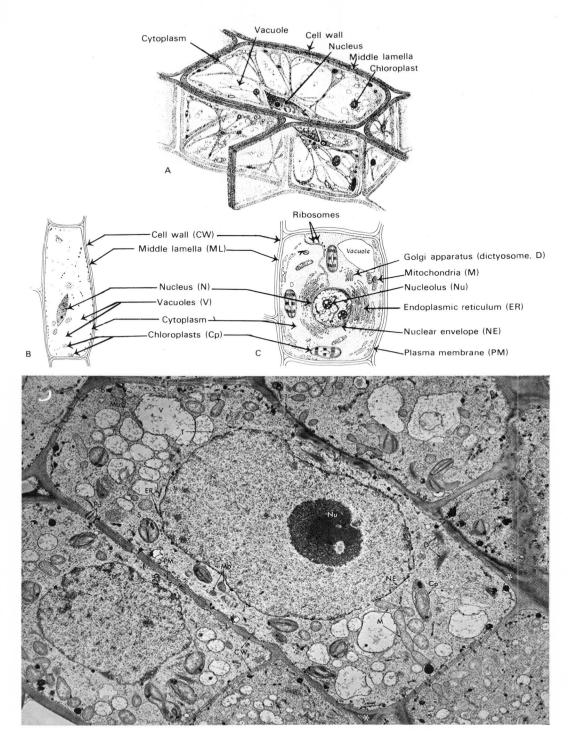

Fig. 2-9. Eukaryotic plant cells. A, three-dimensional view; B, light microscope image; C, artist's diagram of electron microscope image. Letters in parentheses after labels in B and C refer to letters in D, electron photomicrograph of plant cell. (A and B after K. Esau: *Plant Anatomy*. John Wiley & Sons, Inc., New York, 1933. C after Braomet: *The Living Cell*. Copyright © 1961 Scientific American, Inc., all rights reserved. D from M. C. Ledbetter and K. R. Porter: *Introduction to the Fine Structure of Plant Cells*. Springer-Verlag, New York, 1970.)

There was a time when a eukaryotic cell was regarded by some as merely a bag containing a nucleus and various loosely mixed chemicals, but this theory is no longer held. To the discoveries made with the light microscope, the electron microscope has added an almost incredible amount of detail and additional unsuspected structures.

The mitochondrion is one of the easiest cellular organelles to see because of its size and activity. Under a light microscope mitochondria appear as rods or threads which (in living cells) stain greenish-blue with Janus Green B. They can be observed unstained in living cells by the use of special techniques such as dark field and phase contrast microscopy. Long mitochondria wiggle about as though they were independent creatures—in fact, they resemble rod-shaped bacteria. Under an electron microscope each mitochondrion is revealed as composed of a double membrane, the inner one forming folds called **cristae** that extend into the interior.

Biochemical analysis shows that mitochondria are the "powerhouses" of cells. The enzymes of aerobic respiration are located in the walls of the mitochondria. The final energy-giving steps in respiration take place here, yielding so-called "packaged energy" in ATP molecules. The ATP is then available for muscular contraction, secretion, or the synthesis of new complex molecules, such as proteins.

A second conspicuous feature of the cytoplasm is the **endoplasmic reticulum**. This system of membranes is found in all eukaryotic cells, but is especially abundant in cells that synthesize protein. For example, lymphocytes, the white blood cells that make protein antibodies, are rich in endoplasmic reticulum. On the outer side of the endoplasmic reticulum in such cells are minute granules of very uniform size. These are the **ribosomes**, the sites of protein synthesis. It is here that the genetic code is translated into the sequence of amino acids in protein dictated by the sequence of code words in a strand of RNA. It is at the ribosome that the amino acids are linked together to form new protein. Not all ribosomes are located on membranes in the endoplasmic reticulum. Some are free in the cytoplasmic matrix, at times in groups termed polyribosomes.

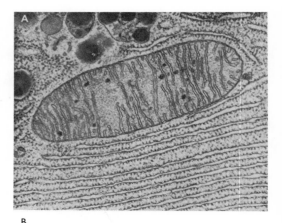

Fig. 2-10. Mitochondrion. A, photo as seen under an electron microscope, × 64,000. (Micrograph of bat pancreas by K. R. Porter. From D. W. Fawcett: *The Cell.* W. B. Saunders Co., Philadelphia, 1966.) B, diagrammatic reconstruction.

A third universal organelle is the **Golgi complex**, also called a **dictyosome**. This is a series of three, four, or more curved membranes stacked like a pile of saucers. The membranes are much like those of the endoplasmic reticulum but lack ribosomes and hence are smooth. Usually the Golgi complex lies near the nucleus. In Figure 2-12 it can be seen that the ends of membranes of the Golgi complex appear to be giving off droplets. The Golgi "apparatus," as it is often called, apparently secretes material, and is in fact well-developed in secretory cells. In dividing plant cells, droplets of calcium pectate which coalesce to form the middle lamella separating the two daughter cells may be secreted by the Golgi apparatus. It also seems to secrete components of the cell wall.

A fourth cytoplasmic structure is the **lysosome**. Lysosomes are small, rounded vesicles varying in diameter from about the width of a

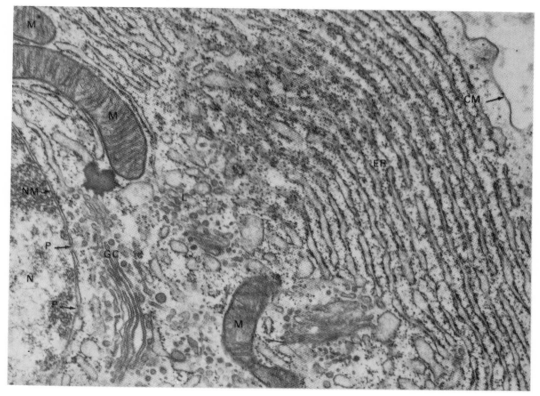

Fig. 2-11. Animal cell (human lymphocyte from blood plasma) as photographed under an electron microscope. ER, endoplasmic reticulum. Note dot-like ribosomes on membrane surface. CM, cell membrane; GC, Golgi complex; L, lysosome vesicles; M, mitochondrion; N, nucleus; NM, nuclear membrane, P, pore through nuclear membrane, × 30,000. (Photo courtesy James A. Freeman. From G. B. Moment: *General Zoology,* 2nd ed. Houghton Mifflin Co., Boston, 1967.)

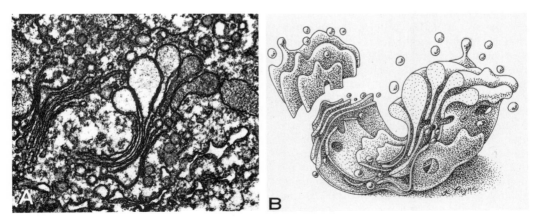

Fig. 2-12. Golgi complex. A, as seen under an electron microscope, × 29,000. (From H. H. Mollenhauer and W. G. Waley: An observation on the functioning of the Golgi apparatus. Journal of Cell Biology, *17:* 223, 1963.) B, diagrammatic reconstruction.

mitochondrion to much smaller. Lysosomes apparently contain lytic (digestive) and other enzymes which become active when the lysosome membrane is ruptured. Since these enzymes digest the cell, lysosomes are popularly called "suicide bags." Apparently they serve to digest unhealthy or injured cells and return their raw materials to the rest of the organism. In some cases at least, lysosomes seem to be produced by the Golgi apparatus.

Photosynthetic organelles, or **chloroplasts**, are found in all eukaryotic cells that are capable of converting light into chemical energy. These green, most often elliptical structures, like the mitochondria, are self-replicating. Under the highest magnification available with the light microscope the green chlorophyll pigment appears to be concentrated in discrete regions called **grana** which are embedded in an optically clear matrix called the **stroma**. The electron microscope has revealed details of the ultrastructure of chloroplasts and their highly organized membranes, or lamellae. Each granum is made up of a stack of lamellar (plate-like) subunits thought by some experts to be like flattened sacs, and referred to as **grana lamellae**. The stroma, which appears granular in the electron microscope, also contains layered membranes called **stroma lamellae**. A more detailed description of chloroplast structure can be found in Chapter 15.

An array of organelles under the collective term **plastids** can contain chlorophyll, pigments other than chlorophyll, or no pigments at all. A variety of chromoplasts, yellow, orange, or red in color, found in the cells of flower petals, certain fruits, and carrot roots, contain carotenoid pigments (xanthophylls, carotenes, and related compounds). The colorless **leucoplasts**, common in the cells of plant tissues that store food reserves, often contain starch grains. Other leucoplasts function as storage organelles for fats or oils. All of the plastids (chloroplasts, chromoplasts, and leucoplasts) are thought to arise from small and structurally simple precursors called **proplastids**. Both proplastids and mature plastids can divide and are randomly distributed between daughter cells during mitosis.

Many cells in both plants and animals possess structures involved in motion and transport called **microtubules**. Such structures have been

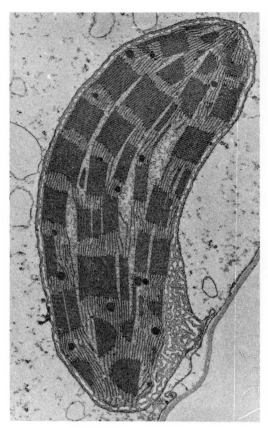

Fig. 2-13. Chloroplast from maize (Indian corn) under an electron microscope. Original magnification × 17,000. (From L. K. Shumway and T. E. Weirer: The chloroplast structure of iojap maize. American Journal of Botany, *54:* 744, 1967.)

known for a decade or so but are now being discovered in many hitherto unsuspected places. An electron microscope is required to see them. Most motile and many fixed eukaryotic cells possess thread-like cytoplasmic extensions called **cilia** (if short) or **flagella** (if long) which are capable of whip-like motion. Both are the same in structure, consisting of a central pair of microtubules (often called fibrils) surrounded by a circle of nine similar microtubules or pairs of microtubules and the whole enclosed by an extension of the cell membrane. At its base in the cell, each cilium or flagellum is connected with a basal granule or body. The electron microscope shows that these are really two very short cylinders of nine microtubules like the flagellum itself, but

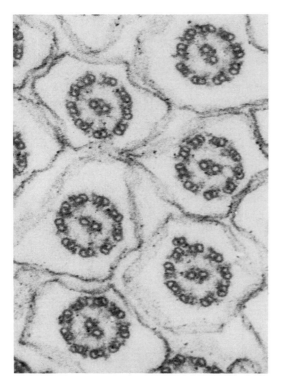

Fig. 2-14. Cross section of cilia in an annelid showing the circle of nine double microtubules (fibrils) plus two central ones. As seen under an electron microscope. (Courtesy Bjorn Afzelius. From G. B. Moment: *General Zoology,* 2nd ed. Houghton Mifflin Co., Boston, 1967.)

without any central tubules. These basal bodies usually lie at right angles to each other and it is supposed that one is the daughter of the other. The present guess is that these basal bodies may be centers which produce the cilia.

Closely similar and perhaps identical structures called **centrioles** are found close to the nuclear membrane in most animal and in a few plant cells. They play an essential role in mitosis. During cell division microtubules extend from the centrioles to a special attachment point on each chromosome. All these microtubules together form the mitotic spindle which is necessary for the separation of the chromosomes into the two daughter cells. Once again the actual mechanics of all this is unknown as are the factors which trigger mitosis. The centrioles duplicate themselves during cell division.

Microtubules have been investigated in developing plant cells by several workers. There is no certainty about their role, but it seems that the tubules are somehow concerned with transporting polysaccharides from the site of synthesis in the Golgi apparatus to the thickening cell wall. So far no such delivery function has been suggested for developing animal cells.

It now appears that microtubules are found wherever cytoplasmic motion takes place. The long radiating cytoplasmic extensions of freshwater and oceanic protozoans, which make them look like sunbursts, consist of more or less rigid extensions of as many as 200 microtubules. These are arranged in two beautifully regular parallel spirals as seen in cross section. In various amoebae the pseudopods (as the locomotory extensions are called) contain microtubules. The evidence to date indicates that motion is somehow produced by the microtubules sliding past eath other.

MEMBRANES IN AND AROUND CELLS

Why all these membranes? It sometimes seems as though life were mostly a membrane phenomenon. No topic in the whole field of cell structure has caused as much recent controversy as the nature of cell membranes.

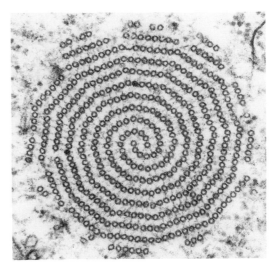

Fig. 2-15. Microtubules seen by electron microscope: cross section of a ray-like pseudopod of a heliozoan, a freshwater protozoan. × 40,000. (Courtesy L. E. Roth, Y. Shigenaka, and D. J. Pihlaja.)

There is no controversy, however, about their importance. Virtually everything that enters or leaves a cell must pass through the cytoplasmic membrane which surrounds it. It is this membrane which separates the living from the nonliving. If the cell membrane of a frog's egg, for example, is ruptured in calcium-free water, the membrane cannot repair itself and the granular cytoplasm of the cell will flow out and drift away. Many hormones produce their effects by doing something to the cell membrane. There is good evidence that at least some hormones act directly on the genes, but even these hormones have to pass through the cell membrane and through the nuclear membrane as well. The function of that nuclear envelope is still much of a mystery. It is provided with pores large enough to see under an electron microscope and therefore large enough for good-size molecules such as proteins and nucleic acids to pass through. The fact that the nuclear membrane is an absolutely constant feature of all eukaryotic cells indicates that its functional significance is great.

Photoreceptors, whether in the eyes of a man or the chloroplasts of a green plant, are composed of layers of membranes. Within cells, the mitochondria and the Golgi apparatus are built of membranes, and much of the cytoplasm is crowded with the membranes of the endoplasmic reticulum adjacent to which lie the ribosomes where proteins are synthesized.

In addition to all this, a nervous impulse passes along a nerve in the form of a self-propagating wave of change in the surface membrane of the nerve fiber. Although we are admittedly only at the beginning of knowledge in this area, all the psychopharmacologically active molecules, whether dangerous artifical mind-benders or the normal hormone of emotion, adrenaline, seem to produce their effects by acting on the surface membranes of nerve cells at the synapse where an impulse from one nerve cell is transmitted to another.

Interest in membranes has grown steadily. A breakthrough came when J. F. Danielli and H. Davson proposed a model of what a cell membrane might be like. As is usually the case, their idea is based on earlier work, in this instance extending back to Benjamin Franklin. By the time of Danielli, the Nobel Prize-winning physicist Langmuir had shown that a drop of oil on the surface of water spreads out into a monomolecular layer. This indicated that cell membranes do not have to be many layers of molecules thick and might even be only one layer in thickness. The model proposed by Danielli and Davson, a very reasonable one that is the basis for all of today's conflicting ideas, is that cellular membranes are like a sandwich four molecules thick. The two outer layers are protein; the two inner layers are phospholipid molecules. The phospholipid, which is the most common lipid found in cells, helps make an effective barrier between the cell's environment which is aqueous and the cell's interior which is also aqueous. The protein covers provide some tensile strength. After all, a cell membrane needs to be stronger than a soap bubble.

Under an electron microscope all cellular membranes, whether on the surface of the cell or in part of an intracellular organelle, appear much the same: there are two dark lines very precisely separated by a clearer layer. The whole membrane is about 90 Å thick. Although measurements vary a bit depending on methods of fixation and other factors, this is about right for a membrane composed of four layers of molecules. In the early 1960's J. D. Robertson drew together all the facts and coined the term "unit membrane" to designate this basic pattern of membrane structure.

One function of intracellular membranes may be to separate different and possibly incompatible reactions within the cell. Mitochondria, for example, are permeable to some substances but not to others. It should be noted that both the outer membrane and the inner one which forms the folds or cristae of a mitochondrion are unit membranes. The inner one can now be seen under the newer electron microscopes to be thinner than the outer one, so all unit membranes need not be precisely the same.

Another important role of cellular membranes appears to be the ordering of enzymes in sequential functional assemblies. Many metabolic reactions require a series of enzymes to act in a particular sequence. The interior of a cell is an enormous space compared with the size of a molecule. If sequential reactions are to take place with reasonable speed, the necessary enzymes need at least to be near each other, if not lined up. Hence, it is not surprising that

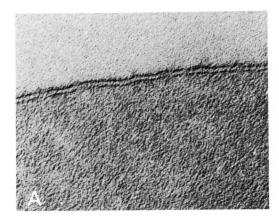

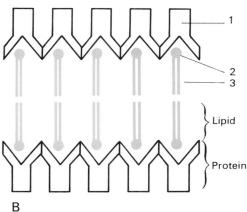

1. Protein which may be in discrete globules

2. Hydrophilic end of phospholipid molecule

3. Hydrophilic chains of phospholipid molecules

Fig. 2-16. Cell membrane. A, under electron microscope, × 275,000. (From J. D. Robertson: Unit Membranes. In *Cellular Membranes in Development,* ed. by M. Locke. Academic Press, New York, 1964.) B, diagram where green represents the lipid layer.

many enzymes now appear to be associated with membranes. The best understood case is the mitochondrion, where the enzymes of aerobic respiration are located.

The function of a cell membrane, in regulating everything which enters or leaves a cell, has been known for decades. However, the passage of materials through cellular membranes has turned out to be a highly complex phenomenon. Despite an enormous amount of work, the whole process is still far from being fully understood.

At least it is clear that some of the properties of cell membranes are primarily physical and dependent on the structure of the membrane. Such physical properties do not require the continuous expenditure of metabolic energy; thus poisons and anesthetics do not interfere with them. The osmotic properties of cell membranes fit into this category. For example, if a solution of protein or sugar in water is separated from pure water by a non-living membrane which is permeable to water molecules but not to sugar or protein, *i.e.,* a **semipermeable membrane**, osmosis will occur. Water will pass into the sugar or protein solution and make it more dilute. This is a fact. The common explanation is that the thermodynamic activity of the water molecules is less in the solution than it is in pure water. This is a theory, although there is good evidence for it.

A simple model to show what happens to cells under different osmotic conditions can be made by filling a cellophane bag with the concentrated sugar solution, tying it tightly closed and placing it in a beaker of water. The result? The bag will swell and ultimately burst. On the contrary, if a bag of water is placed in a beaker of concentrated sugar solution, it will shrink.

Comparable phenomena can easily be seen in living cells. If the surrounding medium has a greater osmotic concentration than the cell interior, it is said to be **hypertonic** and will cause the cell to lose water and shrink, a phenomenon called plasmolysis. If the medium has a lower osmotic concentration, it is said to be **hypotonic** and will cause the cell to take up water, swell, and burst. Such cell breakage is called **lysis**. if the osmotic concentration of the medium is the same as that of the cell, it is said to be **isotonic** or **isosmotic**.

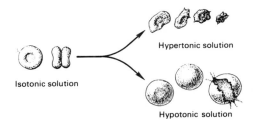

Fig. 2-17. Diagram of red blood cells in iso-, hypo-, and hypertonic solutions.

Osmotic pressure can be measured by filling a thistle tube with sugar solution, covering it tightly with cellophane, and inverting it in pure water. As the sugar solution becomes diluted by water entering through the semipermeable cellophane, the solution will be forced up the tube. The more concentrated the sugar, the greater the osmotic pressure will be, and the higher the solution will climb. The pressure can be measured directly by determining the height of mercury necessary to prevent a rise of solution in the tube. Osmotic pressures can be very great—in some cases as much as 200 atmospheres, or almost 3,000 pounds per square inch.

Osmotic pressure depends not on the actual weight of material dissolved in a given volume of solvent, but rather on the number of molecules or ions. Therefore, for substances that do not dissociate in solution, the osmotic concentration or the potential osmotic pressure of a solution is described in terms of **molarity**. Molar solutions of different substances have the same number of molecules regardless of the size and weight of the individual molecules. A 1.0-molar solution of sucrose has an osmotic pressure of 22.4 atmospheres and is just about isotonic with a sea urchin egg. The cytoplasm of most plant cells has a somewhat lower osmotic concentration.

Dialysis is a phenomenon closely related to osmosis. In fact, both are special forms of diffusion. For instance, dialysis occurs when large molecules of protein in aqueous solution with small salt molecules are separated from pure water by a semipermeable membrane. The salt molecules can pass through such a membrane, whereas the protein cannot. The result is that the proteins can be separated from the unwanted salt or other small molecules.

So far we have considered circumstances where the materials passing through membranes have been passing from regions of greater to regions of lesser concentration, following the laws of diffusion. Cells, or more specifically their membranes, have the ability to move molecules in the reverse direction against a concentration gradient. That this activity requires the continuous expenditure of metabolic energy is indicated by the fact that such **active transport**, as it is often called, is inhibited by metabolic depressants and poisons such as cyanide.

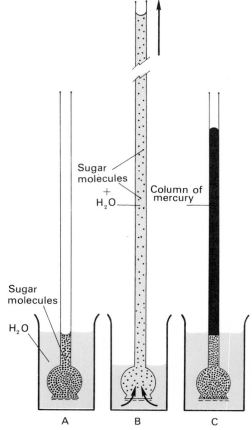

Fig. 2-18. Apparatus to demonstrate osmostic pressure. Green represents water.

The mechanisms by which cells concentrate materials against a concentration gradient are believed to involve special enzymes called **permeases** although the evidence is far from complete. A very simple model can be visualized as follows. Molecule A, outside the cell, unites with a carrier molecule B, within the membrane. Complex AB moves through the membrane and on its inner side breaks into A, which is released into the cell, while B remains in the membrane and eventually picks up another A.

pH: ACIDITY-ALKALINITY

Living cells are extremely sensitive to pH, *i.e.,* the acidity or alkalinity of their environment. pH is measured on a scale which extends

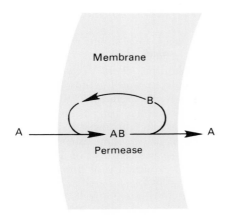

Fig. 2-19. Diagram of permease action in active transport across a membrane.

from 0 to 14, and represents the negative logarithm of the hydrogen ion concentration of a solution. A pH of 1.0 represents an acidity equivalent to that of 1/10 normal hydrochloric acid. A pH of 7.0 is neutral, while a pH of 13 represents an alkalinity like that of 1/10 normal sodium hydroxide. Both internal and external pH are important to cells because most enzymes function only within very restricted ranges of pH.

The pH of a solution is best measured by an electric pH meter. However, many dyes, including natural plant pigments, assume different colors at different pH's. Consequently they can be conveniently used as indicators. Litmus, a pigment obtained from certain lichens, and anthocyanin, the purple pigment of red cabbage, are such indicators.

CELL SIZE

In addition to the basic nuclear-cytoplasmic organization of all eukaryotic cells, they are also essentially alike in size. The enormous differences in bulk between a mouse and an elephant, or between a moss and a redwood tree, are due to differences in numbers of cells, not in cell size. The sperms and eggs, as well as the cells of the liver, skin, brain, and other organs are the same size within very small limits.

Some very potent factors must set both the upper and the lower limits of cell size. Apparently the upper limit is set by the volume of cytoplasm that can be serviced by a single nucleus. Another factor is the geometrical fact that as size increases, volume increases roughly as the cube of the radius of the cell, while surface increases as the square. This means that the relative amount of surface, through which all food and oxygen must enter the cell and through which all wastes must leave, becomes less and less in proportion to the amount of cytoplasm as volume increases. The lower limits of cell size have already been discussed in Chapter 1, where it was pointed out that a certain minimum number of molecules are required to construct a eukaryotic cell with membranes and at least one chromosome, and that the atoms used to construct these essential molecules come only in standard sizes.

Much remains to be discovered about what controls the sizes of cells within the basic limits just described. For example, in animal cells that contain only one set of chromosomes (**monoploid** or **haploid**) instead of the two sets (**diploid**) they normally have, both the nuclei of the cells and the cells themselves are smaller than usual. If cells possess three sets of chromosomes (**triploid**), the nuclei and the cells are larger than normal. Exactly how this regulation is accomplished is not known. Even more puzzling, the triploid animals, although having much larger cells, have fewer cells so that their body size is about normal. The photographs in Figure 2-21 show a portion of the tail fin of each of three larval salamanders with one, two, and three sets. Salamander cells, which are extremely large, are the classic material in which Walter Flemming in the 1870's first described cell division clearly and in detail and coined the word mitosis (*mitos*, thread).

In plants, the relationship between the number of sets of chromosomes in the nuclei and cell size is not so clear-cut as in animals. In some plants, for example, certain of the algae, morphologically identical monoploid plants and diploid plants alternate in the life cycle. In higher plants, where the cells of the plant body are normally diploid, **tetraploid** (four sets of chromosomes) forms are known and frequently are horticulturally desirable because they are larger and more vigorous, with larger flowers and fruits. Some that are triploid, including varieties of pears, hyacinths, and tulips, are sterile and must be propagated vegetatively.

CELL DIVISION

It has been just about a century since Rudolph Virchow (1821-1902) announced his famous aphorism, *Omnis cellula e cellula,* "Every cell from a cell," thus correcting the misconception of Schleiden and Schwann. Since his time, a small army of investigators has sought to describe the facts and explain the mechanism of cell division. Such knowledge is important, because cell division is the method by which the continuity of life is maintained and therefore by which heredity is passed from cell to cell. Furthermore, cell division has an important relationship to cancer. Malignant tumors are essentially populations of cells which have escaped from the normal controls of cell division.

In both animals and plants, cells divide in two quite different though related ways. In embryos and in growing tissues new cells are formed by **mitosis**. This is an asexual process in which the chromosomes of the parent cell are duplicated and passed in identical sets to the two daughter cells. The other method is called **meiosis**, from a Greek word meaning to reduce. The result of this kind of division of the nucleus is that the daughter cells have only one instead of the usual two sets of chromosomes. In animals, meiosis occurs in gonads (sex glands) and results in the formation of sperms and eggs. In plants meiosis usually occurs during the formation of spores. Despite much work, often ingenious and penetrating, very little is known about the physiology and causation of cell division, and even less about what factors induce cells to divide by mitosis on one occasion and by meiosis on another.

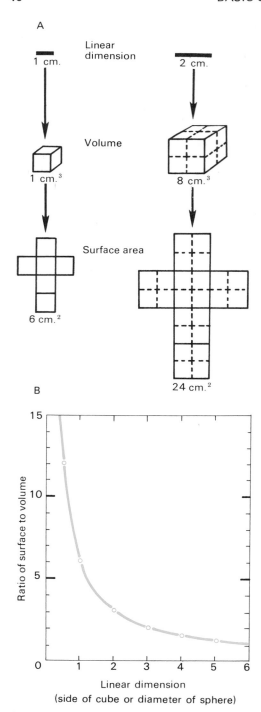

A

Linear dimension

1 cm.

2 cm.

Volume

1 cm.³

8 cm.³

Surface area

6 cm.²

24 cm.²

B

Ratio of surface to volume

15

10

5

0

1   2   3   4   5   6

Linear dimension
(side of cube or diameter of sphere)

linear dimensions vary. For a cube, volume = s^3 (where s = the length of each side) and surface area = $6s^2$. The ratio of surface to volume for a cube = $6s^2/s^3$ or $6/s$. These relationships are shown graphically for cubes of $s = 1$ and $s = 2$ in A. For a sphere, volume = $4/3\ \pi r^3$ or $\pi d^3/6$ (where r = radius and d = diameter) and surface area = $4\ \pi r^2$ or πd^2. The ratio of surface to volume for a sphere = $\pi d^2 \div \pi d^3/6$ or $6/d$. The curve in B is a plot of this relationship. While spherical shape is more characteristic of animal cells and a cube of plant cells, both geometrical shapes have comparable surface to volume relationships which affect the physical limits of cell size.

Fig. 2-20. Diagram of surface-volume relationships in cells. Cubes and spheres share the same mathematical relationship of changing surface to volume ratio as

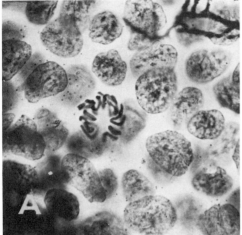

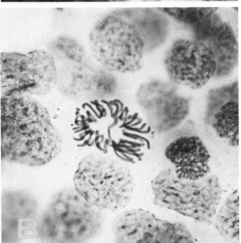

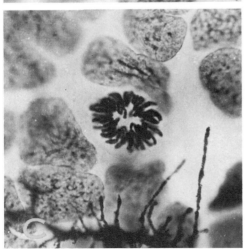

The Events of Mitosis

Interphase

Interphase is the period between divisions, and is the phase in which cells spend most of their lives. The nucleus is surrounded by a membrane within which only a finely granular material plus a nucleolus are visible. In adult animal cells the cytoplasm may be full of some special product of the cell's synthetic activity, perhaps muscle fibers or animal starch, in which case cell division is difficult or impossible and interphase is permanent. Differentiated plant cells often have elaborately thickened cell walls and very special circumstances are required for such cells to divide.

It has now been firmly demonstrated that the duplication of chromosomes takes place during interphase and occurs by the building of a duplicate chromosome (chromatid) alongside the parent chromosome. This conclusion has been reached by feeding cells radioactive thymidine, which is taken up and incorporated into the DNA of chromosomes. All the radioactivity appears on one and only one of each pair of chromatids. If chromosomes reproduced by enlarging and then splitting down the middle, the radioactivity would be found randomly placed in both members of the resulting chromatid pairs (Fig. 2-5).

Prophase

Prophase is the preliminary stage of division, during which the nuclear membrane disappears and the chromosomes become visible, first as long, thin threads and then finally as short, thick threads or rods. It is because of the thread-like appearance of the chromosomes that cell division was named mitosis. Chromosomes in late prophase and subsequent stages of mitosis can be clearly seen in living cells with a

Fig. 2-21. Relationship of the number of chromosomes to cell size in tail fin of a salamander. A, monoploid (haploid) cells; B, diploid cells; C, triploid cells. Note a metaphase group of chromosomes in each case. The dark irregular branching structures are melanin pigment cells. (From G. B. Moment: *General Zoology*, 2nd ed. Houghton Mifflin Co., Boston, 1967. Courtesy G. Frankhauser.)

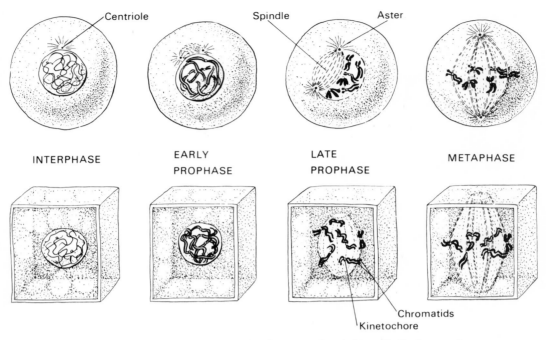

Fig. 2-22. Stages of mitosis in animal cells (top row) and plant cells (bottom row).

phase contrast microscope. In preserved cells they are easily stained with dyes like hematoxylin or various aniline stains. The fact that they can be stained gave them their name (*chroma,* color, + *soma,* body).

It was learned early in the study of cells that the number of chromosomes in every cell of a given species is the same, and it was soon realized that each cell normally contains two similar sets of chromosomes. This means that each chromosome has its own permanent individuality. If the chromosomes in a set from one cell are lined up from the longest chromosome to the shortest and compared with the set from any other cell, provided that it is from the same species, this one-to-one correspondence becomes very evident. In *Drosophila,* the little fruit fly widely used in genetic research, there is always one long chromosome in a set, two middle sized chromosomes, and one short one. Such an inventory of the chromosomes of an organism is known as its **karyotype.**

A conspicuous feature of a chromosome is its **kinetochore** (sometimes called **centromere**). This is a small rounded region that stains differently from the rest of the chromosome and appears to be the kinetic or controlling point in chromosomal movements within the cell. It may be in the middle of a chromosome, near one end, or at some intermediate point. During cell division the kinetochores become attached to the **spindle fibers,** *i.e.,* the microtubules which extend from the poles of the mitotic spindle. It has been well established that the hereditary factors are lined up on the chromosomes in single file and in fixed order.

During early prophase the chromosomes appear as very thin threads which gradually shorten and thicken. In well-preserved material it is easy to see that each chromosome is really double along its entire length. Each member of such a pair is called a **sister chromatid** (although it would be more accurate to call them mother and daughter), and each pair is held together by a single kinetochore.

At the same time that the chromosomes are condensing, the nuclear membrane disappears. In animal cells the centrioles, described earlier along with cilia, duplicate and move toward opposite sides of the cell to form the astral rays of microtubules and a spindle-shaped structure on which the chromosomes become attached at their kinetochores. The star-shaped astral rays are well developed only in large cells such as

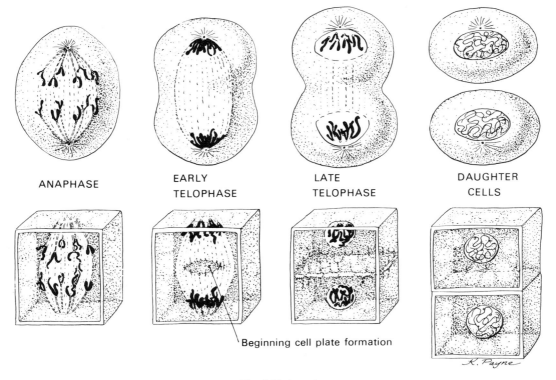

ANAPHASE   EARLY   LATE   DAUGHTER
                 TELOPHASE   TELOPHASE   CELLS

Beginning cell plate formation

K. Payne

Fig. 2-22 (cont.)

those of cleaving eggs. In most plant cells, centrioles and astral rays are absent but microtubules forming the spindle are present.

Metaphase

Following prophase is metaphase, the period in mitosis when the chromosomes have moved to the equator of the spindle. Actually it is only the kinetochore of each chromosome that is on the equator. The rest of the chromosome may dangle at any angle. Each chromosome is still double along its entire length.

Anaphase

The doubling of the kinetochores signals the end of metaphase and the beginning of anaphase. Each of the double chromosomes separates, one chromatid going to each pole of the spindle. The action requires but a few minutes. As the chromosomes move toward the poles, their kinetochores lead the way. It looks as though the kinetochore were being pulled by a spindle fiber. In animal cells, the division of the cytoplasm is initiated during anaphase when the surface membrane begins to constrict between the two poles of the spindle at the level of the equator.

Telophase

In the final phase of mitosis, known as telophase, the cell is divided into two daughter cells. The important point is that as a result of mitosis, each of the two daughter cells has two sets of chromosomes, just as the original cell had. During telophase the nuclear membrane reappears, and the chromosomes lose their stainability as they return to the interphase condition.

In plant cells, **cytokinesis**, or division of the cytoplasm, usually begins at the end of nuclear division. In late telophase droplets appear at the center and later toward the edges of the equatorial plate. Sometimes these pectic substances appear as thickenings of the spindle fibers. The droplets coalesce and form a **cell plate** which divides the protoplasm into two daughter cells. The cell plate later becomes an

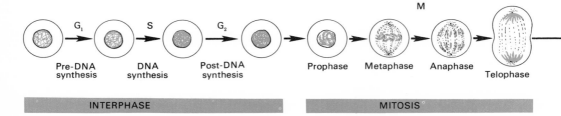

Fig. 2-23. General cell division cycle.

intercellular structure, the **middle lamella**. Plasma membranes are formed at the edge of the daughter protoplasts and become continuous with the remaining segments of plasma membrane of the original cell. The cytoplasm secretes cellulose cell wall material and later, after cell enlargement, additional cellulose is secreted to form secondary cell wall layers.

Another Way of Viewing the Mitotic Cycle

Many investigators of cell division now divide the mitotic cycle into four divisions: M (mitosis), with prophase, metaphase, anaphase, and telophase; G_1, the first gap or growth period; S, the period during which DNA is being synthesized; and G_2, a second gap or growth period. The events occurring and the relative amount of time spent in each period can be learned by feeding a radioactively labeled precursor of DNA such as tritiated thymine and then recording the stage of the cells when it has been incorporated (Table 2-2).

It has been found that the susceptibility of

Table 2-2
Approximate duration of stages in cell division for several organisms

Stage of Cell Cycle	Duration of Stages in Cell Division in			
	Mouse fibro- blasts	Chinese hamster fibroblasts	Pea root tip	*Vicia faba* bean
G_1	2.7[a]	9.0	2.66	5.0
S	5.8	10.0	4.0	7.5
G_2	2,1	2.3	2.5	5.0
M	0.4	0.7	3.0	2.0

[a] Values given in hours.

cells to radiation damage varies markedly with the stage of this cycle. Mitosis is delayed when cells receive relatively small doses of radiation during the G_2 stage. Also, radiation during the S phase can partially or completely block DNA synthesis, although the G_1 and S phases are relatively less sensitive to radiation damage than G_2.

Meiosis

When the facts of mitosis and fertilization became known, particularly that each egg and each sperm carries chromosomes, and that fertilization is the fusion of two such cells, August Weismann (1834-1914) made an important prediction. He foresaw that a stage would be found in the life history of every animal when the number of chromosomes is reduced from the double set or diploid condition to the single set or monoploid (often called haploid) condition. Were this not true, he pointed out, the number of chromosomes would double with each fertilization, *i.e.*, with every generation. This generalization applies to plants as well. There are various times and places where this process, known as meiosis, could conceivably take place. Weismann was nearly blind when he made this prediction and could not himself discover where meiosis does in fact occur. Others investigated this problem and found the following facts.

In animals, meiosis takes place in testes or ovaries during **gametogenesis,** that is, during the formation of the gametes, whether sperms or eggs. This is true of all animals. In plants it is difficult to generalize about where and when meiosis takes place. In the seed plants, the monoploid nuclei are formed in special cells

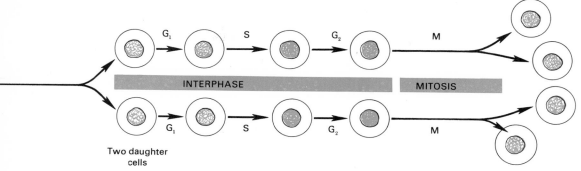

Fig. 2-23 (cont.)

within the male and female parts of the cone or flower. The end products are pollen grains and a cell within the ovary which develops into the embryo sac. Nuclear divisions occur both in pollen grains and the developing embryo sac and both become multinucleate. The monoploid nuclei of the pollen grains and the embryo sac are involved in fertilization.

In all animals and plants meiosis requires two cell divisions. This is just as true between the gills of a mushroom or in a one-celled protozoan as it is in the gonads of a mouse or a man. Neither of these two cell divisions is a normal mitosis, although the cell goes through prophase, metaphase, anaphase, and telophase.

Discussion of the details of meiosis will be postponed until the chapter on genetics, where a knowledge of meiosis is essential for an understanding of the laws of inheritance.

SOCIAL RELATIONS OF CELLS

So far cells have been considered as isolated individuals, but for a multicellular organism one of the most important properties of cells is the ability of individual cells to unite firmly into tissues. In some way cells are able to recognize each other so that they can distinguish not only between themselves and the cells of other species but between cells of different histological types. Thus muscle cells will adhere firmly with other muscle cells, liver cells with liver cells, so that coherent tissues, muscle or liver, are formed.

The instructions which enable cells to do all this are carried by each individual cell, and presumably the cell membrane. Although little is known about the mechanism, a beginning has been made. Many years ago, H. V. Wilson of the University of North Carolina discovered that the scarlet encrusting sponge, *Microciona*, which commonly grows over rocks and oyster

Fig. 2-24. August Weismann, the late 19th and early 20th-century German zoologist who predicted that meiosis would be discovered and held that the organism is the expression of the hereditary determinants in the germ cells and not of the effects of use and disuse. (Courtesy Johns Hopkins Institute of the History or Medicine.)

shells on the Atlantic coast, could be squeezed through very fine silk cloth and the individual cells would be separated but not killed. Such dispersed cells, if permitted to settle in a jar of sea water, will move together and reconstitute minute sponges.

For this behavior to occur, cells must have certain properties. They must be discrete, motile, and possess differential mutual adhesiveness. Obviously, however, there must be much more to the story to account for the great specificity of this behavior. Proteins are known to possess great specificity, being uniquely characteristic of different species and of different tissues within each species. This is clearly shown in the well-known antigen-antibody reactions of immunology. Thus it would seem reasonable to suppose that in some way the proteins of the cell membranes are involved.

Of current medical interest are the problems associated with rejection of transplanted organs. All attempts to transplant the human heart appear to be doomed to failure at this time not because this is a difficult surgical procedure (which it is) but rather because of later complications resulting from the production of antibodies by the recipient. The difficulty lies in the recognition of the transplanted organ by the recipient's immune response system as "non-self" and the resultant production of antibodies. Only when this mechanism, which allows us to fight off foreign disease-causing organisms, is controlled will it be possible to prevent the rejection of transplanted organs.

USEFUL REFERENCES

Buvat, R. *Plant Cells: An Introduction to Plant Protoplasm.* McGraw-Hill Book Co., New York, 1969.

Cohen, S. C. *Are/Were Mitochondria and Chloroplasts Microorganisms?* American Scientist, *58(3):* 281-289, 1970.

DeRobertis, E. D. P., Nowinski, W. W., and Saez, F. A. *General Cytology,* 5th ed. W. B. Saunders Co., Philadelphia, 1970.

Freeman, J. A. *Cellular Fine Structure: An Introductory Student Text and Atlas.* McGraw-Hill Book Co., New York, 1964.

Hendler, R. W. *Protein Synthesis and Membrane Biochemistry.* John Wiley & Sons, Inc., New York, 1968.

Jensen, W. A. *The Plant Cell.* Wadsworth Publishing Co., Inc., Belmont, Calif., 1964.

Jensen, W. A., and Park, R. B. *Cell Ultrastructure.* Wadsworth Publishing Co., Inc., Belmont, Calif., 1967.

Kennedy, D. *The Living Cell: Readings from the Scientific American.* W. H. Freeman and Co., San Francisco, 1965.

Ledbetter, M. C., and Porter, K. R. *Introduction to the Fine Structure of Plant Cells.* Springer-Verlag, New York, 1970.

Swanson, C. P. *The Cell,* 3rd ed. Prentice-Hall Inc., Englewood Cliffs, N. J., 1969.

Wilson, G. B., and Morrison, J. H. *Cytology,* 2nd ed. Reinhold Publishing Corp., New York, 1966.

Molecules and Energy: The Stuff of Life

Life has often been compared to a flame. This ancient analogy is useful because it points to the central fact that life is a steady state that maintains essential constancy of form amid a continually changing flow of matter and energy. However, the flame of life is a very unusual flame, highly structured and complex. In Chapter 2 we considered these complexities as they appear on the cellular level of organization. We will now turn to the key events on the biochemical level, where the matter and energy changes take place.

Modern knowledge of the chemistry of life is both extensive and penetrating, reaching into every dimension of life and to fundamental processes within the cell once thought to be forever beyond human understanding. Such knowledge is also essential for anything but the most superficial comprehension of the ecology of a pond, a forest, or a whole industrial civilization. Consequently, unless some biochemistry is built into our thinking about biological problems of any kind, we will be condemned to remain intellectually stalled in the 1930's or even earlier.

Before beginning a systematic study of the chemistry of life, one should recall from previous chapters some of the areas of human concern where biochemical knowledge is of importance. We can never understand or learn to cope with the intricacies of human behavior without additional information about the chemistry of hormone and drug effects. The same holds true for environmental pollutants. What makes them so terribly dangerous is that contaminants like mercury and DDT have profound effects on enzymes and therefore on the functioning of cells. An understanding of life on the biochemical level requires some knowledge of the relationship of living things to the subatomic and the atomic levels of organization.

THE SUBATOMIC LEVEL

The outstanding fact about events on the subatomic level of electrons, protons, photons, and other wave particles is that they do not seem to follow the rules which hold for the physics and chemistry of our familiar macroscopic world. This peculiar behavior of particles on the subatomic level led to the enunciation of the **quantum theory** by Max Planck in 1900, which states that energy is emitted or absorbed only in discrete units called **quanta**. On the subatomic level, prediction appears to be possible only statistically. Here Heisenberg's **principle of indeterminacy** is applicable. It states that the more accurately you determine the position of a particle the less accurately you can determine its velocity and *vice versa*.

The pertinent question for the biologist is what roles, if any, do events on the subatomic level play in the lives of plants and animals? Clearly quantum events are involved in all aspects of radiation biology from the effects of X-rays in producing mutations to the manufacture of sugar in photosynthesis. A few philosophically minded people think that the indeterminacy principle is important in the functioning of the higher nervous centers. Each reader will have to make up his own mind on this controversial issue but should do so only after studying the nervous system itself.

Electrons are perhaps the best examples of subatomic particles that play a vital role in biological processes, specifically in oxidation-reduction reactions. The loss of electrons is termed **oxidation** and the gain of electrons, **reduction**. Thus, if a substance is an electron donor, it is called a **reducing agent**. If a substance is an electron acceptor, it is an **oxidizing agent**. The emissions of radioactive materials which are so very useful in biological tracer studies in living organisms are all subatomic. Alpha particles are the nuclei of helium atoms (two protons and two neutrons), beta particles are fast moving electrons, and gamma rays are quanta of higher frequency and shorter wavelength than X-rays.

THE ATOMIC LEVEL

On the atomic level, a number of questions—some old, some new—arise. Are living organisms composed of the same chemical elements that are found in the non-living world of rocks and rivers, oceans and clouds, and even the sun and other planets? If we are so made, which elements are involved? Most of the 100-odd or but a few?

The answer to the first question has been clear for many decades. If a plant or animal is killed, dried, and the water that evaporates then measured, it will be found that a very large portion of living matter is water and that the hydrogen and oxygen in this water are the same as any other hydrogen and oxygen. The dried corpse can be burned. Carbon dioxide and water vapor will be given off, revealing a third element, carbon. The remaining ash, on chemi-

cal analysis, yields the following elements in decreasing order of amount: nitrogen, calcium, phosphorus, potassium, sodium, sulfur, chlorine, magnesium, and iron, plus several more such as copper and manganese. It is worth noting that the list represents only a handful of the known elements. Why? No one really knows. The big four as far as quantity goes are oxygen, carbon, hydrogen, and nitrogen, but phosphorus is now known to be so important in both energy-carrying compounds and in nucleic acids that one should really speak of the big five.

The most interesting as well as most important problem in this field today concerns what are called trace elements, or better, **micronutrients**. These are elements essential for survival but required only in minute amounts. For animals, some essential trace elements have been known for many years. Iodine, the lack of which produces goiter, is one. That traces of cobalt are necessary was discovered as a result of an attack of "bush disease" in sheep on Australian ranges in 1895. Since then other micronutrients have been discovered: copper, zinc, and manganese. As seems reasonable for material essential only in minute amounts, these elements are either part of enzyme molecules or behave as catalysts for enzyme action. Copper, for example, is part of the enzyme cytochrome oxidase which is involved

Table 3-1
*Comparison of mass, charge, and penetrating power of naturally occurring
alpha, beta, and gamma radiation and of neutrons*

Kind of Radiation	Nature	Mass	Charge	Penetrating Power and Shielding Needed to Block Radiation
Alpha	Particle composed of 2 protons and 2 neutrons; equivalent to helium nucleus	4	2+	Alpha particles travel only a few centimeters in air; can be stopped by a sheet of paper or skin tissues
Beta	Electron	1/1820	1−	Beta particles travel only about 1 m. in air; can be stopped by several centimeters of wood or a few millimeters of aluminum
Gamma	Energy wave that is part of the electromagnetic spectrum; essentially the same as X radiation	0	None	Gamma rays can travel hundreds of meters in air; several meters of concrete or thick lead shielding needed to block radiation
Neutrons	Occur with protons in nuclei of atoms; not produced by fission of naturally occurring radioisotopes; important product of nuclear reactors	1	None	Neutrons can travel more than 100 m. in air; can be stopped by a meter of water or concrete

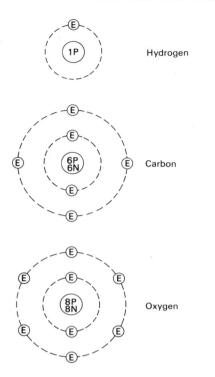

Fig. 3-1. Structure of hydrogen, carbon, and oxygen atoms.

THE MOLECULAR LEVEL— ORGANIC COMPOUNDS

On the molecular level living organisms are characterized by **organic compounds**, or molecules containing the element carbon. Until about a century ago it was believed that all compounds containing carbon could be synthesized only by some unanalyzable vital force in living organisms; hence, the names organic compounds and organic chemistry. This misconception was corrected by a 28-year-old German chemist, Friedrich Wohler. He had no intention of producing an organic compound in his laboratory but he did so in the course of trying to synthesize something else, so his discovery is one of the most important instances of serendipity on record. The substance was urea, $NH_2-CO-NH_2$, a colorless, tasteless, odorless compound which is a waste product of protein breakdown. Of course Wohler did not produce urea the way animals do, but that is not the point. He had killed the old myth and convinced the most skeptical chemists that organic chemicals can be analyzed, synthesized, and understood. Since then enormous numbers of carbon compounds have been synthesized: dyes, drugs, explosives, nylon, hormones, and even nucleic acids.

The carbon atom has a number of properties that make it biologically important. Each carbon atom has four covalent bonds, chemical "links" that enable it to unite with four other atoms by sharing electrons with them. However, it may unite with fewer than four atoms by using two of its bonds to link to another. This is a double bond. The covalent bonds are determined by the number of electrons in the outermost orbit around the nucleus of the carbon atom. They are represented in written formulas by short bars (Fig. 3-2). Each atom in a molecule is represented by the letter symbol for its name: H is hydrogen, C is carbon, Cl is chlorine, and so on. An example of a familiar carbon compound is methane (CH_4) or marsh gas, a waste product of bacteria living in decaying debris on the bottom of quiet ponds and often seen bubbling to the surface. Carbon tetrachloride has a similar structure except that in place of four hydrogen atoms, four chlorine atoms are attached to the carbon. Carbon tetrachloride has been widely used in fire

in oxidative respiration in the mitochondria. This entire subject is one of great practical importance about which much remains to be learned.

The structure of an individual atom bears an interesting resemblance to a minute solar system. There is a central nucleus consisting of a tightly packed group of protons and neutrons having a net positive charge. Revolving around the nucleus are one or more electrons carrying a negative charge and of negligible weight compared with the nucleus. The simplest atom, hydrogen, consists of one proton and one electron in orbit around it. A carbon atom consists of six protons and six neutrons with two electrons in an inner orbit and four in an outer. Oxygen atoms have eight protons and eight neutrons and two electrons in the inner orbit and six in the outer. The chemical properties of an atom are determined very largely by the number of electrons in its outermost orbit. These are the ones involved in valence and shared in covalent bonds.

extinguishers and as a cleaning fluid, although it is highly poisonous. Substituting a hydrogen atom for one of the chlorine atoms produces chloroform.

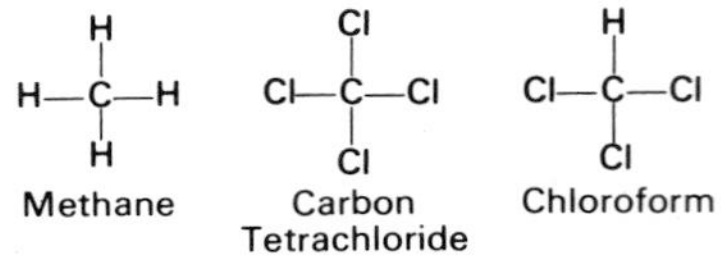

Fig. 3-2. Structure of methane, carbon tetrachloride, and chloroform molecules.

Carbon atoms also have the ability to unite with each other and in this way make extremely large molecules. The carbons may join in short or long single chains, in rings as in benzene, or in many other configurations. Familiar examples include glycerol (also called glycerine), a sweetish lubricant that can be obtained from fats and oils, and ethylene, a sweetish, colorless, highly explosive component of coal gas and a great menace to miners. This compound is also produced by plants and its role in fruit ripening and leaf fall is discussed in Chapter 17. Benzene is obtained from coal tar and is used as a solvent and cleaning fluid and also as the chemical base for the synthesis of thousands of other substances, from dyes to explosives (see formulas).

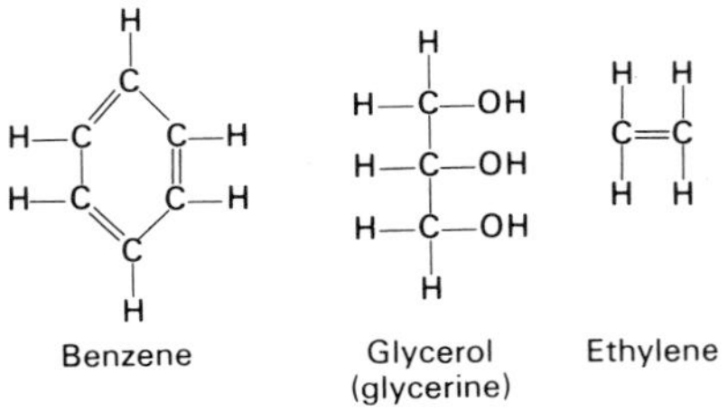

Fig. 3-3. Structure of benzene, glycerol, and ethylene molecules.

The existence of functional groups is of considerable biochemical importance. Functional groups or radicals behave as units in chemical reactions. One of the most familiar is the carboxyl group, COOH, which is also written

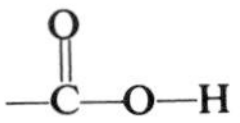

This group may look like a base, but in water it is the hydrogen which is released so that it is in fact acidic. The carboxyl group is present in amino acids and fatty acids. The amino group, $-NH_2$, is basic and is found in amino acids and other important compounds. It behaves like ammonia, which forms ammonium hydroxide with water: $NH_3 + H_2O \rightarrow NH_4OH$. Thus, the amino group is alkaline (Fig. 3-4). Other important groups are the methyl group, the highly reactive keto, and aldehyde groups:

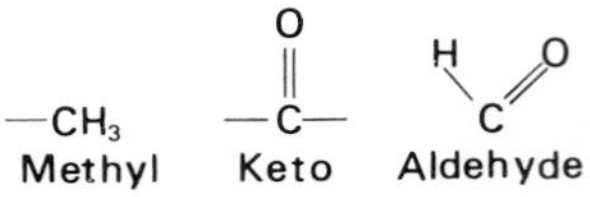

CARBOHYDRATES—THE BUILDING BLOCKS AND FUEL OF LIFE

The familiar sugars and starches are **carbohydrates**. A simple sugar is composed solely of carbon, hydrogen, and oxygen, e.g., $C_6H_{12}O_6$, glucose. Chemically, a carbohydrate often is defined as a simple sugar or a substance which yields simple sugars by hydrolysis, that is, by splitting the compound with the addition of water. Carbohydrates may also be defined as those substances with the general formula $C_x(H_2O)_y$. The values of x and y may range from 3 to several thousand.

If the values of x and y are low, between 3 and 7, the sugar is called a **monosaccharide**. Two monosaccharides linked together comprise

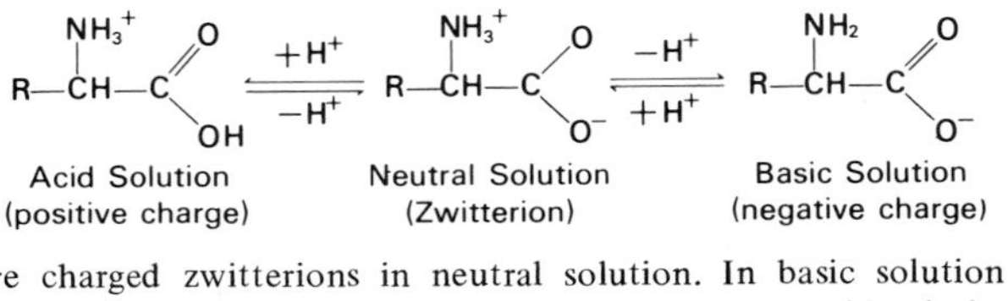

Fig. 3-4. Amino acids are charged zwitterions in neutral solution. In basic solution the positively charged ammonium ion loses a proton to become the uncharged amino group; in acid solution the negatively charged carboxylate ion gains a proton to become the uncharged carboxyl group.

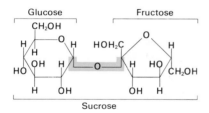

Fig. 3-5. The two monosaccharides, glucose and fructose, joined by a glycoside linkage (green) form the dissacharide, sucrose.

a **disaccharide**, *e.g.,* sucrose, our common table sugar. More than two monosaccharides joined together form a **polysaccharide**. The link that holds one sugar molecule to another is known as a glycoside linkage (Fig. 3-5). Such joining of similar units is called **polymerization,** an extremely important process that we will meet frequently since it is the way proteins and DNA, as well as the whole array of carbohydrate-related materials, are made.

Monosaccharides, *i.e.,* simple sugars with three carbons $(C_3H_6O_3)$, are called triose sugars or merely **trioses**. Those with four carbons are **tetroses**; with five, **pentoses**; and with six, **hexoses**. Glucose, a simple hexose, $C_6H_{12}O_6$, is a highly important sugar because it is the structural unit from which many more complex sugars as well as the starches and cellulose are built.

Another extremely important sugar is a pentose called ribose and the related sugar deoxyribose; deoxy- because it has less oxygen than ribose. While ribose fits into the general formula, $C_x(H_2O)_y$ (*i.e.,* $C_5H_{10}O_5$), deoxyribose $(C_5H_{10}O_4)$ does not. Both of these five-carbon sugars are important constituents of nucleic acids: deoxyribose in the DNA of the chromosomes, and ribose in the RNA of the ribosomes and other components of a cell's machinery for making proteins.

Sugars, especially glucose, are commonly and correctly called the "fuels of life," but it would be incorrect to think that they are only fuels. As just mentioned, the pentoses ribose and deoxyribose form essential parts of nucleic acid molecules. Sugars in general have a marked ability to combine with other molecules and other sugars. The combination of a sugar and a nonsugar is called a **glycoside**, no matter what kind of sugar is involved. If the sugar is glucose,

the term glucoside can be used. Many glycosides have a strong action on the heart. The non-sugar part of these molecules is usually a steroid similar to the sex and adrenal hormones. An exception to this generalization is digitalis, the best known glycoside, where the non-sugar part is digitonin.

The non-sugar part of a glycoside may be an amino acid or simply an amino group, $-NH_2$. It may be recalled that proteins are built up of amino acid chains. An amino-containing glycoside is called a glucosamine (if the sugar is glucose) or simply an **amino sugar**. Amino sugars are polymerized, *i.e.,* united into long chains, to form chitin. This is the highly protective skeletal material of insects, lobsters, and related animals. Interestingly enough, chitin is present in the cell walls of certain molds. Glucosamines are also important in the connective tissues of vertebrates.

Sugar molecules are asymmetrical, *i.e.,* they can exist in two forms that are mirror images like right and left gloves or shoes. This remarkable property of carbohydrates is also characteristic of amino acids. The chemical behavior of the molecule depends not only on the kind and number of atoms in it but on their arrangement (configuration). One of Pasteur's first discoveries, made when he was a very young man, was that simple organic compounds like sugar (he was actually working with a simple organic acid, tartaric) are asymmetrical and that two kinds of crystals can be sorted out under a microscope. As working with model atoms will show convincingly, such asymmetry is inevitable whenever a carbon atom is at-

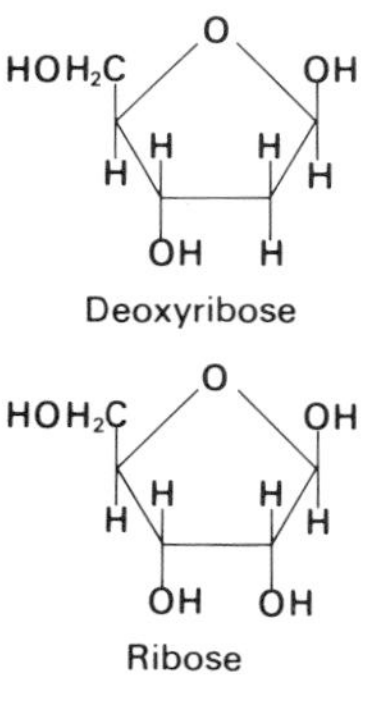

Fig. 3-6. Structural formulas of deoxyribose and ribose, important constituents of nucleic acids.

tached to four atoms that are each different. Such pairs of molecules are called **isomers**. Since they will rotate the plane of polarized light in a clockwise or counterclockwise direction according to their asymmetry, they are commonly called dextro- or levorotary optical isomers. Such isomers are extremely difficult for a chemist to separate but cells do so readily. Virtually all naturally occurring sugars are dextro-sugars, but why this should be is unknown. It may be a mere accident of evolution; it may have some important significance. In any case, digestive enzymes can digest dextro- but not levo-sugars.

By more polymerization, that is, by sugars being linked to other sugars and still other compounds, a vast array of important substances are produced by plants and animals. What is the significance of all this polymerization? First, the process of forming glycoside linkages removes water (dehydration synthesis) which means less oxygen, and that means that the molecule is left richer in energy. Second, these big polymers have biologically useful properties that simple sugars lack, *e.g.*, great tensile strength in fibers, protective qualities in gums and waxes, and lubricating ability in various sorts of mucus. Third, making one large molecule from many small ones (as in making one molecule of starch from many molecules of glucose) reduces the osmotic concentration of the cell in which this polymerization occurs. Ordinarily, green plant cells carrying out photosynthesis rapidly convert their surplus sugars into starch.

Starch and **glycogen**, which is often called animal starch, are two of the more familiar simple polysaccharides. Glycogen is found not only in the vertebrate liver and the cytoplasm of protozoa, but also in blue-green algae and bacteria, so it is produced by both prokaryotes and eukaryotes and is presumably of very ancient origin. Plant starches such as those from potatoes, rice, etc., differ from glycogen. Glycogen molecules are composed of many-branched chains of glucose while plant starch molecules consist of both linear and branched portions. Cellulose, a principal component of wood and plant fibers, is a highly indigestible and apparently unbranched glucose polymer. Many of the gums which are produced by plants after injury are heteropolysaccharides. So also is the ground substance of cartilage, chondroitin, and hyaluronic acid, the "glue" which holds animal cells together. Heparin, a powerful anticoagulant found in the liver and other vertebrate organs, is a mucopolysaccharide.

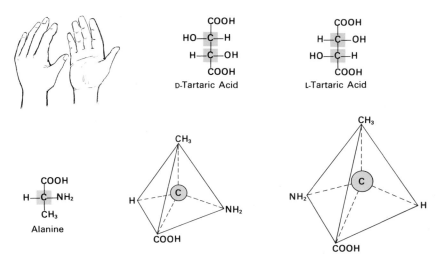

Fig. 3-7. Stereoisomers have the same structural relationship as right and left hands or as an object and its mirror image. Asymmetric carbons in structural formulas are indicated in green.

LIPIDS

Lipids are a heterogeneous group including the fats, oils, steroids, waxes, and related compounds. They are so varied that a satisfactory definition is difficult to formulate. In general, lipids are composed of the same three elements as carbohydrates—carbon, hydrogen, and oxygen, although some also contain phosphorus and other elements. Unlike most carbohydrates, lipids are insoluble in water, but are soluble in various organic solvents such as chloroform, ether, and alcohol. Many lipids, including the common animal fats, can be split into two components, glycerol and fatty acids.

The most obvious function of lipids is energy storage. Since lipids generally have less oxygen per molecule, they carry more energy than starches, but it is less easily available. Migratory birds develop large deposits of fat before they begin their long flights. On arrival hundreds or even thousands of miles from their start, the fat is used up. Hibernating animals also begin the winter with large deposits of fat. This is used partly during the winter and partly when they first emerge and food is scarce. Lipids stored in seeds are utilized during the process of germination.

Lipids are located in the middle layer of both plant and animal cell membranes. In many animals, lipids form heat-insulating layers; the blubber of a whale is nearly a foot thick. Certain vitamins are lipoidal in character and are commonly found associated with other lipids. The sex hormones are lipids, as are cortisone and other hormones. The epidermal cells of plants usually have a coating of wax, also a lipid.

Historically, the chemical nature of fats was understood before that of either carbohydrates or proteins. In fact, a rule-of-thumb knowledge of fat chemistry has been used for centuries in soap making. Fats can be broken down into glycerol (glycerine) and a weak organic or fatty acid. No matter what the fat or oil, the glycerol is always the same although the fatty acids are different for each fat or oil. At the acid end of a fatty acid is a carboxyl group, COOH. From one to over a dozen carbon atoms can be attached to this group. Formic acid, secreted by ants and other insects as a chemical weapon, is the simplest possible organic acid, H—COOH.

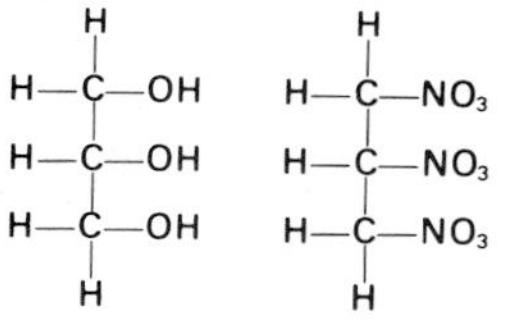

Fig. 3-8. Structure of glycerol and trinitroglycerol.

Acetic acid or vinegar is CH_3—COOH. Stearic acid or beef fat is a veritable train, usually written $C_{17}H_{35}$COOH. Glycerol is a three-carbon compound (see formula). If the three OH groups are removed and NO_3 radicals substituted, the result is trinitroglycerine, the explosive component of dynamite.

The basic reaction by which glycerol and fatty acid are formed from a fat is by splitting the fat with a molecule of water. This process is called **hydrolysis** (*hydro,* water, + *lysis,* to loosen) or hydrolytic splitting. It is one example of digestion, for hydrolysis is what happens to carbohydrates and proteins as well as fats in the digestive tract. The reverse process, by which two chemical substances are combined, is much more complicated and involves a series of steps.

Living things can convert carbohydrates into fats, some of us all too easily. In animal diets, at least two lipids, linoleic and arachidonic acids, are essential. These two organic acids are found in some but not all fats. Fortunately no one needs to worry about this, since it would be difficult to find a diet totally lacking them.

When every carbon in the fatty acid chain is holding two hydrogens the fat or oil is said to be **saturated.** When two adjacent carbons are linked by a double bond, the lipid is said to be **unsaturated.** Butyric and crotonic acids illustrate this contrast. The structural formulas of these two acids are shown in Figure 3-9.

The melting point, which determines whether a lipid is an oil or a solid fat at any given temperature, depends on two factors. The shorter the chain of fatty acids, and the greater their unsaturation, the lower the melting point. Unsaturated fats are thought to be much less easily converted into cholesterol than saturated fats. Although cholesterol is heavily implicated in the cause of heart and other vascular diseases, it should be remembered that all the

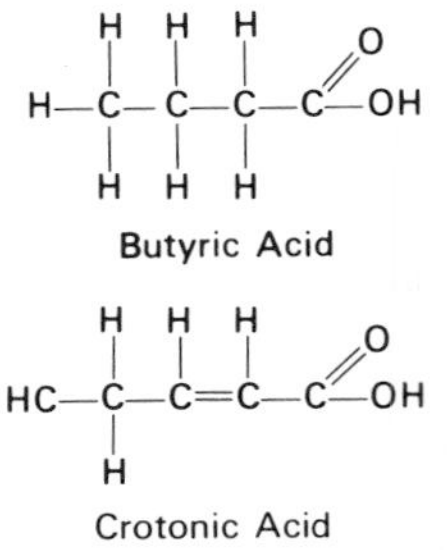

Fig. 3-9. Structure of butyric and crotonic acids.

normally occurring steroid hormones, including the stress hormones of the adrenal gland, adrenaline, as well as cortisone and the like, are derived from cholesterol.

When a lipid has three fatty acid chains, as does stearin or beef fat, it is known as a **triglyceride**. Most of the familiar oils and fats are triglycerides and in most of them not all of the three fatty acids are the same. In fact, each of the three may be different. When a lipid has only two fatty acid chains, it is called a **diglyceride**. These also are important to animals.

We can now take a second look at the lipids in cell membranes. These extremely important lipids are diglycerides that have a phosphate group attached to the glycerol where the third fatty acid might have been. Attached to this phosphate group is usually an additional molecule. In the case of **lecithin**, the most abundant **phosphodiglyceride** in cell membranes, this last group is choline (Fig. 3-10). Choline is a small molecule consisting of a nitrogen atom to which are attached three methyl groups $-CH_3$, plus two carbons with hydrogens. Choline will be discussed again in connection with the conduction of a nerve impulse, which is a membrane phenomenon. The phosphate group of lecithin is an energy-rich group similar to those found in the energy-carrying compounds of respiration. This suggests a role of lecithin in pumping molecules across cell membranes. Thus, like the carbohydrates, the lipids play many roles in the economy of life.

PROTEINS AND AMINO ACIDS

Proteins together with the nucleic acids, DNA and RNA, form the triumvirate of molecular biology. What the genetic code in the nucleic acids spells out are millions of kinds of catalytic proteins, *i.e.,* **enzymes**. Every organism is what its proteins make it because all cellular activities, synthetic, digestive, locomotor, reproductive, and the rest, are either directly controlled by enzymes or by the products of enzymes. The proteins confer species and individual uniqueness. The distinction between self and non-self has its biochemical basis in the proteins characteristic of each organism. The rejection of organ transplants and the identification of blood both depend on proteins. In small amounts they are essential components of the human diet; without them physical and mental deterioration and finally death result.

Many kinds of tissues have high concentrations of protein. Certain seeds such as beans and peanuts are rich sources. A number of animal tissues are constructed largely of protein—especially horns, fingernails, hair, tendons, and feathers. Egg white is almost pure protein and water. Proteins are components of all living cells, giving structure to the cellular membranes as well as constituting important parts of the cellular machinery. Grass contains the proteins which steers build into beef steaks.

Proteins have been recognized for over a century as a distinct group of compounds containing nitrogen in addition to the carbon, oxygen, and hydrogen of the carbohydrates and

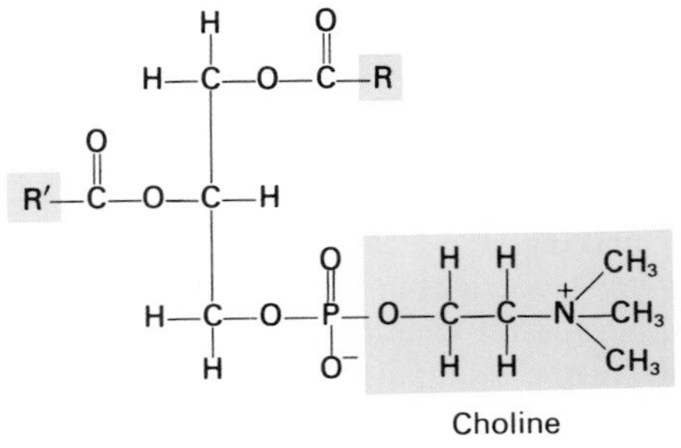

Choline

Fig. 3-10. Structure of lecithin, a phosphodiglyceride with two different fatty acids (R and R'). A phosphate group links choline to the third carbon of glycerol.

lipids. Many proteins also contain small amounts of sulfur and other elements. Over 150 kinds of proteins from a variety of plants and animals have been isolated and purified. Their molecular weights are enormous, ranging from several thousand up to several million. It will be recalled that the molecular weight of water is only 18. All the millions of proteins can be broken down by boiling in acid (or gently by enzymatic digestion) into about two dozen

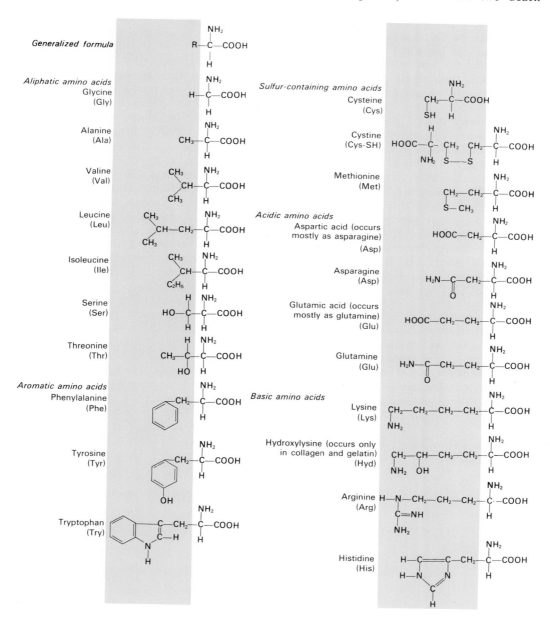

Fig. 3-11. A, generalized formula for an amino acid; B, structures of all naturally occurring amino acids. Green indicates structure equivalent to R.

amino acids. All the millions of plant and animal proteins are made from this same little group of amino acids. They have been compared to the 26 letters of the alphabet which can spell all the words of all the languages of Europe.

All amino acids have a so-called alpha carbon atom to which are attached four other atoms or groups of atoms. There is always a hydrogen, an acidic carboxyl group, —COOH, a basic amino group, —NH$_2$, plus an additional group designated as R in the model amino acid shown (Fig. 3-11).

In the simplest case possible, glycine, the fourth group or R is merely a hydrogen atom. If R is a methyl group, —CH$_3$, the amino acid is alanine. If a phenol ring is substituted for one of the H's in the methyl group, phenylalanine is the result. Phenylalanine is an extremely important substance because it can be made into the amino acid tyrosine merely by adding an OH group to the phenol ring. Tyrosine is used by cells to synthesize thyroxin (the hormone of the thyroid gland), the hormone adrenaline, and the brown, reddish and black melanin pigments of potatoes, skin, hair, and feathers. Tyrosine also forms part of many proteins.

The nutritional value of a protein, whether from meat, fish, peanuts, corn, or any other source, depends on the amino acids it contains. If an animal cannot make a particular amino acid by modifying other amino acids, its diet must provide it. For men and rats, for example, tryptophan (or indolealanine) is essential in the diet or growth cannot take place. Most plants differ from animals in being able to synthesize all the different amino acids that they need.

About 50 years ago a German chemist, Emil Fischer, who devoted much of his life to the study of proteins, discovered how amino acids are joined together into proteins. One can visualize the process by imagining that an —OH is lost from the carboxyl group of one amino acid and an —H from the amino group of another. A molecule of water is released and the two amino acids are left, linked by a bond between the nitrogen of one amino group and the carbon of the carboxyl group of the other. Note that when two amino acids are thus joined into what is called a dipeptide, the resulting molecule still possesses a carboxyl group at one end and an amino group at the other. Consequently, a dipeptide can unite with another amino acid and form a tripeptide, and this in turn with other amino acids until a long **polypeptide** or, more descriptively, a polymer of amino acids, and finally a protein is formed. The linkage is always the same, passing through the alpha carbon to the carboxyl carbon to the nitrogen to the next alpha carbon. Such a bond is called a **peptide linkage.**

The sequence of amino acids in the chain determines the character of a protein. This sequence, termed the primary structure of a protein, is spelled out by the DNA code in the chromosomes, as will be further clarified in the chapter on genetics. The asymmetrical shapes of the amino acids cause a twisting of the long polypeptide chain, usually into a corkscrew-like helix. This helix is called the **secondary**

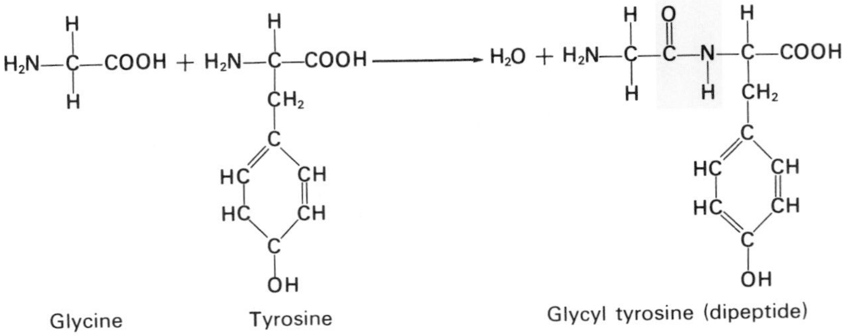

Glycine   Tyrosine   Glycyl tyrosine (dipeptide)

Fig. 3-12. The carboxyl group of glycine and the amino group of tyrosine are involved in the peptide bond (green) joining the two parts of the dipeptide.

structure of a protein. The helical shape is maintained by hydrogen bonds, weak links formed when a hydrogen atom is shared by two other atoms, one of which is usually oxygen. However, few proteins remain as long helical chains. They become folded and bent on themselves in characteristic ways. This **tertiary structure** of a protein is determined by several factors. One of these is the disulfide bonds (−S−S−) that form between the sulfhydryl (−SH) groups of two sulfur-containing amino acids. Since the position of the sulfur-containing amino acids is determined by the genetic code, the folding of any particular kind of protein is determined by heredity, *i.e.*, the structure of DNA.

The primary structure of a protein, *i.e.*, the actual sequence of amino acids, was determined for the first time by Frederick Sanger in Cambridge, England. He used insulin partly because its molecules are relatively small for a protein that can be obtained commercially, and partly because of its great medical importance. The method is both laborious and tricky. It consists of breaking up the molecule by several different methods into pieces of different overlapping lengths, determining the amino acids in each piece, and then finally fitting all the data together into a consistent picture (Fig. 3-13).

The importance of the correct amino acids in the correct sequence has been dramatically shown in sickle cell anemia (see Chapter 7). In this disease the hemoglobin is abnormal; there is a single error in each of the four chains which together make up a hemoglobin molecule. In place of one of the valines is glutamic acid. This is one error out of about 300 amino acids. A consequence of this small change in primary structure is a change in the shape of the red blood cells and in their oxygen-carrying properties.

NUCLEIC ACIDS

The most famous case of an important biological discovery that was neglected for decades is probably Mendel's discovery of the laws of heredity. The discovery of nucleic acids,

which constitute the physical basis of those laws, is another. This emphasizes how difficult

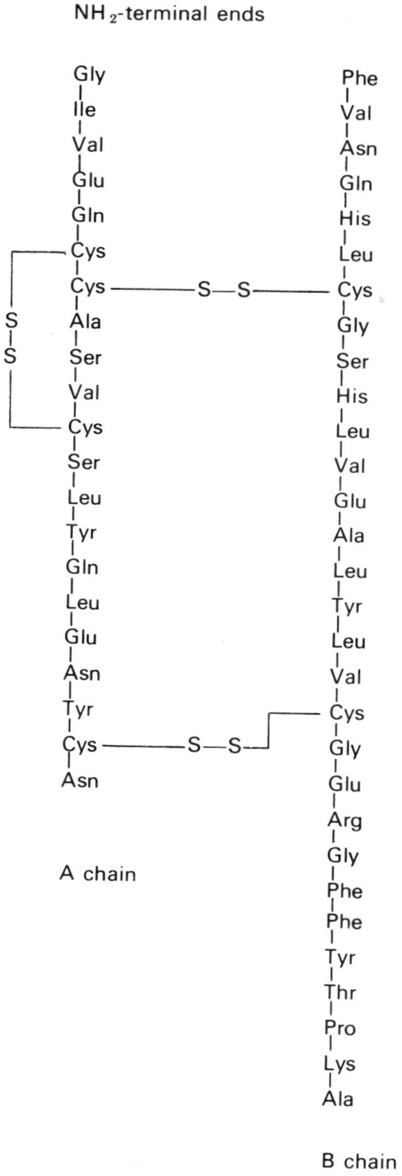

Fig. 3-13. Amino acid sequence of bovine insulin showing disulfide cross linkages. Abbreviations for amino acids are listed in Fig. 3-11.

58 BASIC CONCEPTS

it is to recognize which scientific discoveries are truly important.

An understanding of the molecular structure of nucleic acids, and proof that this structure carries the genetic code from cell to cell and from generation to generation, was achieved within the past two decades. However, a century has passed since Friedrich Miescher began to explore the biochemistry of nuclei. The experimental materials for these early studies were the white blood cells of pus (obtained from wounded soldiers), which have large nuclei (lymphocytes almost lack cytoplasm altogether) and also the heads of salmon sperms which consist of little more than condensed nuclei. Miescher discovered that nuclei were composed almost entirely of an acid (later called nucleic acid). It was unusual in containing not only nitrogen but also phosphorus, a fact so peculiar that his major professor refused to allow him to publish his discovery until after it had been very carefully confirmed more than two years later.

From the 1870's until the 1920's little work was done on nucleic acids. In 1924 Feulgen and others found that the chromosomes could be stained a brilliant red after proper treatment of the deoxyribose located there. This reaction not only afforded a very elegant method for staining but it demonstrated the presence of deoxyribose along the entire length of every chromosome.

Analysis by a variety of chemical and physical methods has now revealed the actual structure of nucleic acids and shown clearly that there are two kinds, deoxyribonucleic acid (DNA) and ribonucleic acid (RNA). DNA is found associated with proteins in the chromosomes. RNA is found in both the nucleus and the cytoplasm. Nucleic acids resemble proteins in that both types of compounds are polymers,

i.e., long chains of small molecules—nucleotides and amino acids, respectively. Each nucleotide consists of a sugar (ribose or deoxyribose), a phosphoric group (PO_4H), and a nitrogen-containing base.

There are only four kinds of nitrogen-containing bases in DNA and also only four common ones in RNA; three of these occur in both. In DNA, two of the bases are purines (adenine and guanine) and two are pyrimidines (cytosine and thymine). In RNA, uracil substitutes for thymine. This fact offers a useful research tool. Radioactively labeled thymine will appear in the DNA in the chromosomes. Labeled uracil will appear in RNA.

When the nucleotides of DNA are linked together, they form two chains with links between them, and resemble a ladder twisted like the stripes on a barber pole. The sides of the ladder are built of alternating phosphoric acid and ribose sugar units. The rungs of the ladder are made of the nitrogen bases, two bases in each rung, one purine and one pyrimidine. In DNA, adenine is always paired with thymine, and guanine with cytosine. The rungs are attached to the uprights of the helical

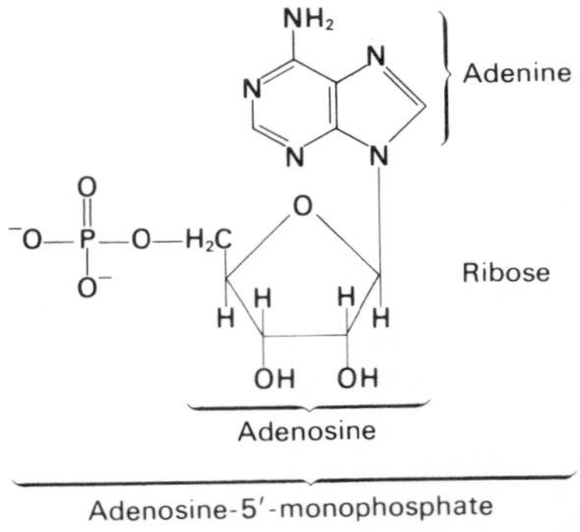

Fig. 3-14. Structure of adenosine-5'-monophosphate, a nucleotide.

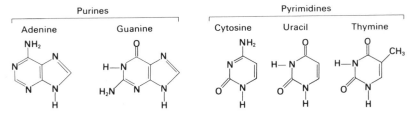

Fig. 3-15. Structures of the nitrogen-containing bases in DNA and RNA.

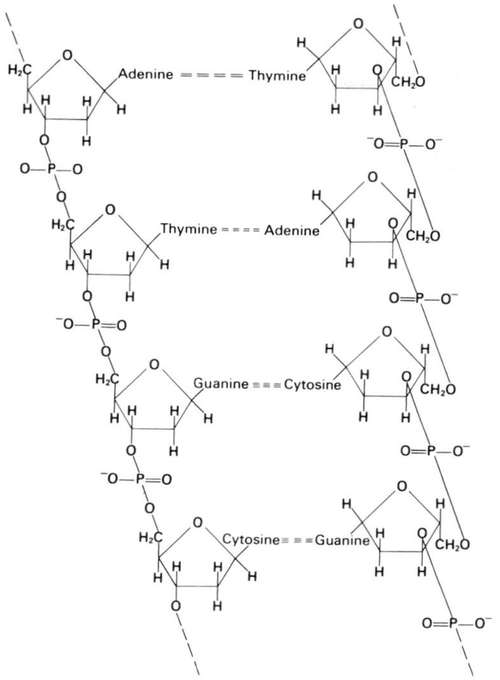

Fig. 3-16. Structure of DNA.

ladder at the sugars. How these four bases constitute a code for making proteins will be discussed in the chapter on genetics.

In living cells the DNA in the chromosome makes more DNA like itself. It also serves as the template for making messenger RNA. The messenger RNA moves out into the cytoplasm where it furnishes the instructions for making proteins, including the most important proteins, enzymes. Thus life is regulated by three key compounds: DNA, RNA, and proteins.

METABOLISM

The proteins, carbohydrates, lipids, and even nucleic acids in a living organism are continually undergoing change and renewal. The old comparisons of life to a flame or to a fountain turn out to be even truer than the ancients suspected. Much of our knowledge of metabolism has come from experiments with radioactive tracers. Tracers have been used not only to unravel complex biochemical processes but also to follow the fate of individual compounds taken up by cells. Tritium, a radioactive isotope

of hydrogen, has been used to label the nitrogenous bases incorporated into nucleic acids. The use of tritiated thymine or uracil has provided information about the replication of chromosomes and the site of RNA synthesis. ^{14}C-labeled amino acids have demonstrated that there is a constant exchange of amino acids between the proteins of which animal tissues are built and the amino acids in the bloodstream. Tracer studies show that the very structure of living things is in flux, with a definite turnover rate. However diverse it may appear, the stream of matter and energy which flows through a living organism is utilized in very few ways. The principle uses of energy are to make big molecules from small ones, to transport material across membranes, or to produce motion.

The sum of all the chemical activities within a living organism is called **metabolism**. Synthetic metabolism such as biosynthesis, which constructs more complex molecules, is known as **anabolism**. Destructive metabolism, such as hydrolysis, breaks up large molecules and is called **catabolism**. The most obvious single aspect of metabolism in animals and ourselves is the mechanical inspiration and expiration of air. Although respiration in this gross mechanical sense has always been the criterion of life, it is only within the last four centuries that anyone has known what breathing does for an animal.

The growth of knowledge about respiration marks one of the most long-continued and important series of achievements in the history of science. In the 17th century a group of young investigators, including Robert Boyle in London, discovered what is still an amazing thing: a breathing mouse and a burning candle both do the same thing to air. By enclosing candles and mice together and separately under glass bell jars, these men were able to show that a candle cannot burn in air in which a mouse has suffocated, nor can a mouse live long in a closed place in which a candle flame has burned out. In 1796 the Dutch physician, Jan Ingen-Housz, published his findings demonstrating that in the dark, plants, like mice and burning candles, remove a component from air that is needed for combustion. In the light, however, adding a plant to the mouse in the bell jar may prevent his suffocation indefinitely.

With the growth of chemistry in the 18th century, Antoine Lavoisier (1743-1794) and Joseph Priestley (1733-1804) proved that in respiration oxygen is taken from the air and carbon dioxide returned to it. By keeping small mammals in confined vessels and measuring accurately both the CO_2 and the heat (*i.e.*, calories) given off, they further proved that animals obey what are now known as the **laws of conservation of matter and of energy.** A breathing guinea pig and burning charcoal give off the same number of calories of heat energy when the same amount of oxygen is consumed or carbon dioxide given off.

The 19th-century researchers refined the discoveries of their predecessors and added some of their own. The simple sugar, glucose, is the material usually burned in a living organism. The over-all equation $C_6H_{12}O_6 + 6O_2 \rightarrow 6CO_2 + 6H_2O$ + energy conceals a vast complexity but needs to be assimilated first. One molecule of glucose combines with six molecules of oxygen to yield six molecules of carbon dioxide and six of water. The energy released is measured in calories (cal.) **A small calorie**, the unit usually used in cellular metabolism, is defined as the amount of heat necessary to raise one gram of

Fig. 3-18. Joseph Priestly, discoverer of the role of O_2 in respiration. (Courtesy Bettman Archive.)

water one degree centigrade (Celsius). (Note that the **large calorie** or kilocalorie is the usual dietetic unit.) At the same time studies were being made on the caloric value of different foodstuffs, extensive knowledge was gained of how oxygen and carbon dioxide are absorbed and transported in the blood, and also about the way respiration is controlled. Many of these facts and principles established in the 19th century are of great importance today in aviation medicine, deep-sea diving, dietary programs, etc.

The achievement of the present century has exceeded the wildest dreams of the 19th. It is nothing less than the successful exploration of the inner workings of cells. The living cell is a complex chemical factory with many diverse and interconnected production lines. The details of respiration in cells are basically the same in plants and animals. This realization has contributed a unity to biology that had not

Fig. 3-17. Jan Ingen-Housz, who showed that plants gave off O_2 only in the light. (Courtesy Bettmann Archive.)

existed before. Knowledge of how oxygen and sugar power these activities, and how they are controlled by enzymes, has led directly to important practical results. This is especially true in understanding how to choose antibiotics and antimetabolites that will block living processes at specific points. It is a large topic to which we will return, but at present it must suffice to cite some basic facts about enzymes.

ENZYMES

Enzymes are of the greatest industrial, medical, and theoretical importance because they control the inner workings of cells. Most poisons are poisonous because they inhibit one or another enzyme; some antibiotics act on enzymes; and vitamins are essential because they form parts of enzyme molecules or cofactors for enzymes. Yet the 19th century understood as little about the inner workings of a cell as a child knows about the inner workings of a juke box. In one case you put in a coin and music comes out; in the other you put in oxygen and sugar, and carbon dioxide, water, and energy come out.

Enzymes as digestive agents have been known in a general way ever since Réné Réaumer (1683-1757) obtained gastric juice

Fig. 3-20. Justus Liebig, renowned chemist who claimed that fermentation was a purely chemical reaction. (Courtesy Bettmann Archive.)

Fig. 3-19. Antoine Lavoisier, established basic laws of respiration before he was beheaded. (Courtesy Bettman Archive.)

from his pet falcon by inducing it to swallow small sponges on strings. The gastric juice thus obtained could soften and dissolve meat. Modern knowledge about enzymes really begins with the famous controversy between Louis Pasteur (1822-1895) in Paris and Justus Liebig (1803-1873) of the University of Heidelberg. Pasteur claimed that alcoholic fermentation was the result of the living activity of yeast cells. Without intact living yeast cells, there was no fermentation. Liebig, a brilliant chemist, claimed that fermentation was a purely chemical process analogous to the rusting of iron. Neither view is entirely wrong nor entirely true. It was Eduard Buchner (1860-1917) who found the answer. He was able to extract from yeast a cell-free juice that had the power of fermentation; that is, it could turn sugar into alcohol and carbon dioxide. He called that active ingredient in his yeast juice an enzyme which means, literally, *in yeast*. The alcohol-producing enzyme in yeast is zymase.

The terminology of enzymes is simple. The material on which an enzyme acts is called its **substrate**. In the case just mentioned, sugar is the substrate. The name of the enzyme is formed by adding *-ase* to the name of the substrate or the activity. Thus an enzyme that acts on lipids is a lipase, an enzyme removing hydrogens, a dehydrogenase. Unfortunately various common enzymes were named before this system arose. Pepsin in gastric juice and ptyalin in saliva are examples.

Enzymes are involved in virtually all the chemical activities of cells, and thus with life itself. There are oxidizing and reducing enzymes, digestive (hydrolytic) enzymes that split carbohydrates, lipids, proteins, and nucleic acids, and synthesizing enzymes that build up these molecules. Enzymes are extremely specific in their activities and particular about the conditions under which they will work. En-

zymes are specific in two ways. First, for a given substrate, many chemical changes might be thermodynamically feasible but only one of these is catalyzed by a given enzyme. Second, enzymes are capable of acting on a limited number of substrates, sometimes only one. Specificity with respect to substrate may involve the linkage between units of a polymer, preference for one over another stereoisomer, or for a specific group within the substrate molecule. For enzymes to be active, proper conditions of temperature and pH are required as well. Since the enzyme molecule is not destroyed when it acts, only very few enzyme molecules are necessary to convert large amounts of substrate to product. A single enzyme molecule can unite or split many thousands of substrate molecules per second.

Most (but not all) enzyme molecules consist of two parts: a large protein portion, the **apoenzyme**, and a small nonprotein part, the **coenzyme** or **prosthetic group**. Many vitamins

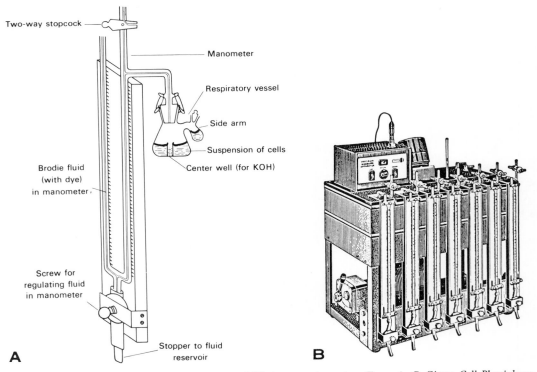

A

Two-way stopcock

Manometer

Respiratory vessel

Side arm

Suspension of cells

Center well (for KOH)

Brodie fluid (with dye) in manometer

Screw for regulating fluid in manometer

Stopper to fluid reservoir

B

Fig. 3-21. A, flask and manometer for the Barcroft-Warburg respirometer. (From A. C. Giese: *Cell Physiology*, 3rd ed. W. B. Saunders Co., Philadelphia, 1968. Courtesy Fisher Scientific Co.) B, Warburg constant temperature water bath showing manometers with vessels attached. A shaking mechanism provides constant movement insuring rapid equilibrium between gas and liquid phases within the reaction vessel.

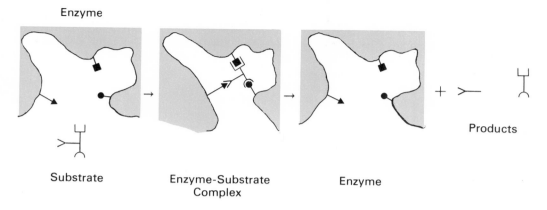

Fig. 3-22. Diagrammatic representation of enzyme-substrate interaction. (Modified from Mallette, Clagett, Phillips, and McCarl: *Introduction to Biochemistry*. The Williams & Wilkins Co., Baltimore, 1971.)

serve as coenzymes or prosthetic groups of particular enzymes. The discovery that enzymes are proteins was the work of James B. Sumner of Cornell University. After 10 years of frustrating but single-minded labor he succeeded in crystallizing urease, an enzyme abundant in the jack bean. Crystallization proved that he had a pure substance. It gave all the standard tests for a protein and, when dissolved, showed urease activity. At first no one believed that it could be true, but today his achievement stands as a landmark in biological science.

The classical method for investigating the action of respiratory enzymes on their various substrates has been to place sliced or finely minced tissue into a Warburg apparatus consisting of glass vessel attached to a manometer, which measures changes in gas pressure. Changes in gas pressure within the flask can be converted to volumes of gas consumed or produced per unit time, *i.e.*, respiration rates. In recent times more elaborate devices such as the oxygen electrode connected to an automatic recorder, or sensitive recording optical devices have been used in studies of specific reactions and intermediates of respiration.

Enzyme reactions can be controlled and the reaction stream guided into one metabolic pathway or another by several methods. Because most enzymes can function only within a very narrow pH range, their action can be blocked by relatively slight pH changes. Furthermore, many, if not all, enzymatic reactions can be made to slow down or even run backwards if the products of the reaction are allowed to accumulate or are added. Many enzymes are irreversibly blocked by combining with various poisons (inhibitors). Some poisons are highly specific but others, such as the heavy metals (lead, mercury, and arsenic), block many enzymes. The environmental consequences of careless release of heavy metals such as mercury and lead into waterways is now being realized.

Theories of Enzyme Action

There is convincing evidence that enzymes produce their results at their surfaces and that there is what the first modern investigator of proteins, Emil Fischer, called a **lock-and-key relationship** between the shape of a particular substrate and the enzyme that acts on it. To be effective, the substrate must fit tightly against the enzyme. It now appears that enzyme molecules are not rigid but bend around their substrates. Hence some prefer to use a **hand-in-glove** analogy. In any case, once molecular intimacy has been achieved, chemical forces cause a large substrate to split or two small ones to unite.

Part of the proof of this theory can be illustrated by succinic dehydrogenase, an enzyme important in respiration. A dehydrogenase is an enzyme that splits off hydrogen from its substrate. In this case, as shown in the diagram, succinic acid fits neatly into the reactive site of the enzyme. Two hydrogen atoms are split off and fumaric acid is formed. The fumaric acid falls away from the enzyme, leaving it ready to repeat the act. If malonic acid, which is very similar to succinic acid, is substituted, the malonic acid is accepted by the enzyme, but since it is not a perfect fit, no reaction occurs. Because its reactive site is occupied, the enzyme is more or less permanently blocked, a situation known as **competitive inhibition**. Its discovery has led to an extensive hunt for such compounds, analogs of substrates which can be used to block unwanted reactions.

Another method of inhibiting enzymes is to give them a "phony" vitamin instead of a "phony" substrate. It will be recalled that many enzymes are really double structures, a large protein plus a non-protein prosthetic group. If a given animal cannot synthesize the prosthetic group, it must have it in its diet. In such a case, the prosthetic group is known as a vitamin. If an animal or bacterium is fed a **vitamin analog**, which the protein part of the enzyme does not distinguish from the genuine vitamin, a nonfunctional pseudo-enzyme is produced. This is the way sulfa drugs work. Luckily, the enzymes of certain disease-causing bacteria are far more seriously damaged than are any essential human enzymes.

Certain metabolic poisons can inactivate enzymes either by blocking the prosthetic group or by changing the shape of the enzyme

molecule so that the active site is modified more or less permanently. Such poisons are called **non-competitive inhibitors**. A basic difference between competitive and noncompetitive inhibitors is that the effects of the former can be overcome by adding excess substrate. With a competitive inhibitor an abundance of substrate molecules can effectively compete for sites on the enzyme molecules, keeping at least some sites unblocked. Inhibitor molecules are not acted upon by the enzyme and therefore, once attached, they block further reaction by that site. Substrate molecules undergo change after attachment to a reactive site, and the products are released, freeing the site for attachment of a new substrate molecule.

METABOLIC PATHWAYS

There have been two massive achievements in this century in the biological sciences. One was the discovery and then the breaking of the genetic code. The other has been the gradual unraveling of the intricacies of cell metabolism. Not only have the general contours of energy flow been sharply delineated, but even the fine details have been traced in an amazingly intricate and extensive system of change and interchange of matter and energy. This new knowledge confers a profound understanding of living processes and their relationship to the non-living world. It also means a vastly increased possibility of control over living processes in many areas—anesthetics, therapeutic medicine, fermentation industries, agriculture, the nervous system, and of animal and human behavior.

At the same time this new knowledge has resulted in a very real simplification. What once seemed a puzzling jumble of unrelated processes now is revealed as a single system. Aerobic and anaerobic respiration, fermentation, and photosynthesis all fit together smoothly. The chief energy-yielding processes are everywhere the same, in bacteria, plants and animals, whether in fermenting yeast or respiring potatoes, in beef hearts, in the flight muscles of honeybees, or in the cilia of a clam.

Anaerobic Respiration—An Overview

The main pathway by which energy is made available for living organisms has two distinct

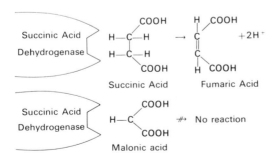

Fig. 3-23. Malonic acid, which is very similar to succinic acid and binds to the enzyme molecule. No reaction occurs, however, and the enzyme remains blocked by this competitive inhibitor.

stages. The first stage does not require free oxygen and is therefore called **anaerobic respiration**. Since sugar is broken down it is also called **glycolysis**, and because alcohol and related substances are end products it is also referred to as **fermentation**, although this term is sometimes restricted to the very final steps of glycolysis. This anaerobic process is also named the **Embden-Meyerhoff glycolytic pathway**, after the men who established its existence.

Glycolysis begins with glucose. In the first step a phosphate group is added. This phosphorylated sugar is then further modified and torn apart by a series of enzymes until it is converted into pyruvic acid. During glycolysis, two molecules (three, in the case of muscle cells) of ATP are produced. It will be recalled that ATP, the nucleotide adenosine triphosphate (Fig. 3-24) is the molecule in which energy is "packaged" and then distributed within cells.

Pyruvic acid, the end product of the Embden-Meyerhoff glycolytic pathway, is highly reactive. It consists of three active groups: a methyl, $-CH_3$, keto, $-\overset{\overset{\text{O}}{\|}}{C}-$, and carboxyl, $-COOH$; connected thus:

$$CH_3\overset{\overset{\text{O}}{\|}}{C}COOH.$$

Pyruvic acid stands at a metabolic crossroads. If no oxygen is present but certain enzymes are, the pyruvic acid is converted into lactic acid (as happens in muscular exercise), or a fatty acid, or some kind of alcohol (as happens in certain yeasts and bacteria), or one of the various other products of the process of fermentation.

Aerobic Respiration—An Overview

A second stage in the release of energy is the **citric acid** or **Krebs cycle**, which receives the products of glycolysis (and often input from other processes) and breaks down these compounds to CO_2. At the very end of the process, the electrons removed by the action of enzymes on several intermediate substrates are transported along a chain of carrier molecules in the mitochondria and finally united with oxygen and protons to form water.

If oxygen is present and also the proper enzymes, then pyruvic acid loses a carbon and CO_2 is released as a waste product. The acetic acid thus formed unites with coenzyme A and enters the citric acid or Krebs cycle. It is at the level of acetic acid and coenzyme A that fat metabolites enter the main energy flow. Residues from amino acids also enter the citric acid cycle at this point, although this is not the only place where they come into it and are used as energy sources. The cycle itself consists of nine different and relatively small organic acids. The two-carbon fragment from pyruvic acid is carried by coenzyme A and enters the cycle by uniting with the last of the nine acids in the cycle, thereby forming citric acid which is the first intermediate in the cycle. As each acid is formed from its immediate predecessor by a series of splittings, coalescences, and modifications, CO_2 is released as waste, more ATP's are produced, and finally the last acid in the cycle is reached. It then unites with more acetyl coenzyme A to form more citric acid and the cycle begins again. The hydrogens from the original glucose are passed along to pyridine

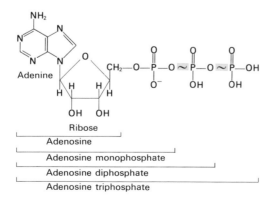

Fig. 3-24. Structure of adenosine triphosphate. Green denotes high-energy bonds.

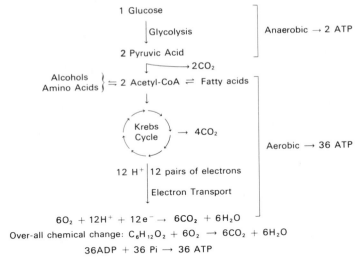

Fig. 3-25. Energy-yielding pathways of carbohydrate breakdown.

nucleotides which pass them along to molecular oxygen with which they form water. The reactions of the Krebs cycle are carried out only by the mitochondria. That all this second or aerobic part of the respiratory pathway occurs in the mitochondria can be demonstrated by breaking cells, separating mitochondria from the other cytoplasmic materials by centrifugation, and then testing the metabolic capabilities of the cytoplasm and of the mitochondria.

The release of energy through the utilization of oxygen and the "storage" of the energy in ATP is called **oxidative phosphorylation**. The relative amounts of energy packaged into ATP molecules by glycolysis and by the aerobic portion of the metabolic pathway are very different. For each molecule of glucose, glycolysis yields only two (or three) molecules of ATP. The aerobic portion yields a total of 36. Actually this varies a bit with the substrate and the organism—yeast cells get 38 ATP's from the aerobic process. The efficiency of the total process is roughly 50 per cent. If one mole (one gram molecular weight) of glucose, which is 180 grams of glucose, is burned to $6CO_2$ + $12H_2O$, about 680,000 small calories of heat energy are released. Making one mole of ATP from adenosine diphosphate + inorganic phosphate stores about 8,000 calories in chemical bond energy. Thus, the two moles of ATP

formed in the glycolytic breakdown of one mole of glucose equal about 16,000 cal. The 36 moles of ATP formed in the aerobic part equal about 288,000 cal. Hence the entire process from glucose to CO_2 and water stores approximately 304,000 cal. packaged as ATP, out of a total of 680,000 cal. released in the complete combustion of one mole of glucose. The energy that is not conserved in ATP is used in breaking down the glucose and producing the ATP's or lost as heat. Evidently there is a packaging charge of over 50 per cent.

Methods and Evidence

Anaerobic Respiration

When you look beneath the dogmatic assertions about glucose uniting with phosphate as the first step in the utilization of sugar, or about a cycle of small organic acids going round and round grinding up acetic acid and producing energy in little ATP bundles, or beneath any of the dozens of other assertions about what is supposed to go on inside the cell, what do you find? How convincing is the whole complicated story, or even any of its parts? How could anyone ever get so much as a finger in a crack to gain access to such deeply hidden events?

A natural starting point is the achievement of

Eduard Buchner (see above). It was he who brought together the oversimplified vitalistic views of men like Pasteur and the equally oversimplified chemical views represented by Liebig. By grinding yeast cells in sharp sand and then filtering and testing the cell-free extract on sugar, he expected to show that such "killed" material could not convert sugar to alcohol and CO_2, but instead he found that it did. This was the beginning. The active agent was not itself living but was produced in living cells.

Buchner died in World War I, but his discovery excited Arthur Harden, then teaching and writing textbooks in Manchester, England. Harden studied the cell-free extracts of yeast and found that they converted sugar into alcohol and CO_2 very rapidly at first and then more and more slowly until action finally stopped. He sought to discover why. After a long search, Harden found that the addition of blood serum or of boiled fresh yeast extract (which would not in itself ferment since it had been boiled) would restore activity. He finally found that the active ingredient in these restorative agents was inorganic phosphate. Fermentation could be virtually at a standstill but then be fully restored by the additon of phosphate. Clearly the enzyme was not wearing out. The surprising thing about this discovery was that sugar, the enzyme zymase, and alcohol do not contain any phosphate. Harden at last discovered that the phosphate combines with the sugar. This is now recognized as the first step in glycolysis.

The discoveries about fermentations in yeasts came to be seen, after some years, to be related to the metabolism of muscle contraction. It became known that muscles contain glycogen, the "animal starch" discovered long ago in livers by Claude Bernard at the Sorbonne. Further, F. G. Hopkins, one of the discoverers of vitamins, showed that a working muscle accumulates lactic acid, an organic acid also produced by various bacteria, notably those which cause milk to sour. It also became known that as a muscle works, it uses up its glycogen.

At this point Otto Meyerhof, at the University of Kiel where Buchner had worked a generation earlier, succeeded in showing first that there is an exact quantitative relationship between the amount of glycogen that dis-

appears and the amount of lactic acid produced just as there is between the sugar used and the alcohol produced in fermentation. Second, Meyerhof showed that in the absence of oxygen this relationship between glycogen and accumulated lactic acid remains the same. In other words, the energy metabolism of muscle in the first stage is anaerobic. When oxygen is again admitted to muscle, the lactic acid disappears and oxygen is utilized. Muscle metabolism thus appeared to consist of two parts: a first anaerobic part, glycolysis, and a second aerobic portion. The working out of the actual steps and enzymes in the anaerobic pathway, of which there are many, between glucose at one end and pyruvic and lactic acids at the other, was largely the work of a husband and wife team, Carl and Gerti Cori. Among other accomplishments they identified the actual structure of the phosphate-sugar compounds which Harden had described only in vague terms.

Aerobic Respiration

Krebs Cycle. What about the aerobic part of respiration? The clues and much of the evidence here were obtained by a remarkable Hungarian, Albert Szent-Gyorgi, who has become the patron saint of Woods Hole, Massachusetts, and a remarkable German, Hans Krebs, who became a professor at Oxford. The final hydrogen runoff was worked out by Otto Warburg, David Keilin, and others. Most recently, the acutal link between the glycolytic pathway and the citric acid cycle has been established by Fritz Lipmann.

Szent-Gyoroyi was interested in investigating the oxygen-requiring part of respiration in muscles. He minced muscle and placed it in a Warburg flask. At first, oxygen was consumed rapidly, but gradually this slowed to a halt. Szent-Gyorgyi speculated that maybe something was being used up, and since it was by then well known that working muscles produce lactic acid, he guessed it might be some compound that was on the pathway between lactic acid and CO_2 and water. After many attempts he finally discovered four substances which restored oxygen uptake. They were all short-chain organic acids—succinic, fumaric, malic, and oxaloacetic. Because any one of the four would work, he concluded correctly not only that they are on the pathway from glycolysis to CO_2 and water, but that they were probably in some kind of a series and could be formed from each other.

At this stage, Hans Krebs, who had been working on the problem of how amino acids are broken down and utilized as sources of energy, left that problem and took up the investigation of aerobic respiration. He found that in addition to the four four-carbon acids (succinic, fumaric, malic, and oxaloacetic) which Szent-Gyorgyi had discovered would restore oxygen uptake in ground muscle, there were also several six-carbon acids, notably citric acid, which would do the same thing. Then, in a series of brilliant experiments, he showed how all these acids fit together in a cycle and that fresh fuel, *i.e.*, acetate formed from pyruvic acid, enters the cycle by uniting with oxalo-acetic acid to form citric acid. From citric acid all the other intermediates in the cycle are

Fig. 3-26. Albert Szent-Gyorgyi, the Hungarian-American biochemist now living in Woods Hole, whose studies of oxygen uptake by minced muscle led to discovery of key intermediates of the Krebs cycle.

formed until finally oxaloacetic acid is reached and the cycle is ready to begin again. The evidence for this is obtained by several methods. Radioactive atoms can be used as labels and these markers can be traced. By the use of appropriate inhibitors such as those described earlier in connection with enzymes, the cycle can be stopped at various points and the substances which accumulate identified; or, various compounds can be fed into the system to determine which ones permit it to proceed. Those that initiate further reactions are probably involved as intermediates.

What is the nature of the actual link between the glycolytic pathway and the citric acid cycle? Pyruvic acid at the end of the anaerobic series has three carbons, but the material entering the citric acid cycle has only two. Fritz Lipmann, working in Germany, Denmark, and fianlly the United States, found the answer to this question. Pyruvic acid loses CO_2, making an acetate group $CH_3\overset{\displaystyle O}{\overset{\|}{C}}O^-$. This two-carbon fragment unites with coenzyme A, forming a compound called acetyl-CoA. The four-carbon oxaloacetic acid unites with the two carbons of the acetate group, forming the six-carbon citric acid. Acetyl-CoA stands at one of the chief metabolic crossroads because derivatives from lipids, amino acids, and other organic compounds enter the citric acid cycle here.

Electron Transport—Terminal Respiration. What about terminal respiration, those steps at the very end of the series involving the dehydrogenases and cytochromes? This is the part which permits the citric acid cycle to keep running by taking care of the hydrogens and their electrons, and, incidentally, produces *three* ATP's per pair of electrons transported. The discovery of this part of the process was largely the work of Warburg and Keilin.

Warburg began to investigate the rate at which tissues fed carbohydrates and amino acids use oxygen. Working with some clever model systems he found that charcoal containing iron (which is true of charcoal made from dried blood) acts as a catalyst speeding up oxygen consumption. Charcoal which lacked iron (charcoal from sugar, for example) failed to do this. He concluded, therefore, that some iron-containing compound was an essential catalyst of oxygen uptake. He went on to show that his iron-containing charcoal plus carbohydrate made an amazingly useful model of cellular respiration. Cyanide in very low concentrations inhibits cellular respiration and also inhibited his model. Anesthetics which inhibit cellular respiration only in high concentrations had the same effect on his models. From such facts Warburg concluded that there is an enzyme containing iron which makes it possible for foodstuffs to be oxidized in the presence of free oxygen. He went on to isolate this enzyme which he called "the respiratory enzyme." It was the first enzyme known to be involved in aerobic respiration. All the others were enzymes of fermentation.

Meanwhile, David Keilin in Cambridge rediscovered a reddish pigment in the wing muscles of the horse botfly and in yeast cells. It is always a safe bet that any substance found in two such widely separated organisms will turn out to be important. Keilin named the substance **cytochrome** (*cyto*, cell, + *chrome*, color). When in the oxidized condition, produced by bubbling oxygen through a solution of the pigment, the cytochrome lost its color. However, if oxygen was excluded and the cytochrome kept in the reduced state, it appeared reddish and showed a characteristic absorption spectrum. If oxygen was readmitted, the bands of the spectrum faded away and the color disappeared. On analysis, cytochrome proved to be a compound similar to hemoglobin—an iron-containing porphyrin attached to a protein.

The really exciting facts came into view when cytochrome was compared with Warburg's respiratory enzyme. They both contain iron. They are both inhibited by cyanide and carbon monoxide. With these poisons cytochrome remains permanently reduced, *i.e.*, reddish, no matter how much oxygen is bubbled through it. Hence, Warburg's enzyme was an enzyme which oxidized cytochrome. Such an enzyme is called a **cytochrome oxidase** because it oxidizes cytochrome in the presence of free oxygen.

In addition, it was found that anesthetics which blocked "respiration," at least oxygen uptake in Warburg's model, also blocked change in the cytochrome keeping it permanently oxidized. Anesthetics then block a different enzyme, one which reduces cyto-

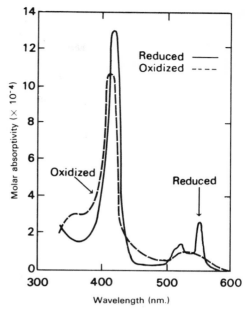

Fig. 3-27. Spectra of oxidized and reduced forms of cytochrome *c*. (After E. Margoliash, Quoted by D. Keilin and E. C. Slater; British Medical Bulletin *9:* 95, 1953.)

chrome. Reduction, it will be recalled, is the addition of hydrogen or electrons. The enzyme, which can be blocked by anesthetics, evidently takes hydrogens or their electrons from something, presumably some breakdown product of a foodstuff, and transfers them to cytochrome. Such an enzyme which accepts hydrogens is commonly called a **dehydrogenase**. The series evidently runs from some substances by means of a dehydrogenase to cytochrome, and from cytochrome by means of an oxidase to oxygen. If hydrogens are somehow being moved along, what can the end product be but water? Subsequent work has shown that there are at least four slightly but significantly different cytochromes which act in series and that there are two dehydrogenases, which stand between them and the Krebs cycle, in this terminal segment of the respiratory pathway.

DETAILS OF THE ENERGY PATHWAY

The results of the work of many investigators in many parts of the world give the following picture of the main energy-yielding pathway.

In glycolysis, the Embden-Meyerhof anae-

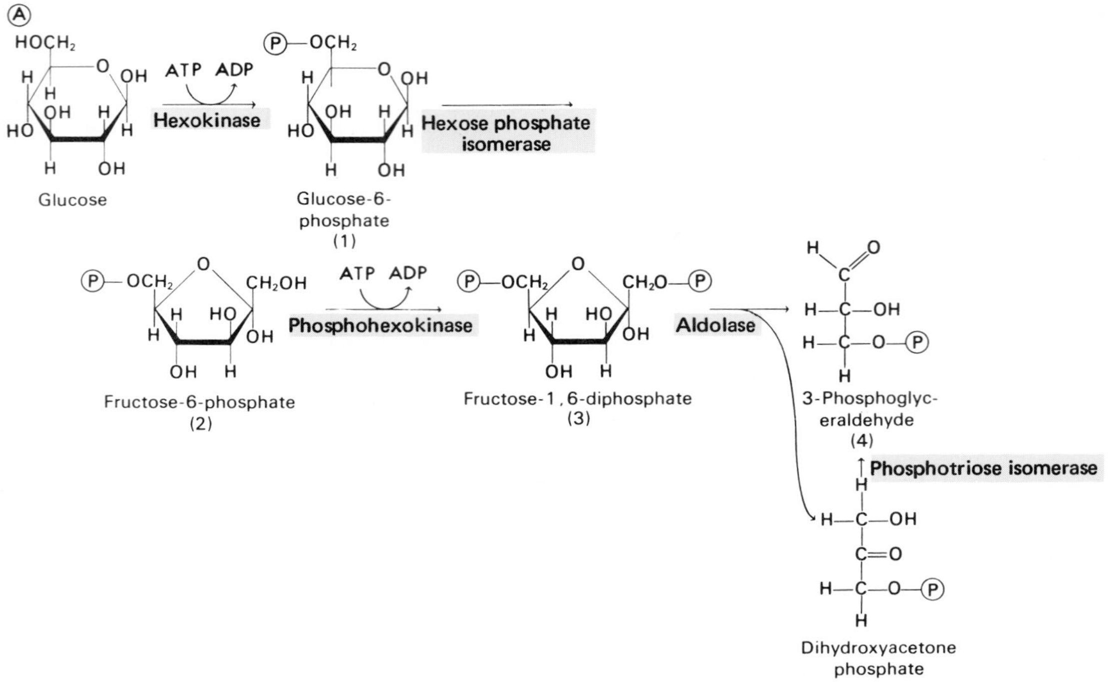

Fig. 3-28. Glycolysis, the stepwise breakdown of glucose to pyruvic acid by a sequence of reactions known as the Embden-Myerhof pathway. Numbers refer to those used in text. Enzymes indicated in gray. Green denotes steps in which ATP is generated, and high energy bonds.

robic pathway, glucose is the usual starting point. It is converted into glucose-6-phosphate (1) by the enzyme hexokinase in the presence of magnesium ions and ATP as a source of energy. Glucose-6-phosphate is so called because the phosphate is attached by the sixth carbon in the glucose molecule. Hexokinase is the enzyme in the brain responsible for breaking down the glucose which provides the energy to read this chapter. Glycogen and starch are other common initial compounds. They are broken down to yield glucose-1-phosphate, which is then changed into glucose-6-phosphate. Glucose-6-phosphate is converted into fructose-6-phosphate (2) to which is added another phosphate, this time to the first carbon, making fructose-1,6-diphosphate (3). The first and third changes require an expenditure of energy which is provided by ATP.

These are the phosphorylating reactions discovered by Harden. Many authors use the term phosphorylation to mean any reaction in which a phosphate group, $-HPO_4$, is added to another compound. Others restrict the term to

those cases where the phosphate group is connected by an energy-rich bond. The context will usually tell which is intended.

Fructose-1,6-diphosphate splits into two compounds, each with a single phosphate, namely, glyceraldehyde phosphate (also called phosphoglyceraldehyde or PGAL (4)) and dihydroxyacetone phosphate These two compounds easily change into each other so there is a triangle here. Moreover, PGAL stands at one of the two chief metabolic crossroads in the entire pathway. By way of dihydroxyacetone phosphate it leads off into fats and alcohols or back from them into the energy-yielding pathway. PGAL can be formed from glucose via the route just outlined, from glycerol, in plants from early products of photosynthesis, and, of course, from dihydroxyacetone phosphate.

Continuing down the main pathway, PGAL is oxidized and gains an inorganic phosphate group to become diphosphoglyceric acid (5). This reaction requires the presence of the coenzyme nicotinamide adenine dinucleotide (NAD) which becomes reduced. Diphosphoglyceric acid is then converted into 3-phospho-glyceric acid (6) with the production of a molecule of ATP. By changing the position of the phosphate group, 2-phosphoglyceric acid (7) is formed. This compound is converted, by the loss of a molecule of water, into a simpler

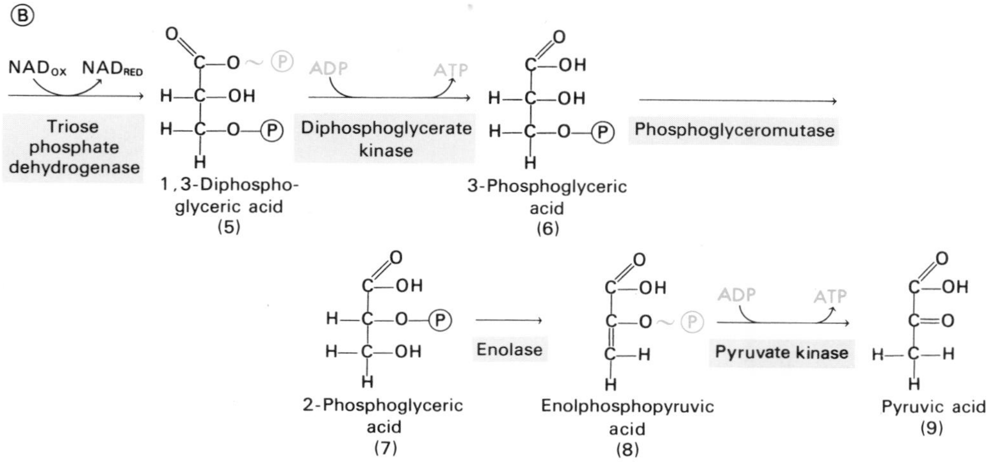

enter the system at this point but also products from metabolized amino acids and lipids.

Coenzyme A was discovered by Fritz Lipmann, who had studied under Meyerhof in Germany. He found that some phosphate bonds have only a low energy yield when broken

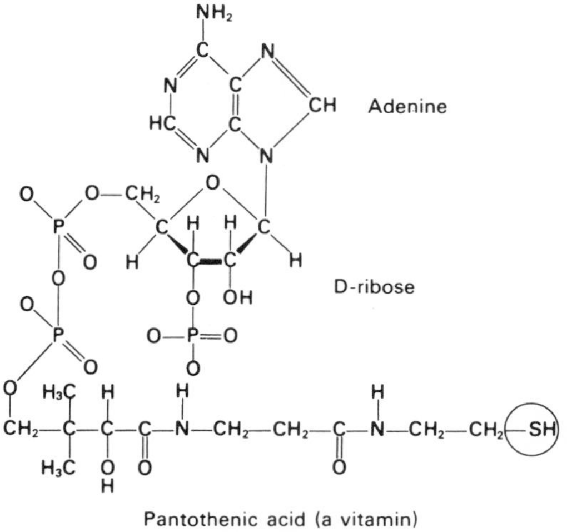

Fig. 3-28 (cont.)

compound, enolphosphopyruvic acid (8), which becomes pyruvic acid (9), plus a second ATP. Each of the steps just reviewed is catalyzed by a specific enzyme.

The molecular structure of these compounds is not very complicated. Glyceraldehyde phosphate consists of the three carbons of glycerol in a row with an aldehyde group, $-\overset{\overset{\displaystyle O}{\|}}{C}-H$, at one end and a phosphate group at the other. The enol form of phosphopyruvic acid is also simple.

Aerobic respiration includes the citric acid or Krebs cycle and terminal respiration (also called electron transport). This portion of the metabolic pathway of energy production begins with the formation of CO_2 and acetic acid from pyruvic acid. The two-carbon acetate group, complexed to coenzyme A, is transferred to the four-carbon oxaloacetic acid in the citric acid cycle to produce the six-carbon citric acid molecule, as previously described. It will be recalled that acetyl-CoA is at the second major metabolic crossroad in the pathway, since not only the breakdown products of carbohydrates

Fig. 3-29. Structure of coenzyme A, carrier of the two-carbon acetate formed from pyruvic acid to the Krebs cycle. When an acetate is attached to the thiol group (–SH) the molecule is called acetyl-CoA.

while others such as those in ATP have a high energy yield. Coenzyme A, which controls the entrance of acetate into the citric acid cycle, turned out to be similar to ATP. It is a very

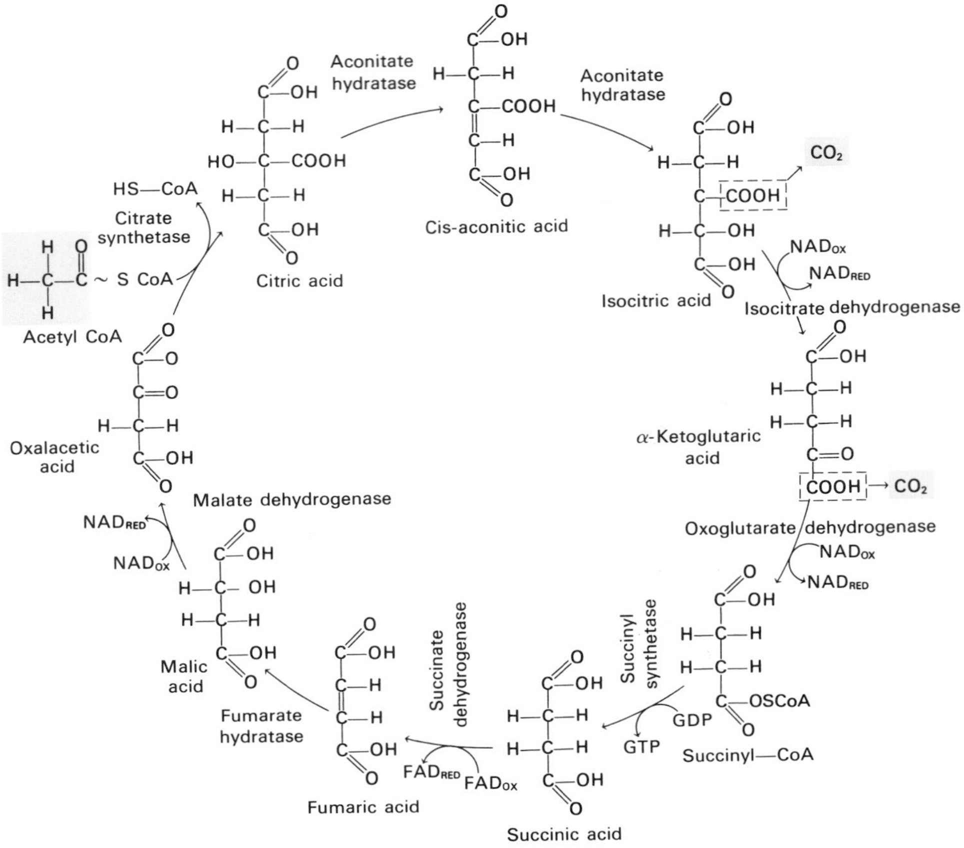

Fig. 3-30. The citric acid (Krebs) cycle. Enzymes indicated in gray. Green denotes CO_2 produced and two-carbon fragment entering cycle.

large molecule, consisting of adenine, ribose, three phosphates, the vitamin panthothenic acid and sulfur-containing thiolthylamine (also called cysteamine). The tiny two-carbon acetyl radical is linked via sulfur. Small wonder the complex is usually called acetyl-CoA.

The citric acid cycle is made up of four six-carbon acids (citric, *cis*-aconitic, isocitric, and oxalosuccinic), one five-carbon acid (α-ketoglutaric acid), and the four four-carbon acids discovered by Szent-Gyorgyi (succinic, fumaric, malic, and oxaloacetic). Once again the structure of the acids is rather simple.

Both α-ketoglutaric acid and succinic acid are notable. α-Ketoglutaric acid is another point at which amino acid breakdown products may enter the system; it is the starting point in the formation of amino acids by transamination, that is, the transfer of an amino group from one molecule to another. Succinic acid forms the

backbone of porphyrin, an important prosthetic group for hemoglobin, cytochrome, chlorophyll, and other compounds.

By a series of dehydrogenations, decarboxylations (removal of CO_2), condensations, splittings, etc., a total of 36 ATP's are produced for every sugar molecule metabolized. Carbon dioxide, water, hydrogen ions, and electrons are given off as wastes.

Terminal Respiration

Let us now consider **terminal respiration** (electron transport). This expression refers to the enzymatic pathway, through which the electrons travel that have been separated from the hydrogen atoms split off in the citric acid cycle. The hydrogen ions and the electrons are finally reunited when they react with oxygen to form water. The first molecules in the electron

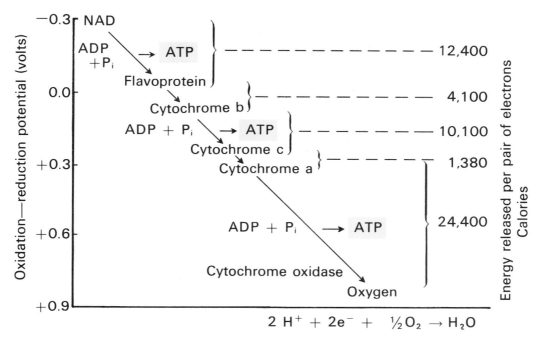

Fig. 3-31. Pathway of electron transport. Intermediates in this pathway can be arranged along a scale of oxidation-reduction potential. Electrons move in a stepwise fashion toward the most electropositive acceptor, oxygen. Energy released in each step per pair of electrons is indicated. In three steps, energy is conserved in the form of ATP.

transport system are dehydrogenases followed by flavins, which are followed by the cytochromes. It is of interest to note that the dehydrogenase NAD is related to the structural units of DNA. Nicotinamide is derived from the vitamin nicotinic acid, one of the substances produced from the amino acid tryptophan. The next kind of molecule in the electron transport chain is flavin adenine dinucleotide (FAD). This flavin is derived from riboflavin, once called vitamin B_2. The adenine in NAD and flavin is the purine found in nucleic acid. The four cytochromes are prophyrin protein complexes with an atom of iron in the center of the molecules. You will recall that hemoglobin is also an iron porphyrin.

As a result of terminal respiration, the citric acid cycle is kept functioning and three ATP's are produced per pair of electrons entering the electron transport system.

ATP (Fig. 3-28) is the key product of both anaerobic and aerobic respiration and is the way energy is packaged and transported within

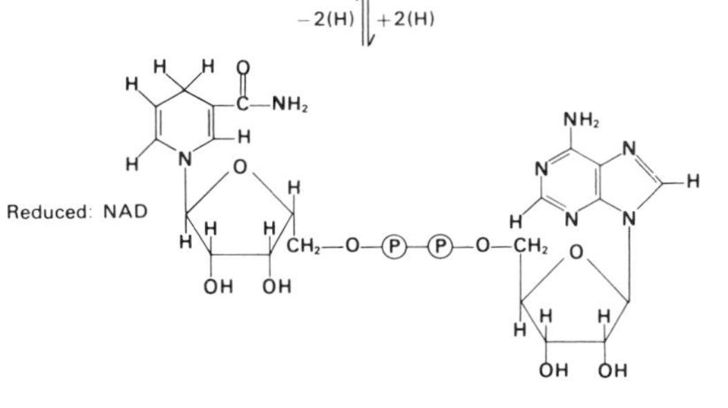

Fig. 3-32. Structure of the coenzyme nicotinamide adenine dinucleotide (NAD). When oxidized, this molecule can act as an oxidant (hydrogen acceptor) and when reduced, as a reductant (hydrogen donor).

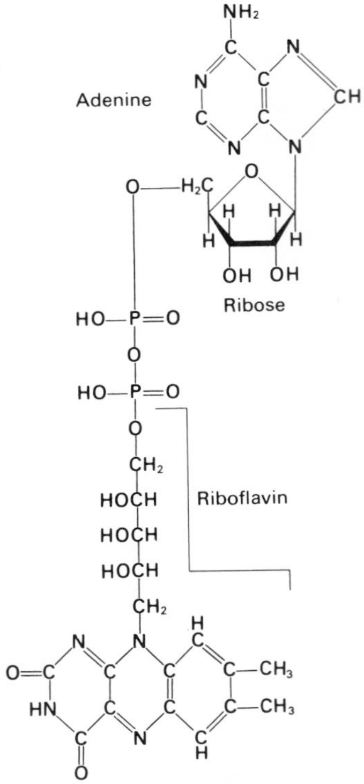

Fig. 3-33. Flavin adenine dinucleotide (FAD). This second electron carrier in terminal respiration has riboflavin as part of the molecule.

cules, the contraction of a muscle, the performance of osmotic work, or the lighting of a firefly.

All of the enzymes of the aerobic portion of carbohydrate metabolism are located within the mitochondria. The enzymes of electron transport are located on the inner membrane, where they are lined up in a way that facilitates these sequential reactions.

No series of discoveries in the whole field of biology has resulted in a more brilliant or more important achievement than the unraveling of the mysteries of cellular respiration. Nor has the solution of any problem of comparable significance ever been to such a marked extent the result of the combined contributions of so many investigators in so many parts of the world.

USEFUL REFERENCES

Haynes, R. H., and Hanawalt, P. C. *The Molecular Bases of Life: Readings from the Scientific American.* W. H. Freeman and Co., San Francisco and London, 1968.

Lehninger, A. L. *Biochemistry.* Worth Publishers, New York, 1970.

Loewy, A. G., and Siekevitz, P. *Cell Structure and Function,* 2nd ed. Holt, Rinehart and Winston, Inc., New York, 1969.

McElroy, W. D. *Cell Physiology and Biochemistry,* 3rd ed. Prentice-Hall Inc., Englewood Cliffs, N. J., 1971.

Stern, H., and Nanney, D. L. *The Biology of Cells.* John Wiley & Sons, Inc., New York, 1965.

cells, regardless of whether the energy will be used in the synthesis of more complex mole-

2
SIMPLE PLANTS AND ANIMALS

Marine Coelenterate from
Contributions to the Natural History of the United States
by Louis Agassiz (1807–1873), father of American zoology.

The Prophyta: Biological Primitives—Blue-Green Algae, Bacteria, and Viruses

The Prophyta constitute an artificial division of the plant kingdom in the sense that no close relationship among them is implied as is the case in other plant groups. Blue-green algae, bacteria, and viruses are a collection of life forms very different from all other living things. Indeed, some biologists think that if life is discovered on another planet in our solar system it will resemble these strange organisms. Of more immediate concern is the hard fact that the Prophyta are of enormous human importance. Blue-green algae are key factors in the pollution of lakes and rivers. Soil bacteria, both the nitrogen fixers, which utilize molecular nitrogen from the air, and the bacteria of decay, play such central roles in ecology that life in anything like its present form and abundance would be impossible without them. Bacteria-caused diseases have wreaked havoc and untold misery on countless millions of people. Other bacteria are very useful, especially in the fermentation industries. Virus diseases of man and of plants are only too familiar.

However, beyond their great immediate importance, the Prophyta present the challenge of the unknown. They have been extremely useful in exploring and solving some very basic problems. Much of today's biochemical genetics is based on investigations of the heredity of the colon bacillus, *Escherichia coli*, and of certain viruses.

The blue-green algae and the bacteria share a unique form of cellular organization differing from the eukaryotic pattern described in Chapter 2 as being almost universal among plants and animals. You will recall that in prokaryotic cells there is no membrane separating the nucleoplasm from the cytoplasm. Prokaryons differ from eukaryons in other respects: they lack the complex membrane-enclosed organelles (such as mitochondria, plastids, etc.) characteristic of eukaryotic cells; they do not exhibit the distinct patterns of mitosis and meiosis observable in eukaryons; only the most primitive kind of sexual reproduction (a form of conjugation) occurs; and finally, the wall materials of prokaryotic plant cells (where present) are chemically different from the cellulose characteristic of most eukaryotic plant cells.

Simpler than the prokaryotic blue-green algae and bacteria, but still considered to be living forms, are the **viruses**. Viruses are neither cells nor free living. They are obligate parasites that can multiply only within living cells. Viruses are considered to be living because they are capable of producing replicas of themselves and because they possess a mechanism of inheritance having the same molecular basis as our own; *i.e.,* information is carried by the nucleic acids.

In the following discussion of the Prophyta we will begin with the blue-green algae, then consider the bacteria and finally the viruses. Perhaps the reverse order would make more sense if we were to emphasize the obvious simplicity of structure of the viruses and the relatively more complex organization and nutrition of the bacteria and blue-green algae. From the viewpoint of evolution, it seems likely, however, that the blue-green algae, the simplest

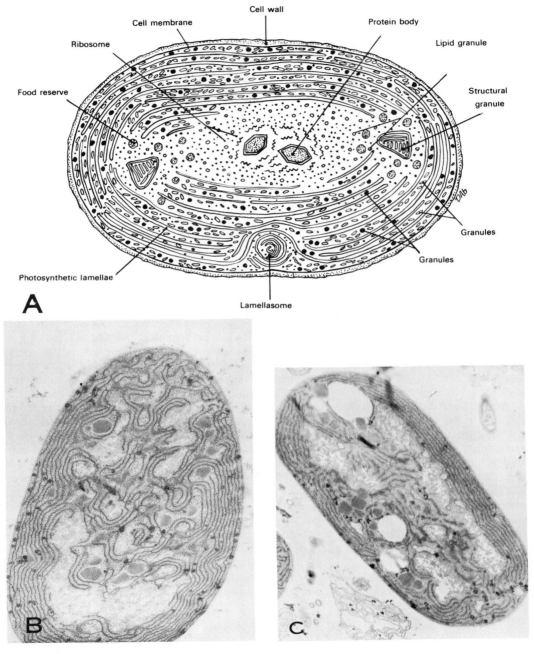

Fig. 4-1. Structural details of photosynthetic prokaryotic cells. A, diagram of structures seen in an electron microscope. B, electron photomicrograph of *Anabena azollae*. × 7,500. (From N. J. Lang: Journal of Phycology, *1:* 127-134, 1964.) C, electron photomicrograph of *Anabena* sp. × 11,500. (From Lang and Rae: Protoplasma, *64:* 67-74, 1967.)

photoautotrophic plants now in existence, must have preceded the largely heterotrophic bacteria and the completely parasitic viruses.

THE BLUE-GREEN ALGAE (CYANOPHYCEAE)

The unique features of the blue-green algae are that they are photosynthetic organisms having a prokaryotic cell structure. Some species have the ability to fix atmospheric nitrogen. Although a few species, such as *Oscillatoria,* exhibit a gliding movement, none is actively motile. Reproduction is asexual, by means of fission, fragmentation of filaments, or the formation of spores.

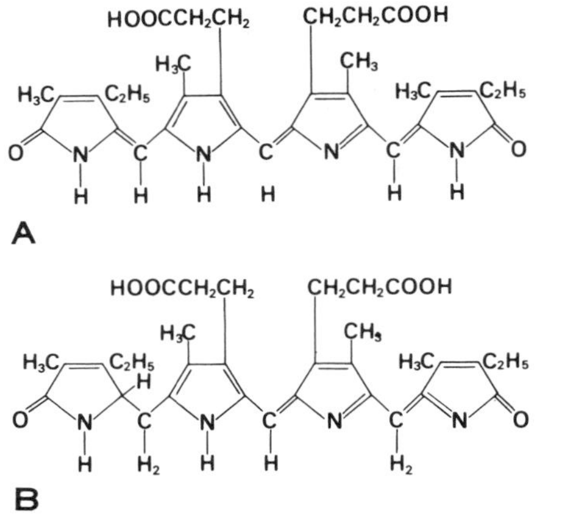

A

B

Fig. 4-2. A, structural formula of the blue pigment phycocyanin; B, structural formula of the red pigment phycoerythrin.

The blue-green algae are widely distributed in nature and can be found in both terrestrial and aquatic habitats. Some species can survive extremes of temperature. They have been found in the Arctic and Antarctic oceans as well as in thermal springs having temperatures up to 85°C. (185°F.). They are frequently found among the free floating species (collectively known as **plankton**) of oceans and lakes. Some blue-green algae live as **endophytes** (within the cavities of other plants), as **epiphytes** (on other algae) or as the algal component of **lichens** (plants which are made up of an alga plus a fungus living together in a state of interdependence called **symbiosis**).

The blue-green algae are **photoautotrophic** organisms, *i.e.,* they can manufacture complex organic molecules by photosynthesis and depend on light as an energy source. The prefix *cyano-* comes from the Greek word *kyanos,* meaning blue, and describes the bluish-green color of many Cyanophyceae. This coloring is due to the presence of the pigment **phycocyanin.** A chemically similar pigment, **phycoerythrin,** is a red pigment found in some of the blue-green algae. Both are water soluble and can be removed from the cells with boiling water. These blue and red pigments, collectively known as the **phycobilins,** are distant relatives of porphyrins such as hemoglobin, the cytochromes, and chlorophyll. Other pigments of the blue-green algae, chlorophyll *a* and the yellow carotenoids, carotene and xanthophyll, are also found in the familiar higher plants. Not all of the blue-green algae are blue-green in

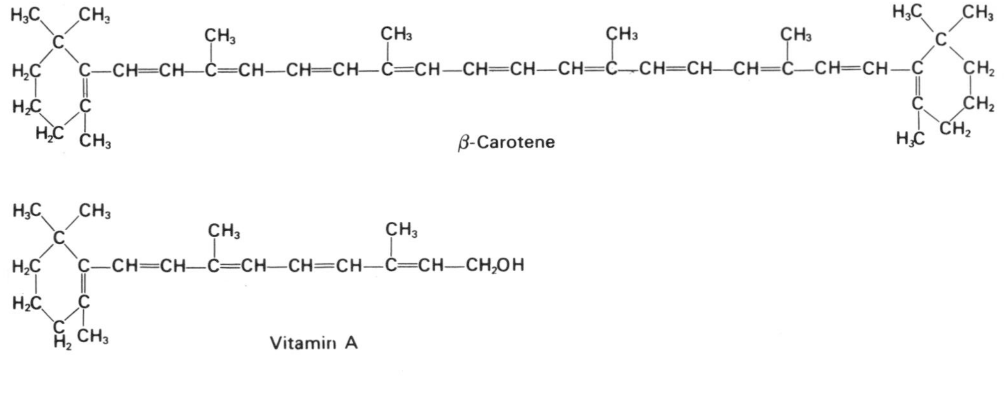

β-Carotene

Vitamin A

Fig. 4-3. Structural formula of β-carotene (top) and vitamin A (bottom). Note that vitamin A is structurally equivalent to half a β-carotene molecule.

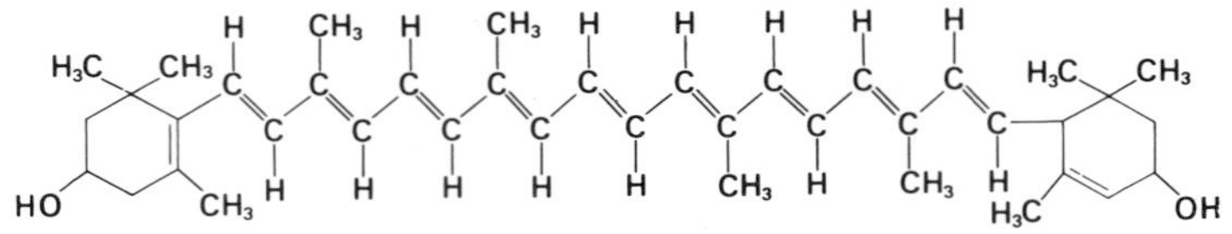

Xanthophyll

Fig. 4-4. Structural formula of xanthophyll. Note similarity in structure of xanthophyll and β-carotene. Both are carotenoids.

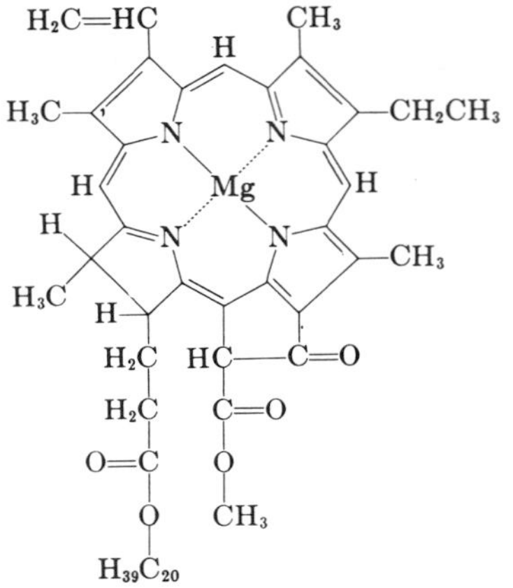

Fig. 4-5. Structural formula of chlorophyll a.

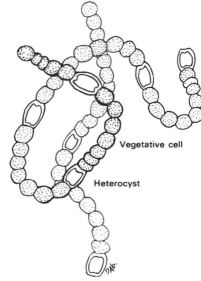

Fig. 4-6. The blue-green alga *Nostoc.* $\times$ 1,100. This member of the Oscillatoriales is able to fix atmospheric nitrogen.

appearance. They may be any of a variety of colors from black to dark purple, brown, or even red, depending on the relative amounts of phycobilins, chlorophylls, and carotenoids in their cells.

As we have already mentioned, some species of blue-green algae can carry on nitrogen fixation. This is a marked advantage because atmospheric nitrogen is universally available on this planet while some soils are so deficient in nitrate and ammonium salts that plants having to obtain their supplies of nitrogen from the soil cannot survive. Their photosynthetic and nitrogen-fixing capacities are contributing factors enabling the blue-green algae to be among the first invaders of areas such as cooled outflows of lava. *Nostoc* and *Calothrix* are two common genera containing species able to fix atmospheric nitrogen.

Individual species of blue-green algae can be unicellular, colonial, or filamentous. The unicellular and colonial forms are thought to be more primitive than filamentous species. There are three orders, the Chroococcales, the Chaemosiphonales, and the Oscillatoriales, which differ in their patterns of cellular organization and in their methods of reproduction.

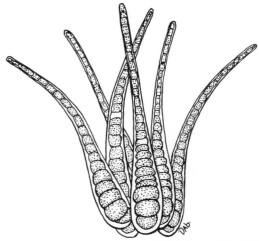

Fig. 4-7. The blue-green alga *Calothrix.* $\times$ 735. This alga, like *Nostoc,* is a nitrogen fixer.

Chroococcales

The Chroococcales are either unicellular or multicellular with cells associated in non-fila-mentous colonies. New plants are formed by cell division or by fragmentation of colonies. Almost all the Chroococcales live in fresh water. The cell walls of all species in this order are composed of two concentric layers: a firm inner layer surrounding the protoplasm and a gelatinous outer layer (the **sheath**). In species of *Gleocapsa* (Fig. 4-8) daughter cells secrete sheaths following division but these are sur-rounded by the stretched sheath of the parent cell. While sheaths provide a mechanism for adherence of cells and therefore for colony formation, colonies can readily be fragmented and the individual cells thus released can continue to grow and divide.

Chaemosiphonales

The Chaemosiphonales are unique in two respects. They grow as epiphytes on other algae and new cells are not produced by a simple division of existing cells. Instead, the protoplast divides to form a series of endospores (Fig. 4-9). After their release, spores float until they reach a suitable new substrate, where they germinate.

Oscillatoriales

The Oscillatoriales are distinguished by a filamentous organization. These are fascinating

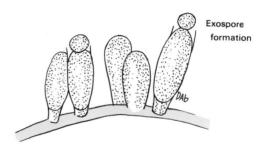

Fig. 4-9. The blue-green alga *Chaemosiphonales* are epiphytes and reproduce by endospore formation. × 3,000.

organisms to observe under the microscope because the filaments often swing from side to side or glide forward and backward. (The name comes from the Latin *oscillare,* to swing.) The filaments of *Oscillatoria* are composed of uniform cells (Fig. 4-10). New colonies can be formed from broken portions of a filament. One group of Oscillatoriales, to which the genus *Nostoc* belongs, develop conspicuously large, special cells called **heterocysts**. There is some evidence that heterocysts can function as reproductive structures. Where spore formation occurs at a specific location on the filament, it always occurs next to a heterocyst. Some relatives of *Nostoc* exhibit structural variations such as branched filaments or tapered rows of cells called trichones (Fig. 4-11).

THE BACTERIA (SCHIZOMYCETES)

Bacteria, the second group of organisms in the Prophyta, are sometimes called the fission fungi (Schizomycetes). These organisms were first observed in the 17th century by Anton van Leeuwenhoek (1632-1723), a Dutch linen merchant, politician, and amateur scientist. This self-educated investigator was a skilled glass blower and metal worker who made lenses as an avocation. He used these to examine many things in his environment—saliva, leaves of plants, and scrapings from his teeth. In such materials he found tiny creatures which he termed "animal-cules" (a word used by the early microscopists for any minute plant or animal). We now know that some of these tiny living things were bacteria.

Bacteria remained curiosities for microsco-pists until the 19th century when, through the work of Louis Pasteur (1822-1895), the signifi-cance of the bacteria in everyday human life

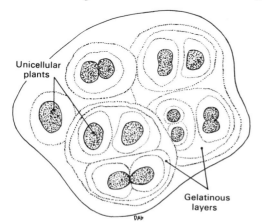

Fig. 4-8. The blue-green alga *Gleocapsa.* × 1,800. Cells in this genus and in all other Chroococcales are surrounded by gelatinous sheaths.

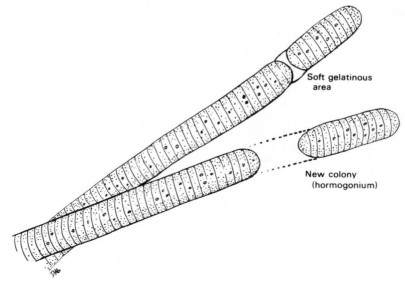

Fig. 4-10. The blue-green alga *Oscillatoria*. × 1,500. Note new colony formed when a portion of the filament is broken off.

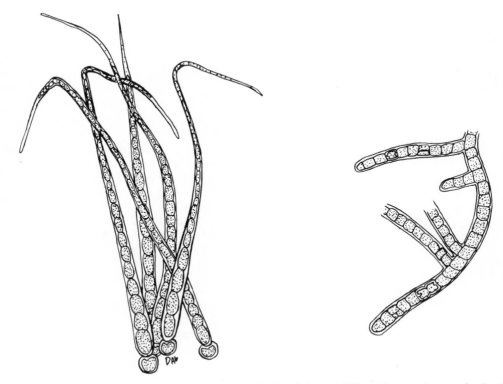

Fig. 4-11. Structural variations in the Oscillatoriales. Left, *Rivularia*, × 1,100, with tapered rows of cells called trichones. Right, haplosiphon, × 900, has branched filaments.

are caused by microorganisms. Seeing the implications of this theory, Joseph Lister, a Quaker surgeon, introduced the first antiseptic techniques into surgery, where appalling numbers of patients were dying of infection. His method was to cover the tissues exposed by the knife with a fine spray of carbolic acid (phenol) to kill the bacteria. It was highly successful in preventing sepsis and thereby saving lives. Still, general acceptance of Pasteur's theory was slow until the work of Robert Koch a decade later. Koch was an obscure German country doctor until, with the encouragement of a professor of

Fig. 4-12. Anton van Leeuwenhoek, early microscopist and observer of what he termed "animal-cules," some of which are now known to be bacteria. (Courtesy Johns Hopkins Institute of the History of Medicine.)

came to be appreciated. A widespread problem in France in the mid-1800's was the souring and spoilage of beers and wines. Pasteur proposed that the "diseases" of these beverages were caused by the growth of contaminating microorganisms. He found that a heat treatment could prevent spoilage (for wine this involved a few minutes at 50 to 60°C.). So long as such heat-treated wines remained sealed, no spoilage occurred. Today pasteurization is commonly used as a means for preserving foods and beverages, from beer to milk.

Pasteur then turned his attention to diseases of man and other living things and in his germ theory of disease argued that these diseases also

Fig. 4-13. Louis Pasteur, originator of the process of pasteurization and proponent of the germ theory of disease. (Courtesy Johns Hopkins Institute of the History of Medicine.)

Fig. 4-14. Robert Koch, whose postulates remain the basis for establishing the relationship between microorganisms and disease. (Courtesy Johns Hopkins Institute of the History of Medicine.)

botany, he showed the world how to culture pure stains of bacteria and enunciated his famous postulates for determining the relationship between a microorganism and a specific disease: (1) The suspected organism must always be present in host organisms exhibiting symptoms of the disease. (2) The organism must be isolated from the diseased host and grown in pure culture. (3) When organisms grown in pure culture are injected into a healthy, susceptible host, the disease must be produced in the host. (4) The disease-producing organisms, when isolated from the experimentally infected host, must be grown in pure culture and shown to be identical with the original isolate.

Bacterial Metabolism

Modes of Nutrition

It is the nature of their metabolism which gives bacteria their importance. The metabolism of most of them is **heterotrophic**, *i.e.*, a source of organically bound carbon is required for growth. This means that the bacteria must obtain the complex organic compounds needed for their nutrition either from tissues of living plants and animals (i.e., they may be **parasitic** or disease producing) or from dead organisms, wastes, or the products of decay (*i.e.*, they may be **saprophytic**).

Although relatively few bacteria possess photosynthetic pigments and are thus capable of autotrophic metabolism in the light, those that are photoautotrophic represent a biochemical diversity not found in any other group of plants. Thus, it is not surprising that these organisms have been studied intensively, not only by investigators interested in the photosynthetic process but also by those concerned with comparative biochemistry and biochemical evolution. Some other autotrophic bacteria obtain energy for the synthesis of complex organic materials from the oxidation of inorganic materials in the dark. Such organisms are known as **chemosynthetic** or **chemoautotrophs**.

The mechanism of photosynthesis in bacteria differs from that in green plants (and even the prokaryotic blue-green algae) in several respects. Chlorophyll *a*, the photoreceptor molecule found in all other photosynthetic organisms, is not present in the bacteria. Instead, some species contain bacteriochlorophyll and others have chlorobium chlorophyll. During photosynthesis, none of the bacteria evolve oxygen. In fact, many of these organisms are strict anaerobes unable to survive in the

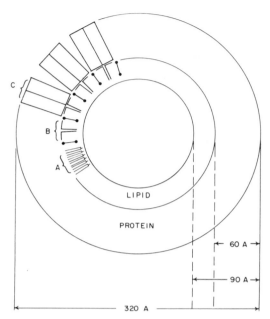

Fig. 4-15. Proposed structure of bacterial chromatophore. Pigment molecules (B) are aligned in a monolayer bounded internally by phospholipid (A) and externally by protein (C).

(From J. A. Bergeron: The bacterial chromatophore. In *The Photochemical Apparatus: Its Structure and Function.* Brookhaven Symposia in Biology, No. 11. Brookhaven National Laboratory, 1958.)

presence of oxygen. Carbon dioxide serves as the source of carbon and is incorporated into more complex molecules during bacterial photosynthesis, provided that an appropriate hydrogen donor is present. In the photosynthetic bacteria, pigments are located in cytoplasmic structures called **chromatophores** (literally, color bearers). These differ from the chloroplasts found in the photosynthetic cells of eukaryotic organisms in that they are neither membrane-enclosed nor self-replicating organelles. However, in all chromatophores the pigments and enzymes are associated with a protein-lipid membrane.

Oxygen Tolerance

Oxygen was discovered by the English nonconformist and writer of political tracts, Joseph Priestley (1733-1804). It is not surprising that it was soon postulated by scientists that oxygen is necessary for all life. About a century later, Pasteur demonstrated that certain yeasts and bacteria can survive in the absence of air. It has since been learned that many common bacteria can not only survive, but in fact require anaerobic conditions.

A knowledge of the oxygen preferences of particular bacteria is of great importance, not only in medicine but also in agriculture, in the fermentation industries, and in the urgent problems of pollution, whether involving sewage treatment, destructive growth of bacteria and algae, or the over-all cycles of matter and energy. Bacteria can be categorized on the basis of their requirements as strict aerobes, facultative organisms able to live either with or without oxygen, microaerophiles, or strict anaerobes.

Those bacteria classified as **strict aerobes** require oxygen. In terms of the metabolic pathways discussed in Chapter 3, aerobes have the enzymatic machinery for complete oxidation of respiratory substrates to CO_2 and H_2O. They utilize molecular oxygen as the terminal hydrogen acceptor: you will recall that hydrogens and electrons removed from intermediates in glycolysis and the Krebs cycle are transferred via dehydrogenases to intermediates (flavins and cytochromes) in electron transport. The reaction with molecular oxygen takes place only in the last step of the electron transport chain.

Some organisms that can utilize molecular oxygen (*i.e.,* they possess the enzymatic machinery of the Krebs cycle and respiratory electron transport) can also survive under anaerobic conditions. Examples of such **facultative** organisms include many common bacteria such as *Escherichia coli* (a common inhabitant of the human intestinal tract) and *Staphylococcus aureus* (an organism which may cause human infections).

Strict anaerobes lack the enzymatic machinery for transfer of hydrogen to molecular oxygen (*i.e.,* they lack cytochromes). *Clostridium tetani,* the organism causing tetanus, or lockjaw, is among the most strictly anaerobic of known bacteria. This organism grows saprophytically on dead tissues and because it is an anaerobe, it can exist in deep wounds. It produces a toxin that causes continuous muscle contraction. The toxins excreted by various *Clostridium* species are among the most poisonous substances known; it takes only 10 micrograms of the toxin of *C. botulinum* to kill a man.

Microaerophilic species grow best at low oxygen concentrations. Such conditions are found in liquids that are not vigorously stirred or below the soil surface. Neither normally aerobic nor completely anaerobic conditions will sustain growth of these species. When grown on solid agar media, the microaerophiles grow best somewhat under the agar surface. This strange requirement of the microaerophiles is due to the fact that in air such organisms produce hydrogen peroxide, but they cannot decompose this toxic product because they lack the enzyme catalase.

Morphology

Individual bacterial cells are either spherical or cylindrical in shape. The spherical forms, called **cocci** (singular, **coccus**), can be further classified on the basis of the ways in which they are grouped. Those clinging together in pairs are called **diplococci**; those forming chains, **streptococci**. In the **staphylococci**, cells cling together in irregular masses, while the **sarcina** form cubical groups of eight cells. The cylindrical forms can be either straight and rod-like (**bacilli**; singular, **bacillus**) or variously curved or twisted. Members of the genus *Vibrio* are much like the bacilli in shape but the rods are curved. Those organisms in which the cells form a rigid spiral belong to the genus *Spirillum,* while the **spirochetes** have cells that are spirally twisted but flexible. *Treponema pallidum,* the agent of syphilis, is a spirochete.

Although the bacteria are visible under the light microscope, they are so small that it is necessary to resort to electron microscopy for most of their structural details to be revealed. Typically, individual cells range from 0.5 to 1.5 μ in diameter. Cylindrical forms as short as 1 μ in length are known, while most are from 2 to 5 μ long with a few as long as 20 to 30 μ. An appreciation of the size of typical bacteria can be gained from the estimate that 9 trillion average sized bacilli could be packed into a 1-cubic-inch volume.

Locomotion

The bacteria exhibit several kinds of movement. **Gliding movement** (similar to that seen in many of the blue-green algae) is found in some of the alga-like bacteria and in slime bacteria. The mechanism of the slow gliding motion of cells in contact with a solid surface is not understood. **Rotary movement** is seen in the spiral cells of spirochetes. This turning motion, which propels the cells, is created by contraction of protein fibers resembling the muscle protein myosin in their properties. **Swimming movements** are possible in those bacteria with **flagella**. Flagella are found in both prokaryotic and eukaryotic organisms, but are structurally different in these two cell types. The flagella of the prokaryotic bacterial cells have no limiting membrane, and have a single bundle of several fibers that are intertwined. Bacterial flagella can be observed in the light microscope only after they are coated with suitable materials to increase their diameter and then dyed. There are several forms and patterns of distribution of flagella in bacterial cells (Fig. 4-17). Flagella are

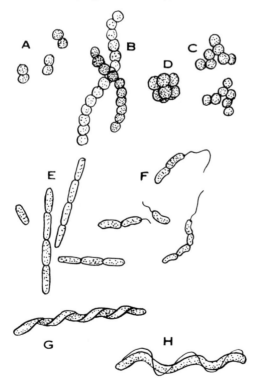

Fig. 4-16. Bacterial morphology. A, diplococci; B, streptococci; C, staphylococci; D, sarcina; E, bacilli; F, vibrio; G, spirillum; H, spirochaete.

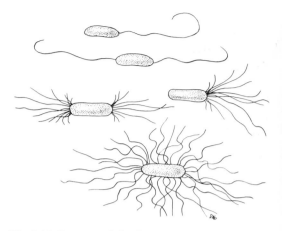

Fig. 4-17. Patterns of distribution of flagella in bacterial cells.

located at the ends (or poles) of cells or over the entire cell surface.

Other Structural Features of the Bacteria

Some of the bacilli have a mass of tiny filamentous **pili** or **fimbriae** on their surfaces. These appendages can be observed only with the electron microscope. Like flagella, they extend from the cytoplasm through the cell membrane and wall and are composed of protein. In the colon bacteria, the pilus is now known to be an organelle involved in conjugation, a mechanism for the exchange of genetic material.

In many prokaryotic cells the cell wall is coated with a hydrated, mucilaginous material called the **capsule**. Chemically, capsules are composed of polymers of sugars or occasionally of amino acids. Because of the distinctiveness of the capsule composition which is unique for each species, it imposes an immunological specificity that is useful in the classification of these organisms. Functionally, the capsule can provide an osmotic barrier around the cell, and it is probably an essential factor in the mechanism of gliding movement. The underlying cell wall, like the surrounding capsule, is highly variable in composition. Typically, the rigid walls of bacterial cells consist of fibers made up of **mucocomplex**, a heteropolymer containing sugars and amino acids as structural units. In some bacteria, layers of lipids are associated with the mucocomplex wall material. The chemistry of the cell wall has considerable influence on the susceptibility of bacteria to

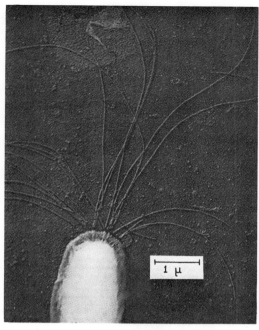

Fig. 4-18. Electron photomicrograph of bacterial flagella in *Spirillum serpens*. Note that flagella pass through the cell wall to the protoplast. (Micrograph by Dr. W. van Iterson. From Biochimica et Biophysica Acta.)

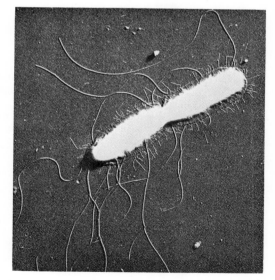

Fig. 4-19. Electron photomicrograph of *Salmonella typhi* × 7,600. This bacillus is dividing. Both flagella and fimbriae can be seen. (Courtesy Prof. P. D. Duguid. From R. R. Gilles and T. C. Dodds: *Bacteriology Illustrated*, 2nd ed. E. & S. Livingstone Ltd., Edinburgh and London, 1968.)

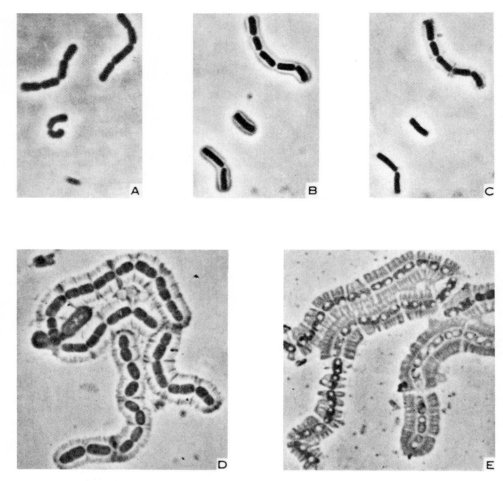

Fig. 4-20. Bacillus capsules as revealed by phase contrast microscopy, × 2,500. Both form and chemical composition of capsules can be determined by antigen-antibody reactions. A, no added antibody; B, with antibody against polypeptide component of capsule; C, same field as B with antibodies against both polypeptide and polysaccharide; D, secondary capsule formed at room temperature after centrifugation, antibody against polysaccharide added; E, same cells as D with polypeptide antibody added. (Courtesy Prof. J. Tomcsik. From the Journal of General Microbiology.)

attack by **penicillin** (an antibiotic) and **lysozymes**. Penicillin acts by inhibiting mucocomplex synthesis in young cells. Those bacteria having other cell wall components (*e.g.*, lipocomplexes) are not completely dependent on the presence of mucocomplex to protect the protoplast from the environment and therefore have less tendency to burst open in the presence of penicillin. The **lysozymes**, enzymes found in egg white, mucous membranes, and skin secretions, attack the bonds linking the structural units of the mucocomplex of the bacterial wall. Thus, in the presence of these enzymes, the cell wall materials are broken down, leaving the protoplast naked and unprotected. Susceptibility of the bacterial cell wall to attack by external agents can be a significant factor in the body's defense against infection by pathogenic bacteria and in the treatment of bacterially caused diseases.

Hereditary Mechanisms in Bacteria: What They Are and What They Mean for Mankind

It had always been assumed that nothing comparable to sex exists among bacteria and that mutation was either absent or extremely rare. Under these conditions a population of bacteria would remain virtually unchanged generation after generation. Consequently, when penicillin and other antibiotics were discovered during and after World War II, it was generally felt that the scourge of bacterial infections in disease and after surgery had been conquered at last. However, resistant strains of bacteria continued to appear, making it evident that mutations do play an important role among bacteria as they do among other organisms. This has meant important changes in the way antibiotics are administered to patients and also a continuing search for new kinds of antibiotics. In addition, bacterial mutations have furnished key information for the understanding of heredity in all living things.

Three different mechanisms exist for the transfer of genetic information (DNA) from one bacterial cell to another. One mechanism, **conjugation**, is more or less comparable to sexuality familiar in the protozoa (Chapter 5). The other two, **transformation** and **transduction**, are not only remarkable in themselves but open up the possibility of developing methods for introducing new and desirable genes into adult plants and animals including man. This whole area of knowledge is in much the same undeveloped state that knowledge of atomic energy was in the 1920's when scientists believed that atomic forces might, perhaps, become available for human use at some time in the remote future. No one expects gene therapy, or genetic engineering as it is sometimes called, to become feasible within the next 10 years, but there are laboratories currently exploring the possibility of curing albinism and diabetes by these methods.

Transformation. Transformation was discovered in the 1920's when it was observed that sometimes one strain of bacteria, if grown in association with a second, can acquire properties characteristic of the second. Two American investigators, Frobisher and Brown, found that nonpathogenic streptococci from cream cheese grown in the presence of streptococci causing scarlet fever could acquire the capability of producing scarlet fever toxin. Other studies by Fred Griffith, an English medical bacteriologist, showed that bacteria could be transformed not only in the presence of other living cells but also by cell-free extracts of dead organisms. Extracts of dead capsule-forming pneumococci could transform a non-encapsualted strain so that cells could now form capsule substances. Later studies by O. T. Avery, Colin McCleod, and Maclyn McCarty at the then Rockefeller Institute in New York showed that the substance responsible for transformation is DNA. Sometimes DNA, formed by replication of the nuclear substance, accumulates in the capsule and can be removed by washing the cells. Transforming DNA can also be obtained by breaking cells and extraction. With DNA so readily extractable and, in some species, produced in excess and secreted into the external layer of bacterial cells, it seems surprising that bacterial populations are so stable. The reason for this stability is that many cells produce and excrete deoxyribonuclease (DNase), an extracellular enzyme which can break down external DNA before it has an opportunity to penetrate into a recipient cell. Also, the presence of a capsule can retard the penetration of DNA fragments, and factors such as pH extremes or salts in the environment can destroy DNA. Once fragments of DNA have entered, they may or may not be incorporated into the DNA strands of the recipient cell. If incorporation occurs, the piece of transforming DNA becomes an integral part of the recipient DNA and can be replicated along with the rest of the nuclear DNA. The fragment of DNA which has been replaced and fragments of transforming DNA which are not incorporated cannot be replicated and are eliminated. Normally, transformation occurs only between related species or strains having DNA's with structural similarities, that is, with similar nucleotide sequences. For altered phenotypes to be observable, the transformed DNA code must be read to form proteins. The resulting protein enzymes must produce products detectably different from those of the original recipient cell, and the transformed cells must multiply.

Those bacterial cells that are "competent" for transformation do not secrete DNase, lack thick interfering capsules, and probably fulfill other prerequisite conditions not yet understood. The penetration of DNA fragments into competent cells is surprisingly rapid and can occur in a matter of seconds. The incorporation of transforming DNA into a cell's genetic material takes from minutes to hours. The **generation time** of bacterial cells (the time required for a single cell to undergo fission or for a population of cells to double in number) also ranges from minutes to hours. The most probable time for the insertion of fragments of transforming DNA is during DNA replication.

The phenomenon of transformation of bacterial cells thus raises the possibility that it may also be possible to transfer DNA from bacteria to plant or animal cells, perhaps even to insert specifically synthesized fragments of DNA into plants or animals to obtain improved varieties. There is evidence that higher plants, in the presence of certain bacteria, can produce an enzyme (choline sulfate permease) which enables their roots to take up a compound that is normally excluded (choline sulfate). A recent paper by the Norwegian, Per Nissen, showed that in order for the changed permeability to occur, bacteria able to produce this enzyme must be in contact with the plant (barley seedlings) and new protein must be synthesized by the plant. Nissen's work raises a number of questions. To what extent do soil bacteria influence the permeability of plant roots? To what extent are interactions of this kind important in bacterially caused plant (or animal) diseases? Can such interactions be used to enhance the yields of agricultural crops?

Conjugation. A second mechanism for changing the genetic potential of a bacterial cell involves direct contact between two cells of the same or closely related species. Following the formation of a **conjugation tube** between the two cells, DNA is transferred from the donor or "male" to the recipient or "female" cell. There is no reciprocal exchange of genetic material in the process and presumably the DNA of the donor cell has already been replicated, or is replicated during conjugation. Thus, a full complement of DNA remains behind in the donor cell. The recombination of characteristics of donor and recipient cells can be observed in

the progeny of the recipient cell. Such a recombination was demonstrated by Tatum and Lederberg in 1948 working with *Escherichia coli*. This organism will grow on a simple nutrient medium. Certain mutant strains require vitamins and amino acids in addition to the minimal medium. One such mutant, requiring added methionine and biotin, and another needing added threonine and leucine, could not grow on media lacking these compounds. After the two strains were mixed and plated onto minimal medium lacking the required compounds, over one hundred colonies appeared. While this does not give us an estimate of the total number of recombining events, it does tell

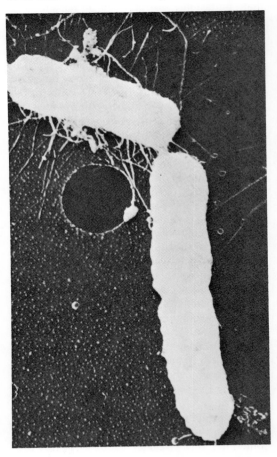

Fig. 4-21. Conjugation in *Escherichia coli*, × 20,000. Genetic material is transferred from donor to recipient cells through the conjugation tube. (From T. F. Anderson and E. L. Wollman: Annales de l'Institut Pasteur (Paris) *93:* 450, 1957.)

us that in this number of cases strains with the synthetic capacity of the wild type were produced.

Transduction. The third mechanism for exchange of genetic information between bacteria was discovered in 1952 by Zinder and Lederberg. Norton Zinder, then a graduate student in Lederberg's laboratory, was assigned the problem of finding out whether strains of *Salmonella* exhibit the kind of exchange of genetic information by conjugation that had been demonstrated in *Escherichia coli*. When variant strains carrying appropriate genetic markers were mixed together and then plated out on media made up so that only recombinant types could grow, colonies appeared in numbers comparable to those observed in experiments with *E. coli*. It appeared that Zinder had found strains of *Salmonella* capable of exchanging genetic information by conjugation. When Zinder and Lederberg repeated these experiments, now separating the two apparently conjugating strains by a porous glass barrier that prevented the two strains of bacterial cells from coming in contact with each other, "recombination" still seemed to occur. Obviously, a mechanism other than conjugation was involved. This mystery was solved when Zinder demonstrated that one strain of *Salmonella* had been infected with a virus (bacteriophage) capable of passing through the glass filter and infecting the second strain.

Viruses can infect bacteria in the same way they attack plant or animal cells. The bacteriophages ("bacteria-eating" viruses) inject their nucleic acid into bacterial cells and cause them to copy the DNA or RNA and protein of the virus rather than their own. Hundreds of new virus particles can be formed in a single bacterial cell. They are released when the bacterial cell ruptures (is **lysed**). Viruses behaving in this way are said to be **virulent**. Viruses do not always cause rapid destruction of the host's cells. They may remain **temperate** for generations of bacterial cells, and during this time there are no observable deleterious effects due to the presence of the virus. In this benign state, phage DNA becomes associated with the genetic material of the host, and pieces of the bacterial chromosome can be incorporated into the nucleic acid of the phage. If the phage becomes virulent, the bacterial cell lyses and

the released phage particles carry with them portions of the nucleic acids of the bacterial cell. Subsequent infection of other cells introduces into the host not only phage nucleic acid but also nucleic acid from the previous host. The recipient cell thus acquires new genetic information as in transformation or conjugation. Thus, transduction, or virus-mediated recombination, which is now a well-documented phenomenon in bacterial systems, presents the possibility of comparable introduction of new genetic material into plants and animals.

Classification of the Bacteria

In classifying the bacteria, a wide array of characteristics is utilized. Many of these (shape, size, color, motility, energy source, etc.) are characteristics equally useful for categorizing other kinds of living things. Because of their small size, the complete classification of bacterial species normally requires much more information than is obtainable from microscopic examination alone. In addition, studies of the nutritional requirements of the organism grown in pure culture and an array of biochemical tests (including tests for the presence of specific enzymes) are utilized by the microbiologists as tools in classification. Approximately 1,500 species of bacteria have been recognized. These have been divided into 10 orders based primarily on morphology and motility.

Spore Formation and Other Peculiarities of Bacterial Reproduction Insuring Survival

A feature of some bacteria that is of much practical importance is their ability to form **heat-resistant spores**, a characteristic that makes these organisms difficult to eliminate by heat sterilization. The process of sporulation results in the formation of a new cell within a pre-existing cell. The newly formed cell, or **endospore**, differs physiologically and chemically from the original cell. After its release, an endospore is capable of surviving for long periods under unfavorable environmental conditions such as extremes of temperature and desiccation, chemicals toxic to vegetative cells, and radiation. When appropriate conditions are encountered, the endospore can germinate to give rise to a new vegetative cell.

Because factors promoting the germination of spores are not completely understood, the spore-forming bacteria pose serious problems for the food preservation industries. Conditions which do not eliminate or destroy all endospores can lead to spoilage of canned foods or transmission of disease-causing organisms or their toxins. Ordinarily, spores germinate only under physiologically normal conditions, *i.e.*, 20 to 37°C., in the presence of adequate nutrients, with pH and oxygen supply appropriate for growth and multiplication of vegetative cells. The activation of spores is promoted by high temperatures (**heat shocking**) or exposure to certain chemicals. The **germination** of activated spores requires the presence of specific nutrients (sugars, salts, and amino acids). Water is imbibed, metabolic activity begins, and nutrients are absorbed. During post-germinative development the spore swells, the wall splits, and an enlarging vegetative cell emerges.

Some Examples of Spore-Forming Bacteria. With the exception of two organisms (*Bacillus anthracis* and *B. thuringensis*), species of the genus *Bacillus* are harmless. Nevertheless, even the harmless species pose problems to the food canning industry and to surgeons and microbiologists who wish to maintain aseptic conditions. *B. anthracis* causes anthrax, a disease of farm animals that is transmissible to man. This organism creates pustules at the point where it enters the body. It can spread throughout the body via the bloodstream, causing further lesions in any organ. After death of the host, *B. anthracis* forms endospores, and animals killed by this organism must be disposed of by incineration to avoid contamination of farm areas and infection of farm workers or of other farm animals. *B. thuringensis* causes a lethal infection of bees. An epidemic of this insect disease can upset plant pollination and cause loss of honey.

In the genus *Clostridium* are over 100 known species of anaerobic spore-formers. Many of the nonpathogenic species are used for industrial fermentations producing a variety of alcohols, organic acids, and vitamins. Others are able to fix atmospheric nitrogen, a property that, combined with their saprophytic habits, makes these organisms important in the decomposi-

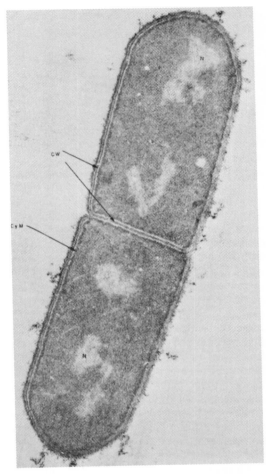

Fig. 4-22. Electron photomicrograph of young vegetative cells of *Clostridium bifermentans,* × 50,000. CW, cell wall; CyM, cell membrane; N, nucleus. (Courtesy Dr. P. D. Walker. From the Journal of Applied Bacteriology.)

tion of dead organisms and the formation of soils. Also included in this genus are the organisms causing lockjaw (*C. tetanus*, mentioned previously), gas gangrene, and botulism (a deadly food poisoning).

Pleuropneumonia-like organisms (PPLO) are pathogenic bacteria without cell walls belonging to the order Mycoplasmatales. They form minute structures called **elementary bodies** or **minimal reproductive units** within their mature cells which are called **large bodies**. The latter are of average bacterial size. The elementary bodies, which are small enough to pass through filters that retain bacterial cells, are released after rupture of the membrane of the large body and thereafter enlarge to form filaments bearing chains of small spheres resembling conidia (asexual spores formed by certain eukaryotic fungi). After release, each spherical body enlarges to form a new large body. Alternative life cycles of the PPLO have been described, including the bacterial type of binary fission and yeast-like budding.

THE RICKETTSIAS AND VIRUSES

These smallest of living things, the viruses and their relatives, differ from the blue-green algae and the bacteria in that they are all intracellular parasites. They cannot multiply outside living cells, and therefore cannot be cultured in chemically defined media as can the blue-green algae and the bacteria. We have placed the rickettsias and viruses together as two orders within the class Microtatobiotes on the basis of their inability to multiply outside of living cells. Some scientists consider these two kinds of life forms sufficiently different in structure to place them in separate divisions of the plant kingdom. The rickettsias have most of the distinctive properties of the bacteria (*i.e.*, they are cells), although in some respects they have characteristics intermediate between the bacteria and viruses. On the other hand, the viruses are unique among living things. The viruses were originally described as invisible, filterable through membranes that retain bacteria, and non-cultivable. However, since the development of the electron microscope in the 1930's, viruses have been observable. With specially prepared filters they can be removed from liquids and, provided with appropriate living cells as culture media, they may even be propagated.

The Rickettsias

All rickettsias are non-motile, intracellular parasites whose cells are smaller in size than bacteria. They are rod-shaped or spherical and multiply by binary fission. These organisms have been cultivated in fertilized eggs (*i.e.*, in developing live chick embryos) and in some tissue cultures.

One family of rickettsias are usually parasitic on arthropods (ticks, fleas, lice, and mites). In the several diseases of humans and other mammals caused by these organisms, the pathogen is transferred to the host via an insect bite. In all rickettsial diseases of man, there is some kind of arthropod-mammal-arthropod chain of transmission in which the arthropod host is not affected by the disease. Rocky mountain spotted fever is transmitted to man from infected ticks. In epidemic typhus, rickettsias are carried from man to man by the body louse. Endemic typhus, a milder disease, is carried by a rat-flea-rat chain of transmission, with the rickettsias transmitted to man by rat flea bites. Q fever, an influenza-like respiratory disease, is caused by a rickettsial parasite of wild animals and ticks. It can be transmitted by ticks to man and domestic animals (goats, sheep, and cattle). The disease can also be transmitted to man by dust, or by contact with infected animals or animal products, *i.e.*, by means other than arthropod bites.

The second family of rickettsias is known as **P-L-T group**, a name derived from the three types of disease that they cause. **Psittacosis** (or parrot fever), a type of pneumonia, is contracted by humans from infected birds. Similar organisms cause **lymphogranuloma venereum**, a venereal disease similar to syphilis, and **trachoma**, a chronic infection of the eye. The P-L-T organisms were for some time thought to be viruses and were called "large" or "mantle" viruses. They are barely resolvable with the light microscope and their properties are in some ways intermediate between those of the viruses and the bacteria. While viruses are dependent on their host cells for energy sources (they use the host's ATP) and for protein and nucleic acid synthesizing systems, the P-L-T organisms can synthesize their own nucleic acids and proteins but cannot form ATP. Thus, they are dependent on the cells of the host for energy. Ordinary bacteria, on the other hand, are generally self-supporting, requiring only carbon energy sources and supplies of inorganic nutrients from the environment.

The life cycles of the P-L-T organisms resemble those of the PPLO forms of bacteria. "Small bodies" and "large bodies" are observable with the electron microscope. Small bodies appear to be much more infective than large bodies and are released only after a latent period.

The Viruses

By the latter part of the 19th century, a great many microorganisms had been discovered and studied. It was generally agreed at that time that the bacteria represented the lower limits of life in terms of size and simplicity of cellular structure. As happens so often in science, an apparently sound generalization was proved erroneous by later discoveries.

The first virus to be discovered is the causative agent of a disease of tobacco plants. In 1892 Iwanowski, a Russian botanist, showed that sap from plants with tobacco mosaic disease could be used to infect healthy plants. Sap from diseased plants remained infective even after filtration through materials known to trap all bacteria. However, it was not until 1935 that the first virus particles were purified. In that year Stanley obtained tobacco mosaic virus (TMV) in crystalline form. This preparation, which remained infective after crystallization, was mostly protein in chemical composition, and for a time viruses appeared to be specialized protein molecules. Several years later it was shown that a constant proportion of nucleic acid is always present, and it is now clear that viruses are complexes of two kinds of molecules: protein and nucleic acid.

Many viral diseases are known, and it is estimated that at least half the infectious diseases of man are caused by viruses. Included in this long list of afflictions are colds, measles, polio, rabies, smallpox, infectious hepatitis, influenza, and probably cancer. The first recognized virus-caused disease of vertebrates, **rabies**, was already under investigation by Pasteur at the time of Iwanowski's discovery of tobacco mosaic virus. By 1900, additional virus-caused diseases were being studied, including foot-and-mouth disease of farm animals, yellow fever in man, and a viral disease of the silkworm.

In 1915, the first observation of virus attack on bacteria was reported by a London bacteriologist, Twort. In his attempts to culture the smallpox virus, Twort's preparations became contaminated with a bacterium, the only thing that grew in his cultures. After a time the bacterial colonies became watery and trans-

Fig. 4-23. Electron photomicrograph of tobacco mosaic virus, × 150,000. (Courtesy Dr. W. R. Allen, Canada Dept. of Agriculture, St. Catherines, Ontario.)

parent (a change described by Twort as a "glassy transformation") and no longer viable. When material from a glassy colony was transferred to a normal colony, it in turn became glassy. The active agent of this bacterial "disease" could not be cultured in any defined medium nor could it induce any change in heat-killed bacterial cells. The agent remained active for long periods, but lost activity after being heated at 60°C. for one hour. Twort did not immediately conclude that he had isolated a bacterial virus. Instead, he proposed three possible explanations for the phenomena he had observed: (1) the apparent bacterial disease is a stage of the life cycle of the organism; (2) the active disease agent is a bacterial enzyme; or (3) the disease agent is a virus that destroys its bacterial host.

In 1917, only two years after Twort published the paper summarizing his findings, there was an independent report of the discovery of bacterial viruses by d'Herelle, a Canadian who worked in many parts of the world. d'Herelle's discovery of a transmissible disease of bacteria began in Mexico where he found that he could start fatal epidemics in populations of locusts by infecting them with bacterial dysentery. In cultures of the bacteria causing this disease, he noted the occasional appearance of transparent patches. Later, at the Pasteur Institute in Paris during an epidemic of human dysentery, d'Herelle prepared cultures of dysentery bacilli from feces of sick men. Once again he observed the occasional appearance of clear spots in cultures incubated on agar in petri plates. He later investigated whether the timing of the destruction of the bacteria in culture was related in any way to the stages of disease in man. Evidence for the presence of the virus (*i.e.*, clear spots) was first observed in cultures of feces collected the day before the first signs of convalescence. Apparently the cultured bacteria and the pathogenic organisms infecting the patient were being destroyed simultaneously by bacterial viruses. In later research, d'Herelle tried to exploit the obvious potential of viruses as agents capable of curing or preventing disease. However, to date no effective therapy based on the destruction of pathogenic bacteria with viruses has been developed. Currently, bacterial viruses and their hosts are used as tools for study of the nature of viruses,

bacterial genetics, and the control of genetic information. They have formed the basis of the new molecular biology.

This bit of biological history should provide insights into the nature of scientific inquiry. Very important discoveries can be accidental. An apparent, immediate, practical application of them can, in the long run, prove to be unworkable or, even if feasible, may be less important than future exploration of new phenomena.

Virus Structure

Viruses are not cellular in structure. Chemically, they are composed of only two kinds of molecules: protein and nucleic acid. DNA *or* RNA but not both makes up the core of the **virion** (virus particle). The presence of only one kind of nucleic acid is a unique feature of the viruses; in all other living things both kinds of nucleic acid are present. The nucleic acid center is surrounded by a protein shell called the **capsid**. The capsid is made up of protein units called **capsomeres**. The nucleic acid core plus its protein capsid is called the **nucleocapsid**. In some viruses an outer limiting membrane (the envelope or mantle) surrounds the nucleocapsid. The envelope or mantle appears to be derived from membranes of the host cell and can contain lipids.

In their over-all morphology, the viruses exhibit a variety of shapes. Some are rod-shaped, others are spherical, cuboidal, or one of many possible polyhedral shapes. Virions vary in size from the large smallpox or vaccinia virus, which is about 0.25 μ long and just resolvable by light microscopy, to 0.020 μ or less. The Japanese type B encephalitis virus has a diameter less than twice that of a molecule of egg albumin protein.

Replication of Viruses

Multiplication of viruses has been studied most extensively in bacterial viruses (the **bacteriophages**) because of the relative ease and rapidity with which their host cells can be cultured. *Escherichia coli*, the colon bacterium most widely used in these studies, has a generation time of only 20 to 30 minutes. A number of bacteriophages parasitic on *E. coli* are known, and they are usually referred to as

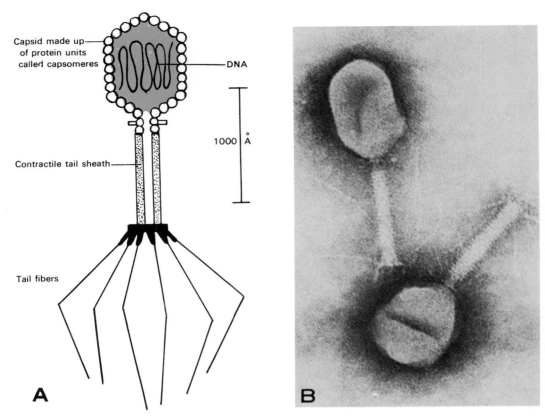

Capsid made up of protein units called capsomeres

DNA

1000 Å

Contractile tail sheath

Tail fibers

A   **B**

Fig. 4-24. T-even phage structure. A, diagrammatic representation; B, electron photomicrograph of *E. Coli* T4 bacteriophage stained with phosphotungstic acid, X 154,000. (From A. J. Rhodes and C. F. van Rooyen: *Textbook of Virology,* 5th ed. The Williams & Wilkins Co., Baltimore, 1968.)

T1, T2, T3, etc. (T stands for type). The T phages are sperm-like in shape with a polyhedral head (the nucleocapsid) and a long tail several head diameters in length. DNA is found at the center of the head. The phage tail is an extension of the protein coating of the head. This hollow tube, enclosed in a retractable sheath, functions in attaching the virus particle to the host cell and in its penetration into the cell.

The first step in the the cycle of penetration, replication, and release of new T phage particles is the **attachment** of the virion to specific **receptor sites** on the cell wall of the bacterium. Following attachment an enzyme at the tip of the phage tail breaks down a portion of the cell wall, and the nucleic acid from the head enters the bacterial cell. The protein shell of the virion's head and tail remain outside the

bacterium. Immediately after **penetration**, the contents of the bacterial cell are not infective and the virus is said to be in **eclipse**, the first phase of the **latent period**. All synthesis of new nucleic acid in the host cell is suppressed, although cellular enzymes and ribosomal RNA already present in the host cell are not destroyed. The viral nucleic acid now becomes dominant. New enzymes called **early protein** are formed which seal the opening in the bacterial cell wall, degrade the cellular DNA, and utilize the breakdown products of the cell's DNA to synthesize new viral nucleic acid. The virion's nucleic acid is replicated many times and utilized by the already existent protein-synthesizing machinery of the host cell to produce the protein units of new virons. Thus, during the eclipse phase, new virus nucleic acids and protein are synthesized.

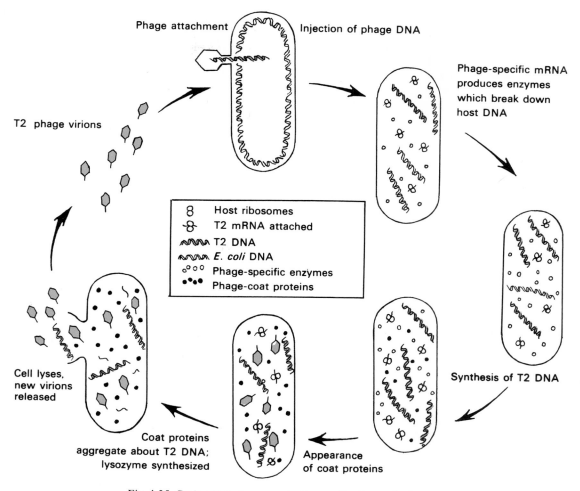

Fig. 4-25. Cycle of T2 phage penetration, replication, and release.

During the period of **maturation** which follows, these component parts combine to form complete new and **infective** virus particles. The appearance of from dozens to hundreds of new virus particles in the host cell signals the end of the eclipse period. The time from penetration to appearance of infective virus particles within the host cell can be as short as 12 minutes (in the case of *E. coli* T2 phage) or as long as many hours. This portion of the viral life cycle is also known as the **vegetative phase**.

The assembly of new virions inside the host cell continues after the end of the eclipse period. In *E. coli* T2, approximately 18 more minutes elapse before the host cell liberates the new virions. Their release initiates a new cycle of infection. One cycle comprises the time from attachment to the host cell to lysis and release of new virus particles.

Virus-Caused Diseases

The kind of viral cycle exhibited by the bacteriophages occurs also when a disease-causing virus invades a receptive non-bacterial cell. This is a key event in the onset of all infectious diseases caused by viruses. We do not always catch all the diseases to which we are exposed, so it would appear that cells can be resistant to invading viruses for any of several reasons, only some of which are understood. A cell may not have appropriate receptor sites (a factor influencing host specificity) or, if receptor sites are present, they could have been altered or destroyed. Conditions such as temperature, pH, the presence of metallic ions, or enzymes from other cells or even other viruses, can change the nature of the receptor site and make a cell resistant to infection. Receptor sites may also be blocked by virions that are defective or inactivated by heat, ultraviolet radiation, or chemical agents. Even active viruses can interfere with or block infection by related viruses. These phenomena form the basis for immunization against certain viral diseases. It should also be evident that vaccination with two live viruses at the same time is not advisable.

Since the last century, vaccines have been developed that utilize our natural defense mechanisms to fight virus-caused diseases. Vaccines are still prepared in much the same way that Pasteur made the first effective vaccine against rabies. Injecting an avirulent or inactivated virus can stimulate the production of antibodies which protect against infection by the virulent form of the virus. The most recently developed vaccines include those against measles and polio.

Although vaccines have been widely used and are an effective means of controlling certain kinds of virus-caused diseases, there are some viral infections that cannot be controlled by vaccination. Examples of the latter include those where the causative agent has not been isolated (*e.g.*, infectious hepatitis and most cancers), others caused by many different kinds of viruses (such as the common cold, which can be caused by any of 100 or more different viruses), or still others that are caused by viruses so unstable that vaccination provides little protection (*e.g.*, influenza). Where vaccination is impossible, one means of treatment is to use chemicals that alter or block the virus reproductive cycle. Effective chemical therapy has been developed for only a few virus-caused diseases such as smallpox, skin lesions resulting from herpes virus, and a single strain of influenza virus.

Interferons. The **interferons**, a newly discovered class of proteins produced by animals in response to viral (and also some non-viral) substances, hold much promise as a new approach to the control of virus diseases. Interferons are produced naturally in all vertebrates in response to virus infection. Their

Fig. 4-26. Left, healthy young tomato plant. Right, tomato plant with spindle tuber disease thought to be caused by a viroid. (Courtesy Dr. T. O. Diener, U. S. Dept. of Agriculture.)

Fig. 4-27. Dr. T. O. Diener (foreground). Ultraviolet absorption profiles and electrophoresis are being used in viroid studies. (Courtesy U. S. Dept. of Agriculture.)

be synthesized in this way. Possibly interferon proteins may some day be artificially synthesized. The alternative approach is to find chemical agents that will stimulate interferon production—either to provide immunity or to help cure a disease after infection has occurred. None of the inducer substances now known such as the synthetic RNA, poly I (a polymer of inosinic-cytidylic acid), or bacterial endotoxins (lipopolysaccharides from the cell walls of certain gram-negative bacteria) can be used because of their undesirable side effects. However, the search for nontoxic inducers continues. Perhaps very soon we may be able to avoid common colds or influenza simply by swallowing an effective interferon inducer.

Classification of Viruses

The earliest attempts to classify the viruses were based on their hosts, the organisms in which they replicate in nature. Viruses are known that infect practically every major group of plants and animals. Thus we can group the viruses into those infecting only a single group of plants or animals, and those having more than one host (some are known that can infect both plant and animal hosts, or two kinds of animals). An alternative but related approach has been the classification according to diseases caused in the host by viral infection. More recent attempts at viral taxonomy (since 1965) have used specific characteristics such as the kind of nucleic acid present (DNA or RNA), the symmetry of the virus particle, presence or absence of an envelope surrounding the nucleocapsid, and size of the nucleocapsid or number of capsomeres. Other factors used include the properties of the nucleic acid present (such as the ratios of bases, sequence, and the number of nucleotides); characteristics of the capsomeres (such as their molecular weight, structure, and antigenic properties); and properties of the capsid (such as number of capsomeres and the reactions of the capsid to chemical or physical agents such as pH or heat).

Viroids. Very recently a Swiss-born investigator in the U.S. Department of Agriculture, Theodor Diener, has discovered particles 80 times smaller than virus particles which produce diseases in tomatoes, potatoes, and perhaps in

production is very different from the production of antibodies which follows vaccination. It appears that practically all body cells can produce interferon. Although synthesis is very rapid (within hours of stimulation), induced interferon disappears within hours or days. Reserve interferon (stored in an inactive, higher molecular weight form) seems to be manufactured by the reticuloendothelial system (see Chapters 27 and 28). The latter form is released in response to certain non-viral inducer substances. Instead of reacting directly with viruses to destroy infectivity as do antibodies, interferon protects cells by inhibiting the replication of viruses. As the yield of new virus particles is thus drastically reduced, the spread of the virus is slowed.

We cannot soon expect viral diseases to be treated with interferons because these antiviral agents are species-specific; thus, to be effective in treating human diseases interferon has to be produced by human cells. Human cell or organ cultures will not help the supply problem because of the very small quantities that could animals and man. He calls these pathogens **viroids**. They are composed only of RNA. By the familiar techniques of electrophoresis through acrylamide gel and density gradient centrifugation, Diener has shown that they have a molecular weight of only about 50,000.

Viroid research is still in its beginning phases and it is too early to conclude that yet another life form has been discovered. We can only wonder whether once again an apparently sound generalization about the ultimate size of living things is about to be challenged.

Biological Significance of Viruses: Evolutionary Products or Precursors?

There are two highly divergent hypotheses about the evolutionary origin of the viruses. According to one view they are considered to be the ultimate in parasites and, therefore, may be the degenerate descendants of organisms that, through their obligate parasitism, gradually lost all their cellular characteristics. The opposite view holds that viruses are the descendants of the first molecular aggregates capable of replication, from which cellular organisms evolved. The array of living forms described in the preceding pages can be thought of as survivors of the various stages of evolution that would have to be postulated to support either of these views. While neither hypothesis can be proven or refuted, each can provide a basis for further speculation.

USEFUL REFERENCES

Bisset, K. A. *The Cytology and Life-History of Bacteria,* 3rd ed. The Williams & Wilkins Co., Baltimore, 1970.

Dubos, R. *The Unseen World.* The Rockefeller Institute Press, New York, 1962.

Frazer, D. *Viruses and Molecular Biology.* The Macmillan Co., New York, 1967.

Frobisher, M. *Fundamentals of Microbiology*, 8th ed. W. B. Saunders Co., Philadelphia, 1968.

Goodheart, C. R. *An Introduction to Virology.* W. B. Saunders Co., Philadelphia, 1969.

Prescott, G. W. *The Algae: A Review.* Houghton Mifflin Co., Boston, 1968.

Stent, G. S. *Molecular Biology of Bacterial Viruses.* W. H. Freeman and Co., San Francisco, 1963.

Watson, J. D. *Molecular Biology of the Gene.* W. A. Benjamin, Inc., New York.

Protozoa: Models of Cellular Life and Agents of Disease

Protozoa are one-celled animals. The variety and the complexity of structure and the diversity of life histories exhibited by these single-celled creatures surpasses the imagination. The human importance of the Protozoa is great in several very different ways. Many are pathogens producing serious diseases of man and domestic animals. The list of protozoan diseases is a long one. Malaria, which afflicts more people than any other single disease, African sleeping sickness, and amoebic dysentery are three familiar ones. Furthermore, the protozoans are extremely useful as tools in basic research. They have been employed extensively in investigating problems of sex, reproduction, and the biochemical side of nutrition. For example, a ciliate *Tetrahymena*, not very different from *Paramecium*, has almost exactly the same nutritional requirements as a man. The Krebs cycle of aerobic respiration takes place in the mitochondria of a protozoan just as it does in the mitochondria of a honeybee or a rat. Occasionally protozoans are of indirect economic importance. For example, the fossil shells of a group of protozoans known as Foraminifera are widely used as guides in drilling for oil. Last, the world of protozoans has a fascination of its own, like the study and collection of stamps or gems.

In the course of hundreds of millions of years, populations of protozoans have evolved in many different directions. There is at least as much diversity among the Protozoa as there is among all the different kinds of birds or mammals. Yet all this extraordinary diversity is achieved within the limitations of size imposed by a single eukaryotic cell. The largest Protozoa, like some of the big ciliated ones and some of the amoebas, escape to a slight degree from the limitation imposed by the volume of

cytoplasm which one nucleus can serve by having several nuclei or a special giant macro-nucleus. Even these large protozoans are still semi-microscopic.

The main difference between the many-celled animals (metazoans) and the protozoans is this: in metazoans, like jellyfish or men, size and complexity are attained by grouping together many cells with individual cells narrowly specialized. Each metazoan nucleus governs a discrete mass of specialized cytoplasm containing muscle fibers, or bearing cilia, or producing hemoglobin, but not doing all three. In the protozoan, one nucleus governs a mass of cytoplasm with many different specializations. Contractile fibers, cilia, and pigment granules may all be present in the same cell.

THE WORLD OF PROTOZOANS

Over 50,000 different species of protozoans have been described, and so very diverse are they that modern protozoologists have been forced to recognize four subphyla. The largest is the **Sarcomastigophora** (*sarkos,* flesh; *mastigion,* little whip; *phoro,* bearing), so called because they bear either pseudopods which are fleshy extensions of cytoplasm or whip-like flagella or, in some species, both. There are two groups in this subphylum. Those species which are primarily characterized by pseudopods, like *Amoeba,* constitute the familiar group called **Sarcodina**. Those species with flagella all or most of their lives are most correctly termed the **Mastigophora** although they are commonly referred to merely as **flagellates**. Here are found many free living pond species and the trypanosomes which cause African sleeping sickness and similar disease in man and many animals. Most of the pathogenic protozoans, including

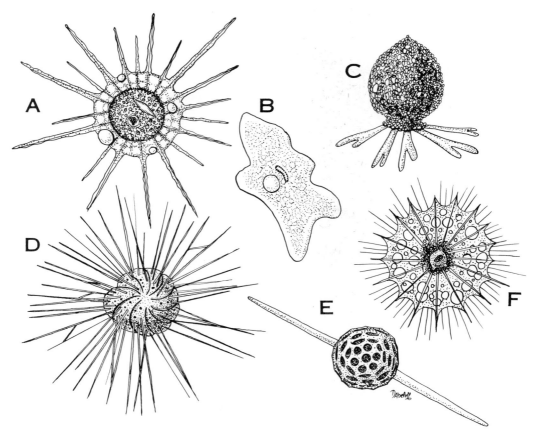

Fig. 5-1. Diversity among Sarcodina. A, a heliozoan from fresh water; B, an amoeba; C, a species of amoeba living within a sand-grain case; D, a marine formaminiferan; E and F, marine radiolarians.

the malarial organism, belong to the subphylum **Sporozoa**, which will be fully discussed in the chapter on symbiosis and parasitism. The third important subphylum is the **Ciliophora** which includes *Paramecium* and thousands of other ciliated forms which almost always possess both a micro- and macronucleus. Representatives of all of these three subphyla have been extensively used in the most basic biological research in all nations where scientific work goes on. The fourth subphylum is very small and unimportant.

BEHAVIOR OF PROTOZOA

The behavior of amoebas, flagellates, and ciliates resembles that of all animals in such fundamental ways that the most basic aspects of behavior can be investigated using protozoans as tools. For example, the same three

aspects of the **stimulus-response situation** are present as in the rat or man.

There must be an adequate stimulus. In the case of light, the adequate stimulus is a more or less sudden change in intensity. If an amoeba moves into a region of brighter light, or if a beam of light is suddenly focused on the advancing lip of a pseudopod, the movement ceases and the animal will move off in another direction. There is a latent period between the application of the stimulus and the first visible response. The stronger the stimulus, the shorter the period. Finally comes the response itself. In the case of a sudden enough increase in light intensity, this is a gelation of the protoplasm which stops the advance of the pseudopod. When subjected to direct electric current, an amoeba will withdraw the pseudopods on the side toward the anode (positive) and move toward the cathode (negative). The amoeboid

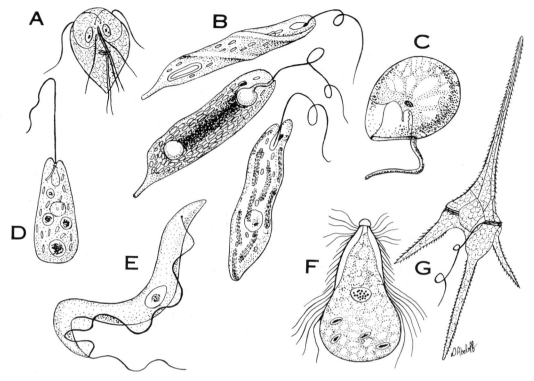

Fig. 5-2. Diversity among Mastigophora (flagellates). A, *Giardia* from human intestine; B, three species of *Euglena;* C, *Noctiluca,* a luminous marine dinoflagellate; D, a freshwater *Paranema;* E, a parasitic trypanosome; F, a wood-digesting flagellate from an insect's gut; G, a marine dinoflagellate.

end of an outgrowing nerve fiber behaves in a similar way.

The response of parameciums to a direct electric current has been much studied ever since the original experiments of Max Verworn in the closing years of the last century. With the current from two 1.5-volt dry cell batteries, it is easy to show that parameciums swim toward the cathode. This behavior is referred to as **galvanotaxis**. The mechanism is puzzling. The beating of the cilia on the side toward the cathode is reversed. This causes the animal to swing around until it heads toward the cathode. In this position only those cilia located at the extreme anterior end beat backwards. The animal consequently moves toward the cathode. However, if the current is made progressively stronger, more and more cilia reverse their effective stroke, and the paramecium swims slowly backwards away from the cathode, *i.e.,* toward the anode. The complexity of this reaction will be realized when it is observed

that if potassium or certain other salts are added to the medium, parameciums swim head first toward the anode.

Except when feeding, the natural state of a healthy paramecium is activity, swimming forward. The rows of cilia are so arranged and the animal is so shaped that a paramecium rotates on its long axis as it moves forward in a corkscrew-like spiral. When it runs into an obstacle—a piece of dirt, the surface of the water, a place where the water is too warm or too cold—a paramecium shows an **avoidance reaction**. It stops, reverses the ciliary beat, backs off, pivots through a moderate angle, and starts off in a new direction. The animal continues in this direction until it runs into something that elicits the avoidance reaction again. This type of behavior was described by Jennings (1880-1940), a life-long student of these animals, as **trial and error**, an unfortunately anthropomorphic term which has stuck with us.

A paramecium appears totally unable to

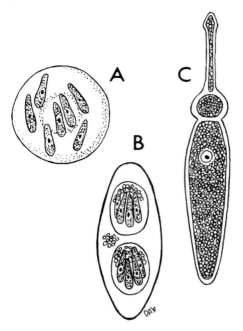

Fig. 5-3. Diversity among Sporozoa. A, malarial organisms within human red blood cell; B, *Isospora,* from human intestine; C, gregarine from an invertebrate gut.

A characteristic reaction of many ciliates, including the parameciums, is to shoot out **trichocysts** in the face of noxious circumstances. Before discharge, a trichocyst appears as a minute, elongate vesicle a short distance under the surface membrane. When discharged by the paramecium and later examined with an electron microscope, the trichocyst is seen to consist of a pointed tip and a long striated shaft. How effective these supposed weapons are is uncertain. They have little effect against a paramecium's arch-enemy, the fellow ciliate, *Didinium.* In fact, the trichocysts have even been observed to backfire into the paramecium itself.

It should be noted that parameciums do not always react to unfavorable conditions by an avoidance reaction. They will readily enter solutions of noxious substances, such as corrosive sublimate or copper sulfate, and quickly die.

Another type of innate response shown by parameciums is **taxis**, the orientation of an organism in a field of force. Parameciums are negatively geotactic and, other things being equal, will swim away from the pull of gravity.

LOCOMOTION OF PROTOZOA

Watching Protozoa dart or glide about has fascinated amateur and professional scientists for generations. This activity is not only a remarkable phenomenon to observe but it involves basic problems including the mechanism of cell division and how our own white blood cells (leucocytes) move.

Amoeboid Motion

There are now two standard theories about amoeboid motion, the flowing motion of a pseudopod. Both are old. Over a century ago Dujardin in France and Schultse in Germany proposed that protoplasmic contractility is a major factor in amoeboid motion. The other theory, developed by American students of Protozoa like Jennings and Mast in the first half of this century, places emphasis on the reversibility of the gel and the sol phases of the colloidal cytoplasm. The electron microscope has recently revealed the presence of short stiff microtubules in the cytoplasm of both plant

learn anything (despite recent efforts to prove the contrary), but the behavior of some of the big stentors, such as *Stentor coeruleus*, can be modified by experience. If an extended stentor with its "foot" fixed to some object is squirted with a gentle current of water, it will jerk down into a rounded position. If, after the stentor has extended again, it is squirted a second time, it will not contract. Perhaps the second response involves some increase in the threshold of stimulation, for a stronger stimulus will produce a vigorous contraction. If the stimulus is repeated several times, the stentor will loosen its foot and swim away.

Parameciums sometimes appear to seek out favorable spots. For example, if one end of a trough of water is heated and the other iced, they will collect in the middle. This comes about because parameciums swim in all directions, and whenever one runs into an area that is too cold or too hot, it turns away. However, it is to be admitted that such more or less blind trial and error methods are also employed by human beings in the face of completely new and strange problems.

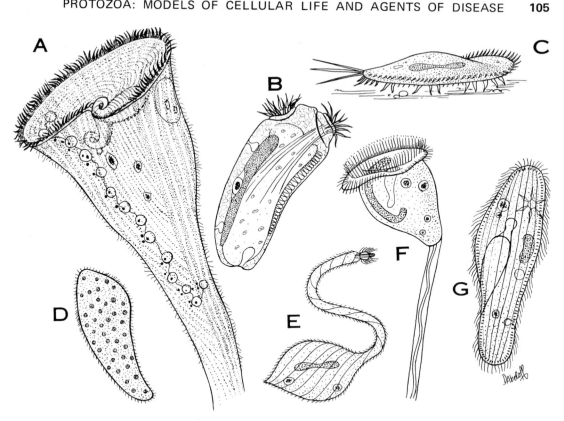

Fig. 5-4. Diversity among Ciliophora. A, *Stentor;* B, *Diplodinium* from cow's stomach; C, a hypotrich from a freshwater pond; D, multinucleate *Opalina* (of questionable relationship) from frog rectum; E, *Lacrymaria,* the tear-drop ciliate; F, *Vorticella;* G, *Paramecium.*

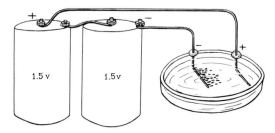

Fig. 5-5. Experimental set-up to demonstrate galvanotaxis in *Paramecium.*

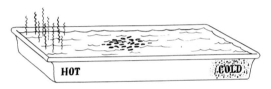

Fig. 5-6. Water trough to demonstrate adaptive result of trial-and-error behavior in *Paramecium.*

and animal cells wherever there is cytoplasmic streaming. These newly discovered structures may furnish the clue which will reconcile these two theories. However, nothing definite is known about the relationship of these organelles to pseudopodial or any other type of cellular motion.

Contractility Theory

Consider first the theory of protoplasmic contractility. If you observe an amoeba like *Arcella,* which lives in a little shell, you will notice that when it moves, a pencil-like pseudopod first flows out to a length of somewhat more than the diameter of the shell. The end adheres to the bottom and the pseudopod contracts and pulls the body of the amoeba and its shell toward the point of attachment. A similar thing can be seen in the heliozoans and

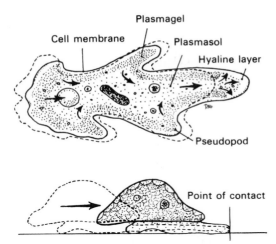

Fig. 5-7. Amoeboid motion diagrammed for *Amoeba proteus* (above) and a thecamoeba, *Arcella* (below).

foraminiferans which send out thin threads of cytoplasm like sun rays.

Added to these old observations are newer lines of evidence. If the long end of a pseudopod is cut off with a thin glass thread, the cut surface will heal, leaving the severed end as a miniature amoeba composed of merely the cytoplasmic membrane *i.e.*, the plasmalemma, enclosing some of the clear hyaline layer of the amoeba. It includes no gel, no granular cytoplasm whatever, and, of course, no nucleus, yet such a fragment will crawl around by amoeboid movement for many hours. The biochemical evidence consists of contractile proteins like those found in muscles that have been extracted from some of the giant amoeboid organisms (see below) which have thousands of nuclei and spread out, sometimes to the size of a half dollar, over damp and decaying wood.

The "Magic Fountain" Theory

The reversible sol-gel theory of amoeboid movement is based on equally valid observations. The advancing pseudopod of an amoeba like *Amoeba proteus* can be seen to have a stream of flowing granular protoplasm in the center of the pseudopod. Tiny crystals, mitochondria, food vacuoles, and the nucleus are carried forward by this stream. Around the edges of the stream and completely enclosing it as a tube or sleeve (the pseudopod is more or less cylindrical) is a continuous layer of protoplasm in the gel phase and in which the granules remain stationary. At the advancing end, the central core of flowing protoplasm spreads out sideways and solidifies, elongating the gel sleeve. Some German biologists have likened this to a "magic fountain" in which the water freezes as it falls and so instead of splashing builds up a collar of ice. At the posterior end of the amoeba the inner surface of the gel becomes fluid and adds itself to the forward flowing stream.

What makes the fountain flow? The answer is not completely known, but part of the answer seems to be the contraction of the plasma membrane. The newly formed end of the pseudopod adheres to some substrate, a glass slide or a leaf. Then a wave of constriction passes backward over the surface, in some way squeezing the contents forward. It may be that in an intact amoeba the gel as well as the plasma membrane contracts. That calcium is essential for amoeboid movement can be easily demonstrated in marine amoebas by placing them in sea water from which the calcium ions have been removed. Amoeboid motion stops. It is worth noting that calcium is also essential for muscular contraction.

Where the contraction waves end at the "tail" of the amoeba they often leave irregular little masses of cytoplasm. Such an aggregation is called a **uroid** (*oura*, tail). They are more prominent in some species of *Amoeba* than in others. They can also be seen in some of the very actively amoeboid white blood cells. From time to time the uroid is completely pinched off and a new one begins to form. Is it justifiable to conclude that both theories are true?

Cilia and Flagella

Ciliates and flagellates both move by the activity of whip-like extensions of their cytoplasm. **Cilia**, which are short and often cover the entire surface of the cell, are found almost throughout the animal kingdom. For example, they cover the gills of oysters and line the tracheas of men. **Flagella**, which are usually few in number and may be as long as the rest of the cell, are more restricted. The tails of all sperms from those of the lower plants to those of man are flagella.

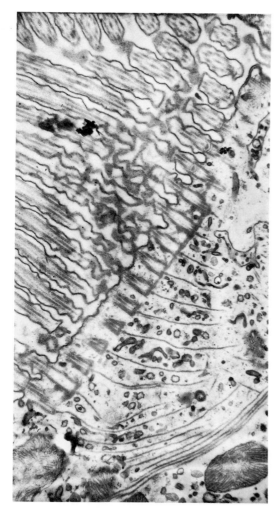

Fig. 5-8. Protozoan ciliation revealed by an electron microscope. Note the row of cilia and their basal bodies (kinetosomes) on the left. A fibril connects adjacent basal bodies, and from each a fibril extends deep into the cell. Near the top of the figure cilia can be seen in cross section, and near the bottom, several mitochondria. (Courtesy H. E. Finley.)

Little is yet known about how either flagella or cilia work. Under a light microscope, it is evident that both arise from a basal granule, the **kinetosome**, in the cytoplasm. A flagellum may rotate like a propeller and pull the animal, or it may wriggle like a writhing snake as it propels the animal. This later action can be observed in the common freshwater flagellate, *Paranema.* In its effective beat a cilium is more or less rigid but appears relaxed on the return stroke. Cilia

are connected to each other by a network of cytoplasmic threads running parallel to and just beneath the cell surface. Under an electron microscope the distinction between flagella and cilia all but disappears. In both, a circle of nine fibrils surrounds a central fibril. The kinetosomes are self-producing organelles and very closely resemble the centrioles in mitosis. There is no more striking indication of the basic unity of all living things than the unity of structure found in cilia and flagella in all animals from the Protozoa to man and in the lower plants which possess motile sperms.

PROTOZOAN NUTRITION AND EXCRETION

The most interesting fact about the dietary requirements of protozoans is their close resemblance to those of human beings. Flagellates, for example, require thiamine (vitamin B_1) just as man does and as do other highly organized animals. Without it carbohydrate metabolism fails and pyruvic acid ($CH_3COCOOH$) accumulates in the blood stream until toxic concentrations are reached and death results. Ciliates also have been shown to have dietary requirements closely similar to those of mammals. Since protozoans derive the energy for living via the anaerobic glycolytic series followed by the further aerobic metabolism of the Krebs cycle and the cytochrome system of the mitochondria, the basic similarities between one- and many-celled animals is only to be expected.

Metabolic wastes like CO_2 merely diffuse through the cell membrane; but excess water, which is taken up by protozoans living in fresh water with its extremely low osmotic concentration, is excreted by a contractile vacuole. How a contractile vacuole works still presents many puzzles. That it requires energy can be demonstrated by exposing a protozoan to a little cyanide, which is a potent inhibitor of the respiratory enzymes of the cytochrome system. The contractile vacuole will slow down and finally stop, after which the animal will swell up and burst.

REPRODUCTION IN PROTOZOA

Reproduction in protozoans has been longer and more intensively investigated than any other aspect of their lives and for a very good reason. In the closing years of the past century several European investigators claimed that a strain of ciliates reproducing by asexual methods could reproduce for but a limited time after which the strain would die out unless they were permitted to reproduce sexually. After sexual exchange of nuclei, the two strains would be reinvigorated and able to undergo another series of asexual generations only to die out again unless sexual reproduction occurred. In direct contradiction were the results of Woodruff with his famous Yale strain of *Paramecium* which appeared to be able to live forever without sex. Also important was the publicity given to Nobel Prize winner Alexis Carrel who claimed to have a strain of chicken heart cells able to survive forever cultured *in vitro*.

Thus there are two very different theories as to the significance of sexual reproduction. One view is that it not only provides the genetic variation, which is the raw material of evolution, but that in some way it is essential for the continued life of a line of cells. The other view is that sex has only one function, the production of variation.

This controversy has recently resumed and is far from settled. No one has ever been able to repeat Carrel's results with normal diploid cells. Moreover, there is good evidence that Carrel's cultures were inadvertently reinoculated with fresh embryonic chick cells from time to time in the chick embryo extract on which his cultures were fed. Within the past several years Hayflick has reported several ingenious experiments which appear to show that there is some kind of a built-in limit to the number of generations cells *in vitro* can undergo. Sonneborn and his students have reinvestigated the question in ciliates and found that many species of ciliates have a characteristic maximum number of possible generations without sexual interchange. Furthermore, different species differ in the number of possible asexual generations.

Sarcodina Reproduction

The common amoebas, *e.g.*, *Amoeba proteus*, so far as is known reproduce only by mitotic cell division and never sexually. There are, of course, the standard stages of cell replication, the M or mitotic stage with the familiar prophase, metaphase, anaphase and telophase, the G_1 or first growth stage, the S stage during which DNA is synthesized, and finally a second growth stage, G_2.

If you ask how certain is it that species such as *Amoeba* never undergo sexual interchange, the answer has to be probably true but not completely certain. After all, bacteria were believed to be completely without sexual phenomena by generations of competent bacteriologists. However, there are many and sometimes complex variations on this theme, including meiosis and fertilization. Amoebas like *Arcella*, for example, have a special problem because of their shells. In *Arcella* the nucleus divides mitotically and half of the protoplasm and one nucleus flows out of the shell, rounds up, and secretes another shell like the first. One daughter cell thus forms a new shell and one keeps the old one. Many amoebas produce rounded cysts with protective walls when conditions become unfavorable. This is true of *Endamoeba histolytica*, the cause of amoebic dysentery.

Among Sarcodina that reproduce sexually, the gametes may be flagellated and there may be an alternation of sexual with asexual generations. This is the case with the Foraminifera, which play an important role in the ecology of the oceans because of their enormous numbers. (Chalk cliffs are composed of the shells of inconceivable myriads of forams.) In the Foraminifera a complex reproductive cycle has been evolved. There are two slightly different body forms. In one, the original chamber of the shell is larger than in the other. Consequently, the two forms are known as **megaspheric** and **microspheric**. The microspheric individuals also known as **schizonts** (a term to be met later in connection with the malarial parasite), are multinucleate. When they attain maximum size they break up asexually into hundreds of tiny amoebas which develop into more schizonts.

A time comes, however, when for unknown reasons some of the offspring of the schizonts do not develop into more schizonts but into megaspheric **gamonts**. These remain uninucleate until they attain maximum size, when the nucleus and cytoplasm divide repeatedly and ultimately form thousands of biflagellate gametes. The gametes from two **different**

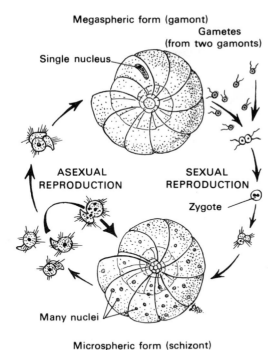

Megaspheric form (gamont)

Gametes (from two gamonts)

Single nucleus

ASEXUAL REPRODUCTION

SEXUAL REPRODUCTION

Zygote

Many nuclei

Microspheric form (schizont)

Fig. 5-9. Sexual and asexual stages in life cycle of a common marine foraminiferan.

gamonts fuse in pairs, forming zygotes, or fertilized eggs, which in turn grow into schizonts.

Ciliate Reproduction

In *Paramecium*, as in other ciliates, there are two nuclei, one very large, the macronucleus, and one (in some species two or more) very small, the micronuclei. In **binary fission**, *i.e.*, asexual reproduction, the macronucleus merely pulls apart without undergoing mitosis. The micronucleus, however, at the same time divides by mitosis. It contains numerous chromosomes. In some strains of *Paramecium caudatum* the haploid number is 18; in others it is 40 or even more. The cytoplasm divides at right angles to the long axis of the body, approximately midway between the two ends. This process requires about half an hour.

Sexual reproduction, *i.e.*, **conjugation**, is more complicated. Two animals come together by their oral groove surfaces. The pairs swim around together for 24 to 48 hours, depending

on the species. In different ciliates the details differ, and even different species of *Paramecium* do not behave exactly alike. The following account applies to *Paramecium caudatum*, often regarded as typical. While the animals are paired, the following events occur: (1) In each animal the micronucleus divides twice, making four haploid micronuclei. This is of course meiosis, as in the formation of any egg or sperm with its single set of chromosomes. (2) Three of the four micronuclei degenerate—how or why is not known. (3) The remaining micronucleus divides by mitosis to give two haploid nuclei which come to lie near the oral region. (4) Each **conjugant**, as the conjugating individuals are called, now extends a cone-shaped protrusion toward the other. These protrusions in the oral region establish a cytoplasmic bridge between the two animals. (5) One of the two haploid nuclei from each conjugant migrates across the cytoplasmic bridge into the other animal and fuses with the stationary micronucleus. The migratory nucleus is comparable to a sperm, the stationary nucleus to an egg. The fused nucleus that results is, of course, diploid and is comparable to the zygote of a metazoan. (6) The conjugants then separate. Each animal is now called an **exconjugant**. (7) In each exconjugant the macronucleus becomes irregular in shape and, during the subsequent divisions of the micronucleus, stretches into twisted skeins and is finally completely absorbed (digested) by the cytoplasm. (8) The diploid micronucleus divides by mitosis three times, giving eight micronuclei. (9) Four of the eight micronuclei in each exconjugant enlarge and grow into new macronuclei. (10) The behavior of the other four micronuclei appears to be related to cell division. Three micronuclei degenerate and the remaining one divides twice by mitosis. Each time this micronucleus divides, the whole animal divides in half, as in asexual reproduction, at right angles to the long axis of the cell. The first division gives two daughters from each exconjugant, each with two immature macronuclei and one diploid micronucleus. The second division yields four granddaughter parameciums from each exconjugant, each with one diploid micronucleus plus one diploid macronucleus. All the individuals produced asexually by mitosis from such an animal are genetically identical.

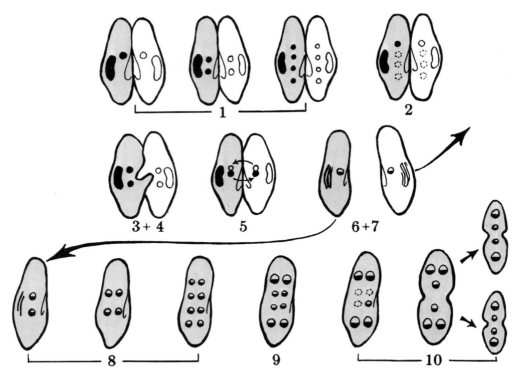

Fig. 5-10. Diagram of conjugation in *Paramecium*. See text for explanation.

These facts make it clear that sexual reproduction in ciliates is basically the same as in the higher animals. There is a cycle of meiosis producing haploid nuclei which fuse to form a diploid zygotic nucleus, and this, after a period of multiplication by mitosis, again undergoes meiosis. The most important genetic difference between a ciliate or other protozoan and a metazoan is that a protozoan lacks the distinction between germ cells and somatic cells. This means that characters acquired by the cytoplasm of a protozoan can be transmitted to subsequent generations because the cytoplasm of the parents is passed on directly to the offspring. Furthermore, during conjugation there is for a brief time a cytoplasmic bridge between the two ciliates so that an interchange of cytoplasmic particles may take place.

Ciliate Mating Types and Scientific Discovery

In investigating the effects of sexual reproduction and hence its meaning in Protozoa, there was one all but insuperable obstacle. Neither parameciums nor other ciliates could be brought together and mated the way most animals can. Groups of parameciums could be mixed together easily enough, but except on rare occasions no conjugation took place. Occasionally, however, and for no known reason, there would be what protozoologists used to call an "epidemic of conjugation." At such times researchers would work around the clock preserving animals in various stages of conjugation.

A new era in the study of sex in ciliates was made possible by the discovery of mating types. The way in which this discovery took place illustrates the accidents that Pasteur claimed happened to "prepared" minds. One winter weekend, Tracy Sonneborn, an American protozoologist, happened to pour two cultures of parameciums together and was astonished to find that they immediately began to clump together, the first stage of conjugation. Mixing other samples of the same two cultures produced the same result. Here at long last appeared to be two sexually differentiated strains, male and female, plus and minus, or whatever one chose to call them. But on

Monday morning the experiment failed to work. Subsequent trials at intervals during the week also failed to produce conjugation. Then on the next weekend conjugation again occurred when the two strains were mixed. The answer was now evident. The university was saving fuel over the weekend and the laboratory cooled off. For conjugation to occur, this species requires a fairly low temperature.

It is now known that every species of *Paramecium*, and of other ciliates so far tested, can be divided into two or more mating types. Strangely enough, there may be more than two interreacting mating types within a species, as well as mutually exclusive groups of mating types. Conjugation will take place only if the conditions of temperature, light, etc., are right. For example, *Paramecium bursaria,* the species which is green due to symbiotic algae, conjugates most readily in the middle of the day, and never before sunrise or after dark. If an animal belongs to mating type A, it can conjugate with any member of type B or type C, but not with another member of A. Likewise a member of type B can conjugate with a member of type A or C but not with one of type B. *Paramecium bursaria* has six mutually exclusive mating type groups. In group 1 there are four mating types that have been called A, B, C, and D; in group 2, there are eight mating types. A member of any one of these can conjugate with a member of any of the other seven in group 2, but not with another member of its own mating type or any of the four mating types in group 1 or with any of the types in the remaining four groups. Of the six mutually exclusive groups of mating types, groups 1, 2, and 3 are from the United States, 4 and 5 are from Russia, and 6 is from England, Ireland, and Czechoslovakia.

In Conclusion

The only certain conclusion that can be reached is that the question of the life spans of **clones** of cells, that is, cells derived from a single progenitor by asexual reproduction, is not settled. It is now known that there was a double catch in that Yale strain of *Paramecium.* First they turned out all to belong to the same mating type so that conjugation is impossible for them anyway. More interesting, it has been found that parameciums undergo periodic nuclear reorganization every so many generations without any nuclear interchange between two individuals. This process is called **autogamy.** The nuclear events are identical for the most part with those seen in conjugation. The macronucleus disintegrates. The micronucleus divides into four haploid nuclei, three of which disintegrate. The remaining nucleus divides by mitosis, but instead of one remaining within the animal and the other acting as a sperm and entering another, these two nuclei fuse, forming a new diploid micronucleus.

If cells actually do have a limited number of possible cell divisions between fertilizations, this fact sets upper limits to growth and longevity. What may cause the progressive aging of a clone is quite unknown. Old clones of ciliates carry more lethal genes than young clones. Perhaps this is also true of clones of metazoan cells growing *in vitro*. No mechanism is known whereby fertilization will rejuvenate a strain of cells.

HOW DO CILIATES MAKE THEMSELVES?

A topic of both basic importance and great fascination remains to be discussed. In normal asexual reproduction where one ciliate divides into two or in regeneration after part of a ciliate has been cut off, how are the new organelles, cilia, oral grooves, contractile vacuoles, etc., made? How does an animal make itself, or, if you prefer, how is the code in the genes translated into the structures of the organism?

In parameciums, for example, the first sign that division is imminent is a duplication of the cilia in the ventral rows near the mouth. The area of duplication spreads in a band around the middle and subsequently out to the ends of the animal. Then a new contractile vacuole is formed immediately anterior to each of the old ones. Thus each new animal will have a new anterior contractile vacuole and a second-hand posterior one. The anterior one of the new animals gets the old mouth, the posterior one a new mouth that seems to be budded off from the old gullet. The cilia certainly seem to be self-duplicating as are the mitochondria and also the symbiotic algae or chloroplasts in those species which possess them. The evidence is

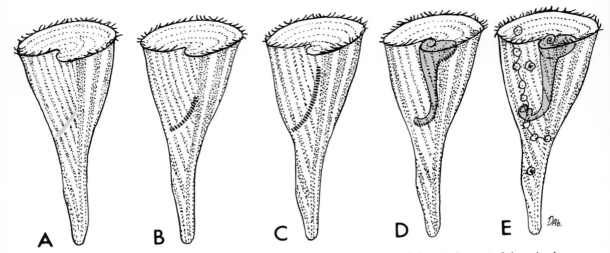

Fig. 5-11. Regeneration of the oral groove in *Stentor* after amputation of the anterior part of the animal.

accumulating that since these structures all carry their own DNA, they are probably the descendants of prokaryotic algae and bacteria which became intracellular **symbionts**. If so, then the problem becomes one of how their duplication is correlated with that of the "host" ciliates or other protozoans and how they come to take up their proper positions within the host cell. The contractile vacuoles seem to arise *de novo*. In some species during asexual division into two individuals, most organelles such as mouth, oral groove, and contractile vacuoles, but not mitochondria, are absorbed and new ones formed in each daughter cell. In other species the organelles are divided between the two new cells, each of which synthesizes some new structures. In ciliates known as **hypotrichs** (see Fig. 5-4C), after injury most or all of the organelles are shed or absorbed and new ones arise. The new structures appear first in a small "regeneration field" and later move out to their definitive positions. Perhaps it is significant that the regeneration field lies close to the macronucleus.

The most extensive investigations of these problems have been done on the giant ciliate, *Stentor*, by Paul Weisz on the East Coast and Vance Tartar on the Pacific. But the investigators are few—which seems strange, because the problems are basic, the techniques no more difficult than in other areas of biological research and, as both Tartar and Weisz point out, the animals are extremely beautiful.

USEFUL REFERENCES

Buetow, D. E. *The Biology of Euglena.* Academic Press, New York, 1968.
Chen, T. T. (ed.) *Research in Protozoology.* Pergamon Press, Inc., New York, 1967.
Curtis, H. *The Marvelous Animals.* Doubleday Natural History Press, New York, 1968.
Hall, R. P. *Protozoan Nutrition.* Blaisdell Publishing Co., New York, 1965.
Jahn, T. L., and Jahn, F. F. *How to Know the Protozoa.* Wm. C. Brown Co., Dubuque, Iowa, 1949.
Kudo, R. R. *Handbook of Protozoology,* 5th ed. Charles C Thomas, Publisher, Springfield, Ill., 1966.
Tartar, V. *The Biology of Stentor.* Pergamon Press, Inc., New York, 1961.
Wolken, J. J. *Euglena: An Experimental Animal for Biochemical and Biophysical Studies,* 2nd ed. Appleton-Hawthorn Books, New York, 1967.

Breakthrough to Bigness: The First Metazoans— Coelenterates, Flatworms, and Sponges

One of the major events in the development of life on our planet, a revolution fully comparable to the appearance of photosynthesis or the conquest of the land, was the emergence of multicellular animals, the **Metazoa**. If animal life had remained on the one-cell level of organization seen in the Protozoa, none of the familiar animals could have evolved and man himself would be unknown. Something like the *Paramecium* is the most that could have been expected.

THEORIES OF METAZOAN ORIGIN

The orthodox theory of how life achieved this new level of complexity was developed by Haeckel in Germany and Huxley in England and is still dominant today. It holds that the first metazoans were **colonial flagellates** similar to *Volvox*, which is a hollow sphere of flagellated cells. Push in one side of this hollow ball and you get a two-layered blind sac resembling the gastrula stage of the embryos of higher animals and also the two-layered body plan of many of the coelenterates. Thus, it is said, the first metazoans were coelenterates like *Hydra*.

The very serious problems with this prevalent theory have long been ignored by the zoological "establishment." Briefly, the flagellates involved are clearly phyto-flagellates, that is, plants producing chlorophyll, cellulose, and starch. They are haploid while animals are diploid. Colonial flagellates are freshwater forms while coelenterates, with only two or three exceptions out of over 10,000 species, are marine.

Moreover, one very large group of coelenterates, the sea anemones and corals, clearly possess three major layers of cells and are thus **triploblastic** like the higher animals. In the words of Libbie Hyman, until her recent death one of the world's outstanding authorities on invertebrate animals: "If one frees oneself from outworn theories dating from Haeckel, one sees at once that there is no essential difference between a cross section of a sea anemone commonly called 'diploblastic' and a cross section of a flatworm commonly called 'triploblastic.' "

The early stages of sea anemone development are **bilateral** and the adults are bilateral internally. These facts strongly suggest that they have evolved from a bilateral group originally. If so, the coelenterates such as the hydras and jellyfish with only an inner and an outer layer of cells separated by a layer of cell-free jelly (**mesoglea**) are specialized descendants from sea anemone-like forms and have lost that middle cellular layer just as modern horses have lost most of the toes possessed by their remote ancestors. However, to quote Hyman again, Haeckel won the argument, "overwhelming his opponents by a flood of expository literature."

The rival theory originated by a Cambridge zoologist, Adam Sedgwick, at the turn of the century has been revived and developed by the Yugoslavian Jovan Hadzi. It holds that metazoans arose from **multinucleate ciliates** which gave rise to simple, gutless (acoelous) flatworms. The points of correspondence between the simplest flatworms and the multinucleate ciliates are both numerous and very striking.

Both live in the ocean where life first arose and where the most primitive forms of animals are to be expected. Both are about 1 mm. long, which is gigantic for a protozoan, but minute for a metazoan. Both are free living heterotrophs, ingesting solid food and dependent on other organisms for complex organic foodstuffs. Furthermore, both are anatomically very similar, with external ciliation and tiny protective vesicles just below the body surface. These are the trichocysts of the ciliates and the rhabdites and sagittocysts of the turbellarian flatworms like planaria. Both lack a gut as well as a body cavity. Food is ingested through a mouth at or near the anterior end and then digested intracellularly.

In many species of primitive gutless flatworms, cells are either absent or few in number with large areas of cytoplasm containing many nuclei unseparated by cell membranes. Recent electron microscope studies of some species show intracellular membranes forming complex interdigitations which resemble the endoplasmic reticulum characteristic of metazoan cells.

Last, as a prominent London zoologist, G. R. de Beer, has emphasized, the ciliates possess "the indispensable requisite for ancestors of the metazoa" in their habit of conjugation with exchange of gamete nuclei rather than complete fusion of male and female organisms as in flagellates. Conjugation in ciliates resembles the mating of multicellular animals in a way the cell fusion of the flagellates does not. Meiosis occurs during gamete formation and fertilization results in a diploid organism in marked contrast to what happens in flagellates where meiosis occurs immediately after fertilization and results in a haploid organism. With many of the simple flatworms copulation is very simple. Each partner injects sperms anywhere into the mesoderm of the other. In some highly specialized ciliates (such as the ophryoscolecids) the migrating nucleus actually resembles a sperm complete with flagellum. All that is lacking in some multinucleate ciliates to make their sexual reproduction essentially metazoan in character is the enclosure of the stationary or female nucleus in its own special cytoplasm to form an egg cell.

It is worth noting that the body plan of a

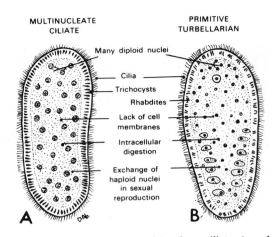

Fig. 6-1. Comparison of a multinucleate ciliate, A, and a primitive gutless flatworm, B.

jellyfish and of a free living flatworm are closely similar (Fig. 6-2). In cross section they are identical in terms of topology (the branch of mathematics which concerns changing shapes). Noteworthy also is the existence of marine (polyclad) flatworms which have a central mouth, a many-branched gut radiating out from the center of the body, and a row of hollow tentacles around the mouth as in sea anemones. If such a flatworm assumed a sedentary life, it would bear a striking resemblance to an anthozoan coelenterate like a sea anemone.

The present evidence is undeniably impressive that the great leap from one-celled protozoa to many-celled metazoa took place when a multinucleate ciliate passed over the very thin line separating it from a simple flatworm. This is the most plausible theory so far advanced, but it remains unproven.

COELENTERATES (CNIDARIA)

For many people jellyfish, sea anemones, corals, and sea fans typify animals without backbones. For the biologist the coelenterates have a special interest. They exhibit in simple form most of the fundamental features and problems of all the higher animals up to and including the social insects, the antisocial octopus, and the big-brained mammals. Furthermore, coelenterates have been important in the development of biological theories of wide application and are still frequently used in research.

There is a story (relevant here) about E. B. Wilson, the first great American investigator of cell structure, who taught invertebrate zoology at Columbia University for many years. He began his course in the fall with the protozoa but when May came and the course was drawing to a close, the class would only be finishing their study of the coelenterates. Professor Wilson felt that if a student had thoroughly studied the protozoa and the coelenterates, he had met all the important problems in animal life and then could, if he was proficient, work out a knowledge of any of the other groups on his own. This is not the outlook of the authors of this book, but Professor Wilson had a point.

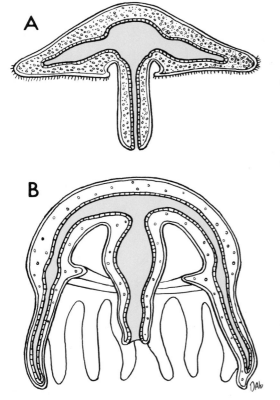

Fig. 6-2. Diagrams of structure of a flatworm, A, and a jellyfish (medusa), B. Green represents the digestive cavities.

Superficially the coelenterates appear very unlike one another. Contrast a jellyfish with a coral or a simple hydra with a sea fan. Yet all the members of this diverse phylum can be defined as **externally radially symmetrical** (the symmetry of a wheel) animals which produce **nematocysts**. Nematocysts, microscopic capsules containing a poisonous thread, give the sea nettle and the Portuguese man-of-war their venomous stings. With the exception of some closely similar structures in some of the ciliate protozoa, no other animals produce nematocysts. For this reason some authors refer to the coelenterates as Cnidaria (cnida is a synonym for nematocyst). All members of the phylum possess a blind sac or gut cavity, called the **coelenteron**, which has a mouth surrounded by **tentacles** but lacks an anus.

Several other features of the coelenterates are usually emphasized, but it is important to remember that they are all either far from universal or are problematical in some way. Over a century ago two Norwegian investigators, Michael Sars and Johann Steenstrup, made the amazing discovery that the eggs of a jellyfish, *i.e.,* of a **medusa**, do not develop into more jellyfish but into a sea anemone-like form called a **hydroid**. This **alternation of generations** is found in most members of two of the major groups of coelentereates, the **hydrozoa** (hydra-like), including small jellyfish, and the **Scyphozoa** (*scyphos,* cup), so called because of the transverse division of their hydroid stage into what resembles a stack of cups, which includes the large jellyfish. The third group, **Anthozoa** (*anthos,* flower), which includes all the sea anemones, corals, and sea fans, never produce anything resembling a medusoid (jellyfish) stage. Medusae are either male or female and their gametes, eggs or sperms, are shed into the sea water. The hydroid forms reproduce asexually, budding off baby jellyfish. Both the jellyfish and the hydroid are diploid, *i.e.,* they possess two sets of chromosomes.

Many books claim that coelenterates are organized on the level of tissues, groups of similar cells, rather than of organs. However, many of the larger jellyfish have well-developed **eyes** with the lens, retina, and pigmented eye cup. Certainly such structures are organs as are tentacles. What coelenterates lack are the

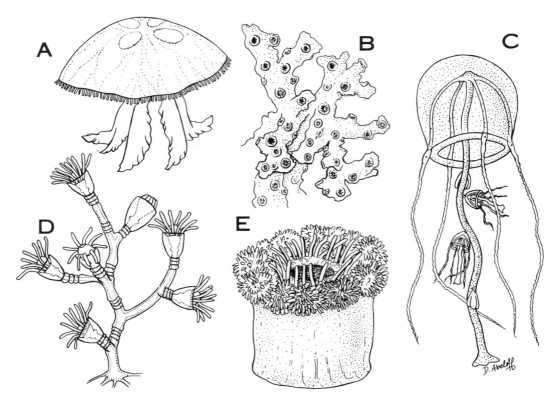

Fig. 6-3. Diversity of form among coelenterates. A, Medusa, a free swimming jellyfish; B, coral; C, a medusa reproducing asexually by budding jellyfish from the manubrium (proboscis); D, a colony of hydroids; E, sea anemone.

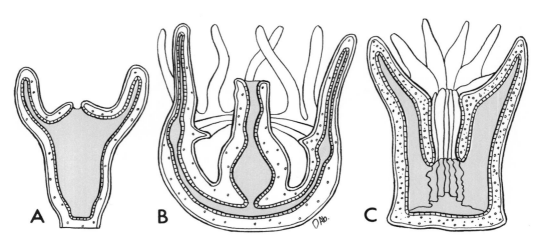

Fig. 6-4. Unity of body plan among coelenterates. A, hydroid; B, medusa; C, sea anemone.

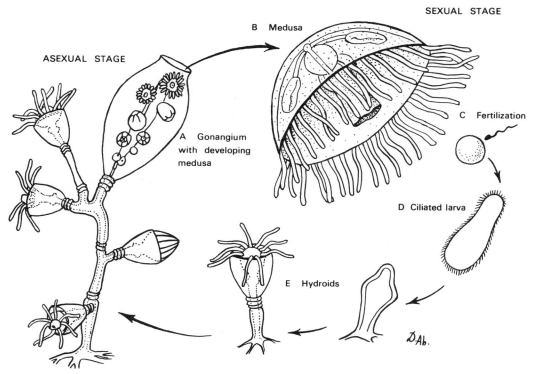

Fig. 6-5. Alternation of generations between asexual stage hydroid on the left and the sexual stage medusa on the right.

complex systems of organs found in the higher animals.

Hydra and Some Fundamental Problems

Coelenterates were introduced to mankind in the 18th century through the highly original experiments on hydra done by Abraham Trembley, whose work astonished his contemporaries. Even the great Voltaire was so dumbfounded that he questioned Trembley's accuracy. Fielding, Smollett, Goldsmith, and other writers of the time ridiculed him. Karl Ernst von Baer, often called the father of embryology, said some years after Trembley's death that his work had led to a revolution in physiology. He certainly raised basic questions about the genesis of animal form and the ability of animals to regenerate lost parts. Many of these questions have not yet been answered, and it is not surprising that *Hydra* continues to be a widely used research animal in the 1970's.

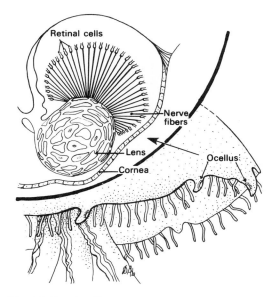

Fig. 6-6. Eye (ocellus) of a medusa. In the enlarged cross section note the retina and lens.

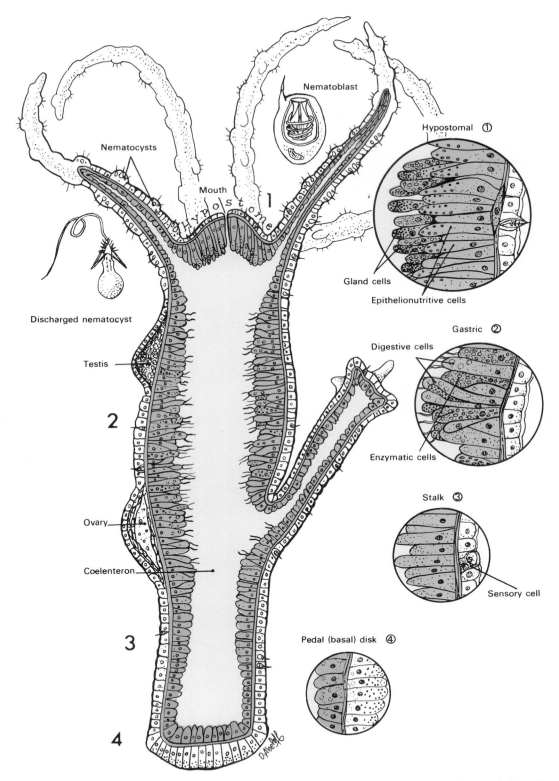

Nematoblast

Nematocysts

Mouth

Hypostome

Hypostomal ①

Gland cells

Epithelionutritive cells

Discharged nematocyst

Testis

Digestive cells

Gastric ②

Enzymatic cells

2

Stalk ③

Ovary

Sensory cell

Coelenteron

3

Pedal (basal) disk ④

4

Fig. 6-7. Diagrammatic longitudinal section of a hydra. Endoderm and digestive cavity (coelenteron) shown in dark and light green, respectively.

Trembley's Experiments and What They Show about Science

In 1740 Trembley discovered the green hydra which seemed to him like a little "water herb" that had the power of movement. Was it an animal or a plant, perhaps similar to the then recently discovered sensitive plant? To answer this question he cut his hydras in half. If the basal part could regenerate a new head with tentacles, then it must be a plant because it was well known that decapitation kills animals. Trembley was amazed at the results: "Who would have thought a head would have returned to it!" Yet Trembley concluded that the hydra was an animal nonetheless. Any experimenter who is not careful to design his experiments so that they can answer the questions he asks can easily find himself in Trembley's dilemma.

Trembley's second experiment is equally instructive because it began a series of investigations which extend over three centuries and illustrates the fact that science is both cumulative and open-ended. The question was very simple, or so it appeared. Can the ectoderm cells which cover the outside of the body assume the form and digestive functions of the endoderm cells lining the gut? To answer this question Trembley turned his hydras inside out and held them that way by pushing a pig's bristle through the animal at right angles to the long axis of the body (Fig. 6-8). A knot was

tied in each end of the bristle so the hydra could not slip off the end and so turn itself right side out again. Under these stringent conditions Trembley found that after several days the outer layer of such hydras had become as clear and smooth as the original one and inner layer had assumed the normal digestive characteristics. At the time everyone was convinced that Trembley had demonstrated that the outer and inner layers could transform into each other according to their position.

Trembley's work was reinvestigated after more than a century had passed because of the emergence of new facts and new ideas, specifically the **germ layer theory** that all animals are composed of three primary layers of cells: **ectoderm, mesoderm,** and **endoderm.** In the rigid form in which this theory was then held, it was thought to be impossible for one to turn into the other. Realizing that either the theory was wrong or Trembley was mistaken, Engelman, professor at Utrecht near where Trembley had carried out his experiments, repeated them without getting the same results. He thereupon rashly concluded that Trembley must have been less than honest since the results he had reported were impossible. At this point the problem was picked up by Nussbaum, a German zoologist who confirmed Trembley's results but showed that the ectoderm cells of an inside-out hydra do not transform themselves into endoderm. Instead they migrate to their proper position on the outside of the hydra,

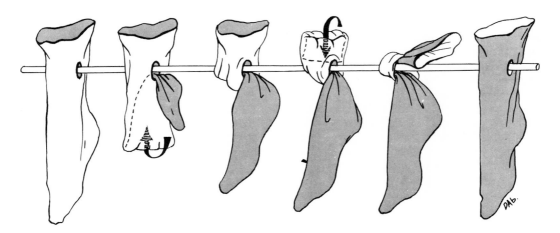

Fig. 6-8. Diagram illustrating Ishikawa's explanation of Trembley's inside-out hydra experiment.

mostly by passing out through the mouth and through the holes made by the silver wire on which the animal had been transfixed.

Further insight into this strange problem was attained by a Japanese investigator who had the good sense to stay up all night and watch his hydras. Ishikawa, who had come to Freiburg to work with Weismann, showed that a transversely transfixed hydra could turn itself right side out without individual cell migration, tearing itself, or getting off the wire. This seemingly impossible feat is accomplished by pushing the basal disk up and out through the space around the transfixing wire.

More recently an investigator in Iowa, Roudabush, has repeated this experiment with a modern twist, statistical analysis. Of 60 hydras turned inside out on a single day, 21 decomposed, 18 turned themselves right side out in the way Ishikawa had reported, and 21 underwent reorganization by the migration of ectoderm cells as first seen by Nussbaum.

But the questions of what makes a hydra turn itself right side out, or why the cells of this or any other multicellular animal are so different from each other, although all are derived from a single fertilized egg, or how an animal can regenerate a "head" end or any other part of its body, remain unanswered.

Hydras in Nature

Hydras are not difficult to find attached to submerged leaves or sticks in ponds and lakes. They can be seen with the naked eye, although a hand lens is helpful. Hydra are carnivorous and eat a variety of small organisms. Like other kinds of animals, there are several species of hydras, each living in some special ecological situation. The green hydra, *Chlorohydra viridissima*, so named by Peter Pallas (a German zoologist who lived in Russia) is notable for the symbiotic green algae living within its endoderm cells. This fact makes more plausible the theory that the chloroplastids of higher plants are the descendants of intracellular symbiotic algae. No one knows how this intimacy of a hydra and an alga began or why the algal cells are limited to the endoderm. The most common hydra in the northern hemisphere is *Hydra oligactis*, a brown hydra also named by Pallas in 1766. This hydra can be identified because

when fully extended the basal third is much thinner than the proximal two-thirds adjacent to the mouth and because it is the only hydra which can paralyze and eat fish fry.

Hydra as a Metazoan

Cellular Differentiation. Viewed under a dissecting microscope, hydras appear to be composed of only one or two kinds of cells but if crushed or sectioned, properly stained, and studied under a compound microscope, the diversity of specialized cell types characteristic of higher animals becomes clear. The ectoderm over the surface of the animal forms an **epithelium**, as any layer of cells covering a surface is called. Here can be seen relatively simple, columnar epithelial cells as well as muscle cells, nerve cells, reproductive cells, mucus-secreting cells, and special cells called **nematoblasts** which manufacture the nematocysts. The endodermal lining of the coelenteron is also composed of epithelial cells, some with flagella which extend into the gut cavity and produce a current in the fluid there. Many of the cells that have amoeboid inner ends ingest and then digest food particles. Hydras are carnivores and eat all manner of small organisms. There are also muscle and nerve cells and glandular cells which secrete digestive enzymes. Digestion is thus both intracellular as in amoebas and other protozoa and extacellular as in the higher metazoa.

Scattered through both ectoderm and endoderm are small undifferentiated cells which are supposed to be embryonic and replace those lost from wear and tear. Between these two layers of cells is the noncellular jelly, **mesoglea**.

Nematocysts. Because nematocysts are so important for protection and food capture and in the accurate identification of all coelenterates, these stinging capsules deserve a close examination. Several types of nematocysts can be seen in hydra and other coelenterates, each supposedly with a different action: some are poisonous, and others wind around the prey. No one knows what induces a cell to produce a nematocyst, much less some particular type. It is known that the **nematoblasts** (cells producing a nematocyst) of hydra arise from small interstitial cells and only in the more distal portions of the body cylinder, never in the basal parts of the body or the tentacles or hypostome. They migrate for unknown reasons from their sites of origin by passing through the mesoglea and endoderm into the coelenteron, where they are carried to the tentacles by the coelenteric circulation.

Each nematoblast bears a pointed thread-like process of cytoplasm, the **cnidocil**, which projects out from the cell surface and is believed to act as a trigger which sets off the discharge of the nematocyst. The nucleus of the nematoblast is pushed to one corner of the cell by the nematocyst, which can be seen under the high power of a microscope to contain a coiled thread and, in some cases, barbs. In discharging, the nematocyst thread turns itself inside out like the finger of a glove everted by air pressure.

The nature of the stimulus necessary for discharge of nematocysts is very uncertain. Mechanical stimulation seems to be involved, yet certain commensal or ectoparasitic ciliates can be seen to run over the cnidocils and bend them down without inducing a discharge. Chemical stimuli are also believed to be involved, yet the juice of *Daphnia*, a favorite food of the hydra, is without appreciable effect.

What are the powerful poisons in nematocysts which can produce redness of the skin,

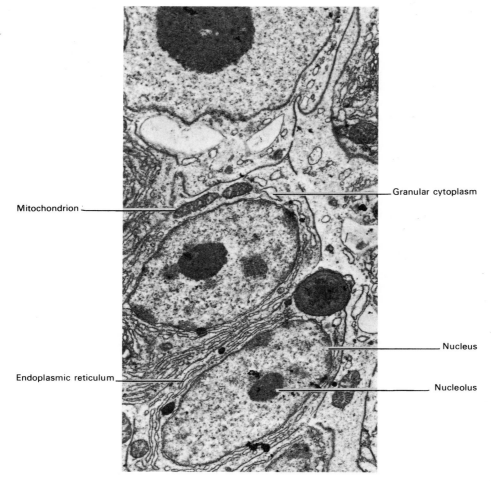

Fig. 6-9. Ectoderm cells of hydra under an electron microscope. Note the nucleus, nucleolus, mitochondria, and endoplasmic reticulum. (Courtesy A. Hess.)

swelling, pain, and even paralysis in man and death in many animals? Paper chromotography and other modern methods have shown that nematocysts contain a number of pharmacologically active substances. **Serotonin** (5-hydroxytryptamine) is abundant in a variety of coelenterates, especially in the tentacles where the nematocysts are concentrated. Serotonin is a powerful constrictor of blood vessels. It is found in the human brain and in abnormal amounts is associated with mental disease. In high concentrations it is a powerful pain producer. An ammonium compound called tetramine and various toxic proteins which block the junctions between nerve cells are also found.

Muscles and Nerves. It is the neuromuscular system which confers on multicellular animals their most important characteristic, namely their complex behavior. The basic features of this system appear in stark outline in hydra. Muscles can only contract; they cannot push. Consequently, muscles have to be arranged in **antagonistic pairs** or sets. In hydra the muscle cells in the ectoderm are T-shaped with the cell body containing the nucleus situated in the epithelium and with a long narrow extension containing the contractile fibrils pressed against the mesoglea. These fibrils run parallel to the long axis of the body extending from apex to base. Their contraction shortens the hydra. Similar **epitheliomuscular cells** occur in the endoderm, but their fibers run at right angles to the long axis. Hence, their contraction causes the hydra to become long and thin.

Basically similar sets of antagonists, flexors and extensors, make movement possible for worms, insects, mollusks, and men. For such pairs of muscles to function at all, a special action of the nervous system is essential, because when the flexors contract, the extensors must relax in a precisely coordinated manner or the animal would simply become rigid. This complex phenomenon is known as **reciprocal inhibition** (see Chapter 32).

With proper staining it is clear that coelenterates (like hydra, for example) possess a **network** of nerve cells throughout the body. Concentrations of neurons making nerve cords run out the "spokes" of a jellyfish disc or encircle the edge of the bell. Each nerve cell is in functional connection with other nerve cells

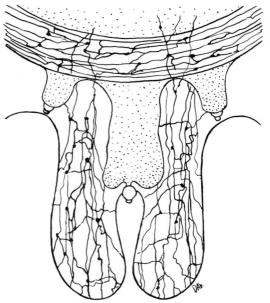

Fig. 6-10. Nerve fiber patterns at the edge of a medusa.

by **synapses**, where the end of a nerve fiber comes into very close juxtaposition either with another nerve fiber or with the body of another nerve cell. Thus the basic structure of this system is the same as in the other animals from flatworms to man and the insects. Many coelenterates lack sense organs, although some of the very active deep-sea jellyfish have both **eyes** and **statocysts**, *i.e.,* pits or vesicles containing minute granules, so they have both some kind of vision and a sense of up and down.

Reproduction. Hydras reproduce both by asexual budding and sexually by eggs and sperm which develop in the ectoderm close to the mesoglea. Some species are hermaphroditic, while in others the sexes are separate. This presents the problems of sex determination, of the stimulus or conditions which bring on the change from asexual to sexual reproduction, and the change from ordinary mitosis to that special type of cell division called meiosis which results in gametes each with a single set of chromosomes.

The Problems of Alternation of Generations and Polymorphism

The egg of a jellyfish develops into a ciliated larva about the size and shape of a paramecium.

Called a **planula**, it swims freely for a while, disseminating the species, and then settles on some object where it grows into a hydroid. For the smaller kinds of jellyfish this hydroid generation is a branching system of stolons through which runs the coelenteron. The colony is thus a kind of biological commune in which all individuals, *i.e.*, the hydranths with their tentacles, share a common digestive tract. The individuals are usually of two sorts, nutritive with tentacles which catch prey, and reproductive which bud off baby jellyfish asexually. Because the whole colony has developed from a single egg and therefore has the same chromosomal composition, all the jellyfish budded off by a single colony will have the same sex. The planulas of the big jellyfish develop into a single hydroid which may be only a few millimeters long. These are the forms which undergo a series of transverse constrictions to form what resembles a pile of cups, each one of which breaks off and grows into a jellyfish, leaving the hydroid to grow a new set of jellyfish. This is one reason why it is so hard to control the sea nettle jellyfish which make swimming so unpleasant in many marine bays and estuaries.

Within this general pattern of alternating hydroid and medusoid generations is much variation. Edward Forbes discovered long ago in that old pirate lair, Penzance Bay, a kind of jellyfish producing other jellyfish by asexual budding. In his own words, "Imagine an elephant with a number of little elephants sprouting from his shoulders and thighs!" (Fig. 6-3C). At the other extreme are species of jellyfish which live far out at sea and skip the hydroid stage by shedding eggs which develop directly into jellyfish. The familiar hydra also lacks the medusa stage.

What are the adaptive advantages of alternation of generations? Which came first in evolution—hydroid, medusa, or something intermediate? No convincing answers are available.

Polymorphism (*poly*, many, + *morphe*, shape) is most easily seen in some of the hydroid forms which grow as a network of branching stolons over a rock or old shell. The "individuals" which grow up from the stolons are variously differentiated. They may be nutritive, with tentacles and mouth; reproductive, budding jellyfish or shedding eggs but lacking tentacles or mouth; defensive, reduced to long flexible stalks with a knob full of nematocysts at the end; or protective, with cones covered with a hard, limy coat. The adaptive significance of this polymorphism is easy to see, but what determines that a given individual will grow into one and not another type of hydroid is quite unknown. In some coelenterates such as the beautiful but highly poisonous Portuguese man-of-war, *Physalia pelagica,* the structures are so complicated that they have baffled generations of biologists. In some ways *Physalia* appears to be a single medusa in which various parts, the manubrium or proboscis, the swimming bell, etc., have multiplied like some Hindu god with many arms or tongues. In other ways the Portuguese man-of-war appears to be a colony of individuals each so highly specialized that it performs only a single function. Alternation of generations can be regarded as an aspect of polymorphism because there are two or more different body forms for a single species in both cases. Both phenomena are also aspects of the general biological problem of **embryonic differentiation**. It is not harder, or easier, to understand how the same set of genes can produce both a jellyfish and a hydroid than it is how they can produce both brain cells and liver cells.

Anthozoa

The third group of coelenterates, the Anthozoa, is comprised of corals, sea anemones,

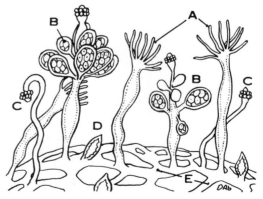

Fig. 6-11. Polymorphism in a colonial hydroid. Four types of individuals can be seen: A, nutritive; B, reproductive; C, defensive; D, protective; E, stolons.

and sea fans, which superficially resemble the familiar hydra, except for their greater size and lack of the medusoid stage in their life cycle. They differ from hydra in having a tubular pharynx leading into a gullet lined with epidermis, a coelenteron longitudinally divided by vertical partitions which in cross section resembles a spoked wheel without a hub, and a much thicker mesoglea elaborated into a fibrous connective tissue.

Throughout history coral has attracted man's attention, whether as a polished, decorative stone or as a navigational hazard. Coral and its relatives were generally regarded as plants until

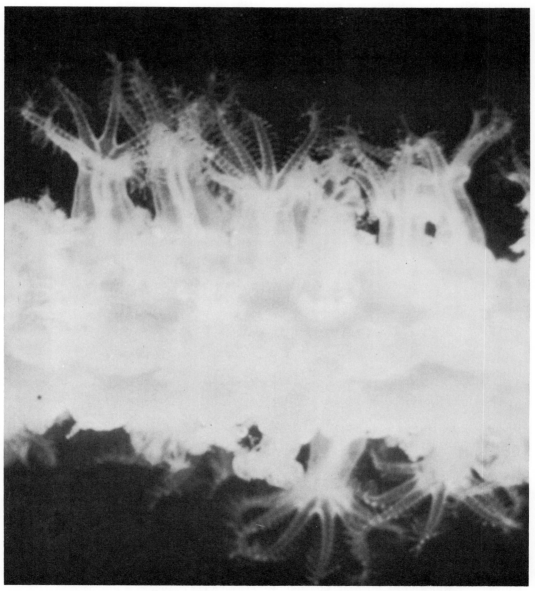

Fig. 6-12. Living coral hydroids. *Leptogorgia*, a relative of red coral. (Courtesy William H. Amos. From G. B. Moment: *General Zoology*, 2nd ed. Houghton Mifflin Co., Boston, 1967.)

the early part of the 18th century when Peysonnel, a Marseilles physician, demonstrated that corals were really animals. Careful examination of a piece of dried, bleached coral reveals within it myriad minute chambers previously occupied by generations of individual polyps whose epidermal cells secreted the hard, limy protective skeletal structure. At night these fragile, carnivorous polyps extend their tentacles to capture zooplankton. Polyps of different species are similar but their secreted skeletons have an amazing diversity of form and color.

Coral reefs, such as the nearly 1,200-mile-long Great Barrier Reef of Australia, are composed of the skeletal accumulation from colonies of polyps which built upon one another generation after generation. This is an impressive example of how in time a lower animal can significantly alter the face of the earth.

THE FLATWORMS (PLATYHELMINTHES)

Flatworms or Platyhelminthes (*platy,* flat, + *helminthos,* worm) are a moderate-sized phylum of about 6,000 species. Included are many harmless forms, some of which have proved useful in research on animal growth. Others are now being used to investigate the possibility that learning can be transmitted from one animal to another by extracting nucleic acid from a trained individual and feeding it to an untrained one. Included also in this phylum are a host of parasitic species which cause diseases in man and in many animals, both domestic and wild. The lung, liver, and blood flukes, and also the tapeworms, are in this group.

All flatworms are characterized by a thin, flat body with no coelom or body cavity, anus, circulatory, respiratory, or skeletal system. Although one small group of tiny marine genera do not have a gut, this structure is present in the rest as a straight rod-shaped tube, a three-branched system, or a many-branched system. There are three primary layers of cells, ectoderm covering the body, endoderm lining the gut, and a rather loose mesoderm filling the space between.

Most flatworms are hermaphroditic. Some parasites, such as tapeworms and liver flukes,

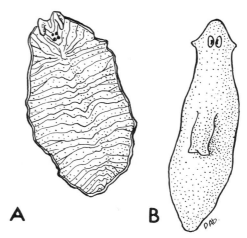

Fig. 6-13. Free living flatworms. A, marine; B, freshwater.

are hermaphroditic, while the highly successful parasitic blood flukes and nematodes are not. The adaptive, evolutionary advantages and disadvantages of hermaphroditism are obscure.

The free living flatworms, the **turbellarians,** have been recognized as a separate group of animals since the time of the American Revolution. These are the flatworms so useful in research. The parasitic flatworms fall into two groups, the flukes or **trematodes,** and the tapeworms or **cestodes.** Both have been a scourge from ancient times.

Turbellarians

Turbellarians are **free living,** non-parasitic flatworms. Externally they are ciliated, especially on the ventral and lateral surfaces. It is from the turbulence of this ciliated surface that they derive their name. The ciliated epithelium covering the body of the worm contains minute rod-shaped vesicles called **rhabdites.** The rhabdites resemble the trichocysts of *Paramecium* and are believed to have a protective function, but this has never been proved. Like most other flatworms, the turbellarians are hermaphroditic; however, unlike the trematodes and cestodes, they have very simple life histories. Some also reproduce asexually by fission, dividing transversely several times until a chain of individuals is produced.

Turbellarians are famous for their spectacular powers of **regeneration,** which not only enable the body to grow a new head but enable a left half of the worm to regenerate a new right half, and *vice versa*. As a result, turbellarians have been extensively used in the experimental analysis of animal development.

Locomotor Systems

Flatworms keep their slithery ribbon shape by virtue of a mass of specialized mesoderm cells that fill the space between the ciliated epithelium of the body surface and the centrally placed, three-branched intestine. These irregularly shaped mesodermal cells that give support are known as **mesenchyme**. Mesenchyme cells are widespread in animals. They permeate and strengthen virtually every organ in the human body.

The gliding motion of flatworms is due to the coordinated beating of the cilia on the ventral surface. The structure of the cilia appears to be essentially the same as in the protozoans. The mechanism of their coordinated beat is very little understood.

Turning and similar motions of turbellarian flatworms are made possible by using two antagonistic sets of muscles. Just under the epidermis is a layer of transverse muscle fibers encircling the animal. Under this is a layer of longitudinal muscles extending from head to tail. Antagonistic pairs of muscles have been seen earlier in the coelenterates, and are found throughout the animal kingdom.

Digestion and Excretion

The three-branched gut of the common turbellarian flatworms leads into the muscular pharynx, a conspicuous tubular structure lying, when retracted, in a pharyngeal cavity. In feeding, the pharynx is protruded through the mouth which is located in the midline on the underside of the body. Particles of food, small animals, and, in the laboratory, bits of liver, meat, or egg yolk are sucked up by the pharynx and then these are passed along into the intestine, where much, and perhaps all, of the digestion is intracellular, *i.e.*, within the cytoplasm of the gut cells. Digested food passes from cell to cell at least in part by diffusion.

As in many of the lower invertebrates, the excretory **apparatus** of the turbellarians consists of tubules ending in **flame cells**, sometimes called flame bulbs. In the parasitic trematodes, to be discussed later, the patterns of arrangement of these excretory tubules and bulbs are an important means of accurate species identification. In most free living flatworms there is an interconnecting row of such tubules along each side of the body. Each flame cell consists of a single cell bearing a long waving tuft of cilia resembling a flame. The cytoplasm of the cell is drawn out into a long tubule around the "flame." This tubule connects with others, commonly in groups of three, which ultimately lead to the exterior via two minute dorsal excretory pores.

The ciliary flame pushes (perhaps fans is a better word) water down the tubule and creates a slight negative pressure within it. As a result, water from the surrounding tissues is drawn into the tubule. As with most excretory systems, including those of vertebrates like ourselves, the primary function is the elimination of excess water. This is indicated by the relative lack of flame cells in marine flatworms, which live in an environment having an osmotic pressure high enough to prevent much water from entering the tissues.

Reproduction

In the simplest turbellarians, the reproductive system consists of little more than paired groups of egg and sperm cells and a protrusible organ, the **penis**, to inject sperms through a temporary break in the "skin" directly into the mesenchyme of another individual.

In the common turbellarians a complete set of reproductive structures of both sexes is present in each individual. A short distance behind the eyes lie a pair of **ovaries**. From each ovary there extends posteriorly an ovovitelline duct which connects with many small **yolk glands**. The two ovovitelline ducts join near the hind end of the body and then enter a cavity known as the **genital antrum**. From near this point extends a long tube, usually called the **vagina**, ending in a sac, the **copulatory bursa**. The male reproductive system consists of two rows of **testes**, each with its own **vas deferens** running parallel to the ovovitelline duct on the same side of the body. Near the posterior end, each vas deferens has an expanded portion where sperms are stored. The vasa deferentia join within the muscular penis, which protrudes into the genital antrum.

In reproducing sexually, each worm inserts its penis into the genital antrum of the other and deposits sperms in the copulatory bursa. The sperms soon pass from the bursa up the ovovitelline ducts to the ovaries, probably by muscular action of the ducts. **Fertilization** takes place as the eggs leave the ovary. As the eggs pass down the ovovitelline ducts, they are joined by large numbers of yolk cells. Fertilized eggs and yolk cells collect in the genital antrum, where droplets from the yolk cells form a protective proteinaceous capsule around the mass of yolk cells and the zygotes, which number from two or three to a dozen. The egg capsules of many species are stalked.

The **cleavage** of the eggs is known as spiral because of the precise spiral pattern in which the early cell divisions take place. In fact, so precise is this pattern that in flatworms, annelids, mollusks, and other protostomes, every cell up to the 24th cleavage has a special name and a fixed destiny.

Regeneration

The remarkable powers of regeneration of the flatworms have long fascinated modern investigators. If a flatworm is cut in half crosswise, the head end forms a new tail, and the hind end forms a new head and pharynx. More remarkable, if a flatworm is cut in half lengthwise, the right half will regenerate a new left half and the left half a new right. Moreover, flatworms appear virtually immortal. If starved, they absorb their reproductive structures, and if further starved simply grow smaller and smaller. It is important to note that the size of the cells does not undergo reduction. The diminution is one of cell number. As in a starving man or rat, the nervous system seems the last to be

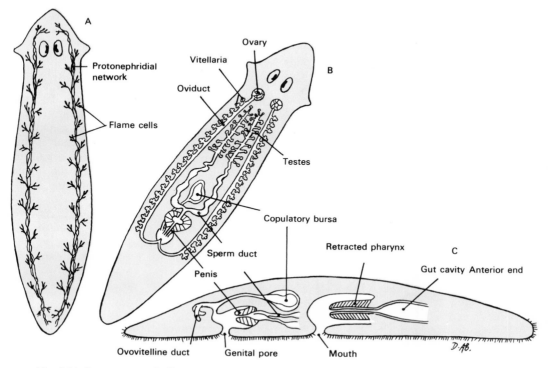

Fig. 6-14. Organ systems in flatworms. A, excretory; B, reproductive; C, reproductive and digestive.

affected. When food is again available these worms grow to adult size with the proliferation of new cells.

Such powers of regeneration present great and far-reaching scientific problems. How is it possible for animals to do all this? When a flatworm is cut in half transversely, how does the cut surface of the anterior end "know," so to speak, that it should grow a new hind end? How does the cut surface of the hind end know that it should regenerate a new head and pharynx? Where do the cells come from that form these new structures?

Nervous System

The two most conspicuous sense organs of the planarian *Dugesia* are the **auricles**, little ear-like points on the sides of the head, and the **eyes**. The auricles are not ears, but olfactory organs. If the auricles are removed, a flatworm can no longer locate food by detecting its odor in the water. The eyes consist of a pair of black pigment cups, each located against the medial margin of a circular area devoid of pigment, so as to give the worms a cross-eyed look. The black pigment is melanin as in all animals, including man. The light-sensitive cells dip into these cups with their photosensitive ends pointing away from the source of light. This amounts to an inverted retina, found elsewhere only in vertebrates and a few mollusks.

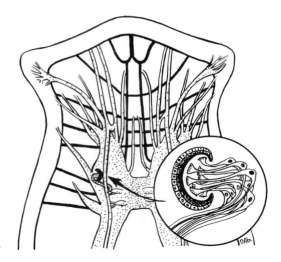

Fig. 6-16. Brain, nerves, and eye of a flatworm. Inset shows right eye cup enlarged.

What is the function of the melanin pigment cups into which the light-sensitive cells dip? Without such shields, light from every direction, except perhaps from directly behind the worm, would reach the light-sensitive cells. Hence, the worm could tell whether it was in the light or the dark, but would have no way of knowing from what direction the light was coming. The opaque cup of the right eye shuts out light rays from all directions except from the animal's right. The left eye acts in the same way. Purely by virtue of their anatomy, these eyes are highly selective and can be stimulated by light from a single direction only. This makes it possible for a worm to turn toward or away from a light source. Such are the simple anatomical beginnings of discriminating, intelligent behavior.

The **brain** is a two-lobed structure close to the eyes. From it a dozen or more nerves extend out into the head and two, or in some species four, main nerve cords pass toward the tail.

Behavior

The behavior of turbellarian flatworms has been extensively explored by American, German, Russian, and Japanese investigators. It is now being reinvestigated in connection with the search for the biochemical basis of learning.

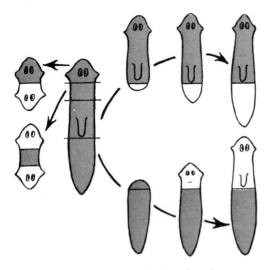

Fig. 6-15. Regeneration in a planarian.

This behavior is of two basic types. The first is a rather general response, usually an increased "random" motion following a generalized increase in some stimulus. For example, a general increase in illumination will cause a quiescent flatworm to glide actively around its dish. Such a non-directed response to light is termed **photokinesis**. A similar response, **thermokinesis**, can be produced by heat.

The second type of reaction is highly specific and directed. As explained above, the anatomy of a flatworm's eyes is such that the right eye can be stimulated only by light reaching it from the animal's right side, and *vice versa*. With this equipment, plus, of course, a nervous system and some muscles, a flatworm will turn away from a source of light. Such a directed reaction in response to an impinging stimulus is termed a **taxis**, in this case a **phototaxis**.

It has been shown that flatworms can learn a simple T maze and can be conditioned to respond in a certain way to a particular stimulus. Incredible as it seems, worms will "remember" what they have been taught after their heads have been removed and new ones regenerated. This is in partial contrast to earthworms, where the amputation of the first several segments, including the brain, does not erase learning until a new brain is regenerated. When that happens, the worm has a "clean" brain and must learn again. McConnell and others have even found that flatworms fed on the brains of trained individuals learn faster than controls. Although investigators in a number of laboratories failed to repeat these amazing results, more recently very careful workers have obtained positive results.

Trematodes and Cestodes

Trematodes (flukes) and cestodes (tapeworms) are known to have been serious and widespread parasites of man and his domestic animals from the most ancient times. However, their often complex life histories, some knowledge of which is necessary for any effective control, have only been worked out in very recent years. Who would have thought that a parasitic worm living in the bloodstream of people in warm parts of the world such as the Nile Valley and Southeast Asia passed a phase of its life history in aquatic snails and that the disease was contracted by wading or washing in the water where these snails lived? Tapeworms present equally surprising life histories and habits. Both types of parasitic flatworms have the same basic structure as is seen in the free living planarians but with special modifications for parasitic life. All these things will be discussed in Chapter 33.

CTENOPHORES

Ctenophores (the *c* is silent) are among the most common marine animals but also among the least familiar. They look like jellyfish without the streaming tentacles. Their ecological importance may be considerable because they eat other animals, especially larvae, but it is largely unstudied. Esthetically they are among the most beautiful creatures known. A ctenophore, as the poet Thomas Grey wrote in an only slightly different connotation, is a "gem of purest ray serene." These are living gems, found not in "the dark unfathomed caves of ocean" but in the dazzling brilliance of the sunlight on the high seas. Examined alive, their bodies possess the incredible glistening transparency of the finest glass.

The first question most often asked about ctenophores is why they are not coelenterates. Zoologists did not generally recognize the distinctions until about 1900. There are, however, several clear-cut and basic differences. Ctenophores never produce nematocysts, which are characteristic of all coelenterates. The chief locomotor system of ctenophores consists of eight vertical rows of comb-like plates of cilia, hence the name of the phylum (*ctenos*, comb, + *phoros*, to bear). The type of cleavage of the egg and the early development of the embryo are both very different from that found in the coelenterates. The cleavage follows a complex and precisely determinate pattern, as it does in flatworms, annelids, and mollusks, but not in coelenterates. The embryo does not form a planula, but develops directly into a small edition of the adult. There is never anything corresponding to a hydroid stage, and the swimming adult is very different from a medusa in its bodily organization. There is no circular band of muscle fibers around the margin of a bell. The body is not umbrella-

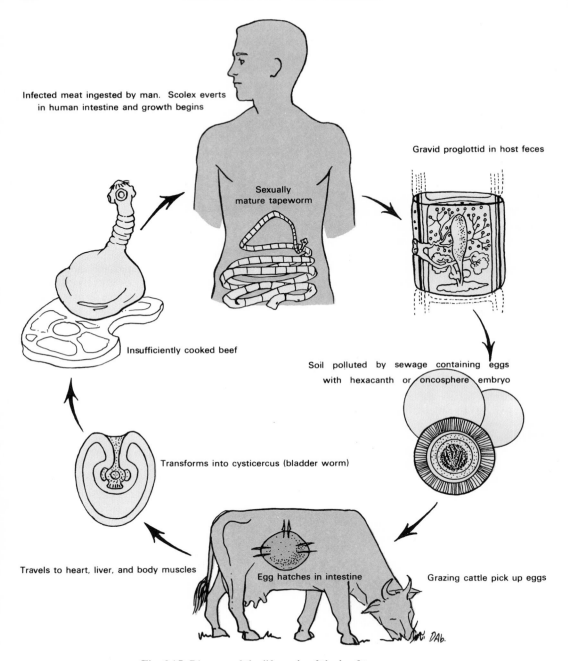

Infected meat ingested by man. Scolex everts in human intestine and growth begins

Gravid proglottid in host feces

Sexually mature tapeworm

Insufficiently cooked beef

Soil polluted by sewage containing eggs with hexacanth or oncosphere embryo

Transforms into cysticercus (bladder worm)

Travels to heart, liver, and body muscles

Egg hatches in intestine

Grazing cattle pick up eggs

Fig. 6-17. Diagram of the life cycle of the beef tapeworm.

shaped, and there is no proboscis, *i.e.*, no manubrium. Instead, the body is generally egg-shaped. The reproductive system of ctenophores is very different from that of jellyfish. Ctenophores are hermaphroditic, while in coelenterates the sexes are usually separate. In fact, the more one knows about a ctenophore, the less it seems like a coelenterate.

Ctenophores may be defined as biradial animals without nematocysts and having eight meridional rows of ciliated **comb-plates**.

It seems strange that a group of animals like the ctenophores, which are so common in the oceans in all parts of the world, should be represented by so few species (there are scarcely 100 in the phylum). This is a real problem for the student of evolution.

Structure

A typical ctenophore is spherical or lemon-shaped. The clear jelly of the body consists mostly of water with some protein. The mouth is at the lower or **oral pole**, and a complex jewel-like sense organ and anal pore are at the opposite **aboral pole**. The **sense organ** consists of a very small calcareous particle supported like a precious stone on four equally spaced tufts of cilia. The whole is enclosed in an astrodome-like statocyst.

From this sense organ eight approximately equally spaced rows of **comb-plates** extend down along eight meridians almost to the mouth. Each comb-plate is made up of a row of large cilia. It is the beating of these cilia that moves the ctenophore through the water and which also produces the shimmering rainbows visible when the animal is removed from the water. Extending from opposite sides of the ctenophore is a pair of long extremely fine and **fringed tentacles**. These are equipped with peculiar adhesive units called **lasso cells** and known only in ctenophores. They capture small creatures of the plankton.

The mouth leads into a straight pharynx which extends upward almost to the sense organ. In the upper portion a "stomach" extends in the plane between the tentacles, and a system of **gastrovascular canals** leads out from it and underlies the eight rows of comb-plates.

Eight elongate gonads, each divided lengthwise into ovarian and testicular halves, lie under each of the eight rows of comb-plates. The eggs undergo a precisely determinate cleavage and the larval form is merely a small rounded version of the adult. There is never any asexual reproduction or alternation of generations.

Luminescence

Of all the animal phyla, ctenophores are easily pre-eminent in luminescence. To anyone familiar only with the light of fireflies and glow worms, the lighting arrangements of ctenophores are truly a revelation.

It has become evident that there is a vast array of light-producing animals: protozoans like *Noctiluca*, jellyfish, hydroids, sea pens, deep-sea squids, various shrimp, brittle stars, marine worms, many fish, various insects, and even some earthworms. Luminescence has been intensively investigated, however, in only three organisms—luminescent bacteria, a tiny but very brilliant Japanese crustacean called *Cypridina*, and the common North American fireflies *Photinus* and *Photuris*.

Of what use is the light to each or all of these diverse organisms? In the fireflies and some marine worms it is known to be a mating signal. In some odd deep-sea fish, the light-producing organs may serve as lures to attract prey or as mating signs, but in many luminous animals like ctenophores it is difficult to imagine a function for the light.

The anatomy of light production in ctenophores is simple. The light is produced most intensely in the walls of the gastrovascular canals that underlie the eight vertical rows of comb-plates. The luminous tissue is either that

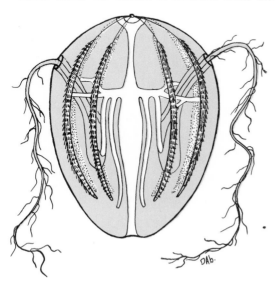

Fig. 6-18. Structure of a ctenophore, *Mnemiopsis*, a common U.S. east coast species (life size).

of the gonads themselves or cells immediately covering the gonads. Since the segmenting eggs of ctenophores are luminous, it seems possible that the gonads are directly involved.

The physiology of **light production** by ctenophores presents some puzzles. Ctenophores do not emit their light continuously but in brilliant flashes upon stimulation. The emission of light is under nervous control and in some way inhibited by external light. A ctenophore acclimated to daylight will not luminesce until it has been in a dark room for 5 to 40 minutes, depending on the intensity of the light to which it has been subjected. Injured but still living ctenophores picked up on the beach in bright sunlight will luminesce brilliantly when touched with a pencil if they are placed in a glass of sea water and allowed to remain in a lightproof closet for half an hour.

Modern knowledge of the biochemistry of light production in animals began with the work of Rene Dubois in 1885 on luminous clams. He found that if the luminous organs are immersed in hot water, the light is extinguished permanently, but ground-up luminous organs will glow for a time. In brief, neither a hot water nor a cold water extract will glow indefinitely, but if they are mixed, even long afterwards, light will reappear. An extract made in hot water evidently contains some heat-stable substance capable of emitting light under the proper conditions. This substance was named **luciferin** by Dubois. The cold water extract evidently contains an enzyme capable of oxidizing luciferin with the production of light. Dubois named it **luciferase**. (Enzymes are inactivated permanently by heat.) A formula can be imagined as follows:

$$\text{Luciferin} + O_2 \xrightarrow{\text{luciferase}} \text{oxyluciferin} + \text{light}$$

By suitable procedures the oxyluciferin can be reduced to luciferin again. Modern research has

Fig. 6-19. Living ctenophores luminescing. (Courtesy George G. Lower. From G. B. Moment: *General Zoology,* 2nd ed., Houghton Mifflin Co., Boston, 1967.)

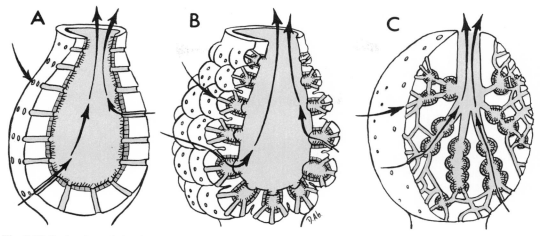

Fig. 6-20. Body plan of the three major types of sponges. A, very primitive sac or ascon sponge; B, sycon type; C, highly evolved leucon sponge such as the commercial sponge.

shown that this scheme will fit all known cases. In fireflies the color of light is controlled by the species contributing the luciferase. *Photinus* emits an orange light and *Photuris* a yellow one. In reciprocal mixtures of luciferin and luciferase between fireflies of these two genera, the color of the light corresponds to the enzyme used.

Luciferase has all the usual chemical and physical properties of an enzyme and thus is presumably a protein. Luciferin is a relatively small molecule with a molecular weight between 250 and 500. Recent work shows it to be an aldehyde associated with a flavin nucleotide, a type of substance concerned with cell respiration. Not surprisingly, **ATP** (adenosine triphosphate), the phosphorus-containing compound essential for the transfer of energy in muscle contraction and in secretion, is also essential for the production of animal light—so deep-set is the unit of living processes on the biochemical level. Bioluminescence is used as a sensitive assay for ATP. Dried firefly abdomens can be purchased for this purpose from most biochemical supply companies.

SPONGES

Although sponges have been articles of commerce for many centuries, it cannot be claimed that they are very important animals economically, scientifically, or esthetically. A few species injure oysters by boring into their shells, and some bright red marine species are poisonous to the human skin, but for the most part sponges are harmless. They are generally admitted to be an evolutionary blind alley and unrelated to any other animal group except perhaps some flagellates.

Over 2,000 years ago Aristotle believed sponges to be animals, but many competent zoologists down to modern times regarded them as plants. This is surprising because one of the most obviously animal-like traits of sponges was discovered in George Washington's day: currents of water continually stream into and out of a living sponge. The final blow to the plant theory of sponges was delivered by a student of the father of American zoology, Louis Agassiz. In the 1860's this student, James Clark, discovered that sponges have flagellated collar cells closely resembling a well-known group of protozoans. Professor Agassiz found this hard to believe and tried to dissuade the student from putting his discovery into print. The professor turned out to be wrong. Although most sponges are marine, both green (from symbiotic algae) and brown species are fairly common in fresh water.

Typical Structure

The characteristic features of sponges are **flagellated collar cells** and **spicules**. The former consist of cells possessing a single long flagellum surrounded at its base by a thin cytoplasmic collar. These cells either line a central cavity or line small chambers in the body of the sponge. The beating flagella draw a current of water in through pores on the sides of the sponge, past collar cells (which capture food particles) and out through a main opening, the osculum.

The skeletal spicules are easily the most notable thing about sponges. They are of importance to science in one way only: they drive home the bewildering complexity and amazing creative diversity so characteristic of life. In the words of Adam Sedgwick:

> The spicules of sponges in their diversity, symmetry, and intricacy of their form, in the perfection and finish of their architecture, constitute some of the most astonishing objects in natural history. In view of them it is impossible to regard the sponges as low in the scale of evolution. Such finish and such perfection of structure can only have been reached as a result of a long process of evolutionary changes.

And then he puts the real problem:

> While it is pretty clear that the main function of the skeletal structures [spicules] is to give support and protection to the sponge body, it is by no means easy to give explanations of the diversity and complexity of form they present.

No sane person, except a specialist in sponges, would think of memorizing all the different kinds of spicules that exist in sponges, but a close look will reveal the magnitude of the problem they present for the student of evolution. Chemically, spicules are either calcium carbonate or silicon dioxide. Spongin, a sulfur-containing protein, forms part or all of the structural support of many sponges which may or may not also have spicules.

Sponges reproduce sexually or asexually. Gametes are formed from both the general mesenchyme cells of the body and from the flagellated collar cells. A single individual may produce both sperms and eggs. These gametes are typical of all animals, each sperm with head, midpiece, and tail. In asexual reproduction among freshwater species, small but rounded

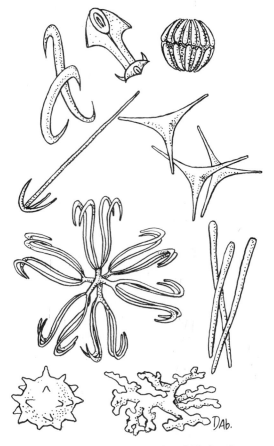

Fig. 6-21. Some of the multitude of kinds of sponge spicules.

bodies called **gemmules** are formed. Visible to the naked eye, they consist of a solid core of amoeboid cells surrounded by a tough coat which protects the inner cells from drying up. Sponges can be cut into pieces and each piece will grow into a whole individual. The most remarkable discovery ever made about sponges was that of H. V. Wilson in 1907. He found that the common scarlet *Microciona* can be squeezed alive through fine silk bolting cloth of the kind used in mills to sift flour. Many of the cells are killed by this treatment, but those that survive will creep together and organize themselves into little sponges if allowed to settle in a dish of sea water. Almost nothing is known about what properties of cells enable them to do this. (It has been found, however, that the cells of various other animals in addition to sponges have this ability.) Nor is it known for certain whether the collar cells, epithelial cells, gland cells, and other specialized cells of the new sponge are derived in each case from corresponding specialized cells in the original sponge or whether the specialized cells of the old sponge, once they are squeezed through the silk cloth and have become amoeboid, are truly dedifferentiated, and hence are able to form any of the cell types necessary to build the new sponge. This is a question of great importance for the theory of development and gene action.

Recent experimenters have discovered that it is possible to induce cells from two different species of sponge to unite into a single individual. This happens only under special circumstances. How does a cell "know" the species to which it belongs? And what makes a cell "forget"?

USEFUL REFERENCES

Bronsted, H. V. *Planarian Regeneration.* Pergamon Press, Inc., New York, 1969.

Cloud, P. E., Jr. Pre-metazoan Evolution and the Origins of the Metazoa. In *Evolution and Environment*, ed. by E. T. Drake. Yale University Press, New Haven, 1968.

Dougherty, E. C. (ed.) *The Lower Metazoa.* University of California Press, Berkeley, 1963.

Hadzi, J. *Evolution of the Metazoa.* Pergamon Press, Inc., New York, 1963.

Harvey, E. N. *Bioluminescence.* Academic Press, New York, 1952.

Hyman, L. H. *The Invertebrates: Protozoa Through Ctenophora. II. Platyhelminthes and Rhynchocoels.* McGraw-Hill Book Co., New York, 1951.

Lenhoff, H. M., and Loomis, W. F. *The Biology of Hydra and of Some Other Coelenterates.* University of Miami Press, Coral Gables, 1961.

Pennak, R. W. *Fresh-Water Invertebrates of the United States.* The Ronald Press Co., New York, 1953.

3

THE STREAM
OF LIFE

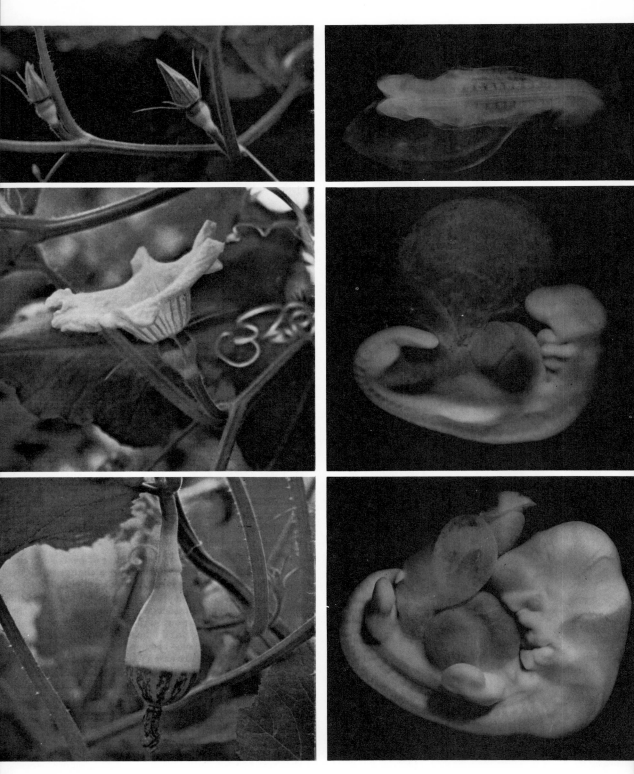

Photos of plants above courtesy of
Chevron Chemical Co., Ortho Division,
200 Bush St., San Francisco, CA. 94120

Genetics

7

The science of genetics is entering a new and, in fact, revolutionary stage of its development. Knowledge of the principles has been achieved by the study of inheritance in garden peas, in the fruit fly, *Drosophila,* and in microorganisms, notably in the mold, *Neurospora,* and in various bacteria. As a result, the gene has become the central concept in biological science. This fundamental knowledge is now being combined with modern techniques for growing human cells outside the body in plastic dishes and the recently acquired insights into the biochemistry of normal and abnormal metabolism to build a true science of human genetics. Cells of organisms as far apart as mouse and man are being hybridized in cultures. These methods are used to map the positions of genes on specific human chromosomes, an accomplishment which is prerequisite for any sound understanding of human heredity.

Not only is such knowledge of great human interest theoretically but it is also of great practical importance, especially as an aid to alleviating the great burden of human genetic disease. The National Foundation–March of Dimes organization estimates that the annual cost of caring for victims of Down's syndrome (Mongolian idiocy), a disease caused by an extra chromosome, is far in excess of a billion dollars in the United States alone. No such figure can express the grief of parents nor the sheer disruption such tragedies inflict on families. Approximately 100 genetic diseases are now well identified, including phenylketonuria, sickle cell anemia, Tay-Sachs disease, sexual anomalies, the self-mutilation Lesch-Nyhan

syndrome, and royal porphyria. Some are common, and some are very rare, but all are debilitating to some extent, and most of them include severe and permanent mental retardation. Happily the new interdisciplinary science of human genetics offers a real hope of significantly improving the human condition. However, before discussion of these issues, the science of genetics itself must be examined, for an understanding of it is essential not only to an understanding of human heredity, but also for any complete insight into any other biological problem from photosynthesis to ecology.

IN THE BEGINNING

The value of taking a hard look at some of man's ancient beliefs is two-fold. Many of them, even though surely false, are still prevalent, and others raise basic and persistent problems. Moreover, genetics illustrates well the important but often forgotten fact that science is both cumulative and open-ended.

To begin with the real classics, the ancient Greeks, Hippocrates (460(?)-377 B.C.), botanist and the "father of medicine," taught that the embryo is formed from a swarm of "seeds" or particles which come from all parts of the body of the two parents and are carried in the reproductive fluids. Thus heredity is particulate. The great philosopher-scientist Aristotle, a younger contemporary of Hippocrates, attacked both parts of this theory and claimed instead that the embryo is formed not by particles but out of the reproductive fluids by the action of a vital formative force which he

139

called "entelechy," a word revived in modern times by Driesch. These fluids, which form the link between parents and children, Aristotle claimed, are not formed all over the body but only in the reproductive organs.

Thus two questions came into clear focus. Is heredity due to particles or to fluids? In other words, is there such a thing as a "blood relative"? Does the hereditary material, whatever it may be, arise from all parts of the body or only from the reproductive organs? This second question obviously involves the question of whether or not the effects of use and disuse, or of anything which happens to the parts of the body, can influence heredity.

ENTER GREGOR MENDEL

Modern genetics began with Mendel, not at the time he published his results, over a century ago, but after they lay completely neglected for nearly 40 years. Why this happened is a complex story but it is important to remember that Mendel's discoveries, like all scientific advances, had antecedents. Mendel himself was widely and actively interested in many branches of science. One of the keys to his success was that he used in his basic experiments a plant, the garden pea, that had been used by many others in studies of heredity. As early as 1823, Thomas Knight had confirmed still earlier reports of dominance and recessiveness and the appearance of ancestral types. Although neither Knight nor any of the others who cross-bred peas noticed regular laws, the necessary basic information about the techniques of breeding pea plants had been obtained.

In addition, several workers, including Charles Darwin, who had studied hybrid pigeons, had stated the problem of heredity very clearly. In his epoch-making book, *Origin of Species,* Darwin wrote:

> The offspring from the first cross between two pure breeds is tolerably and sometimes (as I have found with pigeons) quite uniform in character, and everything seems simple enough: but when these mongrels are crossed one with another for several generations, hardly two of them are alike, and then the difficulty of the task becomes manifest. . . . The slight variability of hybrids in the first

Fig. 7-1. Gregor Mendel, the monk who discovered the laws of heredity for all living things by the use of simple arithmetic and Isaac Newton's binomial theorem. (Courtesy Johns Hopkins Institute of the History of Medicine.)

generation, in contrast with that in succeeding generations, is a curious fact and deserves attention.

The final factor which may have given Mendel the clue to his discovery was his interest in beekeeping. Mendel was a contemporary and almost a neighbor of Johann Dzierzon, the most famous beekeeper of all time. Dzierzon crossed German with Italian bees and found that in the following generation half the drones were German, half Italian. Mendel was thus alerted to the possibility of finding definite ratios.

Why the importance of Mendel's analysis was not appreciated at once is also a complex story. Chromosomes were not taken very seriously by biologists, most of whom were primarily interested in evolution, and the facts of meiosis had not been worked out even though Weismann (see Chapter 8) had predicted that meiosis must take place. Another factor was that Galton and his school had not yet shown the importance of applying statistics to the study of variation.

In 1900 three experimenters eventually rediscovered Mendel's basic laws. They were de Vries in Holland, Correns in Germany, and von Tschermak in Austria. Mendel's paper was found almost immediately afterwards. (It had been mentioned in a book by the renowned American botanist, Liberty Hyde Bailey). At about the same time, de Vries (1848-1935) discovered **mutations**, *i.e.,* sudden changes in heredity. Once alerted to mutations, biologists began to find them in many animals and plants.

MENDEL'S FUNDAMENTAL DISCOVERIES

Segregation: Mendel's First Law

Mendel's experiments, lasting eight years, included the cross-fertilization and raising of many hundreds of plants and the counting of some 8,000 peas. He succeeded where others had failed first, because he simplified his problem and considered single pairs of characters, tall *vs.* short plants, or wrinkled *vs.* smooth seeds, instead of thinking about the whole complex organism. Second, he had the patience to use statistics.

To illustrate the principles discovered by Mendel, consider a cross between two special kinds of chickens, a so-called "splashed white" and a black. If two such chickens are crossed, 100 per cent of the first generation, the F_1 or **first filial generation**, will be neither splashed white nor black but a slaty blue, the "blue Andalusian" (Fig. 7-2). If one of these hybrid blue chickens, either a rooster or a hen, is mated with a black chicken, 50 per cent of the offspring will be black and the other 50 per cent blue. None will be splashed white. These are the facts.

What do these facts mean? Mendel's explanation was simplicity itself. It has since come to be called the **law of segregation**. Each individual produced by sexual reproduction possesses a double set of hereditary factors, one set from each parent. The hereditary make-up of a purebred black chicken can be represented as ●●, and the hereditary constitution of a purebred white chicken as ∞. The hereditary constitution of the first generation cross between the two is then ●○. This produces the intermediate "blue" feathers.

What happens when a black rooster, ●●, is crossed with a blue hen, ●○? Obviously the black rooster can contribute only a factor for black, represented here by ●, to his offspring. The blue hen, however, can contribute either a factor for white, ○, or a factor for black, ●. In other words, all the sperms will carry a factor for black, and one half of the eggs will carry a factor for black and one-half will carry a factor for white. An egg carrying a factor for black, fertilized by a sperm carrying a similar factor, forms a pure black individual, ●●. Con-

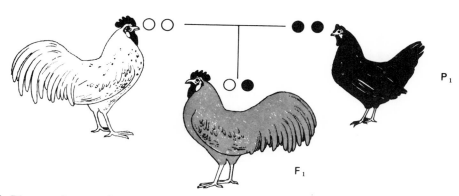

Fig. 7-2. Diagram of a cross between a purebred "splashed white" and a purebred black chicken yielding a "blue" Andalusian.

sequently, 50 per cent of the offspring will be black. The 50 per cent of the eggs which carry a factor for white will also be fertilized by "black" sperms. This 50 per cent of the eggs will thus form the intermediate blue chicks, ●○.

This situation can be diagrammed by the construction of a grid in which the different kinds of eggs are represented along one side, and the different kinds of sperms (in this case only one kind) along the other side. Within the squares the possible crosses are then written. These represent zygotes or fertilized eggs (Fig. 7-3).

Suppose you attempt to breed a race of blue Andalusians by crossing two blue chickens. What will be the result? There will be two kinds of sperms as well as two kinds of eggs. Again the possible combinations can easily be seen by placing the two kinds of eggs along one side of a grid and the two kinds of sperms along the other. This is actually a graphic form of multiplication, to find all the kinds of products. Each kind of egg can then be united with each kind of sperm. It is at once evident that there are four possibilities. The egg and sperm may both carry a factor for black, or both may carry a factor for white. In the third and fourth possibilities, the sperm may carry a factor for black and the egg a factor for white, or *vice versa.* You can thus predict that the result of crossing two blue fowl would be 25 per cent black, 50 per cent blue, and 25 per cent white. You could predict further that if the black offspring are mated to other black chickens, all of their offspring will be black, that the white segregated out of the cross of two blues will also breed true, but that the blues if mated together will again give offspring in a 1:2:1

ratio. This cross has been made repeatedly and the results always agree with prediction.

There are a number of other points to bear in mind. Fertilization takes place at random. In this case a sperm bearing a factor for black is just as apt to fertilize an egg bearing a factor for white as it is to fertilize an egg with a factor for black.

Terminology of Genetics

At this point let us introduce some modern terminology to facilitate discussion. The hereditary factors are now called **genes**. An individual receiving similar genes for a given trait, *e.g.,* white feathers, from each parent is said to be **homozygous** for that trait. An individual receiving different genes for a given trait is said to be **heterozygous** for that trait. All blue Andalusians are heterozygous for feather color. All black sheep must be homozygous for wool color because black is **recessive** to white in sheep. If a sheep bore the gene for white, the sheep would be white even though it also carried a gene for black. In such a case a gene for white is said to be **dominant**. Thus white sheep may be either homozygous or heterozygous for color. A heterozygous individual is called a **carrier** because it carries a gene for the recessive trait but does not show it.

The different forms of a gene at the same position (locus) on a specific chromosome are called **alleles**. Thus the genes for black and for white wool in sheep are alleles. In many known cases there is a series of alleles, as in some of the genes for eye color in *Drosophila.* Because alleles must be at the same position or locus on a specific chromosome (otherwise they would

Fig. 7-3. Genetic grid representing the possible results of a cross between "blue" Andalusian and a purebred black chicken.

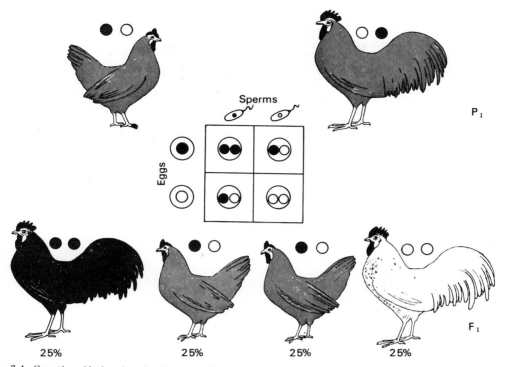

Fig. 7-4. Genetic grid showing the four possible results of a cross between two blue Andalusian chickens. In reality there would be an equal number of males and females in each of the three sorts of offspring.

not be called alleles), any individual diploid animal or plant can carry a maximum of only two alleles of a given gene.

In symbolizing genes it is customary to capitalize the gene symbol for a dominant trait and put its recessive partner or allele in small letters. For example, *a* symbolizes a recessive gene for albinism and *A*, the corresponding dominant allele, the gene for normal pigment. Thus an albino man has the genetic constitution *aa* and is homozygous for albinism. A normal individual may be either homozygous, *AA*, or heterozygous, *Aa;* in the latter case he is a carrier.

Crosses of the kind just described, in which the two individuals differ with respect to a single pair of genes, are said to be **monohybrid** crosses.

Individuals that look alike are said to belong to the same **phenotype** (*phainein*, to show), whether they are genetically the same or not. If they are genetically the same, they are said to belong to the same **genotype** (*genos,* race).

Thus all albinos belong to the same genotype *aa*. All individuals with normal pigmentation belong to the same phenotype but may be of either the homozygous genotype *AA* or the heterozygous genotype *Aa*. The complete **haploid (monoploid)** set of chromosomes characteristic of the cells of any individual constitutes that individual's **genome**. In other words, all the genes carried by a gamete are defined as a genome. However, this term is also commonly used to refer to the entire genetic complement of an organism, *i.e.,* the genes in the diploid set of chromosomes of a particular individual.

Probability and the Product Law

A moment's thought will show that in any cross involving Mendel's law of segregation, the results depend on chance, that is, on **probability**. If 50 per cent of the sperms carry a gene for albinism and 50 per cent carry a gene for normal pigmentation, then there is a 50-50 chance that any particular egg will be fertilized by a sperm carrying the gene for albinism. It will soon be seen that probability also applies to Mendel's second law (see below), as well as to sex determination, and to the genetics of populations in general.

Although it is not yet possible to control the kind of sperm which will fertilize a given egg, it is possible to make some useful predictions about the sex of unborn children, the likelihood of a given marriage producing an albino child, and many other matters. These predictions are based on very simple laws of probability and have an extremely wide application not only in theoretical genetics but also in human life.

Fractions are used to express probabilities. Thus, if the chance of an event happening is one in two, its probability is said to be ½. For example, the probability of "heads" when a coin is tossed is ½. Likewise the chance of getting "tails" is ½. A dice cube has six sides; consequently, the chance that a "four" will land uppermost is 1/6, and so for each of the other sides.

The law of probability which is most important for an understanding of the workings of heredity is the **product law**. It is very simple and can be simply stated. The probability that two independent events will coincide is the product of their individual probabilities. If two coins are tossed simultaneously, the chance that two "heads" will land uppermost is ½ × ½, or ¼. This means that in a sufficiently large series of such double throws, two "heads" will appear in 25 per cent of the cases. Any skeptic need only get two coins and try for himself. It always is true; but you must do it often enough to eliminate the vagaries of chance in small numbers.

Apply this now to a specific problem. A man and his wife are both normal but each is known to carry a gene for albinism because each had one albino parent. One-half of the man's sperms carry a gene for albinism and so do one half of the woman's eggs. Thus the chance that any particular child will be an albino is ½ × ½, or ¼, or one in four. Every time they have a child there is one chance in four that an albino will be born. What is the chance for a homozygous child with normal pigmentation? Again ½ × ½, or a one-in-four chance. Thus 25 per cent of the

children from such marriages will be homozygous albinos and 25 per cent will be homozygous normals.

What about the other 50 per cent of the children? The probability that a sperm carries an *A*, an egg carries an *a,* and the zygote (fertilized egg) *Aa* is therefore ½ × ½ or ¼; the reciprocal combination of an egg with *A* and a sperm with *a* and the resulting zygote *aA* is again ½ × ½ or ¼. Thus 25 per cent of the offspring will be carriers for albinism, having received the gene for that trait from their mothers; 25 per cent will be normal but carriers, having received the gene for albinism from their fathers. Thus 50 per cent of the children will be normal in pigmentation but carriers. Clearly this is the familiar 25:50:25 or 1:2:1 Mendelian ratio.

Suppose that the first child of a couple is an albino. What is the probability that their second child will also be an albino? Neither the second egg nor second sperm months or years later has any way of knowing what the first egg or sperm was like. Therefore the second child is an independent event and the chance of it being an albino is again ¼. However, by applying the same product law to two independent events, each with a probability of only ¼, it is clear that the probability that both children will be albinos is only one in sixteen: ¼ × ¼ = 1/16.

As Mendel himself pointed out, the results of a monohybrid cross can be predicted by use of the binomial theorem usually represented by the familiar equation $(p + q)^2 = p^2 + 2pq + q^2$. Here p represents the frequency of one gene and q the frequency of its allele. Thus if 0.5 of the gametes in a given cross carry gene p and 0.5 carry its allele, gene q, then 0.25 of the progeny will be pp, 0.50 will receive both p and q, and 0.25 will be qq. The use of this formula is basic in population genetics.

Independent Assortment: Mendel's Second Law

Mendel's second law comes into play when the two individuals in a cross differ with respect to two, three, or more pairs of genes. Such crosses are called **dihybrid, trihybrid**, etc. For example, black in rabbits is due to a dominant gene, *B,* and white to its recessive allele or partner gene, *b*. Also, short hair is due to a dominant gene, *S*, and long hair to its allele, *s*. What happens when a homozygous black, short-haired rabbit is crossed with a homozygous white, long-haired one? All the offspring in the first generation will look alike, *i.e.*, will be of the same phenotype, black and short-haired. Their genetic constitution, or genotype, will be heterozygous for both traits, *BbSs*.

When two of these heterozygous rabbits are crossed, the genes inherited from each parent separate in the germ cells of the offspring without any influence on each other. This is the normal segregation of Mendel's first law. Independent assortment means that the way one pair of genes segregates into the germ cells is independent of the way another pair does. In the present illustration one-half of the gametes, either eggs or sperms, will receive a gene for black, *B*, and one-half will receive its allele, the gene for white, *b*. But whether a given gamete gets gene *B* or *b* has nothing to do with whether

or not it gets the gene for short or for long hair. This is determined simply by chance, so that one-half of the gametes that get the gene for black will get the gene for short hair and the other half will get the gene for long hair. This results in four kinds of gametes in equal numbers, symbolized as *BS, Bs, bS, bs*. To predict the results of a dihybrid cross, the four possible types of sperms are written along one side of a square, and the four possible types of eggs along the other. Within the squares appear the possible genotypes resulting from the cross.

It can be seen from the grid in Figure 7-5 that there are only four phenotypes present: black, short-haired; black, long-haired; white, short-haired; and white, long-haired; and that they occur in a ratio of 9:3:3:1. The ratio of the genotypes is very different. For instance, although 9/16 of this generation are of the black, short-haired phenotype, only 1/16 are of the genotype *BBSS*, which is homozygous for both traits. Notice also that only one out of 16

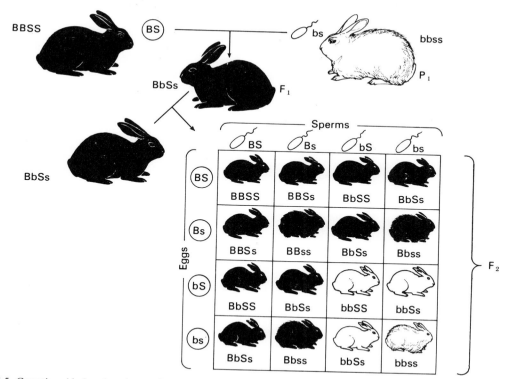

Fig. 7-5. Genetic grid showing the results of a dihybrid cross between two rabbits each heterozygous for both hair length and color. Note that only 1/16 or 6.5 per cent are double recessives.

is the double recessive phenotype—long-haired and white—and that in this case there is only one genotype, *bbss*, that can give this particular phenotype.

It is important to remember in genetic prediction that fertilization is random. When it is said that one-half or 1/16 of the offspring will be of a particular sort, what is really meant is that there is one chance in two or one chance in 16 that the offspring will be so. The genetic ratios predicted are actually realized only in large samples.

CHROMOSOMES AND HEREDITY

Mendel knew nothing of chromosomes, but after the rediscovery of his laws in 1900, several lines of evidence converged to prove that the unit factors Mendel had talked about, the genes, as we say today, are located on the chromosomes. This evidence lies in the precise and extensive parallelism between the behavior of the units of heredity and the behavior of the chromosomes. It is a parallelism so detailed and so extensive that there is no room for doubt that the chromosomes are the very stuff of heredity and hence provide the physical basis of evolution itself. The parallelism runs all through the cycle of fertilization and meiosis, as well as in particular aspects of it, *e.g.,* sex determination and linkage. Furthermore, abnormalities of chromosome behavior are followed by abnormalities of inheritance.

Fertilization and Meiosis: The Cycle

The facts of fertilization in themselves provide some of the most cogent and obvious evidence for the chromosomal theory of heredity. The male contributes equally with the female to the heredity of the offspring, yet the only physical contribution of the male is the head of a sperm. What is the head of a sperm? Microscopic examination of sperm formation in the testis of any animal will reveal that the head of a sperm is little more than a condensed packet of chromosomes. Consequently, it follows that chromosomes are the physical bearers of inheritance. Once inside the egg, the sperm head gradually swells up into a nucleus which ultimately fuses with the egg nucleus.

It was discovered that the number of chromosomes in every cell of a given animal is always the same: 46 in man, 8 in *Drosophila,* the fruit fly, and 36 in chickens. Except for some minor exceptions, this generalization is also true for plants, although many cultivated varieties are polyploid rather than diploid.

Each chromosome has its own permanent individuality as indicated by its size, shape, and the position of its **kinetochore** or centromere, the place where it is attached to the spindle during mitosis. The visible complex of chromosomes, their number, sizes, and shapes, characteristic of any species or individual, is called the **karyotype**, which means, literally, *nuclear type.*

Every cell in an animal has two sets of chromosomes, one set derived from the sperm and one set from the egg. Yet the number of chromosomes characteristic of any species remains the same, generation after generation. How can this be? It was this question which led Weismann to predict that there must be some time when the number of chromosomes is reduced by half, for otherwise they would double in number with every generation. In animals, this reduction, known as **meiosis** (*meiosis,* to diminish), occurs during the formation of gametes, specifically in the two final cell divisions in the formation of sperm or an egg. In the flowering plants, meiosis takes place in the flower. Pollen grains have haploid nuclei as do the specialized structures within the female pistil, called embryo sacs. In fertilization, a sperm nucleus originating in the pollen grain fuses with the egg nucleus within the embryo sac. The resulting zygote develops into an embryo located within the seed, which is released from the parent plant.

When finally unraveled, the over-all facts of fertilization and meiosis turned out to be rather simple. In meiosis the double set of chromosomes found in all the somatic or body cells is reduced to a single set in each sperm or egg. In fertilization the double or diploid set is restored by the fusion, into the nucleus of the fertilized egg, of a single or monoploid set of chromosomes from the sperm with a single set in the egg.

To understand heredity, it is a great advantage to understand in more detail than in

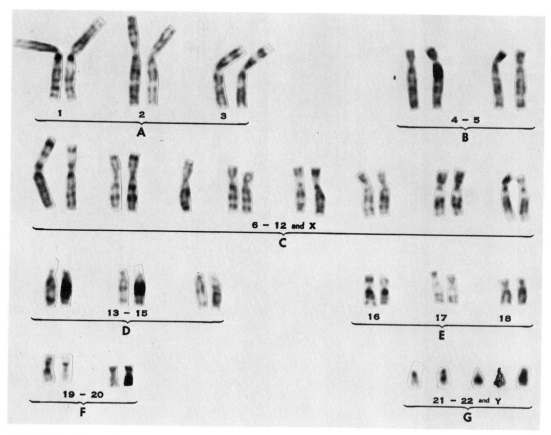

Fig. 7-6. The human karyotype of 23 pairs of chromosomes (Giemsa stain) after brief pretreatment with M8 urea. Each pair is identified by observing its size, the position of the kinetochore (at the constriction), and the banding. The unpaired chromosome between pair 7 and pair 8 in the C group is the X chromosome. The Y chromosome follows pair 22. In the boy from whom this karyotype came, part of one chromosome of pair 21 has been translocated onto the upper arm of one chromosome of pair 6, giving it three instead of the normal two bands, and leaving one of the 21's with only one band. (Courtesy Dr. D. S. Borgaonkar.)

the previous account exactly what happens during the process of meiosis. These details are very much the same in all animals and plants, which is of course the reason why the laws of heredity are the same throughout the living world. In all forms of life meiosis requires two successive divisions. It begins with a single diploid cell and ends with four monoploid (haploid) cells. Two divisions are required to segregate all the pairs of genes because of the occurrence of crossing over (see below) which is about as universal as meiosis itself and which permits genes located on the same chromosome pair to undergo recombination.

Like ordinary mitosis, meiosis begins with a prophase, during which the chromosomes gradually stain more darkly and become shorter and thicker. As in mitosis, a good microscope will show that each chromosome is really double along its entire length, and hence must have duplicated at some time before prophase began.

The first important difference between mitosis and meiosis occurs in late prophase. In mitosis the two sets of doubled chromosomes, one from the male and one from the female parent, pay no attention to each other. However, in meiosis the corresponding chromosomes of paternal and maternal origin come to

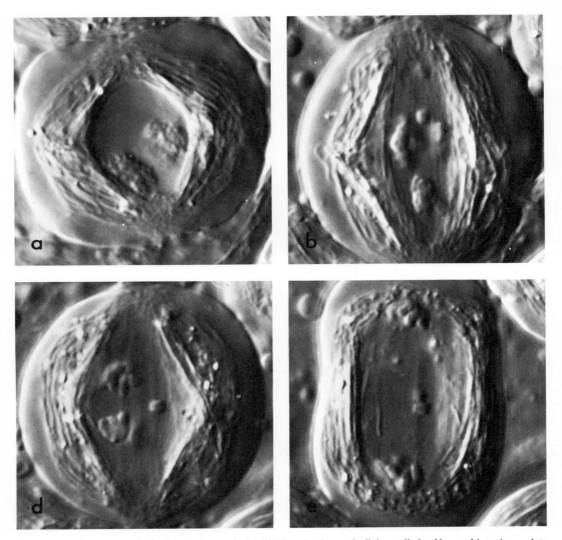

Fig. 7-7A. Chromosome behavior during meiosis, division 1, shown in living cells by Nomarski optics. a, late prophase; b, prometaphase; c, metaphase; d, anaphase; e, telophase; f, cytokinesis. The thread-like structures on each side of the cell are mitochondria. (Courtesy J. R. LaFountain, Jr. From J. R. LaFountain, Jr.: Spindle shape changes as an indicator of force production in crane-fly spermatocytes. Journal of Cell Science, *10:* 79, 1972.)

MEIOSIS: Division I

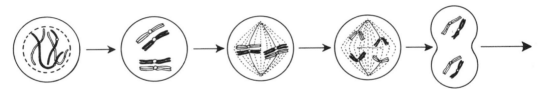

Early Prophase I   Mid-Prophase I   Metaphase I   Anaphase I   Telophase I

Fig. 7-7B. Diagram of meiosis. The green cells are haploid (monoploid).

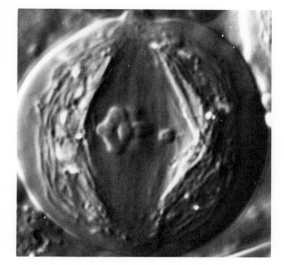

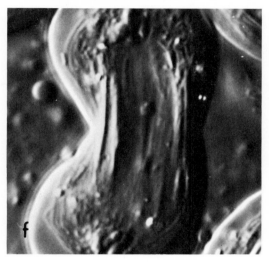

Fig. 7-7A (cont.)

lie side by side, closely aligned at every point from one end to the other. Thus the largest chromosome in the paternal set lies alongside the largest in the maternal set, and so on down to the two smallest chromosomes. This pairing of corresponding or **homologous** chromosomes is called **synapsis**. Synapsis does not occur in mitosis.

Since in meiosis the chromosomes have doubled themselves during the previous interphase, there are really two maternal and two corresponding paternal chromosomes that come to lie side by side in synapsis. The two maternal chromosomes are still held together by their kinetochore or spindle fiber attachment point, and the same is true of the paternal pair. Such a synaptic group is called a **tetrad**. The four chromosomes still held together by their two kinetochores are usually called **chromatids** until they separate. In a species which has three chromosomes in a set, there will, of course, be three tetrads. In the human organism, which has 23 chromosomes in a set, there are 23 tetrads in the prophase and metaphase of the first meiotic division. In other words, there are four complete sets of chromosomes at the first metaphase of meiosis. This is precisely enough to provide each of the four resulting sperms (or four eggs) with one set of chromosomes. To do this, two cell divisions are required.

At the first meiotic division the two kinetochores for every tetrad separate and move to opposite poles of the spindle, each pulling its two chromosomes (more accurately, chromatids). This obviously affords a physical basis for Mendel's law of the segregation of hereditary factors in the formation of sperms or eggs. If, for example, a paternal chromosome carried a

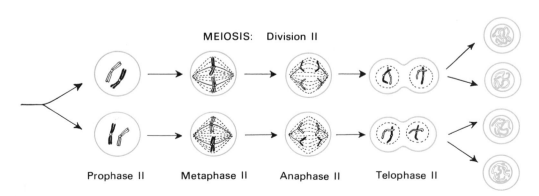

MEIOSIS: Division II

Prophase II   Metaphase II   Anaphase II   Telophase II

Haploid Cells

Fig. 7-7B (cont.)

149

factor for red hair, and the homologous maternal chromosome carried the allele for black hair, one half of the gametes would receive the chromosome carrying the gene for red, the other half the chromosome carrying the gene for black. Said in another way, Mendel's unit factors are present in pairs in the adult. So are chromosomes. Mendel's unit factors separate from each other during reproduction. So do the individual chromosomes in each pair during meiosis.

The way one pair of maternal and paternal chromosomes separate after synapsis is independent of the way any other pair separate. This means that almost no two eggs or no two sperms will have exactly the same assortment of maternal and paternal chromosomes.

At this point the chromosomes are still double and held together by their kinetochores, *i.e.,* the points of spindle fiber attachment. The kinetochores now duplicate themselves. This second division separates the doubled chromosomes so that each of the resulting cells has one chromosome from each of the original tetrads. There is now a total of four haploid cells, each with one complete set of chromosomes.

Genetic Variation—Role of Meiosis and Fertilization

The primary origin of genetic variation is **mutation**, that is, permanent change in a gene. In addition, all organisms which reproduce sexually possess a built-in mechanism to guarantee continual variation by the formation of ever new combinations of the genes via meiosis and fertilization. The way any particular pair of synapsing chromosomes segregates is entirely independent of the way any other pair does. This means that, on the average, every egg and every sperm receives a thoroughly mixed set of chromosomes partly of paternal and partly of maternal origin. What is the chance that a sperm (or an egg), in an organism with only three chromosomes in a set, might receive only paternal chromosomes? Using the product law for the coincidence of two or more independent events, you can easily calculate this chance. For each pair of synapsing chromosomes there is a probability of ½ that the paternal member of the pair will enter a

particular sperm. Consequently the chance that all three paternal chromosomes will enter the same sperm is $\frac{1}{2} \times \frac{1}{2} \times \frac{1}{2} = (\frac{1}{2})^3 = 1/8$. In other words, one in every eight sperms of such a species will contain only paternal chromosomes. By the same reasoning it follows that one in every eight sperms will receive only maternal chromosomes. And of course six of every eight sperms (or eggs) will contain a mixed set. The total number of possible kinds of sperms is thus 2^3, or 8.

The probability that a given human sperm will carry only paternal chromosomes is $(\frac{1}{2})^{23}$ or one chance in 8,388,608. This is another way of saying that every man and every woman can produce over 8,000,000 kinds of sperms or eggs! In fact, crossing over, to be discussed later, yet increases this staggering number.

Fertilization then enters the picture to compound the amount of variation already vouchsafed by meiosis. In a given mating, if there were 8,000,000 kinds of sperms and only two kinds of eggs, there would obviously be 16,000,000 possible kinds of zygotes. However, since any woman can theoretically produce over 8,000,000 kinds of eggs, the total possible kinds of zygotes any human couple can produce is over 8,000,000 times 8,000,000, without counting crossing over. Such is the biological basis of human individuality.

Sex Determination

Natural

In the early years of the present century American cytologists (specialists in the study of cells) made a peculiar discovery about chromosomes in certain insects where chromosomes are favorable for study. In females all the chromosome pairs match perfectly. In males, however, there is one pair called X and Y *chromosomes* which does not match even though the chromosomes come together in synapsis. During meiosis it can be observed that females always possess two X chromosomes, the males an X and a Y. After meiosis every egg will carry an X chromosome, but half of the sperms will carry an X and half a Y. The two kinds of sperms will be produced in equal numbers because at synapsis for every X chromosome there is a Y.

This means that 50 per cent of the eggs will be fertilized by a Y-bearing and 50 per cent by an X-bearing sperm. The result is equal numbers of XX female-producing zygotes and XY male-producing zygotes. In most animals males and females are produced in approximately equal numbers. In some species, perhaps in most, a differential mortality begins before birth and continues long afterward, so that in different age groups the sex ratio varies somewhat to one or the other side of the 50:50 ratio. In man and insects sex is determined by the nature of the sperm, and the sex of an individual is determined at the instant of fertilization.

Artificial Sex Determination

If some way could be found to separate X-bearing from Y-bearing sperms, it would have immediate practical applications in animal husbandry, where artificial insemination is widely practiced. There is some evidence that separation of X-bearing or Y-bearing sperms is possible by the methods of differential centrifugation and electrophoresis. The two kinds of sperms differ slightly in density and in their migration in an electric field, and some observers claim that X-bearing sperms appear under a phase contrast microscope to have slightly larger and more elongated heads. Some have expressed fears least the use of such knowledge would upset the sex ratio in human populations, but there is good evidence that most people desire both boys and girls.

Even though the sex of offspring cannot yet be controlled, it is possible to make some useful predictions about the sex of the unborn. This is done by the use of the product law already discussed. Since all human eggs carry an X chromosome, while 50 per cent of sperms carry an X and 50 per cent carry a Y, the chance of any particular child being a boy is ½, and likewise the chance of its being a girl is ½. In a family of three children, what is the probability that all three will be boys? Applying the product law, $½ \times ½ \times ½ = (½)^3 = 1/8$. This means that if a survey were made of families with three children, it should be found that in 1/8 of the families all three children are boys. It also means that in 1/8 of the families all three

of the children should be girls and in 6/8 of the cases there should be a mixture of boys and girls.

X-LINKED GENES

When a gene is located on the X chromosome, it is said to be X- or **sex-linked** because its inheritance follows the transmission of the X chromosome. The best known X-linked genes in man are those for red-green color blindness and for hemophilia (a faulty clotting mechanism of the blood that results in excessive bleeding, even from a scratch). The gene for white eyes in *Drosophila* is similarly a sex-linked one.

A male has only one X chromosome. Therefore he can carry only one gene for such traits. His one X chromosome may carry a gene for red-green color blindness or a gene for normal vision. But his Y chromosome has no corresponding part. Consequently, even though the genes for white eyes and for color blindness are recessive, they will produce their characteristic effects in a male. For example, if a man has a gene for hemophilia on his X chromosome he is bound to be hemophilic because there is no possibility of the recessive gene being counteracted by a dominant gene for normal blood clotting on a second X chromosome. This type of inheritance is illustrated in the case of white eyes in *Drosophila* (Fig. 7-8).

Of course a female with her two X chromosomes will carry two genes for each sex-linked trait, one gene on each of the X chromosomes. She may be either homozygous or heterozygous. If she is homozygous, she may carry two genes for normal or two genes for white eyes (if she is a fruit fly) or two genes for color blindness (if she is a woman). If she is heterozygous she will carry, of course, one gene for normal eyes and one gene for the abnormal eye trait involved.

A female who is heterozygous for a recessive sex-linked trait is known as a **carrier**. She will be normal, but will transmit a gene for the recessive trait in 50 per cent of her eggs. Whether or not the trait appears in her children will depend on whether they are boys or girls and, in the case of the girls, whether or not the father has a recessive or dominant gene on his X chromosome.

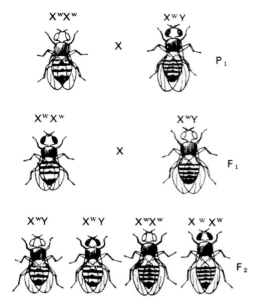

Fig., 7-8. Diagram illustrating the inheritance of genes linked to an X chromosome, often called sex-linked inheritance, in *Drosophila*. W represents a gene for red eyes and w a gene for white eyes.

Autosomal Linkage

Just as Weismann predicted meiosis, so a young graduate student named Sutton at Columbia University predicted **linkage**. He did it on the eminently reasonable grounds that there are many more hereditary factors than chromosomes and that hence each chromosome must carry a group of genes. The group of genes on a single chromosome cannot assort independently at meiosis but has to pass as a unit into the same gamete. In other words, the genes tied together in the same chromosome cannot follow Mendel's second law, the law of independent assortment. To assort quite independently, genes must be on different chromosomes of the set.

There are, consequently, as many linkage groups in any organism as there are chromosomes in its set. So Sutton predicted, and so it has turned out. In *Drosophila* there are four chromosomes in a haploid set, one large, two middle-sized, and one small. Likewise there are four linkage groups of comparable sizes. After linkage was first established in *Drosophila* by T. H. Morgan and his students, it was found to exist in all animals and plants investigated. In humans there are 23 chromosomes in a set and hence 23 possible linkage groups. Twenty-two of the chromosomes are called **autosomes**, and one the sex chromosome, either X or Y. So far, only five autosomal linkage groups plus the sex-linked group have been identified; only about two dozen genes are involved. In mice the situation is somewhat better known, for at least 13 linkage groups have been identified involving about 50 different genes.

Crossing Over

It was learned early in the original work on linkage in *Drosophila* that a phenomenon called **crossing over** occurs. This takes place between the two members of a pair of homologous chromosomes. In the words of T. H. Morgan: "Linkage and crossing over are correlative phenomena, and can be expressed by numerical laws that are as definite as those discovered by Mendel."

For example, on the long second chromosome of *Drosophila* are located the mutant genes for star eyes, black body, purple eye color, dachs (very short) legs, vestigial wings, plexus veins, and speck (black dot at base of wings). Suppose a fly had received a number 2 chromosome with the normal alleles from one parent and a number 2 chromosome with these mutant alleles from the other. Then according to the usual events of linkage, the offspring of this fly will either get all these mutant genes and none of their normal alleles, or else all the normal alleles and none of the mutant genes.

However, in a predictable percentage of cases some offspring will get the genes for speck, plexus, vestigial, and normal alleles for the other mutants, while other offspring will get the normal alleles for the genes first mentioned and the mutant genes for dachs, purple, black, and star. The new combinations now remain as firmly linked as had the old.

Examine now a specific instance more closely. Let b symbolize a recessive gene for black body and B its normal allele, and let v symbolize a recessive gene for vestigial wings and V its normal allele. Then, if a fly, homozygous for the recessive genes, $bvbv$, is crossed with one homozygous for the normal alleles, $BVBV$, all the F_1 generation flies will be heterozygous $BVbv$ and appear normal. When these heterozygous individuals are mated, cross-

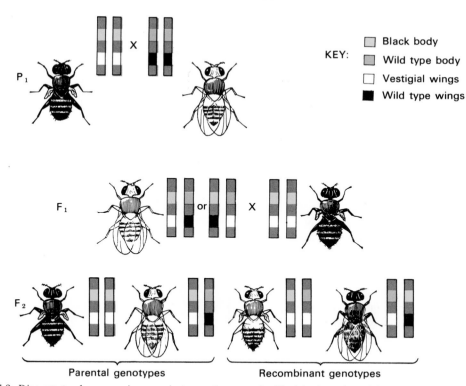

KEY:
☐ Black body
▨ Wild type body
☐ Vestigial wings
■ Wild type wings

Parental genotypes   Recombinant genotypes

Fig. 7-9. Diagram to show crossing over between the genes for black body and vestigial wings and those for wild type body and wings in *Drosophila*.

ing over shows itself. This can most easily be seen if one of the phenotypically normal heterozygotes, a female, is bred to a homozygous recessive, vestigial winged, black male. In the heterozygous female parent, one number 2 chromosome carries both recessive genes b and v while the other number 2 chromosome carries their normal alleles B and V. Thus, with complete linkage, 50 per cent of her eggs will be bv and 50 per cent BV. When mated with a homozygous recessive male, all of whose sperms must carry b and v, as a grid will show, 50 per cent of her offspring would be expected to have normal wings and pigmentation, and 50 per cent to have black, vestigial wings. Breeding experiments, however, show that only 83 per cent of the offspring of such a cross belong to either of these two types. The other 17 per cent are recombinations in which either vestigial is combined with normal body pigmentation, or normal wings with black pigmentation. It is significant that these two new combinations appear in equal numbers.

This result can be explained on the assump-tion that in the formation of 17 per cent of the gametes the homologous chromosomes in the heterozygous female have exchanged parts, so that gene V is now on the same chromosome as gene b, and gene v on the chromosome with B. Direct visible evidence of such crossing over can be seen in meiosis. During synapsis the tightly paired chromosomes become twisted; when separation occurs, the chromosomes have ex-changed parts. Healing appears to be perfect, for once the new combination is formed the chromosomes are as stable as before.

Crossing over produces new combinations of traits. This is important both in natural evolution and in producing desirable new types of domestic animals and plants.

Mapping Chromosomes

Crossing over makes it possible to actually map the locations of genes on a chromosome. It is reasonable to assume that the farther apart two genes lie, the greater the likelihood of a break (with crossing over) between them.

Assume also, and this appears to be true, that crossing over is equally probable at any point along the chromosome. Conversely, the closer together two genes are, the less likely it is that crossing over will occur between them.

Suppose that the percentage of recombinations, *i.e.*, crossovers, between one pair of alleles, say *Aa*, and another pair, say *Bb*, is 5 per cent. It can be said that the two alleles are located 5 arbitrary units apart. Suppose now that crossing over is determined between the *Aa* alleles and a third pair of alleles, *Cc*, and this is found to occur in 15 per cent of the cases. Evidently locus *A* and locus *C* are 15 units apart (Fig. 7-10). There is no way of telling from these data whether locus *B* and locus *C* are on the same side of locus *A* or on different sides. This question can be answered by determining the percentage of crossing over between *B* and *C*. If they are on the same side of *A*, then the percentage of crossing over would be only 10 per cent between them. If, however, *B* and *C* are on opposite sides of *A*, then the crossing over should be the sum of the individual values of distances from *A*.

Very detailed chromosome maps have been constructed by this method, using a large number of genes, not only in the fruit fly and maize plant, but in the chicken, the mouse, the fungus *Neurospora*, and other organisms.

In addition, there is both logical and direct visual evidence of the linear order of genes on chromosomes. The fact that in cell division chromosomes duplicate themselves and pull apart longitudinally rather than breaking in half transversely argues that their important constituents are arranged in a linear series.

Chromosomes, as seen in most cells, are tiny irregular rods. However, about 75 years ago an Italian investigator, Balbiani, discovered that the chromosomes in the salivary glands of flies, gnats, and mosquitos are gigantic in comparison with chromosomes in most cells. He also noticed that these **giant chromosomes** are banded. Beginning in 1930, various workers in this country and in Germany reinvestigated the bands which Balbiani had described long before the rediscovery of Mendel's laws. It soon became obvious that the bands were not haphazard but were constant in number, thickness, and position on any particular chromo-

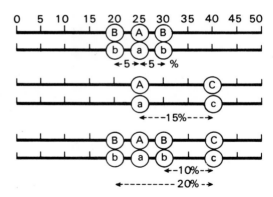

Fig. 7-10. Diagram illustrating the method of mapping the position of genes on a chromosome by comparing crossover frequencies.

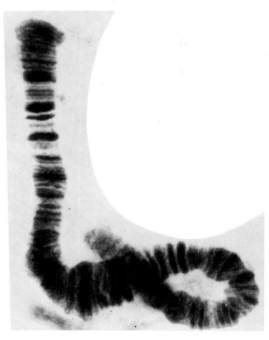

Fig. 7-11. Giant salivary gland chromosomes of a larval insect showing banding. Light microscope. (Courtesy Hans Laufer.)

some. From this fact it was possible to show that specific bands correspond to the position, or locus, to use a more technical term, of particular genes. An abnormality in any particular band is invariably correlated with a particular abnormality in the animal. Thus the chromosome maps constructed from crossover data have been validated by visual evidence.

A new method for discovering which chromosome carries a certain gene is to hybridize normal cells from one organism with cells carrying a biochemical defect from another. For instance, when cells from a normal mouse and a man with some defective gene which fails to produce an enzyme required for some metabolic process are grown together under the proper conditions, the cells hybridize, producing cells with both mouse and human chromosomes. Such cells will be able to carry on the process which the man's cells could not. However, in the course of time chromosomes from these cells will be lost in a random way so that the mixtures of man and mouse chromosomes are different in different cells. These cells can be plated out singly to produce clones descended from single cells. By comparing the chromosomes present with the biochemical abilities of the cells, it is possible to determine which chromosome carries the defective gene and, of course, its normal allele.

MUTATION

Chromosomal and Gene Mutations

Mutations are sudden and relatively permanent changes in the hereditary material. These changes can sometimes be seen under the microscope as changes in the chromosomes; other mutations are changes in the genes themselves.

At the chromosome level, mutations are of four general types. All are caused by the breakage and incorrect rejoining of chromosome fragments. They can be seen in human chromosomes after irradiation.

Deletions are losses of larger or smaller pieces of chromosomes. If too much of a chromosome is lost so that some essential enzyme is not formed, then the mutation will kill the organism or perhaps not permit development even to start. Such a mutation is called **lethal**. A well-known deletion is "notch" in *Drosophila;* this causes a small abnormality in the wings and is lethal when homozygous. This abnormality can be identified cytologically as a small section missing from the third chromosome.

Duplications are repetitions of a portion of a chromosome. These also can be seen in the salivary gland chromosomes. A well-known example is "bar" eye in *Drosophila.*

Inversions are cases where part of a chromosome has been rotated 180° so that the genes, instead of running *ABCDEF,* run *ABEDCF.* Such events produce a wide variety of complications, such as loops, when homologous chromosomes attempt to pair during meiosis. Some inversions are lethal.

Translocation is the term given to mutations in which part of one chromosome becomes permanently attached to the end of a nonhomologous chromosome. Translocation is sometimes called "illegitimate crossing over."

The most interesting and important mutations are those which occur at the gene level and represent mistakes in gene duplication. As will be seen shortly, this means an error in the duplication of a sequence of purine and pyrimidine bases in the DNA.

Mutations have been found to affect just about every part of an organism. In *Drosophila* they change the size, shape, and color of the body; they alter the size, shape, color, and bristles of the eyes. There is even a mutant gene which abolishes eyes. Other genes produce truly revolutionary changes, converting mouth parts into legs, legs into wings, wings into halteres, and halteres into wings, or abolishing wings completely. Some mutations do not produce anatomical results but change physiological or biochemical abilities such as resistance to

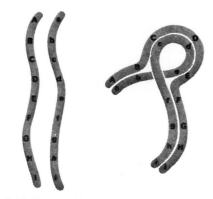

Fig. 7-12. Diagram showing the results of the inversion of a piece of a chromosome when pairing of homologous chromosomes takes place during meiosis.

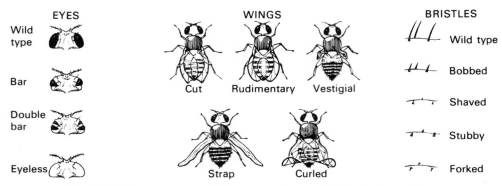

Fig. 7-13. Various mutations affecting eyes, wings, and bristles in *Drosophila*.

poisons or various nutritive capabilities. Such non-morphological mutations are commonly called biochemical mutations. All mutations are, of course, really biochemical, because all the anatomical effects are the result of some change in the enzymatic activity of the cells during development.

Rate of Mutation

The spontaneous rate of mutation is very low and varies from one gene to another. In *Drosophila,* millions of which have been scrutinized in laboratories all over the world, the mutation rate at any one locus on a chromosome may be as high as 1 in every 1,000 gametes or as low as 1 in over 1,000,000 gametes, depending on the gene. The gene mutation that causes ancon (very short legged) sheep has been reported only twice in almost 200 years. Examination of 11,600 flies turned up a mutation which causes yellow body only three times, while a gene for echinoid eyes appeared 18 times.

Records of 94,075 babies born in Copenhagen hospitals showed 10 chondrodystrophic dwarfs. This is the kind of dwarf in which the head and body are normal in size but the arms and legs are abnormally short. The trait is inherited as a dominant. Two of these children had a dwarf parent but the other eight did not; that is, they represent new mutations. These facts show that the gene involved mutates to produce chondrodystrophy approximately once in 11,759 births. Since each child is the result of the fusion of two gametes, an egg and a sperm, and since this trait is a dominant, this means that one out of every 23,518 human gametes in Copenhagen carried this mutated gene.

Artificially Induced Mutations

H. J. Muller, an old student of T. H. Morgan of *Drosophila* fame, then at the University of Texas, and L. J. Stadler, of the University of Missouri, discovered that X-rays will greatly increase the rate of mutation in animals and in plants. It has since been found that ultraviolet light and alpha, beta, and gamma radiation will also produce mutations. Because most mutations are deleterious, an increase in the number of mutations in the human population would be a disaster. Thus, clouds of radioactive dust drifting around the world could easily be ruinous for all mankind. Radiation has been used by plant breeders to obtain mutations in the hope of finding desirable ones from which to start improved varieties of domestic plants.

Many chemicals produce mutations. Such **mutagenic agents** include nitrogen mustards, various cancer-producing substances, alkylating agents, epoxides, caffeine, etc.

Randomness of Mutations

Mutations are random in two different senses: (1) they cannot be predicted except in a statistical way; and (2) they are not related to the needs of the organism. At least 99 per cent of mutations are harmful; this is not surprising when it is remembered that living species are well adapted for their modes of life. Consequently, any random change in the developmental blue print is all but certain not to be an improvement. Whether or not a mutation is advantageous or disadvantageous depends on the environment. For example, mutations reducing wing size so that an insect cannot fly will be a very serious, perhaps fatal, disadvantage in most environments; however, on a tiny windswept oceanic island it will be a life saver. Many such islands support a variety of flightless insects. Similar mutations have occurred and persisted in laboratory cages where they do not affect survival.

There is a sense in which mutations are not random. The kinds of mutations which are possible depend upon the chemical properties of DNA.

GENES

What Are Genes?

The word "gene" was first defined as a unit factor of heredity which followed Mendel's laws. It was coined without reference to chromosomes, frankly as a verbal tool to make it easier to talk and think about heredity. It is in this sense that we have been using the term. However, after T. H. Morgan and his students had mapped the positions of many genes on specific chromosomes by the use of crossover rates, it became fashionable to define a gene as the smallest unit of heredity not divisible by crossing over. Others preferred to define a gene as the smallest unit of mutation. Because each gene controls the formation of an enzyme, or at least a polypeptide chain, others define a gene as the smallest functional unit of a chromosome: **one gene, one polypeptide.**

Obviously all these definitions are very similar, although none is precisely like any other. Each is useful under slightly different circumstances, so that none can be called wrong. This situation will doubtless continue until all the biochemical details of gene structure and action become known. It is to this topic that we now turn our attention.

DNA, deoxyribonucleic acid, is formed something like a twisted ladder, *i.e.,* a **double helix.** The sides of the ladder are composed of linked and alternating molecules of **phosphoric acid** and **deoxyribose,** a five-carbon sugar. The rungs connected to the sugars carry the genetic information. They must if the code really is written in the DNA, because only the rungs vary as you climb the twisted ladder.

These rungs are made up of only four kinds of molecules. Two, **adenine (A)** and **guanine (G),** are purines, which are two-ringed compounds containing nitrogen; the other two, **cytosine (C)** and **thymine (T)** are pyrimidines, which are similar to the purines but have only one ring. **RNA,** found both in the nucleus and in the cytoplasm, is the same as DNA except that the sugar is ribose instead of deoxyribose and **uracil** takes the place of thymine.

The structural formulas of the nitrogenous bases of nucleic acids, *i.e.,* the alphabet of the genetic code, are shown in Figure 7-14.

Two molecules, one purine and one pyrimidine, make a rung. Since the length of the rungs does not vary, the pattern of a purine matched with a pyrimidine remains constant. Two pyrimidines would make a short rung that would not fit. Two purines would make a rung too long. So it comes about that, due to the shape of the molecules, adenine is always matched with thymine by hydrogen bonds, and guanine with cytosine, also by hydrogen bonds. From this it follows that if the nitrogen bases on one side of the ladder are (in order) adenine, guanine, guanine, thymine, thymine,

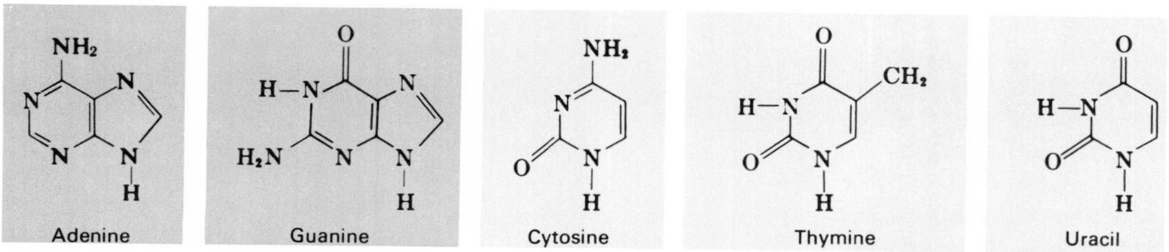

Fig. 7-14. Structural formulae of the five "letters" of the nucleic acid alphabet. (Purines dark green; pyrimidines light.)

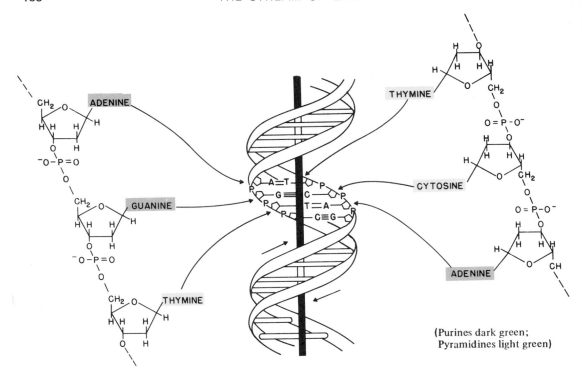

Fig. 7-15. A segment of DNA showing pairs of nucleotides.

thymine . . . , then the corresponding bases on the other side will be thymine, cytosine, cytosine, adenine, adenine, adenine . . . .

There is a slightly different but also useful way of looking at DNA structure. A single phosphoric acid group, plus the sugar, plus the purine or pyrimidine, constitute a **nucleotide**. Thus, when the double helix, or twisted ladder, separates lengthwise into two separate single strands of nucleic acid, each strand will consist of a series of nucleotides just as a protein consists of a series of amino acids.

The Code

There is much evidence, some presented in an earlier chapter and some to come later, that genes preside over the synthesis of enzymes, *i.e.*, proteins. This is not very surprising, since cells are what their enzymes make them, and genes control enzymes. Both nucleic acids and proteins are long chains of smaller units, nucleotides in one case, amino acids in the other. Consequently, it would seem probable that the sequence of bases in the DNA

represents, or codes for, the sequence of amino acids in the proteins.

Since there are four bases, there is a four-letter alphabet. There are about 20 amino acids to specify. Consequently, one-letter words will not suffice because with four letters, A, T, G, and C, there could be only four one-letter words. How many two-letter words can be made from four letters? The first letter of each word might be A, T, G, or C. That makes four possibilities. The second letter might be any one of the four in each case, AA, AT, AG, AC, TA, TT, TG, etc. This makes 4 × 4, or 16 words. This is still not enough. With three-letter words the possibilities are 4 × 4 × 4, or 64. This is enough and perhaps too many, unless some words are merely nonsense or perhaps synonyms or even punctuation.

One of the first questions about a three-letter word system is whether the words can overlap. For example, if *ABC* spells a certain amino acid, and *BCD* spells another amino acid, would *ABCD* spell them both or would you have to write *ABCBCD*? How can this question be put to a test? One way is to note that if the code is an overlapping one, a change (*i.e.*, a mutation) in one letter (one nucleotide) would affect two adjacent amino acids in the protein. The fact is that may cases are now known where a mutation changes a single amino acid but, so far, none has been found where two adjacent ones are changed—something which would happen with most mutations if the code were an overlapping one. Furthermore, if the code were overlapping, it would mean that a given amino acid in a protein would always have one of only four amino acids following it, which is not true. For example, if *ABC* were the code word for glycine, then glycine could only be followed by whatever four amino acids were coded by *BCA, BCB, BCC,* and *BCD.*

To summarize, the code consists of three-letter words made from a four-letter alphabet. It spells out the sequence of the amino acids in the proteins. These three-letter "words" are commonly referred to as **triplets** or **codons.** Whether the code is universal, the same for bacteria, plants, and animals, and how the code can be broken will be discussed in a later section of this chapter.

How Is the Code Duplicated?

In the life cycle of dividing cells, the chromosomes become duplicated and then in mitosis the duplicates separate, one passing to each pole of the spindle. The conditions and specific stimulus, if indeed there is one, which result in chromosome duplication and mitosis are unknown. However, Watson and Crick have proposed a theory, supported by considerable evidence, as to the way in which DNA duplicates. The weak hydrogen bonds, which hold the two strands of the double helix together, break and the strands separate. Each then acts as a model or template for the formation of a new chain of nucleotides complementary to itself along its entire length. The result is two double helices, each identical to the original. In each, one strand will be from the old original double helix and one will be new (Fig. 7-17). In the words of Watson and Crick:

Now our model for deoxyribonucleic acid is, in effect, a pair of templates, each of which is complementary to the other. We

Fig. 7-16. J. D. Watson (left) and F. H. C. Crick with their model of the double helix. (Courtesy Bettman Archive.)

imagine that prior to duplication the hydrogen bonds are broken and the two chains unwind and separate. Each chain then acts as a template for the formation on to itself of a new companion chain, so that eventually we shall have *two* pairs of chains, where we only had one before. Moreover, the sequence of the pairs of bases will have been duplicated exactly.

Much still remains to be learned. It is clear that this is an energy-using process and also that an enzyme, DNA polymerase, is essential. One of the ways of proving that chromosomes duplicate in this manner, *i.e.,* by serving as models which induce a new structure to form alongside of themselves, is to feed cells before division with radioactive thymine. After the formation and separation of the new chromosomes, one of each homologous pair has all the radioactivity and the other member of the pair has none.

How Is the Code Translated into Proteins?

It will be recalled that DNA not only makes more DNA but also, with the help of the proper enzymes, produces messenger RNA, which then moves out of the nucleus into the cytoplasm.

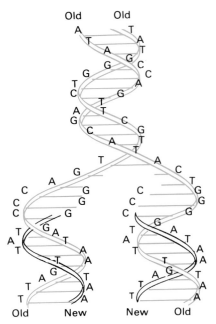

Fig. 7-17. The DNA double helix in the act of self-duplication.

The formation of messenger RNA from its template DNA is referred to as **transcription**. The formation of proteins from the code in the messenger RNA is called **translation**.

Three types of RNA can be identified in the cytoplasm by differential centrifugation and other methods. The first type is **messenger RNA**, usually written mRNA. The messenger RNA occurs in long strands or chains, as Watson and Crick would say. It is the equivalent of a long computer tape carrying coded instructions. A second type is **ribosomal RNA** (rRNA). This makes up about 60 per cent of the structure of ribosomes; the rest is protein. A third type of RNA is soluble or **transfer RNA**, (sRNA or tRNA). Transfer RNA consists of relatively short pieces of RNA.

Translation of the genetic code is merely a somewhat more informative way of saying **protein synthesis**. It occurs as follows. **Ribosomes** become attached to the starter ends of the long strands of mRNA and move along these strands, reading the code and producing proteins by adding one amino acid after another. Series of ribosomes lined up along mRNA can sometimes be seen under an

electron microscope. Such a series of ribosomes is commonly referred to as a **polysome** or polyribosome.

At the same time that the ribosomes are moving along the mRNA strand, specific enzymes in the cytoplasm activate amino acids. Adenosine triphosphate (ATP) complexes with an amino acid. This compound, specifically the amino acid part, then becomes attached to one end of the tRNA molecule.

For a time, transfer RNA was thought to exist in the form of an elongated twisted U or hairpin. More recent evidence indicates that the shape of the tRNA molecule is more like a cloverleaf (Fig. 7-19).

The tRNA molecules, each carrying an amino acid which corresponds to the code it bears, move through the cytoplasm, presumably by

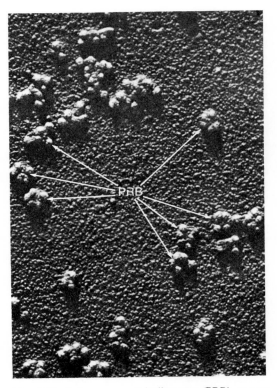

Fig. 7-18. Polysome or polyribosome (PRB) as seen with an electron microscope (×150). (Courtesy Dr. A. Rich. From W. A. Jensen and R. B. Park: *Cell Ultrastructure*. Wadsworth Publishing Co., Inc., Belmont, Calif., 1967.)

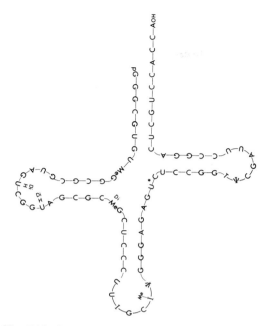

Fig. 7-19. Structure of transfer RNA (tRNA). (From R. W. Holley *et al.*: Structure of ribonucleic acid. Science, *147:* 1464, 1965.)

Brownian movement, and thus come into contact with ribosomes on the mRNA.

Each ribosome, as it passes along the series of triplets in the mRNA, accepts at each position only the appropriately coded tRNA. The tRNA molecule for methionine cannot get functionally tied up with a ribosome until the ribosome reaches a place along the mRNA which reads for methionine. When this happens the methionine-carrying tRNA moves into place on (or perhaps in) the ribosome and the tRNA previously there falls free minus its amino acid. Its amino acid, whatever it was, becomes attached to the methionine carried by the newly arrived tRNA. In this way a chain of amino acids linked by peptide bonds is built up, forming a polypeptide chain. Thus the code is translated into the primary structure of a protein, *i.e.,* its characteristic sequence of amino acids.

It seems reasonable to suppose that the secondary structure—the twisted helical form of the protein molecule—would show itself as soon as the polypeptide chain forms. Likewise, there is no reason to suppose that the forces which

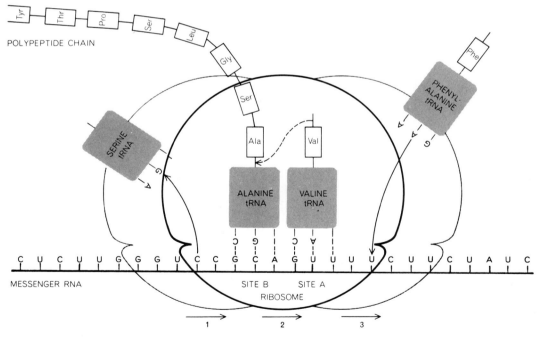

Fig. 7-20. Diagram of protein synthesis as the ribosome and the mRNA "tape" move along each other while the transfer RNA (tRNA) moves in with activated amino acids. (Reprinted by permission from F. H. C. Crick. The genetic code: III. Scientific American, *215:* 55, 1966.)

produce the tertiary structure—the various foldings of the protein molecule—would wait to act until after the full length of the primary structure is complete. Consequently, when the finished protein finally drops free as its ribosome comes to the end of the mRNA ribbon, it would be held in its characteristic folds and bends by its sulfhydryl (—S—S—) and other bonds.

Is the Code Universal?

Is the genetic code the same for microorganisms, plants, and animals? Since it has been firmly established that the proteins of all living things are composed of the same 20 amino acids, and that the nucleic acids of all living things contain the same four (five, counting uracil) purines and pyrimidines, it would be hard to believe that the code would be different. However, plausibility does not constitute proof. The best evidence supporting the idea that the code is the same in all organisms has been obtained by mixing amino acid-charged tRNA from one source (perhaps bacteria), with mRNA from a second source (e.g., from hemoglobin-producing cells of rabbits), and ribosomes from yet a third source; the result is always determined by the kind of mRNA present—in this case rabbit hemoglobin will be formed.

Cracking the Code

One way of discovering what the code actually is, i.e., how to spell the word for any specific amino acid in the three-letter purine-pyrimidine words, is to make your own messenger RNA. This quite spectacular achievement was first accomplished by Marshall Nirenberg working at the National Institutes of Health. If the mRNA has no nucleotides except those carrying uracil, it will cause the synthesis of a polypeptide chain of repeating phenylalanine units. Thus, UUU codes for the amino acid, phenylalanine. Similarly, mRNA carrying only adenine will produce, in a test tube, when mixed with activated tRNA charged with all 20 naturally occurring amino acids, the necessary activating enzymes, and ribosomes, a polypeptide chain consisting of repeating units of only

one amino acid, lysine. Thus, AAA codes for lysine. Three cytosines code for the amino acid proline. By changing the percentages of the different purines and pyrimidines in the mRNA, it has been possible to deduce the code words for all 20 of the naturally occurring amino acids. With different percentages of different mixtures of the four bases built into the mRNA, different amino acids are incorporated into the protein synthesized. For example, if there is no guanine in the mRNA, the amino acid glycine is not incorporated into the polypeptide formed. Therefore the triplet for glycine must contain at least one G (guanine). Messenger RNA with nothing but guanine will not call for glycine, hence there must be some code letter in addition to G. In fact, there has to be twice as much guanine as there is of some other base. Therefore the triplet for glycine has two G's plus one of the other bases. By this kind of analysis it has been found not only that all 20 of the amino acids can be spelled in triplet form, but that the code is redundant, i.e., there is more than one way to spell most of the amino acids. Thus UUU and CUU both correspond to phenylalanine.

GENES IN ACTION

Genes and Enzymes

"One gene, one enzyme," or at least, "one gene, one polypeptide chain" (for there are cases known where two genes are required to produce the complete enzyme), is a central concept of modern biochemical genetics. The idea, however, that genes produce their effects by producing, or failing to produce, specific enzymes was proposed by a physician and biochemist at Oxford, Sir Archibald E. Garrod, during the first decade after the rediscovery of Mendel's laws of heredity. As happened with Mendel's discovery of the elementary laws of inheritance and MacMann's discovery of the respiratory pigment cytochrome, no one paid much attention to Garrod's book, *Inborn Errors of Metabolism,* even though he was Regius Professor of Medicine at Oxford. The reasons for this complete neglect are complex but the basic one is surely that biologists were not intellectually ready to incorporate any of these ideas into the general body of their thinking. (After the battle, it is always much easier to see what the general should have done.)

Garrod studied a hereditary disease called **alkaptonuria**, which still serves as a model of how genes work and how the point at which they act can be discovered. The chief symptom of this rather benign disease is a blackening of the urine after exposure to air. This happens because the urine contains an abnormal constituent, alkapton or homogentisic acid. In normal individuals, the amino acid phenylalanine from food is incorporated into body proteins and also converted into a number of other products, including homogentisic acid. This is then converted into CO_2 and water (see Fig. 7-26).

In the families of people with alkaptonuria, Garrod showed that the disease behaves like one of Mendel's recessive factors (genes). He went on to argue that the accumulation of homogentisic acid which results in its being excreted by the kidneys is due to the absence of a specific enzyme. The missing enzyme is called homogentisic acid oxidase because it can oxidize homogentisic acid, not into a black pigment but into simpler products which in turn are converted into CO_2 and water.

The reasoning is simple. Within the body one compound, say some amino acid or amino acid precursor, is converted into some other compound via a series of steps. If a mutation changes a gene so that it can no longer form the enzyme responsible for some specific conversion of, say, A to B, then A will accumulate in the body and bloodstream and may be excreted in the urine.

This mode of thinking about how genes express themselves did not appear again until about 1935 when Boris Ephrussi and George Beadle undertook to investigate the biochemistry of the formation in *Drosophila* of various eye pigments which were known to be under the control of specific genes. The difficulties of this work led Beadle and E. L. Tatum to turn to *Neurospora*, the pink bread mold. Actually it is more orange than pink, and is found growing wild on sugar cane. B. O. Dodge, of the New York Botanical Garden, had long urged T. H. Morgan to work with *Neurospora,* which he claimed was better than *Drosophila*. Morgan

took some strains of *Neurospora* with him when he left Columbia University for the California Institute of Technology. It was there that Beadle and Tatum obtained this organism, and then the study of biochemical genetics really got under way.

The advantages of *Neurospora* are several. It can be grown readily in pure culture in test tubes and on media of known chemical composition. This means that the mold can synthesize for itself a large number of compounds from very few simple raw materials. The asexual spores, **conidia,** can be irradiated or otherwise treated to produce mutations. When germinated, the growth from these conidia can be crossed with wild strains. After crossing, fruiting bodies are produced which contain cigar-shaped capsules, each holding eight haploid spores. There are two adjacent spores for each of the usual four products of meiosis. Hundreds of these haploid cells can be planted singly in tubes with a complete medium containing a rich variety of vitamins, amino acids, and other substances. They can grow and produce conidia even if they carry mutations which deprive them of one of the enzymes necessary for some essential step in the synthesis of a necessary metabolite. After a good growth has been obtained, each strain can be tested in minimal medium. If it can no longer grow on it the way its ancestor did, then a mutation has occurred, depriving it of some particular enzyme. Because *Neurospora* is a haploid organism, its phenotype reveals its entire genotype, and any change in its genetic constitution is immediately obvious. The next step is to test its ability to grow on media lacking one after another of the various vitamins, amino acids, and other growth substances to discover exactly where the biochemical lesion is located.

Clearly the "one gene, one enzyme" hypothesis fits the facts. It is the biochemical basis of mutational effects in organisms as widely separated as man and mold.

Multiple Factors and Gene Product Interaction

Genes themselves probably never interact with each other directly, but their products can

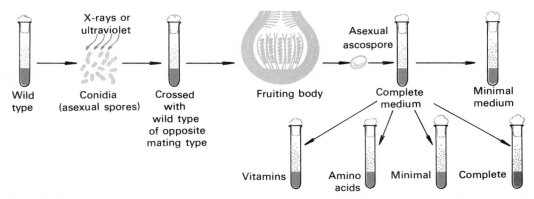

Fig. 7-21. Diagram of the basic experimental design for discovering biochemical mutants in *Neurospora,* the orange sugar cane mold commonly called pink bread mold. The two left media on the bottom row lack vitamins and amino acids.

interact in a variety of ways. A number of instances are known of what is called **epistasis.** In these cases a gene in one locus overrules the effect of a gene in a different locus, perhaps even on a different chromosome. Since this reaction is not between alleles, *i.e.,* alternative forms of the same gene at a given locus on the same chromosome, it differs from ordinary dominance.

A well-known case of epistasis occurs in dogs where the dominant gene *B* gives a solid black coat; the recessive allele *b* gives a brown coat when in the homozygous *bb* condition. However, there is another gene, *I* (for inhibitor), on a different chromosome which prevents pigment formation except in the eyes. To have fur of any color whatever, a dog must be homozygous *ii* for this inhibitor gene. Thus a dog may be homozygous for dominant black and yet be white because of the inhibitor gene *I* which masks the effect of the genes at the other locus. This *I* gene is different from the gene for albinism which prevents the formation of melanin, even in the eyes.

Another type of gene product interaction occurs in all cases where many genes influence a single trait. Body size, intelligence, skin color, eye color, and numerous other general characteristics fall into this class.

How Are Genes Turned On or Off?

One of the great paradoxes of modern biological science is the accepted fact that the cells of a multicellular animal or plant have the same diploid set of chromosomes and therefore the same genes, and yet these cells become differentiated in many very different directions. Clearly there must be some way in which certain genes are turned on in certain cells while other genes are activated in other cells. Or perhaps it should be said the other way around: most of the genes are kept permanently inactive in most cells but the genes that are inactivated are different in different kinds of cells.

The control of such action of one gene by other genes has been known for many years. The cases of repressor genes in *Drosophila* and position effects in corn (maize) also show that somehow genes can affect the action of each other. However, it was not until the work of F. Jacob and J. Monod in 1960 that a plausible model backed up with convincing evidence was proposed to explain how the activity of genes is regulated.

It should be noted first that enzymes are of two general types. There are **constitutive enzymes** which are formed by cells whether or not their substrate is present, at least in detectable amounts, and there are **inducible enzymes,** formed when and only when their substrate is present.

As early as the beginning of this century it was observed that certain enzymes of microorganisms are produced only in the presence of the specific substrates on which these enzymes act. For example, the colon bacterium, *Escherichia coli,* grows on rather simple media containing minerals and carbohydrate, and ferments maltose when this sugar is present as the carbon and energy source. If these bacteria are transferred to a medium where lactose is substituted for maltose, they do not begin to ferment lactose immediately, but are able to do so only after a lag period. During this lag period there is an **induced** production of the enzymes needed to break down this substrate. During such **enzyme induction**, the enzyme which initially is not present in appreciable amounts increases by as much as many thousand-fold.

Escherichia coli exhibits yet another kind of regulated enzyme synthesis where a product of metabolism can control the amount of enzyme formed. If glucose and lactose are both present in the medium, glucose will be metabolized first and the fermentation of lactose will begin only after depletion of glucose. The reason for this is that the enzyme necessary for lactose fermentation (β-galactosidase) is an induced enzyme that is formed only in the presence of lactose. However, in addition, the presence of glucose suppresses the formation of this enzyme. Suppression can be caused by a nutrient or by a metabolic product near the end of a sequence of steps in which the enzyme is involved. The molecule responsible for suppression of enzyme synthesis is called the **repressor**, and this kind of suppression of enzyme synthesis is called **feedback repression.**

Induced synthesis of enzyme insures that enzymes can be formed when specific substrates are available. Feedback repression guarantees that overproduction of enzyme can be avoided. Inducers and repressors can act antagonistically in regulating enzyme production. There are of course certain enzymes that are not subject to regulation by induction or feedback repression. Such constitutive enzymes are always present. It is worth noting that the kinds of enzymes which can be induced by substrates are limited by the hereditary constitution, i.e., the genome of the organism.

The Operon

In 1961, the French bacteriologists Jacob and Monod published a paper summarizing their ideas about bacterial synthesis of inducible and repressible enzymes. This paper laid the groundwork for our understanding of the molecular basis of induction and repression, ideas which seem to be applicable not only to the bacteria, but also to higher plants and animals.

According to the **operon** theory proposed by Jacob and Monod, information (*i.e.,* structural genes) for the synthesis of one to several enzyme molecules comes under the control of a contiguous segment of the DNA molecule called the **operator.** The operator plus the closely linked sites under its control is called an **operon.** The operator can be in one of two states: open or closed. When open, all segments of the DNA molecule linked to the operator are functional and used to synthesize messenger RNA; in turn the polypeptides or proteins coded for in this operon are synthesized. In the closed state of the operator, no mRNA is synthesized in any segment of the DNA molecule under the control of the operator. The operator itself is under the control of the product of a **regulator gene** that is located outside the operon. Two kinds of repressors are produced by regulator genes. One kind can shut off the operator but also can be neutralized by an inducer substance (a nutrient, hormone, etc.) so that it no longer represses the operator. Another kind of repressor must combine with a specific metabolite, the **corepressor,** in order to be effective, *i.e.,* able to inactivate the operator gene. In the absence of the corepressor, the product of the regulator gene is not inhibitory and the operator remains in the open state.

There can be a change (or a mutation) of the regulator gene just as there can be mutations in portions of the DNA coding for specific proteins or enzymes. If the regulator gene becomes inactivated by mutation and no longer forms a product which can repress its operator, then the operator can no longer be repressed and remains permanently in its open state. Under these conditions one or several enzymes will always be produced and will appear to be constitutive. Also, a change in the operator can modify its susceptibility to repression. The information under the control of the operator can become permanently available and the associated structural genes would give rise to constitutive synthesis of enzyme. The general scheme by which inducers may regulate gene action can be diagrammed as shown in Figure 7-22.

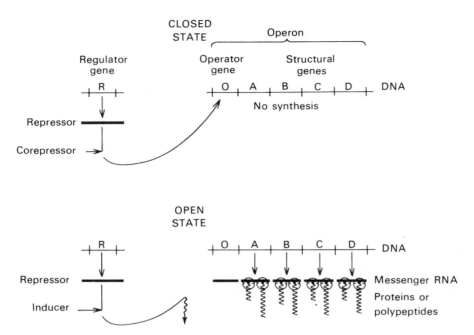

Fig. 7-22. Diagram of the operon theory of gene action.

HEREDITY VERSUS ENVIRONMENT

The General Problem

Which is more important, heredity or environment? This question refuses to die. The answer depends on cases, even on your point of view, because the development of any organism is like the flight of an arrow. At every point the trajectory of the arrow is the result of two forces. One is the initial propulsion imparted by the spring of the bow. This may be compared to heredity, specifically the set of genes in the fertilized egg. The other force is the environmental force of gravity, modified by wind; and this may be compared to the action of regular and random environmental forces on the course of development.

In one sense the genes reign supreme. We are human and not starfish, or even anthropoid apes, precisely because of the kind of fertilized eggs we developed from and for no other reason. In the teeming waters of the ocean the fertilized eggs of thousands of kinds of animals and plants develop: fish, worms, sea urchins, medusae, clams, and the myriad kinds of algae.

The environment may be the same over hundreds of cubic miles of sea, but the organisms that come from these different kinds of eggs are drastically different.

In some minor aspects as well, the genes are supreme. If a man is born without a gene for normal pigmentation he will be an albino regardless of environment. The blood groups to which a person belongs are determined by his genes and by them alone.

However, in another sense environmental forces are equally important. Genes do not function in a vacuum. Only within a narrow range of temperature found only on a planet within a certain distance from a sun is it at all possible for genes to form and function. In innumerable lesser ways the environment controls the actual as contrasted with the potential behavior.

In *Drosophila* it has been possible to produce copies of many of the well-known mutants by manipulation of the environment, treating the embryo with heat or various chemicals at a certain stage of its development. In wild flies, up to 75 per cent of the individuals can be converted into phenocopies of the "curly" mutant by heat. In man, the persistent exposure to sunlight of an individual with a genotype producing a very light skin can cause conversion to the phenocopy of an individual of a genotype which in a relatively sunless environment produces an equally dark hue.

In summary, what is inherited by fruit fly, man, or any other creature is a capacity to respond in certain ways or a lack of ability to respond. As far as the external environment is concerned, there is a whole spectrum of cases, from genotypes that produce their characteristic effects in all known environments to genotypes that are very sensitive to environmental factors. Albinism is in the first class, melanin production during sun tanning in the second.

Are the Effects of Environment and Activity Inherited?

The belief that effects of use and disuse and of the environment are inherited is extremely ancient but still persists. Perhaps this is because it offers hope of being able to influence the hereditary nature of domestic animals and plants and even of man himself.

Shortly after the turn of the century an American zoologist, Castle, and his colleagues tried a new way to get at this problem. They removed the ovaries from a purebred white guinea pig and ingrafted ovaries from a black animal. Later this white female with ovaries from a black animal was mated to a white male. The offspring were all black. In no case was it possible to detect any modification of the black inheritance even though the ovaries from the black female were actually within the white female and were necessarily nourished by her bloodstream.

After the *Drosophila* work was well advanced, T. H. Morgan pointed out that some of the observations of his school really gave the old theory of the inheritance of acquired characteristics the *coup de grace*. If a fly with very small vestigial wings, which of course cannot be used in flight, from a long line of flies all with vestigial wings, is crossed with a normal fly, all the first generation will have normal wings, just as large and functional as

ever even though half of their ancestors had never used their wings for many generations.

Methods for Genetic Improvement

A method of heredity control that does produce improved breeds of animals and plants is the application of selection alone or, better, in conjunction with outcrossing and inbreeding. This is the method that has given such spectacular results with corn, now one of the mostly widely cultivated crops from the plains of Iowa to the Ukraine.

Hybridization

The development of improved varieties of grains, especially corn, wheat, and rice, and their impact on the methods of agriculture and the availability of food supplies, has been a phenomenon of the past half century. The development of hybrid corn caused revolutionary changes in U. S. agriculture, and was a significant factor in the recovery of Europe following World War II. In the past decade, the introduction of new varieties of rice in Asia has had even more profound effects on the health and economy of nations.

Hybridization of corn is certainly not new. *Zea mays* is a hybrid in nature because cross-pollination ordinarily occurs. The American Indians traditionally planted different varieties of corn in the same fields because they were aware that this led to increased yields and greater vigor. The literature on corn hybridization goes back to the early days of the American colonies. In 1716 Cotton Mather, who is better known for his sermons and witch-hunting, wrote about his observations of the crossing of corn varieties in nature. It was not until the middle of the next century, however, that controlled experiments on the hybridization of corn were initiated by Charles Darwin. Darwin compared corn plants that were the progeny of self-pollinated *vs.* cross-pollinated individuals. He observed that crosses between unrelated varieties resulted in progeny that were often more vigorous and produced a higher yield than either parent strain, a phenomenon referred to as **hybrid vigor**.

Darwin corresponded about his experiments with the American botanist, Asa Gray. One of Gray's students, William Beal, working at Michigan State University, wrote the next chapter in this tale by attempting to obtain improved corn varieties through hybrid vigor. Although unsuccessful, Beal made a very significant contribution to the ultimate attainment of his objective: the method for obtaining hybrid plants. He grew two varieties in the same field and made certain that no other corn was grown nearby. All developing tassels (male flowers) were removed from one variety. Corn kernels developing in the ears of these emasculated plants all resulted from fertilization of the ovules by pollen from the second variety of corn planted in the field. Unfortunately, in Beal's attempts to obtain better varieties of corn, the yields from the resulting hybrids were not sufficiently improved to justify all the extra work involved.

The factor overlooked in Beal's experiments was revealed at the beginning of this century by the work of George H. Shull, of Princeton. Shull's contribution came from his attempts to obtain pure lines (*i.e.,* inbred varieties) of corn. Shull's objective was to obtain better and more uniform corn strains by inbreeding, the so-called "pure line" approach. His corn varieties did indeed become more uniform after several self-pollinated generations, and he developed, with continued inbreeding, a number of pure lines of remarkable uniformity. Some of these pure lines were greatly inferior in yield to the original stock; others were moderately inferior and some were about the same.

Shull's next research objective was a study of the inheritance of the number of rows of kernels, and he proceeded to make crosses between his pure lines. The resulting hybrids were indeed uniform but, most importantly, were greatly superior in vigor and yield to the parental strains. Thus, without intending to do so, Shull had developed a method for obtaining improved strains of corn with high yield. This approach involved two steps: (1) inbreeding to obtain the best pure lines and (2) crossing two pure lines to obtain a hybrid of vastly improved productivity.

By the 1920's one final step was added to Shull's classically simple approach to corn improvement by Edward M. East and Donald Jones of the Connecticut Agricultural Experiment Station. This was to obtain seeds from a double cross, thus combining desirable traits from four inbred strains and making large amounts of seeds for agricultural use from scarce single crossed seed. From a few bushels of single-cross seeds, several thousands of bushels of double cross seeds can be obtained (Fig. 7-23).

By the early 1930's commercial production of hybrid corn had begun and by 1950 more than 65 million acres of hybrid corn were grown in the United States (more than 75% of the total corn acreage). It was introduced in the U.S.S.R. by Nikita Kruschev. Research aimed at the improvement of corn strains still continues. Much of the current effort is to introduce characteristics such as stiff stalks (which make mechanical harvesting easier) or improved nutritional content of the kernels (for higher protein and vitamin content). The greatest impact of the story of hybrid corn has been, of course, on the yield per acre, which has enabled U. S. agriculture to consistently produce excess crops for export to parts of the world where critical food shortages exist.

Selection

In the past half century there has been wide application of genetic knowledge for improving strains of plants and animals that result in increased agricultural production. With a relatively unimproved group of cows, chickens, or other animals to begin with, rigorous **selection** of the best individuals results in hereditary improvement. This is the way most of the current breeds of farm animals have been produced. By this method, improvement is usually rapid at first, but a plateau is reached after a variable number of generations.

Progeny Testing

To achieve further results, a method originated over a century ago by the great French plant breeder, Louis de Vilmorin, has been applied with success. This is the **progeny test**. It has been especially successful in dairy herds and in chickens. A bull obviously cannot be judged

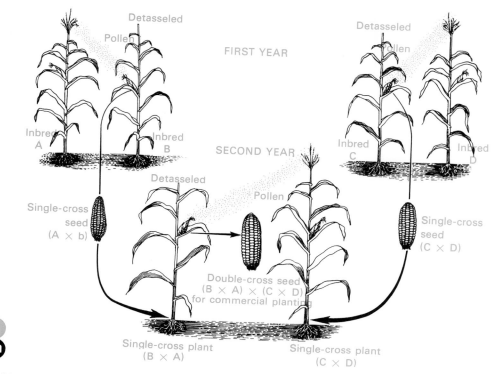

Fig. 7-23. Diagram of the Shull-East-Jones method of producing dramatic improvement in corn by double hybridization.

on his milk production. It is the milking record of his daughters that reveals whether or not he carries genes that make for high milk productivity.

Excellent results from using the progeny test on chickens have been achieved in agricultural experiment stations in several states. The older method of selecting hens for breeders on the basis of their own egg-laying records was followed until for several years no further improvement was obtained. Then the selection of the hens and roosters to be used as breeders was based on the egg production record of their offspring. This new method of selection gave marked additional improvement.

Heterosis

Outcrossing (or **cross-breeding**) is an extremely valuable method for improving varieties of animals as well as plants. Outcrossing is used for two purposes. First, it may combine in one animal desirable traits from various breeds. A notable case is the Santa Gertrudis cattle of Texas, which have been produced by crossing Herefords and shorthorns with Brahman cattle from India. The resulting hybrid has some of the superior beef qualities of the European breeds combined with the resistance to heat, drought and insects of the Indian cattle. The second generation after any such cross shows a considerable variability, as would be expected from the segregation and recombination of genes. Consequently, in the second and later generations, selection must play an important part in the elimination of undesirable traits. Some of the later generations might equally well combine the poor beef qualities of the Indian cattle with the susceptibility to drought and heat of the European breeds.

Cross-breeding, as shown by hybrid corn, not only produces new combinations of traits from which the desirable ones can be selected. It also produces hybrid vigor or **heterosis.** Heterosis was defined by Gowen as "the evident superiority of the hybrid over the better parent in any measurable character such as size, general vegetative vigor, or yield." The explanation of hybrid vigor is still obscure. Part of it may be due to heterozygosity of harmful recessive genes. Some think that for unknown reasons heterozygosity is in itself beneficial. Others believe that in certain strains there just happen to be genes for "vigor" which have a complementary reinforcing action when brought together.

Less is known about hybrid vigor in animals than in plants, although in *Drosophila*, heterosis, as defined above, has been shown to exist to a very marked degree in lifetime egg production and to lesser degrees in various other traits.

The hybrid mule is a much hardier animal and able to thrive on poorer food than either the donkey or the mare, its parents. The lack of fertility in mules is not due to any fault in reproductive anatomy or physiology but to a failure in chromosome behavior during meiosis. In synapsis, the horse chromosomes do not pair properly with the donkey chromosomes. The eggs or sperms produced do not have complete sets of chromosomes and hence do not give rise to viable offspring.

Inbreeding

The value of **inbreeding** is based on the fact that inbreeding produces homozygosity. This is a point made by Mendel in his original paper. He established what perhaps should be called Mendel's third law, namely that under conditions of self-fertilization or close inbreeding, the proportion of homozygous individuals becomes greater and greater and the proportion of heterozygotes becomes less and less. It is easy to understand why this should be so, because with close inbreeding the homozygous individuals will produce only homozygous offspring, whereas only half the offspring of the heterozygotes will be heterozygous. In the 10th generation of self-fertilized peas, Mendel calculated that the offspring from a monohybrid cross would be in the ratio of 1,023 homozygous dominants to 2 heterozygotes, to 1,023 homozygous recessives. In other words, homozygotes to heterozygotes would be in a ratio of 1,023:1. Of course the effect could not be so

extreme with inbreeding in animals, but with brother-sister and father-daughter crossing, similar results follow.

The first great practical result of inbreeding is that the resulting homozygosity "stabilizes" the desirable traits so that animals breed true. This is the method that has been used in the production of most present-day breeds of domestic animals. The first step may or may not be a series of crosses, but there is always selection of the animals having the desirable genes, and this is usually followed by inbreeding to make the strain homozygous.

A second useful fact of inbreeding is that homozygosity reveals desirable and undesirable recessives. Such genes can then be either retained or eliminated. In corn breeding, which set the pattern, inbreeding produces a number of strains, some very poor. Only the best inbred strains are subsequently used in the crossing that gives the spectacular hybrid vigor.

CONCERNING THE HUMAN GENE POOL

The grand total of all the genes, good, bad, and indifferent, possessed by the human race is unknown but certainly enormous. Scientific knowledge of this vast treasury of potentialities is only just beginning, but at long last a true information explosion has begun. The study of human heredity will always be hampered by great difficulties. Compared with *Drosophila* or mice, man is a slow-breeding animal. Moreover, a single female does not produce 75 to 100 offspring, nor can specific crosses be made to order for genetic purposes. Many of the most important and interesting human traits are behavioral and therefore governed by cultural as well as genetic influences. Added to these basic difficulties are several minor ones. Occasionally, individuals are not completely honest in reporting the facts. Furthermore, a number of human genes either vary in the degree to which they are expressed in different individuals or lack 100 per cent **penetrance**. In other words, the same gene may manifest itself more strongly in one person then in another, or in some cases it may not produce any detectable effect. The genes for hyperextension of the thumb and for lack of a complete set of permanent teeth fall into this category.

The Blood Groups

It is now over half a century since Karl Landsteiner, in reinvestigating the old problem of blood reactions, discovered that the blood serum of some people would cause an agglutination, *i.e.,* clumping, of the red cells of certain individuals but not of others. Following this discovery it became evident that all people fall into one of four blood groups, now called O, A, B, and AB. One who belongs to group O has blood serum (the fluid part of his blood minus the clotting fibrin) which carries proteins known as **antibodies** that will clump the red cells from any person belonging to group A, B, or AB. If a man belongs to group A, his serum carries antibodies that will agglutinate cells from a member of groups B or AB. If he belongs to group B, his blood will clump cells from members of groups A and AB. Finally, if a man is of group AB, his serum will not clump blood cells of A, B, or O type. A necessary corollary of the presence in group O serum of antibodies against both A and B cells is that the anti-B antibody in A serum and the anti-A antibody in B serum will not clump group O red blood cells.

In transfusions it is important to use blood of a type matching that of the recipient. In extreme emergencies group O blood can be transfused into persons belonging to the other three groups because the O cells cannot be agglutinated by the anti-A or anti-B antibodies present in the blood of the recipient, and at the same time the antibodies in the O serum are so diluted that they do not clump the recipient's cells.

The inheritance of the ABO factors is simple. Three alleles, all located at the same or nearly the same position, produce the four blood groups. Gene I^A (abbreviated from isoagglutin A) yields A cells and anti-B antibody in the serum. Its allele I^B produces B cells and anti-A antibody in the serum. When a person receives gene I^A from one parent and gene I^B from the other, he is of the genotype $I^A I^B$ and belongs to group AB. Gene I^O, if homozygous, puts a man or woman in group O.

Subsequent studies have uncovered a dozen or more other blood groups by which people can be distinguished. The best known of these are the MN system and the Rh or rhesus factor

Fig. 7-24. Karl Landsteiner, physician and protein chemist. (Courtesy Johns Hopkins Institute of the History of Medicine.)

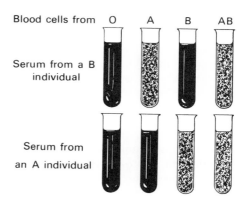

Fig. 7-25. Test for blood groups of the ABO series. Blood cells from O, A, B, and AB individuals are mixed with serum from a B individual which contains an antibody clumping cells carrying the A factor, and from an A individual whose serum will clump red blood cells carrying the B factor.

blood group. In these systems there are normally no corresponding antibodies in the serum, but only antigens on the red cells. All are inherited according to Mendelian rules. The rhesus factor was discovered first in the blood of the macaque or rhesus monkey, *Macaça,* which is the short-tailed, brown monkey of Southern Asia used in medical research. The medical importance of the rhesus factor lies in the fact that if a Rh negative (Rh⁻) person receives a transfusion of Rh positive (Rh⁺) blood, the Rh⁻ person forms antibodies against the Rh⁺ erythrocytes or red cells. On a subsequent transfusion with Rh⁺ cells, the foreign Rh⁺ cells will be clumped, and that can lead to serious blocks in the vascular system. Furthermore, if an Rh⁻ woman becomes pregnant with an Rh⁺ child, some Rh positive factors may pass through the placenta into the blood of the mother. She will then produce antibodies against the Rh⁺ cells. These antibodies from the mother will pass through the placenta into the fetal bloodstream. Consequently, in a subsequent pregnancy with an Rh⁺ child, this child may develop a hemolytic disease called **erythroblastosis fetalis** in which the red blood cells are destroyed.

Although about 15 per cent of people of European ancestry have Rh⁻ blood, which means that about 12 per cent of all marriages are between Rh⁻ women and Rh⁺ men, fortunately very few of these marriages result in erythroblastic babies. This is partly because some women do not make Rh⁺ antibodies at all or fast enough to be very harmful, and partly because many of their husbands are heterozygous for the Rh factor. Rh⁺ is a simple dominant. One-half of the sperms of a heterozygous man carry a gene for Rh⁻. A child resulting from such a sperm would be homozygous for the Rh negative factor and would be unharmed by the mother's Rh⁺ antibodies, even if abundant. Consequently, instead of 12 per cent of marriages yielding one or more erythroblastic children, only about one-half of 1 per cent do so. Erythroblastic babies, if not too badly affected, can be tided over the crisis by transfusion with about 500 ml. of Rh negative red cells which are immune to the antibodies the baby has received from its mother. It is now possible to prevent the mother from forming destructive antibodies by administering to her an Rh immunoglobulin (Rho-GAM) which suppresses the maternal immune system. There will remain some cases of sensitization caused by less common blood factors and by incompatibilities in the ABO system.

Blood group genes are especially useful to anthropologists interested in the historical interrelationships of the various races of mankind. This is because, unlike skull shape or tooth wear or skin color and many other traits, blood groups (1) are not changed by differences in food, climate, or other environmental influences; (2) are inherited according to very simple Mendelian laws; and (3) are sharply defined, "all-or-none" characteristics in marked contrast to the blending nature of traits like skull shape or hair structure; (4) can even be determined in prehistoric mummies and bones of other ancient human remains.

All three of the ABO blood group genes have been found in all racial groups, but their frequencies vary widely. Gene I^A, for example, is almost completely absent among the Indians of Central and South America, but is very common among the Blackfeet and other tribes centering in Montana and among the natives of northern Norway.

As far as blood transfusions are concerned, the only thing that matters is that the two individuals concerned belong to the same blood group. Thus natives of Sweden, China and Africa can safely exchange bloods if they are of the same blood group, while two brothers cannot if they belong to different blood groups.

Genetic Diseases

Phenylketonuria (PKU) is caused by a recessive autosomal gene which blocks the conversion of phenylalanine into tyrosine. This is a very simple reaction, merely the addition of an —OH group (hydroxylation) to the phenol ring. Since tyrosine (*p*-hydroxyphenylalanine) is a normal precursor for adrenaline, melanin, and thyroxine, it might be thought that PKU sufferers would lack adrenaline, and be albinos with goiters. This is not the case because tyrosine is an amino acid found in a normal diet. In the disease, if the ingested phenylalanine is not hydroxylated into tyrosine, it and its derivatives, phenylpyruvic acid, etc., accumulate in the blood and cerebrospinal fluid, and appear in the urine.

Although some PKU cases are only slightly

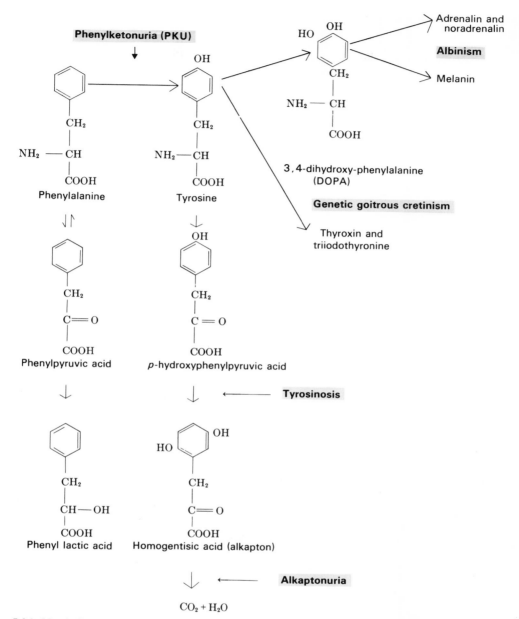

Fig. 7-26. Metabolic pathways leading from the amino acid phenylalanine to the hormones adrenaline and thyroxine and the pigment melanin together with the pathological results of enzyme deficiencies at specific points. (From G. B. Moment: *General Zoology*, 2nd ed. Houghton Mifflin Co., Boston, 1967.)

retarded, most are feebleminded or worse. Medical opinion holds that brain damage can be avoided by feeding only enough phenylalanine to provide for essential protein building. The results of this treatment are extremely hard to assess because of the variability of untreated cases. A rough diagnosis can be made with the "diaper test" because phenylpyruvic acid will turn $FeCl_3$ green. This procedure is being replaced by the more reliable but very laborious Guthrie test.

Sickle cell anemia is so called because the red blood cells shrink into rough crescents when deprived of oxygen (as when placed on a microscopic slide). In heterozygous individuals the disease causes a more or less debilitating anemia, especially if the individual lives at high altitudes. It is characterized by occasional crises of swollen and very painful joints and spleen. The homozygous condition is invariably fatal. This gene is found among people of African descent and in certain groups in Greece and India. It appears to have been preserved by natural selection because it confers a degree of immunity to malaria. The sickle cell trait causes an error of only two wrong amino acids in the primary structure of human hemoglobin protein. There is no known cure, but urea, an agent which tends to keep proteins in solution, is a palliative.

Porphyria is a derangement of porphyrin metabolism in which the urine is the red color of Port wine and during acute attacks the patient suffers abdominal cramps, nausea, sweating, restlessness, and great mental confusion. The most famous case in history was George III of England, who suffered recurring attacks from 1765 until his death at the age of 81 in 1820. Not surprisingly, this disease was rated as insanity. It may have been a factor in the mismanagement of the North American colonies which resulted in the American Revolution. More interesting is the fact that before it was known that he suffered from porphyria, George III's recurring periods of madness were very carefully studied by an American psychiatrist who concluded that the king suffered a manic-depressive psychosis and wrote that:

> Self-blame, indecision and frustration destroyed the sanity of George III. A vulnerable individual, this unstable man could not tolerate his own timorous uncertainty [and] broke under the strain. [Had the king] been a country squire, he would in all probability not have been psychotic.

In recent years it has been found that various infections, even rather mild ones, may bring on an attack in a person with porphyria, and that various drugs such as barbiturates or sulfonimides always do. Many of the royal relatives of George III, including his remote cousin Frederick the Great and his ancestor Mary, Queen of Scots, are now known to have had porphyria. One is reminded of the way the X-linked gene for hemophilia was spread around Europe among the descendants of another queen, Victoria.

In addition to phenylketonuria, there are now recognized several dozen genetic conditions which cause mental impairment of some sort. Down's syndrome (Mongolian idiocy), the Lesch-Nyhan syndrome, and even extra X or Y

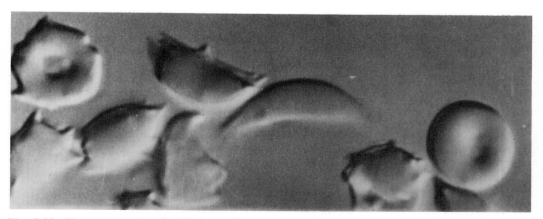

Fig. 7-27. Photomicrograph of sickled red blood cells. Normal cell is at far right. (Courtesy National Foundation.)

chromosomes in women or men. It is significant that a 1965 study by Reed and Reed showed that 5 million out of the 6 million cases of mental retardation in the United States had either a retarded parent or a normal parent with a retarding sibling.

The Lesch-Nyhan syndrome is one of the most dramatic instances where a single gene, with a well understood mode of inheritance, produces a profound and tragic behavioral abnormality. This disease is X-linked, and so is inherited like hemophilia or red-green color blindness. So far it has been seen only in boys. These unfortunate children have an uncontrollable tendency for self-mutilation. Unless their hands are kept bandaged, they chew off their fingers. Before their fingers are destroyed, if not prevented by wearing padded gloves, they will sometimes destroy their noses piece by piece. Physically these patients appear normal but their muscular coordination is very poor, often spastic; and they are mentally retarded. They are afflicted with a marked deficiency in the enzyme xanthine oxidase, a much studied enzyme which converts hypoxanthine to xanthine and then to uric acid. Whether this biochemical lesion is causally related to the behavioral abnormality and mental retardation is unknown.

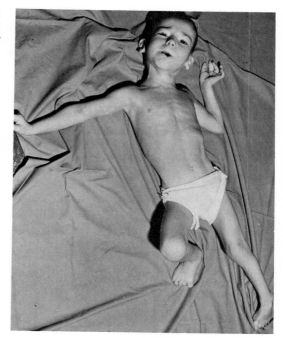

Fig. 7-28. Victim of the X-linked Lesch-Nyhan syndrome. Note the spastic posture and that part of the lower lip has been chewed away. (Courtesy W. L. Nyhan.)

Down's syndrome or **Mongolian idiocy** is now known to be associated with **trisomy** of chromosome 21. Why having three instead of the normal two of these chromosomes should produce the stolid, somewhat fat individuals with rounded face and head, stubby hands and body build, mental retardation, and (fortunately) a friendly disposition is quite unknown. The condition has nothing whatever to do with Mongolian ancestry and occurs in all human races. The chance of it occurring rises sharply in mothers over 35, although it can occur in the children born to women of any age.

Tay-Sachs disease is found mostly among people of Jewish ancestry, especially those originating in certain parts of Eastern Europe. It is a tragic disease that is always fatal after a child has developed normally for a year or so.

All these hereditary diseases represent a great genetic load for the human race, more costly in personal tragedy and in money badly needed for other purposes than the vast majority of bacterial diseases. As such, genetic ailments are a public health problem of the first magnitude. Fortunately this situation can now be vastly alleviated by a new technique for early diagnosis of genetic defects, **amniocentesis**, followed by biochemical and cytological analysis of the cells so obtained (Fig. 7-29). The cells sloughed off by the fetus into the amniotic fluid can be grown *in vitro* and their chromosomes studied. In this way Down's syndrome can be detected long before birth. The metabolic defect causing Tay-Sachs disease and glucose-6-phosphate dehydrogenase (G-6-PD) deficiency (favism) can be detected, as can the enzyme deficiency in Lesch-Nyhan disease.

After amniocentesis and analysis of the fetal cells, parents can be informed as to whether they can expect a normal child. In cases like Lesch-Nyhan disease it is difficult to understand how any humane society would do anything but recommend the immediate termination of pregnancy. With some of the other

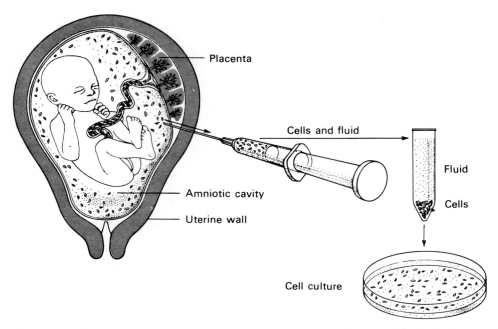

Fig. 7-29. Diagram showing technique of amniocentesis and subsequent plating out of fetal cells for growth *in vitro.*

diseases the situation may seem different but it is a noteworthy fact that few people who think parents should be forced to bring an abnormal child into the world are themselves willing to donate the large sums of money and the devotion required to support such tragic malformations of human life.

Sex Chromosomes and Human Abnormalities

Altogether there are about 60 genes now known to be located on the human X chromosome. In addition to the genes for red-green color blindness and hemophilia are the genes for childhood (Duchenne type) muscular dystrophy, glucose-6-phosphate dehydrogenase (G-6-PD) deficiency, total color blindness, a blood group termed Xg, a lack of gamma globulin essential as antibody against bacteria, and numerous others.

The Lyon Hypothesis

Male mammals have only one X chromosome while females have two. Since this is so, you would expect females to have twice as much of the product of all the genes on the X chromosome as do males. This is not the case. It has been established for some years that women do not have any more of the anti-hemophilia globulin, for example, than do men. The same is true for an abnormal hemoglobin due to a gene on the X chromosome.

Several mechanisms that would result in "dosage compensation" giving this result can be imagined. Several investigators hit upon what appears to be the correct explanation at about the same time, but the clearest statement and most convincing evidence was presented by Dr. Mary Lyon of Harwell, England, an explanation commonly known as the **Lyon hypothesis.** According to this theory, early in mammalian development (probably in the stage when the egg is becoming implanted on the wall of the uterus) one or the other of the X chromosomes in each cell of a female blastula becomes inactivated. These inactive X chromosomes are the **Barr bodies** which can be seen adhering to the nuclear membrane of many cells, whether from cheek epithelium, nerve ganglia or liver in females. They are the "drumsticks" attached to the irregular shaped, "polymorphic" nuclei of

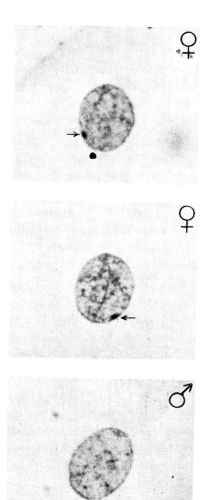

Fig. 7-30. Human male and female epithelium cells from the lining of the cheeks, showing Barr bodies. (From K. L. Moore and M. L. Barr: Smears from the oral mucosa in the detection of chromosomal sex. Lancet, 2: 57, 1955.)

single color but a heterozygous female is mottled with some patches of fur of one and some of the other color. A similar situation occurs in tortoise shell cats. Males fail to show such mosaics.

Another instance is the population of erythrocytes in the blood of a female heterozygous for G-6-PD deficiency. It is possible by a special staining technique to distinguish red cells which have this enzyme from those which do not. As expected, heterozygous women show both types of cells in approximately equal numbers.

Dosage compensation, however, is not always complete. If it were, XXY men would be normal rather than suffering from Klinefelter's syndrome.

Sex Anomalies

You will recall that in humans (and fruit flies) sex is determined by the sex chromosomes. XX individuals are females, while XY individuals are males. During meiosis mistakes can occur which result in gametes with missing or extra chromosomes. When such deletions or duplications occur, any resulting zygotes will have the wrong number of chromosomes. Abnormal numbers of X or Y chromosomes can influence not only the sex of the individual but also fertility. In an insect such as *Drosophila*, where all XY zygotes develop into males, XO zygotes, which have no Y chromosome because of its loss during early development of the egg, also develop into normal though sterile males. Any XXY zygote develops into a normal appearing and fertile female.

The situation in humans is somewhat different. Human XO zygotes do not develop into males but into females. Such girls develop normally until the age when puberty should occur. But menstruation does not take place and there is neither development of breasts nor of axillary or pubic hair. Such a condition is called **Turner's syndrome**. At first it was thought that the trouble might be a failure of the pituitary gland to secrete its normal gonad-stimulating hormone, but it was soon shown that there is ample gonadotropin from the pituitary in the blood of afflicted individuals. Biopsy, however, revealed that the gonads, *i.e.,* the ovaries, were virtually absent.

An egg which contains two X chromosomes

the white blood cells of a woman. Which X chromosomes becomes inactive is apparently a matter of chance, but once one of the two is committed to inactivity, all of its descendants follow suit and that X chromosome remains inactive. The evidence for the Lyon hypothesis is convincing and the results striking, if hardly surprising. Lyon based her theory on X-linked coat color in mice. A homozygous female has a plus a Y chromosome develops into a man with **Klinefelter's syndrome.** Such men appear more or less normal but are sterile because spermatogenesis does not occur.

The facts of Turner's and Klinefelter's syndromes make it evident that sex determination in human beings (and probably all other mammals), while basically determined by an X and Y chromosomal mechanism as in insects, nevertheless has some important differences. The Y chromosome, which appears to be without any function in insects except to exclude the presence of a second X, in man carries genes which have a positive male-producing action. It is of interest here to point out that a very similar situation exists in the dioecious plant *Melandrium,* where a Y makes the plant a "male," *i.e.,* a pollen producer only, whether one or two X's are present.

Eggs with abnormal numbers of chromosomes can be produced by various agents—radiation, extremes of temperature, and certain chemicals. Any derangement of meiosis in which an X or Y chromosome is lost, or the X chromosome fails to separate properly, could lead to the sexual abnormalities just discussed.

Recently a number of men with two Y's and an X have been found in penal institutions to which they had been committed because their extremely aggressive and even belligerent natures made them unacceptable as members of society. It is possible that the XYY condition of these unfortunate men has no causal relation to their behavior, but the evidence accumulated so far suggests such a relationship. Many more cases will have to be examined, especially among men outside of penal or mental institutions, before a firm answer can be given to the question of causality.

A BIOLOGICAL LOOK AT RACES

Racial differences clearly involve genetic differences, but they also involve so many cultural factors and so many unknowns that a scientific discussion of race is difficult. At the outset, value judgments must be set aside as irrelevant to a scientific study. That Swedes have big ears is a statistical fact which can be confirmed by going to Sweden and measuring people's ears. To say "big ears are beautiful" is to express a value judgment and is outside the sphere of the natural sciences.

Physical measurements are easiest to obtain and compare, and they form the classic criteria for racial distinctions—skin color, skull shape (whether long, medium, or round), hair type, and, more recently, the frequencies of various blood groups. An important conclusion can be drawn from these physical measurements, which is: "race" is a very fuzzy term. There are many genes shared in common by different races. Although it is useful to talk about yellow (which, when translated into Mongoloid, includes the American Indians), black, and white races, there are no sharp boundaries. Moreover, there are "races" which defy inclusion in even such a broad classification. The "hairy Ainu," the aborigines of Japan, in many traits resemble Asiatics but in others Europeans. The aboriginal Blackfellows of Australia are as black as the blackest Africans and resemble them in other respects, yet they possess traits which link them to the Caucasians. There are different gene pools but how different the gene content of these great collections of genes is remains unknown.

It is certain that in many traits—height, physical strength, resistance to disease, hairiness, and others, including intellectual abilities—there is a great overlap among the three major races of man. Intelligence is extremely difficult, perhaps impossible, to measure within one group and far more so between individuals of diverse groups; but it does not require any elaborate psychological testing to know that the smart, the average, and the stupid are no monopoly of any one race. When individuals of all races vary among themselves in height, weight, body build, and a myriad of other traits, it is impossible to believe there is no variation in intellectual endowment. The only real question is averages, and there seems to be no satisfactory way to measure these.

One of the most persistent questions asked biologists concerns **racial crossing.** There are three classic biological objections that have been raised against racial crossing. The first is that harmonious adaptive patterns will be broken. A favorite example is the lightly pigmented skin and narrow nostrils of northern peoples compared with the heavy pigmentation and broad nostrils characteristic of many equatorial races. It has turned out to be harder to prove that these differences are truly adaptive than you might suppose, but it is a plausible idea that can be accepted for the sake of argument. Imagine a man with fair skin and a wide nose or one with dark skin and narrow nostrils. Such men would resemble Socrates and Gandhi and hosts of others who appear not to have suffered at all from their supposedly disharmonious condition.

The second objection, which is similar to the first, is that disharmonious combinations of traits will result. The classic example is in the work of C. B. Davenport on crosses between Europeans and Africans in the West Indies. The arms of these crossbreeds were said to be too long for their legs. The fact is that their arms were on the average 1.1 cm. (not quite ½ inch) longer than in the original white group and 0.6 cm. longer than in the original blacks! Certainly a negligible effect and quite possibly due to errors of measurement, diet or even to a touch of hybrid vigor. Surely everyone has known mongrel dogs which showed neither physical nor behavioral deficiencies; in fact, many are smart, vigorous, and cooperative without being subservient.

The third objection raised against racial crossing is that if one race is superior, its qualities will be diluted. This objection depends on a value judgment. How can anyone decide which race is superior? One suggested criterion is similarity to the great apes; the more ape-like, the more inferior. The apes have very thin lips and wavy hair, so by these traits blacks are clearly superior to whites. If body pigmentation is the test, then the Europeans are superior to Africans. If general hairiness is the test, then the Asiatics are first, the Negroes second, and the Europeans and the hairy Ainu last. Who can regard such traits as of any great importance?

Another approach is to consider genes for traits that are universally regarded as harmful. There is a list of hundreds: albinism, hemophilia, Down's syndrome, diabetes, Huntington's chorea, deaf-mutism, porphyria, etc. But the vast majority of these genes have been identified in all racial groups. Presumably the relative frequencies of these genes vary somewhat from race to race but that certainly provides no basis for judging one group as a whole superior to another.

Brain size is sometimes taken as a criterion. Within very broad limits, intelligence is correlated with size. Frogs have very small brains and no one is surprised that their intellectual abilities are limited. Great caution must be exercised, however, in extrapolating such observations to mankind. Microcephalic idiots exist in all races. But it must be remembered that much more than mere size is important. Cuvier had one of the largest human brains ever recorded and his contributions to biological science were great, but not nearly so important or original as those of the far smaller-brained Lamarck. Anatole France, a writer commonly regarded as the equal of Voltaire during his lifetime, possessed one of the smallest of recorded human brains. Remember that in size and complexity the brain of the porpoise is equal to man's. There is some evidence that average brain size is greatest in Central Asia and becomes less as you move out to the far corners of the world. Perhaps potential intellectual capacity is greatest in Central Asians and less in the rest of us. Perhaps, but no one knows. Quite possibly no one ever will.

Human racial crosses turn out well from the biological point of view even though there are sometimes sociological problems of a severe kind. Cross-cultural marriages often have rough going. One of the most dramatic racial crosses studied was between the Australian Blackfellows, who had one of the most primitive cultures ever seen, with no written language and no use of metals, and people from highly industrialized England. The offspring are healthy (no disharmonious gene combinations), capable, and do well at school and college. The most famous racial cross is probably that between the mutineers on *H.M.S. Bounty* and their Polynesian wives. Their descendants have lived in virtually complete isolation on Pitcairn and Norfolk Islands in remote parts of the Pacific. One human geneticist reported the results of this cross as deplorable, but earlier and several more recent visitors to Pitcairn have returned with moving pictures and other documentation showing that the Pitcairn Islanders are in actuality "robust, long-lived, smart, and ingenious," and in addition endowed with wit and friendliness. Certainly not inferior in anybody's language!

A specific question often put to biologists is how much gene exchange there has been between races in North America. By arbitrary convention, anyone with any sign of African genes is called Negro or black. From the biological point of view this becomes more and more absurd as the proportion of genes of European origin increases. Even when the number of African and European genes are present in equal numbers, to call a man a black is as arbitrary as calling a zebra a black horse with white stripes or a white horse with black stripes.

Casual observation makes it obvious that there has been racial crossing on a vast scale. Careful calculations based on the familiar Hardy-Weinberg principle (discussed in detail in Chapter 35) indicate that approximately 20 percent of the gene pool of American blacks is of non-African origin. This is a general average. Of course in some individuals it is much more and in others, less. Two individuals who happen to have the same proportion of European to African genes may have this proportion consist of rather different genes. The genes for skin color and hair type, for example, are not necessarily inherited together.

Finally, a biologist should warn against falling into the trap of thinking that cultural inheritance is transmitted via chromosomes. There are no genes which carry instructions for speaking English or Urdu, Swahili or Japanese. It is in no way necessary to possess DNA which originated in Greece to be an heir to the cultural heritage of Athens. That belongs to all mankind regardless of race. A biologist who thinks of the commonality of all living things based everywhere on the double helix of deoxyribonucleic acid finds it easy to remember the words which Socrates uttered so long ago: "As for me, it is not the Athenians, nor even the Greeks, that are my brothers, but all mankind."

What of Eugenics?

Many people have been interested in eugenics, the science of "being well born." In fact, this concern extends far back into history.

Plato and other Greeks of his age discussed it. The citizens of Sparta, that stern and bleak prototype of the dictator state, practiced a crude form of race improvement by exposing all weak or deformed infants to the elements to die.

Two basic issues are involved in eugenics. The first concerns what is biologically possible and what is impossible or extremely difficult to do. These are questions of fact and are a normal part of science. The second group of questions concerns value judgments. What traits are desirable in human beings? These questions begin in biology but extend far beyond the province of ordinary science.

What is biologically possible and what could be done without great social change depends to a large extent on whether or not a particular gene is dominant or recessive, and on whether the intent is to reduce the number of harmful genes to the lowest possible level or to increase the number of desirable ones. Consider first so-called **negative eugenics**, *i.e.,* reduction in human load of harmful genes. There are many which produce results so sad and so horrible that it is impossible to believe any sane person would wish them to be perpetuated. As noted 700 years ago by St. Thomas Aquinas, some of the most frightful kinds of idiocy are hereditary. Sickle cell anemia and Down's syndrome are mild ailments compared with the Lesch-Nyhan disease or dozens of others. The only real question is one of method. The ancient Spartans would have exposed afflicted infants on a hillside. The Nazis would have used gas chambers.

In our own times humane methods and potentially very effective ones are becoming available. If a deleterious gene is dominant, it can be completely eliminated, except for rare mutations, in a single generation by preventing afflicted individuals from reproducing. Minor surgery tying off the sperm ducts or oviducts makes the conception of offspring impossible. By the use of amniocentesis and appropriate testing of the cells obtained, even afflicted individuals, if heterozygous, can be assured of normal children.

The problem with completely recessive genes or dominant genes which are not expressed until after the age of childbearing is more complex. Huntington's disease (chorea) illustrates the latter type. It begins in middle age with involuntary twitching of face and limbs which slowly becomes more and more severe. In the end there is degeneration of the brain and complete dementia. In a freely intermarrying population, preventing the afflicted homozygous recessives from having children will do little to reduce the number of the recessive genes in the gene pool. For example, there are five albinos in every 100,000 people. To produce these five on the basis of random matings requires 1,420 heterozygous carriers. Thus, removing the 10 recessive genes present in the five albinos would reduce the number of genes for albinism in the population from 1,430 to 1,420, a negligible amount. It is with the recessive genes that amniocentesis, testing of the cells for the presence or absence of such genes and, if indicated, termination of pregnancy, offers our chief hope.

What are the possibilities for **positive eugenics**? The desire that children should come into the world sound in body and mind is universal. It is when increased numbers of superior children are demanded that eugenics runs into a difficulty far transcending the proper bounds of biology. Who is to judge what the really desirable traits are and in what proportions they are desirable?

Under a ruthless dictatorship, either of a single man or of a group of so-called philosopher-kings, it would be possible to produce as diverse and fantastic breeds of men as has already been done with dogs and pigeons. The methods are the same as those that have produced the various breeds of domestic animals—using mutations, cross-breeding, selection, inbreeding, and more selection. As explained in the chapter on animal development, it would be possible by the new methods of nuclear transplantation to produce hundreds and even thousands of individuals all with identical genotypes.

A world without Shakespeares, Pavlovas, Beethovens, Gandhis, Edisons, Einsteins, and opera singers would surely be a poor place. Yet who would care to face a world composed solely of such people? Or who would advocate producing a standard model human being so all men would be as alike as identical twins? There is even evidence that certain afflictions can serve as gadflies to achievements of great benefit to the human race. Homer was blind, Edison deaf, Steinmetz crippled, Byron clubfooted. Clearly, our aim should be a golden mean of rich human diversity, perhaps not very different from what we now have but free of the present burden of genetic defects.

USEFUL REFERENCES

Crick, F. H. C. The genetic code. III. Scientific American, *215:* 55-62. Scientific American Offprint No. 1052, 1966.

Dunn, L. C., and Dobzhanshy, T. *Heredity, Race, and Society.* Mentor Books, New American Library, New York and Toronto, 1952.

Hartman, P. E., and Suskind, S. R. *Gene Action.* Prentice-Hall Inc., Englewood Cliffs, N. J., 1965.

Hayes, W. *The Genetics of Bacteria and Viruses,* 2nd ed. John Wiley & Sons, Inc., New York, 1968.

Lawrence, W. J. C. *Plant Breeding.* St. Martin's Press, New York. 1968.

Lynch, H. T. *Dynamic Genetic Counseling for Clinicians.* Charles C Thomas, Publisher, Springfield, Ill., 1969.

McKusick, V. A. *Human Genetics,* 2nd ed. Prentice-Hall Inc., Englewood Cliffs, N. J., 1969.

Mead, M., Dobzhansky, T., Tobach, E., and Light, R. E. (eds.) *Science and Concept of Race.* Columbia University Press, New York, 1968.

Mendel, G. *Experiments in Plant Hybridization.* Reprinted by Harvard University Press, Cambridge, Mass., 1967.

Watson, J. D. *The Double Helix.* Atheneum, New York, 1968.

Animal Development

The birth of a child is not only one of the most deeply moving of human events but it also forces on our minds our basic unity with the rest of the living world and leaves us wondering about the deep mysteries of the origin and continuity of life. What is the nature of the hereditary link between two parents and their offspring? How can a highly complex organism endowed with speech and conscious purpose possibly develop out of a minute speck—a fertilized egg? Anything like complete answers to these questions lies far in the future. Meanwhile, a significant amount of knowledge has been achieved. Within the egg is the code, written in sequences of purines and pyrimidines, which spells out the instructions for making a new human or a new starfish. The central modern problem in development is to discover how the information in the egg is translated into the adult; how the word, if you will, becomes flesh.

In the pursuit of this major problem, many satellite problems are also under investigation. Some carry important implications for mankind. The way is now open to produce hundreds or thousands of identical, *i.e.*, monozygotic, men or women. Whether anything of the sort is desirable is a question extending far beyond the boundaries of the natural sciences. For other forms of animal life there are obviously instances where many individuals with the same genome would be desirable. The methods? It is now possible to remove a nucleus from a cell of the intestine and transplant it into an enucleated egg which can then develop into a second animal with a genetic construction identical with the donor of the nucleus. Since, in any individual, there are

thousands of gut cells which could provide nuclei and any number of eggs, very large numbers of individuals are obtainable. The eggs with the donor nuclei would have to be placed in the uterus of a suitable female host, but this feat has been accomplished routinely in mice, rabbits, and other mammals.

The regeneration of new organs, which avoids the problems of organ transplants, has long been under investigation in Europe, North America, and Japan. Many of the lower vertebrates can regenerate new feet, new legs, or new lenses for the eye. Why are mammals unable to do this? Some of the new discoveries about development have been unexpected. Who would have anticipated, for example, that the development of the brain in mammals is permanently affected by male sex hormones soon after birth?

REMOVING THE ROADBLOCK TO MODERN KNOWLEDGE

The work of Fabricius at Padua on the chick and of his famous student, William Harvey, discoverer of the circulation of the blood, are often taken as the starting points of modern embryology. Harvey's work, published in the middle of the 17th century, illustrates the bewildering confusion then current. Like many others, the learned Harvey believed that mares could become pregnant not only by mating with stallions but also, according to ancient tradition, from breathing in the air on hilltops at certain seasons. He seriously discussed the comparison between the conception of an idea by the brain and the conception of an embryo in the uterus and argued for a basic similarity.

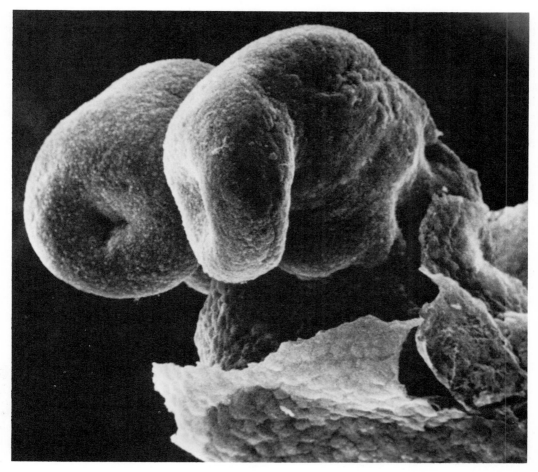

Fig. 8-1. Embryonic form of the structures which become the most complex object known to science, the mammalian brain. From a hamster as seen by a scanning electron microscope. (Kindness of Robert E. Waterman.)

Amid all this confusion, Harvey made important advances. In a truly royal experiment on the king's deer, doubtless what we would today call "government-sponsored research," Harvey showed that the popular and age-old idea that the embryo is formed by coagulation could not be true. Harvey took some female deer very shortly after mating and divided them into two groups. One he left as controls. These gave birth to fawns at the end of the normal time, eight months. Harvey sacrificed the does in the other group at intervals and examined their uteri. At no time, least of all soon after mating, was there any coagulating mass in the uterus. In Harvey's words, "After coition there is nothing at all to be found in the uterus, more than there was before." These experiments disposed of the time-honored theory that the embryo is formed by coagulation of blood and seminal fluid.

Gametogenesis and Gametes

The process by which gametes, whether sperms or eggs, are produced is called **gametogenesis**. It has two aspects. One is the formation of gametes, either motile sperms or yolk-laden eggs. The other concerns the reduction of the chromosomes from two sets to a single set, *i.e.,* meiosis, in the sperm or egg.

The mechanisms underlying both meiosis and

gametogenesis are very obscure. Both vitamin E (tocopherol) and vitamin A (a carotene derivative) are essential for gametogenesis, especially in males. A sufficiently high level of pituitary hormone is also essential. If the pituitary gland is removed, spermatogenesis stops in males, and in females all ova above a certain size degenerate.

Spermatogenesis and Sperms

Spermatogenesis, the formation of sperms, takes place in the **testes**. The testes vary in number, position, and shape according to the kind of animal. In many invertebrates and in fish and salamanders among vetebrates, the testis is composed of box-like compartments often called lobules. All the sperm cells in a given lobule develop more or less simultaneously, so they will all be approximately in the same stage of development. Such a testis is

relatively easy to study, and much of the pioneer research on spermatogenesis was done on the testes of insects for this reason. In the higher vertebrates, from frogs to men, sperms are produced in the **seminiferous tubules**. The stem cells, dividing by mitosis, lie around the periphery of such a tubule, and cells in the various stages of meiosis are crowded into the center.

The structure of mature sperm of either a vertebrate or an invertebrate varies little from species to species. These "little traveling libraries of genetic information" invariably consist of a head, a midpiece, and a tail.

The **head** of a sperm is the condensed nucleus of the spermatid. The tip of the head is covered by a cap-like membrane called the **acrosome**, which plays an important role in the penetration of the egg.

The **midpiece** of the sperm is notable chiefly because it contains mitochondria, sometimes

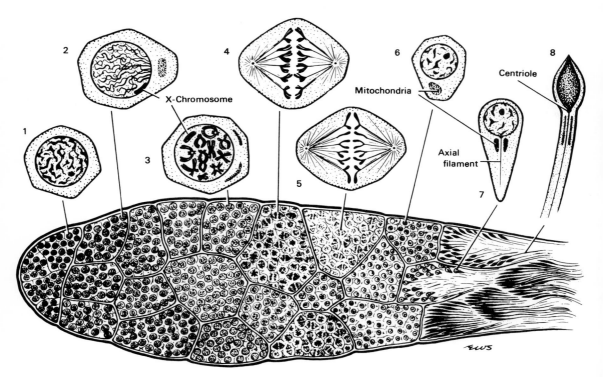

Fig. 8-2. Longisection through the easily studied testis of a grasshopper. Note how all the cells in a compartment are in the same stage. 1, spermatogonium; 2, early primary spermatocyte; 3, primary spermatocyte with tetrads (synapsis); 4, metaphase of first meiotic division; 5, metaphase of second meiotic division (separation of **kinetochores** and production of haploid cells); 6, spermatid; 7 and 8, developing sperms.

several wound in a spiral around a central axial filament. The midpiece may be a mere dot or longer than the head, depending on the species. Where the head and midpiece join, there is a pair of **centrioles** similar to those found at the base of any flagellum or cilium. Evidently, the function of the midpiece is to furnish the energy for the action of the flagellum. In many species of animals the midpiece enters the egg cytoplasm along with the head and contributes one or both of its centrioles which form the astral rays of the first cell division of the egg.

The **tail** has the same basic flagellar structure in all animals, from sponges to mammals. There is a pair of central **fibrils** surrounded by a circle of nine fibrils, the whole surrounded by a membrane. In some kinds of sperm a beautiful undulating membrane extends the length of the tail. This can readily be seen with a compound microscope in the living sperms of salamanders.

Oogenesis and Eggs

Oogenesis, the development of ova, takes place in **ovaries** which, like testes, vary in number, shape, and position according to the

kind of animal. In frogs, the two ovaries are hollow sacs each made of a thin double membrane; in higher vertebrates the ovary is solid. Within it are blood vessels, nerves, and developing eggs. Each egg is surrounded by **follicle cells,** which are important in supplying the egg cytoplasm with yolk. Surrounding the egg cell, and separating it from the follicle cells, is a yolk or **vitelline membrane.** In mammalian eggs this is a thick conspicuous structure. As a developing frog's egg increases in size, the follicle cells stretch out flat and thin. In mammals the follicle cells remain with the egg at ovulation and are not dispersed until fertilization. The smallest, least developed egg cells are known as **oogonia.** The larger ova undergoing meiosis are called **oocytes.** A notable feature of the mammalian ovary is that all the oocytes that will ever be present are present at birth. In the human ovary the number is estimated to lie between 200,000 and 400,000.

During their gradual growth in the ovary, eggs acquire a polarity. The yolk accumulates

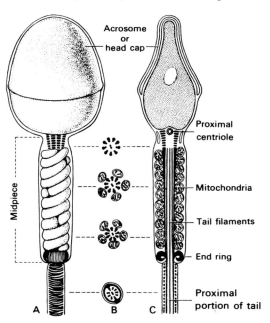

Fig. 8-3. A typical mammalian sperm as seen in the electron microscope. A, side view; B, cross sections; C, edge view in section. (After D. W. Fawcett.)

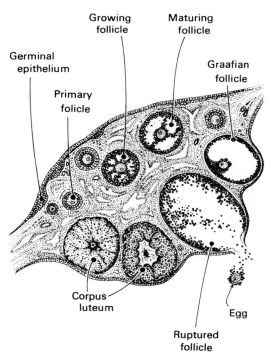

Fig. 8-4. Mammalian ovary in section showing enlarging Graafian follicles and ovulation.

more densely around one pole of the egg, which is known as the lower or **vegetal pole**, because the old naturalists believed it was primarily vegetative and plant-like in function. The opposite pole of the egg contains the nucleus and much more cytoplasm. This pole floats uppermost and is called the **animal pole**. The follicle cells not only provide yolk but also nucleoproteins, which are absorbed by the ovum and are ready to be built into new chromosomes during the rapid series of mitoses in the cleavage stages which follow fertilization.

When the primary oocyte with its giant nucleus, *i.e.*, the germinal vesicle, undergoes the first meiotic division, the division of the cytoplasm is so unequal that the egg seems to be extruding a tiny sphere. This minute cell is called the **first polar body**. It is, of course, a secondary oocyte, the egg itself being the other secondary oocyte. The first polar body seldom, if ever, divides. The egg undergoes a second meiotic division, extruding a **second polar body** containing a single set of chromosomes. The egg, now an **ootid**, also contains a single set of chromosomes.

Types of Eggs

Eggs differ in two important respects. The most far-reaching distinction is between the **spirally cleaving, mosaic eggs** characteristic of flatworms, annelids, mollusks, and many other groups of animals and the **radially cleaving, regulative eggs** found chiefly in echinoderms and vertebrates. This distinction will be discussed more completely in the section on cleavage of eggs.

Eggs also differ in the amount and disposition of yolk, a fact which makes a big difference in how development proceeds. Those in which the yolk is fairly uniformly distributed throughout the cytoplasm, as in the human or starfish egg, are known as **isolecithal** (*isos*, equal, + *lecithos*, yolk). These eggs are all small. Egg cells in which the yolk is concentrated toward one pole (or end), as in frogs and birds, are called **telolecithal** (*telos*, end, + *lecithos*, yolk). Frogs and salamanders have moderately telolecithal eggs; reptiles, birds, and some very primitive egg-laying mammals of Australia such as the platypus, have extremely telolecithal eggs. In these cases the actual ovum or egg cell is commonly called the yolk of the egg. Only this yolk is formed in the ovary of the bird or reptile. It is a single cell and is the true ovum. The white of the egg and the shell are secreted by the oviduct around the true ovum as it passes to the exterior. This enormous ovum is inert except for a small disk of living cytoplasm containing the nucleus and lying on the top of the yolk. The third type of egg is the **centrolecithal** found in insects and other arthro-

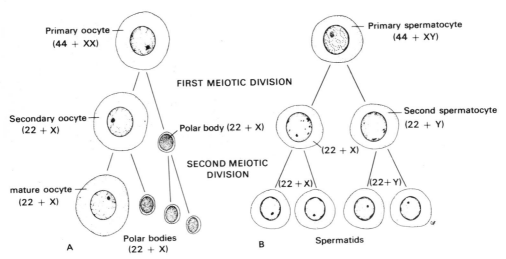

Fig. 8-5. Polar body (polocyte) formation in the formation of human eggs and sperms. (From J. Langman: *Medical Embryology*, 2nd ed. The Williams & Wilkins Co., Baltimore, 1969.)

pods. It may be round, oval, or even sausage-shaped, but the living cytoplasm always forms a thin layer over a centrally placed yolk.

OVULATION

Ovulation, the process of extrusion of the egg from the ovary, is, in vertebrates at least, under the control of the pituitary gland which is located on the underside of the brain. Vertebrates in general can be induced to ovulate by injecting anterior pituitary glands or their extracts. This method is used to cure certain types of sterility in women but has to be administered with great precision; too little extract has no effect, and too much produces multiple ovulations and therefore multiple births of as many as 10 infants which have almost no chance of survival.

In many animals, especially birds but also various mammals such as deer, ovulation is also controlled by the number of hours of light per day to which the animal has been exposed. In most rodents and in the primates, including man, ovulation follows an internal rhythm, with a four- or five-day cycle in the mouse, a 28-day cycle in the human, and a 32-day cycle in the baboon. In a very few mammals, such as cats and rabbits, ovulation occurs a short and definite time after the stimulus of mating. Rabbits, in fact, can be made to ovulate by stamping near their cages. In all these cases, the pituitary is believed to be involved, stimulated in some way by light, mating, or internal rhythms.

FERTILIZATION

The location of the egg when it is fertilized and becomes a **zygote** depends largely on the environment in which the adults live. Most, though by no means all, aquatic animals have external fertilization in which both eggs and sperms are simply poured out into the water to meet there. Many species of sharks and other live-bearing fish are, of course, exceptions. Terrestrial animals such as insects and mammals are forced to have internal fertilization by the simple fact that neither their sperms nor their eggs can live or move on dry land. In reptiles and birds fertilization takes place at or near the upper end of the **oviduct**, before the albumen

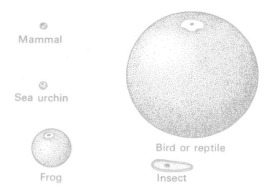

Fig. 8-6. Types of eggs, according to the amount and distribution of yolk. Mammalian and sea urchin eggs are isolecithal, frog and bird telolecithal, and the insect egg centrolecithal.

and shell are secreted around the yolk, the actual ovum. In mammals, likewise, the sperms are carried to the upper ends of the oviducts, or **Fallopian tubes** as they are known in human beings, by muscular contraction of the ducts.

Results of Fertilization

There are three major results of fertilization. First, fertilization stimulates the egg to begin its development. The second major result of fertilization is the restoration of diploidy, since the sperm brings its set of chromosomes into the egg. Diploidy in itself seems to be a source of vigor, but the main advantage of introducing a second set of chromosomes is an evolutionary one, the production of variation, of new kinds of individuals which result from new combinations of the genes. This is one of the main driving forces of evolution, and what sex is really all about. Third in mammals and many other animals but not all, the sex of the new individual is determined by whether the sperm carries an X or a Y chromosome.

In addition to the two primary results of fertilization there are a number of lesser ones having to do with the activation of the egg. The most obvious is the almost immediate appearance of a **fertilization membrane** within which the egg is free to rotate. This membrane prevents other sperms from entering the egg. The fertilization membrane also serves to hold together and protect the **blastomeres** as the first cells into which the egg cleaves are called. In the case of the human egg, where the first polar body is given off before and the second after fertilization, the first polar body lies outside and the second inside the fertilization membrane. The fertilization membrane appears to be formed directly from the vitelline membrane, which tightly surrounds the egg during its ovarian development, by the absorption of water and its collection immediately under the membrane. This lifts the membrane away from the cytoplasm.

Artificial Fertilization

This term refers to the introduction of sperms into the female reproductive tract by mechanical means. It has been used successfully in cases of human sterility. It is a common practice in certain branches of animal husbandry. Pedigreed lambs have been produced in Idaho from sires living on the U. S. Department of Agriculture farms in Beltsville, Maryland. Cows are frequently bred by this method because it saves the expense, trouble, and danger of keeping a bull, and allows wide use of sires that are known to transmit desirable milking qualities to their daughters.

The longevity of functional sperms has been much studied, especially in connection with artificial insemination. If chilled and kept in thermos jars, mammalian sperms can be kept about a week. If kept anaerobic at approximately $-70°C.$ in "deep deep-freeze," sperms seem capable of indefinite existence. This is the method of "sperm banks" where human sperms are deposited.

Microanatomy of Fertilization

Many of the details of fertilization were first clearly seen during the summer of 1875, which Oscar and Richard Hertwig spent on the shores of the Mediterranean Sea at Naples. There they discovered that the eggs of starfish and sea urchins are so transparent that the major events of fertilization can be seen in the living egg. In these animals the head and midpiece of the sperm enter the cytoplasm of the ovum. The midpiece gives rise to a star-shaped aster, or division center, which soon becomes double. The head of the sperm moves toward the egg nucleus and gradually becomes transformed into a nucleus itself, lying pressed against the egg nucleus.

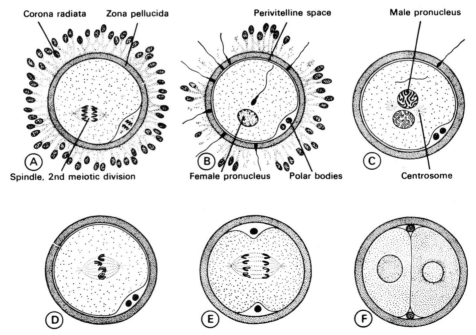

Fig. 8-7. Microanatomy of fertilization, completion of meiosis, and first cleavage in a mammal. (From J. Langman: *Medical Embryology,* 2nd ed. The Williams & Wilkins Co., Baltimore, 1969.)

There is no compelling evidence that animal sperms are attracted toward eggs. Swimming appears to be random. The first visible event in the actual fertilization process is the extrusion of the acrosome into a longer or shorter tubular thread or vesicle which penetrates the jelly around the egg and makes contact with the cytoplasmic surface of the ovum. The **acrosome reaction** is almost certainly in response to a glycoprotein, **fertilizin**, which diffuses from the egg. The response of the egg is the formation of a cytoplasmic fertilization cone which engulfs the head and midpiece of the sperm. In some species, including the human, the second polar body is given off and in most the egg membrane swells away from the egg.

Physiology and Biochemistry of Fertilization

Despite nearly a century of hard work in many marine laboratories since that summer when the Hertwigs discovered what favorable material for research echinoderm eggs are, we still cannot answer the question of what fertilization does that breaks the block to cleavage and subsequent development of an egg. We do not know the nature of the block. Nor do we know the similarity or dissimilarity between the stimulus of fertilization and the normal stimulus for ordinary cell division. There are at least two main possibilities. Entrance of the sperm might somehow inactivate or remove an inhibitor. Or it might trigger the activity of some enzyme or cofactor or even add a factor essential in minute amounts.

In addition to the elevation of the fertilization membrane and the disintegration of granules at the egg surface, other events are known to occur very soon after fertilization. The **permeability** of the eggs changes. The egg becomes impermeable to other sperms but much more permeable to certain ions. Potassium and calcium exchange between the egg cytoplasm and the surrounding water increases about 15-fold. Radioactive phosphorus, ^{32}P, is taken up by the egg over 100 times faster after fertilization.

In some species the rate of **respiration** greatly increases immediately after fertilization. This is true of sea urchin eggs and was at first supposed to be the key aspect in activating development.

Then it was discovered that in some species, such as *Chaetopterus,* a marine annelid, and in various mollusks, the rate of respiration falls after fertilization. Still more puzzling, if sea urchin eggs are fertilized immediately after they are shed and while they are still healthy, they show no appreciable change in respiration. It was once thought that unfertilized eggs respired through a non-cytochrome system. But this too has been shown to be untrue as a generalization.

There is also a marked rise in **protein synthesis**. This can be demonstrated by various methods such as measuring the rapid increase in uptake of labeled amino acids. It would be logical to suppose that immediately after fertilization the zygote nucleus begins to pour out messenger RNA into the cytoplasm. However, Ethel Harvey showed that enucleated fragments of sea urchin eggs can be stimulated to cleave parthenogenetically without any nucleus whatever. They never develop beyond a ball of cells. Moreover, mRNA can be extracted by the usual methods from unfertilized eggs. Transfer RNA and ribosomes are also present in unfertilized eggs, along with amino acids.

Since all the protein-forming machinery is already present before fertilization, the possibility exists that one or more of the parts of this apparatus are present in inactive form. This hypothesis has been tested by several people in different laboratories and notably by Alberto Monroy of the University of Palermo. He measured the incorporation of radioactively labeled amino acids into proteins, *in vitro,* with a mixture of ribosomes from unfertilized eggs and mRNA from adult liver. The result was invariably negative. No protein formation took place. However, with ribosomes from liver and mRNA from unfertilized eggs, protein synthesis did occur. This clearly indicates that the ribosomes of unfertilized eggs are, somehow, inactivated or masked. If ribosomes from unfertilized eggs are treated with proteinases, there is some evidence that they can be rendered active. This would indicate that before fertilization the ribosomes are covered with an insulating coat of protein. This theory is supported by the fact that activity of proteolytic enzymes increases after fertilization. If the theory is correct, a question to be answered is: how does the sperm release the activity of the ribosomes?

Parthenogenesis

Many investigators have thought that **parthenogenesis**, the development of an egg without fertilization, would reveal how fertilization removes the block to development. Parthenogenesis was first extensively studied years ago by Jacques Loeb and others at the laboratory at Woods Hole on Cape Cod. They and their successors have found that a large variety of treatments will stimulate unfertilized eggs to develop. Adding weak organic acids such as butyric acid or various fat solvents such as ether, alcohol, or benzene, temperature shock by exposure to either heat or cold, osmotic changes which can be produced by sugar or urea, ultraviolet light, and the prick of a needle (especially if previously dipped in blood plasma), all will stimulate the development of unfertilized eggs. Among the most interesting of the more recent experiments are those of John Shaver, who showed that injection of small amounts of the granular, presumably mitochondrial and ribosomal, fraction obtained by centrifuging homogenated adult tissue will produce parthenogenesis, while the clear supernatant fluid will not. The trick in all these experiments is to have the concentrations or dosage and the timing precisely right.

Parthenogenesis has been artificially induced in many organisms, including sea urchins, marine annelids, frogs, and rabbits. The resulting animals become normal looking adults. Natural parthenogenesis has long been known to occur in rotifers, certain small crustaceans, aphids, and honeybees (where it gives rise to males).

So many different procedures will result in artificial parthenogenesis that one can only conclude that an egg is set up to begin cleavage and embryo formation much as a muscle cell is set up to contract. A stimulus can elicit the only response of which the cell is capable. So far, the study of various possible stimuli has not furnished the clue to the old problem of how a sperm breaks the block to development.

CLEAVAGE OF THE ZYGOTE

Cleavage is the name given to the series of cell divisions which divide the fertilized ovum, or zygote, into successively smaller and smaller cells until first a mulberry-like group of cells, the **morula** and then a hollow ball called the **blastula** is formed. The chromosomes separate by ordinary mitosis, so that each new cell is diploid. The cell divisions follow each other rapidly; hence cleavage must be a period of rapid synthesis of nucleic acids and proteins to form new chromosomes and new protoplasm.

Anatomically, depending on the amount of yolk, cleavage differs markedly in different kinds of animals. In eggs with little yolk, cleavage is total, or **holoblastic**, and completely divides the egg successively into 2, 4, 8, 16, etc., separate cells. Isolecithal eggs, such as those of echinoderms and mammals, have this type of cleavage. So also do the moderately telolecithal eggs of frogs. In the extremely telolecithal eggs, such as those of reptiles and birds where there is only a tiny disk of protoplasm on top of an enormous yolk, cleavage does not divide the entire ovum into separate cells. Cleavage here involves only the thin disk on top of the egg, and is termed partial or **meroblastic** cleavage (*meros,* part).

In mammals, cleavage takes place as the egg passes down the oviduct. This requires about 4½ or 5 days in all mammals regardless of size. On the fifth day, whether in mice or women, the egg is a hollow blastula, called the blastocyst in mammals, which enters the uterus and becomes more or less deeply implanted on the uterine lining.

There are two patterns of cleavage in the animal kingdom. In echinoderms and vertebrates cleavage is **radial** and also **indeterminate**. It is called indeterminate because there is only a very general relation between the position of any particular cell formed during cleavage and the specific tissues it will form in the embryo.

In contrast, in marine annelids and mollusks, etc, the first two cleavages result in four equal cells, but the third time the cells divide they each give off a small cell in a clockwise direction, as seen from above. The large cells later give off a second quartet of cells, this time in a counterclockwise direction. At about the same time the first quartet of small cells divides in a spiral direction. In this way an elaborate and extremely precise jewel-like design of cells is produced which is exactly the same in every blastula, and in which the ultimate destiny of

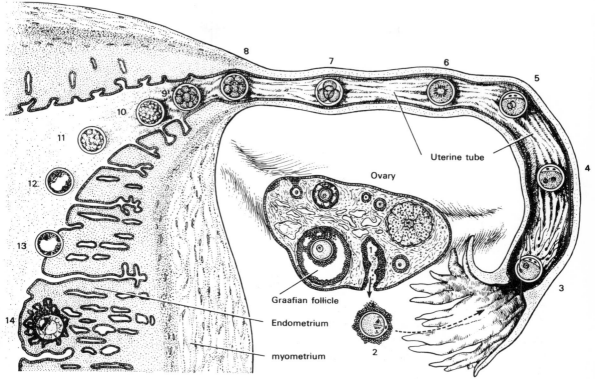

Fig. 8-8. Early stages of human development. Ovulation, fertilization, cleavage, and implantation in the endometrium (lining) of the uterus.

every cell can be precisely foretold. Consequently, such cleavage is termed **spiral** and also **determinate**. Both types of cleavage are illustrated in Figure 8-9.

NUCLEUS OR CYTOPLASM?
THE PARADOX OF DIFFERENTIATION

Is It the Nucleus?

As soon as it became clear that every cell, whether a brain or liver cell, carries exactly the same double set of chromosomes, a very awkward but fundamental question arose. If the chromosomes really carry the hereditary factors which determine all the physical traits of the body, why are not all the cells exactly alike? How can they become differentiated into so many very different sorts of cells? Despite great advances in knowledge there is still no satisfactory answer to this question.

August Weismann, the blind biologist and founder of the germ-plasm theory that an organism is the expression of its hereditary material (germ-plasm), proposed that the answer to this question is that the nuclei of different kinds of cells carry different hereditary determinants (we would say genes). Specifically he held that during the cleavage of the egg the hereditary factors for the different parts of the body are sorted out into different cells. At the end of cleavage certain cells contain only genes for the right arm or for skin, or red blood cells, and that is why they become a right arm or skin or blood cells.

About this time a very gifted experimenter, Theodor Boveri, discovered that in the cleavage of the egg of the nematode worm, *Ascaris,* the chromosomes in the different cells actually do become different. Only the germ cells which will form the future eggs and sperms retain all the chromosomes, a fact which is still not understood. Other investigators found that if you separate the first two blastomeres (as the

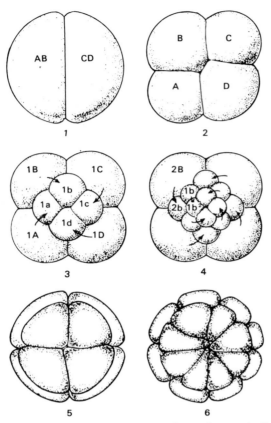

Fig. 8-9. The two basic types of egg cleavage (cell division) in the animal kingdom. 1 and 2, first two cleavages similar in both; 3 and 4, subsequent cleavages in the spiral and determinate type of annelids and mollusks; 5 and 6, subsequent cleavages in the radial and somewhat irregular type of mammals and echinoderms.

cells of a cleaving egg are called) of the egg of an animal which has determinate cleavage, then two half-embryos develop. Each of the first four cells, if separated, forms a quarter-embryo. This type of development is commonly called **mosaic cleavage.** It seemed to support Weismann's differential chromosomal segregation theory.

But at about this point in the argument a speculative zoologist, Hans Driesch, showed that if you separate the first two cells of a starfish egg by shaking in calcium-free sea water, two whole embryos will form. In his flamboyant phrase, each cell is a **"harmonic equipotential system."** His experiment certainly argues against the idea that nuclei become different from each other in genetic content during cleavage so that all the genes for the left half of the body get into the left of the first two blastomeres, and the genes for the right half into the right blastomere. Hans Driesch himself became so excited by his experiment that he left his laboratory forever and became a philosopher. He proposed that the development of an embryo is due to an "entelechy," which is something like Henri Bergson's vital force, and which "carries its purpose within itself." No mechanical or truly scientific explanation of embryonic development is possible, he claimed. After all, one cannot cut a typewriter or other machine in two and expect each half to behave as an harmonic equipotential system and regulate into two perfect small typewriters. Therefore, at the very least, these early nuclei must all be equivalents and cannot be responsible for differentiation.

Is It the Cytoplasm?

If all the nuclei are equivalent, then it would seem that cytoplasmic differences must somehow be responsible for differentiation.

The cytoplasm of uncleaved eggs commonly shows its heterogenous character by differences in pigmentation and cytoplasmic constituents in different regions of the ovum. Perhaps these cytoplasmic substances are organ-forming, acting either directly or in some roundabout way by influencing genes. To test such a hypothesis, eggs can be centrifuged. If this is done in a solution of approximately the same density as the eggs, the eggs will be gradually pulled apart. The lighter components of the cytoplasm will move to the upper (centripetal) pole of the egg and the heavier to the lower (centrifugal) pole. By repeating this process with each of the separated halves of the egg, sea urchin eggs can be neatly separated into four quarters. Each quarter can be fertilized and thus tested for developmental potentialities. This experiment was carried out by Ethel B. Harvey.

The lightest, uppermost quarter contains mostly a clear fluid plus a few oil droplets, together with the egg nucleus. The second is a mitochondrial quarter which contains all, or at least all the detectable, mitochondria. The third and fourth are, respectively, a yolk-filled quarter and a pigment quarter. Upon fertilization all four quarters will develop into ciliated swimming larvae, although, of course, all but those derived from the clear first quarters are haploid. More remarkable, the clear quarters, which lack mitochondria, yolk, pigment, and whatever other substances are localized in the other three quarters, develop into approximately normal larvae with skeleton, gut, and pigment spots. The yolk quarters also form larvae with skeleton, gut, and pigment, but with a less normal body shape. Other centrifugation experiments on the eggs of this and other species of echinoderms make it certain that the visible substances in the egg of sea urchins are not a direct cause of differentiation.

Thus we are left with the paradoxical conclusion that neither the nucleus nor the cytoplasm is responsible for initiating differentiation. Yet differentiation obviously does occur and chromosomes can be shown to control the development of the specific traits of an animal.

Modern Answers

Several avenues of escape from this dilemma of development are open. Theodor Boveri showed that in *Ascaris* a loss of chromosomal material during any particular cleavage depends on the cytoplasm in which the nuclei come to lie. Then E. G. Conklin, of Princeton and Woods Hole, showed in a long series of largely overlooked experiments with the eggs of marine relatives of vertebrates called ascidians that abnormalities can be produced by centrifuging the cytoplasm to abnormal positions, provided that the centrifugal force is sufficiently strong and applied at the right time. Conklin's line of work is now being pursued by T. C. Tung in Communist China but the results are not available. It may turn out after all that there are substances in the cytoplasm which derepress or activate specific genes.

Recent work with actinomycin, which blocks the transcription of new RNA from DNA, and puromycin, which blocks the synthesis of proteins, has thrown new light on the differences between regulative and determinate eggs.

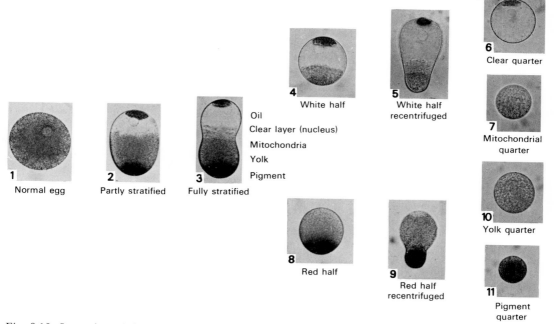

Fig. 8-10. Separation of the cytoplasmic components of a sea urchin egg into four parts by centrifugation. (Kindness of Ethel B. Harvey.)

Puromycin blocks development in both,—therefore, both require protein synthesis for this process. Actinomycin blocks development of a regulative egg such as that of a sea urchin but not that of a determinate egg like that of an ascidian. Such eggs continue to develop into normal tadpoles in actinomycin but do not metamorphose into adults. Evidently new RNA is required for making the adult body but the egg already has all the instructions ready as mRNA for the proteins needed to form a larva.

Recently Briggs and King and others in the United States and abroad have been investigating the problem of whether or not the nuclei become different during cleavage or are interchangeable, a problem that Pflüger had investigated in 1884 by squeezing frog eggs between glass plates. The new method, a very difficult one, is the **transplantation of nuclei** from various parts of frog embryos and even tadpoles in different stages of development into activated but enucleated uncleaved eggs. With proper care, the host eggs with foreign nuclei will develop in as many as 80 per cent of the cases. The advantage of the new method is that it allows the testing of nuclei from much more differentiated stages than did the old Pflüger technique. The first results of the new method quickly confirmed Pflüger's conclusion that all the early cleavage nuclei are interchangeable. However, when older and older donors are used, the resulting embryos show more and more abnormalities. Nuclei taken from such maldeveloping embryos and retransplanted into second and third generations of enucleated eggs have given clones of embryos showing similar abnormal development. The actual meaning of this is very puzzling because more than a few cases have been found where nuclei from older embryos gave entirely normal development. Furthermore, work by Gurdon at Oxford with the African toad *Xenopus* has given large numbers of normal tadpoles with nuclei from differentiated epithelial cells of the tadpole intestine. By this method large numbers of tadpoles can be obtained, all with the same genome because all were produced by nuclei from the same donor intestine. It is this technique combined with intrauterine implantation into host females that would make possible large numbers of identical mammals, including humans.

The problem of the developmental equivalency of nuclei can be tested also by fusing two eggs. Unless the nuclei are equivalent, such a fusion would produce a monstrous creature with a mixture of parts of two animals. The most extensive and exciting work of this kind is now being done by Beatrice Mintz in Philadelphia. She has developed a method for fusing two mouse eggs in the two-cell stage. When replaced in the uterus of a properly prepared

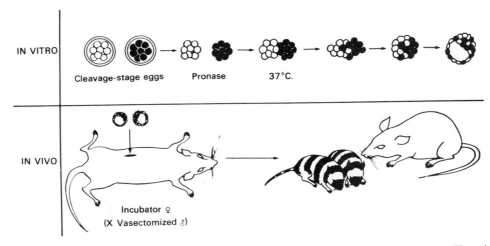

IN VITRO   Cleavage-stage eggs   Pronase   37°C.

IN VIVO   Incubator ♀ (X Vasectomized ♂)

Fig. 8-11. Two quadriparental mice produced by the fusion of two eggs in early cleavage stages. The mice are normal except for the unexplained zebra-like stripes when eggs from white and black mice are fused. (Kindness of Beatrice Mintz.)

host mother, such fusions develop into whole normal but tetraparental mice. If one egg was destined to become an albino and the other pigmented, the resulting mouse will have zebra-like bands of black and white.

Study of over 1,000 mice produced by such fusions of fertilized eggs from parents carrying different genic alleles have revealed that there are no immunological antagonisms developed, such as would have occurred had the two eggs developed into separate individuals. When the egg of a strain which always develops cancer is fused with one which does not, the cells of the cancerous strain become cancerous while adjacent cells of the immune strain remain healthy. It is also apparent that many tissues in an embryo are developed as a clone—descendants of a single cell.

Experimentation during later stages of development has revealed much about how differentiation is induced, but before discussing these discoveries, we will consider normal embryo formation.

BUILDING THE EMBRYO

Gastrulation and the Germ Layers

Gastrulation is the process that converts a ball of cells, the blastula, into an embryo. After fertilization, it is the most crucial stage in development. A gastrula stage occurs in most metazoans, but the following account will deal only with echinoderms and vertebrates. The details of gastrulation are very different in different kinds of animals, although the basic process is similar in all, as are the results.

After gastrulation three layers of cells, called germ layers, can be distinguished. The **ectoderm** covers the embryo externally and will form the external layers of the skin and skin derivatives like hair, hoofs, and nails, plus the entire nervous system. The **endoderm** forms the gut and will become the epithelial lining of the entire alimentary canal and also of all the structures derived from the gut in the course of development, such as the lungs, liver, and

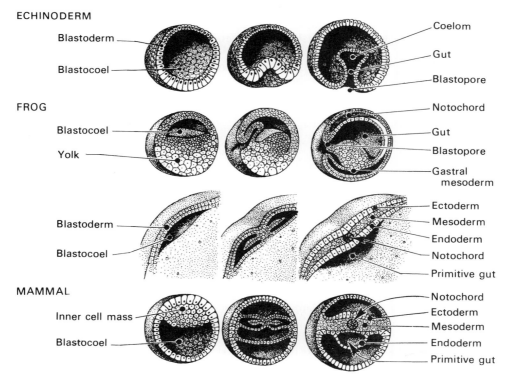

Fig. 8-12. Gastrulation and the formation of germ layers in four types of animals.

pancreas. The **mesoderm** lies between ectoderm and endoderm and is itself split into two layers. The **parietal** or somatic layer presses against the ectoderm. The **visceral** or splanchnic layer of mesoderm presses against the endoderm of the gut. The cavity between parietal and visceral mesoderm is the **coelom.** From the mesoderm are formed muscle, skeleton, vascular system, and the connective tissues which hold the body together.

In the eggs of starfish and sea urchins, where there is very little yolk, gastrulation begins with the inpocketing of the cells around the lower or vegetal pole. It looks as though a thumb had pushed in one side of an old hollow rubber ball. The cavity of the blastula, *i.e.,* the **blastocoel,** is slowly obliterated. The new cavity formed (after the hypothetical thumb is removed) is the primitive gut or enteron and is therefore called the **archenteron** (*archaios,* ancient, + *enteron,* gut). The opening into the archenteron is called the **blastopore.** The archenteron is the origin of the endoderm. The mesoderm is formed as a pair of hollow buds from the inner end of the archenteron.

In a frog's egg there is so much yolk that the cavity of the blastula is very small, and consequently gastrulation of the type just described is impossible. The egg "cheats," as it were. The archenteron pushes into the blastula cavity from one side. Cells move down from the animal hemisphere and roll in over the edge or **dorsal lip** of the **blastopore** to add themselves to the wall of the archenteron. The dorsal lip is the "organizer" of the embryo.

In the telolecithal eggs of reptiles and birds, where the yolk is enormous, gastrulation takes place by the inpushing of cells along a line called the **primitive streak.** At the anterior end of the primitive streak is a pit beside a hillock of cells. This is Hensen's node and corresponds to the dorsal lip of the blastopore. At the end of gastrulation the embryo consists of three layers of cells, a layer of endoderm against the yolk, a layer of mesoderm on top of it, and a layer of ectoderm over the mesoderm. The three are commonly adherent along the **notochord,** which extends anteriorly from Hensen's node and lies in the mesoderm. It is the forerunner of the backbone.

In mammals gastrulation occurs in the same way as in birds and reptiles, clearly an inheritance from our reptilian ancestors. The blastula of placental mammals, such as man, is a hollow sphere called the **blastocyst.** It is the blastocyst which becomes implanted on or, in higher mammals, into the lining of the uterus, an event which occurs on the fifth or sixth day of development. A solid mass of cells lies at one pole of the blastocyst. Two cavities form in this mass, and in the plate of cells separating the two cavities, the primitive streak is formed and gastrulation takes place. The upper cavity becomes the **amnion,** a membrane which surrounds the embryo until birth. A third cavity, the **primitive gut,** pushes into the lower cavity, which is obliterated.

Neurulation

Neurulation is the process by which the primitive nervous system is formed. Immediately after gastrulation a broad strip of ectoderm overlying the notochord and the mesoderm adjacent to it begins to thicken. This strip of ectoderm, the **neural plate,** forms the brain and spinal cord in a most remarkable way. The edges roll up to form a groove, and by the meeting and fusion of the edges of the groove, a tube is formed. The anterior part of this tube enlarges into the brain, and the posterior part becomes the spinal cord. In these early stages the embryo is called a **neurula.**

The motor (efferent) nerves grow out from the brain and the spinal cord and make connections with their appropriate muscles and glands. This remarkable fact was deduced from the study of sectioned embryos by two rather isolated investigators, Santiago Ramon y Cajal in Spain and Wilhelm His in Switzerland. However, the scientific world continued to hold to the prevailing theory that nerves are formed by the coalescence of long chains of cells until Ross G. Harrison discovered that cells can be grown outside the animal's body and was able to demonstrate the outgrowth of nerve fibers from isolated pieces of embryonic spinal cord. Harrison's famous experiment at Johns Hopkins was the beginning of animal tissue culture *in vitro* as a usable technique.

Along the edges of the neural folds is a narrow strip of tissue called the **neural crest.** The cells of the neural crest separate from both the surface ectoderm and the neural folds as

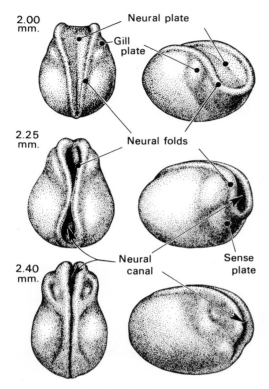

2.00 mm.

Neural plate

Gill plate

2.25 mm.

Neural folds

2.40 mm.

Neural canal

Sense plate

Fig. 8-13. Neurulation in a frog. Compare with Figure 8-1, neurulation in a mammal.

they are closing, and migrate down over each side of the body between the ectoderm and the underlying mesoderm. The neural crest cells develop into at least four different important tissues; although this has been hard for anatomists and embryologists to believe, it has been abundantly proved by microsurgical studies in which parts of the neural crest have been removed and transplanted to other sites in the embryo. The neural crest cells form the dorsal sensory ganglia and hence the sensory (afferent) nerves to both brain and spinal cord. They become the pigment cells, both the melanocytes of a fish or a frog and the pigment-forming cells in the feathers of birds and the hair of mammals. They form the cartilage and bone of the so-called visceral skeleton which comprises the jaws and gill arch skeleton of fish and the jaws and hyoid bones of higher vertebrates. They form the medulla of the suprarenal (adrenal) gland and related endocrine structures. In connection with the nerves,

some of the neural crest cells become Schwann cells which line up along the nerve fibers and, by winding around them, myelinate them, *i.e.,* cover them with a special kind of protective sheath which insulates one fiber from another and increases the speed of the nerve impulses.

Gill Slits

In the most anterior part of the gut in all chordates, **gill slits** develop. These are paired openings on the side of the throat from the exterior into the pharnyx or throat itself. At the most anterior and ventral part of the gut, the mouth breaks through. This stage, in which pharyngeal gill slits appear, is called the **pharyngula**. As pharyngulas, vertebrates of widely separate groups—birds, mammals, fish— resemble each other more closely than in any other stage, for the eggs differ as markedly as do the adults. In this stage a simple diagram will do for them all. The heart is located forward, not far behind the mouth. Blood is pumped forward and then up to the dorsal side of the gut, through six pairs of aortic arches. In fish and in larval frogs and salamanders, the aortic arches send blood vessels out into gills, where the blood vessels become thin-walled capillaries through the walls of which oxygen and carbon dioxide are exchanged with the surrounding water. The capillary circulation in the gills makes a handsome sight when brilliantly illuminated under a binocular dissecting microscope. In mammals all the gill slits grow closed except the first pair, which form the Eustachian tubes of mammalian ears. The hands and feet grow out as paddle-shaped structures which then develop fingers or toes.

The relationship between a frog embryo and a chick in the neurular or pharyngula stage can be visualized by cutting a chick embryo off from the underlying yolk and pulling the edges together ventrally. Since the days of Fabricius and Harvey, the chick embryo has been a favorite object of study because of its ready availability and ease of handling. A 72-hour chick embryo corresponds to a human embryo at about the end of the first month of intrauterine life. Both are in the gill slit stage. It is in the early stages during the first three months that human embryos are most susceptible to deleterious influences, such as the virus

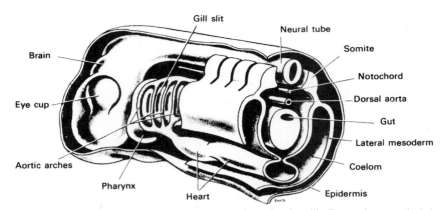

Fig. 8-14. Diagram of basic structure of a vertebrate embryo in the gill slit or pharyngeal cleft stage. This corresponds to a chick of 72 hours incubation or a human embryo at the end of the first month. (For "eye cup" read "eye vesicle.")

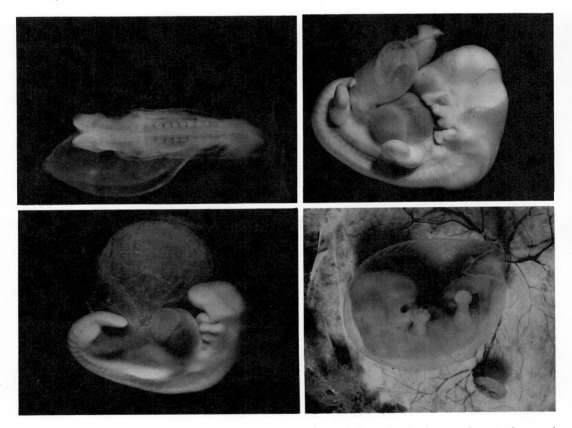

Fig. 8-15. Human embryos at approximately three, four, five, and six weeks. At three weeks note the neural folds at the left and the somites on either side of the midline. Gill slits are visible at four and five weeks and the rudiments of fingers at six. An amnion surrounds the six-week embryo. (Kindness of the Carnegie Institution of Washington.)

of rubella (German measles) or the products of thalidomide. However, there is no stage in which the unborn cannot be damaged.

EMBRYONIC INDUCTION— TURNING ON GENES

Investigators seeking the forces that produce gastrulation have made little progress, but in the effort important facts have been learned about the causes of differentiation, which visibly begins during gastrulation. The pioneer work was carried out on amphibian embryos by Warren Lewis in Baltimore and Hans Spemann in Freiburg. Briefly, they found that the dorsal lip of the blastopore is in some unknown way the "**organizer**," or better, the **evocator**, which calls forth the development of an embryo. If a frog blastula is constricted with a fine hair into two separate halves, two embryos, indentical twins form, but only if the constriction cuts through the part of the blastula that will form the dorsal lip of the blastopore. If the blastula is so constricted that one-half gets all the dorsal lip material, that half, and that half only, will form an embryo. The other half remains a mere ball of cells.

More dramatically, if the dorsal lip is cut out and implanted in another embryo by injecting it into the cavity of the blastula with a fine pipette, the transplanted dorsal lip will be pushed down against the ventral belly wall during gastrulation and will there induce or evocate a secondary embryo in the host tissue (Fig. 8-16). Specifically, the cells with lie over the notochord and its adjacent mesoderm become induced to form neural plate. These inducing tissues are found as the dorsal lip of the blastopore in the early stages of gastrulation. In birds and mammals, where gastrulation occurs along the primitive streak which thus corresponds to the blastopore, no one was surprised when it was discovered that Hensen's node is the essential evocator or organizer of bird and mammalian embryos.

The dorsal lip is the primary evocator without which no embryo forms. There are also **secondary evocators**. For example, the eye is formed as an outgrowth from the brain. Where this outgrowing **optic vesicle**, as it is called, comes into contact with the skin of the embryonic face, a lens forms in the embryonic facial ectoderm. If the optic vesicle is removed before it touches the skin, in most species of animals, no lens forms. The optic vesicle is therefore a secondary evocator.

How does the evocator work? Does it have to be living? The answer to the second question is no. Subsequent work has shown that a wide variety of substances, when soaked up in agar jelly and imbedded in a blastula, will evocate a secondary embryo. These include: extracts of adult brain or other tissues; various pure chemicals (especially phenanthrenes, which are chemicals resembling both the sex hormones and adrenal cortical hormones); and a number of cancer-inducing agents. Hypotonic salt solution will also cause ectoderm explanted into a glass dish to form neural tissue. The situation is comparable to that in fertilization and artificial parthenogenesis. So many agents will produce the effect that all one can say is that, as with muscular contraction, many agents can trigger the reaction.

Recent experiments suggest that evocators are diffusible chemicals. Clifford Grobstein separated the inducing tissue from the inducible tissue by cellulose ester membranes of different thickness and porosity. Even if the pores in such membranes are small enough to prevent any cytoplasmic contact (about 0.1 μ in diameter), induction can still take place. For example, kidney tubules are induced in mesoderm by substances diffusing from a region of adjacent neural tube. Twitty and Niu have shown that tissues grown *in vitro* exude inducing molecules into the medium. The work of Yamada in Japan strongly suggests that these inducer substances are proteins; other experiments suggest that they are mRNA.

Possible Biochemical Mechanisms

It will be recalled that the functional genetic unit consists of an operator gene and one or more structural genes which the operator controls. The structural genes are the templates for the formation of messenger RNA. Also, the operator gene itself is under the control of a regulator gene which keeps the operator gene repressed by a repressor substance, probably a protein. According to this Jacob and and

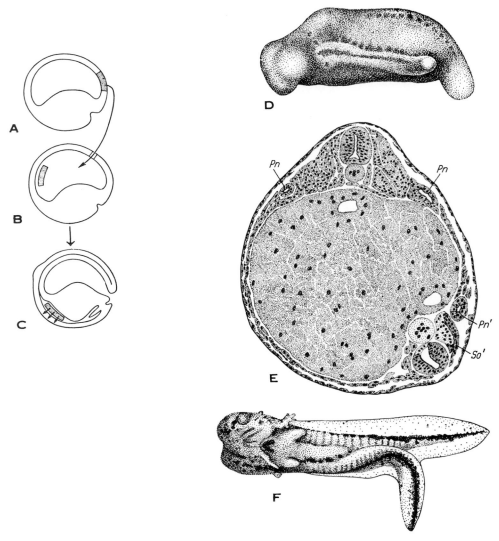

Fig. 8-16. Inductive action of dorsal lip of the blastopore using a salamander gastrula. The dorsal lip is placed inside the blastocoel of the host and is pushed against the lower wall of the egg during gastrulation. (From A. Kühn (Translated by R. Milkman): *Lectures on Developmental Physiology,* 2nd ed. Springer-Verlag, New York, 1971.)

Monod model of gene action and its control, genes become activated, *i.e.*, derepressed, by specific derepressor, *i.e.*, inducer, substances. These may be proteins or other things that combine with or otherwise counteract the repressor substance. If embryonic inducers act in this way, they are acting essentially at the gene level.

It is also possible that embryonic inducers act at the level of proteins by what is called **end product feedback inhibition.** There are a number of well-worked-out cases in microorganisms where the entire biosynthetic pathway is known from substance A to B to C to D to E, each step controlled by its own enzyme.

In such cases it has been found that if the end product E is added to the medium in which the organisms are growing, the cells no longer form E. Further investigation has revealed that it is not the final step from D to E that is blocked by the presence of excess E but the enzyme governing the first step, A to B, that is blocked. Here clearly is another mechanism that could function during induction of embryonic differentiation. Does it?

Meryl Rose, an investigator with wide experience with animals from different phyla, has proposed that as each tissue is differentiated it exudes some kind of substance which inhibits the differentiation of additional cells of the same kind—a "don't-make-any-more-of-me" substance. Such a substance would explain the limitation of a given tissue, although it does not explain how the genes controlling the next tissue to differentiate become derepressed (activated).

A number of striking cases are known where a continuously acting influence of one structure on a second is essential to maintain a differentiated condition. If the retina is removed from the eye of a salamander, the black pigmented cells just back of the retina lose their pigment and become embryonic. The stages in this process can be readily seen if salamanders are sacrificed at intervals after removal of the retina. The formerly pigmented cells undergo mitosis and form a new layer of cells lining the big vitreous chamber of the eye. The cells on the inner surface of this layer differentiate into a new retina, and those on the outer surface form a new pigmented layer. This demonstrates that the state of differentiation of at least the pigmented cells requires the continuous presence of some influence from the retina. It also shows that differentiation is not an irreversible process.

A similar kind of relationship is seen in the fiddler crab, where there is a fantastic difference in size between the two big claws (chelipeds). If the larger of the two chelipeds is removed, then in the subsequent molts the smaller remaining limb develops to a huge size while the position of the original large claw is occupied by a very small one. In all such cases it is easy to believe that the original structure gave off some kind of inhibitory material which prevented the development of a similar structure on the opposite side of the body. However, there is no proof.

Sequential Gene Action

Recent discoveries about the sequence in which different specific proteins appear during development have brought some of the problems just discussed into a sharper focus. Each protein, or each polypeptide part of a protein composed of more than one chain, is the expression of a single gene. Therefore, since the proteins found in later stages of development are different from those found earlier, the genes responsible must be different. The three best understood cases are seen in hemoglobin, in lactate dehydrogenase, and in the visual pigments and liver enzymes of a metamorphosing frog tadpole.

It has been known for nearly a century that fetal hemoglobin is different from the hemoglobin of an adult. Adult hemoglobin is easily denatured by acid or alkalis but fetal hemoglobin is not. Modern analyses show that the normal adult hemoglobin molecule is composed of four polypeptide chains, two of alpha hemoglobin and two of beta. The normal fetal hemoglobin molecule is composed of two alpha chains and two gamma chains. During human development the total production of hemoglobin continues at a steady rate, but the synthesis of the alpha chains falls while that of the beta chains increases.

What is the adaptive meaning of this change? Fetal hemoglobin has a greater affinity for oxygen than does adult hemoglobin. Consequently, it is able to acquire oxygen from maternal hemoglobin on the other side of the placental barrier. This is true of the monkey, sheep, cow, and mouse, as well as man.

Eventually the structural gene for alpha hemoglobin becomes completely repressed. The interesting question is: how does the gene for the alpha polypeptide get repressed while the gene for the beta chain becomes derepressed? There are at least three possibilities. The lower oxygen concentration in the fetus may be responsible. Severe anemias due to various diseases and to bleeding, all of which would lead to oxygen deprivation, also lead to the appearance of fetal, i.e., alpha, hemoglobin even in fully mature adults.

It may be that hormones play an important role. It is known that some hormones act at the gene level. Furthermore, women with certain types of ovarian tumors produce small amounts of fetal hemoglobin. Another suggestion is that the precursor cells for erythrocytes begin with some cytoplasmic repressor for the beta gene, which becomes diluted or disappears as the precursor cells proliferate. At that point the structural gene for beta hemoglobin begins producing its mRNA. Such mRNA not only furnishes the instructions for the synthesis of beta chains but also may act as a repressor for the alpha gene.

Lactate dehydrogenase (LDH) is an enzyme which converts lactic acid to pyruvic acid (CH_3 CHOHCOOH to $CH_3COCOOH$). It is also a tetramer of four polypeptides. Two genes are involved in its synthesis. One regulates the production of an A and the other a B polypeptide. Since they unite by chance, five possible kinds of LDH are possible, one composed of four A's, one of four B's, and three of combinations of A and B chains, thus: A^4B^0, A^3B^1, A^2B^2, A^1B^3, and A^0B^4. Such different enzymes all having the same function are known as **isozymes**. This set of five is favorable for study because they can be easily separated by electrophoresis. LDH1, which is A^0B^4, is the most negative and therefore migrates the fastest and forms the band farthest from the starting point. LDH5 is the least negative and forms the band closest to the origin. LDH5 is found in all embryonic tissues while LDH1 is characteristic of adult tissues except for muscles, which keep LDH5.

Clearly gene activity changes here as it does in the case of hemoglobin, and it is possible that oxygen concentration may be a controlling factor. In any case, LDH5 is more efficient at low oxygen tensions than LDH1, and fetal tissues and muscles are characterized by lower concentrations of oxygen than most adult tissues.

Changes in gene activity are also known to occur during the life history of the frog. The visual pigment in a tadpole is porphyropsin (as in fish). In the adult it becomes rhodopsin as in terrestrial vertebrates. In the tadpole, waste nitrogen simply diffuses out through the gills as ammonia, while in the frog it is converted into urea by enzymes in the liver. Clearly genes are active in the adult frog which were repressed in the tadpole. Again it is easy to believe that

hormones are responsible. The "turning on" of specific genes by the activity of known hormones can be seen in certain insects; we will now turn to this topic.

Visible Gene Activation

It has been known for some time that in the nuclei of salivary gland cells and of cells in various other tissues of the larvae of flies and gnats are **giant-sized** and **banded chromosomes**. Moreover, the bands correspond to specific gene loci. Several years ago W. Beerman discovered that various bands show puffs or swellings, and that specific bands become puffed in a defnitite sequence during the development of the larva. In a cross between two species of gnats (*Chironomus*), one species forms special granules in certain cells of the salivary gland near its duct; the other does not. Conventional genetic experiments showed that the ability to secrete these granules is inherited as a simple Mendelian dominant and that the gene responsible is located adjacent to the kinetochore of one of the chromosomes. Cytological studies showed that in the cells which had these granules—and only in those cells—there is a conspicuous puff adjacent to

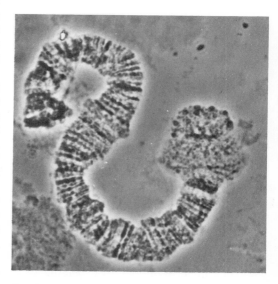

Fig. 8-17. Chromosome in action. Bands and a puff in a salivary chromosome of a gnat, *Rhynchosciara*. "Squash" preparation, × 765. (Courtesy of Mary T. Handel and the Oak Ridge National Laboratory.)

the kinetochore. Most significantly, Beerman then showed that in heterozygous individuals only the chromosome carrying the dominant allele shows a puff in that locus. What more convincing evidence could there be that puffing represents gene activity?

More convincing evidence has been presented by Ulrich Clever. Also using *Chironomus,* Clever showed that injections of the insect steroid hormone ecdysone into the larva will result in pupation much sooner than would otherwise occur. Normal pupation in this gnat is preceded by some very characteristic puffs in specific chromosomes. The same puffs appear within two hours after ecdysone injection, and a long time before any other signs of impending metamorphosis can be detected. Histochemical stains and radioactive labeling reveal that the puffs are sites of messenger RNA synthesis. Moreover, actinomycin D, which is known to inhibit the synthesis of mRNA, inhibits the formation of these puffs. Hence the puffs apparently correspond to the loops of the lampbrush chromosomes found in the germinal vesicles of oocytes. Precisely what this newly formed material does has not been established, but it is clear that steroid hormones can act directly at the gene level.

In higher organisms it may be that whole "batteries of genes" are derepressed together and that gene derepression, which is another name for differentiation, may be a much more complex process in higher organisms than in bacteria on which the current Jacob-Monod model of gene activation is based. Alkaline proteins known as histones, which adhere to chromosomes, may play a key role in the repression and derepresion of genes.

SEX DIFFERENTIATION

Both male and female vertebrate embryos begin development with the same set of reproductive primordia. In mammals a pair of **gonads** is located on either side of the midline near the kidneys, and two pairs of ducts, the **mesonephric** or **Wolffian ducts** and the **Mullerian ducts**, extend from near the kidneys posteriorly to connect with the exterior via the common urogenital opening. In addition there are the external genitalia. These consist of a median **genital tubercle** immediately anterior to the urogenital opening, on either side of which is a **genital fold**. If the zygote from which an embryo develops contains one *X* and one *Y* chromosomes, *i.e.,* if it is a male, the gonads will develop into testes and descend from the abdomen into a scrotum, the genital tubercle will grow into a penis, each Wolffian duct will develop into a vas deferens which will conduct sperms to the penis, the Mullerian ducts will degenerate and virtually disappear, and the genital folds will form the scrotum.

If the egg carries two *X* chromosomes and develops into a female, the gonads become ovaries and remain abdominal although they move much farther down toward the pelvis than they were originally placed, the Wolffian ducts degenerate and virtually disappear, the Mullerian ducts develop into the oviducts (Fallopian tubes of human anatomy), the uterus, and vagina. The genital tubercle becomes the clitoris and the genital folds become the labia majora.

The differentiation of sex is a very complex affair with considerable medical importance and much variation in mechanisms from one kind of animal to another. Sex determination as distinguished from the process of sex differentiation during development has been discussed in Chapter 7.

In some vertebrates and invertebrates, notably insects and crustaceans, many sexual differences are controlled by local gene action within the tissues rather than by a circulating hormone. The classical and best studied case is **gynandromorphism** in *Drosophila,* where one-half of the body is male and the other half is female. Cytological study of the chromosomes of such animals shows that the female side has the normal *XX* composition but the male side has lost one of the *X*'s and is *XO*, which is the same thing as an *XY* in this fly. In the gynandromorph *Drosophila* shown in Figure 8-18, the remaining *X* on the male side carried recessive genes for white eyes and for miniature wings. The normal female side must have had an *X* carrying the normal dominant alleles for these characteristics.

In placental mammals present evidence indicates that **Wiesner's one-hormone theory**, first proposed in 1934, is correct. According to this theory female structures develop autono-

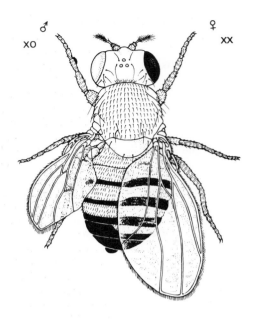

♂
XO

♀
XX

Fig. 8-18. Gynandromorph in *Drosophila*. The *X* chromosome on the male carried the recessive genes for white eyes and miniature wings while the other *X* carried their normal dominant alleles. Note the male sex comb on the left front leg and the male type abdomen on that side only.

mously, and male structures develop under the influence of male sex hormone. If male genital primordia from a mouse or rat fetus are cultured *in vitro* with embryonic testes or with testosterone (male sex hormone), the Wolffian ducts continue to grow while the Mullerian ducts degenerate. If the testes or testosterone is omitted, the Wolffian ducts degenerate. If female genital primordia are explanted *in vitro*, the female ducts continue to develop with or without ovaries. In other experiments normal and castrated female fetuses and castrated males all developed essentially alike, furnishing further support for Wiesner's non-hormonal theory of female sexual differentiation.

Additional support comes from a study involving freemartins. **A freemartin** is a female calf which had a male calf as a twin and which has undeveloped or maldeveloped reproductive organs although the male calf is normal. F. R. Lillie showed long ago that in such cases the placentas of the twins are so closely joined that there is a vascular interconnection permitting blood—and consequently hormones—from the male twin to reach the female.

Finally, the fact that female sex hormone passes through the placenta and reaches the

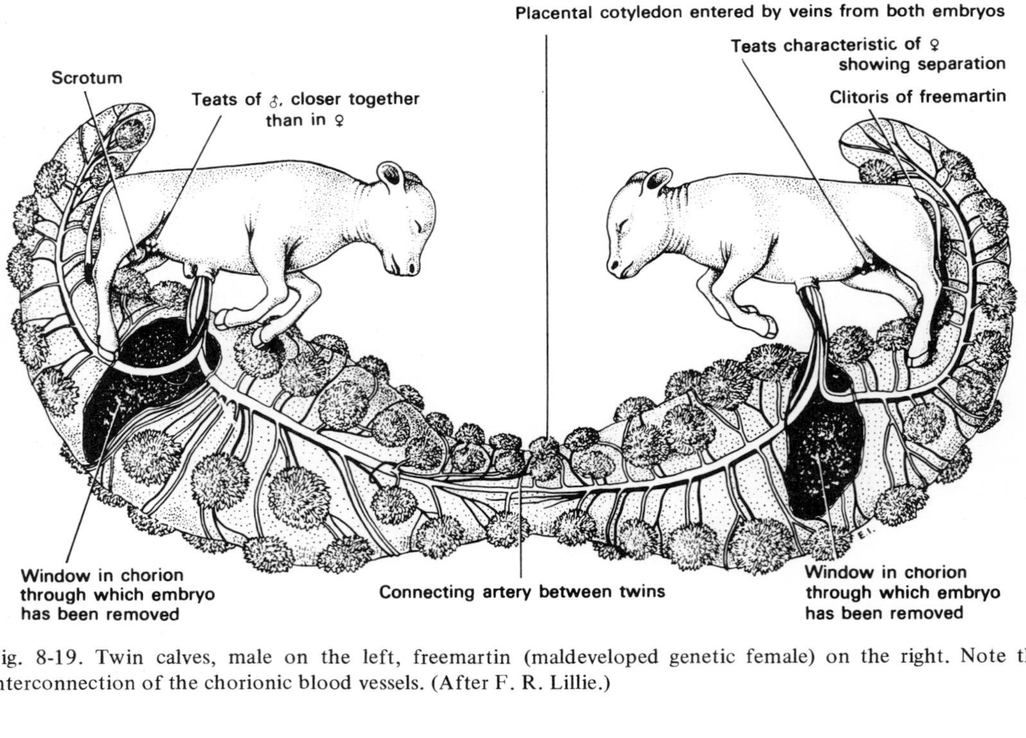

Placental cotyledon entered by veins from both embryos

Teats characteristic of ♀ showing separation

Clitoris of freemartin

Scrotum

Teats of ♂, closer together than in ♀

Window in chorion through which embryo has been removed

Connecting artery between twins

Window in chorion through which embryo has been removed

Fig. 8-19. Twin calves, male on the left, freemartin (maldeveloped genetic female) on the right. Note the interconnection of the chorionic blood vessels. (After F. R. Lillie.)

developing fetus from its mother virtually necessitates a system whereby the male develops under the influence of its own hormones while the female develops without hormones. If female sex hormone were necessary to insure that a genetic female developed in that direction, then, since female sex hormone is always present in placental transmission from the mother, only females would be possible.

The development of the secondary sexual characteristics that appear at sexual maturity, such as differences in hair, voice, etc., are clearly under hormonal control in both sexes of mammals. Whether various other sex differences, such as general bodily conformation, are or not is an open question. The well-confirmed existence of true gynandromorphs in certain strains of mice where the sex ducts and other structures are male on one side of the body and female on the other presents some interesting problems. It is perhaps possible that mammalian gynandromorphs can be explained by the loss of a chromosome as in insects, plus the operation of Wiesner's theory.

One recently discovered and important fact is that the **hypothalamus** (the portion of the brainstem immediately above the pituitary gland) behaves differently in adult males and females and that this difference is imprinted into the hyopthalamus of the newborn mammal by male sex hormones very soon after birth. A single injection of as little as 1.25 mg of testosterone into a female rat two or three days old produces permanent sterility, apparently by producing a continuum of hormone production instead of the cyclical one characteristic of the female. Since it is the hypothalamus which signals the pituitary to release gonad-stimulating hormones at the appropriate intervals, the investigators, Barraclough and Gorski at U.C.L.A., tested the idea that the lack of rhythms of activity in the hypothalamus was responsible for lack of ovulation by stimulating the hypothalamus electrically. Such stimulation produced ovulation.

MEMBRANES

The membranes which surround an egg, it will be recalled, are of two different origins. The vitelline membrane is secreted by the follicle cells and probably also by the cyto-

plasm of the egg. It is formed while the egg is still in the ovary and becomes transformed into the fertilization membrane immediately after fertilization. Various other membranes are secreted around the egg by the oviduct during its passage to the exterior of the body. In a frog these are proteinous coats which swell to form a jelly as soon as the eggs reach the water. In the chick the first membranes secreted by the oviduct are several layers of protein, commonly called the white of the egg. The layer next to the yolk itself is drawn out into two whitish strings called **chalazas** which help keep the yolk properly oriented within the shell. Around the albumenous layers, the lower portion of the oviduct secretes two very thin but somewhat leathery shell membranes, and finally, a calcareous shell.

With reptiles, birds, and mammals, the embryo itself grows four membranes of living tissue which surround and protect the embryo as well as mediate its nutrition, respiration, and excretion. These four embryonic membranes make it possible for reptiles and birds to lay their eggs on dry land, and for mammals to be effectively viviparous, that is, to bear living young.

The membrane closest to the embryo is the **amnion,** which completely encloses the embryo within a fluid-filled space—its own private pond in which it passes through the gill slit stage, although the gill slits are never used for respiration. Occasionally a baby is born with a bit of the amnion on its head. Termed a "caul," it is counted good luck in folklore. It will be noted later that it is the amnion which gives the name **amniotes** to the three groups of vertebrates possessing it.

The second membrane, the **yolk sac,** grows out from the belly side of the embryo and encloses the yolk from which it absorbs food. A yolk sac membrane forms in the higher mammals also, even though there is no yolk to enclose.

The third membrane is called the **allantois** (*allas,* sausage), because in many animals it becomes an elongate cylindrical structure. The allantois grows out from the hind gut, so that the cavity of the hind gut is continuous with the cavity of the allantois. In reptiles and birds it is highly vascular and lies under the shell, serving as a lung. The allantois stores fetal urine

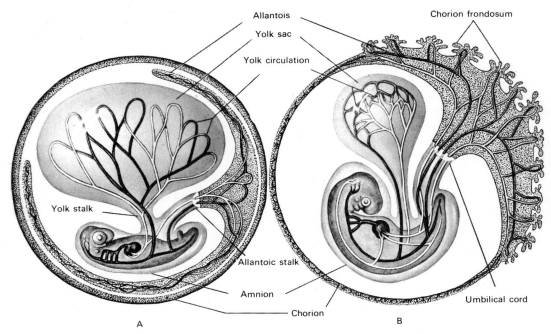

Fig. 8-20. Comparison of fetal membranes of a bird, A, and a mammal, B.

in mammals with poorly developed placentas, and plays a major part in the formation of the placenta in other mammals.

The **chorion** is the most external of the four membranes. In reptiles and birds it presses against the shell membrane. In mammals it develops directly from the blastocyst wall and presses against the lining of the uretus. Where the allantois fuses with the overlying chorion, the blood vessels sprout out into root-like extensions which fit tightly into the lining of the uterus. The region where these chorionic extensions grow out forms the **placenta**, which is the only place where actual exchanges of materials between mother and fetus occur.

In carnivores, like cats and dogs, the chorion is an elongate sac. The chorionic villi develop a placenta that encircles the sausage-shaped chorion like a cigar band. In deer and cows, many small rounded placentas are scattered over the chorion. In primates there is a single disk-shaped placenta; the lining of the uterus proliferates, forming part of the structure.

It is important to note that even in cases, as in man, where the maternal tissue lining the uterus is extensively digested away, forming

crypts into which maternal blood empties, there is no direct exchange of blood between fetus and mother. The fetal blood is enclosed within capillaries which are themselves enclosed within the various tissues of the villi. Only by some breakage, which rarely occurs, is there direct mixing of bloods. Nevertheless, a host of materials are continually exchanged via the placenta between fetal and maternal blood-streams. Indeed, that is the whole point of the placenta. Carbon dioxide, oxygen, digested food, and metabolic wastes all pass through, as do antibodies. The odor of garlic, though not of onions, passes through, so that if a woman chews garlic shortly before her baby is born, it will come into the world with garlic on its breath, *i.e.*, excreting it from its bloodstream via the lungs.

Fetal Membranes and Twins

Monozygotic, or genetically identical, twins bear a special relationship to the fetal membranes. One-egg twins develop from a single ovum in which there are two centers of development in the blastocyst. Then, since the chorion develops directly from the blastocyst, one-egg twins are always enclosed within a single chorion and have more or less fused placentas. Each twin is provided—or rather provides itself—with its own amnion and usually with its own yolk sac.

Fraternal twins, whether boys, girls, or a boy and a girl, result from simultaneous fertilization of two separate eggs. Each is enclosed in its own chorion and provided with a separate placenta. Unfortunately from the diagnostic point of view, occasionally a pair of fraternal twins become implanted so close together on the wall of the uterus that their placentas and even their chorions tend to fuse.

The cause of fraternal twins, fraternal triplets, or any higher number of fraternal sibs is multiple ovulation. To produce a pair of fraternal twins, two eggs must be ovulated at approximately the same time. There is evidence that fraternal twinning tends to be inherited. A woman can inherit this tendency to ovulate two or even more eggs at a time from either her mother or her father. On the other hand, one-egg twins are produced by certain factors within the fertilized egg. Consequently, the male as well as the female parent might contribute the factor which makes for the appearance of two centers of development in a single egg. What factors actually cause this are uknown, but it has been found in many organisms from coelenterates up that if the original growth center is inhibited, then one or more secondary growth centers will arise. If the inhibition is removed or overcome, multiple buds will result.

DEVELOPMENT OF THE NERVOUS SYSTEM

Brain and Neural Circuitry

The development of the vertebrate nervous system can be conveniently divided into three phases. First is the induction of the neural plate by the underlying chordamesoderm. The chordamesoderm is derived directly from the primary evocator or organizer of the vertebrate embryo, the dorsal lip of the blastopore. Immediately following this induction comes formation of the neural groove and neural tube.

The second phase is the conversion of the neural tube into the five basic subdivisions of the brain, plus the spinal cord. The five brain subdivisions are the same in all vertebrates from fish to man and form comparable structures in the adults. As soon as it is formed, the anterior end of the neural tube develops three hollow swellings—**forebrain**, **midbrain**, and **hindbrain**. The cavities become the various ventricles of the adult brain and remain open into each other and into the lumen of the spinal cord. The forebrain develops into the right and left cerebral hemispheres and, immediately posterior to them, the **diencephalon** from which the optic vesicles grow out on either side. Dorsally the diencephalon forms the pineal gland or third eye in many reptiles. The sides become the thalamus and hypothalamus which contain centers controlling hunger, thirst, pleasure, sleep, etc. From the midventral surface the posterior lobe of the pituitary gland grows out.

The midbrain becomes a visual center and is where the optic lobes of the frog brain are located. In mammals this visual center to which nerve fibers go which come directly from the eye is called the optic tectum. The hindbrain forms the cerebellum and the medulla oblongata which narrows into the spinal cord.

Neural Circuitry and the Development of Behavior

During the later stages in the development of the nervous system, a neural circuitry of enormous complexity and incredible precision is spun out between the hundreds of millions of cells in the system. The nerve fibers which compose this system are longer or shorter extensions of single nerve cells. A group of nerve fibers running close together and parallel within the brain or spinal cord constitutes a **fiber tract**. When a group of nerve fibers run together outside the central nervous system, they form a **nerve**. Ever since the descriptive investigations of Ramon y Cajal and His and experiments of Ross Harrison, it has been clear that nerve fibers grow out from their cell bodies and make connections with their end organs— muscle, gland, sense organ, or (within the nervous system) other nerve cells.

The relationship of the development of

behavior to the development of this neural circuitry is a highly important and also highly controversial area where biology and psychology overlap. One persistent question is the relative importance of cytological maturation of the nervous system compared with learning in the appearance of any specific behavior.

G. E. Coghill and L. Carmichael early showed that the various swimming movements in a developing salamander larva appear in a fixed sequence, the same in every individual, and that these motions are correlated with the development of the appropriate connections within the nervous system. The larva first bends only into a C, then later into an S. Swimming becomes more and more perfect as the neural connections become more completely formed. The growth of the nerve fibers concerned was traced in histological sections stained with silver,

which blackens nerve fibers. Coghill also showed that if an embryo salamander is kept under an anesthetic until control animals have developed into swimming larvae, the anesthetized animal swims immediately on being removed from the anesthetic and does so without passing through the usual awkward stages which look like learning. Flying in many species of birds—pigeons, swallows, and cliff-nesting sea birds, among others—is not learned but appears fully developed the first time the bird attempts it, even if the birds are raised with their wings bound to their bodies.

The Central Problem

A central problem remains which presents many unanswered questions. Does the developing nervous system grow as a diffuse, equipo-

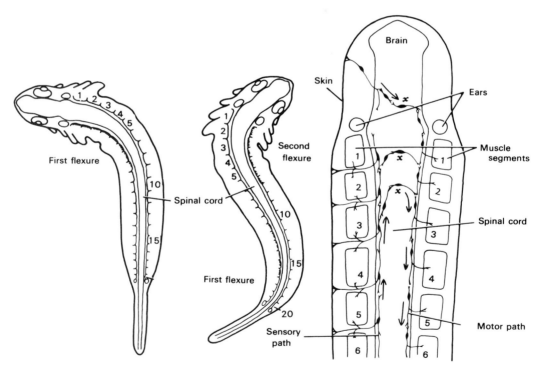

Fig. 8-21. Coordinated development of neural circuitry and behavior in a salamander. Observe that any stimulus applied to the left side of the body, either on the head or in the muscle segments, will pass over to the right side, causing the animal to bend away from the stimulus. Because the crossover nerves, x, are only at the anterior end of the body, the flexure away from the stimulus will begin at the head and pass toward the tail. Since there is a sensory pathway on the right side (not shown), contraction of the muscles on that side will send an impulse forward which will cross over at the head end to the left side and there initiate a second flexure. (After G. E. Coghill.)

tential nervous network on which function, *i.e.,* conditioned reflexes or some other type of learning, imposes reaction patterns? Or does the growth of nerve fibers within the central nervous system establish genetically programmed neural circuits which, in turn, result in preordained behavior patterns such as the precise kind of web a spider will spin or the elaborate coordination of muscular action essential to produce skillful flight? It should be noted that the second alternative does not preclude the possibility that the same nervous system could also grow circuits which provide the physical basis for learning. How do the outgrowing nerve fibers know where to grow and what connections to make? How does the brain know which muscle or gland a certain nerve fiber innervates or from what spot on the retina, for example, a certain nerve fiber comes?

Here are some of the facts. They give only partial answers but they are a beginning. In embryonic and even larval salamanders and frogs, muscles can be transplanted to unnatural sites on the body. A whole leg can be grafted into the socket from which an eye has been removed. In such cases nerves which would normally never see a leg will make functional connections with the strange muscles. Very commonly such transplanted muscles will contract synchronously with the corresponding muscles in the normally placed leg. Facts such as these have led to the conclusion that nerves merely innervate the first muscle they meet and are guided in their outgrowth purely by physical topography. The explanation of the synchronous contraction with the corresponding muscles in the normally placed limb is very obscure. In the higher vertebrates, especially in mammals, scrambling nerves and muscles leads to serious malfunction. The meaning of these facts remains to be clarified.

Highly specific behavior in the growth of nerve fibers from specific parts of the retina to specific parts of the brain (in the optic tectum, of course) has been shown by R. W. Sperry at Cal Tech in fish and by Leon Stone at Yale in salamanders. In the common pond salamander, *Diemictylus* (*Triturus*) *viridescens,* it is possible to transplant the eyes with a return of vision because the low oxygen requirements of a cold-blooded animal permit revascularization before the death of the eye. The new nerve fibers grow into the brain from their cell bodies located in the retina. It is necessary only to make sure that the cut ends of the optic nerve are held together; otherwise the regenerating fibers grow out of the cut end and merely form an irregular mass.

If a salamander is anesthetized and one eye cut out, rotated 180°, and replaced, when vision returns, that eye will see everything upside down and backwards. If both eyes are thus rotated the animal will live indefinitely but must be hand-fed because it never learns not to reach up for food placed below it or reach down in trying to grab food which is, in fact, dangled above it. If the left eye is grafted into the place of the right eye but not rotated, the salamander will see backwards but not upside down with that eye. If that left eye is rotated 180°, then the salamander will see upside down but not backwards with that eye. In other words, it will jump up to get food presented under its nose but lurch forward to get a piece of meat held a short distance in front of it. These facts indicate that nerve fibers coming from each particular spot on the retina either carry some kind of label (which the brain can recognize) as to their place of origin in the retina, or else that in regeneration each nerve fiber somehow grows back to its proper location in the brain.

In fish, which are favorable material for such investigations, Sperry showed that nerve fibers from the eye actually do grow back very precisely to their original terminal sites in the brain. How do they do this? Perhaps they are guided by some kind of chemical gradients.

AGING

The fact of aging is obvious to all, yet nothing certain is known about its causes. It may very well be that it is controlled in different kinds of animals by entirely different mechanisms. What determines the time when teeth will erupt, or when the thymus gland undergoes atrophy, or sexual maturity occurs, or menopause takes place? Why should a rat have a potential life span of three years, being the equivalent of a man in his 90's at 36 months, while a flying squirrel having basically the same size, anatomy, and physiology easily lives six or seven years? The life span of *Drosophila* can be lengthened by a factor of three by low temperature, or cut in half by raising the temperature. However, despite some intensive studies, no one knows how temperature affects the mechanisms that control aging.

There are many theories about senescence, most of them easily disposed of. Metschnikoff held that aging was due to poisons absorbed from intestinal and other bacteria. But animals raised in a sterile environment grow and age like others—or even die sooner. Steinach in Germany proposed that aging is caused by changes in the reproductive glands. But castrated animals grow and age at almost the same rate as do normal ones. Bogomolets in the U.S.S.R argued that changes in the connective tissue system are mainly responsible, but his theory is so vague it can scarcely be proved or disproved. Others have suggested dietary factors and changes in the calcium or water content of the cells.

In many vertebrate tissues—heart, liver, and others—a curious yellowish-brown pigment, lipofuscin, a complex of lipoids and proteins, accumulates with age. Various enzymes change during a life span. But whether this is merely a concomitant of aging, or is a cause of aging, or what its significance is, remains unknown.

A now popular theory, which has been under discussion for over a decade, holds that senescence is due to the accumulation of random errors in the replication of DNA during cell division until so much "noise" is built up in the genetic information system that an adequate number of functional cells cannot be maintained. X-irradiation accelerates aging, a fact which supports this theory. However, the dosages which produce mutations and those which produce aging often differ, the chemical mutagens tested do not produce senescence, and in ciliates lethal mutations are extremely rare in young clones but become very frequent in old ones or in young nuclei introduced into old cytoplasm. Semi-starvation greatly prolongs the life of mice and rats but it is not known to retard mutations.

Perhaps the clue to the problems of aging will be found by studying the programmed death of blocks of cells which occurs during development. Even when transplanted to a

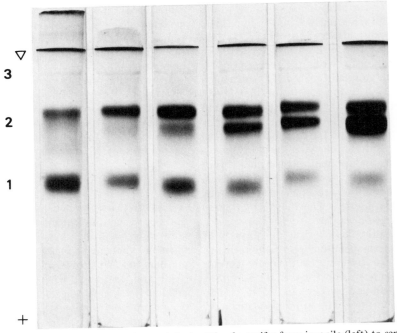

Fig. 8-22. Changes in enzymes (acid phosphatases) during life of a rotifer from juvenile (left) to senescent (right) revealed by electrophoresis through acrylamide gel followed by staining. (Kindness of N. Meadow.)

presumably favorable environment, these cells die, although surrounding cells remain healthy. J. Saunders in the United States and L. Amprino in Italy have been investigating this strange phenomenon in chicks. Recently, others have found the same thing in mammals but no one can even guess what the "death clock" is. It will be recalled from Chapter 5 that ciliates and very probably metazoan cells have a limited clonal life span. If this is a fact then it will influence the maximum attainable age.

No one can predict how close we are to an understanding of the mechanisms which control aging. There may be a very simple answer; but then, there might not be.

USEFUL REFERENCES

Arey, L. *Developmental Anatomy of Man,* 7th ed. W. B. Saunders Co., Philadelphia, 1965.

De Haan, R. L., and Ursprung, H. (eds.) *Organogenesis.* Holt, Rinehart and Winston, Inc., New York, 1965.

Ebert, J. D. *Interacting Systems in Development.* Holt, Rinehart and Winston, Inc., New York, 1965.

Hamilton, H. L. Lillie's *Development of the Chick.* Henry Holt & Co., Inc., New York, 1952.

Markert, C. L. and Ursprung, H. *Developmental Genetics,* Prentice-Hall, Inc., Englewood Cliffs, N. J., 1971.

Patten, B. M. *Early Embryology of the Chick*, 4th ed. Blakiston, Philadelphia, 1951.

Rugh, R. *The Frog: Its Reproduction and Development.* P. Blakiston's Sons & Co., Philadelphia, 1951.

Strehler, B. L. *Time, Cells, and Aging.* Academic Press, New York, 1962.

Plant Development

9

The question, "How does a plant grow?" can be asked, and answered, in many ways. When the question is asked in the sense of "How well?" it assumes vital importance to mankind, because all animals, including ourselves, are completely dependent on plants to provide the base for food chains, to supply the essential raw materials for products as diverse as lumber and drugs, and to maintain adequate levels of oxygen in the air we breathe.

When we ask "What happens?" or "What is formed?" a number of chapters, in this text are needed to describe, in even the briefest way, the profusion of developmental patterns that have evolved since the origin of plants on this planet. Here we will limit the consideration of these questions to the key events common to all plants and then concentrate on the dominant plants in our terrestrial environment, the seed plants.

Until relatively recently, studies of development were limited to the questions "What happens?" and "What is formed?" It was not until the last decade, when an understanding was achieved of how DNA in the nucleus of a cell controls the synthesis of specific proteins in the cytoplasm, that questions about development could be posed in a more sophisticated way: "How is it that one kind of DNA can produce an alga while another can give rise to an oak tree?"

Development of a plant (or an animal) is an orderly sequence of events, predictable for any species. It can be thought of as the product of reading the genetic code. The basic information needed to make a new plant is present in the nucleus of each cell. In some ways, the genetic code and the plant which is formed from its set of instructions are analogous to the written musical score of a symphony and the performance of it. In both cases a number of variables can influence the quality of the product. The performance of a symphony orchestra is influenced by factors such as the expertise of the players, the quality of the instruments, the accoustical properties of the hall, and the correct proportion and variety of instruments and musicians. In other words, the product is variable and dependent on factors other than the musical score. Plants can develop in ways that fall far short of their genetic potential, just as the performance of a musical score may leave much to be desired. A number of external factors must be present and available in the correct proportions for the vigorous growth and development of plants. These factors include appropriate temperature, moisture, minerals from the soil, oxygen and carbon dioxide from the atmosphere, and light. The influence of these external factors will be discussed in detail in the chapters on plant metabolism (Chapters 15-18).

LIFE CYCLES: READOUTS OF THE GENETIC CODE

The life cycles of plants present what appears to be a bewildering diversity. What should be recognized is that, in spite of the apparent complexity of the living things around us, there is a common underlying theme. The life cycle of any plant from the simplest alga or fungus to a large and complex conifer or flowering plant can be summarized by means of a diagram (Fig.

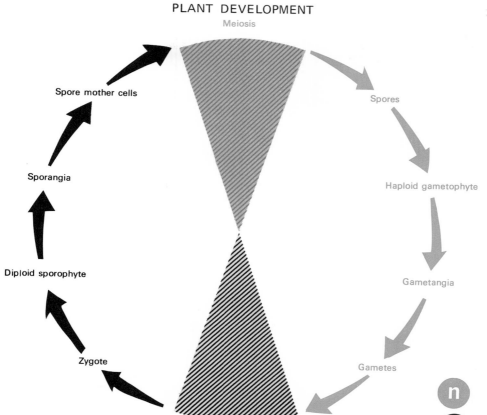

Fig. 9-1. Generalized plant life cycle. The sequence of events in the life of any plant can be represented by appropriate variation of this basic scheme.

9-1). Once a few key events are recognized, it becomes relatively simple to understand all the diverse life cycles of plants.

The two most significant events that occur are **meiosis**, a specialized pair of cell divisions in which haploid cells are formed, and **fertilization,** the fusion of nuclei which restores diploidy. A widespread phenomenon in plants is the separation of these key steps in the life cycle. In many plants they occur in separate and independent plant bodies.

The plant body composed of haploid cells is called the **gametophyte**. As it reaches maturity, the gametophyte plant produces gametes in specialized structures called **gametangia**. In most plants, two kinds of gametes are formed which differ in size and structure. The smaller,

frequently motile, gametes are called sperms and are produced in male gametangia or **antheridia** (singular, antheridium). Larger, usually non-motile, gametes called eggs are produced in the female gametangia or **archegonia** (singular, archegonium). Archegonia and antheridia can be formed on the same gametophyte plant or on separate plants. Fusion of haploid sperm and egg nuclei in the process of fertilization results, as in animals, in a single diploid cell called the **zygote**. This cell is the beginning of the diploid phase of the life cycle. From it, a diploid organism called the **sporophyte** develops. When the sporophyte matures, structures called **sporangia** are formed. In the sporangia, special cells called **spore mother cells** are produced. These cells undergo meiosis and the resulting haploid cells, **spores**, are released. Spores, on development, form new haploid gametophyte plants.

Among several aspects of the life cycles of plants that are different from the circumstances found in animals is the already noted phenomenon of separate haploid and diploid phases with morphologically different and independent gametophyte and sporophyte generations. This life style, which is apparent in the mosses and ferns, has been given the name **alternation of generations**, an unfortunate choice of terminology because the same term is used by zoologists when referring to the morphologically distinct, but both diploid, phases in certain coelenterate life cycles.

The alternation of haploid and diploid phases in the life cycles of plants insures a separation in time of the key steps of meiosis and fertilization. There is a general evolutionary tendency in plants for the gametophyte phase

of the life cycle to become reduced in size and duration. In the most highly evolved plant group, the familiar seed plants, the haploid portion of the life cycle is reduced to a few specialized cells.

Life Patterns in the Seed Plants

The seed plants dominate our environment with a vast host of species—the grasses and maple trees, sunflowers, pines, and orchids. They form the basis of our agriculture and forestry. The plants that surround us are in the diploid (sporophyte) phase of their life cycles. The haploid or gametophyte stages have been reduced to such an extent that haploid cells occur only as the immediate products of meiosis in parts of specialized reproductive structures, **flowers** or **cones**. The key events in the life cycle of a flower-forming seed plant (an angiosperm) are summarized in Figure 9-2. New

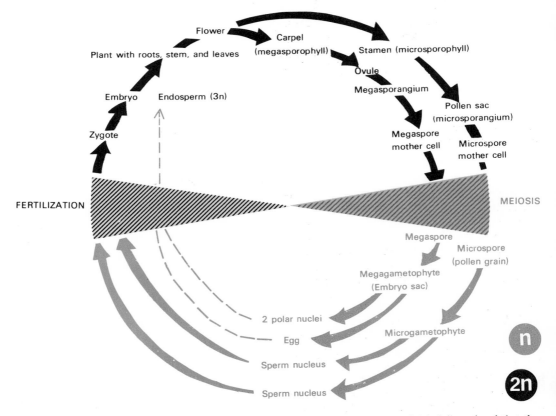

Fig. 9-2. Life cycle of a flowering plant. Note that the basic scheme seen in Figure 9-1 is followed and that there is an alternation of haploid and diploid stages.

diploid plants develop from **seeds**. In each seed there is an **embryo** plus food reserves, frequently in a specialized tissue called the **endosperm**. The entire structure is covered with a protective **seed coat** formed from tissue of the sporophyte flower. After seeds are formed, there is often a period of arrested development **(dormancy)** before further growth of the new sporophyte plant takes place. Once dormancy

has ended and appropriate conditions of temperature, moisture, oxygen, and light prevail, **germination** begins. Factors such as gravity and light assure the proper orientation of the young diploid plant during its early development from embryo to seedling. Vegetative growth continues until internal and external factors are appropriate for the onset of reproductive development, or flowering. For detailed de-

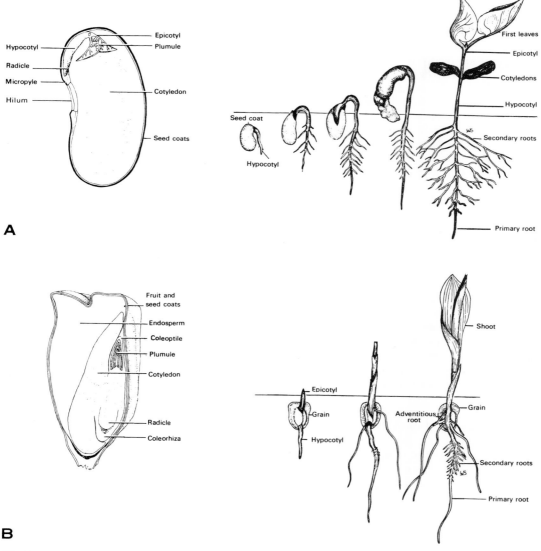

Fig. 9-3. Seed structure and germination. A, bean; B, corn. (From V. A. Greulach and J. E. Adams: *Plants, An Introduction to Modern Botany.* John Wiley & Sons, Inc., New York, 1967; and H. J. Fuller and O. Tippo: *College Botany.* Henry Holt & Co., Inc., New York, 1954.)

scriptions of life cycles in the seed plants, see Chapters 13 and 14.

THE STRUCTURE OF SEED PLANTS— THE PRODUCT OF DEVELOPMENT

A basic difference between plants and animals is that animals tend to grow "all over" while in plants new parts are added only in restricted regions called **meristems**. However, before the organization and function of meristems are discussed, we will have to examine the basic structural patterns of the seed plants.

It is to the distinct advantage of the botanist that plants are relatively simple in organization with far fewer organs to be named and recognized than in animals. There are only six major parts of a flowering plant: roots, stems, and leaves (vegetative organs); and flowers, seeds, and fruits (reproductive organs). The simplest way to describe a seed plant is to say that it is made up of a vertical axis plus appendages. That portion of a plant normally above the ground is called the **shoot**, and that below the soil level called the **root**. The vertical axis of the shoot is the **stem** and the appendages found on the stem are the **leaves**. The usually broad, flat portion of the leaf, the **blade**, is attached to the stem by a stalk termed the **petiole**. A region of the stem where leaves are attached is called a **node**. For every species there is a definite pattern of leaf attachment or **phyllotaxy**. Portions of a stem between nodes are called **internodes**. The angle between the petiole of a leaf and the stem is called an **axil**. This is where **buds** are formed. A bud can develop into a branch of the stem with its own attached leaves, or can develop into a flower, or both.

Stems have three functions: support, conduction, and storage. The rigidity and strength of stems is the result of thickened and reinforced walls of the cells found in stem tissues.

Cells of the **vascular** (or conducting) tissues are arranged in very definite quantities and ways within stems. Herbaceous plants, which frequently complete an entire life cycle in one growing season, contain relatively less vascular tissue in their stems than do trees living for many years. There are two kinds of conducting tissues: the **xylem**, making up the woody portion of a tree trunk; and **phloem**, found in

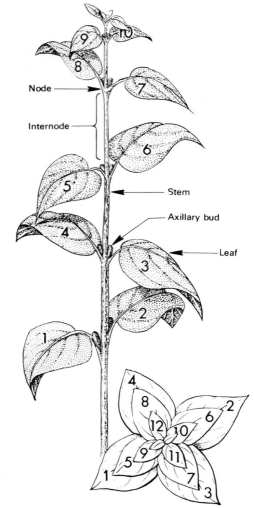

Fig. 9-4. Structural organization of a shoot showing stem, leaves, nodes and internodes, and axillary buds. Top view shows phyllotaxy with leaves numbered to indicate order of formation.

the bark region of trees. In herbaceous plants, xylem and phloem are located together in vascular bundles embedded in the soft, relatively undifferentiated inner tissues of the stem. Two patterns of vascular bundles characteristic of many herbaceous species of flowering plants are illustrated in Figure 9-5. In the **monocots**, such as corn and lilies, vascular bundles are scattered at random through the central **ground tissue**. In the **dicots**, such as sunflowers and beans, strands of conducting tissue are usually found in a ring around a central **pith**. Within each strand, xylem cells are located on the inside and phloem cells toward the outside of the stem. Beyond the ring of vascular bundles is a second layer of relatively thin-walled, undifferentiated cells called the **cortex**, and at the outer surface of the stem we find an **epidermal layer**. Frequently the epidermal cells of both stems and leaves secrete a waxy coating (cutin) to form the cuticle, which helps to retard water loss.

The arrangement of tissues in woody stems is somewhat more complicated than in herbaceous stems. Here the xylem and phloem are generally laid down as concentric layers that constitute hollow cylinders running lengthwise within the stem (Fig. 9-6).

Primary growth results in the formation of new stems or lengths added to old stems or roots. The tissues laid down during the addition of such new plant parts are referred to as **primary tissues**. Primary growth is responsible for establishing branching patterns of the shoot and root and is the product of terminal growing regions called **apical meristems**. **Secondary growth**, on the other hand, brings about a thickening of already formed stems or roots. The regions of actively dividing cells responsible for secondary growth are layers of cells called the **vascular cambium** and the **cork cambium**.

The vascular cambium is a single layer of cells located at the interface between the xylem and the phloem. The products of mitosis in this type of cambial cell are a new cambial cell plus either a secondary xylem or a secondary phloem cell. The accumulation of a new ring of secondary xylem cells inside the vascular cambium pushes the cambium and all tissues beyond it toward the outside. Thus, a stem or

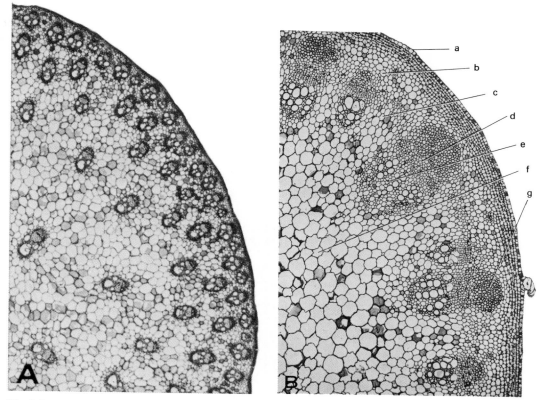

Fig. 9-5. Arrangement of vascular bundles in herbaceous stems. Left, monocot pattern: cross section of a corn stem. (Courtesy Carolina Biological Supply Co.) Right, dicot pattern: cross section of a sunflower stem. a, epidermis; b, cortex; c, pith ray; d, pith; e, primary phloem; f, primary xylem; g, cortical fibers. © General Biological Supply House, Inc., Chicago.)

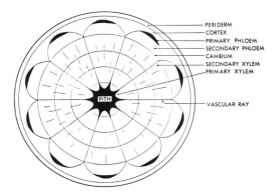

Fig. 9-6. Diagrammatic cross section of a young woody stem at the end of three years' growth. Note three annual rings. (From H. J. Fuller and O. Tippo: *College Botany*. Henry Holt & Co., Inc., New York, 1954.)

root increases in diameter by adding layers of secondary xylem and phloem. New secondary tissues are formed each growing season. Particularly in the secondary xylem, which makes up the bulk of the wood in tree trunks, the seasonal increments are readily seen in cross section. Relatively larger cells are formed early in the growing season while cell size later in the season is smaller because environmental factors are less favorable for growth. These seasonally formed layers, or **annual rings**, are visible in most wood products finished so that the grain is visible.

The outermost layers of the stem, the phloem, cortex, and epidermis, are obviously stretched during secondary growth as new layers of cells are laid down on the inside. Eventually the epidermis and cortex can actually crack open. This stimulates some of the exposed cells of the cortex to differentiate into cork cambium. Cork formed by division of cells in the cork cambium provides a waterproof protective layer for the outside of the stem.

Phloem tissues are located outside of the vascular cambium and, like all other tissues in the bark region, are subject to great stress and the pressure of outward movement. Only the innermost portion of the secondary phloem forms a continuous ring. Older secondary and primary phloem form discontinuous patches in

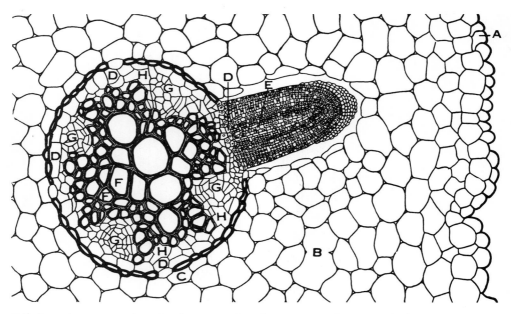

Fig. 9-7. Root in cross section showing young lateral root penetrating outward through the cortex and epidermis. A, epidermis; B, cortex; C, endodermis; D, pericycle; E, branch root originating from pericycle; F, xylem; G, phloem; H, parenchyma of stele. (From H. J. Fuller and O. Tippo: *College Botany*. Henry Holt & Co., Inc., New York, 1954.)

Fig. 9-8. Root tip of radish seedling showing root hairs and root cap (× 15). (From T. P. O'Brien and M. E. McCully: *Plant Structure and Development.* The Macmillan Co., New York, 1969.)

the bark region.

There are many structural similarities between roots and stems. Both can exhibit secondary as well as primary growth. Products of the vascular and cork cambium in roots are comparable to those described in stems. Several fundamental differences exist, however, including an absence of pith (instead there is a core of primary xylem at the center of roots) and the absence of leaves (and nodes). The mode of origin of secondary roots also differs from that of stems. In contrast to the new branches of the shoot which arise from buds formed in the axils of leaves, lateral roots arise internally from the **pericycle**, a layer just inside the cortex. Developing lateral roots actually penetrate outward through the cortex and epidermis (Fig. 9-7).

Roots not only anchor plants in the ground, but also serve as the port of entry for water and dissolved minerals from the soil. These necessary raw materials enter the root system through the thin-walled extensions of epidermal cells of young roots called **root hairs** (Fig. 9-8). Water and dissolved minerals diffuse from cell to cell from the epidermis through the cortex and **endodermis** (a layer of cells just inside the cortex) to the cells of the vascular tissue which serves to carry water and dissolved materials to all parts of the plant.

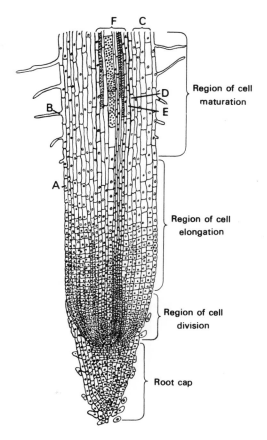

Fig. 9-9. Longitudinal section of a young root of barley. A, epidermis; B, root hair; C, cortex; D, endodermis; E, pericycle; F, differentiating conducting tissues of stele. (From H. J. Fuller and O. Tippo: *College Botany.* Henry Holt & Co., Inc., New York, 1954.)

APICAL MERISTEMS: ORGANIZATION AND FUNCTION OF THE REGIONS OF GROWTH RESPONSIBLE FOR ELONGATION

As we have already mentioned, primary growth (or increasing length) of stems and roots results from the activity of the apical meristems. The apical meristem of the root is somewhat simpler in organization than its counterpart in the shoot. It is covered by a protective mass of cells called the **root cap**. New cells are formed in a region of cell division located just behind the tip of the root. Once new cells are formed, they undergo elongation and finally differentiation. The elongating,

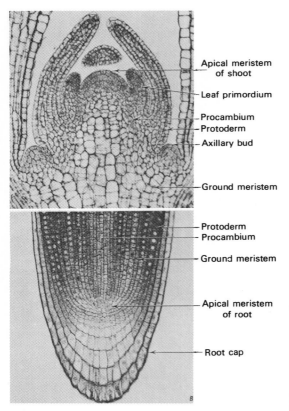

Apical meristem
of shoot

Leaf primordium

Procambium
Protoderm
Axillary bud

Ground meristem

Protoderm
Procambium

Ground meristem

Apical meristem
of root

Root cap

Fig. 9-10. Apical meristems of shoot and root, longitudinal sections. A, flax seedling shoot tip, ×120 (From J. E. Sass: *Botanical Microtechnique*, 3rd ed. The Iowa State College Press, Ames, 1958.) B, root tip, ×130 (From K. Esau: *Anatomy of Seed Plants.* John Wiley & Sons, Inc., New York, 1960.)

newly formed cells provide the impetus to push the tip of the root through the soil. The growth of a root is analogous to the unrolling of a ball of yarn. The ball of yarn moves forward as it unwinds, leaving the unwound yarn stationary behind it. The older portions of a root system are in fixed position in the soil with only the tip portion pushing forward as a result of new growth.

The products of shoot growth are new stem and new leaves. The apical meristem of the shoot is surrounded by young developing leaves, the youngest located closest to the apex with the larger expanding primordia attached beneath and surrounding the younger, more apical leaves.

In most seed plants the shoot apex consists of two regions, an outer mantle of one to several layers of cells (called the **tunica**) plus an inner core of central tissue made up of larger dividing cells (the **corpus**). While the terms tunica and corpus describe a structural pattern common to apical meristems of many flowering plants, they are of little help in explaining the workings of these structures or the manner in which new plant parts are formed. There is some evidence that the summit area of the shoot apex is meristematically inactive. It is the subterminal and peripheral zone that produces the leaf primordia and gives rise to cortex and vascular tissue of the stem. A third region, located below the apical zone and inside the peripheral region where leaf primordia are formed, is thought to give rise to cells which mature into pith.

Although the structural patterns of apical meristems have been identified and named, there are still many unanswered questions about their functioning. What factors control the orientation and frequency of cell divisions? What controls the pattern of initiation of leaf primordia? Speculative answers to some of these questions are available. To a great extent the location of cells determines their potential. For instance, internal cells are under very different physical restrictions and pressures from those in the epidermal, or surface, layers. There is some evidence that the distribution of metabolites and regulators of growth becomes non-uniform in the appex, and that their pattern of accumulation may control the inception of leaf primordia. Existing leaf primordia may also influence the positioning of

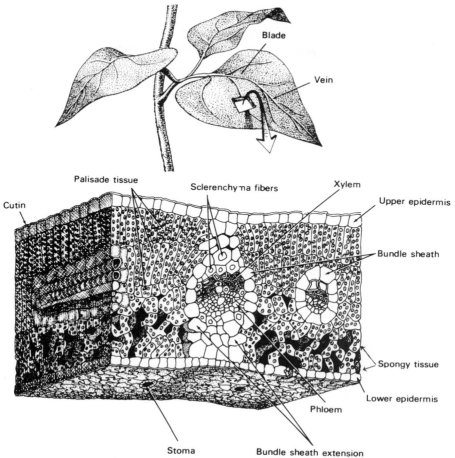

Fig. 9-11. Internal structure of a leaf.

additional primordia. Certainly the genetic code is a major factor in determining the patterns of leaf attachment.

GROWTH VS. DIFFERENTIATION

We are all familiar with the changes in morphology (form) that occur during the development of most organisms. Just as tadpoles are very different from frogs, the early developmental stages of most plants differ in structure and function from the adult. While growth increases the size of an individual, the process of differentiation results in the internal and external modifications of the organism which make a maple tree different from a fern. We can consider differentiation at the levels of cells, tissues, and organs. At every level, form is closely related to function. For example,

chloroplasts, the green cellular organelles that are the site of photosynthesis (a process requiring light), are found in the parts of plants located above ground, primarily in the cells of the leaves but also in the outer layers of young stems (the regions most easily exposed to light). The form of leaves is well adapted to their primary function as the food-synthesizing organs of the plant. The broad, flat surface is appropriate for absorption of radiant energy from the sun, and the relative thinness allows quanta of incident light to penetrate to the inner layers of cells where the chloroplasts are found. As a further aid to function, there are openings called **stomates** in the leaf's surface layers through which the gases involved in photosynthesis and respiration can be exchanged with the atmosphere. Leaf veins provide the means for importing to the leaves the water and minerals absorbed by the roots of the plant and for exporting the products of photosynthesis. The veins of the leaves are made up of cells and tissues specialized for transport of water and dissolved substances, and they are connected to the conducting tissues of the stem and root.

On the cellular level, differentiation can result in formation of specialized subcellular organelles, such as the chloroplasts, or in structural modifications such as the thickenings found in cell walls of supporting and vascular tissues. In tissues, differentiation can result in the association of various kinds of specialized cells such as the xylem and phloem, conducting cells found in the veins of a leaf and in the vascular strands of the stem.

On the level of organs, the unequal rates of cell division in different directions result in the characteristic form of the organ. The characteristic shapes of a maple leaf, of flower petals, or of fruits such as an apple or a squash are results of differences in rates of cell division (along their various axes) during development.

An important fact to keep in mind about growth and differentiation is that they occur simultaneously in practically all organisms. Thus it is difficult to study just one of these two aspects of development. There is, however, a group of organisms in which the processes of growth and differentiation are separated in time. These organisms, the slime molds, are fungi (see Chapter 10) that have served as the experimental material for many classical studies of development. With these organisms, cell multiplication can be studied without the complication of cell differentiation and *vice versa.*

The Slime Molds—Plants That First Grow and Then Differentiate

Dictyostelium discoideum, a species of slime mold studied extensively by Kenneth Raper and John Tyler Bonner and their respective collaborators, can be grown in culture and its life cycle is completed in a few days. During vegetative growth there is a multiplication of amoeboid cells which feed on bacteria. On depletion of the food supply, **aggregation** begins. At the onset of this stage, cell growth and division come to an end. A streaming of the

cells toward centers of aggregation is accompanied by a change in cell shape from the typical amoeboid form characteristic of the vegetative stage to a longer and thinner form. The direction of cell movement is believed to be in response to a specific chemical, called **acrasin**, secreted by cells at the centers of aggregation. Sudden pulses of movement have been observed during this stage, and such pulses may be related to fluctuations in acrasin synthesis. As more and more cells stream in, the mass of cells mounds up at the center. This marks the beginning of **migration**, a time of movement. The mound of cells (called a **pseudoplasmodium**) consists of from 1,000 to 200,000 cells enveloped in a noncellular sheath. Initially compact and erect, the cigar-shaped mass later bends over, elongates, and begins to move, leaving behind a streak of slime. The direction of movement tends to be along light and temperature gradients, toward higher light intensities and warmer temperatures.

Raper has demonstrated that if food (*i.e.*, bacteria) is added during the migration stage, the pseudoplasmodium disintegrates and individual cells return to the vegetative stage until

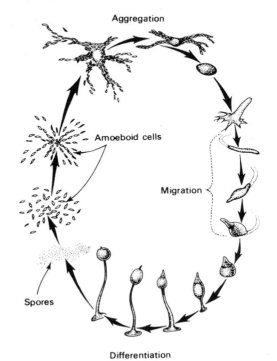

Fig. 9-12. Life cycle of *Dictyostelium discoideum.*

the bacteria have been consumed. Therefore, it appears that conditions favoring growth prevent aggregation.

Differentiation begins at the end of migration when the pseudoplasmodium again becomes vertical in orientation and forms a stalk with a round mass of spores at the top. This entire structure is called the **sporocarp**. Potentially any one of the free living amoeboid cells of the vegetative stage is capable of differentiating into any cell of the mature sporocarp. Whether a given cell will eventually be a stalk cell or a spore is determined by its position in the pseudoplasmodium. The cells of the aggregation center form the anterior part of the pseudoplasmodium and finally stalk cells. Cells reaching the pseudoplasmodium late in aggregation become the posterior cells during migration and ultimately become spore cells. Anterior cells forming the stalk enlarge, become vacuolated, and secrete a cellulose sheath. Posterior cells move up outside the basal cells of the stalk. Those at the top and on the outside of the rising stalk eventually differentiate to form spores.

In the slime molds cell multiplication, migration, and differentiation are separated in time, and they can be studied as isolated phenomena. Slime molds still present a host of unanswered questions, however. What are the factors which cause certain cells to become aggregation centers? What is there about relative cell position that enables certain cells to become stalk cells, while others become spores? Answers to such questions may someday be obtained by students now reading this book.

Another Approach to the Separation of Growth and Differentiation

The slime molds are not the only experimental material available to botanists who wish to study differentiation in the absence of growth. The advent of the nuclear age has made available new tools for the investigation of biological phenomena. At the Oak Ridge National Laboratory, biologists are studying the effects of radiation on living organisms. Haber, Foard, and their co-workers have studied the development of seedlings from seeds receiving sufficient gamma irradiation to completely inhibit DNA synthesis, mitosis, and cell divi-

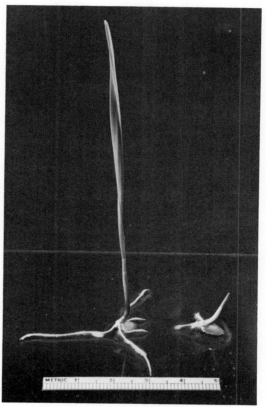

Fig. 9-13. Wheat seedlings 9 days after the beginning of germination. Left, unirradiated control; right, "gamma plantlet" from grain that had been exposed to 800 kr of ^{60}Co gamma radiation. The gamma plantlet has grown by expansion of cells within the embryo without cell division and DNA synthesis. (From A. H. Haber, W. L. Carrier, and D. E. Foard: Metabolic studies of gamma-irradiated wheat growing without cell division. American Journal of Botany, *48:* 431-438, 1961.)

sion. Grains of wheat can still germinate following exposure to 800 kiloroentgens of gamma radiation. The resulting seedlings (called **"gamma plantlets,"** Fig. 9-13) can grow in the sense that cells present in the embryo at the time of irradiation can enlarge and differentiate. Thus, botanists have utilized radiation as a tool and have taken advantage of the fact that cell division is relatively radiation-sensitive, while cell expansion and differentiation are relatively radiation-resistant. The gamma plantlets developing from gamma-irradiated wheat have provided experimental materials for the study of many aspects of the differentiation and metabolism of young seedlings in the absence of cell multiplication.

ENVIRONMENTAL CONTROL OF PLANT GROWTH AND DEVELOPMENT

All aspects of the environment, from climate to man-made smog, affect not only the development but also the very survival of plants. Climate, a collective term for many factors including rainfall and temperature extremes, is influenced by latitude, altitude, and geographical features such as lakes and oceans. Climate has a great influence on the kinds of plants that are native to or can be introduced into any given area. We acknowledge this when we associate palm trees with tropic islands or hemlock and spruce trees with northern forests. Climate is the prime factor determining whether agriculture is feasible. Regions of limited rainfall or seasonal extremes of temperature are usually unproductive and not used for farming or grazing. Unexpected fluctuations in climate (too much or too little rainfall, abnormally cold summer weather, etc.) can affect agricultural yields and ultimately the cost of commodities in the local supermarket.

Temperature

All plants respond to temperature and usually grow best within a limited temperature range. Frequently plants will grow more rapidly under conditions of warm days and cool nights than they will under conditions of constant warm temperature. This response to diurnal fluctuations in temperature is called **thermoperiodism**. An example of such effects of temperature on growth is shown in Figure 9-19. Under conditions of constant temperature, growth rates for the stems of tomato plants were highest at 26.5°C. However, when tomatoes were grown at this optimum temperature during the day but at lower temperatures during the night, stems elongated even more rapidly. Vegetative growth is not the only developmental response exhibiting thermoperiodism. Fruit set in many garden plants from peppers to squash is improved under conditions of lower night temperature, and thermoperiodism can influence such varied responses as leaf shape and flowering.

Seasonal fluctuations in temperature also play an important role in plant life cycles. Buds and seeds of many species remain dormant until subjected to a period of cold temperature. In some kinds of grains normally planted in the fall (*e.g.*, the so-called "winter" varieties of rye and wheat), a cold period during vegetative growth is required for flowering. These varieties are usually planted in the fall and harvested the following year. If such varieties are planted in the spring, they will grow but not flower. After seeds of winter varieties are soaked in water and then exposed to low temperature (0 to 10°C.) for a period of a few hours to several days, they can be planted in the spring and will flower in the same season. This process is called **vernalization** and is an agricultural practice used extensively in Russia. Many biennials, such as carnations, which normally form flowers only in their second season of growth, can be made to bloom in their first season if subjected to a period of low temperature. The requirement for cold prior to reproduction in many plants is comparable to the situation in some insects which will not emerge from their pupal stage unless subjected to a period of cold.

Basically, temperature affects plants through its control of the rates of individual biochemical reactions. Within the physiological

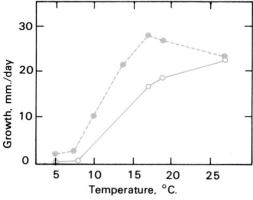

Fig. 9-14. Effects of temperature on growth of tomato plants. Greatest growth occurs when night temperatures are lower than day temperatures. ●- - -●, day temperature of 26°C. alternating with night temperature indicated; ●—●, constant temperature (day and night) as indicated. (After A. C. Leopold: *Plant Growth and Development.* McGraw-Hill Book Co., New York, 1964.)

range of temperatures that a species can tolerate, an increase in temperature will generally increase the rates of cellular metabolic processes just as rates of all chemical reactions are increased by increased temperature. The adverse effects of temperatures which are too low (at the freezing point of water or below) or too high (in excess of about 40°C.) generally can be explained on the basis of their damaging effects on cellular components. Freezing temperatures can result in the formation of ice crystals which rupture cell membranes. Elevated temperatures can denature proteins and thus destroy the capacity of enzymes and membranes to function normally.

Moisture

The prevailing conditions of rainfall and soil moisture determine which species can survive in any given area. Plants exhibit wide differences in their tolerance for unusually dry or wet conditions, and the selection of appropriate varieties plays an important role in determining the success or failure of agriculture.

All plants require water and are dependent on their environment for its adequate supply. In many respects, plants act like wicks in their removal of moisture from the soil. The movement of water is a one-way process from the soil, through plants, to the atmosphere (see Chapter 16). Although relatively little of the water moving through a plant becomes incorporated into cellular components, water is necessary for a great many metabolic processes, from the first steps in the germination of seeds and spores (which involve an imbibition of water) to hydrolysis of stored foodstuffs (a splitting of complex molecules through addition of water) and synthesis of nutrients in the process of photosynthesis (for which water is a raw material).

Atmosphere

The gaseous mantle which now surrounds the earth is a product of many billions of years of chemical and biological evolution. Three component gases in air (nitrogen, 78.1 per cent; oxygen, 20.9 per cent; and carbon dioxide, 0.03 per cent) play major roles in plant metabolism and are therefore essential for growth and differentiation.

Except for the nitrogen-fixing bacteria, plants are not able to utilize atmospheric nitrogen. Thus the nitrogen required for synthesis of all amino acids, nucleic acids, porphyrins, and a wide range of other essential constituents of cells must be absorbed from the soil solution in the form of soluble nitrate or ammonium ions.

The maintenance of a constant level of 20.9 per cent oxygen in our atmosphere is a consequence of the balance between utilization in all oxidation reactions requiring oxygen and the single process which adds significant amounts of molecular oxygen, namely green plant photosynthesis (see Chapter 15). With the exception of anaerobic forms such as some of the bacteria, plants require oxygen to remain alive. Early in the process of seed germination, after imbibition of water has initiated the hydrolysis of food reserves, there is a sharp increase in the rate of oxygen uptake. This occurs because respiratory metabolism requires oxygen for the complete oxidation of foodstuffs and the release of chemical energy. Under conditions where oxygen is restricted (this can happen in waterlogged soils following unusually heavy rains), seeds sometimes fail to germinate. Roots are the plant organ most likely to be subjected to anaerobic conditions in nature. In flooded soil, air spaces are filled with water and the root systems can no longer function normally. Most of us have observed trees killed by the damming of streams by beavers or the creation of man-made lakes.

Carbon dioxide, even though a minor component of air, is constantly being added to our atmosphere as a product of the combustion of fuels and food. It is an essential raw material for the process of photosynthesis, the basis for all food chains, and under normal conditions is the most likely factor to be limiting in this process.

Within the past several decades we have become increasingly aware of the presence of man-made atmospheric pollutants and their effects on the biosphere. We are only beginning to understand the nature of damage to plants and animals caused by automobile exhaust and industrial fumes. It is becoming clear, however, that plants can be harmed by automobile exhausts just as humans are. In many forested areas around the Los Angeles basin, some pine species are being killed by ever spreading automobile-generated smog. In many ways, plants are a good indicator of the general "health" of our atmosphere.

Soil Factors

Soil provides a medium in which the root system anchors a plant and absorbs water and essential minerals. Soils vary greatly in their composition, consistency, and their capacity to support plant growth. Agricultural practices such as crop rotation, contour plowing, addition of chemical fertilizers, irrigation, and the like help to maintain fertility and productivity of agricultural lands and make possible the sustained high yields necessary to feed our growing world population. The essential mineral elements which plants must obtain from the soil are discussed in Chapter 16. For now it will suffice to point out that, if the soil is deficient in any of the essential minerals, plant growth will be impaired and productivity will be reduced.

Even when essential minerals are present in adequate amounts, other factors, such as pH of the soil, can control their availability and thus influence plant growth. Soils may be acid, neutral, or alkaline. Some plants that grow under a variety of pH conditions actually can indicate, through responses such as flower color, the approximate pH of the soil. The familiar hydrangeas respond in this way with red flowers in alkaline soils and blue flowers in acid soils. In nature, acid-loving plants (azaleas, rhododendrons, and the like) often flourish in forest communities, while entirely different species such as grasses populate soils which are neutral or alkaline.

Although the preferences of various plant species for a particular soil pH are not readily explainable, the reasons for the deleterious effects of extremes of pH are quite obvious. Among other effects, pH can control the solubility (and hence the availability) of essential minerals. Iron, which is needed for chlorophyll synthesis and is a necessary component of a number of enzymes, including cytochromes and catalase, tends to form in-

Fig. 9-15. A, effects of air pollution. Young grafted white pine plants. Left, after 7 months' exposure to polluted air; right, after 7 months in pollution-free air. (Courtesy U.S. Dept. of Agriculture.) B, plant varieties differ in resistance to air pollution. Left, Greenpod 407 bean, a variety resistant to air pollution. Note uninjured foliage and good yield. Right, Tempo bean, a variety which is very sensitive to air pollution. Note damaged foliage and poor yield. Both of the plants in this photograph were grown under the same conditions in unfiltered air. (Courtesy Robert K. Howell, U. S. Dept. of Agriculture.)

soluble salts at pH levels above neutrality. Extremes of pH can also denature proteins and thus influence not only membrane permeability but also the status of cellular proteins. Fortunately, soil pH can be modified by appropriate use of fertilizers. Lime (predominately calcium carbonate) is frequently applied to make soils less acid. Calcium ions in limestone serve not only as an essential mineral but also as a factor in the release of other essential elements which may be bound to soil particles.

Another factor influencing soil fertility is its consistency. Clay soils contain much fine-grained material; they drain poorly and tend to become very hard when dry. The addition of organic matter (leaves, decomposing plant materials, peat, etc.), inorganic materials such as sand or "vermiculite," or synthetic soil conditioners can make clay soils loose, crumbly, and porous. These conditions are necessary for proper aeration, percolation (drainage) of water, and penetration by roots.

Light

Light is an essential factor for the sustained growth and normal differentiation of most plants. In addition to its essential role as an energy source in photosynthesis (see Chapter 15), it has profound effects on such varied aspects of development as leaf and stem morphology, pigment synthesis, direction of growth, and time of flowering.

Some Facts about Light

Before we discuss any specific aspects of light-dependent development (**photomorphogenesis**), let us review some basic facts about its physical nature and how light can promote

chemical change in living systems. Without sunlight no life would exist on earth. Thus, a knowledge of light is basic to any complete understanding of the total ecology of our planet. Light is usually defined as the visible portion of the electromagnetic spectrum of radiation. Thus the term light, by definition, involves human perception of one part of a continuum including cosmic rays, gamma and X-rays, ultraviolet and visible light, infrared (or heat) waves, and radio waves. These types of electromagnetic radiation differ in wavelength. Those wavelengths which influence plant development extend beyond what our eyes can see to longer wavelengths in the far red and near infrared and to shorter wavelengths in the ultraviolet region. The units most frequently used for measuring wavelengths of light are the angstrom unit ($1\ \text{Å} = 10^{-8}$ cm.), and the nanometer (nm.) or millimicron ($m\mu$) (1 nm. or $1\ m\mu = 10^{-9}$ M., 10^{-7} cm., or 10 Å). Light from the sun or from an incandescent lamp contains a mixture of wavelengths which we perceive as white light. When light hits an object, the component wavelengths can be absorbed, reflected, or transmitted.

Colored materials absorb some wavelengths while transmitting or reflecting the rest. In order for light to be absorbed, a pigment must be present. This light absorption by a pigment

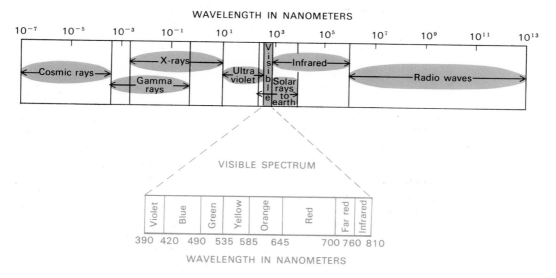

Fig. 9-16. The electromagnetic spectrum. Note that light, the visible portion of the spectrum, is a very narrow band in this continuum.

(called the **photoreceptor**) is always the initiating step in photochemical or photobiological processes. The pattern of light absorption by each pigment is unique and this **absorption spectrum** can be measured with a spectrophotometer. Colored substances can be identified and distinguished from similarly colored compounds on the basis of their absorption spectra. Absorbance increases with increased concentration. This relationship has been formulated as the **Beer-Lambert law** and can be utilized to measure the concentration of colored compounds.

In addition to wave properties which allow us to describe light in terms of wavelength and **frequency** (number of repeating waves per unit time), electromagnetic radiations also have the properties of particles. In an apparently paradoxical way, light, and also ultraviolet, gamma and X radiation, etc., is absorbed or emitted in discrete packets called **quanta** (singular, quantum). A quantum of visible radiation is called a **photon**. The energy per quantum is inversely proportional to its wavelength.

A fixed amount of energy is associated with each quantum of electromagnetic radiation, and the amount of energy can be calculated from the following relationship formulated by Planck early in this century.

$$\text{Energy per quantum} = h\nu = \frac{hc}{\lambda}$$

where h = Planck's constant, and

$$\nu = \text{frequency} = \frac{\text{velocity } (c)}{\text{wavelength } (\lambda)}$$

Both h (Planck's constant = 6.625×10^{-27} erg-sec.) and c (the velocity of light = 2.99×10^{10} cm.sec.$^{-1}$ are constants. The only variable is the frequency.

Thus, short wavelength radiations such as ultraviolet or gamma rays contain more energy per quantum than longer wavelengths such as visible or infrared radiations. This relationship has important implications because the absorption of a high energy quantum can result in the input of considerably more energy, with greater potential for cell damage or useful work, than a low energy quantum.

In order for light to affect a chemical or biological system, it must be absorbed. This obvious fact, recognized since the early 1800's, is known as the first law of photochemistry or the **Grotthus-Draper law**. The so-called second law of photochemistry was not formulated until the early years of the present century. This **Stark-Einstein law of photochemical equivalence** asserts that the absorption of a single quantum of light by a photoreactive molecule can result in change in one molecule. This implies that the **quantum yield** (or the number of molecules undergoing chemical reaction relative to the number of quanta absorbed) should be equal to unity for all strictly photochemical processes. This is rarely the case, however, and measured yields vary from a small fraction of the theoretical value of one to several million molecular changes per absorbed quantum. The reason for such apparent divergence from the theoretical values, especially in biological systems, is the fact that a single light-dependent step often initiates a long series of biochemical dark reactions. The product measured can be many steps removed from the initiating light-dependent reaction. Quantum yield studies can be of considerable importance, however, for they can contribute to our understanding of the mechanism of light-dependent processes. High quantum yields indicate that the light-dependent step may initiate a chain reaction. Quantum yields of less than one indicate that several photochemical reactions may be needed to produce the change being measured.

In terms of over-all efficiency, most photobiological processes are very inefficient. Foremost among the many reasons for such inefficiency is the fact that some wavelengths are readily absorbed by the photoreceptor while others tend to be transmitted or reflected. The absorption spectrum of the photoreceptor tells us something about the relative probability that a given wavelength will be absorbed. Obviously only those wavelengths likely to be absorbed will be effective in promoting a light-dependent process. The relative response to light plotted as a function of wavelength is called an **action spectrum**. Usually the action spectrum of a light-dependent process corresponds closely to the absorption spectrum of the photoreceptor. This correspondence is used to determine what

pigment is the photoreceptor of a light-dependent reaction.

Some Examples of Light-Controlled Plant Development

Many kinds of seeds can germinate in the
dark and can grow for a limited time at the
expense of their food reserves, but such
dark-grown seedlings are abnormal in form and
pigmentation, with long, weak stems and small,
unexpanded, yellow leaves. Plants exhibiting
such symptoms are said to be **etiolated**. We can
observe etiolation to some extent in houseplants or in garden plants grown in deeply
shaded areas. At least moderate intensities of
light are needed for normal chlorophyll synthesis and leaf expansion, while moderate to
high intensities are necessary to inhibit stem
elongation. Plants grown in adequate light have

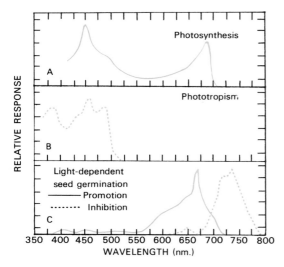

Fig. 9-17. Action spectra of several important light-
dependent plant processes. A, photosynthesis. The
photoreceptor, chlorophyll, is a green pigment. The
most effective wavelengths are in the blue and red
regions of the spectrum. B, phototropism, or the
bending growth toward light, is promoted by blue
wavelengths. The photoreceptor for this response is
not known. C, light-dependent seed germination is one
of many plant responses mediated by phytochrome.
Red light promotes, while far red light inhibits
germination. (After R. B. Withrow (ed.): *Photoperiodism and Related Phenomena in Plants and
Animals.* American Association for the Advancement
of Science, 1959.)

shorter and stronger stems than those grown at
lower light intensities.

Plants exhibit many other readily observable
light-dependent developmental responses. When
houseplants are kept near a window, growth of
the shoot tends to be toward the source of
light. Such a growth response to unilateral
illumination is called **phototropism** and is
caused by an asymmetric distribution of the
plant hormone auxin (See Chapter 17).

Other developmental responses are controlled by the periodicity of illumination, *i.e.,*
the relative length of day and night in each
diurnal cycle. The best known example of
photoperiodism is the control of flowering (see
Chapter 18).

There are numerous light-dependent phenomena involved in the germination of seeds
and the early development of seedlings. Some
varieties of seeds (such as lettuce and certain
grasses) require light for germination to occur.
Later in seedling development when a part of
the seedling emerges from the soil, light is
required for its hypocotyl to straighten, a
change which moves the developing shoot into
a vertical position (Fig. 9-18). In both of these
responses the quality of light (*i.e.,* its wavelength or color) is critical. Wavelengths in the
red portion of the spectrum are most effective
in promoting, whereas longer wavelengths (in
the far red region) can inhibit these responses
or even counteract the effects of an earlier
exposure to red light. Promotion by red and
inhibition by far red light are characteristic of a
wide variety of responses in plants mediated by
the **phytochrome** pigment system. The action
spectra for these responses indicate that the
photoreceptor can exist in either of two
interconvertible forms: one, known as P_R (the
red absorbing form of phytochrome), is a blue
pigment with an absorption peak at approximately 660 nm. It can be converted by red
illumination to P_{FR} (the far red absorbing form
of phytochrome) with an absorption maximum
at 710 to 730 nm. The far red absorbing form
is converted back to P_R by slow dark reactions
or, more rapidly, by exposure to far red light.

The recognition that many light-dependent
developmental phenomena in plants are mediated by the phytochrome pigment system came
from studies in numerous laboratories in many
parts of the world. However, the pioneering

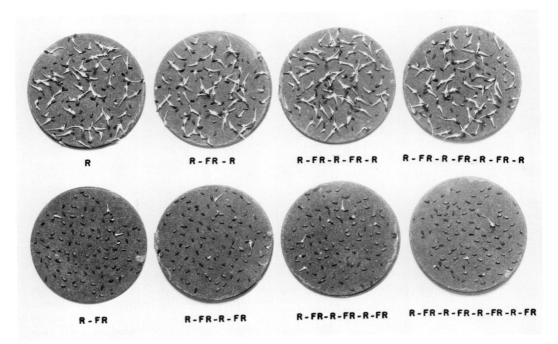

Fig. 9-18. Effects of light on the germination of lettuce seeds. Red light (R) promotes, while far red light (FR) inhibits germination. When exposed to red and far red light in sequence, it is the quality of the light used last that determines the response. (Courtesy Harry A. Borthwick, U. S. Dept. of Agriculture.)

studies on action spectra and the first successful isolation of phytochrome were accomplishments of scientists at the United States Department of Agriculture Plant Industry Station in Beltsville, Maryland. The pioneer work on photoperiodic induction of flowering was begun by Garner and Allard. Careful studies on action spectra were contributed by Borthwick and Hendricks, and finally the phytochrome pigment was isolated by Siegelman and Butler. Among the responses now known to be under control of phytochrome are day length-dependent flowering, light-dependent seed germination, production of anthocyanin pigments in fruits and leaves, stem elongation, leaf expansion, sleep movements of leaves, and the straightening of seedlings as they emerge from the soil.

Gravity

Everyone knows that plant shoots grow up, and roots grow down. The environmental factor controlling this phenomenon is the earth's gravitational field; the growth response to gravity is called **geotropism**. The mechanism of geotropism, like that of phototropism, involves the asymmetric distribution of the plant hormone, auxin (see Chapter 17). Plant physiologists have long questioned the effects on plant morphology of an absence of gravity. The original ingenious approach was to tie potted plants to a rapidly turning water wheel. This was done by Thomas Knight, an 18th-century horticulturalist and forerunner of Mendel in cross-breeding peas. The later 19th century approach was to use a simple instrument called a *clinostat*. Such experiments established that a plant can grow quite normally in a horizontal position if it is rotated continually around the horizontal axis. Recently, with the advent of the space age, it has become possible to test the effects of much reduced gravitational fields on developing seedlings and excised plant shoots by monitoring their response while in orbiting satellites. After retrieval, orbited materials can be compared to controls maintained under normal conditions on earth. An unexpected result of space experiments has been evidence that weightlessness does not appear to impair the development of seedlings. If anything, a slight stimulation of growth has been observed.

INTERNAL CONTROL OF DEVELOPMENT—THE CONCEPT OF TOTIPOTENCY

So far we have considered the influence of external, *i.e.*, environmental, controls over plant development: light, soil, water, tempera-

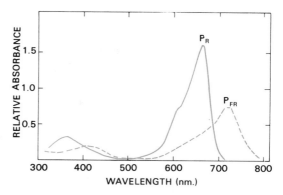

Fig. 9-20. Action spectra for photoconversion of phytochrome. The red-absorbing form of phytochrome (P_R) can be converted by red light to the far red absorbing form (P_{FR}). Conversion to P_{FR} of P_R can occur by slow dark reactions or, more rapidly, by exposure to far red light.

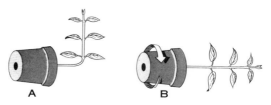

Fig. 9-21. A plant placed in a horizontal position soon turns upward (A), but if rotated in a horizontal plane such a plant will continue to grow normally (B).

ture, and the like. Plants respond to all these factors but the manner of any plant's response depends on its genetic make-up. Obviously no amount of nutrients or water will convert a daisy into a sunflower. Often the responses of a plant are mediated by hormones the plant synthesizes. There is another and, in a sense, more primary way in which the nature of plant development is subject to internal controls. This is the master plan, the information in the genetic code which dictates the patterns of development and the sequence of events in the life cycle. In the nucleus of each cell there is a complete set of instructions for the plant.

Evidence that the Nucleus Controls Cellular Differentiation

One of the most convincing and informative organisms, either plant or animal, in which to

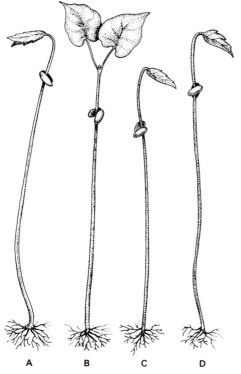

Fig. 9-19. Effects of short periods of light on development of dark-grown bean seedlings. A, 10-day-old seedling maintained in continuous darkness; B, seedling grown 10 days in the dark and then exposed to 2 minutes of red light; C, seedling exposed to 2 minutes of red plus 5 minutes of far red light; D, seedling exposed to 5 minutes of far red light. Red light promotes leaf expansion and straightening of the stem. Far red light reverses the effect of red light, but alone has no effect. (After R. J. Downs, in R. B. Withrow ed.): *Photoperiodism and Related Phenomena in Plants and Animals.* American Association for the Advancement of Science, 1959.)

investigate the role of the nucleus is the remarkable *Acetabularia,* a single-celled marine alga. This plant has cells 5 to 9 inches long that differentiate to form a long stalk anchored by a rhizoidal (root-like) base and an umbrella-shaped cap. Because of their unusual capacity for regeneration and the fact that *Acetabularia*

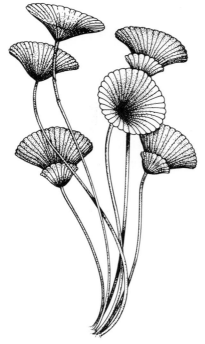

Fig. 9-22. *Acetabularia,* a single-celled marine alga (×2).

can be cultured in the laboratory, these single-celled plants have been utilized for many studies of regeneration, the influence of the nucleus on cell development, and the interaction between the nucleus and the cytoplasm. If the cap of an *Acetabularia* cell is cut off, the basal segment can regenerate a new cap. Likewise, an apical segment (*i.e.,* stalk plus cap) can form a second cap at its base if a sufficiently long segment of stalk is present. In this case, regeneration occurs in the absence of a nucleus, so apparently a nucleus does not always have to be present for regeneration to occur. However, it is clear from other experiments that substances formed in the presence of a nucleus are required for cap regeneration. There is enough of such material already present in the cytoplasm for an enucleate stalk to regenerate a cap, but not enough for a second one to be regenerated if the first formed cap is removed. On the other hand, stalks with rhizoids (*i.e.,* with a nucleus) can repeatedly form new caps.

Transplantation and Grafting Experiments—Further Evidence of Interaction between Nucleus and Cytoplasm

After several seasons of growth, an *Acetabularia* cell forms reproductive structures called cysts within the adult cap. A prerequisite to cyst formation is the proliferation of nuclei in

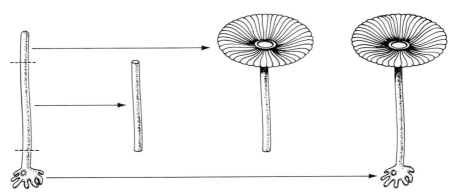

Fig. 9-23. Regeneration in *Acetabularia.* When a young *Acetabularia* cell is cut into three pieces, the middle section remains alive but cannot grow. The apical piece forms a cap and the basal piece (which contains the nucleus) develops into a mature cell with base, stalk, and cap.

the rhizoid and the migration of nuclei and chloroplasts to the cap. Removing a cap just before division of the nucleus delays **division until a new cap is regenerated**.

It is possible to graft together apical and basal segments of *Acetabularia* from two individuals of the same species or from individuals of two different species by pushing together the cut ends of the two segments and thus bringing their cytoplasms into contact. From such manipulation it has been learned that grafting an old cap onto a young rhizoid induces premature division of the nucleus. The opposite combination (a young cap on a mature rhizome) delays division of the nucleus. Thus it appears that there is something within the cytoplasm that can control division of the nucleus.

Another kind of grafting experiment clearly shows the control exerted by the nucleus over development elsewhere in the cell. It is possible to graft together cell segments from two different species of *Acetabularia* that differ in cap morphology. When a capless stalk from *A. mediterranea* is grafted onto a short basal piece of *A. crenulata* stalk with its nucleus-containing rhizoid, a new cap will regenerate. The shape of the new cap will be intermediate between the two species, indicating an interaction between cytoplasmic substances in the *A. mediterranea* stalk and the *A. crenulata* nucleus in the rhizoid. However, if the new cap is removed, the second regenerated cap is of the *A. crenulata* type. Apparently the cytoplasmic cap-forming substance originally present is used up during formation of the first cap; so, when the second cap is formed, its morphology is determined by cap-forming substance of the *A. crenulata* type. This indicates that synthesis is now under the total control of the *A. crenulata* nucleus in the rhizoid. The reciprocal graft (*A. crenulata* stalk on *A. mediterranea* base) similarly produces an intermediate type cap at first; afterwards all new caps are of the type corresponding to the nucleus present in the base.

The mixing of two types of cytoplasm that occurs when two *Acetabularia* species are grafted together also introduces into a single cell the chloroplasts and other subcellular organelles of the two species. The functioning of these organelles in the absence of a nucleus or in the presence of the nucleus of another species poses many interesting problems for physiologists and biochemists.

Although the nature of the cap-forming substance (or substances) is not known, it is probable that messenger RNA's synthesized in the nucleus are involved as the templates for synthesis of the enzymes and structural proteins utilized in the formation of new cap material.

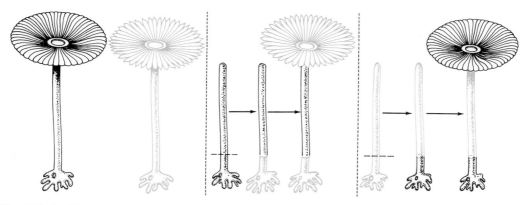

Fig. 9-24. Grafting experiments with different species of *Acetabularia* demonstrate the effect of the nucleus on development. *A. mediterranea* is shown in black, *A. crenulata* in green. When a young *A. mediterranea* is cut in half and the top is grafted onto an *A. crenulata* rhizoid, the grafted cell will develop a cap characteristic of *A. crenulata*. A young *A. crenulata* stalk grafted onto an *A. mediterranea* rhizoid forms an *A. mediterranea* cap. The type of cap formed corresponds to the kind of nucleus in the rhizoid.

Evidence of Totipotency from Tissue Culture

In multicellular organisms only a fraction of the information in the DNA of each cell nucleus is utilized. Cells and tissues differ not only with respect to the details of their structure but also in their complement of enzymes and chemical products. It has long puzzled students of development how an organism's cells can develop so differently from the same set of instructions. The position of any cell within the plant appears to have a large effect on its developmental pattern because the position of a cell determines the conditions under which it develops. In plants, however, individual cells appear to retain great developmental potential. This can readily be demonstrated when stem cuttings form new roots, or when new plants form from leaves.

The term **totipotency** has been used by both zoologists and botanists to express the idea that the nucleus of any cell contains a complete set of genetic information for the organism. Therefore, an entire plant or animal can, at least in theory, be regenerated from a single cell provided that conditions inducing multiplication and differentiation can be provided.

Evidence supporting the concept of totipotency comes from studies of the growth of plant tissues in culture. Many plant scientists have worked to develop media in which isolated plant parts can be cultured. F. C. Steward and his collaborators at Cornell University studied the conditions necessary for proliferation of pieces of mature phloem tissue from carrot roots. Small explants of carrot phloem were removed aseptically and transferred to media containing minerals plus a carbohydrate source and necessary growth factors. Under appropriate conditions of temperature and aeration, cell division occurs in these explants; the tissue increases in size and weight. New cells differ from those removed from the carrot root, however. The undifferentiated and unorganized mass of new cells is called a **callus**. Steward found that for prolonged growth it is necessary to add special growth factors to the medium. The liquid endosperm of coconuts was found to be a convenient source of these factors, now recognized to be a class of plant hormones

called the cytokinins (see Chapter 17). In time, differentiation can be observed within the callus. Growth centers of lignified elements that resemble the cells of vascular tissues form. Further growth and organization leads to the formation of roots. If a rooted callus is transferred from the original liquid medium to agar and added plant hormones are carefully controlled, buds can be induced which give rise to shoots. Thus, in culture, a sequence of events including proliferation of cells, dedifferentiation of the cells of mature tissues, and redifferentiation of callus cells with initiation of new roots, shoots, and finally entire new plants, is possible.

Even more intriguing, however, are the events following isolation of single cells from the mass of carrot callus. Such single-cell isolates divide and follow a pattern of development which closely parallels that of an embryo developing from the zygote. Thus, removed from the normal restraints imposed by the

Fig. 9-25. F. C. Steward, whose pioneer studies on regeneration of plants from cultured cells and tissues support the concept of totipotency. (From Annual Review of Plant Physiology, 22, 1971.)

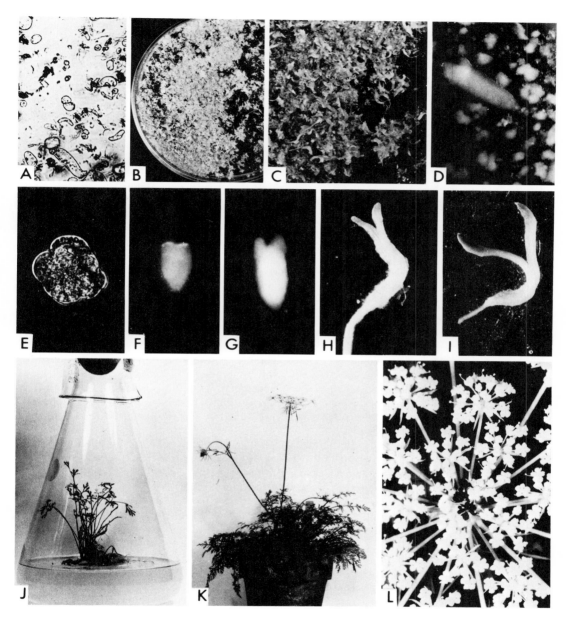

Fig. 9-26. Development of carrot plants from free cell suspensions. A, carrot cells grown in liquid medium; B, growth of embryoids from carrot cell suspension transferred to agar plate; C, higher magnification of embryoids shown in B; D, growth of globular masses in liquid medium. These were derived from a cell suspension similar to that shown in A. One torpedo stage of embryo is seen. E, globular mass of cells developed form free cell suspension; F, heart-shaped embryoid; G, torpedo-shaped embryoid; H, young plantlet developed from embryoid; I, older plantlet; J, carrot plant developed from embryoid growing in agar medium; K, carrot plant in flower after about six months' growth starting with free cell suspension; L, close-up of inflorescence of *Daucus carota* plant grown from cells of embryo origin. (From F. C. Steward: Growth and Development of cultured plant cells. Science, *143:* 3601, 1964.)

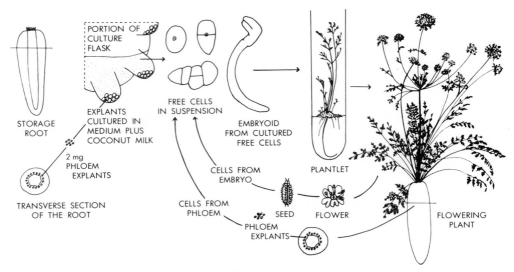

Fig. 9-27. Carrot cells to new carrot plants. Diagram showing sequence from mature phloem explants or isolated embryos to proliferating cultures, free cells in suspension, embryoids, plantlets, and mature carrot plants. (From F. C. Steward: Growth and development of cultured plant cells. Science, *143:* 3601, 1964.)

surrounding cells within a tissue, and provided with nutrients normally available to the zygote, developmental events occur which ordinarily only follow fertilization. First, polarity is established, and then the growing cell mass goes through a filamentous stage. Later, recognizable parts of a plant embryo (cotyledons, epicotyl, and hypocotyl) are formed, and ultimately an entire plant can be regenerated from a single diploid cell.

Other recent studies have indicated unusual regenerative powers in tissue cultures derived from parts of the carrot plant other than mature phloem in the root. Wetherell and his associates at the University of Connecticut have shown that immature embryos of the wild carrot can be grown in tissue culture. If such developing embryos are disintegrated, entire new embryos can be regenerated from individual cells or group cells. Thus, from a single embryo it is possible to obtain literally thousands of genetically identical plants.

Support for the Concept of Totipotency from Molecular Biology

Except for cells which are the products of meiosis, special polyploid cells in the endosperm, and interphase cells about to divide,

every cell of a seed plant contains the same amount and kind of DNA. Evidence that the complete genetic complement is retained in differentiated cells and can be utilized under appropriate conditions has been reviewed in the description of tissue culture experiments in the preceding section. However, differentiated cells differ greatly in the amounts and kinds of proteins and other cellular constituents normally present. Thus, it is obvious that much (if not most) of the genetic information available to an individual cell is ordinarily inert or **repressed**. Clearly, there must be some mechanism that determines in which cells and at what times during development a particular gene will be active (**derepressed**) and can be used as a template for synthesis of its characteristic messenger RNA. As the plant embryo develops from the zygote, cells begin to differ from one another and to acquire the characteristics of the specialized cells of the adult organism. Within the nucleus of each cell there is a program which determines the proper sequence of repression and derepression of genes necessary for orderly development. This program for the sequence of gene activity must itself be a part of the genetic information, since the course of development and final form are inheritable.

We must turn to molecular biology for

insights into the mechanisms which control the repression and derepression of genes. It is now possible to break open cells and to separate the nuclear material from other cell components. The isolated chromatin thus obtained can be used as a template for synthesis of messenger RNA. The product of this DNA-dependent RNA synthesis can then be used in an *in vitro* protein-synthesizing system as a template for protein synthesis. The proteins formed can be compared with those normally synthesized by cells from which the chromosomal DNA was isolated. These procedures have been followed by James Bonner and Ru Chih Huang in their studies of the protein globulin which is synthesized and stored in the cotyledons of germinating peas. This globulin is not found in other tissues of the pea plant.

The first step in Bonner and Huang's experiments was the isolation of chromosomes from pea cotyledon cells. They did this by grinding tissues in such a way that the cells were broken but their nuclei remained intact. Because the chromosomes are heavier than other cellular organelles except for starch grains, they can be isolated by centrifugation.

Once purified chromatin is obtained, it can be used as a template for messenger RNA synthesis. The necessary ingredients for this *in vitro* synthesis of mRNA are the four nucleotide structural units of RNA, purified chromatin (which provides the DNA template), and the enzyme RNA polymerase (which joins together the structural units for the RNA product). The messenger RNA synthesized in this step can then be used as a template for protein synthesis. For *in vitro* protein synthesis, the

following components must be present: ribosomes, amino acids, transfer RNA, activating enzymes, ATP, and messenger RNA. (For a detailed summary of how these components work, see Chapters 3 and 7.) The extent of amino acid incorporation into protein can most readily be followed by labeling the amino acids with radioactive carbon (^{14}C).

Following the procedures just outlined, Bonner and Huang prepared chromatin from pea seedling cotyledons and also from apical bud tissues. They utilized the chromatin from these two sources as the templates for RNA synthesis, used the mRNA's thus formed as the templates for protein synthesis, and finally compared the kinds of proteins formed, starting with chromatin from these sources. Their measure for the total amount of protein synthesized was the amount of radioactivity from labeled amino acids incorporated into protein isolated from the reaction mixture. By then isolating and measuring the radioactivity of individual proteins from the mixture of products, they could determine the relative amounts of individual proteins that had been synthesized.

Their experimental results (summarized in Table 9-1) indicate that isolated chromatin can function in the same manner as it does in the nuclei of intact cells. Genetic information that is inert in the intact nuclei of cells of the apical meristem of the pea seedling (*i.e.*, information needed for globulin synthesis) remains repressed in the chromatin isolated from this source. On the other hand, the significant amounts of globulin that are synthesized from the genetic information available in the chro-

TABLE 9-1
Protein synthesized from messenger RNA generated by two kinds of pea plant chromatin[a]

Template for RNA Synthesis	Total Protein Synthesized, *cpm*[b]	Globulin Synthesized, *cpm*[b]	Globulin, % of Total Protein
Apical bud chromatin			
Expt. 1	15,650	16	0.10
Expt. 2	41,200	54	0.13
Cotyledon chromatin			
Expt. 1	8,650	623	7.2
Expt. 2	6,500	462	6.9

[a] From J. Bonner: *The Molecular Biology of Development.* Oxford University Press, London, 1965, p. 9.
[b] Radioactivity in counts per minute.

matin isolated from cotyledonary tissues correspond to the amounts synthesized in the corresponding intact tissue.

The obvious question raised by these findings is: What is the mechanism by which part of the genetic information remains inert (or repressed)? Bonner and his associates were aware of the fact that, in isolated chromatin, large amounts of protein are bound to DNA. A major portion of this bound protein consists of histones. It was found that bound proteins can be separated from DNA by dispersing isolated chromatin in 4-molar cesium chloride. The high ionic strength of the solution disrupts the ionic bonds linking the protein to DNA. DNA can then be separated from protein by centrifugation. The resulting deproteinized DNA can still support messenger RNA synthesis.

After removal of histones, the DNA's isolated from two different pea tissues have equal capacity to support globulin synthesis. (Table 9-2). However, the percentage of globulins produced from deproteinized cotyledonary chromatin is reduced because the genes for synthesis of many other proteins have been derepressed. While these experiments support the hypothesis that the histones bound to DNA can act as repressors, they do not prove that all repressors of gene activity are histones. Furthermore, we are left with the question of how repressors (histones, other bound proteins, or whatever) are selectively removed so that appropriate bits of genetic information become available at appropriate times in development. Certainly the removal of histones by subjecting isolated chromatin to high salt concentrations is a technique that cannot be used by living cells.

Mechanisms of Derepression

There is increasing evidence that hormones act as derepressors of genes. The classic definition of a **hormone** is that it is a specific chemical synthesized in one organ and affecting another organ. Generally, very small amounts of hormones have very large effects. How are such amazingly amplified effects achieved?

It is now generally agreed that the application of a hormone to its target organ derepresses previously repressed genes which control the production of specific enzymes. Thus,

hormones cause the production of specific new species of messenger RNA. Once again, it is appropriate to ask how such derepression is brought about. One possible mechanism is that a hormone binds with a specific protein repressor (presumably histone), changing the conformation of that protein so that the resulting hormone-protein complex no longer can act as a repressor. (For more complete accounts of hormonal derepression of genes, see Chapters 17 and 30.)

In addition to genes which can be derepressed by the presence of specific small molecules, i.e., hormones, there are also genes which are known to be regulated by specific metabolic products. Such genes can be turned on (or off) by the presence or absence of a product of enzymatic activity. For instance, in higher plants the enzyme nitrate reductase is not formed so long as an adequate level of nitrogen in the form of ammonium ions is available. This is clearly an example of the repression of genetic information by the product (NH_4^+) of an enzyme (nitrate reductase).

Mechanisms of Phytochrome Action. Although the mode of action of phytochrome is still not completely understood, there is considerable evidence that P_{FR} (the far red absorbing form) is the effector molecule which can trigger many varied responses. Although at first it was difficult to imagine how the small changes in a few pigment molecules could be amplified to produce such large and obvious changes in the whole plant, there is now good evidence that two basic mechanisms exist which make amplification possible. The first operates through differential gene activation, a mechanism proposed by Hans Mohr. His studies of phytochrome-induced anthocyanin synthesis in mustard seedlings indicate that P_{FR} promotes the synthesis of mRNA specific for enzymes involved in formation of these blue or red pigments. Phytochrome-mediated pigment synthesis can be blocked either by actinomycin D (which inhibits DNA-dependent RNA synthesis) or by puromycin (which inhibits mRNA-dependent protein synthesis). An amplification of such hormonal response can readily be understood if we consider that a number of mRNA's can be synthesized from the information in one activated gene, a number of enzyme molecules can be synthesized per mRNA, and each enzyme molecule can catalyze a given reaction many times.

A second mechanism of phytochrome action involves the control of membrane permeability. Some responses are too rapid to be explained by the multiple steps in gene activation and protein synthesis which require a minimum of several hours to be observed. Changes such as the movement of leaflets of *Mimosa pudica* (the sensitive plant) are evident within times as short as 5 seconds. When these plants are irradiated with far red light and then transferred to the dark, their leaflets remain open for many hours; but the leaflets close rapidly when exposed to red light prior to the dark period. Such immediate responses are most readily explained in terms of changes in membrane permeability which are known to occur with great speed.

TABLE 9-2

Protein synthesized from messenger RNA generated by pea plant chromatin after removal of histone[a]

Template for RNA Synthesis	Total Protein Synthesized, *cpm*[b]	Globulin Synthesized, *cpm*[b]	Globulin, % of Total Protein
DNA of apical bud chromatin			
Expt. 1	15,200	60	0.40
Expt. 2	14,200	72	0.51
DNA of cotyledon chromatin			
Expt. 1	5,600	22	0.39
Expt. 2	60,000	314	0.52

[a] From J. Bonner: *The Molecular Biology of Development.* Oxford University Press, London, 1965, p. 23.
[b] Radioactivity in counts per minute.

Fig. 9-28. Effects of light quality on sleep movements of *Mimosa pudica* leaves. Leaflets normally open in the light and close in the dark. Exposure to red light (R) prior to darkness promotes closure while far red light (FR) prior to darkness delays closure. Effects of red and far red in sequence are comparable to their effects on lettuce seed germination. The quality of the last light period governs the response. (From J. C. Fondeville, H. A. Borthwick, and S. B. Hendricks: Leaflet movement of *Mimosa pudica* L. indicative of phytochrome action. Planta (Berlin), *69:* 357-364, 1966.)

Both the quality of light (*i.e.*, color) and the duration of illumination are important in responses such as the photoperiodic control of flowering. Seasonal changes in the vegetation of the temperate zones are triggered by changes in day length (see Chapter 18).

Haploid Plants

In discussing the topic of totipotency we emphasized that any somatic cell of a diploid plant (and apparently any animal, also) contains all the information needed to produce the entire organism. But one can raise serious questions about the necessity for all this DNA. Do diploid organisms really require duplicate sets of information? Recent work indicates that somatic cells of plants may possess more information than they actually need. Nitsch and Nitsch at Gif-sur-Yvette in France have recently succeeded in growing haploid tobacco plants from single pollen grains. The ability to produce a haploid plant at will has certain practical advantages. As you recall, experimental organisms such as the pink mold, *Neurospora crassa,* have been widely used in genetic studies precisely because they are haploid: induced mutations are readily visible; genotype and phenotype are the same. Thus, if haploid higher plants could be produced at will, mutations could be detected immediately and the desired changes in their genotypes preserved. The means for doubling chromosome numbers are available, for example, applying colchicine or by the process of **endomitosis**, a doubling of chromosomes without nuclear division. Callus cultures tend to undergo endomitosis when grown on special media. Thus, completely homozygous individuals, a rarity among plants, can be obtained.

The secret of growing haploid tobacco plants from pollen grains is to remove stamens at the appropriate stage of flower development (stage 2 as shown in Fig. 9-31), and to culture them on a suitable medium. After three to four weeks, embryos and plantlets emerge which can be transferred to a simpler medium. Once a root system has formed, the young plants can be transferred to pots where they will grow to maturity and flower. One strange (and perhaps reassuring) aspect of the development of em-

bryos from pollen grains is that their pattern of development follows precisely the same steps seen in the normal development of the embryo plant from a zygote.

AGING, SENESCENCE, AND DEATH

Old age is a predictable phase in the life of any multicellular organism, except those meeting accidental death. Associated with aging and senescence is a loss of vigor and deterioration of cells and tissues. In annual plants, sensescence usually follows the development of the reproductive structures (flowers, fruits, and seeds), and death of the organism is precipitated by changes in climate, such as the first frost. The underlying causes of aging in either animals or plants are unknown. Consequently, an intensive search for clues is under way in laboratories throughout the world.

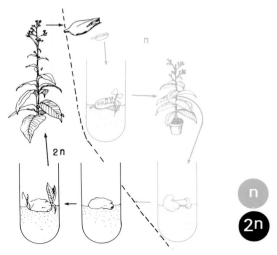

Fig. 9-29. Homozygous plants from cultured anthers. Flower buds excised from a diploid plant are grown in culture. The resulting haploid plants are transferred to pots. A piece of haploid stem or petiole is transferred to a synthetic medium stimulating callus growth. Doubling of chromosomes occurs during callus development. Diploid callus cultures are transferred to a medium that stimulates bud formation and give rise to diploid plants which are homozygous. (From J. P. Nitsch: Plant propagation at the cellular level, a basis for future developments. Proceedings of the International Plant Propagator's Society, vol. 19, pp. 123-132, 1969.)

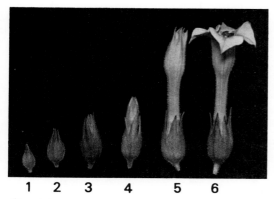

Fig. 9-30. Stages in the development of tobacco flowers. Meiosis occurs in the microspores at stage 2 and stamens excised at this time yield the greatest numbers of haploid plantlets. (From J. P. Nitsch: Haploid plants from pollen. Zeitschrift für Pflanzenzüchtung, *67:* 3-18, 1972.)

Juvenile vs. Adult Morphology

Although senescence and death are the final phases in the life cycle, they are not the only changes associated with the aging process. There is a constant turnover in cellular constituents from the time of fertilization in animals, and from the time of germination in seed plants. Furthermore, plants undergo changes in morphology as they grow that are comparable in many ways to the morphological changes so familiar in animals. We can thus recognize certain shapes and proportions as characteristic of juvenile or adult forms. Such differences in shape are readily observable in leaves formed at different stages of development. Some examples of these differences in leaf shape in ivy are shown in Figure 9-33. As leaves are products of the apical meristem of the shoot, it follows that there must be changes in the pical meristem preceding changes in leaf morphology. An even more drastic change occurs in this region when vegetative growth stops and the onset of reproductive growth begins. With the onset of flowering, massive changes in the shape and organization of the apical meristem itself can be observed. Can we then conclude that the meristem itself ages?

There are two views concerning this question. According to the first view, meristems do not age. Juvenile or adult leaves, or the initiation of flowering, result from a balance of factors such as supply of mineral nutrients, carbohydrates, nitrogenous substances, and growth factors. This view implies that the direction and extent of meristematic activity are controlled by chemical and/or hormonal signals originating elsewhere in the plant and can be significantly influenced by environmental factors. According to the second hypothesis, meristems do age during the development of plants, so that an apical meristem can be classified as juvenile, adult, or senescent. An extreme interpretation of the latter view would predict an unalterable sequence of development set by the genetic code of the organism. Plant development actually follows a middle road, with the pattern set by heredity but modified by internal and external factors. If apical meristems are isolated in tissue cultures or as cuttings, they retain their characteristic organization and continue to form new appendages (leaves) characteristic of the species. When isolated surgically from adjacent tissues, but left in contact with the plant, an apical meristem can proceed to form a normal shoot even though connected with the rest of the plant only through undifferentiated pith cells.

As we have already stated, changes in apical meristems can be detected by changes in the products of their activity. Leaves formed at various stages of development can differ in their morphology. At the end of the last century, the

Fig. 9-31. Haploid tobacco plantlets emerging from anther cultured on a synthetic agar medium containing salts and sucrose. (From J. Nitsch: Haploid plants from pollen. Zeitschrift für Pflanzenzüchtung, *67:* 3-18, 1972.)

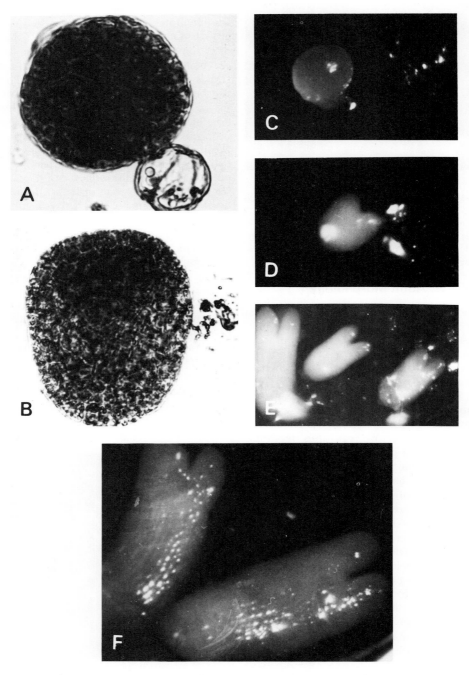

Fig. 9-32. Stages in embryo development from pollen grains. A, spherical mass of cells and empty integument of pollen grain; B, later view of developing cell mass showing polarity; C, globular stage of embryo development; D, heart-shaped embryoid; E and F, early and later torpedo stage. (Courtesy C. Nitsch, Laboratoire de Physiologie Pluricellulaire, Centre National de la Recherche Scientifique, Gif-sur-Yvette, France.)

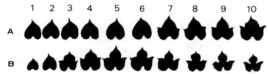

Fig. 9-33. Changing morphology of ivy leaves formed at successive nodes. A, leaves formed at first 10 nodes of a plant grown in the shade; B, leaves formed at first 10 nodes of a plant grown at a higher light intensity. (From D. Briggs and S. M. Walters: *Plant Variation and Evolution.* McGraw-Hill Book Co., New York, 1969.)

German botanist Goebel used the term **heteroblastic development** to describe the situation where juvenile and adult leaves are strikingly different in shape. Such changes in leaf shape from node to node have been used as a measure of physiological age. Krenke, another German botanist whose works were published early in the 1900's, regarded successive nodes as units of physiological time. Leaves formed at successive nodes often differ in shape, and flowering normally follows a predictable succession of changes in leaf morphology.

The ideas of Goebel and Krenke have been used in studies of aging in the sunflower (*Helianthus annuus*). This species lends itself very well to studies of plant aging because its development is not only heteroblastic but also determinate, *i.e.,* it develops a definite predictable number of leaves. The latter trait is an unusual feature in plants, which are normally **indeterminate** in their growth patterns (*i.e.,* most plants form an unpredictable number of appendages during vegetative growth). In the sunflower, however, the extent of vegetative growth, *i.e.,* the number of leaves formed, is controlled by photoperiod. Thus, by maintaining a constant day length (13 hours), it is possible to control the number of leaves formed (14 to 16 pairs) prior to a shift in the activity of the single apical meristem to reproductive development.

Thus, a sunflower's apical meristem can be assigned a physiological age based on the number of leaves it has formed. Under conditions of controlled photoperiod, the remaining vegetative life expectancy of the meristem is predictable because its total life expectancy is fixed, and the number of leaves and primordia already formed can be counted.

Experiments can then be designed to answer the specific question: What effect would a changed biological environment have on the life expectancy of an apical meristem? The experimental approach to such a question is relatively simple because sunflowers can be grafted. Using this very old horticultural technique, the apical meristems of seedlings have been grafted onto physiologically older plants. By counting the number of leaves formed by a **scion** (the grafted seedling), it is possible to determine whether the older stock has influenced the rate of aging of the scion. The results clearly show that transfer to a physiologically older environment can accelerate the aging of an apical meristem.

The sum of the leaves on the stock plus scion always adds up to approximately the same total. A seedling grafted onto a physiologically old stock already in bud will form only two or three pairs of leaves before flowering (compared to 14 or 16 pairs normally formed by a sunflower plant). Accompanying such early flowering is a compression of the pattern of changes in leaf shape: morphological changes occur in fewer nodes (or units of physiological time). It is therefore obvious that the rest of the plant can influence the functioning of an apical meristem.

Although grafting provides a means for transferring young apical meristems to older environments, the nature of the chemical signal transferred from old tissues that triggers changes in the developmental patterns of physiologically young scions is as yet unknown. When such experiments are done in reverse and the old apical meristems are grafted onto seedlings, there is practically no influence of the stock on the scion. In other words, while exposure to older tissues can accelerate aging of young meristems, young tissues apparently are unable to influence the rate of aging of older meristems.

Senescence

At any point in its life cycle, a higher plant is a mixture of young and old, living and dead components. In perennial plants such as the trees, life spans of several hundred years are not uncommon, yet there is an annual production of new tissues and organs. In deciduous species, new leaves are formed at the beginning and shed at the end of each growing season. In evergreen species such as the conifers, leaves persist for two to several seasons, but some leaves are replaced each year. If we examine the stems of older trees, we find that large amounts of the central xylem and also of the outer bark tissues are dead, so that only relatively minor regions on either side of the vascular cambium are made up of living cells.

In annual plants, even while the entire bulk of the organism is composed of living tissues, we find a mixture of young, newly formed tissues, older organs, and aging parts. Leaves at the base, formed early in vegetative growth, are obviously older physiologically than recently formed leaves near the apical meristem. We can expect to find differences in morphology, composition, and metabolism related to plant age and leaf position.

Most studies of aging in plants have been concentrated on changes in single tissues or organs after removal from the plant. Such isolated plant parts, or **explants**, do deteriorate with time and undergo chemical and physical changes comparable to those occurring in intact plants. The most obvious of these changes are the decomposition of chlorophyll and hydrolysis of proteins.

Leaf Abscission

Among the most interesting studies of aging phenomena using explants have been the investigations of factors controlling **leaf abscission**, the separation of leaves from the stem in autumn. The loss of leaves in the fall is usually preceded by the formation of an abscission layer across the base of the petiole. The weak, thin-walled cells of the abscission layer soon disintegrate, leaving only the strands of conducting tissues in the petiole to support the leaf. Many factors are known to influence the formation of the abscission layer. Abeles, Rubenstein, and their co-workers at the former U. S. Army Biological Laboratories in Frederick, Maryland, have shown that all factors promoting leaf abscission have one effect in common: they induce the production of ethylene. Furthermore, they have shown that the introduction of ethylene gas during incubation of petiole explants can accelerate abscission. The mechanism by which ethylene promotes abscission appears to be comparable to the way in which hormones control development. Abeles and Holm have shown that ethylene promotes the synthesis of new kinds of messenger RNA and that inhibitors of RNA and protein synthesis can counteract the promoting

Fig. 9-34. A, freshly cut explant of cotton includes section of stem and bases of two petioles. B, explant of cotton after formation of abscission layer and drop of petioles. Ethylene promotes changes associated with abscission. (Courtesy F. B. Abeles, U. S. Dept. of Agriculture.)

effects of ethylene on abscission. Thus, ethylene can be classed with other hormones which act as specific derepressors of genetic information. (For further discussion of ethylene and other plant hormones, see Chapter 17.)

Delay of Senescence by Cytokinins

There is a whole class of naturally occurring plant hormones, the **cytokinins** (see Chapter 17), that have the general effect of slowing the changes in plant tissues characteristic of senescence. A knowledge of the effects of these hormones has already been applied in the preservation of vegetables such as spinach and broccoli after harvesting. You will recall that the cytokinins are now known to be the mysterious components in the coconut milk used by Steward and others to supply the factors needed for continued cell division in plant tissue cultures. The cytokinins are essential throughout the life of a plant beginning with the earliest cell divisions in the developing embryo.

In the development of plants, there is a constant interplay of internal factors (genes and hormones) and the essential ingredients of plant life provided by the environment (water, soil, light, atmospheric gases, and climate). The earliest studies of plant development were descriptive, dealing primarily with establishing the sequence of events in the life history of a given species. Once the predictability of development and life cycles was recognized, more penetrating questions could be asked: What is the basis for this predictability and for the orderliness of development in living things? How can the patterns of development be modified? In recent decades the ideas and tools of molecular biology have brought new insights to studies of plant development, and of animal development as well. In the decades ahead we can look for a better understanding of the ways in which the blueprint for an organism provided by its genetic code is read and carried out.

USEFUL REFERENCES

Bonner, J. *The Molecular Biology of Development.* Oxford University Press, London, 1965.

Cutter, E. G. *Plant Anatomy: Experiment and Interpretation,* Part I: Cells and Tissues (1969); Part II: Organs (1971). Addison-Wesley Publishing Co., Reading, Mass.

Galston, A., and Davies, P. J. *Control Mechanisms in Plant Development.* Prentice-Hall Inc., Englewood Cliffs, N. J., 1970.

O'Brien, T. P., and McCully, M. E. *Plant Structure and Development: A Pictorial and Physiological Approach.* The Macmillan Co. and Crollier-Macmillan Canada, Ltd., Toronto, 1969.

Steward, F. C. (ed.) *Plant Physiology, A Treatise,* Vol. VIC: Physiology of Development: From Seeds to Sexuality. Academic Press, Inc., New York, 1972.

Torrey, J. G. *Development in Flowering Plants.* The Macmillan Co. and Crollier-Macmillan Canada, Ltd., Toronto, 1967.

4

THE WORLD OF LIFE: SPORE-BEARING PLANTS

The More Complex Algae and Fungi

10

The algae and fungi are often ignored, yet their importance can hardly be exaggerated. The photosynthetic algae are the primary producers in all aquatic food chains, not only in freshwater lakes, ponds, and streams, but also in the oceans. If the algae were to disappear from the earth, animal populations would be exterminated soon thereafter, because all animals are ultimately dependent on the algae, either as their direct beneficiaries within food chains or as utilizers of atmospheric oxygen. Marine algae contribute far more than any other group of plants to the maintenance of constant levels of atmospheric oxygen. They are responsible for as much as 90 per cent of all photosynthesis on this planet.

In contrast to the photosynthetic algae, the fungi are not like most plants in their nutritional habits: all are heterotrophic, that is, dependent on their environment for an energy source in the form of complex organic molecules. Many are saprophytic, playing an important role in the breakdown and decay of dead organisms; others are parasites. The fungi assume a variety of roles in the interaction of living organisms—some beneficial and others deleterious. Some fungi cause plant, animal, and human diseases. Hundreds of millions of dollars of damage is done annually to wheat and other crops as well as to trees. Human diseases caused by fungi range from athlete's foot to blastomycosis, which causes stubborn ulcers on the skin, in the eyes, or in the lungs. Some fungi, such as *Penicillium,* serve as sources of a variety of antibiotics used in the treatment of bacterial infections. Certain species of fungi are edible; still others are sources of hallucinogenic drugs.

It is unfortunate that so few people bother to look closely at the algae and the fungi because, whether microscopic or macroscopic, these organisms reveal a diversity of form and color rivaling any other living things in beauty.

CLASSIFICATION OF THE ALGAE AND FUNGI

In some systems of classification the algae and fungi are grouped together in the division Thallophyta, or thallus-forming plants. A justification for this grouping is the characteristic simplicity of the algae and fungi, which have none of the usual organs (roots, stems, and leaves) found in the higher plants. An opposite approach to their taxonomy that is presently fashionable places the numerous large groups of algae and fungi in separate divisions of the plant kingdom. We have chosen a middle course. The prokaryotic blue-green algae (Cyanophyceae) and bacteria (Schizomycetes) have been grouped with the akaryotic viruses in the division Prophyta (see Chapter 4). All the eukaryotic algae have been grouped together in the division Algae, and the seven generally accepted groups of algae have been assigned subdivision status. All eukaryotic fungi have been grouped together in the single division Fungi, with the four generally recognized categories assigned subdivision status (Tables 10-1 and 10-2).

THE ALGAE

Algae range in size and complexity from microscopic unicellular forms to macroscopic multicellular organisms 30 meters or more in length. Unicellular species are found in all subdivisions except the brown algae. Many of the smaller species (both unicellular and colonial) float freely near the surface of their aquatic environments. Small, free living plants and animals floating or feebly swimming in water and carried by currents are collectively called **plankton**.

Reproduction and Life Cycles of the Algae

The algae exhibit great diversity not only in size, form, and patterns of growth, but also in

TABLE 10-1
Division Algae

Subdivision	Common Name
Chlorophyta	Green algae
Euglenophyta	Euglenoids (sometimes classified as animals)
Charophyta	Stoneworts
Phaeophyta	Brown algae
Rhodophyta	Red algae
Chrysophyta	Yellow-green algae, golden-brown algae, and diatoms
Pyrrophyta	Cryptomonads and dinoflagellates

TABLE 10-2
Division Fungi

Subdivision	Common Name
Myxomycophyta	Slime molds
Phycomycophyta	Water molds and black bread molds
Ascomycophyta	Yeasts and brown, green, and pink molds
Basidiomycophyta	Rusts, smuts, and mushrooms

their means of asexual and sexual reproduction. Obviously cell division in the unicellular algae and fragmentation in colonial and multicellular forms can result in new organisms. In all forms of asexual reproduction the offspring remain identical to the parent in genetic constitution.

Sexual reproduction is known to occur in species representing all subdivisions of the eukaryotic algae. All modes of sexual reproduction involve the fusion of specialized cells called **gametes**. Gametes, especially in the motile, unicellular algae, may look identical to vegetative cells. More frequently, however, gametes differ morphologically from vegetative cells and, in multicellular forms, are produced in specialized structures called **gametangia**. The two fusing gametes may be the same in size and morphology or different. Male gametes (sperms) are produced in gametangia called **antheridia**, and female gametes (eggs) are produced in **oogonia**. In some species, both kinds of gametes are produced by the same plant, while in others the two kinds of gametes are formed by different individuals.

The zygote, formed by fusion of gametes in the process of fertilization, may be free floating or fixed within the body of the parent alga. The zygote is always a diploid cell because it results from the fusion of two haploid gametes. Its formation marks the beginning of a new individual. A second crucial step in any life cycle is the process of meiosis, a specialized cell division in which the chromosome number is halved (see Chapter 7).

Because of the great diversity in their life patterns, the algae have provided clues about what the first plants might have been like and

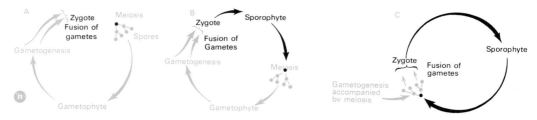

Fig. 10-1. Patterns of life cycles in algae. A, haploid gametes fuse to form a diploid zygote which immediately undergoes meiosis. A gametophyte plant develops from the haploid spores. B, true alternation of generations. Haploid gametophyte plant produces gametes which fuse to form a zygote. A diploid sporophyte plant is formed from the zygote. Meiosis occurs during sporogenesis. Gametrophyte plants develop from haploid spores. C, meiosis occurs during gametogenesis. Fusion of gametes produces a diploid zygote from which a new sporophyte plant develops.

how higher plants may have evolved. As we noted at the beginning of this chapter, the algae represent an "artificial" division of the plant kingdom in the sense that these groups of organisms are not closely related. The various algal subdivisions are currently thought to have evolved independently of one another. Collectively they can be regarded as representing a level of development ranging from the primitive to one almost approaching that of the vascular plants in complexity. We see in the algae many features found in higher plants and evolutionary tendencies considered by botanists to be advanced. The algae have a eukaryotic cellular organization and most have cellulose cell walls. Multicellular algae exhibit a specialization of cell structure related to function, complex branching patterns, and external differentiation into organs resembling the roots, stems, and leaves of higher plants. Their life patterns and modes of reproduction range from the simplest asexual replication to sexual reproduction involving specialized and complex reproductive structures. In the algae we find the phenomenon of alternating haploid and diploid phases within the life cycle. This alternation of generations is a recurrent theme that will be encountered again in all other divisions of the plant kingdom.

A frequently asked question is whether more highly evolved plants originated from alga-like ancestral forms. It is now generally agreed that higher plant groups evolved from ancestral stocks related to the present-day green algae (Chlorophyta).

The Green Algae (Subdivision Chlorophyta)

The green algae are both numerous and widely distributed. About 5,000 species have been described. They are found in the oceans (as plankton and also as benthic, including intertidal, organisms); in freshwater lakes and streams; on and in the soil; and on the moist surfaces of rocks and trees. Some species are **epiphytes**, that is, they exist attached to the surfaces of other plants or animals. A few species are **endophytes**, living within certain protozoa, coelenterates, and sponges.

The Volvocales

The Volvocales are a group of green algae usually found in fresh water. Their vegetative cells are motile and propelled by flagella. Some species are unicellular; others are multicellular colonial forms. A representative unicellular genus, *Chlamydomonas*, is found in soil and aquatic habitats. All species in this genus have two anterior flagella and most have a single cup-shaped chloroplast (Fig. 10-2).

Colonial Volvocales (for instance, *Pandorina, Eudorina,* and *Volvox*) exhibit a number of evolutionary advances characteristic of complex multicellular organisms. The simplest colonial algae are no more than aggregates of cells that fail to separate after mitosis. In an organism like *Volvox,* however, we find that the component cells are not all identical and that they exhibit some differentiation or specialization related to function. Certain enlarged cells near

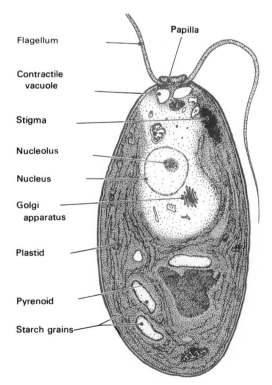

Flagellum

Papilla

Contractile vacuole

Stigma

Nucleolus

Nucleus

Golgi apparatus

Plastid

Pyrenoid

Starch grains

Fig. 10-2. *Chlamydomonas,* a green alga representative of the unicellular Volvocales. Cellular structure as seen in the electron microscope (×6,000).

the equator are capable of forming new colonies; others give rise to gametes. Furthermore, there is evidence of cytoplasmic connections between adjacent cells. Such cytoplasmic connections could provide a mechanism for the transfer of stimuli and would be of obvious advantage for coordinated movement of the colony as a whole. There is indeed coordinated movement in *Volvox*. Its name is derived from the Latin *volvere* (to roll), and the graceful rolling motion of this colonial organism through its aqueous environment is readily observable.

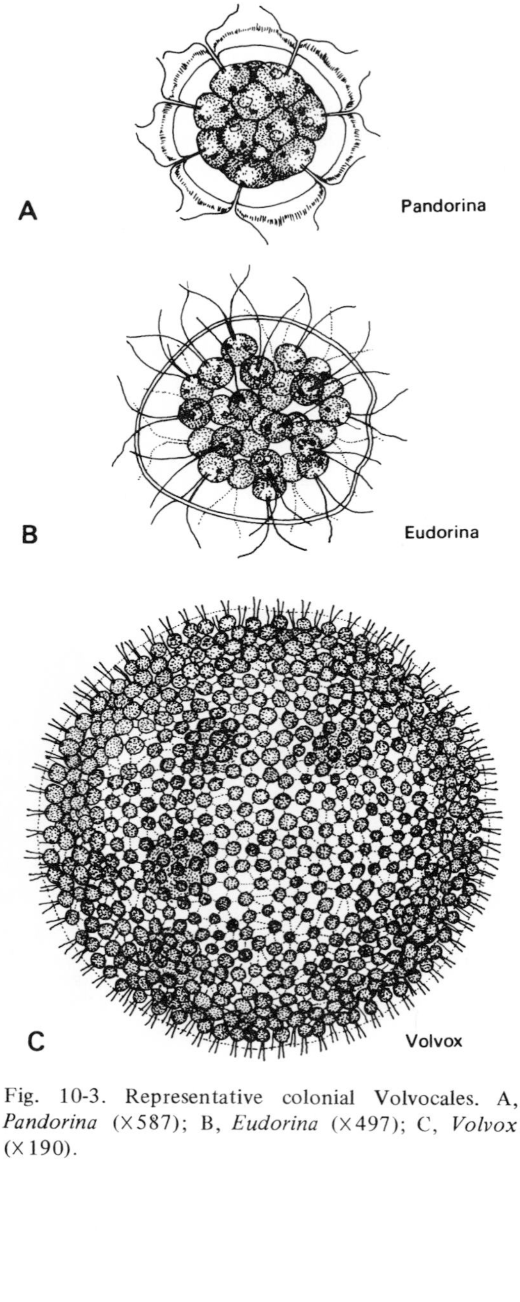

Filamentous Forms

The Ulotrichales and Oedogoniales are rather similar groups of green algae. These are mostly filamentous freshwater forms with cells united end-to-end in simple or branched filaments. The basal cell of each filament (or **holdfast** cell) is modified to anchor the plant. Individual cells contain one nucleus and a single chloroplast (sometimes with unusual morphology). In *Ulothrix*, the chloroplast has the shape of an open ring, whereas in *Oedogonium* it is composed of a network of joined strands encircling the protoplasm. In these, as in most other algal cells, we find **pyrenoids**, or centers of starch formation, in the chloroplasts.

Both *Ulothrix* and *Oedogonium* can reproduce asexually by fragmentation of their filaments or by spore formation; however, their methods of sexual reproduction differ. In *Ulothrix*, identical gametes produced by different filaments fuse to form the zygote. After a resting period, the zygote divides meiotically to form four haploid spores. The spores become attached to the substratum and new multicellular filaments are produced by cell division. In *Oedogonium*, male and female gametes differ in size and shape. The motile sperm enters the oogonium through a pore in the wall of this enlarged cell. After fertilization the zygote is released from the parent filament. The first division of the zygote is meiotic and four haploid spores are produced which develop into new haploid plants.

"Sea Lettuce"

The Ulvales are a group of green algae that exhibit an alternation of generations. In *Ulva*, or "sea lettuce," a littoral marine alga growing attached to rocks and piers in the intertidal zone, a haploid, gamete-producing plant (the **gametophyte**) and a diploid, spore-producing plant (the **sporophyte**) follow each other and together comprise one turn of the life cycle. The gametophyte and sporophyte plants are identical in appearance.

Fig. 10-3. Representative colonial Volvocales. A, *Pandorina* (×587); B, *Eudorina* (×497); C, *Volvox* (×190).

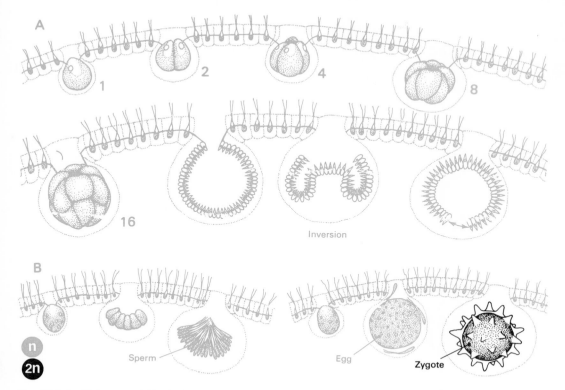

Fig. 10-4. In *Volvox* specialized cells near the equator of the colony are capable of forming new colonies or of giving rise to gametes. A, new colony formation (×950). Numbers indicate cells present in stages shown. B, gametogenesis (×866).

The diploid, sporophyte stage of *Ulva* reproduces by means of spores. These spores have four flagella and are haploid cells formed by meiotic division of special cells of the sporophyte thallus. After release, the spores develop via an unbranched filamentous stage into one of two kinds of haploid gametophyte plant. Half the spores from a given sporangium develop into gametophyte plants that produce small biflagellate male gametes, and the other half develop into those producing large biflagellate female gametes. After fertilization, the diploid zygote develops into a sporophyte thallus, completing the life cycle.

Mermaid's Wineglass

The Siphonales are mostly marine forms having rather large, tubular plant bodies. *Acetabularia* (commonly called mermaid's wine-glass) is a unicellular plant 5 to 9 cm. long, consisting of a long stalk with an umbrella-like cap at the top anchored by a rhizoidal base that contains a single nucleus (Fig. 10-7).

Because of its large size, its unusual capacity for regeneration and the fact that it can be cultured in the laboratory, *Acetabularia* has been utilized for many studies of regeneration and influence of the nucleus on cell development (see Chapter 9). It is possible to graft together apical and basal segments of *Acetabularia* from two individuals of the same species or from individuals of two different species by pushing together the cut ends of the two segments so that their cytoplasms are brought into contact.

Development of a new *Acetabularia* plant begins with the fusion of two haploid biflagellate gametes. The zygote enlarges and the rhizoidal base forms first. Next, the elongating

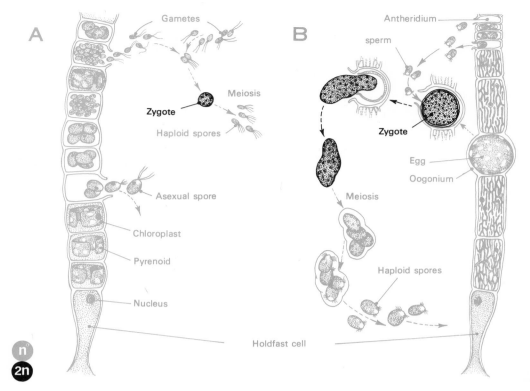

Fig. 10-5. Filamentous green algae. A, *Ulothrix*. Following fusion of gametes, zygote divides meiotically to form haploid spores which give rise to new haploid filament. B, *Oedogonium*. Note that male and female gametes differ in size and shape. Following fertilization zygote is released from oogonium and divides meiotically to give rise to four haploid spores.

stalk forms and finally a whorl of gametangia forms at the apex. These fuse to form the cap. The entire differentiated structure is one cell with a single nucleus located in one of the lobes of the rhizoidal base. *Acetabularia* lives for several years. In the fall a crosswall is formed above the base of the rhizoid. The stalk and cap disintegrate while the rhizoid persists and sends up a new stalk the next spring. Fertile gametangia are usually formed only after several seasons. At this time the nucleus (still in the rhizoid) undergoes many divisions. The resulting thousands of secondary nuclei move up the stalk. They carry with them the green chloroplasts which had been located in the rhizoid and stalk. When the nuclei, chloroplasts, and accompanying cytoplasm have reached the chambers of the cap, cysts are formed, each with a single nucleus. A single cap can form as many as 15,000 cysts. Meiosis occurs in the first division of the uninucleate cyst. During maturation, haploid nuclei within each cyst divide repeatedly, giving rise to as many as 1,800 binucleate, chloroplast-containing gametes per cyst. After their release from the mature cyst, gametes fuse. The diploid zygote develops into a new single-celled individual, forming first a rhizoid, then a stalk, and finally a cap.

Spirogyra and Desmids

Spirogyra, a freshwater green alga, has a long spiral chloroplast coiling around the periphery of the protoplast. Its close relative *Zygnema* has cells with two star-shaped plastids on either side of the central nucleus (Fig. 10-8). These plants reproduce asexually by fragmentation of filaments or sexually by **conjugation**. During conjugation, the cells of two adjacent filaments grow outward. An opening forms at the point where extensions from two cells meet, forming

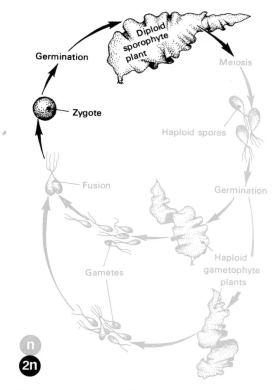

Fig. 10-6. In life cycle of *Ulva*, morphologically identical haploid and diploid stages alternate.

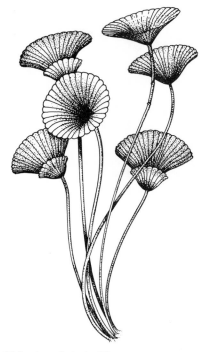

Fig. 10-7. *Acetabularia*. The cap, stalk, and rhizoidal base are parts of a single cell (×2).

a **conjugation tube**. Thereupon the contents of cells in one or both filaments become amoeboid. In those species where only one gamete (male) is active, fertilization takes place in the female gametangium. In those species where both gametes are active, fertilization occurs in the conjugation tube. Usually all cells within a given filament become gametes at the same time, and during conjugation they all migrate or remain stationary. A ladder-like structure forms as conjugation tubes connect cells of two parallel filaments. The zygotes ultimately released from the parent filament usually do not begin further development until the following spring. The zygote undergoes meiosis prior to germination. Three of the resulting haploid nuclei disintegrate and the fourth serves as the nucleus for the first cell of a new haploid filament.

The desmids, a group of freshwater organisms related to the filamentous *Spirogyra* just described, exhibit a variety of exotic shapes, many with bilateral symmetry (Fig.

10-10). Desmids may be solitary or may exist in filamentous or amorphous colonies. Reproduction is asexual (by cell division) or sexual (by conjugation). Their chloroplasts are of the spiral and star-shaped types. The desmids are thought to have evolved from more complex filamentous forms.

Evolutionary Trends in the Green Algae

The green algae illustrate several evolutionary trends that will be seen again in groups of higher plants. The primordial form from which all groups of green algae are thought to have evolved probably resembled the biflagellate, unicellular *Chlamydomonas*. The grouping together of such organisms probably resulted in the evolution of colonial forms resembling *Volvox*. Loss of flagella in aggregating cells could have led to stationary colonial forms. Other branches of the evolutionary tree culminated in the tubular Siphonales, the several filamentous groups of algae, and the flat, sheet-like thalli of the Ulvales. It is significant that green algae are the simplest plants exhibit-

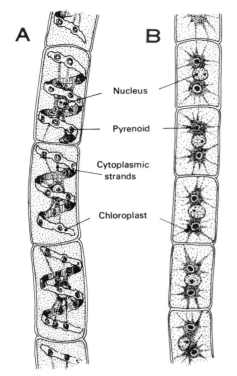

Nucleus

Pyrenoid

Cytoplasmic
strands

Chloroplast

Fig. 10-8. A, *Spirogyra* (×700); B, *Zygnema* (×360).

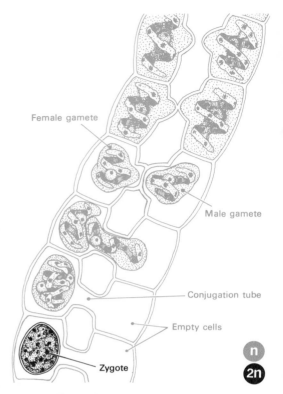

Female gamete

Male gamete

Conjugation tube

Empty cells

Zygote

Fig. 10-9. Conjugation in *Spirogyra.*

ing the phenomenon of alternation of generations, a life pattern that is encountered throughout the plant kingdom. The metabolic similarity of the green algae to higher plants (with identical photosynthetic pigments and products) provides further compelling evidence that the green algae represent the evolutionary base from which the higher plants evolved.

The Euglenoids
(Subdivision Euglenophyta)

The euglenoids, organisms claimed by both botanists and zoologists and often used in research, can be described most simply as unicellular organisms exhibiting some animal-like and some plant-like characteristics. Their motility, the absence of rigid walls, their ability to ingest some materials, and the presence of contractile vacuoles make them resemble many Protozoa. On the other hand, the presence of photosynthetic pigments located within chloroplasts is definitely a plant-like characteristic. The dilemma can be illustrated by the fresh-

water forms, *Euglena* and *Phacus* (Fig. 10-11). Although they are photosynthetic organisms, they differ from other chlorophyll-containing plants in that they require organic nutrients (usually vitamins) from the environment. Aside from their vitamin requirement they need only minerals when grown in the light. In the dark they must also be provided with a source of carbohydrate. Whether these organisms ought to be regarded as plants that have acquired animal-like characteristics or as animals that have acquired a degree of photosynthetic self-sufficiency is a moot question.

This is a rather small group of organisms with only 355 species belonging to approximately 25 genera. Because the photosynthetic euglenoids have plastid pigments (chlorophylls and carotenoids) identical to those found in the green algae and higher plants, it is tempting to include them among the green algae. However, the absence of a cellulose cell wall, accumulation of **paramylum** rather than starch as a product of photosynthesis, and the presence in some

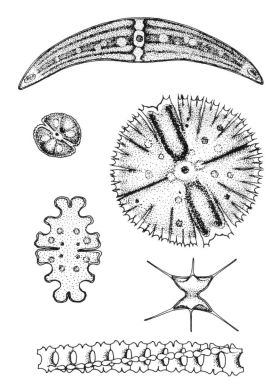

Fig. 10-10. Desmids exhibit great variety in their morphology (×400).

species of only a single flagellum are characteristics setting the euglenoids apart from the green algae.

Species of *Euglena* multiply by cell division. They can survive unfavorable conditions by forming thick-walled resting cells or cysts. These spindle-shaped organisms are propelled by a single anterior flagellum inserted in a tubular groove resembling the gullets found in the Protozoa. Each *Euglena* cell has a single central nucleus and many small ellipsoidal chloroplasts. The elastic plasma membrane in *Euglena* permits variations in cell shape.

Species of *Phacus* have a rigid plasma membrane (**periblast**) and, as a consequence, cell shape is fixed. Cells are often flattened and in some species may be slightly curved. Both *Euglena* and *Phacus* possess a pigmented, cup-shaped eyespot near the anterior end of the cell. This eyespot provides part of the mechanism by which these organisms are able to orient themselves with respect to light, a response called **phototaxis**. The other part of

the phototactic machinery is a light-sensitive photoreceptor located at the base of the flagellum. The eyespot acts as a shield that can cast a shadow on the photoreceptor when light comes from behind the eyespot. In *Euglena*, the eyespot is oriented so that the photoreceptor is continually illuminated as long as the cell is moving toward the light. If the direction of incident light is changed so that the shadow cast by the eyespot falls on the photoreceptor, *Euglena* turns so that the photosensitive material is again illuminated. Staying in the light is clearly an adaptive advantage for photosynthetic organisms.

The Stoneworts (Subdivision Charophyta)

The stoneworts are notable chiefly because their cells are so large that these organisms have been widely used in studies of membrane permeability and uptake of specific chemicals from the environment. In some species the cell volume is sufficiently large so that concentra-

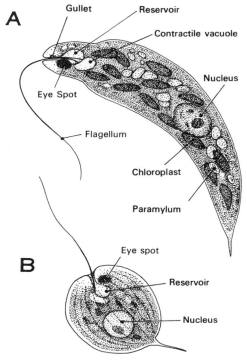

Fig. 10-11. Two representative euglenoids. A, *Euglena* (×900); B, *Phacus* (×1,150).

tions of absorbed substances in individual cells can be measured. The name stonewort (brittle-wort is also used) refers to the layers of calcium carbonate frequently accumulated by these algae. These are multicellular forms with an erect, branched structure. The vertical axis is divided into distinct nodes and internodes, each node bearing a whorl of branches of limited growth. The axis grows by formation of new cells at the apex. The plant is anchored to the substrate by branching filaments called **rhizoids.** Reproductive structures are borne on the plant's lateral branches. In many stonewort species complex multicellular male and female gametangia are formed on the same plant.

Two representative stoneworts found in many lakes and ponds are *Nitella* and *Chara* (Fig. 10-12). In these genera individual cells can be several millimeters in length.

The Brown Algae
(Subdivision Phaeophyta)

The brown algae owe their name and their color to the presence of a golden-brown carotenoid pigment, **fucoxanthin,** that is localized within the chloroplasts. The brown algae are all macroscopic and multicellular, literally the giants of their kind. Some of the ocean kelps attain lengths comparable to the height of many trees. They are found in great abundance in colder, shallow, offshore ocean waters and in the intertidal zone along rocky coastlines. The littoral species can withstand periods of desiccation and are regularly exposed to the atmosphere during low tides. Although the plant body is always attached to the bottom, the reproductive cells are motile and have two laterally inserted flagella of unequal length.

All of the brown algae are photosynthetic but they do not accumulate starch, the usual carbohydrate storage product of photosynthesis. Instead they accumulate **laminarin,** a related polysaccharide containing glucose. They also accumulate the relatively insoluble hexose, **mannitol,** and fat droplets.

Individual cells in the brown algae have prominent nuclei. During mitosis, centrosomes and astral rays appear. These structures, norm-ally seen during the division of animal cells, are

Fig. 10-12. Stoneworts. A, *Chara* (×½); B, *Nitella* (×½).

not usually found in plants. A cellulose cell wall is present and may be covered with **algin,** a polysaccharide composed of mannuronic acid residues. Algin can be removed by alkaline extraction and is one of the few algal products of commercial importance. It is used as an emulsifying agent in many products, including ice cream and puddings.

Ectocarpales

Although the brown algae are best known for their unusually large size (such as the kelps already mentioned) the first group that we will consider, the Ectocarpales, are all small algae living as epiphytes on larger marine algae. Members of the genus *Ectocarpus* are common along the Atlantic coast of the United States, where they are frequently found in the interti-dal zone growing as epiphytes on *Fucus,* another brown alga. The *Ectocarpus* plant body is composed of branched filaments differentiated into a flattened portion attached to the host and an erect tuft-like branching system. Reproductive structures are formed at the ends of short lateral branches (Fig. 10-13).

Kelps

The Laminariales, better known as **kelps**, are of interest because of their large size and complexity of structure. The sporophyte stage, often several meters long, is differentiated into a **holdfast**, a **stipe**, and a **blade**. The holdfast anchors the plant and can be either disk-shaped or made up of a system of forked branches resembling roots. The unbranched stipe, or vertical axis, can be cylindrical or flattened. The flattened blade is attached to the stipe laterally or terminally. Kelp blades can be

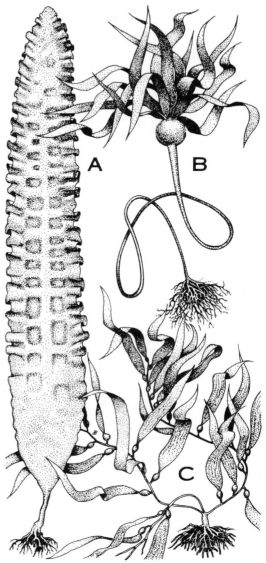

Fig. 10-14. Representative kelps. A, *Laminaria* (×1/5); B, *Nereocystis* (×1/30); C, *Macrocystis* (×1/15).

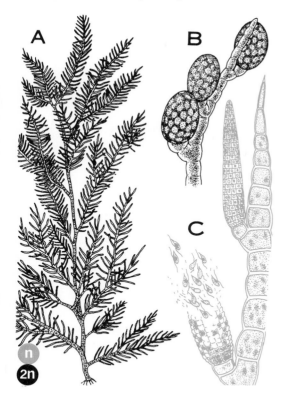

Fig. 10-13. The brown alga *Ectocarpus*. A, tuft-like branching system (×1); B, enlarged section of sporophyte filament with sporangia (×540); C, enlarged section of gametophyte filament with gametangia (×540).

simple or split longitudinally, with smooth or convoluted surfaces.

For a time the kelps were harvested commercially as a source of minerals such as potassium and iodine. The discovery of terrestrial deposits of these minerals a century ago made such mining of the sea unprofitable. However, numbers of kelp species are still harvested by the Japanese for a staple food called *kombu.*

Life cycles in the kelps resemble, in some ways, the pattern found in the ferns (where spores are formed in special structures borne on the undersurfaces of the leaves and gametophytes developing from the spores are small compared to the sporophyte stage). In species of *Laminaria,* clusters of sporangia develop on surfaces of the blades late in the growing seasc n. Motile spores released at high tide swarm for a time, and then become rounded and secrete a wall. On germination, a small multicellular gametophyte plant forms. Of the 32 or 64 spores released from a single sporangium, half develop into male and the rest into female gametophytes. Both kinds of gametophytes are small, but they can be distinguished either on the basis of the kind of sex organs formed or, prior to fruiting, by cell diameter, which is twice as large in the female gametophyte cells as in the male. Stages in the life cycle of *Laminaria* are illustrated in Figure 10-15. The diploid sporophyte plant body is large, complex, and perennial, while the haploid gametophyte is much reduced and relatively short-lived. The attachment of the sporophyte to the gametophyte during its early development is yet another characteristic of the kelps that is similar to the situation encountered in the ferns.

Fucus and Other Rockweeds

The Fucales are marine algae found attached to rocks in the intertidal zone. They have flattened thalli that are repeatedly branched into two equal parts. Such **dichotomous** branching can be observed in many diverse species of plants.

The rockweeds, unlike the giant kelps and the small epiphytic Ectocarpales, do not exhibit an alternation of haploid and diploid phases in their life cycles. They are like animals in being diploid with meiosis occurring during the formation of gametes. For some species, tides are a factor in the release of gametes. In these species, gelatinous masses of gametes are extruded during low tide as the thallus dries. As the tide returns, the gametes are released.

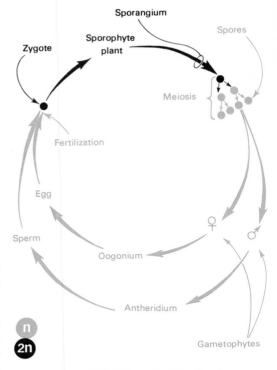

Fig. 10-15. Life cycle of *Laminaria.*

Immediately after fertilization, the zygote begins to develop into a new sporophyte plant.

The Red Algae (Subdivision Rhodophyta)

The red algae or sea mosses (Rhodophyta) are mostly marine, although there are some freshwater species. Their delicate and beautiful forms make them the darlings of amate ir seaweed collectors and their flavor the favorites of European and Oriental gourmets. Several species of Rhodophyta are the only algae that have been regarded as edible by Europeans. Irish moss (*Chondrus crispus*) is used to prepare a pudding type of dessert. *Porphyra,* used in soup by both Europeans and Japanese, is actually a cultivated crop in Japan. *Rhodymenia,* also called **dulse** or **sea kale,** is also used in soups or can be dried and eaten. Perhaps the most widespread commercial use of the red algae is the harvesting of species in the genus *Gelidium* as a source of **agar,** a solidifying agent

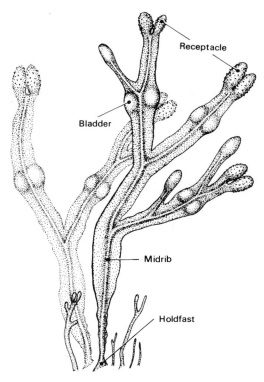

Fig. 10-16. The brown alga *Fucus* (×½).

to absorb the light that passes through the shallow water. Because quanta of light absorbed by the accessory pigments can be transferred to chlorophyll and utilized in photosynthesis, the red algae are able to survive in habitats unsuitable for other algae that require wavelengths of light available only near the surface of the water.

Most of the red algae are multicellular, although there are two genera with unicellular representatives. A unique feature of the Rhodophyta is that their gametes do not have flagella. In most species, morphologically similar gametophyte and sporophyte stages alternate in the life cycle. *Porphyra* (*porphyra* is the Greek word for purple) has an irridescent, purple-brown thallus, resembling the green algae, *Ulva,* in morphology. It grows attached to rocks in the intertidal zone. The flat thallus is one or two cells in thickness and cells are embedded in a gelatinous matrix. Porphyra can reproduce asexually by means of non-motile spores that form from vegetative cells within the haploid thallus. After liberation, spores germinate and develop into a filamentous stage that produces a new haploid thallus. Sexual reproduction also occurs regularly.

used in the preparation of media for laboratory culture of bacteria, fungi, and algae. In some coastal areas the red algae are used agriculturally as fodder, particularly for sheep.

The coralline algae, or those **Rhodophyta** that accumulate calcium carbonate, play an important role in land building. It has been estimated that their contribution to reef formation is in fact greater than that of the coral animals, which usually get all the credit for this activity.

In addition to the usual photosynthetic pigments, a red pigment, **phycoerythrin,** is found in the plastids of these plants. In some species there is also a blue pigment, **phycocyanin.** We have previously encountered these pigments (known collectively as **phycobilins**) in the Cyanophyceae, or blue-green algae (Chapter 4). There is an ecological advantage in the array of accessory pigments found in the red algae. These algae are usually found in deeper water than the green algae. The phycobilins are able

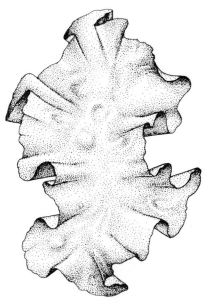

Fig. 10-17. The red alga *Porphyra* (×½).

Yellow-Green Algae, Golden-Brown Algae, and Diatoms (Subdivision Chrysophyta)

There are three groups of Chrysophyta: the yellow-green algae, (Xanthophyceae), golden-brown algae (Chrysophyceae), and diatoms (Bacillariophyceae). The large amounts of carotenoids present in their plastids cause the characteristic colors of these algae. Food reserves accumulating from photosynthesis are never stored as starch in these organisms but instead as oils or as **leucosin**, an insoluble carbohydrate. In the Chrysophyta are found groups of algae comparable in many ways to certain green algae. In both subdivisions there are organisms with unbranched filaments, flagellated or non-flagellated single cells, and forms with coenocytic (multinucleate) cells.

The Yellow-Green Algae (Xanthophyceae)

The yellow-green algae are usually found in fresh water, although a few species are marine. Some are aerial and can be found growing on tree trunks, walls, other plants, or soil. In many genera the cells have walls made up of two overlapping halves, much like the bottom and cover of a box. In filamentous genera, the bottom half of the wall of each cell is joined to the top half of the wall of the cell below, forming an H-shaped wall segment in which the crossbar of the H forms the crosswall of the filament.

The Golden-Brown Algae (Chrysophyceae)

Cells of the golden-brown algae may be either motile and flagellated or immobile; naked or with a cell wall; solitary or in colonies. These predominantly freshwater species are most abundant when the water is cool. Golden-brown algae are animal-like in some respects. There are amoeboid species that ingest solid foods. Reproduction is asexual by cell division in unicellular species or by proliferation from fragments of colonial species. In some genera, a unique kind of spore called a **statospore** (or **cyst**) may form. Such cysts have walls made up of two overlapping halves with a small pore closed by a plug. The silica-containing walls of the statospores are often ornately sculptured. On germination of the statospore, the plug separates from the rest of the wall, the spore moves out of its enclosure by amoeboid movement, developing flagella during or after this migration.

The Diatoms (Bacillarophyceae)

The **diatoms** are usually single-celled organisms but sometimes they form filaments or colonies. Found in all fresh and salt waters, they exhibit a multitude of forms. Among the 170 genera there are over 5,000 known species with rod, disc, wedge, triangular, and other assorted shapes. Vegetative cells lack flagella and have walls with overlapping halves, sometimes called **valves**. Walls are composed of pectic materials impregnated with silica and usually are ornamented with bilaterally or radially symmetrical markings.

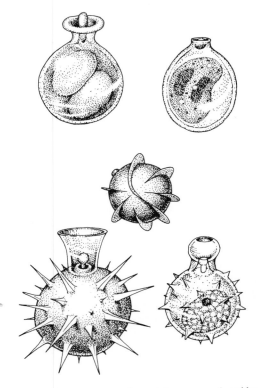

Fig. 10-18. Representative statospores of golden-brown algae. Walls are often ornately sculptured.

The silicaceous materials of the cell walls of the diatoms frequently remain intact after cell death and decay. Deposits of these empty walls can accumulate on ocean and lake bottoms. Great deposits of fossil diatoms, known as diatomaceous earth, have been discovered in various parts of the world. Although these are remains of marine species, they are now found inland as a result of geological changes in the earth's crust. The thickest known deposits (about 3,000 feet thick) are in the Santa Maria oil fields of California. Other deposits have been found in the southeastern United States, in Russia, and in Spain. These deposits are mined and are used in a number of industrial processes. Diatomaceous earth has been used to filter liquids in sugar and oil refineries. Formed

into bricks, or as a loose powder, it is a very effective insulating material for use in boilers and blast furnaces. Another use has been as an abrasive in toothpastes and in metal polishes. Diatomaceous earth has also been used as an additive to concrete and as a lightweight building material.

The diatoms reproduce asexually by cell division. Each daughter protoplast forms half of the cell wall and uses as the other half one of the valves of the parent cell. As each of the new valves is secreted inside an old one, the daughter cells are slightly smaller than the parent cell. Progeny might therefore be expected to become smaller and smaller, but this does not occur. The valves apparently can increase in size slightly. Furthermore, during spore formation, protoplasts escape from the cell walls and increase in size. Diatoms formed from such **auxospores** have the maximum size characteristic of the species. Auxospores can also act as gametes. Under these circumstances, meiosis occurs before the release of the protoplasts and diploidy is restored when the protoplasts unite to form a zygote.

The classification of the diatoms is based largely on the nature of their wall markings which are either radially or bilaterally symmetrical. The Centrales, or centric, diatoms are radially symmetrical, whereas the Pennales, or pennate, diatoms are bilaterally symmetrical (Fig. 10-20).

Cryptomonads and Dinoflagellates (Subdivision Pyrrophyta)

The cryptomonads and dinoflagellates resemble the Euglenophyta in having both green and colorless species and in being claimed by both zoologists and botanists. Most of the pyrrophyta are unicellular and biflagellate. Cell walls, when present, contain cellulose and those species that are photosynthetic accumulate starch.

The dinoflagellates are unicellular marine plankton often extremely abundant and brilliantly luminescent. Some, such as *Gonidium* and *Gonyaulax*, periodically become so abundant that their pigments color the coastal waters yellow, red, or olive-green. The factors controlling the sudden appearance of such "blooms" called "red tides" are not known.

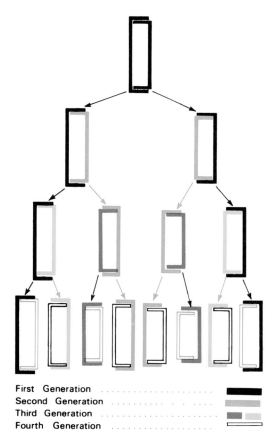

First Generation .... ▬▬▬
Second Generation .... ▬▬▬
Third Generation .... ■ ▬
Fourth Generation .... ▭

Fig. 10-19. In successive cell divisions, diatoms secrete new valves inside an old one derived from parent cell. Thus, daughter cells become smaller.

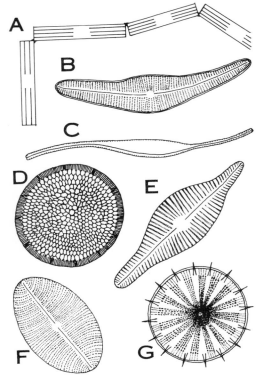

Fig. 10-20. Representative diatoms. A, B, C, E, and F are bilaterally symmetrical or pennate; D and G are radially symmetrical or centrate.

Even during periods when the numbers of dinoflagellates are much reduced, they are ingested by shellfish that are somehow immune and can isolate the poison produced by these algae. Thus shellfish that feed on certain dinoflagellates can become hazardous to humans. Victims of dinoflagellate poisoning can be paralyzed (indicating that this is a nerve poison). Unfortunately, no antidote is known. Although the poison is of no apparent advantage to the dinoflagellates, it can be a factor in producing blooms. The poison kills fish which then disintegrate, releasing nutrients which can promote the production of more dinoflagellates. Many primitive peoples had elaborate taboos regulating how and when fish and shellfish could be eaten. Indians who lived along the Pacific coast of North America posted guards to prevent the taking of fish and shellfish during parts of the year when the likelihood of poisoning was the greatest.

Most dinoflagellates have cell walls made up

of interlocking plates which may be covered with spines or ridges. Frequently there is a groove, called the **girdle**, circling the central part of the cell. Dinoflagellates are either autotrophic or heterotrophic. Those species without plastids are dependent on their environment for a source of food.

Most of the cryptomonads are motile single cells but a few form immobile colonies. *Cryptomonas* is a typical genus that resembles the euglenoids in cellular organization. An animal-like characteristic of this genus is that the cell surface near the base of the flagella is invaginated to form a large gullet. A contractile vacuole appears to discharge its contents into the gullet. *Cryptomonas* has two plastids oriented longitudinally at the edge of the cytoplasm.

THE FUNGI

In contrast to the nutritionally independent or autotrophic algae, the one most distinguishing feature of the fungi is their lack of chlorophyll and their heterotrophic nutritional

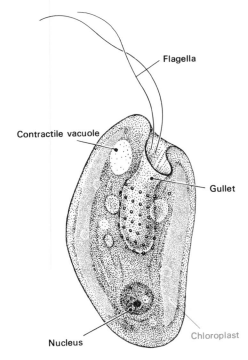

Fig. 10-21. *Cryptomonas*, a unicellular cryptomonad resembling the euglenoids.

status. In other respects, the fungi are as diverse in structure, physiology, and life patterns as the algae. Like the algae, the subdivisions of the fungi are thought to represent independent lines of evolutionary development and their origin is unknown. The fungi may have been derived from algae that lost their photosynthetic capacity. Some similarities between the algae and fungi in their vegetative and reproductive structures support this hypothesis. On the other hand, many experts consider at least one fungal subdivision, the slime molds (Myxomycophyta) to be more closely related to the simplest animals, the protozoans.

The Slime Molds
(Subdivision Myxomycophyta)

Slime molds have little in common with the rest of the plant kingdom. Their most plant-like characteristic is the production of spores that resemble those produced by other fungi.

The vegetative body of slime molds, the **plasmodium**, is frequently a naked mass of protoplasm containing many nuclei. It can grow to many square centimeters in area, creeping slowly over damp rotting logs or other material by a flowing amoeboid movement in response to factors such as light, temperature, and food supply. Most of slime molds are **saprophytes**; that is, they obtain nutrients from dead and decaying organisms in their environment. A few species are parasitic on other fungi or on higher plants.

Life cycles in the slime molds generally follow a pattern in which a period of vegetative growth precedes the formation of complex and beautiful reproductive structures. This unusual separation of growth and differentiation has made the slime molds very useful as experimental organisms for studies of these aspects of development (see Chapter 9).

There are three classes of slime molds: the Myxomycetes, the Acrasiomycetes and the Plasmodiomycetes.

Myxomycetes

Physarum is a genus typical of the Myxomycetes. The amoeboid plasmodia of *Physarum* are composed of naked multinucleate protoplasm and are colored (yellow or blue). Their mode of nutrition is partially saprophytic.

Under unfavorable conditions the plasmodia can contract to form dormant **sclerotia** which are able to survive periods of extreme temperature or desiccation. Plasmodia re-form from the sclerotia when conditions are again favorable. In *Physarum,* as well as other slime molds, light (particularly the blue portion of the spectrum) is required to induce the formation of sporangia. Some fruiting bodies typical of

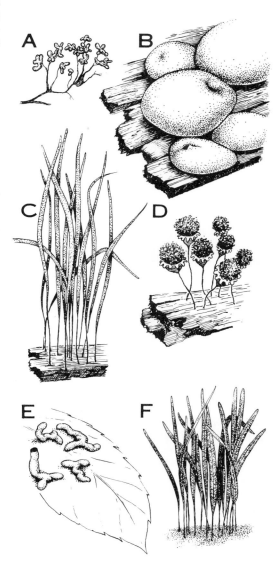

Fig. 10-22. Fruiting bodies of representative slime molds. A, *Physarum polycephalum;* B, *Lycogala* sp.; C, *Stemonitis* sp.; D, *Hemitrichia;* E, *Physarum alpinum;* F, *Stemonitis splendens.* All × 6.

these organisms are illustrated in Figure 10-22. These slime molds are often found on lawns and compost.

Acrasiomycetes

Slime molds belonging to the class Acrasiomycetes are abundant yet rarely seen in nature because of their inconspicuous structures. The genus *Dictyostelium* found in this group is well known because of its use as a research tool. Its life cycle and use in studies of growth and differentiation were discussed in Chapter 9.

Plasmodiomycetes

In the Plasmodiomycetes, we find the only slime molds that are parasites. Their usual hosts are higher plants, although species parasitizing algae and other fungi are known. They may cause abnormal enlargement of the host cells (**hypertrophy**) and abnormally rapid cell division (**hyperplasia**). Little is known about the substances produced by the parasite and released into host cells or how they cause enlargement of infected parts of the host. It would seem that this plant-fungus system could provide information potentially useful in studies of tumors and cancers.

Species that are of economic importance are the organisms causing the "powdery scab" disease of potatoes and "club root" disease of cabbage. The latter parasite is located within the root cells of the host plant during most of its life cycle. When the host plant dies and its roots disintegrate, haploid resting spores that can remain viable for many years are released into the soil.

The Algal Fungi (Subdivision Phycomycophyta)

The algal fungi (Phycomycophyta) resemble the green algae in their structural organization and modes of reproduction. In fact they have been regarded by some experts as green algae that have lost their chlorophyll. The parasitic forms of algal fungi are among the most destructive parasites of crop plants: mildews and blights can affect large agricultural areas while others can cause spoilage of fruit and vegetable crops in transit and in storage.

An example of the potential impact on human affairs of a plant disease is the potato blight caused by a species of Phycomycophyta. The Irish famine of the mid-19th century resulted from crop destruction by this organism. Two circumstances resulted in unbelievable misery and disruption. First, economic conditions forced the population of Ireland to rely on easily grown, productive, and filling potatoes for their staple food; second, the potato crops were almost completely destroyed by late blight in 1845 and 1846. Between 1845 and 1860, Ireland lost almost a third of its population as a direct consequence of this plant disease. A million people died from starvation or from diseases complicated by malnutrition. Another million and a half people emigrated, many of them to the United States.

Like the green algae, most of the algal fungi live in water or in soil. They are either saprophytic or parasitic. Some species that are obligate parasites on seed plants spend their entire life cycles within the host.

The algal fungi include some single-celled organisms but most are multicellular with plant bodies made up of masses of filaments called **hyphae**. The vegetative plant body is called the **mycelium**. All of the algal fungi can reproduce asexually by means of spores. Aquatic species form motile spores bearing cilia or flagella. Terrestrial species form spores lacking cilia or flagella which are dispersed by air currents. During a typical life cycle, the growing mycelium produces sporangia in which asexual spores are formed. After their release, these spores germinate to form hyphae that develop into a new mycelium. Such a cycle of asexual reproduction can occur repeatedly during the growing season.

Chytrids

The chytrids are a unicellular group of algal fungi. They lack a true mycelium, but in some species, rhizoidal extensions of the base of the cell may be extensive enough to superficially resemble a true mycelium. Chytrids are found in the water (where they may be parasitic on algae) and in the soil (where they are parasitic on higher fungi and on seed plants). *Rhizophydium,* one of the simplest unicellular chytrids, grows parasitically on the green alga *Spirogyra*.

Blastocladiales

A second group of algal fungi, the Blastocladiales, like the chytrids are found both in water and soil. Some are obligate parasites on insects. Others are parasitic on worms and other small animals. *Allomyces* grows saprophytically on dead insects or other animal remains. The plant body in *Allomyces* is a true mycelium consisting of branched rhizoids (that attach the fungus to its substrate), a trunk-like portion, and many side branches on which the reproductive structures are formed. The hyphae are coenocytic (lacking cell walls between nuclei) and cell walls contain chitin, a rare component of plant cells but a plentiful constituent of the insects on which these fungi live.

Saprolegniales

A third group of algal fungi, the water molds, (Saprolegniales) are abundant forms that can be either saprophytic or parasitic on fish and

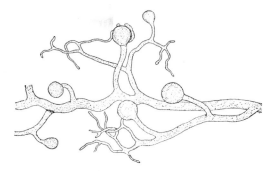

Fig. 10-24. The water mold *Saprolegnia* (×107). (From G. M. Smith: *Cryptogamic Botany,* Vol. I: Algae and Fungi. McGraw-Hill Book Co., New York, 1938.)

amphibians. Species such as *Saprolegnia*, a common parasite on freshwater fish, can cause considerable damage in fish hatcheries or in specimens kept in aquariums.

Peronosporales

A fourth group of algal fungi, the Peronosporales, are destructive parasites of agricultural plants. These fungi cause damping off of seedlings. The white rusts and downy mildews exhibit an unusual specialization in their parasitism. A given species usually can infect only a small group of closely related host plants, and some species can be distinguished only in the hosts that they infect. For each host there is a specific **biological form**, morphologically identical to other strains of the same parasitic species but differing in its capacity to infect the various potential hosts. The infection of host plants occurs when spores germinate on the wet surfaces of leaves and send out germ tubes that penetrate the host cells and absorb nutrients from their protoplasts.

Mucorales

The Mucorales, also called black molds because of the presence of melanin pigments in their spores, are mostly terrestrial saprophytes found on dead organic matter such as decaying plant or animal remains and dung. Some species are economically important because they grow on fruits, vegetables, and other foodstuffs such as bread (*e.g., Rhizopus nigricans*). Many of

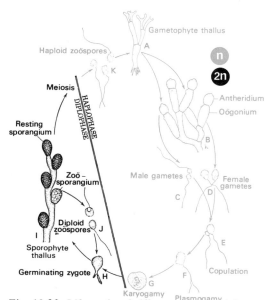

Fig. 10-23. Life cycle of *Allomyces*, an algal fungus. (From C. J. Alexopoulos: *Introductory Mycology.* John Wiley & Sons, Inc., New York, 1952. Artwork by Mrs. Sun Huang Sung.)

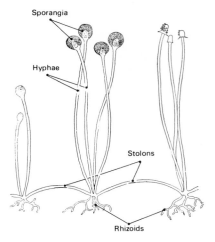

Fig. 10-25. The black mold *Rhizopus nigricans* (×25). (From G. M. Smith: *Cryptogamic Botany,* Vol. I; Algae and Fungi. McGraw-Hill Book Co., New York, 1938.)

these algal fungi play an important role in the decay of plant and animal bodies and their wastes. They thus aid in maintaining soil fertility as well as in the cycling of chemical components of the biosphere. There has been some use of these organisms in industrial fermentations to produce alcohols, fumaric acid (used to make resins and plasticizers) and lactic acid (used in preparing hides before tanning), and as a mordant for dyes in the textile industry.

Yeasts, Molds, and Mildews (Subdivision Ascomycophyta)

There are 30,000 species of yeasts, black molds, and green molds. One universal characteristic of the Ascomycophyta is the formation of a reproductive structure called the **ascus,** a sac-like hypha in which eight **ascospores** develop. Another universal characteristic is the absence of motile cells.

These fungi range in size from microscopic single-celled organisms (the yeasts) to macroscopic forms. Some are parasites, but most are saprophytic. The black molds (*Aspergillus*) and green molds (*Penicillium*) can cause damage to food, leather, and fabrics. Others cause many diseases of agricultural crops, timber, and ornamental plants. Some of these plant diseases (such as chestnut blight and Dutch elm disease)

have been so severe and widespread that they have practically eliminated certain plant species. Although they cause a number of plant diseases and some animal diseases, such as the respiratory ailment aspergillosis, most of the Ascomycophyta are beneficial to mankind. Records of the earliest civilizations indicate that yeasts were used to ferment alcoholic beverages. This was undoubtedly the first use of microorganisms in the processing of plant materials for human consumption. The leavening of bread, a second traditional use of the yeasts, originated much later, about 6,000 years ago, in Egypt. Certain species of *Penicillium* are used in the production of cheeses such as Roquefort and Camembert. In the 1940's, the antibiotic properties of some fungal products were discovered. Since that time, a large industry has been developed for the preparation of penicillin and other antibiotics. Other fermentation industries based on these organisms produce a wide array of products including alcohols, organic acids, and vitamins. No discussion of the beneficial uses of these organisms would be complete without mentioning the edible morels and truffles which are considered by gourmets to be without equal in flavor.

The Ascomycophyta reproduce both sexually and asexually. These two distinct reproductive phases are sometimes referred to as the **perfect stage** (when sexual reproduction occurs and asci are formed) and the **imperfect stage** (when asexual reproduction occurs by means of spores called **conidia**). Additional means of asexual reproduction include fragmentation, fission, and budding.

Yeasts

The yeasts are unicellular Ascomycophyta that usually do not form a mycelium. In the Schizosaccharomycetes, found on grapes and other fruits, the vegetative yeast cells are haploid and the zygote, formed when two vegetative cells come in contact, is the only diploid cell in the life cycle. In the Saccharomycodes it is the vegetative cells that are diploid and the ascospores, which function as gametes, are the only haploid cells. A third pattern is found in the Saccharomycetes, a genus including brewer's yeast. Here, both haploid and diploid cells reproduce asexually by budding.

Haploid cells fuse to form diploid cells that ultimately are converted into asci containing four haploid ascospores. On germination, each ascospore can give rise to a population of haploid cells.

The usefulness of yeasts in the brewing, baking, and fermentation industries has already been noted. A few yeasts are parasitic on higher plants and several can cause human diseases. In one such chronic infection, called blasto-mycosis, the skin and central nervous system are attacked.

Blue, Brown, and Green molds

The Aspergillales, described variously as blue, brown, or green molds, are the organisms responsible for moldiness and deterioration of leather, foodstuffs, and clothing in warm, humid climates. The two common genera, *Aspergillus* and *Penicillium*, may be either exceedingly harmful or beneficial to mankind. Species of *Aspergillus* cause decay and spoilage of many foodstuffs, deterioration of fabrics and leather, and certain human and animal diseases. But these organisms are also widely used in the manufacture of alcohols and organic acids. Species of *Penicillium* cause decay of fruits and bread and discoloring and deterioration of paper, lumber, and leather. However, *P. roque-forti* and *P. camemberti* are responsible for the characteristic flavors of cheeses. Another spe-cies is the source of the antibiotic **penicillin**. This bactericidal drug has been used in the treatment of many infectious diseases including pneumonia, syphilis, and ear and throat infec-tions.

Sexual reproduction is a rare occurrence in both *Aspergillus* and *Penicillium* and propaga-tion is usually by means of conidia. The conidial stage of *Aspergillus* is illustrated in Figure 10-26.

Pink Molds

The best known genus of the Sphaeriales (pink molds) is *Neurospora*, often called pink bread mold, although the species frequently used in research is orange and grows wild on sugar cane. Species in this genus have been used in genetic studies in part because they grow and reproduce rapidly, but especially because they

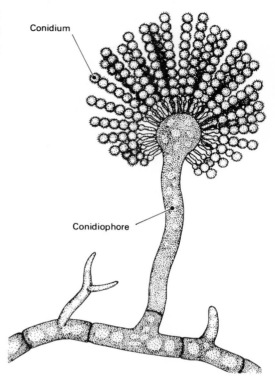

Fig. 10-26. Conidial stage of *Aspergillus*, an ascomy-cete. Conidia are asexual spores (× 340).

are haploid. Consequently, genotype and phenotype are always the same and any mutation is immediately expressed. Wild type *Neurospora* can be grown in the laboratory on a relatively simple, chemically defined medium containing mineral salts, sugar, and the vitamin biotin. Certain biochemical mutants cannot grow on this simple "minimal" medium because they have lost the ability to synthesize certain enzymes. Such mutants can grow if the products of these missing enzymes are added to the minimal medium. For example, mutant *Neurospora* strains unable to make the enzyme tryptophan synthetase (and therefore unable to grow on minimal medium) will grow as rapidly as the wild type in the presence of added tryptophan. Biochemical mutants can be char-acterized by determining what supplement must be added to the medium for growth to occur. Many of our basic ideas about the genetic control of enzyme synthesis have come from studies with this organism.

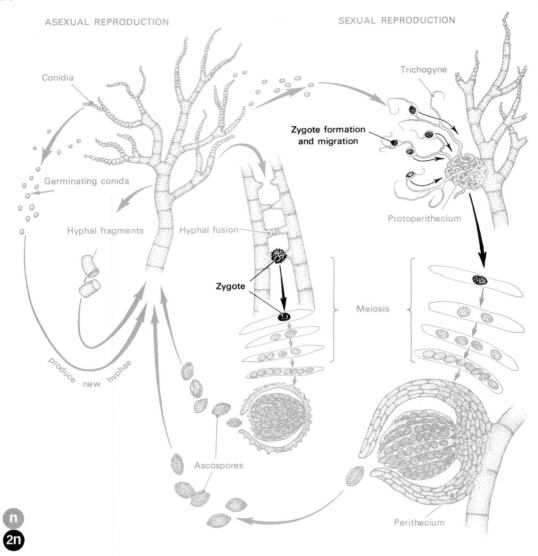

Fig. 10-27. Life cycle of *Neurospora*.

The mycelium of *Neurospora* consists of a mass of branched hyphae. Conidiophores that resemble vegetative hyphae are borne on the upper parts of the mycelium and produce chains of pink conidia. Although *Neurospora* is **hermaphroditic,** *i.e.,* both male and female gametangia are formed on the same mycelium, it is self-sterile. Sexual reproduction occurs only when two strains of different mating types are brought together. When this occurs, the hyphae fuse and form a mycelium called a **heterokaryon** which contains nuclei of both strains.

At times conidia can act as male gametes. Such conidia, coming in contact with a hypha serving as the female parent, will contribute their nuclei to heterokaryon formation and set into motion the formation of perithecia. Perithecia are dark colored because of the presence of melanins. They contain many asci, each developing eight ascospores. Young asci have two nuclei, one descending from each of the parent strains. Fusion of these two nuclei after their characteristic prolonged association in the dikaryon is followed by three divisions during which meiosis occurs. At maturity, the eight haploid ascospores from each ascus are discharged from the flask-shaped perithecium through a small aperture at its apex. On germination each ascospore can form a new mycelium that will grow vegetatively and produce conidia. Although such mycelia are capable of prolific asexual reproduction by means of conidiospores, sexual reproduction will occur only if the mycelium comes into contact with a second mycelium of appropriate mating type. Four of the eight ascospores of each ascus are of one mating type while the other four give rise to mycelia of the opposite type. In genetic studies the ripe perithecia are broken open under a dissecting microscope and each ascospore within an ascus is isolated and grown separately so that the genotype of the resulting mycelium can be ascertained. The use of *Neurospora crassa* in genetic studies has been discussed in Chapter 7.

Powdery Mildews

The powdery mildews (Erysiphales) are parasites whose hyphae invade and absorb nutrients from the epidermal cells of the leaves and stems of flowering plants. They are named for the mealy or powdery patches they form on the surfaces of infected parts. Among the higher plant species attacked by these fungi are lilacs, roses, and cereal grains. The powdery mildews are obligate parasites that can exist only on specific host species. Fortunately, infection by these mildews causes relatively little tissue damage.

Cup or Sac Fungi

The most conspicuous feature of a fourth group of Ascomycophyta, the cup or sac fungi (Pezizales), is the presence of a large, fleshy, often brightly colored fruiting body. Most are saprophytic, living on dead wood, soil, humus, and dung. Their fruiting bodies may be cup-shaped or sponge-like structures borne on stalks. The best known genus is *Morchella*, which contains the edible species known as morels.

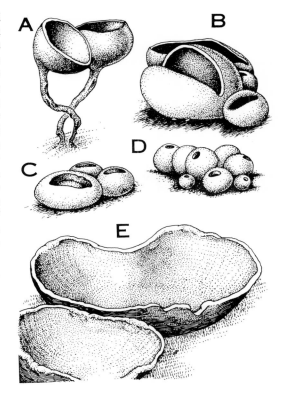

Fig. 10-28. Representative cup and sac fungi. A, *Sclerotinia* (life size); B, *Peziza* (×3); C, *Humaria* (×5); D, *Pyronema* (×6); E, *Sarcoscypha* (×3).

Hypocreales

The Hypocreales are a large and heterogeneous group of Ascomycophyta, many of which have brightly colored perithecia with soft or waxy walls. In this order is the genus *Claviceps* which is parasitic on grasses and certain cereal grains. *Claviceps purpurea* causes a disease called ergot which results in rye grains becoming enlarged, purple in color, and filled with hyphae. The fungus synthesizes a number of alkaloids, toxic to animals and man, which cause a condition called **ergotism**. Cattle can be poisoned by grazing on grasses infected with *Claviceps*. In the past, it was not uncommon for humans to die from ergot poisoning in areas where rye bread was a staple part of the diet. Although modern methods of preparing rye flour have reduced the incidence of ergotism, mass poisonings still occur occasionally.

Truffles

Another minor order of Ascomycophyta, the Tuberales, are commonly called truffles. These edible fungi grow beneath the soil surface. They are highly prized delicacies in Europe and are gathered for market with the help of trained dogs or pigs. These animals are able to detect them by their superior sense of smell and are taught to dig them out of the ground.

Mushrooms, Toadstools, Puffballs, Bracket Fungi, Smuts, and Rusts (Subdivision Basidiomycophyta)

The Basidiomycophyta are divided into two classes: the rusts and smuts. (Hemibasidiomycetes) and the mushrooms and puffballs (Holobasidiomycetes). These most advanced fungi

Fig. 10-29. White pine tree infected with blister rust. Currants and gooseberries serve as the alternate host for this disease-causing fungus. Elimination of all currants and gooseberries has been the method of control in areas where white pines are an economically important crop. (Courtesy U. S. Dept. of Agriculture.)

produce a special reproductive structure called the **basidium**. This club-shaped cell usually forms four **basidiospores**.

Among the Basidiomycophyta are many harmful as well as useful species. The rusts and smuts cause millions of dollars of damage annually to corn and wheat crops. Some of the Basidiomycophyta attack trees or destroy wood products. This type of destruction is especially troublesome in the humid tropics. Certain mushrooms are edible and the cultivation of one species, *Agaricus campestris*, has become a large industry in parts of the United States. Certain wild species have superior flavor but their collection for human consumption is something best left to experts because a number of poisonous species closely resemble edible ones.

In addition to toxic substances, certain mushrooms contain hallucinogenic compounds. Some of the ancient Indian civilizations of Mexico and the southwestern United States used these hallucinogenic mushrooms in their religious rites. The active principles in some fungi have been analyzed chemically, for example, psilocybine and psilocine. Our understanding of the biological effects of such naturally occurring compounds on the mind and body is still fragmentary.

Hemibasidiomycetes

Rusts. The rusts (Uredinales) are parasites that form reddish-brown spores on the stem and leaf surfaces of vascular plants. One peculiarity of the rusts is that frequently two unrelated host plants are successively parasitized. For example, the wheat rust *Puccinia graminis* alternately infects a cereal grain and the common barberry, and both hosts must be present for completion of the life cycle. The cedar-apple rust alternates in its life cycle between cedars and apple or hawthorne trees. White pine blister rust parasitizes the white pine during part of its life cycle and wild gooseberries or currants during other stages. Any of these rusts which require two hosts can be controlled by systematic elimination of one of the alternate hosts.

The wheat rust fungus *Puccinia graminis* illustrates well the complexity of life cycles in the rusts. During the summer, hyphae of this

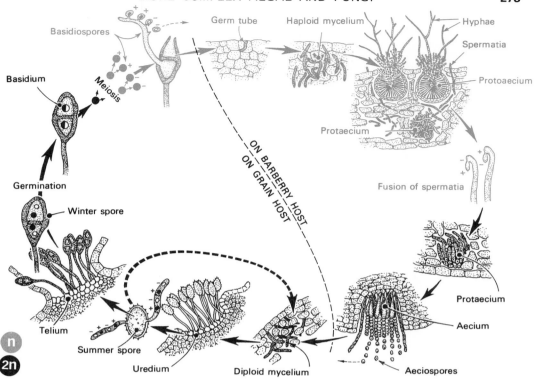

Fig. 10-30. Life cycle of the wheat rust *Puccinia graminis*.

fungus invade the stem and leaf tissues of wheat plants, producing surface blisters in which are formed many unicellular summer spores. These are carried by wind and can infect other wheat plants. Late in the growing season, the hyphae form two-celled winter spores that are resistant to low temperatures and remain dormant within the wheat straw. In the spring, the winter spores germinate to form basidia, each of which produces four basidiospores. The basidiospores are also carried by wind and can germinate only on the leaves of the common barberry. The resulting hyphae form flask-shaped gametangia on the upper surfaces of the barberry leaves. Small cells called **spermatia**, formed within the gametangia, ooze out onto the leaf surfaces in small droplets of nectar that attract insects. The spermatia formed on a given hypha are either plus or minus mating type and, after being carried by insects to the surface of another infected leaf, can fuse only with spermatia of the opposite kind. Following fusion, a new mycelium forms which produces small cup-like structures on the lower surfaces

of the barberry leaves. These structures, called **aecia** (singular, aecium) form **aeciospores** that, in turn, are carried by wind to young wheat plants which they infect. Hyphae within infected wheat plants produce summer spores during the growing season, and thus the cycle is repeated. Wheat rust can be effectively controlled in cold climates by eliminating barberry plants, thus removing the host required for germination of the over-wintering spores. This means of control is not effective in mild climates, however, because the summer spores can survive and reinfect young wheat plants during the following spring..

Smuts. The smuts (Ustilaginales) are parasitic on seed grains, including corn, oats, barley, wheat, rice, and sorghum. The corn smut *Ustilago zeae* causes enlarged black outgrowths, especially of the ovary tissues (ears). These sooty masses of spores are what give smuts their name. The mass of spores, remains of the sterile hyphae, and host cells finally rupture the epidermis of the host, whereupon the spores are released and carried by wind. They can germi-

Fig. 10-31. Smut-infested corn. (Courtesy U. S. Dept. of Agriculture.)

nate immediately or remain dormant until the next growing season. On germination, meiosis occurs and a basidial hypha is formed that divides to form four linearly arranged cells, each with a haploid nucleus. The resulting basidiospores can germinate if they fall on the epidermis of leaves of appropriate developing host plants. Germination tubes penetrate the leaf surface and form intracellular mycelia. Mycelia of two appropriate strains must infect the same host tissue for fusions of hyphae to take place. After fusion occurs, the binucleate mycelium enlarges and a gall-like growth is produced by host cell enlargement within the infected region. The protoplasts of the short-celled hyphae then round up, their walls disintegrate, and each protoplast secretes a thick wall. The resulting spores are spherical and frequently have wall markings character-istic of the species.

Jelly and Ear Fungi. Two minor orders of the basidium-forming fungi are the jelly fungi (Tremellales) and ear fungi (Auriculariales). Neither group is of economic importance but

they provide examples of the diversity of form observable among the fungi. Fruiting bodies of the jelly fungi vary in shape from crust-like to stalked, and from cushion-like structures with a wrinkled surface or leaf-like folds. Some species of the genus *Tremella* are used as food by the Chinese. The ear fungi are a diverse group including saprophytes and parasites on plants and on insects. Fruiting bodies vary in size and complexity from rather simple web-like masses of hyphae to larger gelatinous, leathery struc-tures.

Mushrooms, Puffballs, and Shelf Fungi (Holobasidiomycetes)

The best known kinds of fungi—the mush-rooms, puffballs, and shelf fungi—belong to the second class of Basidiomycophyta, the Holo-basidiomycetes. It is not even necessary to leave the city to see these forms. They are frequently found growing in lawns of city parks and on golf greens. Most of the mushrooms and their

Fig. 10-32. Ear and jelly fungi. A, *Auricularia*, an ear fungus (×½); B, *Tremella*, a jelly fungus (life size).

relatives are saprophytic, deriving their nutrients from decomposing organic matter such
as leaves, dead bark, and dung.

Mushrooms and Toadstools. The gilled fungi
(Agaricales) are the order to which both edible
"mushrooms" and inedible "toadstools" belong. The portion of the fungus that we
normally see above ground is its fruiting body,
or **basidiocarp**, which appears after development of a mycelium below the surface. Usually
at a particular time of year (according to the
species), localized areas of hyphal growth form
masses of compacted hyphae called "buttons."
Mature fruiting bodies are formed from these
button stages by uptake of water. The resulting
basidiocarp consists of a stalk and a broad,
unbrella-like cap. Thin plates of radially arranged hyphal tissue, called **gills**, are found on
the undersurface of the cap. The basidia are
formed on the surface layer of the gills where
they are found mixed with sterile cells. Many
billions of spores are released from the gills of a
large mushroom during the several days of
discharge. The pattern of spore discharge can be
demonstrated by placing a mushroom cap on a
white piece of paper and covering it with a bell
jar. After several days a distinctive spore print
can be seen on the paper.

Basidiospores can germinate immediately
after liberation, forming a mycelium of uninucleate cells. In the commercial cultivation of
mushrooms, such mycelia inoculated into compost mixtures of soil, leaf mold, and manure are
sold as mushroom "spawn."

The best known of the edible mushrooms is
Agaricus campestris (Fig. 10-33). A genus of
poisonous forms, *Amanita*, unfortunately has
numerous representatives that closely resemble
the edible *Agaricus*. Today, there is growing
interest in collecting edible mushrooms in the
field. Many superstitions have flourished about
ways to distinguish between edible and poisonous forms. Some say that silverware or onions
will be blackened by toxic species; others give
generalizations about the color of the gills.
Unfortunately, none of these rules has any
scientific basis and the only safe procedure is to
learn to identify individual edible species.

Pore Fungi. In contrast to the gill fungi, the
pore fungi, or Polyphorales, have many tiny
tubes, or pores, instead of gills on the underside

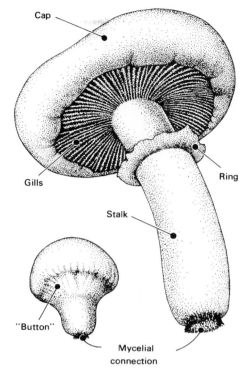

Fig. 10-33. Young "button" and mature stages of
edible mushroom, *Agaricus campestris* (life size).

of the fruiting body. Basidia are formed in the
tissues around the pores and the basidiospores
are released through these openings. Some pore
fungi are found in soils but most live as
parasites or saprophytes on the trunks of trees,
forming shelf- or bracket-like outgrowths. In
species growing on trees, the vegetative mycelium penetrates the bark and wood by
excreting digestive enzymes. As tissues of the
host are broken down, nutrients are released
that are absorbed by the fungus. The vegetative
mycelium usually develops extensively within
the host tissues before fruiting bodies are
developed. Once formed, the basidiocarp may
be annual or perennial. Pore fungi can cause
costly damage to forest timber and stored
lumber.

The pore fungi whose fruiting bodies remain
closed during spore development (Gastero-
mycetes) include puffballs and stinkhorns.

Puffballs often grow to great size. The

fruiting bodies of the giant puffball, *Calvatia gigantea*, can be as large as 50 to 100 cm. in length. This species is frequently found in lawns or grassy fields. All the puffballs are edible and are best for eating when the center is white (before spores form). At maturity, the surface of the fruiting body becomes dry and leathery and the center is filled with powdery spores.

The stinkhorns, or Phallales, include many beautifully shaped and colored fungi that are usually not greatly admired because of their very unpleasant odor. The odor is caused by the autodigestion of part of the fruiting body wall. The resulting foul-smelling gelatinous matrix attracts flies and other insects, which help to

disseminate the spores that stick to their mouth parts and bodies.

Origin of the Basidiomycophyta and Their Relationship to Other Fungi

The Basidiomycophyta resemble the Ascomycophyta more closely than any other group of fungi. The structure of their hyphae and the development of asci and basidia have many similarities. Most experts agree that the Basidiomycophyta evolved from the Ascomycophyta.

The origin of the Ascomycophyta and indeed of the fungi in general is still open to question. The groups of organisms that have been suggested as ancestral forms from which the fungi could have evolved include the Actinomycetes, a group of filamentous bacteria, the Algae (particularly the Rhodophyta, or the red algae), and the Protozoa. It has even been suggested that the fungi evolved independently from any other group of plants or animals.

Fungi Imperfecti

The fungi imperfecti are organisms known to reproduce only asexually. Because their sexual stages are either unknown or nonexistent, they cannot be assigned to any of the groups of fungi already described. Even though they are unclassifiable, the fungi imperfecti are of considerable importance to man. Many are parasitic on plants, animals, and humans. They cause a variety of human diseases including ringworm, athelete's foot, and an array of skin and pulmonary infections.

LICHENS

There are about 15,000 species of lichens. In these symbiotic associations of an alga plus a fungus, the fungus is dependent on its algal partner for photosynthetic products, while the alga benefits from the water and possibly mineral nutrients absorbed by the fungus. The algal components of lichens (usually either blue-green or green algae) can be grown in the absence of their fungal partners. However, the fungal components of lichens (usually Ascomycophyta) apparently cannot survive in the absence of the appropriate alga.

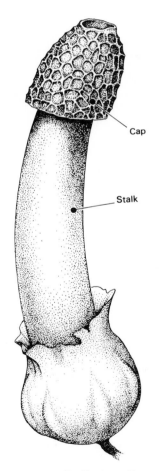

Cap

Stalk

Fig. 10-34. A stinkhorn, *Phallus impudicus*, representative of the Phallales (life size).

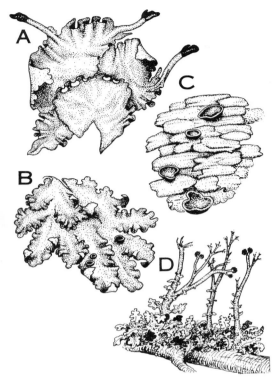

Fig. 10-35. Lichens. A, *Peltigera,* and B, *Parmelia,* both foliose lichens (×½); C, *Xanthoria,* a crustose lichen (×2); Ð, *Cladonia,* a fruiticose lichen (×½).

Lichens are found on the surfaces of rocks, trees, soil, and wooden structures. They can be quite colorful, with gray-green, white, yellow, orange, yellow-green, brown, and black forms known. Lichen bodies exist in several shapes. **Crustose** lichens are hard, flat forms often growing in the surfaces of rocks. **Foliose** lichens

are flattened, leaf-like forms and **fruiticose** lichens are usually erect, branched structures. The best known fruiticose lichen is "reindeer moss," *Cladonia rangifera,* a lichen that is the principal food of the reindeer, caribou, musk ox, and other animals inhabiting the tundra. Although sometimes used as fodder for animals, most of the lichens have a bitter taste and have very rarely been used as food for humans. There are reports of Iceland "moss" being eaten by Arctic explorers and it is thought that the biblical "manna" may have been a lichen, *Lecanora esculenta,* that occurs in desert areas.

Lichens can reproduce by fragmentation or by special bodies called **soredia**. These are disc- or cup-shaped structures containing one or more algal cells surrounded by fungal hyphae. The algal and fungal components of lichens can also reproduce independently but it seems unlikely that such means are important in the propagation of these symbiotic forms..

USEFUL REFERENCES

Alexopoulos, C. J. *Introductory Mycology.* John Wiley & Sons, Inc., New York, 1971.

Bold, H. C. *Morphology of Plants,* 2nd ed., Harper & Row, Publishers, Inc., New York, 1967.

Christensen, C. M. *The Molds and Man: An Introduction to the Fungi,* 3rd ed. McGraw-Hill Book Co., New York, 1965.

Dawson, E. Y. *How To Know the Seaweeds.* Wm. C. Brown Co., Dubuque, Iowa, 1956.

Doyle, W. T. *Non-Vascular Plants: Form and Function.* Wadsworth Publishing Co., Inc., Belmont, Calif., 1964.

Prescott, G. W. *The Algae: A Review.* Houghton Mifflin Co., Boston, 1968.

Webster, J. *Introduction to Fungi.* Cambridge University Press, London, 1970.

Conquest of the Land: Liverworts and Mosses

11

In emerging from the primeval seas where they arose, living organisms faced a variety of problems peculiar to existence on the land. The first prerequisite was protection from the ultraviolet radiation of the sun. This was provided by a shield of ozone that absorbed most of this lethal radiation. Since ozone is formed from atmospheric oxygen it could not have formed until after photosynthesis appeared and algae became abundant in the ocean. Before that time the atmosphere of our planet was anaerobic, without free oxygen.

The aqueous environment in which life originated is relatively unchallenging. There is an ever-present supply of water, a requirement for all life; temperatures are relatively constant, changing slowly with the seasons; the aqueous medium provides dissolved nutrients; water is buoyant and thus eliminates the need for elaborate supporting structures; and gametes can be released without the need for protective coverings to guard against desiccation.

REQUIREMENTS FOR TERRESTRIAL LIFE

Living things inhabiting terrestrial environments have developed special ways of coping with conditions such as the constant danger of desiccation, restricted availability of water and nutrients, rapid shifts in temperature, a hostile environment for gametes and young embryos, and a greatly increased gravity problem. It is worth noting that the first organisms to survive on land must have been plants. Before animals could follow, the continents had to be populated with the vegetation that is basic to all food chains.

If we examine terrestrial plants, we can readily see the kinds of modifications that make it possible for them to survive. Land plants are generally more massive, with a smaller surface relative to volume than their aquatic ancestors. Greater bulk contributes both supporting and water storage tissues, while reduced surface retards water loss. Almost all land plants are anchored to the soil, which supplies both water and inorganic nutrients. Exposed surfaces are covered with a protective waterproof layer, often made of waxy material. An exchange of gases with the environment must take place through the outer layers of the aerial parts of land plants because they must not only import CO_2 for photosynthesis and O_2 for respiration, but also release the gaseous byproducts of these processes. Gas exchanges are facilitated, and rates of water loss can be regulated, by pores (actually valves known as stomates), in the surfaces of the aerial parts of land plants.

A protected embryonic development is another feature that sets terrestrial plants apart from aquatic plants. In fact, the presence or absence of this adaptation has been used as the basis for dividing the plant kingdom into two subkingdoms: a largely aquatic group which includes all the eukaryotic algae and fungi discussed in Chapter 10, and a primarily terrestrial group called the **Embryophyta**, which includes all other plants from the relatively simple mosses to the most complex seed plants.

The single outstanding characteristic of the Embryophyta, development of the embryo within the female sex organs, provides the same kind of protective and nutritional advantages that we find in the mammals. A layer of cells in the outermost part of the reproductive structures protects against desiccation of the developing gametes and the zygote that results from fertilization. Another universal characteristic of the Embryophyta is that the female gametes, or eggs, are non-motile.

The more highly evolved groups of Embryophyta have a complex water-conducting system that also provides rigidity and support. The presence or absence of conducting tissues (xylem and phloem) is the basis for dividing the embryo-forming plants into two groups: the nonvascular and the vascular plants. In this chapter we will consider the two divisions of nonvascular plants: liverworts (Hepatophyta) and mosses (Bryophyta). These are considered to be not only the most primitive of the groups of plants adapted to existence on land, but also the simplest of the embryo-forming plants.

LIVERWORTS AND HORNWORTS (DIVISION HEPATOPHYTA)

An important (but not universal) characteristic of the liverworts and hornworts (Hepatophyta) is the flattened thallus of the gametophyte stage that resembles the lobed structures of an animal liver. In all, about 9,000 species of liverworts and hornworts have been recognized. They are usually found in very moist habitats growing on soil, rocks, or the bark of trees.

Root-like **rhizoids**, found on the ventral (under) side of the gametophyte thallus, are of the simplest possible structure, *i.e.,* unicellular,

as opposed to the multicellular rhizoids of the mosses. Reproductive structures develop either terminally on the gametophyte thallus or on the dorsal (upper) side.

These plants exhibit a clear alternation of generations in which the gametophyte stage is free living and dominant. The sporophyte is always attached to the gametophyte and largely dependent upon it for water, mineral nutrients, and, in some cases, even the complex organic compounds produced by photosynthesis.

The Liverworts (Hepadopsida)

Gametophyte liverwort plants are flat, dichotomously branched, and prostrate. Although simple in appearance, there is a surprising degree of internal differentiation of the cells. In *Marchantia,* the thallus has a thickened central midrib where cells are elongate, lack chloroplasts, and have thickened cell walls. Surface cells elsewhere on the thallus are arranged in a hexagonal pattern which can be seen without magnification. With a hand lens, it is possible to see the pore located at the center of each hexagonal area. This pore is the opening from an internal chamber lined with photosynthetic cells containing chloroplasts. Although the apertures of the pores cannot change, they provide for a rapid exchange of gases between cells within the thallus and the atmosphere. Thus in function they appear to be comparable to the stomates found on the leaves of higher plants. Layers of cells ventral to the air chambers usually lack chloroplasts. On the

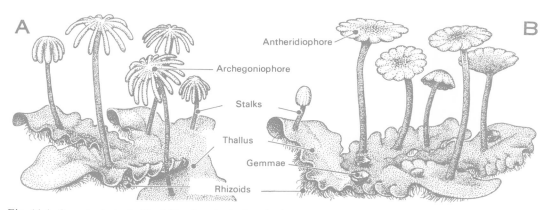

Fig. 11-1. Gametophyte thalli of *Marchantia.* A, female thallus with archegoniophores (life size); B, male thallus with antheridiophores (life size).

lower surface of the thallus there are two or more rows of scales and rhizoids. The latter are root-like in function, anchoring the plant and absorbing water and nutrients from the soil.

Marchantia plants grow apically, forming new branches only at their tips, while the posterior end disintegrates. Thus the plant moves along the ground, continually dying behind and growing ahead. New and separate plants are formed when disintegration of the posterior end reaches a branch point. A second, more elaborate means of asexual reproduction is also found in *Marchantia*. Multicellular structures called **gemmae** can develop within cup-shaped outgrowths on the dorsal surface of the thallus. Mature gemmae are shaped like notched or indented discs and are attached to the gametophyte by a short basal stalk. They possess no dorsiventral differentiation and most of their cells contain chloroplasts. When mature, gemmae are released and dispersed by wind or water. Symmetry becomes fixed when the gemmae settle on a suitable substrate. At this time, rhizoids develop from colorless cells on the surface next to the soil. Then meristems (regions of cell division located in the two indentations) begin to develop into two new thalli and the central portion of the gemma

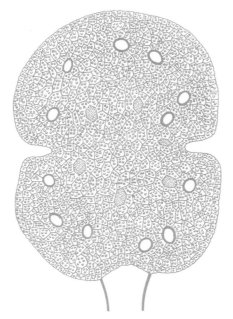

Fig. 11-2. Gemma of *Marchantia* (× 257).

begins to disintegrate, thereby forming two new gametophyte plants. Gemmae are prevented from germinating while still attached to the parent thallus by auxins synthesized in the apical growing regions of the vegetative plant. Auxin (indoleacetic acid) is a hormone involved in many aspects of plant development including tropisms (bending growths in response to environmental factors such as gravity and light), shoot elongation, and apical dominance (see Chapter 17). In these lowly, primitive plants the same hormone is present that controls the growth of flowers and trees. Once again the unity and economy in design of living organisms can be observed.

The gametophyte thalli of *Marchantia* are either male or female. Thus, during sexual reproduction, the structures in which gametes are formed (the **gametangiophores**) are borne on separate plants. These unique, umbrella-shaped, gamete-producing branches appear to be modifications of the dichotomously branching vegetative thallus; they form rhizoids in grooves along their stalks; and on the upper surfaces of the caps of mature gametangiophores there are air chambers similar to those on the dorsal surfaces of the vegetative thalli.

Archegonia, the structures in which eggs are formed, develop on the dorsal surfaces of female gametangiophores. They are flask-shaped, with enlarged round bases and long slender necks. A single egg develops in the base of each. During maturation, the upper surface of the cap of the female gametangiophore expands rapidly, so that at maturity the archegonia are located, necks pointing downward, on the lower surface. Cells of the neck canal tend to disintegrate when the egg matures, forming a sticky fluid environment through which the male gametes must swim to reach the egg. Numerous sterile outgrowths on the undersurface of the cap of the female gametangiophore separate the rows of archegonia.

Antheridia, the structures in which male gametes are formed, are borne on the dorsal surface of the cap-like top of the male gametangiophores. They are located in radial rows below the surface. At maturity, a narrow canal develops which connects the chamber in which each antheridium is housed with the surface. Thousands of biflagellate **antherozoids**

are formed in each antheridium and are discharged in a gelatinous mass as free swimming motile male gametes. These gametes are attracted toward specific chemicals, particularly certain proteins, potassium salts, and organic acids such as malic acid. Thus, the entrance of the male gametes into the neck of the archegonium is probably a chemotactic re-

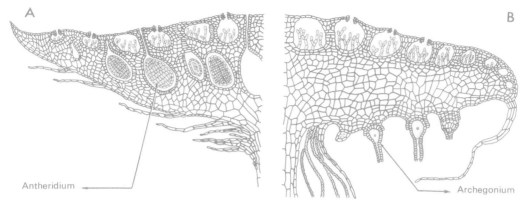

Fig. 11-3. Detailed structure of gametangiophores of *Marchantia*. A, male antheridiophore (×111); B, female archegoniophore (×111).

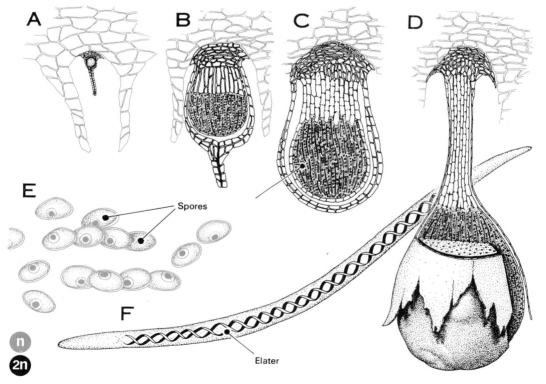

Fig. 11-4. A through D, stages of sporophyte development in *Marchantia*. Note that diploid phase develops within the venter of the archegonium and remains attached to the female gametangiophore (×12½). E, enlarged view of spores (×700). F, enlarged view of elater (×700).

sponse. This contrasts with what is known about animal fertilization where chemical sperm attractants probably do not exist. Union of the antherozoid and egg not only produces a diploid cell, the zygote, that marks the beginning of the sporophyte generation, but fertilization also provides the stimulus for cells surrounding the base of the archegonium to proliferate. The resulting protective sheath formed from female gametophyte cells surrounds the developing sporophyte until it is nearly mature.

The diploid sporophyte develops within the **venter**, or base, of the archegonium. Usually only a few of the eggs produced in a female gametangiophore develop into sporophytes following fertilization because of their nutritional and spatial needs. During early development, cells at the base of the sporophyte enlarge and penetrate the base of the venter. The resulting **foot** anchors the developing diploid plant and facilitates the absorption of nutrients. Two additional parts soon differentiate—a cylindrical **seta** and a terminal **capsule** or **sporangium**. Cells within the capsule are initially all alike but later differentiate into elongate cells and rows of spherical cells. These spherical cells function as spore mother cells and undergo meiosis, producing haploid spores. The elongate cells, or **elaters**, form spiral thickenings in their cell walls. Their twisting movements assist in spore dispersal in response to changes in moisture content of the atmosphere.

When mature spores have been formed, the seta elongates, pushing the capsule out through its protecting layers of gametophyte tissue. The outer layer of the capsule splits open **(dehisces)**, exposing masses of spores and elaters. The haploid spores ejected by the coiling and uncoiling of the elaters are carried by air currents. Spores falling on a damp soil surface germinate and form new male or female gametophyte plants. Of the four spores formed from a single spore mother cell, two develop into male and two into female gemetophyte plants.

Although cells of the *Marchantia* sporophyte contain chloroplasts, and the diploid stage is therefore at least potentially photoautotrophic, it relies completely on the gametophyte for water and inorganic nutrients. The extent to

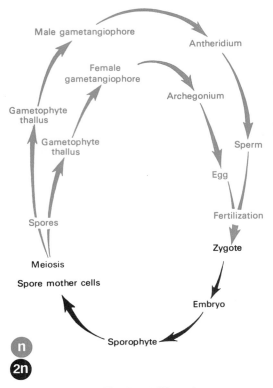

Fig. 11-5. *Marchantia* life cycle.

which organic nutrients are also provided by the haploid phase is uncertain. In all probability the utilization of complex organic substrates provided by the gametophyte plant is essential at least during the early development of the sporophyte.

Horned Liverworts (Anthoceropsida)

Anthoceros is the best known genus of horned liverworts (Anthoceropsida). The dark green, thallose gametophytes of *Anthoceros* inhabit moist soil or rocks in the temperate zone and superficially resemble the non-leafy liverworts. However, there is no regular dichotomy in the horned liverworts and their only distinctive morphological feature is a slender, cylindrical sporophyte.

This insignificant and readily overlooked group of plants is of interest because of a feature ordinarily associated with flowering plants, the control of sexual reproduction by photoperiod. The formation of gametes is

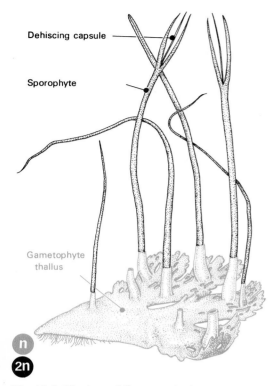

Dehiscing capsule

Sporophyte

Gametophyte thallus

n

2n

Fig. 11-6. The horned liverwort *Anthoceros;* gametophyte thallus with attached sporophytes (× 1½).

to changes in atmospheric moisture. Dehiscence continues as additional spores mature, a process that continues for some time. As in the liverworts, *Anthoceros* produces male and female spores in equal numbers.

Asexual reproduction by progressive growth and death is less frequent in the horned liverworts than in the liverworts. Gemmae are not formed, but during prolonged desiccation marginal thickenings may form on the thallus where the superficial cells develop a waterproof corky layer. These structures have been called "tubers" and develop into new gametophyte plants when favorable conditions again prevail.

MOSSES (DIVISION BRYOPHYTA)

There is much confusion about the term "moss" among non-botanists. The term is most often applied incorrectly to various algae, especially species such as *Protococcus,* which is found on tree trunks in moist areas, to "Irish moss" (a marine red alga), and to "reindeer moss" (a lichen composed of an alga and fungus living together symbiotically). The mosses are generally inconspicuous and of little direct human importance. The dense growths of sphagnum moss that occur in some boggy areas provide one of the only bryophyte "crops" harvested commercially. The early New England settlers used this sphagnum in its dried form to line cradles and to provide an absorbent lining for diapers. Today this species, which can absorb and retain large quantities of water, is sometimes used by nurseries as a packing material for trees and shrubs. When dried, it is sold for use as a mulch and soil conditioner. Deposits of **peat**, sphagnum that has accumulated and decomposed under the soil surface are also used as mulch. In countries such as Scotland and Ireland peat still has limited use, as it has had in the past, as a fuel.

Although of relatively minor importance to the human economy, the mosses can be of considerable significance in the natural environment. These small plants are frequently found growing on the trunks of trees, on rocks, and on poor soil. Today, as they were millions of years ago, these are often "pioneer" species invading areas disturbed by earthmoving equipment or areas where most other plant species cannot survive. Some kinds of mosses, the

initiated in the fall of the year at a time of decreasing day length. Both male and female gametangia are formed in pits on the upper surface of the thallus on the same haploid plant. The fusion of gametes produces a zygote which is the first cell of the sporophyte phase. A lobed foot is formed at the base of the sporophyte and its cylindrical stalk is surrounded by tissues formed from the upward growth of the gametophyte thallus. The sporophyte grows for several weeks by continued meristematic activity of cells at its base. The cells in its outer layers are photosynthetic, and stomates resembling those found in higher plants are found in the epidermis. The long slender capsule that forms at the apex of the sporophyte begins to dehisce at its tip, forming two longitudinal slits. As the outer layers of the capsule curl back, spores and multicellular elaters attached to the central column become exposed. The dissemination of spores is assisted by movements of valves and elaters in response so-called "copper mosses," can flourish in soils having concentrations of copper or antimony too high to be tolerated by most plants. Because of their ability to survive extreme conditions and to form dense stands where other plants are unable to grow, the mosses sometimes appear to have very restricted habitats. However, rather than reflecting unusual requirements, this indicates unusual tolerance. We do not see mosses in more favorable environments because they have been crowded out by their more complex and better adapted descendents. With about 600 genera and 14,000 species, the mosses are more numerous and have a broader distribution than the liverworts. They can easily be found in woods, even in lawns on acid or poor soils, and can be the dominant form of plant life in alpine and arctic habitats.

In the mosses, as in the liverworts, the gametophyte phase dominates the life cycle. However, two stages are evident in the moss gametophyte. Following germination of the spore, a simple filamentous or thallous structure called a **protonema** is formed which later gives rise to upright, leafy branches called gametophores that bear the sex organs. Usually the protonemata disintegrate once the erect branches are formed. Other morphological characteristics help set apart the mosses and liverworts. Mosses have multicellular rhizoids, their gametangia are borne on longer stalks, and, because of a lack of elaters in the sporophyte phase, complex mechanisms usually exist for opening the mature capsules and releasing spores.

The Peat Mosses (Sphagnopsida)

The peat mosses are ordinarily restricted to waterlogged areas with acid soil. In spite of this apparently restricted distribution, they can form massive growths several feet thick. Such accumulations on the surface of bogs are often sufficiently thick to walk on and to provide a solid substratum for the invasion of species of land plants. These are the mosses that form peat.

In *Sphagnum,* a representative peat moss, the adult gametophyte plant has an upright axis with whorls of branches. Its "leaves" are made up of two kinds of cells, large colorless cells

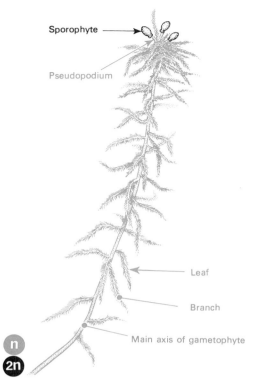

Fig. 11-7. Gametophyte of the peat moss *Sphagnum* with attached sporophytes (life size).

with thickened walls and spiral markings that die as they reach maturity and small photosynthetic cells derived from and located on either side of the larger dead cells, which are porous and can store huge quantities of water. An abundance of colorless cells produces the pale green color of the mature plants.

Vegetative, or asexual, reproduction in *Sphagnum* results from the disintegration of older parts, which causes branches to separate and form new individual plants. *Sphagnum* species may be either monoecious or dioecious. Antheridia are formed in the axils of leaves near the tips of lateral branches that are inserted close to the apex of the mature gametophyte. Antheridial branches are often readily distinguishable by their pigmentation which can be red, purple, or brown. The male gametes have two flagella of equal length and, as in the liverworts, an aqueous environment is needed for the motile male gametes to reach the egg following their release from the antheridia.

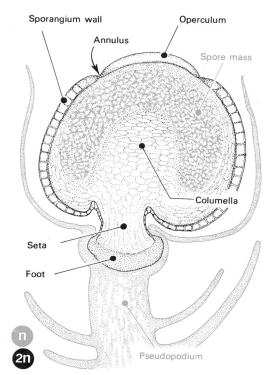

Fig. 11-8. Detail of *Sphagnum* sporophyte (×30).

The zygote formed in fertilization develops into a sporophyte having the same parts (**foot, seta,** and **capsule**) seen in the liverworts. A peculiarity in *Sphagnum* is that the capsule is raised above the gametophyte plant, not by elongation of the seta, but rather by enlargement of a portion of the female gametophyte where the archegonia were formed. This leafless axis is called the **pseudopodium** (an unfortunate word choice because of the very different meaning in zoological terminology). The developing capsule is covered with leaves of the female gametophyte and the remains of the archegonium until just before spores are released. At maturity, spores are shed explosively when the upper part of the capsule, the **operculum,** is torn loose. Spores can be scattered through distances as great as 10 cm. by this explosive discharge. Spores germinate, forming first a filamentous and then a thallose protonema. Finally, a leafy shoot, or gametophore, is formed and the protonema disintegrates.

The Granite Mosses (Andreaopsida)

The granite mosses (Andreaopsida), as their common name implies, are found on the surfaces of rocks. They are rarely over 1 cm. in height and are very restricted in their distribution. *Andrea,* a genus of tiny black or dark brown mosses found on rocks in cold climates, is limited to arctic and alpine regions.

The True Mosses (Mnionopsida)

The Mnionopsida, or "true mosses," are the most numerous and widely distributed of the Bryophyta, with three orders, over 600 genera, and some 14,000 species. Although most species are restricted to moist habitats, some can be found in very dry places and others are able to grow while submerged in aquatic environments. Terrestrial species are found in soil, on rocks and the bark of trees, and on wood. Some of the true mosses are perennial, forming dense growths each year over large areas. Annual species of the temperate zone usually are seen developing during the fall and winter.

As in other bryophytes, there are two distinct phases in development of the gametophyte in the true mosses. An inconspicuous, branching, filamentous protonema develops that can spread over an area more than 40 cm. in diameter. Some of the branches of this protonema penetrate the substratum, covering exposed soil and preventing erosion. Cells of the superficial branches are photosynthetic and contain lens-shaped plastids. The shape of the

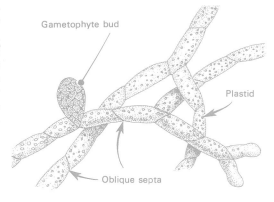

Fig. 11-9. Moss protonema (×80).

plastids and the oblique end walls of cells in the filament make the moss protonemata readily distinguishable from terrestrial green algae living in the same habitat.

The more conspicuous stage of the true moss gametophyte is the erect, leafy adult. Leaves are usually only one cell in thickness except in the central midrib region. They are attached to the axis in spirals, although in some species they tend to lie all in one plane. The vertical axis is usually simple, but within this group we find the first zonation of differentiated cells in concentric rings. Such an arrangement of specialized cells appears to represent rudimentary vascular and supporting tissues.

Asexual reproduction in the true mosses occurs by means of vegetative propagation from almost any part of the gametophyte (regenera-

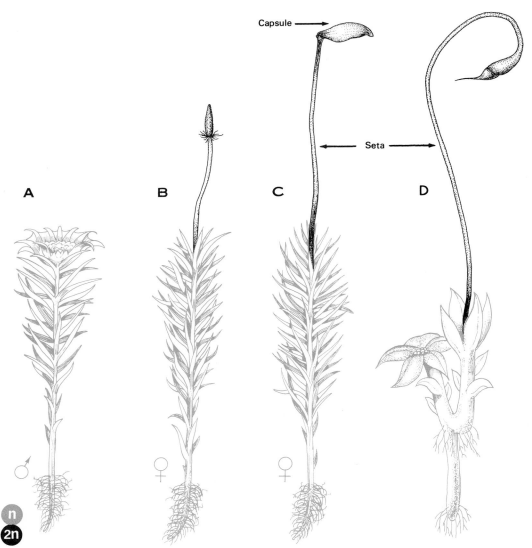

Fig. 11-10. The "true mosses." A, *Polytricum,* male gametophyte; B, *Polytricum,* female gametophyte with young sporophyte; C, *Polytricum,* female gametophyte with mature sporophyte; D, *Funaria,* female gametophyte with mature sporophyte. All × 2½.

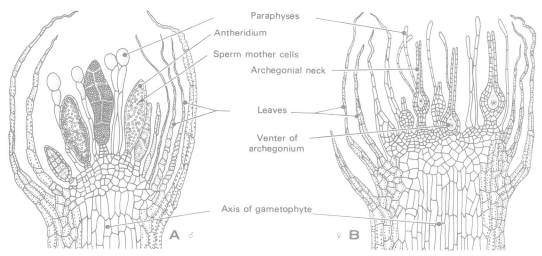

Fig. 11-11. Details of moss gametangia. A, male gametophyte with terminal antheridia (×80); B, female gametophyte with terminal archegonia (×80).

tion of an entire new plant is possible from leaf, stem, or rhizoid) or by production of gemmae. Some species seem to reproduce only by asexual means as the sporophyte phase is almost never in evidence. During sexual reproduction in monoecious species, gametangia are borne on the same or separate gametophores; many dioecious species are also known. There is a tendency for gametangia to be clustered and interspersed with many sterile hairs (**paraphyses**). The clusters of gametangia are often surrounded by a whorl of leaves. Sex organs can be formed either terminally on erect axes or laterally along creeping forms.

The true moss sporophyte develops immediately after fertilization, forming a foot, elongate seta, and a capsule. An unusual aspect of sporophyte development is the formation of a persistent protective layer that surrounds the capsule. This layer is made up of cells originating in the venter of the archegonium, and it appears to be essential for normal development of the capsule. The developing sporophyte has chlorophyll-containing cells and is at least potentially photosynthetic. However, it receives its supply of inorganic nutrients and water from the gametophyte by absorption through the foot, and it is probable that the gametophyte provides some complex organic nutrients as well. The sporophyte is little more than a slender cylinder until the elongation of the seta

is complete. At this time, development of the capsule begins. The apical portion forms a sterile, cap-like lid called the **operculum** which

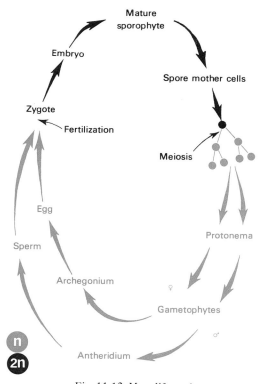

Fig. 11-12. Moss life cycle.

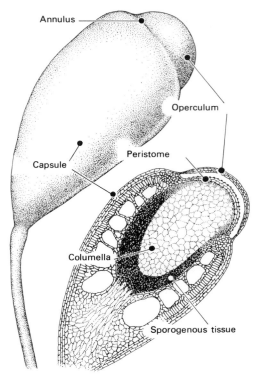

Fig. 11-13. Sporophyte capsule of the moss *Funaria* (× 45).

separates from the rest of the capsule when the ring of cells in the **annulus** at its base dries out. Under the operculum is a structure called the **peristome** (either a ring of tooth-like structures or a porous plate) that functions in spore dispersal.

EVOLUTIONARY ADVANCES AND SIGNIFICANCE OF LIVERWORTS AND MOSSES

Students of evolution are fascinated by the liverworts and mosses because of their obvious advances over the algae, their intermediate status between organisms adapted to existence in water and those adapted to terrestrial life, and the pattern of their life cycles in which alternation of generations achieves increased importance. In the liverworts and mosses we find not only that the gametophyte and sporophyte stages in the life cycle are distinctly different, but also that the haploid, gamete-producing plant is dominant, while the diploid,

spore-producing phase is nutritionally dependent on the sexual gametophyte.

The liverworts and mosses differ from other land plants in the absence of vascular, or conducting, tissues. Their lack of xylem and phloem, the specialized tissues involved in water and food transport, is probably a factor accounting for their relatively small size. Another factor influencing their size is a dependence on water as the medium through which the motile male gametes must travel to the egg before fertilization can occur.

Although primitive in many ways, the liverworts and mosses exhibit a number of advances over their algal ancestors including gamete-producing structures, gametangia, surrounded by protective tissues; airborne spores; an increasing tendency for the diploid sporophyte to be photosynthetic; and the development of structures resembling the stems and leaves of higher plants.

It is interesting to note that the presence of sex chromosomes in plants was demonstrated for the first time in a species of liverwort (*Sphaerocarpus.*) In both the liverworts and mosses, the gametophytes tend to be dioecious, *i.e.,* with separate male and female plants. In *Sphaerocarpus* the nuclei of female gametophyte cells, which, of course, are haploid, contain one large *X* chromosome while male gametophyte cells have a small *Y* chromosome. In addition to the *X* or *Y* sex chromosome, there are seven autosomes. Not only do we encounter sex chromosomes in these primitive plants, but also phenomena that we associate with the more advanced seed plants: photoperiodic control of reproduction and control of development by the widely distributed plant hormone auxin.

A general trend of evolution in the plant kingdom is toward greater size and independence of the sporophyte with an accompanying reduction in the gametophyte. Does the number of sets of chromosomes dictate the pattern of development? How does it influence cell metabolism? We should remember that there are organisms such as the common alga "sea lettuce," or *Ulva*, with morphologically identical haploid and diploid plant bodies. In such plants, there is no evidence of biochemical differences between vegetative cells of the gametophyte and sporophyte. The real difference in algae with morphologically identical haploid and diploid plant bodies occurs when the gametophyte follows the instructions in the genetic code for producing gametangia and gametes, and when special cells of the sporophyte follow the instructions to undergo meiosis and form spores.

There are special circumstances when gamete-producing and spore-producing plants having the "wrong" chromosome number are formed. We find examples of such diploid gametophytes and haploid or polyploid sporophytes in the liverworts and mosses. Under some conditions a tiny piece of a sporophyte plant can develop into a diploid **protonema** (the initial phase in the development of a haploid plant from a spore). This gives rise to diploid gametophyte and, following fertilization, to tetraploid sporophytes. Also, structures that look like the sporophyte can develop from the egg without fertilization. The resulting or-

ganism, although like the sporophyte phase in appearance, is haploid or monoploid. This type of development from an unfertilized egg is called **parthenogenesis**, a phenomenon found not only in plants but also in some groups of animals.

USEFUL REFERENCES

Bell, P. R., and Woodcock, C. L. F. *The Diversity of Green Plants.* Addison-Wesley Publishing Co., Reading, Mass., 1968.

Bold, H. C. *Morphology of Plants,* 2nd ed. Harper & Row, Publishers, Inc., 1967.

Bold, H. C. *The Plant Kingdom.* Prentice-Hall, Inc., Englewood Cliffs, N. J., 1964.

Conrad, H. S. *How to Know the Mosses and Liverworts.* Wm. C. Brown Co., Dubuque, Iowa, 1956.

Doyle, W. T. *Nonvascular Plants: Form and Function.* Wadsworth Publishing Co., Inc., Belmont, Calif., 1964.

Harris, R. M. *Plant Diversity.* Wm. C. Brown Co., Dubuque, Iowa, 1969.

Primitive Vascular Plants: Psilophytes, Club Mosses, Horsetails, and Ferns

The presence of a single structural feature—vascular tissues with their specialized conducting cells, xylem and phloem—has enabled some land plants to attain great size and, in a sense, to inherit the earth. In this chapter we will consider those vascular plants that are thought to be primitive and which may have contained the ancestral stocks that gave rise to the most highly evolved plant groups, the cone-bearing and flowering plants. The latter comprise the great forests of today and are the basis of the agriculture upon which our civilization depends. There is an abundant fossil record indicating that the psilophytes, club mosses, horsetails, and ferns have a long history on earth, and that in past ages they were far more dominant forms of plant life than they are today. Extensive forests of club mosses that attained a height of over 30 meters once existed. Coal was formed from their compressed remains.

The groups discussed in this chapter share with all other vascular plants the presence of conducting tissues which often also add rigidity and support. Furthermore, we find in the club mosses, horsetails, and ferns ample evidence of increased size and independence of the diploid, or sporophyte, stage in the life cycle. The vascular plants (sometimes referred to as the Tracheophyta) are the giants of the plant kingdom. Here, too, are the largest numbers of species, nearly 80 per cent of the known kinds of plants.

Why do we bother to describe plants that are rare, comprise a minor part of the vegetation around us, and are of little value economically?

Botanists study the simpler vascular plants in part because, like the mountain climber who is challenged to scale new peaks, they are there. From an evolutionary point of view, they are studied because they represent remnants of the past that provide clues about how living things became what they are today. Furthermore, we should not disregard large segments of the living world because they have no obvious practical importance according to our current knowledge of them. After all, who could have predicted a half century ago that the lowly molds would yield antibiotics or become the tools of the new molecular biology? We should bear in mind that most of the plant kingdom represents unexplored territory for the chemist and pharmacologist. It is conceivable that a fern may someday provide a cure for cancer or a club moss an unexpected miracle drug.

LIVING FOSSILS: PSILOPHYTES (DIVISION PSILOPHYTA)

Because of their simple structure and an abundant fossil record, the psilophytes have for some time been regarded as the group from which all other vascular plants must have evolved. Although such a view is indeed appealing, it cannot be accepted as undisputed fact. *Rhynia,* one of the best known of the ancient psilophytes, can be studied only from the fossil record. This plant consisted of a **rhizome** (underground stem) with **rhizoids** (root-like filaments) plus leafless aerial stems that branched dichotomously into two equal parts (Fig. 12-1).

Fossil Psilophytes and the Telome Theory

Rhynia and its relatives are of interest to botanists because, according to the telome theory proposed by Zimmermann in 1930, branching systems of the *Rhynia* type are viewed as the basic structure from which leaves evolved. A **telome** is nothing more than the terminal segment of a branching axis. If this segment happens to bear a sporangium, it is called a fertile telome; if no sporangium is present it is called a sterile telome or phylloid (leaf-like). Beginning with the simple *Rhynia* type branching axis, Zimmermann proposed that only a few basic types of modifications were needed in order for leaves to be evolved. These changes are: (1) **overtopping**, in which the branching becomes unequal; (2) **planation**, in which the lateral branches become restricted to a single plane; and (3) **webbing**, in which there is a fusion of the separate telomes into a dichotomously veined and flattened lamina (Fig. 12-2). It will probably never be possible to prove that the leaves of higher plants originated in this fashion but the fossil remains of psilophytes illustrate at least the stages of overtopping and planation. Although the telome theory has been popular with students of plant evolution for close to a half century, the discovery of fossils of other major groups of vascular plants that are as old or older than those of the psilophytes indicates that these ancient plants, although simple, may not be as primitive as they seem. Instead, they might represent an evolutionary blind alley.

Living Psilophytes

The two living genera in the division Psilophyta are rare and of practically no consequence in commerce or human affairs. In these two genera, *Psilotum* and *Tmesipteris*, only three species are known. Yet an abundant fossil record indicates that they were extensively distributed in the Devonian and Silurian periods of the Paleozoic era (about 360 to 400 million years ago).

Psilotum is the more common of the two living psilophyte genera. The two known

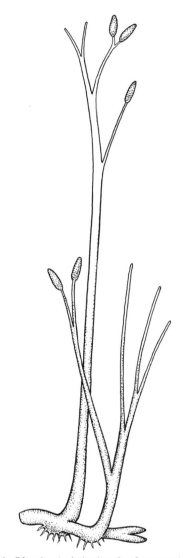

Fig. 12-1. *Rhynia*. Artist's sketch of restoration from fossil remains, ×1/3. (After Kidston and Lang: Transactions of the Royal Society of Edinburgh, Vol. 52, Part IV, 1921.)

species, *P. nudum* and *P. flaccidum,* are widely distributed in subtropical regions including Florida, Bermuda, and Hawaii. *Psilotum* often grows as an epiphyte on the trunks of trees, but is also found growing on soil. *P. nudum* thrives under greenhouse conditions and is included in the collections of many botanical gardens.

Although these plants have no roots, rhizoids aid in the absorption of water and minerals. It is interesting to note that the rhizoids provide a route of entry for fungi that invade the rhizome and live symbiotically within this subterranean stem. This association between a fungus and a vascular plant resembles the mycorhizal roots found in many seed plants where the fungus is thought to aid in absorption of mineral nutrients from the soil. The symbiotic

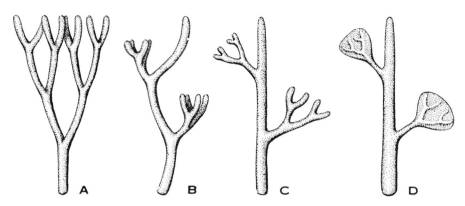

Fig. 12-2. The telome theory of the evolution of leaves. A, early stage of dichotomous branching as found in fossil psilophytes; B, unequal branching or "overtopping"; C, "planation" of smaller branching systems; D, "webbing" produces a flat leaf-like structure with dichotomously branching veins. (From A. S. Foster and E. M. Gifford, Jr.: *Comparative Morphology of Vascular Plants.* W. H. Freeman & Co., San Francisco, 1959.)

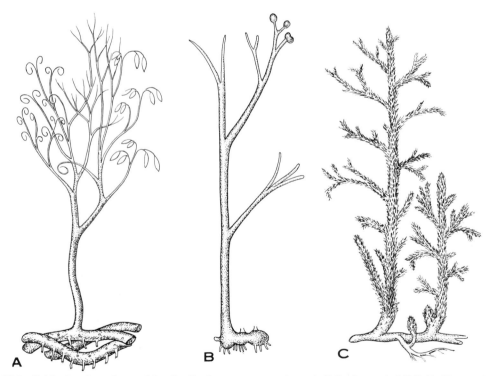

Fig. 12-3. Habit sketches of several fossil psilophyte reconstructions. A, *Psilophyton* (×1/10); B, *Horneophyton* (×2/3); *Asteroxylon* (×1/15).

Psilotum and *Tmesipteris* share many structural features with the fossil members of this division. Among these are the dichotomous branching patterns of the shoots, the absence of true leaves (the leaf-like appendages of the shoot lack vascular tissues), sporangia that are located terminally on branches of the shoot, and the presence of strands of vascular tissues in the stem. Although roots are absent, the rhizomes (horizontal, fleshy stems) with rhizoids function in the absorption of nutrients and water. Because of the strikingly close similarities between living and fossil forms, present-day psilophytes are generally considered to be the most primitive of vascular plants. An alternate point of view is that the living psilophytes are products of evolution in which there has been a progressive reduction or loss of advanced structures. The fossil record does not help us to choose between these two extreme views because no fossil remains more recent than those from Devonian times have yet been discovered.

Fig. 12-4. *Psilotum nudum.* Aerial branches of sporophyte plant and its subterranean rhizome (×¾). (From H. C. Bold: *Morphology of Plants*, 2nd ed. Harper & Row, Publishers, Inc., New York, 1967.)

fungi are certainly advantageous to *Psilotum* plants growing as epiphytes. The breakdown of organic compounds of the host brought about by the fungus releases bound mineral nutrients essential for the photoautotrophic green plant. The second genus of living psilophytes, *Tmesipteris,* has a very limited distribution in Australia, New Zealand, New Caledonia, and some other islands in the western Pacific. Like *Psilotum, Tmesipteris* often grows as an epiphyte.

Fig. 12-5. *Tmesipteris*. Habit sketch (× 1/3).

GROUND PINES, CLUB MOSSES, SPIKE MOSSES, AND QUILLWORTS (DIVISION LYCOPSIDA)

Plants in the division Lycopsida (also called Microphyllophyta), like the psilophytes, achieved their greatest prominence in the earth's flora during past ages, when many species grew to the size of trees and formed vast forests. These forests formed the basis for deposits of coal—hence the name Carboniferous for the era when the lycopsids flourished.

At the present time this division of the plant kingdom is represented by fewer than a thousand species that tend to be small and an inconspicuous part of the vegetation. The ground pines, club mosses, and their relatives are more advanced than the psilophytes, since their diploid sporophyte stages have roots, stems, and true leaves. Many species have rhizomes, and adventitious roots can form on such modified stems. Leaves tend to be small, with a single vein connecting to the vascular tissues of the stem.

Five orders of lycopsids have been recognized. Of these, only three have living representatives: the ground pines and club mosses (Lycopodiales), the spike mosses (Selaginellales), and the quillworts (Isoetales). The other two orders include extinct species known only through their fossil remains.

The Ground Pines and Club Mosses (Lycopodiales)

Many common species of lycopods found in the forests of the United States are known as ground pines because of their superficial resemblance to young evergreens. The smaller species are also known as club mosses, an unfortunate descriptive name because these plants are not closely related to the mosses discussed in Chapter 11. A major characteristic of the lycopods which sets them apart from the bryophytes is the fact that the diploid sporophyte phase of the life cycle predominates.

The human usefulness of ground pines and club mosses is at best modest. In New England these plants are gathered during the Christmas season for use in wreaths and table decorations. They are indeed attractive and long lasting but their growth is so slow that extensive use would

soon exterminate them. Another use of these plants is in a classic method for measuring the wavelength and frequency of sound waves devised by Kundt in 1866. This method used ground pine (*Lycopodium*) spores because of their small and uniform size. Standing sound waves in a glass tube set these small haploid cells into vibration. They settle at the nodes of the sound waves (locations of minimum displacement). The distance between heaps of spores is equal to half the wavelength of the sound waves in the gas that fills the tube. Beyond such modest use, the lycopods provide diversion for those who are interested in the rare living things in our environment and for scientists interested in the evolution of plants.

The order Lycopodiales contains both living and fossil genera. One living genus, *Lycopodium*, with about 200 species, is worldwide in its distribution and has been found from the Arctic to the tropics, while the other, *Phylloglossum*, has but a single species found in parts of Australia, New Zealand, and Tasmania.

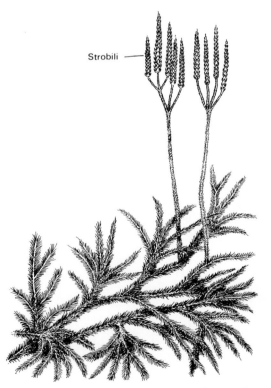

Strobili

Fig. 12-6. *Lycopodium.* Habit sketch of sporophyte plant (× 2/3).

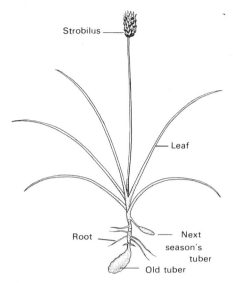

Fig. 12-7. *Phylloglossum*. Habit sketch of sporophyte plant (×1.½).

The spore-bearing plants consist of upright branches originating from rhizomes, specialized prostrate stems growing at or just beneath the soil surface. In all of the Lycopodiales, only one kind of spore is formed in sporangia borne on leaves. Such spore-bearing leaves or **sporophylls** are either located at any point on the stem or they may be restricted to the tips of specialized branches forming a **cone** or **strobilus**. The tiny spores are disseminated by wind. Spores may remain dormant for some time (up to several years). Such postponed development appears to be related to the presence of surface markings which affect their wettability. The rougher the spore surface, the less permeable it is to water and the more likely that spores will be washed into the soil. On germination, tiny egg- or carrot-shaped gametophyte plants are formed which develop at or below the soil surface. Except for those at the soil surface, they are colorless and non-photosynthetic with a nutrition that is saprophytic and largely dependent on the fungi that invade their cells and make available nutrients from decaying organic matter in the soil. Droplet-shaped sperms with two or three flagella are formed in spherically shaped antheridia located at or just below the upper surface of the prothallus. Archegonia are sunken and connected to the surface by the neck. A single egg

is formed in the venter at the base of each archegonium. At maturity the neck canal cells disintegrate, forming a passage to the egg. Sperm cells appear to be chemically attracted to the archegonia. Following fertilization, the zygote remains within the venter. As the young sporophyte plant grows, it eventually breaks through the surface of the haploid prothallus. During its early development the sporophyte is dependent on nutrients absorbed from the gametophyte.

The club mosses can reproduce asexually as well as sexually. Structures similar to the gemmae formed by liverworts and mosses develop on the tips of young stems. When these fall to the soil surface they develop into new sporophyte plants. Also reminiscent of the liverworts and mosses is the tendency for older parts to die and disintegrate to branch points, giving rise to new and independent plants each season.

The Spike Mosses (Selaginellales)

Although *Selaginella* is the only living genus within this order, over 700 species are known

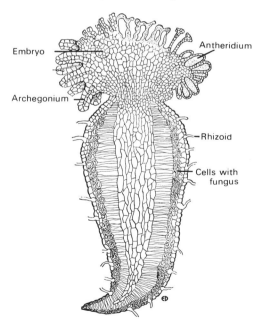

Fig. 12-8. *Lycopodium*. Longitudinal section of gametophyte plant (×20). (From H. J. Fuller and O. Tippo: *College Botany*. Henry Holt & Co., Inc., New York, 1954.)

and they are widely distributed. Generally these tiny plants prefer moist and shady habitats. They lend themselves well to cultivation in terrariums and are widely used as ornamental plants in greenhouses. One well-known desert species of spike moss is the resurrection plant which curls up when dry and unfolds when watered.

The Selaginellales differ from the Lycopodiales in that they form two kinds of spores (**microspores** and **megaspores**) which develop into male and female gametophytes. The sporangia are located in cones formed at the tips of fertile branches and are borne singly in the axils of specialized leaves called **sporophylls**. In a microsporangium, many microspore mother cells undergo meiosis, each producing four microspores. In a megasporangium only one megaspore mother cell undergoes reduction division to form four megaspores. In some species microspores and megaspores germinate within the sporangia, resulting in a developmental pattern which could be regarded as a prototype of the situation found in the seed plants. The gametophyte stage of the life cycle is much reduced and the structure that is shed from the megasporangium is an embryonic sporophyte plant. In other species both microspores and megaspores are released and fall to the ground, where they germinate to form tiny male and female haploid plants. Here, too, the gametophytes develop within the spore wall just as they do in species where spores germinate within sporangia attached to the diploid parent plant. Biflagellate sperms (antherozoids) are formed in the male gametophyte and at maturity the male haploid plant consists of a mass of antherozoids confined within the microspore wall. The motile gametes need a layer of water to reach a female plant. The preference of the spike mosses for moist habitats is at least partially explained by the need for water as a medium for the motile male gametes.

The Quillworts (Isoetales)

The quillworts (Isoetales) are quite different in appearance from their lycopod relatives, resembling more the rushes that belong in the monocot class of flowering plants. The quillworts are either aquatic or found in areas

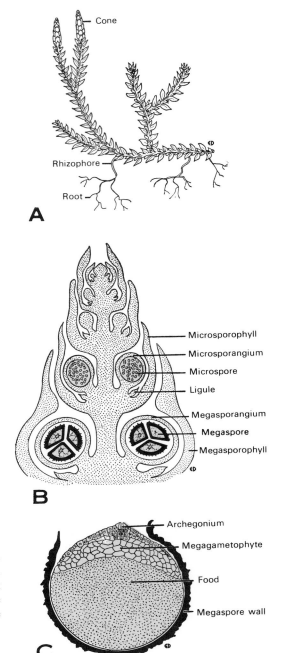

Fig. 12-9. *Selaginella.* A, habit sketch of sporophyte plant (×½); B, diagrammatic longitudinal section through cone showing microsporangia and megasporangia (×25); C, megaspore with internally developing female gametophyte (×50). (From H. J. Fuller and O. Tippo: *College Botany.* Henry Holt & Co., Inc., New York, 1954.)

subject to periodic flooding. There are only two genera: *Isoetes* and *Stylites*.

The sporophyte plant of *Isoetes* consists of a short, unbranched rootstock growing at or just below the soil level that bears a mass of leaves at the top and roots below. These rootstocks are eaten by aquatic animals such as ducks and muskrats. Shoot and root meristems are found at the upper and lower centers of the rootstock. New leaves and roots are produced by the central meristems, while old leaves and roots are pushed toward the periphery. Growth is seasonal. Because old parts die and are shed, the mature sporophyte remains a constant size.

Stylites, the second living genus of the quillworts, resembles *Isoetes* but has a branched rootstock. Its distribution is limited and it has been found only on the shores of glacial lakes of the Peruvian Andes.

HORSETAILS (DIVISION SPHENOPSIDA)

The horsetails (members of the genus *Equisetum*) comprise the only living representatives of their division of the plant kingdom. These are common plants found in many parts of the world with the notable exceptions of Australia and New Zealand. Of the approximately 25 surviving *Equisetum* species, most are found in moist habitats but a few compete successfully in the dry and barren edges of railroad tracks and roadways. The horsetails have true roots, stems, and leaves. Their stems are jointed and hollow and leaves are borne in whorls at the nodes.

The order Equisetales, in which the horsetails are classified, includes, in addition to the living genus *Equisetum,* fossil representatives belonging to the genus *Calamites.* These plants resembled present members of the order except for their size, which was tree-like (up to 30 meters tall). These giant horsetails reached their peak distribution during the Carboniferous era. They were a major component of the forests which ultimately gave rise to the vast deposits of coal that are being utilized to this day. The *Calamites* were able to grow to such enormous size because, unlike present-day *Equisetum,* they had a cambium and were therefore capable of secondary growth. An unusual feature of the secondary tissues in the fossil remains of the *Calamites* is the absence of growth rings. Is this

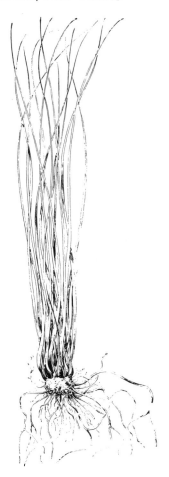

Fig. 12-10. *Isoetes* (quillwort). Habit sketch (×1). (From A. S. Foster and E. M. Gifford, Jr.: *Comparative Morphology of Vascular Plants.* W. H. Freeman & Co., San Francisco, 1959.)

perhaps an indication that the climate in their era was very uniform and without seasons?

Three additional orders of the division Sphenopsida are known only from fossil records of the Devonian and Carboniferous periods of the Paleozoic era. These rather primitive land plants reached the height of their development at a time when primitive reptiles and insects appeared; they were declining when the vertebrates were invading the land. Today only a few remnant species of *Equisetum* remain of these plants that constituted vast forests in past ages. We can only wonder whether excessive grazing by increasing numbers of herbivorous

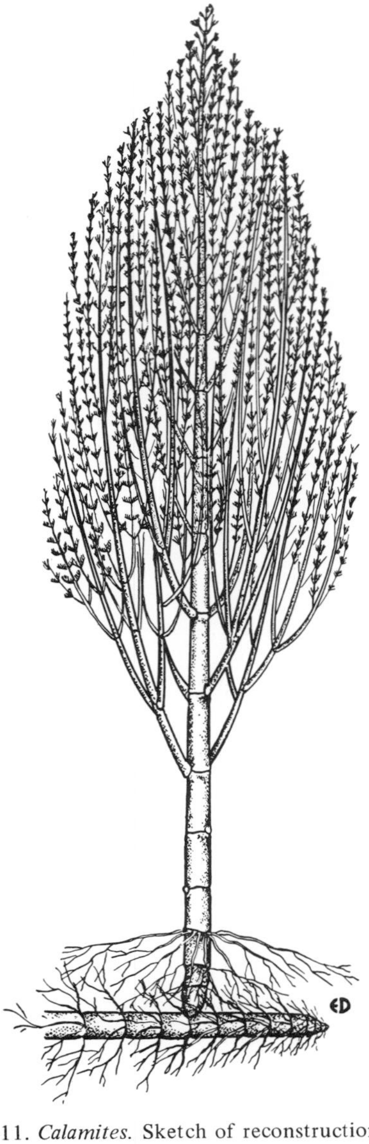

These modified stems grow below the soil surface producing roots at nodes and giving rise to two kinds of aerial stems: unbranched fertile stems with terminal, spore-bearing cones, and bushy, green shoots with whorls of branches arising from the nodes. This bushy appearance is the reason for the name horsetail. Scouring rushes, the other common term for these plants, comes from their coarse texture, due to deposits of silicates in the epidermal cells of

Fig. 12-11. *Calamites.* Sketch of reconstruction showing rhizome, roots, jointed stem and branches bearing leaves of this giant, tree-sized horsetail (×1/100). (From H. J. Fuller and O. Tippo: *College Botany.* Henry Holt & Co., Inc., New York, 1954.)

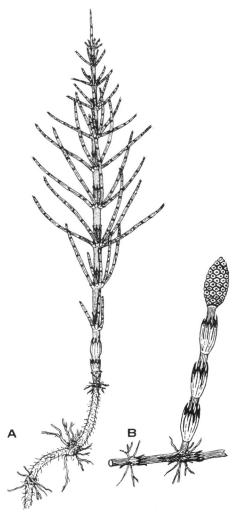

Fig. 12-12. *Equisetum arvense* (a horsetail). A, vegetative shoot (×½); B, fertile shoot with terminal strobilus (×½). (From H. C. Bold: *Morphology of Plants*, 2nd ed. Harper & Row, Publishers, Inc., New York, 1967.)

animals spreading across the land could have contributed to the decline of horsetails.

Present-day species of *Equisetum* are intriguing because they resemble so closely their many extinct fossil relatives. *Equisetum* sporophytes have horizontal, branched rhizomes.

their stems. *Equisetum* species were used by the early settlers in North America to clean pots and pans, and even to scrub floors.

Individual sporangia are borne on modified stems (sporangiophores) in a cone located at the apex of a fertile stem. The sporangiophores are arranged in whorls at the cone surface. Visible on the outside of each sporangiophore is a five- or six-sided plate. The haploid spores produced within the sporangia are unique in two respects: they contain chloroplasts and on their outer surface are four strap-like structures called **elaters**. The elaters are hygroscopic. When wet, they are tightly coiled around the spore surface. As they dry, they expand and straighten, expelling the spore from the sporangium (Fig. 12-13). Spores germinate soon after dispersal. The haploid prothalli, which are small, lobed, and photosynthetic, form gametangia within a month of germination. Archegonia and antheridia resembling those of the lycopods are located on the upper surface of the prothallus. In bisexual plants, antheridia are formed later than archegonia. Sperm cells are twisted much like a corkscrew, with a tuft of flagella at one end.

Following fertilization, the zygote develops within the archegonium. The root of the developing embryo grows down through the parent plant to the soil. Soon after the shoot breaks through the neck of the archegonium, the stem branches. One or more of these branches grow downward to form rhizomes. These underground stems of the sporophyte can penetrate as far as one or two meters below the soil surface. They are undoubtedly a factor

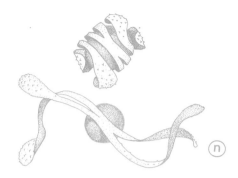

Fig. 12-13. *Equisetum* spores with elaters coiled and extended (×350).

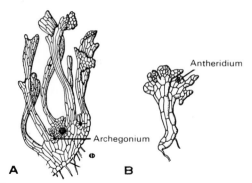

Fig. 12-14. *Equisetum* gametophytes. A, female gametophyte; B, male gametophyte (both approximately ×20). (From H. J. Fuller and O. Tippo: *College Botany*. Henry Holt & Co., Inc., New York, 1954.)

enabling some horsetails to persist in relatively dry habitats such as railroad embankments, sand, and gravel soils where they are often found in abundance. A single branching rhizome can give rise to a whole colony of horsetails.

FERNS (DIVISION PTERIDOPHYTA)

The ferns are a division of the plant kingdom decidedly different from those described earlier in this chapter. An appreciation of their complexity and beauty was expressed by the 19th-century American naturalist, Thoreau, who wrote: "God made ferns to show what He could do with leaves." The Pteridophyta are more closely related to the seed plants than they are to the more primitive vascular plants just described. Ferns share with the seed plants several important characteristics. Both have true leaves with branched veins that are connected to the vascular tissues of the stem. Where strands of conducting tissue enter the leaves, there are **leaf gaps**, or discontinuities left in the otherwise continuous vascular system of the stem. In both the ferns and seed plants, leaves are often large and can be lobed or compound. The sporophyte stages of their life cycles are dominant. An important difference is that in the ferns, the much reduced gametophyte stage is a nutritionally independent and separate plant, while in the seed plants the much more reduced haploid stage is dependent on the dominant sporophyte plant.

Ferns are well represented in the fossil record and comprised a major part of the vegetation in past ages. Even today there are several hundred genera and approximately 10,000 species of ferns among the flora of the earth, indicating that these ancient plants have been able to compete successfully even to the present. Generally the ferns are most abundant in wet, shaded habitats and they are particularly numerous in the tropics. There is great diversity in the Pteridophyta. Some forms are epiphytic, growing attached to the shoots of other plants; others are tree-like, and some species even grow as vines.

Most ferns found in the temperate zones are perennials, with persistent rhizomes that elongate and form new leaves at their anterior end each year, while a portion of the posterior end dies and decomposes. Roots are found along the entire length of the rhizome. Such roots are termed "adventitious," since they originate from stem tissues rather than from the activity of the root apex or as lateral outgrowths of roots. Fern leaves exhibit a wide array of shapes ranging from simple undivided blades to lobes with variously indented margins, to compound leaves with a central midrib (or **rachis**) and two rows of leaflets (or **pinnae**). Compound leaves of the type described are called **pinnate**. Even more complex compound leaves are found in some ferns. In **bipinnate** leaves, each lateral leaflet is further divided into smaller leaflets. Tripinnate and even higher degrees of compounding exist in some species. As developing fern leaves first emerge, many are tightly rolled and this **circinate** structure or "fiddlehead" gradually uncoils outward.

The two classes of ferns, the Eusporangiopsida and the Leptosporangiopsida, differ in their patterns of sporangium development. In the first, sporangium development resembles that found in the lycopods and horsetails in that sporangia are large and at least partially embedded in parent tissues. Large indefinite numbers of spores are produced. In the latter class, sporangia develop from single cells, tend to remain small, and project from the surface of the parent plant. Spores are produced in smaller numbers, usually in multiples of four (often 32 or 64).

Eusporangiate Ferns

Ophioglossum, the adder's tongue fern, and *Botrychium,* the grape fern or rattlesnake fern, are representative of the Ophioglossales class of eusporangiate ferns. The adder's tongue fern looks very unlike most common ferns. It has a single simple leaf borne on a short upright stem and a single fertile spike bearing a terminal cluster of sporangia. The grape fern conforms more closely to how we expect a fern to look; but, like the adder's tongue, it has a single leaf and a single fertile spike. Its leaf is compound with leaflets resembling the leaves of grape vines.

Ophioglossum and *Botrychium* are unique among the ferns in that sporangia develop on a single fertile spike which originates from the petiole of the sterile leaf. Each sporangium may contain as many as 15,000 haploid spores. *Ophioglossum* species are unique in yet another respect. Their chromosome numbers are unusually large. Each diploid cell of the sporophyte plant in *O. vulgatum* contains 512 chromosomes, while *O. petiolatum,* a species found in the tropics with a diploid number of chromosomes in excess of 1,000, holds the record for number of chromosomes among vascular plants.

Gametophyte plants formed on germination of spores are subterranean, non-photosynthetic and shaped like a twisted cylinder. The size of these haploid plants ranges up to 5 cm. in length and 1 cm. in diameter. Like the gametophytes of *Psilotum* and *Lycopodium,* the haploid stages of the eusporangiate ferns are infected with the hyphae of symbiotic fungi. They are nutritionally dependent on the breakdown products of dead organisms in the soil released by the saprophytic fungi for which they are hosts. The gametophytes of *Ophioglossum* are bisexual. Antheridia and archegonia originate from surface cells, but in their development they become at least partially buried within the tissues of the haploid plant. Following fertilization, growth of the new sporophyte plant is surprisingly slow. It usually takes several years for the diploid plant to reach maturity and form spores.

Other eusporangiate ferns are found only in

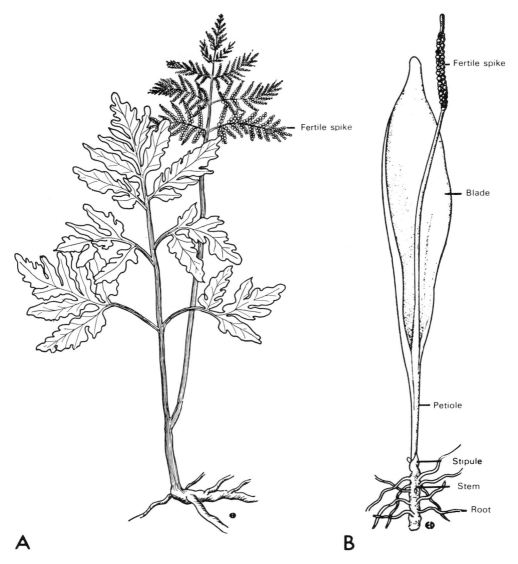

Fertile spike

Fertile spike

Blade

Petiole

Stipule

Stem

Root

A   **B**

Fig. 12-15. Eusporangiate ferns. A, *Ophioglossum* (adder's tongue fern); B, *Botrychium* (grape fern) (both approximately ×1/5). (From H. J. Fuller and O. Tippo: *College Botany*. Henry Holt & Co., Inc., New York, 1954.)

warm climates, on islands in the Pacific, including Hawaii and the Philippines, or in the tropical regions of both the northern and southern hemispheres. While none are native to the United States, six species are found in Mexico, Central America, and the West Indies. Some are large plants with leaves (fronds) up to 10 meters long.

Leptosporangiate Ferns

The leptosporangiate ferns are far more numerous than the eusporangiate ferns, both with respect to population sizes and the number of known genera (about 500) and species (approximately 4,000). Although widely distributed in the temperate zones, they

Fig. 12-16. *Ophioglossum* gametophyte (×20). (From H. C. Bold: *Morphology of Plants*, 2nd ed. Harper & Row, Publishers, Inc., New York, 1967.)

are more abundant in the tropics. In size, they range from tiny aquatic species to the tree ferns. There are three kinds of ferns in this class: the Filicales, or true ferns, and two groups of water ferns, the Marsileales and Salvineales.

True Ferns (Filicales)

In the Filicales sporangia are usually found in clusters on the undersides of the leaves and may be protected by an outgrowth from the leaf surface called the **indusium**. The entire structure is called a **sorus**. Individual sporangia are borne on stalks and their walls are characteristically thin and rather transparent except for a strip of thick-walled cells called the **annulus**. The loss of water from the cells of the annulus causes contraction of their outer surfaces with a straightening of this ring of cells and a rupturing of the wall of the sporangium. These

changes in the annulus result in the ejection of spores.

Polypodium is the most abundant and widespread genus in the Filicales. This common fern is found both in the tropics and in the temperate regions. Individual species have a wide variety of environmental preferences, ranging from the Venus maidenhair which is found on wet limestone cliffs to the resurrection fern which can survive extremely dry conditions. Generally, however, temperate species in the genus *Polypodium* are found in moist woodlands. Their perennial rhizomes form several circinate leaves each year which unroll during enlargement, making the familiar fiddlehead. Sori are located on the undersides of the leaves and are often mistaken for brown bugs or disease spots on the leaves.

Fern Life Cycle

In the life cycle of the fern there is an alternation of haploid and diploid phases in which the diploid sporophyte is large and independent except during the early stages of its embryonic development. After spores are dispersed, they germinate, forming a filamentous structure which grows to form a small, often heart-shaped, prothallus. This haploid gametophyte plant is photosynthetic and absorbs essential mineral nutrients and water from the soil through rhizoids located on its lower surface. Antheridia formed near the rhizoids are spherical and sessile. At maturity they contain many spiral-shaped, multiciliated sperm cells. Archegonia form near the notch of the heart-shaped prothallus and resemble those found in the mosses. Usually the archegonia and antheridia of a given prothallus mature at different times, a situation that ensures cross-fertilization. There is evidence of an antheridium-inducing factor in many fern species. Gametophytes grown in isolation tend not to form antheridia, yet in crowded laboratory cultures they tend to form antheridia precociously. It has recently been shown that the addition of gibberellins to media on which fern gametophytes are grown in culture will sometimes stimulate the development of antheridia. This naturally occurring plant hormone (see Chapter 17) may be the substance responsible for initiating the development of

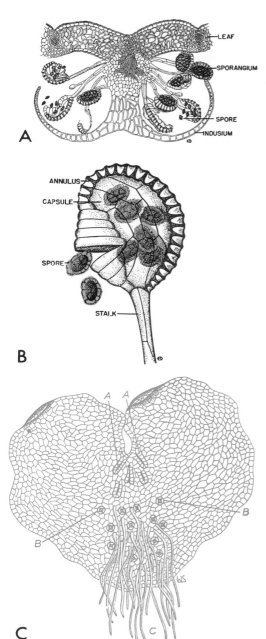

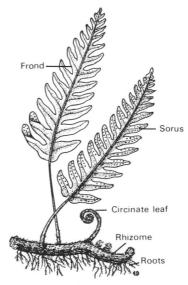

Fig. 12-18. *Polypodium.* Habit sketch of a common leptosporangiate fern (×1/10). (From H. J. Fuller and O. Tippo: *College Botany.* Henry Holt & Co., Inc., New York, 1954.)

Fig. 12-17. Reproductive structures in ferns. A, leaf and sorus in cross section. The indusium covers numerous sporangia (×30). B, single fern sporangium (×200). C, gametophyte plant (prothallus). Lower surface with archegonia (a), antheridia (b), and rhizoids (c), ×6. (From H. J. Fuller and O. Tippo: *College Botany.* Henry Holt & Co., Inc., New York, 1954.)

male gemetangia in ferns. The effects of crowding *vs.* isolation on antheridium formation can be explained either on the basis of a critical concentration being essential or the increased production of gibberellins by older gametophyte plants.

Following fertilization, the diploid zygote develops into an embryo. For a time the young sporophyte is nourished by food and water absorbed through its foot from the gametophyte prothallus. Soon the developing root begins to absorb inorganic nutrients and water from the soil. With the emergence of the first leaf, the developing sporophyte becomes independent of the gametophyte and the prothallus soon disintegrates.

Alternative Life Styles
—Apogamy and Apospory

Although the life cycle just described represents the normal pattern of fern development, alternatives of the sort found in other plant groups ranging from mosses to seed plants do occur. New sporophyte plants can be produced asexually from the rhizome where disintegration of the posterior part up to a branch point automatically separates parts of a single plant into two independent sporophytes. Also, new sporophyte and gametophyte plants can arise without the usually necessary steps of fertilization and meiosis, *i.e.,* by apogamy or apospory.

Apogamy is the formation of a new sporophyte plant from a gametophyte without the fusion of gametes. It occurs naturally in some species of ferns and can be induced in others. By maintaining gametophytes under conditions sufficiently dry so that sperms are unable to swim, fertilization is prevented and apogamy can result. An alternative means is to grow gametophytes on a medium supplemented by sugars (2.5 per cent of the monosaccharides glucose or fructose, or of the disaccharides sucrose or maltose). Apogamy is also stimulated by high light intensity, a factor producing increased rates of photosynthesis and added carbohydrate reserves. It appears that the common stimulus provided by added carbohy-

drates or increased light intensity is the availability of excess respiratory substrates.

Under those circumstances where apogamy occurs naturally, it has been shown that a constant chromosome number is maintained in both the sporophyte and gametophyte phases which, according to species, may be either haploid or diploid. Where apogamy is induced it is likely that the resulting sporophytes are haploid.

Apospory is the production of new gametophyte plants by means other than the production, release, and germination of spores. It too can occur naturally in mosses, liverworts, and ferns, or it can be induced experimentally. Young fern leaves grown in culture tend to form filaments that develop into morphologically normal yet diploid gametophyte plants.

The phenomena of apogamy and apospory

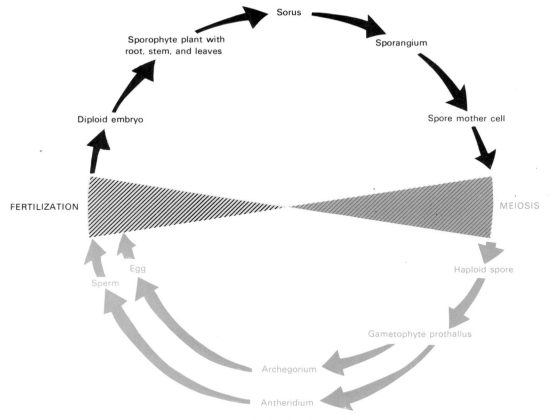

Fig. 12-19. Fern life cycle.

raise very basic questions about the way genetic instructions carried by the DNA in chromosomes are carried out. Something in addition to the presence of a single or double set of chromosomes within the nucleus of a cell must be involved in directing the pathway of development. Many factors appear to play a role, including the cytoplasmic environment of the nucleus, chemical and physical forces exerted by the presence or absence of neighboring cells, and environmental variables such as the availability of nutrients, light, temperature and gravity. In the past, the phenomenon of alternation of generations, illustrated so dramatically by the ferns, provided two sweeping generalizations for botanists: (1) alternating haploid and diploid plant bodies can differ drastically in morphology; and (2) there is an evolutionary tendency for increased size and independence of the diploid sporophyte with reduction in the haploid gametophyte. The exceptions to what normally develops in the presence of one *vs.* two sets of chromosomes provide interesting challenges for interpretation by molecular biologists.

Diversity of Leptosporangiate Ferns

A complete inventory of the ferns would require many volumes and would include a great number of rare species with limited geographic distribution. Obviously such an approach would be of little interest or value to most students. Yet the array of ferns commonly observed in the temperate environments where many of us live provides little indication of the diversity of forms of leptosporangiate ferns which exist in other parts of the world. It is our purpose here to provide some illustrations of this diversity.

Tree Ferns. Tree ferns, found in the tropical and subtropical rain forests of Hawaii, Australia, New Zealand, Central America, and the West Indies, can grow to heights of 20 meters. These unusual plants are popular specimens in botanical gardens but must be maintained in greenhouses if winter temperatures drop below 0°C. The trunk of a tree fern is normally unbranched and the base is covered with a mat of adventitious roots which provide anchorage and support. At the top of the trunk is a crown of pinnately compounded leaves which can be up to 5 meters in length. There are six or seven genera of tree ferns belonging to the family Cyatheaceae. Perhaps the most unusual feature of these large ferns is the absence of secondary growth. There is no cambium to form secondary xylem and phloem, the tissues making up the wood and much of the bark in the seed plants.

Water Ferns. The Marsileales are amphibious ferns found in ponds and ditches. They are usually aquatic but can survive dry periods if the soil remains moist. Rhizomes grow at or just below the water-soil interface. Adventitious roots originating at the nodes of the rhizomes grow downward and anchor the plant to the soil. Compound leaves with four leaflets resembling four-leaf clovers are borne on long petioles. Leaflets may be submerged, floating, or aerial. Sporangia form on specialized lateral branches arising from the lower parts of the petioles, usually after a dry period. These spore-producing structures are at first green and later become brown and hard. The outer wall persists for several years in nature and spores are not released until erosion or bacterial action ruptures this tough stony material. The delayed release of spores has no adverse effect on their germination because they remain viable for 20 to 30 years. Once free, spores germinate and male and female gametophyte plants develop in less than a day. Within a few hours after fertilization, the diploid embryo begins developing. The advantages for survival of a very rapid gametophyte development in the water ferns are obvious. Spore formation occurs when desiccation threatens the sporophyte plants. The release of spores, which can occur several years later, is initiated by flooding, a condition favoring the development of sporophyte plants. Germination, gametophyte development, fertilization, and growth of a new diploid plant proceed rapidly and the sporophyte persists in the vegetative state until drought threatens its survival.

The Saviniales, small aquatic ferns that float on the surface of pools and sluggish streams constitute the third order of leptosporangiate ferns. Most species in the genus *Salvinia* are native to Africa with only one species found in North America. In a second genus, *Azolla,* there

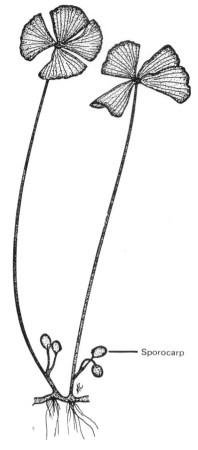

genera are incompletely described on the basis of bits of stem, petiole, or leaf and their growth habits and life histories are therefore largely unknown. The three orders of leptosporangiate ferns just discussed, the Marattiales, the Filicales, and the Salviniales, all include a number of extinct genera known only from their fossil remains. Other orders have no living representatives.

The seed ferns, another group of extinct plants with fern-like appearance and foliage that can easily be mistaken for that of ferns, have been assigned to the division Spermatophyta, the seed plants. This unusual group of plants from which the most highly evolved members of the plant kingdom are thought to have been derived were very abundant during

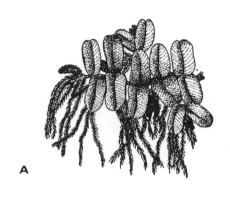

A

B

Fig. 12-20. *Marsilia* (a water fern). Habit sketch (×1). (From Russell: *An Introduction to the Plant Kingdom*. The C. V. Mosby Co., St. Louis, 1958.)

are two species that occur in the United States. These tiny forms resemble the common duckweeds (*Lemna*) that also float freely on the surfaces of ponds but are seed plants.

Fossil Ferns. The ferns, like the psilophytes, club mosses, and horsetails, are abundantly represented in the fossil record. All of these primitive vascular plants which are now minor components of the earth's vegetation were far more predominant in earlier ages. The Carboniferous era has been described as the Age of Ferns and the fossil remains of these ancient plants provide evidence that they were the precursors of present-day seed plants. Many of these now extinct ferns are known only from fragmentary remains. Such so-called **form**

Fig. 12-21. Representative water ferns belonging to the Salviniales. A, *Salvinia natans* (×2); B, *Azolla* (×8). (From R. M. Holman and W. W. Robbins: *Textbook of General Botany For Colleges and Universities*, 4th ed. John Wiley and Sons, New York, 1947.)

the Carboniferous era. In fact it has now become evident that much of the foliage remains of that era which on first examination appeared to be ferns really belonged to the seed ferns. Thus, referring to the Carboniferous era as the Age of Ferns is not strictly correct.

The Usefulness of Ferns

Although ferns are widely grown in greenhouses and shaded gardens because of their attractive foliage, present-day species are of negligible economic importance. There are reports that the rhizomes of some species are edible and there has been some limited use of fern extracts as drugs to treat parasites such as tapeworms and liver flukes. Significant economic usefulness is found only in the remains of these plants of the Carboniferous era that contributed to the coal reserves that serve as

fuel for homes, industry, and electric power plants. Except in parts of the tropics, ferns are now a minor component of natural environments.

USEFUL REFERENCES

Andrews, H. N. *Ancient Plants and the World They Lived In.* Comstock Publishing Co., Ithaca, N.Y., 1947.

Arnold, C. A. *An Introduction to Paleobotany.* McGraw-Hill Book Co., Inc., New York, 1947.

Bell, P. R., and Woodcock, C. C. F. *The Diversity of Green Plants.* Addison-Wesley Publishing Co., Reading, Mass., 1968.

Bold, H. C. *Morphology of Plants.* 2nd ed. Harper & Row, Publishers, Inc., New York, 1967.

Scagel, R. F., Bandoni, R. J., Rouse, G. E., Schofield, W. B., Stein, J. R., and Taylor, T. M. C. *An Evolutionary Survey of the Plant Kingdom.* Wadsworth Publishing Co. Inc., Belmont, Calif., 1965.

5

THE WORLD OF LIFE: SEED-BEARING PLANTS

13

Cone-Bearing Plants and Their Relatives: The Gymnosperms

The plants which clothe the continents today much as the algae populate the seas are the gymnosperms and angiosperms, the two groups of seed plants. You will recall that in discussing the ferns, we noted that they share many of the characteristics of seed plants such as large, true leaves and a life cycle in which the diploid sporophyte stage is dominant. In the seed plants, the diploid plant releases a special reproductive structure, the **seed**, which contains the embryo of a new diploid plant. Female spores are retained within the megasporangia of the parent plant. The female gametophyte develops and fertilization takes place within a sporophyte structure called the **ovule**. The term gymnosperm means "naked seed," a universal characteristic of plants in the subdivision Gymnospermae. This characteristic, in which the megasporangia are borne on the surface of a **sporophyll** (spore-producing leaf), differs from the situation found in the second group of seed plants, the Angiospermae or flowering plants, where the seeds are enclosed within the megasporophylls (megaspore-producing leaves).

SEEDS—KEYS TO SURVIVAL AND SUCCESS

In a real sense, seeds provide a kind of mobility for the species that is not available to an individual plant. The rapid invasion of clearings by all manner of seed plants is so commonplace that we give it little thought. Yet it would be startling to see our lawns invaded by horsetails and ferns rather than dandelions, chickweed, and tree seedlings. The reason why this will never happen is that the more primitive plants lack the unique dispersal mechanism provided by seeds. Seeds have very real adaptive advantages over the spores that are produced by ferns and other less highly evolved plants. They carry an embryo ready to send out roots and shoots and a supply of stored food that provides a capital reserve to nourish the young plant until it is photosynthetically self-sufficient.

The gymnosperm seed is a combination of sporophyte and gametophyte tissues. This structure is the result of fertilization and enlargement of the megasporangium (ovule). The embryonic sporophyte plant that develops from the zygote is embedded in nutritive tissue of the female gametophyte. Surrounding the embryo and its food reserves are outer layers derived from the parent sporophyte plant that harden to form a protective seed coat. Such a seed is derived from three generations: the parent sporophyte, the gametophyte, and the new embryonic sporophyte. This structure, with its built-in food reserves and the capacity to survive periods of low temperatures or desiccation that are hostile to development, is easily transported by means of air currents, water, or animals (see discussion of seed dispersal in Chapter 18).

Evolutionary Origins of Seeds

Although studies of the fossil record may never reveal all the stages in what was probably a long evolutionary development of the seed habit, it is obvious that certain conditions were necessary. The most essential prerequisite was

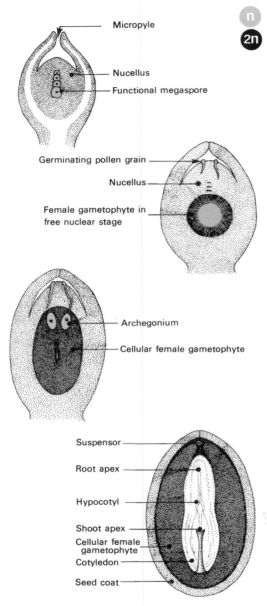

Fig. 13-1. Seed development in gymnosperms. A, megasporangium at the end of meiosis. Of the four haploid cells formed, only one is a functional megaspore. B, free nuclear stage of female gametophyte development. C, archegonia form at the micropylar end of the cellular gametophyte. Pollen grains landing on the surface of the nucellus have germinated and pollen tubes are growing toward the archegonia. D, mature seed containing an embryo surrounded by nutritive tissue of the female gameto-

the production of two different kinds of spores: larger **megaspores** and smaller **microspores**. The larger spores developed differently from the smaller spores, tending to remain where they formed, ultimately giving rise to a structure containing eggs. The smaller microspores continued to be released as in the more primitive plants, but in the seed plants they became pollen grains. In function, the megaspores are female, while the microspores are male. Of course, like the spores of the ancestral ferns, the micro- and megaspores of the seed plants are haploid and are produced in groups of four by the familiar two divisions of meiosis. These conditions certainly did not suddenly arise in the immediate predecessors of early seed plants. More primitive prototypes still exist in some of the spore-forming plants. The spike mosses (Selaginellales) in the division Lycopsida exhibit all three of these conditions and can be regarded as one of nature's early experimental prototypes of the seed habit (see Chapter 12).

One aspect of megaspore development that is unique to the seed plants is the involvement of the surrounding sporophyte tissue in the development of the megasporangium. The integument that forms a protective covering for the developing ovule later serves as the seed coat.

Seed Development in Gymnosperms

The developmental sequence in seed development from a gymnosperm megasporangium is shown in Figure 13-1. Following meiosis, there are within each ovule four haploid megaspores embedded in a structure called the **nucellus** that is equivalent to the megasporangium wall. Only the spore at the base of the ovule undergoes further development. Outside the nucellus is a several-layered integument which has a tiny opening, the **micropyle,** which provides a means for the male gametophyte **(pollen)** to enter the ovule. During development of the female gametophyte, there is a period of nuclear divisions without cell wall formation **(free nuclear stage),** followed by the laying down of walls from the periphery inward. As cell walls are formed, a few individual cells near the

phyte. Outer layers derived from the parent sporophyte harden to form a protective seed coat.

micropylar end become archegonia. Meanwhile pollen grains, *i.e.,* **microspores,** that have been carried by air currents and swept through the micropyle reach the surface of the nucellus and germinate. The pollen tubes grow down through the nucellus toward the archegonia. In the cycads and *Ginkgo* the developing pollen tubes are much like fungal hyphae, penetrating and absorbing nutrients from the nucellus. During fertilization, two motile, flagellated sperms are released from the burst basal end of the pollen tube and penetrate an archegonium, where one of them fuses with the nucleus of the single egg. In the conifers, pollen tubes grow directly downward through the nucellus and release non-flagellated gametes into the archegonium. This sperm-carrying function of the pollen tube exhibited by the more advanced gymnosperms is found in all of the flowering plants (angiosperms).

Following fertilization, there is in most gymnosperms a period without cell wall formation in the development of the embryo just as there is such a phase in the development of the female gametophyte. As the embryo becomes cellular, differentiation occurs with formation of a suspensor and organization of the embryo into shoot (with apical meristem and seed leaves or cotyledons), hypocotyl, and radicle.

Polyembryony, the tendency for a single gametophyte to form several embryos, occurs in many gymnosperms. As in animals, multiple embryo formation may result from the fertilization of more than one egg or from the separation and independent development of cells derived from a single young embryo. Although polyembryony is common during seed development, usually one embryo becomes dominant and the rest are eliminated by the time seeds are shed.

During seed maturation, the nucellus becomes disorganized and persists only as a thin layer near the micropyle. The integument becomes a hard, durable protective shell. Once shed, the seeds of most gymnosperms go through a period of dormancy when germination cannot occur even though appropriate conditions of temperature and moisture may be encountered. The phenomenon of dormancy is a distinct advantage, protecting the progeny of seed plants from initiating development during temporarily favorable conditions in late summer and fall which will be followed by cold temperatures that would quickly kill young seedlings.

MAJOR GROUPS OF GYMNOSPERMS

There is evidence from the fossil record that the gymnosperms are of ancient origin with many extinct representatives. Their fossil remains indicate that our pines, firs, and other gymnosperms may have evolved from fern-like ancestors. These intermediate forms are known as **seed ferns** (the Cycadofilicales, also called Pteridospermales).

According to modern experts on the taxonomy of the gymnosperms, there are four diverse classes of living plants in this subdivision of the seed plants: **cycads** (Cycadopsida, also called cycadophytes); the Ginkgopsida, which only has one living representative in the ginkgo (maidenhair) tree; the conifers (Coniferopsida, also called coniferophytes); and the Gnetopsida, a collection of rather rare and exotic forms.

The cycads, familiar as decorations at formal dances and funerals, and the ornamental ginkgo tree are relics of an earlier age. The cycads are limited in their distribution to tropical areas and are never a dominant part of the vegetation. The single representative of the Ginkgopsida, *Ginkgo biloba,* although native to China, is now widely grown in the United States and Europe as an ornamental tree. It is in the class Coniferopsida that we find large numbers of familar and economically important gymnosperms including the pines, spruces, hemlocks, firs, and cedars. These conifers are worldwide in distribution (although mostly in temperate climates), often forming extensive forests. They dominate a wide subarctic belt called the taiga, which extends across Siberia, northern Europe, Canada, and Alaska. They provide the raw materials for the lumbering industry, paper products, and manufacture of synthetic fabrics and plastics, and a variety of distillation products, including turpentine and resins. In this group of the gymnosperms we find not only the largest plants but also plants with the longest known lifespans. Some of the giant redwoods and the bristlecone pines of the Sierra Nevada have lived for thousands of years.

The Gnetopsida are composed of three

Fig. 13-2. Bristlecone pine (*Pinus aristata*). The wood of these trees can withstand the elements for thousands of years after death. Tree ring records from such ancient remains combined with borings from living trees that are almost 5,000 years old have provided a continuous chronology going back more than 7,000 years. The veteran trees shown here are growing at an elevation of 3,000 meters. (Courtesy U. S. Forest Service.)

orders of very strange plants which are different from other gymnosperms, bear little resemblance to each other, and are strangely distributed in out-of-the-way places such as the Namib Desert of Southwest Africa.

Although they are a diverse group, all gymnosperms share the following characteristics: the dominant sporophyte stage of the life cycle is usually tree-like; secondary growth always occurs, producing the large amounts of conducting and supporting tissues needed; male and female gametophytes are very much reduced and non-photosynthetic so that they are nutritionally dependent on the sporophyte; and following fertilization the embryo develops within the ovule and it is the young sporophyte embryo, plus its surrounding protective maternal sporophyte tissues (the seed), which is shed from the parent plant.

As with many groups of animals, extinction has been a fairly common feature of plant

evolution. The causes are unknown, so we can only record the stark facts. Many gymnosperms are known only from their fossil remains. Two main evolutionary lines have been recognized: the fern-like cycadophytes, which, except for the present-day cycads, are all extinct, and the coniferophytes. The fossil records of both groups extend back to the Paleozic era. The cycadophytes were characterized by fern-like leaves and short, unbranched trunks. Included in this evolutionary grouping are the seed ferns, which are thought to be a major evolutionary link between the ferns and the seed plants, and the Bennettitales, a group of fossil plants resembling present-day cycads, that flourished during the age of dinosaurs in the Mesozoic era. The Coniferophytes include the present-day conifers and ginkgos plus a completely extinct group, the Cordaitales, which had tall, often branched woody trunks and some had needle-like leaves. Cycadophytes and coniferophytes are thought to have originated in the Paleozoic era and evolved along parallel lines.

The Seed Ferns (Cycadofilicales)— Originators of the Seed Habit

The position of the seed ferns in the evolution of the plant kingdom is somewhat analogous to that of the Model T Ford in the automobile industry. Both were highly successful early models that gave rise to lines of successful descendants that are now found in most parts of the world. Seeds insure an efficient means of replication and distribution of the seed plants just as the assembly line and modern transportation techniques insure the proliferation and dissemination of automobiles.

Fragmentary remains of the seed ferns have been found in strata beginning in the early Carboniferous era to the Cretaceous. They reached their maximum numbers and distribution late in the Carboniferous and were among the most important coal-forming plants. Prior to this century, fossil remains of the seed ferns were assumed to be those of true ferns. It was not until 1903 that the English investigators, Oliver and Scott, discovered that seeds were borne on certain fossil plants that had fronds resembling those of true ferns. Since that time, the fossil remains of many seeds and pollen-producing organs have been found attached to

Fig. 13-3. *Lyginopteris oldhamia,* a seed fern (×¼). Reconstruction of this rather small plant from fossil remains reveals a stem only several centimeters in diameter and pinnate leaves no more than half a meter long. (Redrawn from R. F. Scagel *et al.: An Evolutionary Survey of the Plant Kingdom.* Wadsworth Publishing Co., Inc., Belmont Calif., 1965.)

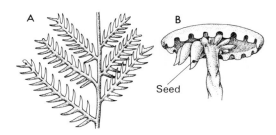

Fig. 13-4. *Lepidopteris,* a seed fern. A, piece of frond (×¼); B, seeds attached to underside of disk-shaped structure (× 2). (Redrawn from R. F. Scagel *et al.: An Evolutionary Survey of the Plant Kingdom.* Wadsworth Publishing Co., Inc., Belmont, Calif., 1965.)

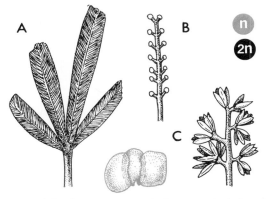

Fig. 13-5. *Caytonia,* a seed fern. A, leaf found associated with fossil *Caytonia* seeds, ×¼; B, seed-bearing axis, ×½; C, pollen-bearing structure, ×½, and single pollen grain, ×600. (Redrawn from R. F. Scagel *et al.: An Evolutionary Survey of the Plant Kingdom.* Wadsworth Publishing Co., Belmont, Calif., 1965.)

fronds. These plants are present as fossils in coal seams and undoubtedly our coal represents vast forests of plants like these. Seeds were borne on leaves and were much like the seeds of present-day cycads in structure. The seed ferns were tree-like with woody stems.

In the Permian period of the Paleozoic era, some 300 million years ago, and the Triassic period of the Mesozoic era that followed the Carboniferous periods, the earth underwent drastic changes in climate. The oceans receded and the land became increasingly arid. Later there was extensive glaciation and areas that had formerly been tropical were covered with ice. As a consequence of these changes, the seed ferns declined and many genera became extinct.

Leaves, seeds, and microsporophylls of the genus *Lepidopteris* have been found in the Triassic remains of eastern Greenland. These plants had twice pinnate, fern-like leaves. Seeds were borne in a circle on the lower surface of a stalked disk about 1.5 cm. in diameter (Fig. 13-4). The Caytoniaceae have been found in strata of the Triassic through the Lower Cretaceous periods of the Mesozoic era, and thus they were contemporaries of the dinosaurs, the first mammals, the insects, and the primitive birds. The fossil remains of their seed-bearing structures are so similar to ovules of the present-day flowering plants that when

Caytonia was discovered it was for a time thought to be a primitive angiosperm.

Cycads

There are two orders of cycads, one extinct (the Bennettitales) and the other with living representatives (the Cycadales). The Bennettitales arose, flourished, and perished in the Mesozoic era, coinciding in time with the dinosaurs. We can only speculate whether their parallel existences indicate a mutual interdependence or a need for comparable environmental factors. Although both the Bennettitales and the Cycadales are thought to have evolved from the seed ferns, these two lines appear to have followed separate paths for most of their

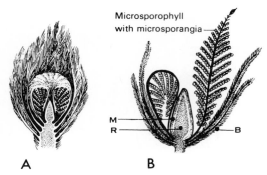

Microsporophyll
with microsporangia

M
R   B

A   B

Fig. 13-6. Bisexual strobilus of *Cycadeoidea*, a fossil cycad member of the Bennettitales. A, early stage of development; B, later stage with unfolding microsporophylls (both ×1/20). Structure labeled *M* is the megasporophyll; *R* is the receptacle; and *B* is a bract. There is a resemblance between this structure and the angiosperm flower—a case of parallel evolution? (Redrawn from H. J. Fuller and O. Tippo: *College Botany,* revised edition. Henry Holt and Co., New York, 1954.)

histories. One intriguing feature of the Bennettitales is the structure of their reproductive organs. Their strobili were bisexual with a general organization very similar to that of a primitive flower (Fig. 13-6). In other respects, the Bennettitales were much like present-day cycads in appearance (Fig. 13-7). Fossil remains of these plants have been found in many parts of the world, including parts of Europe, Asia, and North America. In the United States many specimens have been discovered in Wyoming and in the Black Hills of South Dakota. The latter area is so rich in fossil cycads that a national monument has been established to preserve and protect this deposit.

The order Cycadales includes all nine genera of living cycads. They are limited in their distribution to restricted regions of the tropics and subtropics (Table 13-1). The cycads look very much like palm trees with woody stems that may be short or tall, bearing frond-like leaves at the top. Most of the cycads are less than 2 meters high. *Zamia,* the only genus of cycads native to the United States, has a squat, tuberous stem with a crown of pinnate leaves. These plants are **dioecious**: male and female cones are formed at the center of the crown of leaves on separate sporophyte plants. After their release from the parent plant, Zamia seeds

are able to germinate immediately. The absence of dormancy, a phenomenon important in the survival of seed plants in regions where there are seasonal changes in climatic conditions, is certainly a reason for the very restricted distribution of the cycads in tropical areas of the world.

The Maidenhair Tree (Ginkgo): Survivor of an Earlier Age

The maidenhair tree (*Ginkgo biloba*) is the sole survivor of a class of gymnosperms that was widespread in distribution and relatively abundant in the Mesozoic era. The ginkgo tree was discovered growing in the temple gardens of eastern China and Japan and it seems likely that cultivation since ancient times prevented its extinction. This "living fossil" is now grown

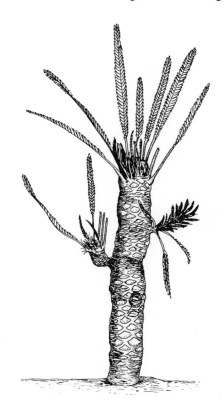

Fig. 13-7. *Williamsoniana.* Artist's sketch of reconstruction from fossil remains, ×1/30. (Redrawn from R. F. Scagel *et al.: An Evolutionary Survey of the Plant Kingdom.* Wadsworth Publishing Co., Belmont, Calif., 1965.)

extensively in North America and Europe and seems to thrive in our cities. The female trees are definitely less desirable as ornamental plantings than are male trees because after their fruits drop to the ground the fleshy outer layers decay, producing butyric acid, a compound that fills the air with the pungent odor of rancid butter and dog excrement. What possible evolutionary advantage the accumulation of this strange but not at all complicated compound provided for the *Ginkgo* is not known. The ginkgos share with the cycads and conifers

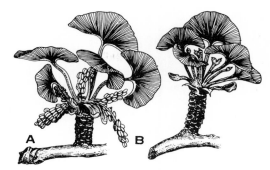

Fig. 13-9. *Ginkgo biloba* L., the maidenhair tree. A, short shoot of male showing young leaves and microstrobili (life size); B, short shoot of female tree showing young leaves and paired ovules borne on stalks (life size).

certain characteristics such as the structure of their spore-forming organs and patterns of gametophyte development. Ginkgo trees reach 30 meters or more in height with trunk diameters of up to a meter. Leaves are wedge-shaped with dichotomously branched veins. They are borne in clusters at the ends of specialized branches called short shoots.

Ginkgo trees are **deciduous**, forming a new set of leaves in late spring that are shed in the autumn. These plants are also **dioecious**, *i.e.,* they are either male or female and form only one kind of cone. The cones, like leaves, are borne at the tips of short shoots. Male strobili contain many stalked sporophylls, each of which bears two microsporangia. Pollen grains are released from the microsporangia at a four-celled stage of male gametophyte development. Ginkgo, like the cycads, is wind pollinated and the pollen tube, which develops after the male gametophyte has been carried by air currents through the micropyle, eventually releases two ciliated sperm cells. The female strobilus consists of a stalk bearing two ovules at its apex. Several such strobili can form at the tip of a short shoot. Within each ovule is a single megasporangium. The development of the haploid megaspore into a female gametophyte closely resembles that of the cycads. An unusual feature of the ginkgo female gametophyte is the presence of chlorophyll, a condition reminiscent of the photosynthetic and nutritionally independent gametophytes found in the lower plant groups. Following fertilization the ovule is transformed into a seed.

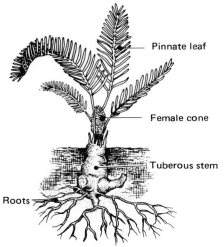

Fig. 13-8. *Zamia,* the only cycad native to the United States, has a short tuberous stem with roots at its base and a crown of pinnate leaves. A female plant is shown (×1/20).

Table 13-1
Distribution of living cycads[a]

Genus	Geographic Distribution
Cycas	Australia, East Indies, India, China, southern Japan
Macrozamia	Australia
Bowenia	Australia
Encephalartos	South Africa
Stangeria	South Africa
Zamia	Florida, West Indies, Mexico, Central America, northern South America
Microcycas	Western Cuba
Ceratozamia	Mexico
Dioon	Mexico

[a] After R. F. Scagel *et al.*: *An Evolutionary Survey of the Plant Kingdom.* Wadsworth Publishing Company, Inc., Belmont, Calif., 1968, p. 472.

Conifers (Coniferopsida)

The conifers are the most familiar and plentiful of the gymnosperms. Although they have been a part of the earth's vegetation since the Carboniferous period of the Paleozoic era and reached their peak during the Mesozoic era, they are still plentiful and form vast forests in many parts of the world.

All conifers are woody plants—a few are shrubs, but most are trees. Their leaves are simple (in contrast to the compound, fern-like leaves of the cycads), tend to be needle- or scale-like, and usually are evergreen, persisting for several years (in contrast to the deciduous leaves of the ginkgo).

Life History of a Typical Conifer—The Pine

Pines are common in the northern hemisphere. The sporophyte stage of the pine life cycle is the tree that we all recognize with its cones and clusters of evergreen, needle-like leaves. The familiar diploid tree is monoecious, producing both male and female cones. Microstrobili (male cones) are smaller than macrostrobili (female cones) and are borne in clusters at the ends of short shoots. They form during the summer, persist over winter covered with brown scales, and in the spring enlarge, releasing their pollen. Each male cone is made up of many spirally attached microsporophylls. Each microsporophyll has a stalk and two microsporangia. In the spring, microspore mother cells undergo meiosis, forming many haploid microspores. Before the pollen grains are released from the microsporangium, two cell divisions occur and the outer wall bulges at two points, producing a characteristic winged structure. The four-celled male gametophytes are carried by air currents, and these tiny, buoyant structures can travel up to hundreds of miles in a stiff wind.

Female strobili are generally much larger than male cones (over half a meter long in some species of pines) and they are borne singly. The

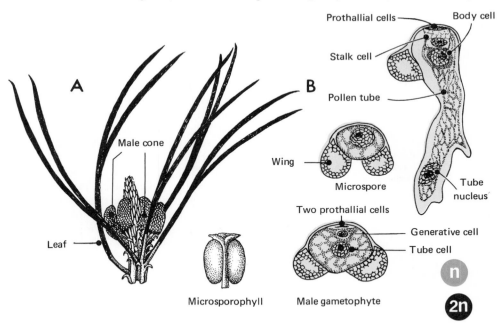

Fig. 13-10. Pine male cones and male gametophyte. A, short shoot bearing cluster of male cones (×½). A single microsporophyll is enlarged at right (×20). B, stages in development of the male gametophyte from the microspore. By the time of its release, the winged pollen grain is a male gametophyte with four cells (left). A nearly mature male gametophyte with developing pollen tube is shown on the right. (From H. J. Fuller and O. Tippo: *College Botany,* revised edition. Henry Holt and Co., New York, 1954.)

general structure is similar to that of male cones with a central axis and spirally attached megasporophylls. Each megasporophyll is attached in the axil of a tiny leaf-like bract, and on its upper surface there are two ovules. Each ovule consists of a megasporangium with its surrounding integument. A single megaspore mother cell undergoes meiosis and forms four haploid megaspores, three of which disintegrate. Note that this is what happens in the production of animal eggs. The remaining haploid cell goes through a period of free nuclear division, followed by cell wall formation and the maturation of a female gametophyte having two or three archegonia. Early in its development, the female cone is a soft, green structure. Hardening and the change in color to brown occur after pollination.

At the time of pollination in the spring, a sticky fluid exuded through the micropyle of the ovule traps the windborne pollen grains. This sticky, sugar-containing substance contracts as it dries, drawing the pollen grains through the micropyle and into contact with the nucellus. At this point the male gametophytes germinate, sending pollen tubes down through the nucellus. While the pollen grains are germinating, the megasporocyte is about to enter meiosis. Development of the female gametophyte is a slow process which is not completed until the next spring. Thus, fertilization occurs about a year after pollination. The male gametophyte that persists for all this time has a branched pollen tube that absorbs nutrients from the nucellus. Its growth is controlled by the tube nucleus. Just a few days before the pollen tube finally reaches the female gametophyte, the nucleus of the body cell of the male gametophyte divides to form two sperm nuclei. As the tip of the pollen tube reaches an archegonium, its cytoplasm and nuclei are discharged into the egg cell. One of the sperm nuclei unites with the egg nucleus to form the zygote and the remaining three nuclei of the male gametophyte disintegrate.

At the beginning of embryonic development, the zygote undergoes a period of free nuclear division, forming nuclei which migrate to the lower end of the single multinucleate cell. There, cell walls are laid down forming four rows of four cells each. Each cell in the bottom

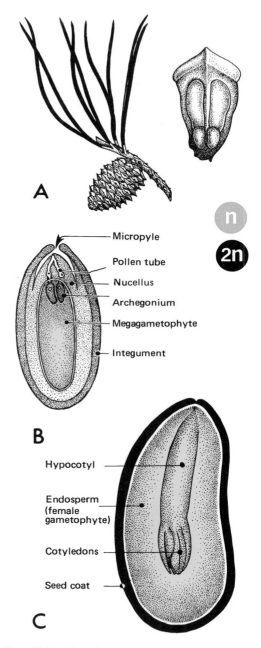

Fig. 13-11. Pine female cone and female gametophyte. A, twig with female cone (X¼). A single megasporophyll enlarged at right shows two winged seeds (X5). B, longitudinal section through pine ovule. Pollen tubes are shown penetrating the nucellus (X50). C, longitudinal section through pine seed (wing not shown), X50.

layer of cells can form an embryo but ordinarily three of the four abort; one embryo continues development while the rest disintegrate. As embryonic development continues the ovule is transformed into a seed. This structure consists of the embryo (differentiated into a hypocotyl, several cotyledons, and an epicotyl) embedded in the endosperm (female gametophyte tissue) and surrounded by a protecting seed coat (derived from maternal sporophyte tissue). In its second year, the cone becomes hard and woody. At the time of seed dispersal, about 18 months after the female cone began to form, the scales curve outward. Pine seeds have wing-like extensions of the integument that assist in their dispersal by wind. Following a period of dormancy, the seeds germinate when appropriate conditions of temperature and moisture prevail. During germination, the radicle (young root) at the tip of the hypocotyl is the first part of the young sporophyte to push through the seed coat. As the cotyledons (or seed leaves) emerge and the seed coat is shed, the young diploid plant begins its existence as a photosynthetic organism. Years later it will reach the stage of maturity when the first cones form and the cycle of life will be repeated.

Conifers of the Past and Present

Two orders of conifers are recognized: the extinct Cordaitales and the Coniferales, which include both fossil and living representatives.

The Cordaitales. The *Cordaitales* are believed to have been the progenitors of the conifers. Most of the fossil remains of this group that have been studied are no more than fragments

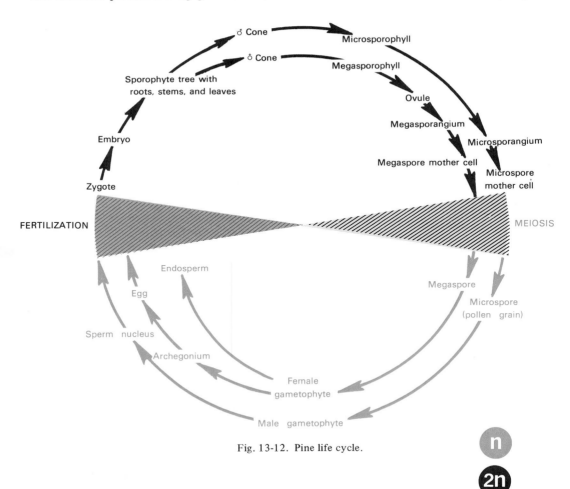

Fig. 13-12. Pine life cycle.

Strobili →

Fig. 13-13. *Cordaites.* Artists's sketch of reconstruction from fossil remains (×1/100). Terminal branches with long slender leaves and strobili were borne on tall trunks. (Redrawn from H. J. Fuller and O. Tippo: *College Botany,* revised edition. Henry Holt and Co., New York, 1954.)

of petrified wood. The best known species is *Cordaites,* one of the tallest trees of its time. Fossils of leaves and strobili attached to stems have been found for this species and reconstructions of this plant have been made (Fig. 13-13).

Cordaites had a long, slender trunk and petrified stems have been discovered that are over 20 meters long. At the top of the trunk, branches bore spirals of narrow, tapered leaves that were up to a meter in length. Strobili were borne in the axils of leaves on outer branches.

The Coniferales. The *Coniferales* are a varied and abundant group with six distinctly different families. Because of their peculiar geographic distribution, with some found only in the northern hemisphere and others only in the southern hemisphere, several of the conifer families seem strangely foreign to all except those who are world travelers or habitués of botanical gardens. In order to introduce those families of conifers which are native to distant parts of the world, Figure 13-15 shows representative species of each family and Table 13-2 lists some of their basic characteristics, including the geographical areas where they are found.

The Gnetopsida

The fourth class of gymnosperms, the Gnetopsida, are a strange collection of plants exhibiting characteristics that can be considered intermediate between the rest of the cone-bearing plants and the flowering plants. They have a very diverse morphology.

The order Welwitschiales, with its single species *Welwitschia mirabilis,* which occurs only in the Namib Desert of Southwest Africa, is like no other plant. Its woody stem, shaped like an inverted cone, is over a meter in diameter at the top and extends below ground as a long

Fig. 13-14. *Welwitschia mirabilis.* A, female plant, one of the largest living specimens, is about 1.5 m. in diameter. B, close-up of megastrobili. Each is about 15 cm. long, C, a large male plant over a meter in diameter. (From C. H. Borman *et al.: Madoqua,* Series II, vol. 1, Nos. 54-62, 1972.)

Table 13-2
Summary of the Coniferales: representative genera, general characteristics, and distribution[a]

Family Name	Number of Genera	Common Names of Representative Genera	General Characteristics	Geographical Range
Pinaceae (*Abietaceae*)	10	Pine, larch, spruce, hemlock, fir	Leaves needle-like and borne in spirals; conical growth forms; monoecious	Most are restricted to the Northern Hemisphere
Taxodiaceae	10	Redwood, bald cypress	Leaves varied: scale-like, flat-bladed, or needle-like; monoecious	Native to China, Japan, Formosa, Tasmania, U. S. (Pacific coast and southeast), and Mexico
Cupressaceae	16	Juniper, cypress, arbor vitae, white cedar	Reduced, scale-like leaves; monoecious or dioecious	Widespread in both Northern and Southern hemispheres
Araucariaceae	2	Monkey-puzzle tree, Norfolk Island pine	Leaves broad to needle-like; whorled; symmetrical branches; monoecious or dioecious	Almost all are native to the Southern Hemisphere, widely distributed as cultivars
Podocarpaceae	7	*Podocarpus*	Leaves needle-like or broader; dioecious	Most are native to the Southern Hemisphere, a few as far north as Central America and the West Indies; cultivated as a shrub in the southern U. S.
Taxaceae	5	Yew, *Torreya*	Female cones absent; ovules borne terminally on short lateral branches; dioecious	Most are native to the Northern Hemisphere

[a] After R. F. Scagel *et al.: An Evolutionary Survey of the Plant Kingdom.* Wadsworth Publishing Company, Inc., Belmont, Calif., 1968, p. 499.

Fig. 13-15. Foliage and cones of representative families of conifers. A, Pinaceae: a, *Pinus* (pine), ×¼; b, *Larix* (larch), ×¼; c, *Keteleeria,* female cone, ×¼; d, *Abies* (fir), needles and erect cones, ×¼. B, Taxodiaceae: a, *Sequoia* (redwood), male cone at tip of twig (top) and female cone (bottom), ×¼; b, *Sciadopitys,* clusters of leaves at branch tips and one female cone, ×¼; c, *Cryptomeria,* pointed needle-like leaves and female cone (right) and male cone (left), ×¼; d, *Cunninghamia,* needle-like leaves and female cone, ×¼. C, Cupressaceae: a, *Biota,* scale-like leaves and cone with opened scales and attached ovules, ×¼; b, *Cupressus* (arborvitae), scale-like leaves and female cones, × 1/3; c, *Juniperus* (juniper), relatively larger leaves on branches and scale-like leaves on cone-bearing twigs (left), female cones with fused bracts and scales (right), life size. D, Araucariaceae: a, *Agathis,* leaves (left) and male cone (right), ×¼; b, *Araucaria,* male cone of *A. araucaria* (left), shoot of *A. excelsa* (right), ×¼. E, Podocarpaceae: a, *Podocarpus,* male cones and leaves, ×¼; b, *Dacrydium,* scale-like leaves, ×¼; c, *Phyllocladus,* leaf-like appendages are modified branches, ×¼. F, Taxaceae: a, *Taxus* (yew), leaves borne in two rows and seeds, ×¼; b, *Torreya,* needle-like leaves and seeds, ×¼. (Redrawn from R. F. Scagel *et al.*: *An Evolutionary Survey of the Plant Kingdom.* Wadsworth Publishing Co., Belmont, Calif., 1965).

taproot. Two enormous leaves persist for the life of the plant (a hundred years or more), growing continually through the activity of their basal meristems.

The Gnetales are woody plants; some are vine-like and others are trees; all have broad, evergreen leaves.

The Ephedrales, shrubs found in the desert areas of the southwestern United States, in some ways resemble *Psilotum* and *Equisetum*. In spite of their rarity and limited geographic distribution, at least one order of the Gnetopsida, the Ephedrales, have been used extensively because of their medicinal properties.

The alkaloid, ephedrine, which is an active vasoconstrictor (causing contraction of blood vessels), is used in the treatment of allergies to relieve nasal congestion. It is extracted from the mahuang plant (*Ephedra vulgaris*), a species native to China and used by the Chinese for over 4,000 years (since 2737 B.C.). The medicinal value of the Ephedrales was also known to the Indians of the southwestern United States and Mexico. These people used extracts of roots and stems to treat urogenital ailments and as a beverage, and the seeds were used to make bread. A species of *Ephedra* known as "joint fir," found in the desert areas of the southwestern United States, is a plant grazed by cattle and sheep and therefore an important component of the rangeland vegetation.

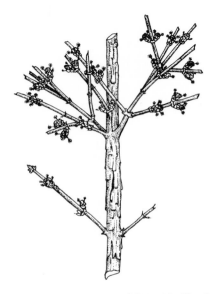

Fig. 13-16. *Ephedra antisyphilitica* M. Woody shoot with microstrobili, ×2/3. (Redrawn from H. C. Bold: *Morphology of Plants*, Harper & Row, Publishers, New York, 1967.)

USEFUL REFERENCES

Bold, H. C. *Morphology of Plants,* 2nd ed. Harper & Row, Publishers, New York, 1967.

Chamberlain, C. J. *Gymnosperms: Structure and Evolution.* University of Chicago Press, Chicago, 1935.

Foster, A. S., and Gifford, E. M. *Comparative Morphology of Vascular Plants.* W. H. Freeman and Co., San Francisco, 1959.

Jaques, H. E., *How to Know the Trees.* Wm. C. Brown Co., Dubuque, Iowa, 1946.

Scagel, R. F., Bandoni, R. J., Rouse, G. E., Schofield, W. B., Stein, J. R., and Taylor, T. M. C. *An Evolutionary Survey of the Plant Kingdom.* Wadsworth Publishing Co., Inc., Belmont, Calif., 1967.

Flowering Plants: The Angiosperms

14

In all the plant kingdom, no group exhibits greater diversity of form or number of species (some 200,000 to 300,000) than the flowering plants. They form the dominant kinds of vegetation over much of the earth's land masses; they are the plants on which our agriculture is based; and there are very few of mankind's activities in which the flowering plants are not somehow directly or indirectly involved. Although they are the most highly evolved plant group, they first appeared over 100 million years ago during the Cretaceous period, when dinosaurs dominated the land. Their long survival and present dominance are due to the great diversity and effectiveness of their adaptations. These include the capacity to survive under a tremendous range of climatic conditions, development of herbaceous forms which can complete their life cycles in a single growing season, and physiological or biochemical subtleties such as germination inhibitors which must be washed out before a seed can sprout. In the angiosperms the seed habit has reached astounding levels of efficiency. There are marvelous adaptations such as parachutes, wings, barbs, sticky surfaces, and the like (see Chapter 18) which ensure a wide dissemination of the embryo plants housed in seeds.

ANGIOSPERM CHARACTERISTICS

Flowers, the reproductive organs of the angiosperms, are used by botanists as key characteristics for their identification and classification. These are parts of the plant which produce seeds and ensure the survival of the species. Some aspects of flower structure, including the colorful and showy petals, nectars, and aromas that entice cooperative insects and other animals to transport the pollen needed for fertilization, also are features which make these structures esthetically pleasing. Evidence of the use of flowers for ornamentation and as motif for art can be found in the artifacts and records of practically all civilizations.

The term **angiosperm** (*angeion*, enclosing vessel, + *sperma,* seed) designates the single universal characteristic of the flowering plants: the enclosure of the ovule, which matures to form a seed, in a hollow structure called the ovary. Aside from this feature and a peculiarity in sexual reproduction where a double fertilization occurs, it is impossible to make generalizations about the flowering plants that are universally true. Although the angiosperms are mostly terrestrial, individual species have preferences for degrees of dryness or wetness in their habitats ranging from those characteristic of xerophytes (like cactus) to hydrophytes, with a few like water lilies being strictly aquatic. In their preferences for or tolerance of other environmental factors from temperature to light intensity, the flowering plants exhibit a diversity as great as they exhibit with respect to moisture. Their range extends from the barren arctic to the lush tropics, from alpine heights to sea level.

HUMAN USE OF THE ANGIOSPERMS

The economic importance of the angiosperms is beyond estimation. Human food

Fig. 14-1. Early painting of a flowering plant from the *Dioskurides*. This book, first prepared about 512 A.D. for the imperial princess Anicia Juliana at Byzantium, contains illustrations of many medicinal herbs which had been described by Dioskurides in the first century A.D. The picture reproduced here is from a facsimile edition of the restored manuscript in the Austrian National Library.

drugs, rubber, resins, dyes, and insecticides. To introduce the variety of special products obtained from the flowering plants, a very abbreviated sampling is provided in Table 14-1.

MAJOR GROUPS OF ANGIOSPERMS

The seed plants are divided into two subdivisions: the gymnosperms already discussed in Chapter 13 and the angiosperms. The flowering plants are in turn divided into two classes: the **dicots** (Dicotyledoneae) and the **monocots** (Monocotyledoneae). These two classes of flowering plants differ in a number of easily recognizable ways. The characteristic difference between monocots and dicots is found in their seeds and seedlings: monocots have but one cotyledon or seed leaf while dicots have two. What the adaptive and evolutionary meaning of this difference can be remains a mystery. The difference certainly began far back in the history of plant life. The dicots are herbs, vines, shrubs, or trees (familiar examples are the dandelions, tomatoes, poison ivy, roses and lilacs, maples, and oaks), having broad leaves with veins in a ribbed or netted pattern, and stems with vascular tissues arranged in a circular pattern, often with considerable secondary growth (*i.e.*, in diameter). The monocots, on the other hand, characteristically have narrow leaves with parallel veins. Their stems have relatively less conducting tissues than those of the dicots. Vascular bundles are scattered through ground tissue at the center of the stem. Because there is no cambium, secondary xylem and phloem are not formed. Familiar examples of the monocots include the grains, grasses, rushes, sedges, lilies, and palms.

The two classes of flowering plants are further divided into subclasses and orders. A complete listing of the angiosperm orders can be found in the classification of the plant kingdom (see appendix). Although the total number of known species of flowering plants is now approaching 300,000, many more may yet be discovered in parts of the world that are still relatively unexplored by trained botanists. Thus we can expect the number of known species to increase as the vegetations of such regions as South America, Africa, and Southeast Asia are examined more closely—in some cases for the

supplies are heavily dependent on this plant group. In many parts of the world grains such as rice, wheat, or corn provide the staple of human nutrition supplemented by fruits and beverages prepared from seeds (coffee and cocoa) or leaves (tea). The extravagant, protein-rich diets of North Americans and Europeans are only a step or two removed from the angiosperms in the food chain. The cattle, sheep, pigs, and fowl that provide the meats in our diets graze on angiosperm foliage or consume their fruits or seeds in the grains that they are fed. In addition to providing foods and beverages, the flowering plants provide fibers that make our clothing (cotton and linen); the hardwoods used to make furniture and as building materials; spices, oils, waxes, perfume, first time. There has been a tendency in recent years to regard taxonomy as a rather dull and unrewarding aspect of the life sciences. In terms of our understanding of evolution, or our search for new crops, drugs, and pesticides, however, plant taxonomy provides the input of basic information. For, after all, a name and a description are needed if a plant species is to be used as the source of a chemical, for introduction as a new crop, or for the genetic engineering involved in the improvement of existing crop varieties.

The Angiosperm Life Cycle

The essential features of the life cycles of flowering plants have been briefly described in Chapter 9 ("Plant Development") and will be dealt with again in Chapter 18 ("Reproduction in Plants"). Here we will concentrate on aspects of the angiosperm life style which are unique to this group of living things.

The most logical place to begin is with the seed, that remarkable structure which is the product of sexual reproduction and the beginning of a new diploid sporophyte plant. Seeds released from the parent plants are transported by such varied agents as wind, water, gravity, and animals. When their period of dormancy has passed and conditions are favorable for germination (appropriate soil, moisture, temperature, oxygen, and light) the further development of the new sporophyte plant begins. As germination proceeds, the root is the first part of the embryo to break through the seed coat. It anchors the young plant in the soil and absorbs water and mineral nutrients needed by the growing embryo. As the shoot emerges and the first leaves expand and become green, the seedling's period of dependence on stored foods obtained from the parent plant ends, and it becomes capable of synthesizing its own organic compounds by the process of photosynthesis (see Chapter 15). Ordinarily the sporophyte plant which develops, whether a tomato plant or a maple tree, goes through a period of vegetative development lasting for a few months to several years before reaching the stage of maturity when reproductive structures are formed. During the phase of vegetative growth, existing plant parts enlarge and addi-

tional new organs (roots, stems, and leaves) are formed. The resulting plant consists of diploid cells.

When conditions are right for reproductive development, flowers form. In these reproductive structures of the angiosperms, microspores (which develop into pollen grains) and megaspores are formed in the male and female parts of the flower, the **stamens** and **pistils**. The gametophyte stage of the life cycle that begins with the formation of haploid micro- and megaspores is very much reduced in all seed plants. In fact, if a further reduction were to occur, they would almost disappear completely. The male gametophyte consists of the pollen grain and the pollen tube that develops as it germinates. This structure contains three haploid nuclei, one **tube nucleus** that guides the development of the pollen tube, and two **sperm nuclei** that function in fertilization. The female gametophyte that develops from the haploid **megaspore** is called the **embryo sac**. This structure located within the ovule is multinucleate. A typical embryo sac has eight nuclei: one **egg nucleus**, two **synergids**, two **polar nuclei**, and three **antipodals**. When the pollen tube reaches the embryo sac, the two sperm nuclei are released and a **double fertilization** occurs. One sperm nucleus fuses with the egg to form the diploid **zygote** from which an **embryo** develops. The other sperm nucleus fuses with the two polar nuclei, producing a triploid **primary endosperm nucleus** from which the **endosperm** tissue that provides reserve nutrients for the embryo is formed. This second fertilization that forms the initial cell of a polyploid nutritive tissue occurs only in the angiosperms. You will recall that in the gymnosperms the endosperm consists of haploid cells derived from the female gametophyte.

In the flowering plants the reduced male and female gametophytes are completely dependent on the parent sporophyte generation. The female gametophyte, the embryo sac, remains embedded within the ovule. Antheridia and archegonia are not formed; the only vestiges of these structures are the egg and sperm nuclei.

In their sexual reproduction the flowering plants certainly represent the epitome of adaptation to life on the land. Male gametophytes (**pollen grains**) are carried by wind,

Table 14-1

Flowering plants useful to man—a sampling of those yielding products of special use

Plant Product	Source Species	Family	Geographical Area Where Grown or Collected for Commercial Use	Comments on Active Components, Parts of Plant Used, and Processing
Beverages				
Coffee	*Coffea arabica*	*Rubiaceae*	South America (Brazil, Columbia); Africa (Ivory Coast, Angola, Uganda); Central America (Salvador, Guatemala, Costa Rica); Mexico; Indonesia.	Caffeine acts as a mild stimulant and diuretic. Seeds are dried and roasted, extracted with boiling water.
Tea	*Thea sinensis*	*Theaceae*	Northeast India, Ceylon, China, Japan, Indonesia	Caffeine acts as a mild stimulant. Leaves are wilted (allowed to ferment) prior to drying.
Cocoa	*Theobroma cacao*	*Sterculiaceae*	Africa (Ghana, Nigeria, Ivory Coast, Cameroons); South America (Brazil, Equador); Mexico; Dominican Republic	Seeds are fermented and then roasted. Extracted materials include chocolate flavoring, cacao butter (an edible fat used to make chocolate candy) and the caffeine-like alkaloid, theobromine, used in making cola beverages.
Cola	*Cola nitanda*	*Sterculiaceae*	Africa, Jamaica	Seeds are rich in caffeine and alkaloids.
Oils, fats, and waxes				
Coconut oil	*Cocos nucifera*	*Palmaceae*	South Pacific	Coconut fruits yield oil and edible coconut "meat."
Palm oil	*Elaeis guineensis*	*Palmaceae*	West Africa, Indonesia, Malaysia	Oil is extracted from outer husks of fruits. Used in industry, foods, and dentifrices.
Linseed oil	*Linum usitatissinum*	*Linaciae*	U. S., Canada, Argentina, India	Oil is pressed from seeds, used in manufacture of paints, oil cloths, and linoleum.
Safflower oil	*Cathamus tinctoria*	*Compositae*	U. S., India, Russia	Oil is extracted from seeds by solvents or pressing; used as cooking oil, to make margarine, manufacture of resins and paints.
Carnauba wax	*Copernica cerifora*	*Palmaceae*	Brazil	Wax, separated from surfaces of wilted leaves, is used in manufacture of floor and automobile polishes, phonograph records, plastics, and cosmetics.
Spices, flavorings, and perfumes				
Black pepper	*Piper nigrum*	*Piperaceae*	Southeast Asia	Dried fruits are ground to produce the spice.
Cinnamon	*Cinnamonum zeylanicum*	*Lauraceae*	Ceylon, India	Dried bark is used directly (cinnamon sticks) or ground to form powder.
Vanilla	*Vanilla planifolia*	*Orchidaceae*	Madagascar	Vanilla beans (seed pods) are fermented and then dried and extracted with alcohol. Vanillin, the active ingredient ($C_8H_8O_3$), can easily be synthesized.

	Scientific name	Family	Region	Uses
Rose oil	*Rosa* spp.	*Rosaceae*	Southern Europe, Asia Minor	Flowers are collected in the late bud stages; essential oils (citronellol, geraniol, nerol, linalol, etc.) are separated by distillation or solvent extraction; used for perfumes.
Industrial oils				
Camphor	*Cinnamomum camphora*	*Lauraceae*	China, Japan	Camphor oil is extracted from the wood of camphor trees; used in the manufacture of liniments and insecticides and formerly in the making of celluloid from nitrocellulose.
Medicines and drugs				
Gum arabic	*Acacia* spp.	*Leguminoseae*	North Africa	Extracts have been used as medicines since the time of Herodotus; presently used as an emulsifying agent in cough syrups.
Belladonna	*Atropa belladonna*	*Salonaceae*	Europe, North America	The alkaloid, atropine, is extracted from dried foliage with solvents such as ether or ethyl acetate; used to stimulate circulation, to dilate pupils of the eyes for optical examination, and as a stimulant for the central nervous system.
Opium	*Papaver somniferum*	*Papaveraceae*	Balkans, Near East, Southeast Asia, China, Japan	Active ingredients in opium are the alkaloids morphine and codeine. Seed capsules are picked by hand; the next day drops of milky exudate are scraped off and dried. Widely used as a pain killer and an addictive drug.
Quinine	*Chinchona* spp.	*Rubiaceae*	Java, Andean highlands of South America	Extracts of the bark of *Cinchona* trees are used to treat malaria. Extracts also contain the alkaloid quinidine, useful as a cardiac depressant.
Reserpine	*Rauwolfia serpentina*	*Apocynaceae*	India	Root extracts have long been used in one treatment of mental illness. Reserpine is chemically similar to serotonin and LSD. Acts as a tranquilizer and lowers blood pressure.
Insecticides				
Pyrethrum	*Chrysanthemum cinaerarifolium*	*Compositae*	Japan, Africa	Dried flower heads are used as a source of pyrethrins and volatile oils. Used in sprays and dusts to protect gardens and stored grains against insect damage.
Rotenone	*Lonchocarpus nicon*	*Leguminoseae*	South America, Far East	Dried roots are extracted or pulverized for use as dusts. Active components are lipophilic aromatic compounds. Nontoxic to mammals—used by South American Indians as a poison for fish used for food.

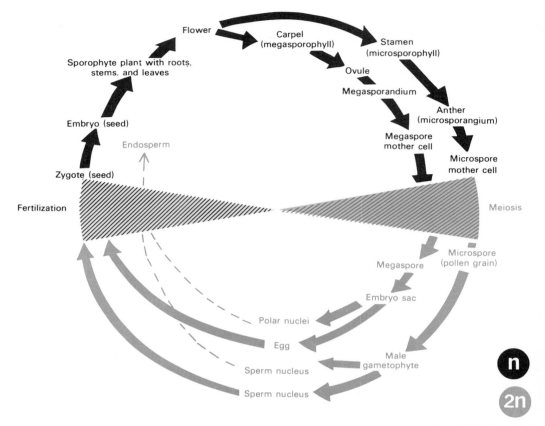

Fig. 14-2. Angiosperm life cycle. (After H. J. Fuller and O. Tippo: *College Botany*, rev. ed., p. 832. Henry Holt and Co., New York, 1954.)

insects, or other animal vectors. Once transferred to the stigma of a flower, pollen tubes develop that carry the sperm nuclei to the embryo sac where fertilization occurs. Nowhere in this sequence is there a need for water as a medium for the transport of gametes.

Once aware of the drastic reduction in the haploid phase of the life cycle in the angiosperms, it is tempting to question whether the ultimate reduction in gametophytes has been achieved by these plants or whether further reduction is possible. Certainly the male gametophyte has reached the limit of reduction: each of the sperm nuclei is involved in fertilization and the tube nucleus is needed to regulate the development of the pollen tube. On the other hand, among the eight nuclei present in the typical embryo sac described, there are five with no obvious function: the

three antipodals and two synergids. If we examine other angiosperm megagametophytes, however, we soon find that further reduction can be found, along with examples of less reduction. There is great variety in the patterns of development with one, two, or four haploid spores taking part in the formation of the female gametophyte. In all, 10 different types of embryo sac development have been recognized with anywhere from 4 to 16 nuclei present in the mature structure. Even in the most reduced case (four haploid nuclei) there is at least one superfluous nucleus within the embryo sac.

Following fertilization the embryo plant develops from the zygote. The nutritive endosperm tissue may remain a structural entity recognizable in the mature seed (as it is in castor beans and corn) or it may be absorbed

by the developing embryo during seed maturation (as it is in peas and beans). As the seed
matures, the ovule wall hardens to become the
seed coat. In many plant species the walls of
the ovary in which the ovules are borne become
fleshy and the resulting fruit (containing one to
many mature ovules, or seeds) is released from
the parent sporophyte plant. Thus we arrive
back at the point where we began to examine
the life cycle of the flowering plants, albeit one
generation later.

Flowers

In reviewing the essential features of angiosperm life cycles, we said little about flower
structure except to mention that male gametophytes (pollen grains) are produced by the
stamens and that female gametophytes (embryo
sacs) form within the ovules which are located
in the **pistils**. Angiosperm flowers exhibit the
most diverse imaginable colors, shapes, and
structural patterns. This diversity is seen in the
number and arrangement of the functional
parts, the stamens and pistils, and also in the
leaf-life sepals and petals.

What is a flower? It seems appropriate that
Goethe, the German poet and naturalist, was
one of the first to propose a theory about the
nature of flowers. In 1790, he suggested that a
flower is a modified shoot, the appendages of
which are morphologically equivalent to leaves.
It is not at all difficult to see the close
resemblance between the sepals or petals of
many flowers and leaves. Look at the sepals and
petals of a water lily: their flattened, blade-like
shapes and venation are visible to the naked
eye. It takes somewhat more imagination to
visualize how a stamen or a pistil might have
evolved from leaves bearing surface sporangia.
Yet, beginning with such a structure, its rolling
inward and fusion with like parts could result
in the hypothetical sequence shown in Figure
14-4, which would have produced the kind of
compound ovary that you can see every time
you slice a tomato. The same kind of evolutionary sequence is thought to have given rise
to the anthers in which pollen grains are
produced. What evidence for this theory exists
is found in the stamens of supposedly primitive
flowering plants of the family Ranales, which
includes the magnolias and buttercups (Fig.
14-5).

Fig. 14-3. J. W. von Goethe, 18th-century German
poet and naturalist, whose concept of the flower as a
modified shoot is still accepted by botanists. (From a
painting by Stieler, in *Goethes Morphologische Schriften*, ed. by W. Troll. Eugen Diederichs Verlag, Jena.)

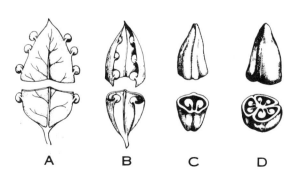

A B C D

Fig. 14-4. Possible stages in the
evolution of the pistil. A, primitive pistil with seeds borne along
the edges of a leaf-like carpel; B,
carpel rolling inward; C, simple
pistil after fusion of edges of
rolled carpel; D, compound pistil
resulting from fusion of three
carpels. (Redrawn from W. H.
Muller: *Botany: A Functional
Approach*, 2nd ed. The Macmillan
Co., London, 1969.)

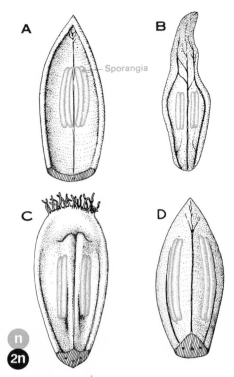

Fig. 14-5. Primitive types of stamens found in the Ranales (buttercup family). Microsporangia are borne on leaf-like structures. A, *Austrobaileya*; B, *Himmantandra*; C, *Dengeneria*; D, *Magnolia*. (Redrawn from Canwright: American Journal of Botany, *39:* 484, 1952.)

The component parts of flowers—the sepals, petals, stamens, and pistils—may be assembled in an endless variety of ways, yet this structure is the least variable part of a plant and therefore the most reliable feature to use in the identification of angiosperm species. Flower parts are attached to an enlarged apex of the flower stalk called the **receptacle**. The outermost appendages, the **sepals**, that open first as a flower bud enlarges and unfolds, are frequently green and are the most leaf-like of the floral parts. Next apically and inward are the **petals**, often colored and showy. Collectively, sepals are called the **calyx** and petals are referred to as the **corolla**; together they form the **perianth**. The petals can play an important role in attracting insects and other animal pollinating agents.

Animals are attracted to some flowers by their colors and the patterns of pigments, some of which absorb ultraviolet light. Patterns due to ultraviolet absorption are visible to insects but not to humans. In other species, particularly those attracting bees or hummingbirds, the lure is nectar produced by glands located near the bases of the petals. In yet other species, pollen carriers are attracted by fragrances and essential oils produced by the petals.

Inside the perianth are the **stamens**, the male structures of the flower. Most stamens have two easily recognized parts: a long supporting **filament** and an enlarged part called the **anther**. Within the anther, the microspore mother cells are formed which undergo meiosis and give rise to tetrads of haploid microspores. These in turn, develop into male gametophytes, the **pollen grains**.

At the center of the flower is the female functional part or **pistil**. At the top of the pistil is a sticky, enlarged surface called the **stigma**, on which the pollen grains land and germinate. The pollen tubes formed by the male gametophyte grow downward through an elongated part of the pistil called the **style** that connects the stigma and the bulbous basal **ovary**. Within the ovary there are one to many **ovules**. Within each ovule there is a megaspore mother cell from which haploid megaspores are formed, and finally the mature female gametophyte, the **embryo sac**. The pollen tubes grow down through the fleshy walls of the ovary to reach the ovules, and penetrate these structures through an opening in their integuments called the **micropyle**. Following fertilization, the embryo sporophyte plant forms within the ovule. A mature ovule is a **seed**. Pistils may be either simple or compound, *i.e.*, made up of one or more **carpels**, the unit of pistil structure. The garden pea (*Pisum sativum*) has a simple pistil composed of one carpel. Familiar species with compound pistils (consisting of two or more fused carpels) are the tomato (*Lycopersicon esculentum*) and the lily.

A given flower may or may not have all the four component parts just discussed. If all are present, the flower is said to be **complete**; if both stamens and pistils are present (regardless of whether there are sepals or petals) a flower is said to be **perfect**. A unisexual flower, missing

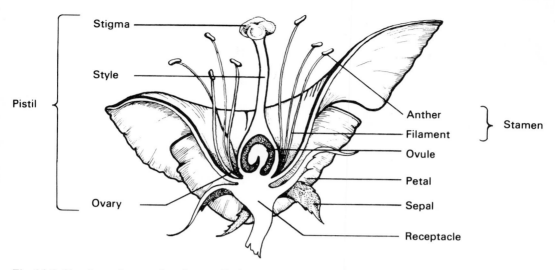

Fig. 14-6. Structure of a complete flower. (Redrawn from W. H. Muller: *Botany: A Functional Approach*, 2nd ed. The Macmillan Co., London, 1969.)

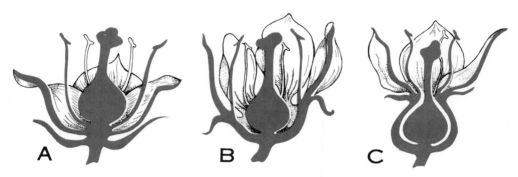

Fig. 14-7. Ways in which floral parts are attached relative to the position of the ovary. A, hypogynous flower with superior ovary; B, perigynous flower in which fused bases of other floral parts form a tube around the ovary but the ovary is still superior; C, epigynous flower with inferior ovary. (Redrawn from W. H. Muller: *Botany: A Functional Approach*, 2nd ed. The Macmillan Co., London, 1969.)

either stamens or pistils, is said to be **imperfect**; a bisexual flower is perfect, but may be **incomplete**.

In addition to variations in the floral parts present in a flower, there are several different ways in which floral parts are attached relative to the position of the ovary. In many flowers the ovaries are **superior**, *i.e.,* attached to the receptacle above stamens, petals, and sepals (as they are in the morning glories and buttercups). Such a flower is said to be **hypogynous**. In other flowers the bases of the stamens, petals, and sepals are fused to form a tube-like structure attached to the receptacle below the ovary. Examples of such **perigynous** flowers (in which the ovary is still superior) are the cherry, plum, and phlox. In a third pattern, the ovary is enclosed within an outgrowth of the receptacle. Such flowers (iris, apple, and dandelion) are said to be **epigynous**. Their ovaries are **inferior** and other flower parts are attached to the receptacle at a level near the top of the ovary.

Finally, in our review of the diversity of flower structure, we must note two further

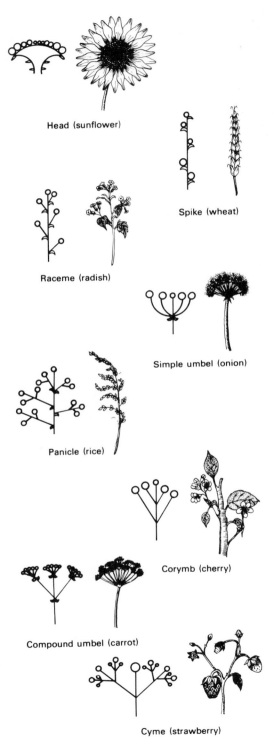

Head (sunflower)

Spike (wheat)

Raceme (radish)

Simple umbel (onion)

Panicle (rice)

Corymb (cherry)

Compound umbel (carrot)

Cyme (strawberry)

patterns: first, the fusion of parts, which gives rise to a variety of tubular and bell-shaped structures; and secondly, the tendency in some species or flowers to be borne in clusters (**inflorescences**).

Evolutionary Trends in the Flowering Plants

Faced with such variety, botanists have devoted much effort to trying to decide which characteristics are primitive and which are advanced so that evolutionary trends within the angiosperms could be established. So little was known about the origins of flowering plants a century ago that Charles Darwin referred to this very important aspect of the evolution of life as "an abominable mystery." Circumstances have not much improved since his time. Because of the fragile nature of flowers, they form a very rare part of the fossil record. Consequently, most of the hypotheses concerning the origins of the angiosperms, their relationships with other plant groups, and their evolutionary trends are based on comparative studies of living representatives. With so little evidence, how is it possible to decide what characteristics are primitive or advanced? Obviously the most primitive flowers should resemble the strobili of the conifers, presumed ancestral forms that have spirally arrayed parts on an elongate axis. It is not difficult to imagine how such an arrangement could be compressed, giving rise to a whorled attachment of floral parts. Also, separate parts would have to be present before they could fuse. Table 14-2 lists a number of floral characteristics, indicating for each what is generally agreed to be primitive and advanced; also, primitive and advanced patterns of growth are compared.

Although there is still little agreement among the experts on plant evolution about the origin of the flowering plants, it is quite probable that they evolved from one of the gymnosperm groups, possibly the seed ferns (Cycadofilicales).

There are two major theories about what living group of flowering plants is most primi-

Fig. 14-8. Some common examples of flowers borne in clusters (inflorescences).

Table 14-2
Evolutionary trends in the angiosperms

Floral Characteristics	Primitive	Advanced
Arrangement of floral parts	Spirally arranged	Whorled
Number of parts	Indefinite, *i.e.,* variable and numerous	Definite and few
Fusion of parts	Separate	Fusion of same or different parts
Symmetry	Radial	Bilateral
Position of ovary	Superior	Inferior
Differentiation of perianth	Sepals and petals similar	Sepals and petals different in color and shape
Solitary *vs.* clustered	Borne singly	Borne in clusters; the most advanced inflorescences are umbels and heads
Bisexual *vs.* unisexual	Bisexual (perfect)	Unisexual (imperfect)
Method of pollination	Insects	Wind
Growth patterns		
Growth form	Trees and shrubs	Herbs
Longevity	Perennial	Annual

Fig. 14-10. Buttercup flower (*Ranunculus*). Note whorl of five petals, numerous stamens, and central pistil composed of many separate carpels, each containing a single ovary. (From P. Jaeger: *The Wonderful Life of Flowers* (translated by J. P. M. Brenan, p. 64. E. P. Dutton & Co., Inc., New York, 1961.)

Fig. 14-9. Inflorescences of the goat willow (*Salix caprea*). These catkins open in early spring. (From P. Jaeger: *The Wonderful Life of Flowers* (translated by J. P. M. Brenan, p. 23. E. P. Dutton & Co., Inc., New York, 1961.)

tive. One proposes that the Hamamelidae, a group in which the willows, oaks, walnuts, and birches are found, are primitive. Their inflorescences (catkins) resemble the cones of gynosperms with many flowers borne in a long spike. Individual flowers are unisexual with wind-borne pollen. Flowers are yellow or green in color and lack petals. According to the second theory, the Ranales (buttercups) are the most primitive. In this group, flowers are borne singly, have numerous parts that are spirally arranged, are bisexual, colored, and are insect-pollinated. If we are to accept the generalizations about evolutionary trends in the angiosperms summarized in Table 14-2, then we should prefer the latter theory because of the obviously more primitive characteristics of flowers in the Ranales.

In spite of the controversy over which order, the willow-oak-birch group or the buttercup-magnolia group, contained the original ancestral angiosperm, it seems clear that one or another of these dicots rather than any monocot is the most primitive. The dicots differ from monocots and resemble the obviously more primitive cone-bearing gymnosperms like the pines in possessing an active cambium. Thus they can grow in diameter, adding annual rings in the woody forms. Even herbaceous dicots share with the gymnosperms the arrangement of their conducting tissues (xylem and phloem) in a ring within their stems. The monocots (lilies, grasses, palms, orchids, and their relatives) are thought to have evolved from some primitive group of dicots; in the process the cambium and one cotyledon were lost and vascular bundles became scattered.

Esthetic Appeal of the Angiosperms

The earth would be a very drab place without the flowering plants. Nothing in our environment provides more variety and color than the seasonal sequence of blooms we take so much for granted; without angiosperms there would be no lawns or gardens as we know them. Amateur horticulture, whether limited to ornamentals grown indoors, in flower boxes or in the yard, or extended to fruits and vegetables cultivated for the table or freezer, is a hobby available to anyone regardless of age, finances, or geography. Their cultivation provides a healthy respite from the stress of our technologically oriented society.

USEFUL REFERENCES

Baker, H. C. *Plants and Civilization*, 2nd ed. Wadsworth Publishing Co., Inc., Belmont, Calif., 1970.

Cronquist, A. *The Evolution and Classification of Flowering Plants*. Houghton Mifflin Co., Boston, 1968.

Cuthbert, M. B. *How to Know the Fall Flowers*. Wm. C. Brown Co., Dubuque, Iowa, 1948.

Cuthbert, M. B. *How to Know the Spring Flowers*. Wm. C. Brown Co., Dubuque, Iowa, 1948.

Eames, A. J. *Morphology of the Angiosperms*. McGraw-Hill Book Co., New York, 1961.

Esau, K. *Anatomy of Seed Plants*. John Wiley & Sons, Inc., New York, 1960.

Jaeger, P. *The Wonderful Life of Flowers* (translated by J. P. M. Brenan). E. P. Dutton & Co., Inc., New York, 1961.

Janick, J., Shery, R. W., Woods, F. W., and Ruttan, V. W. *Plant Science: An Introduction to World Crops*. W. H. Freeman and Co., San Francisco, 1969.

Pohl, R. W. *How to Know the Grasses*. Wm. C. Brown Co., Dubuque, Iowa, 1954.

6

PLANTS AT WORK

Photosynthesis

15

UNIQUE AND UNIVERSAL IMPORTANCE OF PHOTOSYNTHESIS

Photosynthesis is unique among all the metabolic processes that occur in living cells, because it provides the only means for an input of energy from the sun into the earth's biosphere. All living things except for a few obscure chemosynthetic bacteria are ultimately dependent on this process, a complex set of reactions that result in the conversion of light energy into the chemical bond energy of organic compounds.

Photosynthesis occurs only in those cells containing the green pigment chlorophyll. From the substrates carbon dioxide and water, new complex organic molecules are formed and molecular oxygen is released. Accompanying this, the energy of sunlight is stored in the energy-rich bonds of ATP. This compound can later be used as a source of energy for making more cells or reserve foodstuffs.

As intellectually curious animals, we should be appropriately impressed by the biochemical versatility and productivity of green plants. Not only our present well-being, but also much of our history is directly dependent on past photosynthetic productivity. For example, the industrial revolution was possible only because of the availability of vast reserves of fossil fuels, such as coal and oil, which are the partially decomposed products of plants living in past ages. If we look to the future, we realize that long voyages into space will be feasible only if appropriate kinds and numbers of photosynthetic organisms are taken along to recycle wastes and to provide a continuing source of oxygen and food. Photosynthetic organisms are the primary producers that form the basis of all food chains. Furthermore, these organisms are responsible for maintaining our aerobic atmosphere. Photosynthesis is the only source of molecular oxygen and all aerobic organisms are dependent on green plants for its continued availability.

The photosynthetic process has certainly influenced the evolution of all the organisms found on this planet. Aerobic respiration could not have evolved before photosynthetic organisms appeared. Furthermore, atmospheric oxygen made possible the accumulation of an ozone layer which acts as a shield to screen out much of the potentially harmful ultraviolet radiation in sunlight. Without such a screen, plants and animals could not live on the land.

Considering the amount of oxygen utilized in respiration and in the ceaselsss oxidation of materials in our environment, it is evident that the continuing production of oxygen by photosynthesis is essential to the maintenance of life on earth. There have been many estimates of the total amount of photosynthesis occurring annually. No two estimates agree but all are fantastically large. One such estimate, expressing the worldwide rate of photosynthesis in terms of the weight of carbon incorporated into new organic compounds, is given as 16×10^9 *tons* per year. Most of this 16 billion tons of carbon is captured by photosynthesis occurring in the oceans that cover approximately four-fifths of the earth's surface.

CHLOROPLASTS: THE PHOTOSYNTHETIC MACHINES

Chlorophyll, the molecule that captures light energy, is located in a cellular organelle called the chloroplast. There is great variety in chloroplast shape and size, especially in algal cells, where chloroplasts range from stars to spiral ribbons in external appearance (see Chapter 10). Most chloroplasts resemble a football, *i.e.,* are ellipsoidal. The number of chloroplasts per cell can vary from one to 50 or more.

Viewed with a light microscope under low power, chloroplasts appear to be uniformly green. Under oil immersion it is evident that chlorophyll is not uniformly dispersed within the chloroplasts, but is instead limited to certain regions called **grana**, which appear to be scattered through an optically clear colorless matrix called the **stroma**.

As revealed by the electron microscope, the basic structural pattern of a chloroplast is much like a many-layered sandwich. This organelle is bounded by a membrane that, like the unit membranes of the nucleus, mitochondria, endoplasmic reticulum, etc., is made up of double protein and lipid layers. Internally, there are many alternating double protein and lipid layers or **lamellae**. Some of these lamellae appear to extend through the entire length of the plastid, and at intervals there are discrete stacks of membranes much like collapsed balloons. The electron microscope reveals that these stacks of lamellae correspond to the grana visible in the light microscope. These chlorophyll-containing layers are also called **grana lamellae**; the layers extending through the colorless matrix are called **stroma lamellae**.

The positioning of chlorophyll molecules within the grana lamellae is thought to be determined by the chemical structure of chlorophyll. Like the cytochromes and hemoglobin, chlorophyll is a **porphyrin**. Porphyrins are a diverse family of chemicals having as their basic structural feature a **tetrapyrrole nucleus**, a ring structure made up of four pyrrole units (five-membered rings made up of four carbons plus a nitrogen atom) joined together by carbon bridges. Many porphyrin molecules have an atom of a metallic element at the center of the tetrapyrrole nucleus: iron in the case of the cytochromes and hemoglobin, magnesium in chlorophyll. Most of the porphyrins are colored, *i.e.,* they are light-absorbing pigments. They also tend to be water soluble and are usually associated with proteins. Thus, the porphyrin heme plus the protein globin together form hemoglobin. Porphyrins can function as the prosthetic groups for many enzymes (which you will recall are proteins) including cytochrome oxidase and catalase.

Chlorophyll follows family tradition in exhibiting all but one of the characteristics just mentioned. An unusual structural feature, a long-chain hydrocarbon ($C_{20}H_{39}$, a **phytol** group) attached to one of the four pyrroles of its porphyrin structure, makes chlorophyll insoluble in water. In other words, because of the influence of its phytol group on solubility, chlorophyll is **lipophilic**, or soluble in organic solvents. Chlorophyll molecules are located within the lipid portions of the grana lamellae

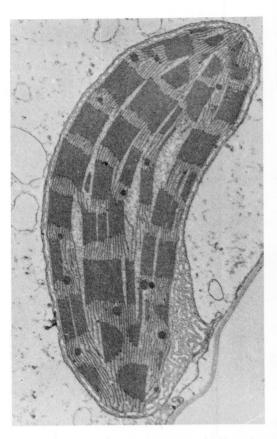

Fig. 15-1. Electron photomicrograph of chloroplast from maize (Indian corn), × 17,000. (From L. K. Shumway and T. E. Weirer: The chloroplast structure of iojap maize. American Journal of Botany, *54:* 744, 1967.)

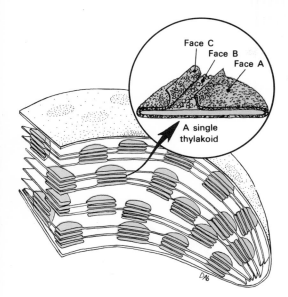

Fig. 15-2. Diagram showing details of chloroplast structure revealed by electron microscopy. Chlorophyll containing grana (colored green) are composed of stacks of vesicles called thylakoids. Freeze-etching techniques have revealed that thylakoid membranes are made up of smooth lipid layers penetrated by protein particles (see inset). Note that grana lamellae may be continuous with stroma lamellae. The plastid is bounded by a double membrane.

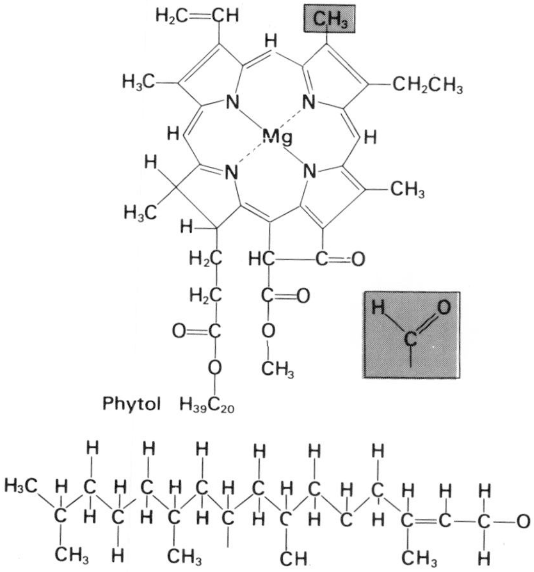

with their tetrapyrrole "heads" aligned along the lipid-protein interface. Associated with the chlorophylls in the lipid layers of the grana lamellae are the yellow carotenoid pigments. These carotenoids and other accessory pigments can also absorb light quanta. Thus they assist in

Fig. 15-4. Structural formula of chlorophyll *a*. Chlorophyll *b* has the same structure except that a –CHO group instead of a –CH$_3$ group is attached to one corner of the porphyrin ring. Structure of phytol (H$_{39}$C$_{20}$) group is shown below.

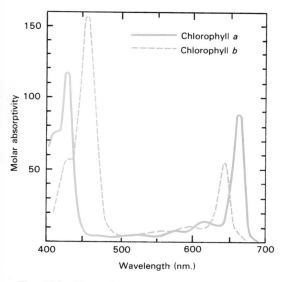

Fig. 15-3. Absorption spectra of chlorophyll *a* and chlorophyll *b* in ether.

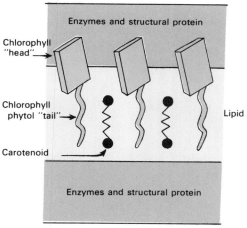

Fig. 15-5. Diagram showing possible orientation of chlorophyll and carotenoid molecules within protein and lipid layers of the grana lamellae.

the transfer of the absorbed light energy to reactive sites within the grana. The protein layers of the chloroplast lamellae contain structural protein and enzymes. Other enzymes are located in the optically clear matrix, or **stroma**.

Evidence that Chlorophyll Is the Light Receptor for Photosynthesis

An appreciation of the uniqueness of green plant tissues—their ability to remove CO_2 from air and to generate oxygen in the light—originates with the earliest chemical studies of gases and the composition of air. Priestley observed that an animal placed in a closed container would soon suffocate, but if a plant were added to the container, the animal could survive.

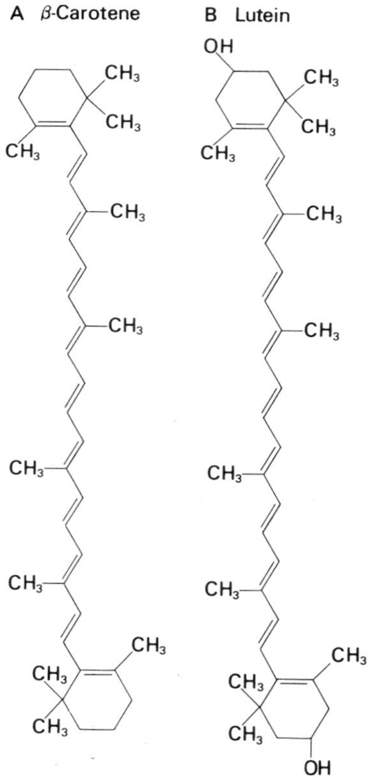

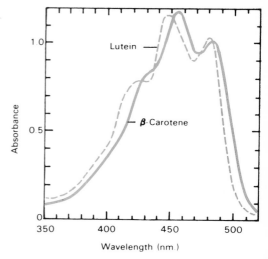

Fig. 15-7. Absorption spectra of two common carotenoids, β-carotene and lutein.

Therefore, he correctly concluded that plants remove a product of the animal's respiration from the air and so, in a sense, purify that air. However, Priestley did not realize why, on later occasions, this kind of experiment was not completely successful. In the 1790's Jan Ingen-Housz, the personal physician to Maria Theresa, Empress of Austria, and pioneer in the use of smallpox immunization, correctly interpreted Priestley's work in terms of the newly discovered gases, carbon dioxide and oxygen, and explained Priestley's later failures by showing that plants require light in order to remove CO_2 from the air and generate oxygen. Ingen-Housz also noted that only the green parts of plants can do this. He was the first to point out the importance of plants in the cycling of inorganic and organic material in nature: organic compounds synthesized from CO_2 and H_2O in green plant photosynthesis are consumed by herbivores; herbivores are in turn consumed by carnivores. CO_2 and H_2O, the products of respiration and of the decay of dead plants and animals, are released into the atmosphere. The only major input into this cycle of carbon compounds is photosynthesis.

Although the importance of green plants was recognized by earlier investigators, it was not until the late 19th century that experimental proof was obtained for the participation of

A β-Carotene   B Lutein

Fig. 15-6. Structure of two common carotenoids. A, β-carotene; B, lutein, a xanthophyll. These accesory pigments are located within the lipid portion of grana lamellae.

chlorophyll and chloroplasts in photosynthesis. This was the achievement in the 1880's of the German investigator, Theodor Engelmann, a physiologist also known for his work on the striations of voluntary muscle fibers. He initiated modern experimental studies of photosynthesis using very simple tools. An old fashioned light microscope and the response of motile bacteria, which move toward regions of greater oxygen concentration, enabled him to detect the site of photosynthetic oxygen evolution. By clever manipulation, he was able to focus a narrow beam of white light on filamentous green algae and diatoms, thereby illuminating only one portion of a cell at a time: a chloroplast, nucleus, or an area of the cytoplasm, etc. By placing the motile bacteria in the aqueous medium surrounding the algal cells and observing their response, it was possible to determine under what circumstances oxygen was being evolved. Engelmann observed that the bacteria clustered around the illuminated cell as close as possible to a chloroplast when it received light. He observed a random distribution of bacteria with no tendency to cluster when parts of the cell other than the chloroplast received light. Engelmann therefore

Fig. 15-8. Theodor Engelmann, the German physiologist who proved that chlorophyll is the photoreceptor of photosynthesis and is also known for his studies of muscle fibers. (Courtesy Johns Hopkins Institute of the History of Medicine.)

concluded that photosynthesis must occur within the chloroplasts.

The most apparent colored component of chloroplasts is the green pigment chlorophyll. How was it possible for Engelmann to demonstrate that this is the **photoreceptor,** *i.e.,* the light-absorbing component, of the photosynthetic apparatus?

As you will recall from the discussion of light effects on plant development in Chapter 9, white light is a mixture of all colors in the visible spectrum. Component wavelengths can be separated by passing a beam of white light through a prism. Projected on a white screen or a piece of paper, the separated wavelengths appear as bands of color blending gradually through violet, blue, green, yellow, orange, and red. The appearance of any colored object is determined by which wavelengths of the visible spectrum are absorbed by the pigments present and which are reflected or transmitted. Chlorophyll appears green because it absorbs poorly the green portion of the spectrum, whereas it readily absorbs the blues, oranges, and reds. In order to determine whether a given pigment, such as chlorophyll, can be the photoreceptor for a light-dependent process, it is necessary to demonstrate that the wavelengths of light absorbed by the pigment in question correspond to those portions of the spectrum that are effective in promoting the light-dependent process. If, for example, we find a pigment that absorbs only blue light but we subsequently find that it is not blue but red light that makes the reaction go, then clearly the blue-absorbing pigment is not the photoreceptor we are looking for.

The relative effectiveness of various parts of the visible spectrum in promoting a process like photosynthesis can be ascertained in many ways. Usually an aspect of rate such as the amount of oxygen evolved or CO_2 absorbed per unit time (in the case of photosynthesis) is measured at wavelengths in all parts of the spectrum with a constant intensity of light quanta used for each measurement. **Quanta** (singular, quantum) are the smallest particles of light energy. (The term **photon** is often used interchangeably with the term "quantum" for visible light.) These rate data are then plotted against wavelength to provide an **action** spectrum for the light-dependent process. Such an action spectrum should closely resemble the **absorption spectrum** of the **photoreceptor**, a pigment, in this case chlorophyll. This is because the absorbing material is in fact the photoreceptor for the process measured. The absorption spectrum of a pure compound is a unique property related to its chemical composition and molecular structure. It is one of the best ways to identify and describe the compound. An absorption spectrum can be determined on an instrument called a **spectrophotometer**, which measures the amount of light absorbed by a dissolved pigment relative to that absorbed by its solvent. The units of optical density or absorbance are used as a measure of the light absorbed.

The basic concepts concerning the relationship between color and absorbed light were known to Engelmann even though sophisticated spectrophotometers were not yet developed. Even without this instrument, however, Engelmann was able to measure the action spectrum of photosynthesis in an elegantly simple way and to compare it with the absorption spectrum of chlorophyll, thereby proving conclusively that chlorophyll is, in fact, the photoreceptor for the photosynthetic process—a major ac-

complishment, indeed. To measure the action spectrum, he placed a prism in the path of the microbeam of white light used in previous experiments and focused the resulting visible spectrum across the chloroplasts of a single algal cell. The motile bacteria in his system clustered in greatest numbers around those sections of the cell where rates of oxygen production were highest. Thus, the distribution of bacteria provided a picture of the relative effectiveness of various wavelengths of light in promoting oxygen production and depicted the action spectrum of photosynthesis. He found that the most effective wavelengths of light are in the blue and red regions of the spectrum. This is to be expected because these are the wavelengths of light most readily absorbed by the photoreceptor, a fact that is easily demonstrated by measuring the absorption spectrum of chlorophyll in a modern spectrophotometer.

Quantasomes: The Machinery for Conversion of Light to Chemical Energy

From the work of Engelmann and subsequent investigators, we know conclusively that chlorophyll is the photoreceptor, but what about other essential components of the photosynthetic apparatus? What are the working parts necessary to convert light energy, carbon dioxide, and water into complex organic compounds? Are these other essential parts located near chlorophyll molecules in the grana of the chloroplasts, or are some in the colorless stroma?

Although the electron microscope has been able to reveal the complexity and internal order of chloroplast structure, other methods and approaches were necessary to answer these questions and to ascertain the chemical composition and functioning of these organelles.

It has been known for some years that intact chloroplasts isolated from broken cells retain the capacity to photosynthesize. More recently, isolated chloroplasts have been broken to determine how small a fragment still retains some photosynthetic capability. From this, it has been learned that the grana can be broken into subunits that can, when illuminated, evolve

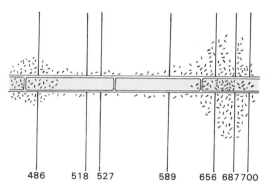

486   518 527   589   656 687 700

Fig. 15-9. Portion of an algal filament (*Cladophora*) in a microspectrum of light. Swarming bacteria cluster around those regions receiving red and blue light. Numbered lines indicate wavelength in nanometers. This was the first action spectrum of photosynthesis. (After T. W. Engelmann: On the production of oxygen by plant cells in a microspectrum. Botanische Zeitung, *40:* 419-426, 1882. Translation of this paper is in *Great Experiments in Biology* (M. L. Gabriel and S. Fogel, eds.), Prentice-Hall Inc., Englewood Cliffs, N. J., 1955.)

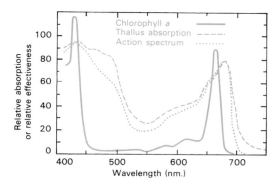

Fig. 15-10. Absorption spectrum of chlorophyll *a*, action spectrum of photosynthesis, and thallus absorption in the green alga *Ulva taeniata*. Note the close correspondence between wavelengths promoting photosynthesis and those absorbed by chlorophyll.

oxygen and store light energy as ATP (although they cannot utilize CO_2). These subunits have been named **quantasomes**.

Chemical analysis (Table 15-1) has shown these small, uniform pieces to have an over-all molecular weight of approximately 2,000,000. They contain two kinds of chlorophylls, four kinds of carotenoid pigments, several plastoquinones (derivatives of benzoquinone, yellow compounds that can be reversibly oxidized and reduced), and a variety of lipids. About half of the weight of the quantasome is made up of protein. Some of the 12 iron atoms present are associated with an iron-containing protein called ferredoxin, while others are part of cytochrome molecules. Copper atoms are found in an unusual green pigment called plastocyanin. It is obvious that the chemical machinery of photosynthesis is highly complex.

Roderick Park and his associates at the University of California at Berkeley, who were the first to characterize the quantasomes chemically, noted that under the electron microscope preparations of isolated chloroplast lamellae showed a repeating structure on the inner surface of the unit membrane. These regular arrays of spherical subunits appeared to be the morphological equivalent of what had been physically isolated and chemically analyzed as the quantasome. More recent work has led to controversy about whether or not the quantasomes can actually be seen under the electron microscope. Even though in doubt as

an identifiable structural entity, the quantasome has been a useful concept in studies of the ultimate physical and chemical functioning subunit of choloroplasts. It is the working part where light energy is converted to chemical energy (*i.e.*, where ATP is generated), and it contains the machinery for oxygen evolution.

Because quantasomes do function in the presence of light energy but cannot utilize CO_2, *i.e.*, are unable to "fix" it by the addition of hydrogen, it seems reasonable to suppose that the actual conversion of CO_2 into carbohydrate takes place in the colorless matrix and not in the chlorophyll-containing grana. Both classical and recent chemical investigations support this conclusion.

Table 15-1
The machinery of photosynthesis: chemical composition of a single quantasome

Fraction	Molecular Weights	
Lipid		
230 chlorophyll molecules		206,400
160 chlorophyll *a*	143,000	
70 chlorophyll *b*	63,400	
48 carotenoid molecules		27,400
14 s-carotene	7,600	
22 lutein	12,600	
6 violaxanthin	3,600	
6 neoxanthin	3,600	
46 quinone molecules		31,800
16 plastoquinone A	12,000	
8 plastoquinone B	9,000	
4 plastoquinone C	3,000	
8-10 α-tocopherol	3,800	
4 α-tocopherylquinone	2,000	
4 vitamin K₁	2,000	
116 phospholipid molecules (phosphatidylglycerols)		90,800
144 digalactosyldiglyceride		134,800
346 monogalactosyldiglyceride		268,000
48 sulfolipid		41,000
? sterols		15,000
unidentified lipids		175,600
	Total lipid	990,000
Protein		
9,380 nitrogen atoms as protein		928,000
2 manganese		110
12 iron, including two cytochrome		672
6 copper		218
	Total protein	930,000
Total lipid + protein		1,920,000

After R. B. Park and J. Biggins: Quantasome: size and composition. Science, *144:* 1009-1011, 1964.

Fig. 15-11. Spinach chloroplast grana lamellae. The crystalline array seen in the electron microscope image of this chromium-shadowed preparation was thought to be structurally equivalent to the quantasomes. (Courtesy R. B. Park.)

EFFECTS OF ENVIRONMENTAL FACTORS ON RATES OF PHOTOSYNTHESIS: WHAT THEY REVEAL ABOUT THE PROCESS

Not only agricultural productivity on land (indeed even the feasibility of agriculture) but also the harvests of fish and other seafood from the oceans depend on a complex of environmental and climatic factors influencing the rates of photosynthesis.

The first systematic studies of the kinetics of photosynthesis (that is, the effects of factors such as temperature, CO_2 concentration, and light intensity on photosynthetic rates) were done early in the present century by Blackman and Matthei. Their studies yielded important insights into the machinery of photosynthesis, including the realization that part of the photosynthetic process can take place in the dark and is therefore independent of light. That portion of the photosynthetic process not requiring light is sometimes called the Blackman reaction. The rationale behind all studies of the kinetics of a process is to vary a single factor at a time and to measure the effects of such variation with all other conditions held constant.

The Effects of Light Intensity

The dependence of photosynthesis on light as a source of energy would lead us to expect that photosynthetic rates should be zero in the dark and rates of oxygen evolution or CO_2 consumption should increase with increasing light intensity. Such is indeed the case at low light intensities. When rates are measured under physiologically appropriate conditions of constant temperature and CO_2 concentration, there is a range of light intensity where rates increase roughly linearly with increasing light (*i.e.,* doubling intensity approximately doubles rate). There is also a light intensity that saturates the photosynthetic process under these conditions. Beyond this saturating intensity further increase in light does not increase the photosynthetic rate. If we measure the intensities of light that saturate photosynthesis in most green plants, we find that they are in the range of 1,000 to 2,000 foot-candles, a figure that is much lower than the maximum light intensity attainable on a clear sunny day in most areas (which can be as high as 10,000 foot-candles). This raises challenging questions. What is the evolutionary explanation of this low saturation point? Did chlorophyll first

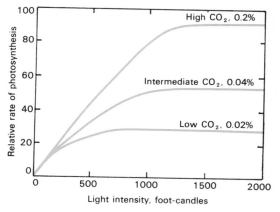

Fig. 15-12. Effect of light intensity on rate of photosynthesis. The intensity needed for and the rate at light saturation depend on the adequacy of other factors. In this graph the effects of light intensity on rates at three different concentrations of CO_2 are shown. (Redrawn from W. H. Johnson and W. C. Steere (eds.): *This is Life, Essays in Modern Biology.* Holt, Rinehart and Winston, New York, 1962.)

evolve in deep or murky water? Do the laws of photochemistry limit a process based on a porphyrin to this level of efficiency? Would it be possible, perhaps in the distant future, for man to devise his own method of photosynthesis based on porphyrin, or on some very different compound, that would have a far greater efficiency?

Another aspect of this problem is seen if we measure the intensity of light required for rate saturation at a higher but still physiologically tolerable temperature. We then find that the light intensity needed for saturation is higher. Thus, if rates are measured first at 10°C. and then at 20°C., both the rate at light saturation and the light intensity needed for saturation are higher at 20°C. than at 10°C. (Fig. 15-13).

At low light intensities, still another phenomenon is evident. Below an intensity called the **compensation point**, the rate of photosynthesis is too low to be readily measurable in intact plant cells. The compensation point is the light intensity needed for the rate of photosynthetic oxygen production to equal the rate of oxygen consumption in respiration. The compensating intensity (roughly 200 to 300 foot-candles) is only a fraction of saturating light.

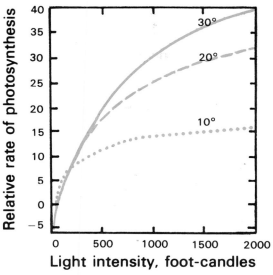

Fig. 15-13. Effect of light intensity on rate of photosynthesis at three different temperatures. In these curves the net rate of photosynthesis is plotted as a function of light intensity. At light intensities below the compensation point, rate of respiration exceeds that of photosynthesis and net rates of photosynthetic assimilation are negative. (Redrawn from W. Noddack and C. Kopp: Zeitschrift für physikalische Chemie, *A187:* 79, 1940.)

Temperature Effects and the Discovery of Dark Reactions

The effects of temperature on photosynthetic rates were the evidence that led Blackman to suggest that photosynthesis is not a strictly photochemical process but rather one that involves both light and dark (or enzymatic) reactions. One characteristic of all strictly photochemical reactions is their temperature independence. However, an effect of temperature on rates of photosynthesis was already noted in our consideration of light intensity effects. This temperature dependence is unusual in a light-dependent process. In fact, it should be impossible.

Blackman found, at a constant low light intensity (above the compensation point but well below saturation), and at constant CO_2 concentration, that temperature has relatively little effect on photosynthetic rates. Thus, at low light intensity, the photosynthetic process

responds in a way expected for a light-dependent reaction (Fig. 15-14). However, at constant high light intensity (at or above saturation levels) with CO_2 concentration again held constant, he found that photosynthetic rates respond to temperature just as most enzymatic dark reactions do. That is, rates increase as the temperature is raised up to 30 or 35°C.; at temperatures above 35°C. rates drop due to the denaturation of enzymatic components of the photosynthetic apparatus. Clearly, there must be an important part of photosynthesis, or, more accurately, the production of carbohydrate from CO_2 and H_2O which is not directly dependent on light.

CO_2 Concentration

Carbon dioxide serves as a substrate, or raw material, for the process of photosynthesis and is the essential ingredient from which the carbon skeletons of all the complex organic components of plant and animal cells are

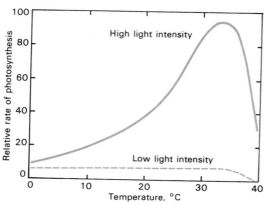

Fig. 15-14. Effect of temperature on rate of photosynthesis. At low light intensities temperature has little effect on rate of photosynthesis. At high light intensities photosynthesis responds to temperature much as enzymatic reactions do. Temperature studies provided the first clear indication that photosynthesis involves dark enxymatic as well as light reactions. (Redrawn from W. H. Johnson and W. C. Steere (eds.): *This is Life; Essays in Modern Biology.* Holt, Rinehart and Winston, New York, 1962.)

manufactured. We would therefore expect that the carbon dioxide level would influence the rate of photosynthesis and that photosynthetic rates would be zero in the absence of this essential substrate.

Increasing CO_2 concentration does indeed increase the rates of CO_2 uptake and of O_2 evolution; this effect depends on light intensity. Saturating levels of CO_2 are reached at about 0.05 per cent in low light intensity, whereas at saturating light intensities, CO_2 concentrations of 0.2 per cent or higher are needed for maximum photosynthetic rates. When we consider that the normal level of CO_2 in our atmosphere is approximately 0.03 per cent, it is evident that this essential substrate is the factor most likely to be rate-limiting in photosynthetic processes on this planet. This knowledge has been used to increase the growth and productivity of plants in greenhouses, where CO_2 levels can be monitored and increased (up to several per cent) during the light period.

As the prevailing 0.03 per cent CO_2 is well below the optimum for all known species of green plants, it is tempting to speculate that photosynthetic organisms may have evolved, and that the photosynthetic apparatus must

have developed to its present specifications, under conditions of considerably higher CO_2 concentrations. Perhaps this is also the evolutionary explanation for the low intensity of light needed to saturate the photosynthetic process: there simply was not enough CO_2 present to make a higher saturation level worthwhile, even though it certainly seems very likely that CO_2 levels were greater during past ages. During the Carboniferous era vast forests of primitive plants developed. From such vegetation deposits of plant material were laid down, subjected to heat and pressure, and later gradually transformed to coal or oil. Over millions of years, most of the earth's available carbon became bound in fossil fuels, in the form of carbonates in rocks (due in part to protozoa and mollusks secreting calcareous shells) or in the form of complex organic molecules in living organisms.

Today, this trend toward diminished atmospheric CO_2 levels may be reversing itself as the continued rapid depletion of coal and oil reserves is accompanied by the return of CO_2 to the atmosphere as a product of combustion. It is debatable whether average temperatures will change with these increasing CO_2 levels. However, we can predict that such changes should result in a general increase in photosyn-

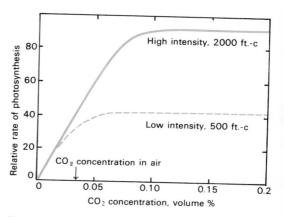

Fig. 15-15. Effect of CO_2 concentration on rate of photosynthesis. Note that the normal concentration of CO_2 in air (0.03%) is well below that needed for saturation of photosynthesis even at low light intensities. (Redrawn from W. H. Johnson and W. C. Steere (eds.): *This is Life, Essays in Modern Biology.* Holt, Rinehart and Winston, New York, 1962.)

thetic rates and possibly in agricultural productivity.

Some Implications of Our Knowledge of Limiting Factors in Photosynthesis

A knowledge of the ways in which limiting factors such as CO_2 concentration, light intensity, and temperature influence photosynthetic rates helps to explain why plant species are distributed as they are. Although plants generally respond in the same way to variations in light intensity, temperature, and CO_2 concentration, individual species or varieties differ with respect to the conditions that they can tolerate. Species growing in alpine regions have adapted to extremes of temperature and to relatively high levels of ultraviolet radiation. Jens Clausen and William Heisey of the Carnegie Institution in Stanford, California, have made intensive studies of native species (including *Mimulus,* or monkey flower, and *Poa,* a grass) able to survive in habitats ranging from sea level to timberline in the Sierras. When the photosynthetic capabilities of these "ecological races" from different altitudes were compared, they exhibited wide variations in their responses to factors such as light intensity and temperature.

Knowledge of the effects of light intensity and temperature on photosynthetic rates can be applied to the question of whether or not it is profitable (or feasible) to grow crops in a given area. The number of days required for a plant species to mature is in part dependent on its photosynthetic productivity. Climatic factors, such as seasonal extremes in temperature, can determine whether agriculture is possible. In marginal regions, a careful selection of plant species or varieties is necessary. Surprisingly good agricultural yields are possible in some areas such as the Matanuska Valley in Alaska, where a very short growing season is compensated for in part by exceptionally long days.

Even though terrestrial plants carry on only a small portion of the total photosynthesis on earth (probably only 1/10th with about 9/10ths performed by marine algae), it has been estimated that land plants assimilate an amount of carbon equal to all the CO_2 in the atmosphere above the continents in less than 10 years. This colossal productivity makes the constant recycling of carbon necessary in order to maintain our food chains and to keep our atmosphere aerobic.

PRIMARY REACTIONS OF PHOTOSYNTHESIS

As Blackman first demonstrated, modern studies confirm that photosynthesis consists of two kinds of reactions: those dependent on light and others that are dark (enzymatic) reactions. The light reactions utilize sunlight to produce energy-rich molecules of ATP, reduced molecules of the pyridine nucleotide NADP, and molecular oxygen. The dark reactions synthesize carbohydrate from CO_2, using ATP as the source of energy and reduced NADP as a source of hydrogens. We will first discuss the light-dependent, or primary, reactions of photosynthesis that take place within the quantasomes.

Current views of the nature of these primary reactions have come from a mass of experimental data that at times seemed very puzzling. It is the obligation of science to provide theories compatible with all experimental observations; in the case of photosynthesis, the observations that we must explain were downright mystifying. A close look at these strange

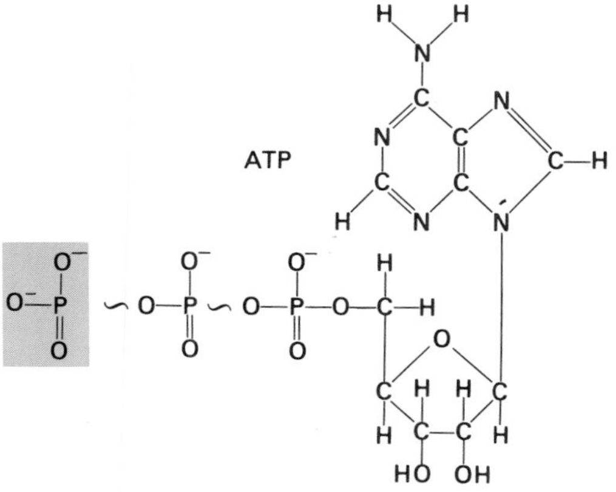

Fig. 15-16. Structure of ATP (adenosine triphosphate). The light reactions of photosynthesis trap the energy of sunlight and store it in the energy-rich bonds of ATP ($\sim$). When the last phosphate group (shown in green) is removed, ADP is formed. (Redrawn from R. P. Levine: The mechanism of photosynthesis. Scientific American, *221*(6): 61, 1969.

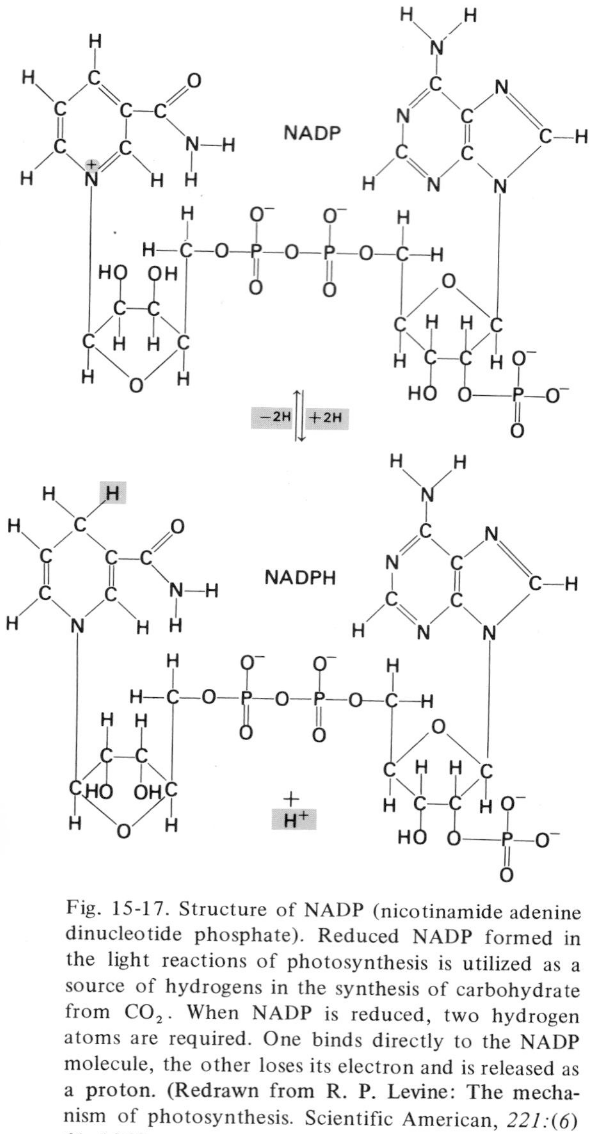

NADP

NADPH

$-2H$ $+2H$

$+$
H^+

Fig. 15-17. Structure of NADP (nicotinamide adenine dinucleotide phosphate). Reduced NADP formed in the light reactions of photosynthesis is utilized as a source of hydrogens in the synthesis of carbohydrate from CO_2. When NADP is reduced, two hydrogen atoms are required. One binds directly to the NADP molecule, the other loses its electron and is released as a proton. (Redrawn from R. P. Levine: The mechanism of photosynthesis. Scientific American, *221*:(6) 61, 1969.

facts will aid a student in understanding our contemporary ideas about photosynthesis as well as show something about the way science achieves knowledge of things once thought unknowable.

First of all, by the 1930's scientists knew that only a small fraction of the chlorophyll molecules present in a chloroplast were capable of doing anything. This became apparent when

Robert Emerson and William Arnold at the Carnegie Institution in Stanford, California, studied the effects on photosynthetic rates of short flashes of light separated by relatively long dark periods. Essentially, their approach was a very ingenious extension of the old Blackman light-dark experiments. Their objective was to provide sufficient light to completely saturate the photosynthetic apparatus, then to follow this by a dark time long enough for all of the products of the light reaction to be used up. They were able to adjust the pattern of light and dark periods by placing a rotating disc between the light source and their experimental algae. The relative length of the light flash could be controlled by the size of the opening cut in the disc, and the number of

Fig. 15-18. Robert Emerson (1903-1959). His experiments with Arnold on yields of photosynthesis in flashing light confirmed the concept of a photosynthetic unit. Later studies on quantum yield and his discovery of the phenomenon of enhancement provided a basis for present views of the two photoreactions in photosynthesis. (From Plant Physiology, *34(3)*: 178, 1959.)

flashes per unit time could be regulated by the speed of the disc's rotation.

Emerson and Arnold worked with their system until they found the conditions that resulted in the maximum rate of photosynthesis per flash. Then the amount of oxygen produced per flash of light and the amount of chlorophyll present in the experimental cells were measured for these conditions. If the photosynthetic apparatus was completely saturated with light and provided with sufficient time in the dark for all subsequent reactions to go to completion, they expected that one O_2 molecule would be produced and one molecule of CO_2 would be consumed for each molecule of chlorophyll. Instead, Emerson and Arnold consistently found that, under optimum conditions, the ratio of O_2 or CO_2 per chlorophyll molecule was in the range of one per several hundred to several thousand. The only possible conclusion they could draw from such observations was that only a small fraction of the chlorophyll present in a chloroplast can participate in a primary reaction. Stated another way, it appears that there are anywhere from several hundred to several thousand chlorophyll molecules per reactive site. This chlorophyll aggregate works as a **photosynthetic unit**. Quanta of light absorbed anywhere within the unit can be transferred from chlorophyll to chlorophyll until one located next to the appropriate enzymatic machinery is reached.

A second peculiarity of the photosynthetic apparatus, that must be explained by any theories concerning its operation, is the observation that action spectra and absorption spectra do not correspond perfectly at long wavelengths. Instead, photosynthetic rates and efficiency decrease in the far red region of the spectrum where there is still considerable absorption of light by chlorophyll. This phenomenon of unexpected decrease in efficiency at wavelengths longer than approximately 680 nm. is called the **red drop**.

The same Robert Emerson who, with Arnold, did the flashing light experiments already described, in later years became Professor of Botany at the University of Illinois. There he made painstaking measurements of the action spectra and the quantum requirements of photosynthesis in algae. During such studies, Emerson observed that small amounts

of short wavelength (blue) light added to saturating amounts of long wavelength light increase rates of photosynthesis far in excess of what he expected from the sum of rates in low intensity blue light plus rates in high intensity far red light. This phenomenon, the stimulating effect of short wavelength light added to long wavelength light, is called the **Emerson enhancement effect**. This strange effect has now been explained and is known to result from the fact that there are two separate light-dependent steps in the photosynthetic process.

The evidence that two light-dependent events occur in photosynthesis came in part from studies of quantum requirements, experiments that point to a third peculiarity of the photosynthetic process. Many investigators, including the German Nobel laureate Otto Warburg, sought to answer the question of how many quanta (photons) of light are needed by the photosynthetic apparatus for the incorporation of one molecule of CO_2 or the evolution of one molecule of O_2. In other words, what is the **quantum requirement** of photosynthesis? We should recall at this point that the amount of energy per quantum of light varies inversely

Fig. 15-19. Otto Warburg, known for his classic studies of respiratory enzymes, was the first to demonstrate that multiple quanta must be absorbed for the incorporation on one CO_2 molecule into carbohydrate and the evolution of one molecule of O_2 in photosynthesis. (Courtesy Johns Hopkins Institute of the History of Medicine.)

with wavelength. Thus, red light, with a wavelength of 660 nm., contains less energy per quantum than does shorter wavelength (450 nm.) blue light (to review these energy relationships, see Chapter 9). We can extend our question about the quantum requirement of photosynthesis as follows: Does the quantum requirement remain the same for all wavelengths of light that are effective in promoting this process?

The quantum requirement for photosynthesis has been measured for all wavelengths of light from the ultraviolet to the infrared. One rather strange thing about these data is that when quantum requirements are plotted as a function of wavelength, the resulting curve looks very different from the action spectrum of photosynthesis: it is a flat line through most of the visible spectrum, increasing sharply in the far red region. Alternatively, these same data can be plotted as **quantum yield**, which for photosynthesis is the amount of O_2 produced or CO_2 consumed per quantum. This is nothing but the reciprocal of the quantum requirement, so when plotted as a function of wavelength, it is a flat line dropping to zero in the far red. The reason that these plots do not resemble action spectra is simple: in measurements of quantum requirements and yields, only the **absorbed** quanta count, but in measurements of action spectra the rates at constant intensities of

incident quanta (whether absorbed or not) are important. You will recall from the absorption spectrum of chlorophyll (which can be viewed as a picture of the relative probability for light quanta of different wavelengths to be absorbed) that blue and red quanta are readily absorbed by chlorophyll while green and far red light quanta are absorbed very poorly. The drop in quantum yield in the far red region (*i.e.,* the need for additional quanta to accomplish a given amount of change) is obviously related to the decreased efficiency of light absorption by chlorophyll molecules in this long wavelength part of the spectrum. The discrepancy between the action spectrum of photosynthesis and the absorption spectrum of chlorophyll is due to the fact that there are two different light-requiring steps in the photosynthetic machinery. One of these involves a form of chlorophyll that can absorb light at longer wavelengths than the other, *i.e.,* their action spectra are different. These two primary reactions will be discussed in detail later.

The important point about quantum data, which we emphasize here, is that at least 8 quanta of light must be absorbed in order for one molecule of O_2 to be evolved. This multiple quantum requirement (most measurements range between 8 and 12 or more quanta per molecule of O_2) was perhaps the best indication that no simple, direct reaction between molecules of CO_2 and chlorophyll could explain the mechanism of photosynthesis.

In addition to the peculiarities of the photosynthetic apparatus already noted (the photosynthetic unit, the red drop, enhancement, and the multiple quantum requirement), there is a fourth and final fact coming from experiments with tracer oxygen that must be explained by any general theory. In the 1940's, Ruben, Randall, Kamen, and Hyde, working at Washington University in St. Louis, posed the question: Where does the O_2 evolved in photosynthesis originate? Both substrates, CO_2 and H_2O, are molecules containing oxygen. Using an isotope of oxygen (with an atomic weight of 18) to label either the CO_2 or H_2O, they ran experiments to determine the origin of the O_2 evolved. Their results are summarized by the two following equations:

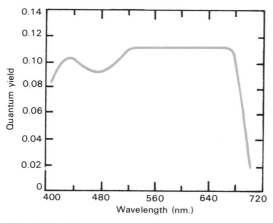

Fig. 15-20. The quantum yield of photosynthesis as a function of wavelength in *Navicula minima*, a diatom. (Redrawn from T. Tanada: American Journal of Botany, *38:* 276, 1951).

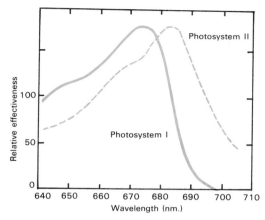

Fig. 15-21. Action spectra of Photosystems I and II of photosynthesis. Because the photoreceptor for Photosystem I absorbs at slightly longer wavelengths, it can be activated by quanta beyond approximately 680 nm. that cannot be absorbed by Photosystem II.

$$C^{18}O_2 + 2H_2^{16}O \rightarrow [CH_2O] + {}^{16}O_2$$
$$C^{16}O_2 + 2H_2^{18}O \rightarrow [CH_2O] + {}^{18}O_2$$

The label appeared in the evolved oxygen only when it was introduced in the water in which photosynthetic cells were suspended. Therefore, the obvious conclusion from these experiments was that the oxygen produced in photosynthesis comes from water.

VAN NIEL'S UNIFYING CONCEPT: PHOTOLYSIS, THE LIGHT-DEPENDENT SPLITTING OF WATER

Until the last decade there was only one theory which could account for at least several of the puzzling observations just discussed—that of C. B. Van Niel, a microbiologist working at the Hopkins Marine Station of Stanford University in Pacific Grove, California. His ideas provided the first theoretical framework for an understanding of the mechanism of photosynthesis and still serve as the basis for research. Van Niel examined an array of photosynthetic microorganisms, including algae and photosynthetic bacteria, and from such comparative studies he concluded that photosynthesis could be viewed as a sequence of reactions between a hydrogen donor and a hydrogen acceptor

(CO_2). A hydrogen donor is a **reducing agent**, that is, a molecule that can contribute hydrogen or an electron to a receptor molecule (oxidant). In the green plants, H_2O serves as the hydrogen donor while alternate compounds serve this function in the bacteria (*e.g.,* H_2S in the sulfur bacteria). Thus for any organism, photosynthesis could be summarized by the general equation:

$$CO_2 + 2H_2A \xrightarrow[\text{chlorophyll}]{\text{light}} (CH_2O) + 2A$$

Fig. 15-22. C. B. Van Niel, whose concept of photolysis, or the light-dependent splitting of water, provided the first theoretical framework for understanding the mechanism of photosynthesis. (From Annual Review of Plant Physiology, *13:* 1962.)

where H_2A stands for the hydrogen donor and (CH_2O) represents carbohydrate. In green plants this equation becomes the familiar:

$$CO_2 + 2H_2O \xrightarrow[\text{chlorophyll}]{\text{light}} (CH_2O) + O_2$$

whereas in the sulfur bacteria it becomes:

$$CO_2 + 2H_2S \xrightarrow[\text{chlorophyll}]{\text{light}} (CH_2O) + 2S.$$

A further generalization made by Van Niel was the suggestion that, in *all* photosynthetic organisms, the function of light is the same. He proposed that light quanta are utilized to split water (a step he termed **photolysis**) into a reduced product, [H], and an oxidized product, [OH]. The reduced product ultimately serves to reduce CO_2 to the level of carbohydrate (CH_2O). Oxygen is evolved in green plants from the oxidized product, and in the bacteria a variety of organic molecules are oxidized by the [OH]. Van Niel's scheme can be summarized as follows:

$$4 \text{ quanta} \rightsquigarrow 4H_2O$$

$$4H \,\vert\, OH$$

$$CO_2 \qquad 4[H] \qquad 4[OH] \xrightarrow[+ 2H_2A]{\text{bacteria}} 4H_2O + 2A$$

$$(CH_2O) + H_2O \xrightarrow[\text{green plants}]{} 2H_2O + O_2$$

This hypothesis can explain a number of the puzzling observations that were made during the past half century. However, it now is clear that light doesn't split water directly but rather is essential to the utilization of H^+ and OH^- ions in the way Van Niel proposed. His scheme is certainly consistent with the requirement for more than one quantum of light per O_2 evolved or CO_2 consumed. A minimum of 4 quanta would be required by Van Niel's scheme, since he postulated that four molecules of water are split, each by 1 quantum of light energy. Any discrepancy between this theoretical requirement of 4 quanta and actual measurements could be explained by a loss of quanta as heat, etc. Van Niel's theory is also completely consistent with the tracer studies indicating that the oxygen produced in photosynthesis comes from water. However, viewed from the critical vantage point of thermodynamics, Van Niel's original picture of the photolysis of water cannot be accepted, and for one very simple reason. There is just not enough energy in a single quantum of visible light to split a molecule of water. Photolysis, using one quantum of light, is thermodynamically unfeasible; using more than one quantum, it becomes a statistically improbable event because it would take too long for a single chlorophyll to be hit by two photons. Even without the insight of thermodynamics, Van Niel's theory has been open to re-examination for the further reason that it does not explain the enhancement effects already described. And, after all, a theory must be consistent with all of the facts. Today this generalization has not been abandoned, but, as so commonly happens with theories, refined with better knowledge of the mechanism by which light energy can separate the components of water.

The Technology of Isolated Chloroplasts: The Hill Reaction

There exists a long history of experimentation on photosynthesis based not on leaves or even on whole cells, but instead on isolated cellular organelles, the chloroplasts. Attempts to separate photosynthesis from the rest of cellular metabolism go back to the last century when Haberlandt in 1888 and Ewart in 1896 observed that chloroplasts isolated from leaves by grinding in water were capable of liberating small quantities of oxygen when illuminated. Later, in 1904 and again in 1925, Molisch reported the evolution of oxygen from preparations of dried leaf powders. Inman, in the 1930's, returned to the kinds of experiments done by Haberlandt and Ewart, grinding fresh leaves to isolate chloroplasts. He demonstrated that the limited ability of his isolated chloroplasts to evolve oxygen in the light could be destroyed by factors such as high temperature or protein-digesting enzymes, and therefore suggested that oxygen evolution is enzymatic.

The experiments of all these pioneer investigators were disappointing, however, because the amounts of oxygen evolved by their preparations during illumination were never more than a small fraction of the photosynthetic rates of the living cells and tissues from which their chloroplasts were isolated. It was not until the late 1930's that the English investigator, Robin Hill, discovered that to sustain high rates of oxygen evolution in the light, a hydrogen acceptor (*i.e.*, an oxidant) had to be added to suspensions of isolated chloroplasts. Using a mixture of hemoglobin and ferric oxalate as the hydrogen acceptor, Hill demonstrated that chloroplasts continued to evolve oxygen in the light until the oxidant (the ferric oxalate mixture) was completely reduced. Although the amount of oxygen evolved was limited by the amount of oxidant initially present, the *rate* of oxygen evolution depended on chloroplast concentration and light intensity. In terms of Van Niel's theory, Hill's chloroplast reaction can be summarized as follows:

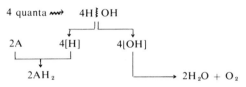

where A = the hydrogen acceptor or oxidant

It is now clear that the mechanism of oxygen evolution by isolated chloroplasts is identical to that followed by photosynthesizing cells. Thus, Hill's discovery provided convincing evidence of the validity of Van Niel's view of the primary reactions of photosynthesis and was a landmark in our knowledge of this process.

In the years that followed Hill's discovery, many additional compounds were shown to be good oxidants for the family of reactions now referred to as **Hill reactions**. The list includes a variety of dyes, ferricyanide salts, quinones, and even molecular oxygen. The term Hill reaction is used for all chloroplast reactions in which a suitable oxidant reacts with a reductant generated by illuminated chloroplasts. Reduction of the hydrogen acceptor, or oxidant, is accompanied by evolution of oxygen. The oxidant essentially substitutes for the reactions involved in CO_2 reduction in intact photosynthesis, and for a while it seemed that the CO_2-fixing system might even be located somewhere outside the chloroplasts. However, as methods were improved it eventually became possible to isolate unbroken chloroplasts and to demonstrate that they are capable of the entire set of reactions of photosynthesis including the synthesis of complex organic compounds. Once chloroplasts have been isolated, they can be broken down further with exciting results. A recent achievement of two Australian investigators, Boardman and Anderson, has been the separation of two distinctly different chlorophyll-containing fractions from broken chloroplasts. Each fraction contains the components of a unique part of the photochemical apparatus. According to current terminology, these two parts are called **Photosystems I** and **II**. They are at the crux of current theories about the action of light in photosynthesis, discussed at length in the next section.

With the perfection of a technology for isolating chloroplasts and individual parts of the photosynthetic apparatus from the rest of the cell, rapid strides have been made toward an understanding of the light-dependent reactions in photosynthesis. One major aspect of the process still poorly understood, however, is the mechanism of oxygen evolution. The only undisputed fact here is that manganese is essential. It has long been recognized that photosynthetic cells deprived of manganese lose their capacity to evolve oxygen, and that the enzymatic machinery for oxygen evolution can be rapidly restored by adding inorganic manganese salts to the medium in which deficient cells are suspended. Unlike the magnesium, which is known to be part of the chlorophyll molecule, or other elements associated with specific components of chloroplasts, the role of manganese is unknown. It is generally agreed, however, that the reactions involved in oxygen evolution are closely linked to that part of the photosynthetic apparatus called Photosystem II. Still, the solution of the mystery of the mechanism of oxygen evolution remains an accomplishment eagerly sought but yet to be achieved.

PRESENT VIEWS ON THE PHOTOCHEMISTRY OF PHOTOSYNTHESIS

Many disciplines have contributed to the current picture of what happens when light is absorbed within a chloroplast. Physicists tell us that light absorption results in the displacement of an electron from the photoreceptor molecule. This displacement is nothing more than movement of the electron to an orbit further from its atomic nucleus. Several options are possible thereafter. First, the electron can fall back into place (that is, back to its "ground state") with the result that its absorbed energy is dissipated as heat. A second possibility is that, as the electron falls back to its ground state, light can be re-emitted either as fluorescence or phosphorescence, and again lost. Finally, the electron can be captured by an appropriate acceptor molecule. Only under the last circumstance can the energy of the absorbed light quantum be conserved and used; that is, converted into potential chemical energy.

Perhaps the most significant progress in human understanding of photosynthesis during the past decade has been the realization that we can account for the strange characteristics noted earlier only by assuming the collaboration of two different primary reactions; that is, two distinct light-dependent steps, carried out by Photosystems I and II.

Energy, trapped anywhere within the aggregate of chlorophyll and accessory pigment molecules called the photosynthetic unit, is transferred to a reaction center. At such a center a chlorophyll molecule has associated with it an electron donor (**reductant**) and an electron acceptor (**oxidant**). The absorption of a quantum of light by the chlorophyll at a reaction center results in the displacement of an electron from the chlorophyll molecule; it is trapped by the acceptor molecule which thus becomes reduced. The chlorophyll is returned to electrical neutrality by receiving an electron from its donor molecule. In doing so, the donor becomes oxidized. Thus, the over-all result of the absorption of a photon by the chlorophyll at a reaction center (*i.e.*, a **primary reaction**) is the formation of a reduced electron acceptor and an oxidized electron donor. We have already encountered donor and acceptor molecules that function similarly to those involved in the photosynthetic apparatus in the electron transport system of the mitochondria (which contains cytochromes; see Chapter 3). As you will recall, such molecules can be described by their oxidation-reduction potential, a quantity

that is either positive or negative and is expressed in volts. Electrons, being negative, move spontaneously from a donor having a negative potential to an acceptor with a less negative potential. As electrons move along a gradient of carrier molecules toward the most electropositive acceptor, energy is released. In both the mitochondria and the chloroplasts, energy released by the movement of electrons along a chain of carriers is stored as ATP. The mitochondria and chloroplasts differ in one very important aspect, however. In the mitochondria, electrons removed during the breakdown of foodstuffs all flow "downhill" to the ultimate electron acceptor, oxygen. In the chloroplasts, two of the steps in electron transport are against the gradient of electrochemical potential, and for each of these two steps, a light quantum, *i.e.*, a photon, is needed. Thus, there is an input of energy which accomplishes the feat of moving electrons "uphill," from water molecules (at a potential of +0.8 volts, which serve as electron donors) to the pyridine nucleotide NADP, nicotinamide adenine dinucleotide phosphate (at a potential of −0.3 volts), which serves as the terminal electron acceptor.

The two energy input steps occur at reaction centers having different action spectra (although a form of chlorophyll *a* is thought to be involved in both) and different electron donor and acceptor molecules. These two kinds of reaction centers are the Photosystem I and Photosystem II already mentioned. In Photosystem II, the primary reaction results in the generation of a strong oxidant and a weak reductant. In Photosystem I, a weak oxidant and a strong reductant are formed. The weak reductant formed by Photosystem II and the weak oxidant generated by Photosystem I are linked together by a chain of carriers along which electrons can move spontaneously.

An electron ejected from the chlorophyll of Photosystem II is transferred from its electron acceptor through intermediates including plastoquinone, plastocyanin, and cytochromes *f* and *b* to the positively charged chlorophyll remaining after removal of an electron from the chlorophyll of Photosystem I. The electron ejected from Photosystem I chlorophyll is trapped by its electron acceptor and transferred via ferredoxin to NADP.

But what about the remaining positively

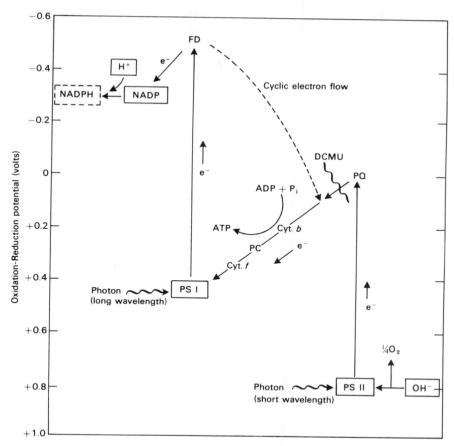

Fig. 15-23. The primary reactions of photosynthesis. The two photoreactions are linked by a chain of electron-carrying intermediates. (See text for discussion.) The following abreviations are used: PQ, plastoquinone; PC, plastocyanin; FD, ferredoxin.

charged chlorophyll II? And how does **ferredoxin** (which accepts electrons from Photosystem I) reduce NADP? After all, protons as well as electrons are needed for the reduction of this pyridine nucleotide. Also, we have said nothing thus far about the involvement of water and CO_2, the recognized substrates of photosynthesis. Light quanta cannot be used to tear apart water molecules by any known mechanism. How, then, does water become involved in these events? Once again we must remind ourselves about a basic fact of chemistry—that a fraction of all water molecules, including those in the chloroplast, are ionized, or dissociated into negatively charged hydroxyl ions (OH^-) and positively charged hydrogen ions (H^+, or protons). These ions provide the means for restoring the photosynthetic apparatus to elec-

trical neutrality following the absorption of light quanta. A negatively charged hydroxyl ion can donate an electron to restore the positively charged chlorophyll II molecule to its original state, while a proton (i.e., an H^+) joins the electron removed during the reoxidation of ferredoxin to reduce the pyridine nucleotide NADP. The hydrogens thus made available from reduced NADP are utilized in the sequence of reactions to be described later, where CO_2 is reduced to the level of carbohydrate. Two quanta of light are required for each electron moved through Photosystems I and II from water (i.e., OH^-) to the point where electrons and protons reduce NADP. It takes four hydrogens (i.e., four protons and four electrons) for the reduction of a single molecule of CO_2 to the level of carbohydrate and four

[OH] radicals for the formation of one molecule of O_2 plus the two molecules of water needed to balance our chemical ledger. These aspects of the chemistry of photosynthesis are summarized in the following equations:

$$4H_2O \rightarrow 4H^+ + 4OH^-$$

$$4OH^- \rightarrow O_2 + 2H_2O + 4e^-$$

$$2NADP + 4e^- + 2H^+ \rightarrow 2NADPH$$

$$2H^+ + 2NADPH + CO_2 \rightarrow 2NADP + H_2O + (CH_2O)$$

Net change: $H_2O + CO_2 \rightarrow (CH_2O) + O_2$

Thus, it appears that the experimentally measured requirement of at least 8 quanta of light for the evolution of one molecule of oxygen from, or the entry of one molecule of CO_2 into, the photochemical apparatus makes sense theoretically. Our present view of the primary reactions of photosynthesis is consistent with tracer experiments indicating that the oxygen evolved in photosynthesis comes from water. It must be emphasized, however, that water molecules are not broken apart directly through the action of light, but rather products of the ionization of water molecules can serve as electron acceptors and electron donors. The energy of light quanta is needed for only one purpose: to move electrons from one energy level to a higher energy level in a molecule of chlorophyll. Electrons thus displaced from their ground state can be removed from chlorophyll if caught by appropriate acceptor molecules.

There is a stepwise loss in energy as electrons move spontaneously along the chain of intermediates connecting the two photosystems. As we have noted, the situation here is similar to that in the electron transport system of mitochondria. Coupled to electron transport in both chloroplasts and mitochondria are reactions in which ATP is synthesized from ADP and inorganic phosphate. In the chloroplast system, this formation of ATP is called **photosynthetic phosphorylation** or **photophosphorylation**. More will be said later about this light-dependent synthesis of ATP.

Primary Reactions of Photosynthesis and the Composition of the Quantasomes

One very comforting aspect of the present view of how photosynthesis works is that it indicates the usefulness of a number of the once strange-seeming components found in the quantasomes. In our discussion of the workings of Photosystems I and II we noted the involvement of chemicals listed earlier as constituents of the quantasomes, including copper in plastocyanin, a green pigment that is not a chlorophyll; quinones such as plastoquinone; iron in the cytochromes and ferredoxin; and pigments such as chlorophylls and carotenoids. An abundance of chlorophyll molecules is needed, of course, with a few of the more than 200 chlorophyll molecules per quantasome located near an appropriate electron acceptor. The carotenoids and other accessory pigments share with chlorophylls the job of transferring absorbed light quanta to reactive sites. The carotenoids also have a protective function in preventing the destruction of chlorophyll molecules by photo-oxidation. In most green cells there is a constant turnover of chlorophyll, with new synthesis more or less compensating for breakdown. In certain mutants of algae, photosynthetic bacteria, and higher plants that lack carotenoid pigments, the photodestruction of chlorophyll is so rapid that it cannot accumulate; such cells are albino in spite of their ability to synthesize chlorophyll.

The oxidants and reductants in the quantasomes include the various quinones. Of these, plastoquinone is located in the electron transport system between Photosystems I and II. Whether this compound serves as the initial electron acceptor for Photosystem II or not is debatable. If not, it is certainly located close to the initial acceptor, since it is possible to demonstrate that plastoquinone becomes reduced very rapidly when chloroplasts are illuminated.

The same sort of uncertainty applies to the role of ferredoxin as an electron acceptor for Photosystem I. Ferredoxin is an iron-containing compound but is not a porphyrin, *i.e.*, it is not related to the cytochromes and chlorophyll. There is evidence that a more electronegative compound is the initial acceptor and that this electron carrier in turn reduces ferredoxin. A rapid reduction of ferredoxin can be demonstrated spectrophotometrically in chloroplast preparations illuminated by light absorbed only by Photosystem I.

Present views of the sequence of interaction of components of the electron transport system in the two photosystem scheme of photosynthesis are based on very sensitive and complex optical experiments in which changes in absorption by individual compounds can be observed in times as short as milli- or even microseconds.

As a component of membranes, the various lipids of the quantasomes are involved in maintaining the structural integrity of the photosynthetic apparatus. It is within the lipid layers of the grana lamellae that the chlorophylls, carotenoids, and quinones are located. The proteins, which make up close to one-half of the total molecular weight of the quantasome, are involved not only in membrane structure, but also as enzymes and as proteins associated with components of the electron transport system such as the cytochromes and plastocyanin.

The components of our scheme that contain iron are ferredoxin and cytochromes b and f. The six atoms of copper per quantasome all appear to be associated with the green pigment plastocyanin, a component of the electron transport chain connecting Photosystems I and II. The two atoms of manganese are not associated with any known component of the quantasomes, but their recognized role in photosynthetic oxygen evolution leads us to assume that they are associated with an enzyme involved in conversion of the oxidized product of Photosystem II to molecular oxygen.

The Two Photosystem Scheme and the Phenomena of Red Drop and Enhancement

Two peculiar characteristics of the photosynthetic apparatus noted earlier in this chapter must now be re-examined in light of the mechanisms we have been discussing: red drop and enhancement. As we have already noted, the photoreceptors of Photosystems I and II are forms of chlorophyll with different absorption and action spectra. Through most of the visible spectrum, light quanta have approximately equal probability of being absorbed by either photosystem. It is at long wavelengths that absorption and action spectra of the two photosystems differ significantly. The effectiveness of Photosystem I extends beyond that of Photosystem II into the long wavelength portion of the spectrum (*i.e.*, in the far red; see Fig. 15-21). Thus, at wavelengths beyond approximately 680 nm., incident light quanta can be absorbed quite readily by Photosystem I but very poorly, if at all, by Photosystem II. The explanation of lower rates and the decrease in quantum yield (or increase in quantum requirement) seen at long wavelengths is a simple one: the absorption of light by system II chlorophyll drops rapidly at long wavelengths, while the absorption of light by system I chlorophyll continues. The differences in the absorption characteristics and in the action spectra of the two photosystems result in a deficiency of light quanta entering system II at long wavelengths. But for the entire process to continue to operate, light quanta must continually be absorbed by both photosystems. The stimulating effect of added short wavelength light is also readily explainable in that it provides the input of light energy needed for Photosystem II, which is rate-limiting under these conditions.

PHOTOPHOSPHORYLATION

The coupling of ATP synthesis to the electron transport chain connecting Photosystems I and II, *i.e.*, photophosphorylation in chloroplasts, and the seeming parallel in mechanism of oxidative phosphorylation of mitochondria have already been noted. Both produce ATP, but ATP synthesis in chloroplasts has certain peculiarities indicating that this light-dependent synthesis is different from anything encountered previously in the mitochondria. To begin with, the entire photosynthetic apparatus need not be operational for photophosphorylation to take place. When an external electron donor is provided, **cyclic photophosphorylation** can proceed, utilizing light absorbed exclusively by Photosystem I. Cyclic photophosphorylation can also take place in the presence of an inhibitor such as dichlorophenyldimethyl urea (DCMU), which inhibits Photosystem II. The "short circuit" around Photosystem I involved in cyclic photophosphorylation can provide needed ATP under conditions where neither the reduction of CO_2 nor the evolution of oxygen can occur.

ATP synthesis also occurs when both photosystems are operational, and under these circumstances it is referred to as **non-cyclic photophosphorylation**. In this case, ATP formation is accompanied by the production of reduced NADP and the evolution of oxygen.

For the complete photosynthetic process, the usual **stoichiometry** (or ratio of reactants and products) is that for each molecule of CO_2 entering the carbon cycle, one molecule of O_2 is evolved, two reduced NADP's are formed, and three ATP's are generated. It has been postulated that more than one photophosphorylation site must exist for such stoichiometry to be possible, but much controversy still exists concerning the exact location of these sites of ATP synthesis. Some light quanta may be utilized in cyclic photophosphorylation, while only the ATP's formed at the same site by non-cyclic photophosphorylation are accompanied by NADP reduction and oxygen evolution. Viewed in this way, it appears that several phosphorylation sites may not be absolutely essential to account for this over-all stoichiometry. The measured quantum requirement for O_2 evolution (or CO_2 consumption) generally exceeds the theoretical minimum of 8 by 4 to 6 quanta. It is likely that some of these extra quanta are utilized for cyclic photophosphorylation.

CARBOHYDRATE SYNTHESIS: THE CALVIN CYCLE

The enzymatic reactions involved in reducing CO_2 to the level of carbohydrates and other complex organic components of plants are thought to take place outside the grana, that is, in the colorless stroma of the chloroplasts. The reactions of CO_2 fixation are dark reactions dependent on reduced NADP and ATP generated in the light reactions which take place in the quantasomes.

Our understanding of the biochemistry of the path of carbon in photosynthesis comes largely from the work of Melvin Calvin and his associates at the University of California at Berkeley. For his achievement in elucidating this complex pathway, Calvin received the Nobel Prize in 1961. Calvin's work demonstrates the importance to scientific progress of the availability of appropriate materials and techniques. These experiments would have been impossible without the existence of a radioactive isotope of carbon, ^{14}C, which was used to label CO_2 entering the photosynthetic apparatus and to trace the intermediates and products formed. Paper chromatography was utilized to separate individual compounds from the complex mixtures isolated from cells and to identify compounds containing the radioactive tracer.

The technique of paper chromatography is based on the differences in solubility of different compounds in various solvents and in the degree to which they tend to be adsorbed on paper. A small drop of sample is applied on one corner of a piece of filter paper and the edge of the paper is dipped into a solvent. As the solvent rises by capillary action, it carries the dissolved molecules with it. Molecules will move in accordance with their size, solubility,

Fig. 15-24. Melvin Calvin, whose investigations of the path of carbon in photosynthesis resulted in elucidation of the complex reaction cycle bearing his name. (Courtesy Johns Hopkins Institute of the History of Medicine.)

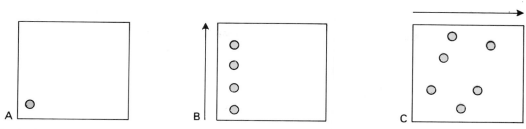

Fig. 15-25. The technique of paper chromatography. A, spot of sample is applied to one corner of a sheet of filter paper. B, separation of components in the sample after development in first solvent. Direction of solvent movement is indicated by arrow. C, further separation of components after development with a second solvent with paper rotated 90° (direction indicated by arrow).

and other characteristics. Further separation can be attained by rotating the paper 90° and "developing" in a new solvent.

Radioactively labeled components can be located by a technique called **radioautography**, which involves placing the dried chromatogram next to a sheet of photographic paper or film. After development, dark spots appear on areas of the film that were in contact with portions of the chromatogram where radioactively labeled compounds were located. Individual compounds can be identified from the location of labeled pure compounds subjected to the same conditions, or by chemical analysis of the spots themselves.

In the early experiments with ^{14}C-labeled CO_2, photosynthesizing cells were exposed to the tracer for relatively long periods of time. As a result, almost every compound in these cells became labeled. Therefore, it was apparent that exposure to the tracer had to be very short if only the early intermediates between CO_2 and carbohydrate were to be labeled. With sufficiently short exposure times, the label appeared in only the earliest intermediates; with longer exposure times more compounds became labeled. By varying the exposure times and by identifying all the labeled compounds, it was possible to piece together the sequence of reactions taking place.

Exposure of cells to tracer and light was done in a flat, round glass container appropriately called a "lollipop." After an appropriate reaction time, the contents of the lollipop (algal cells in a suspending medium containing ^{14}C-labeled CO_2) were drained rapidly into a beaker of hot alcohol. This treatment killed the cells and stopped all reactions instantly. The cells

were then extracted, the extracts concentrated, and the components of the extract were separated and identified by paper chromatography.

Details of the intermediates and reactions discovered by these means are shown in Figure 15-26). There are certain aspects of the cycle that we should all understand and appreciate, although the detailed memorization of most intermediates is a task best left to the organic chemists. CO_2 molecules enter the cycle by attachment to a five-carbon, phosphorylated acceptor molecule, **ribulose diphosphate (RuDP)**. The resulting six-carbon intermediate is very unstable, and breaks into two molecules of **3-phosphoglyceric acid (PGA)** which is the first identifiable labeled compound. PGA is next reduced at the expense of reduced NADP generated in the light reactions to form two interconvertible triose phosphates, **dihydroxyacetone phosphate** and **3-phosphoglyceraldehyde**. At this point compounds have been formed that are familiar as intermediates in the respiratory breakdown of foodstuffs. As you will recall, in glycolysis these two triose phosphates are the products of the action of the enzyme aldolase on fructose 1,6-diphosphate. Their further breakdown to pyruvic acid provides the key substrate leading to the Krebs cycle. Thus, it is evident that very early intermediates of the Calvin cycle can be diverted into the respiratory metabolism of the cell.

The remainder of the cycle is made up of a complex sequence of reactions in which a new acceptor molecule is generated. The regeneration of a new phosphorylated acceptor molecule and the formation of triose phosphates

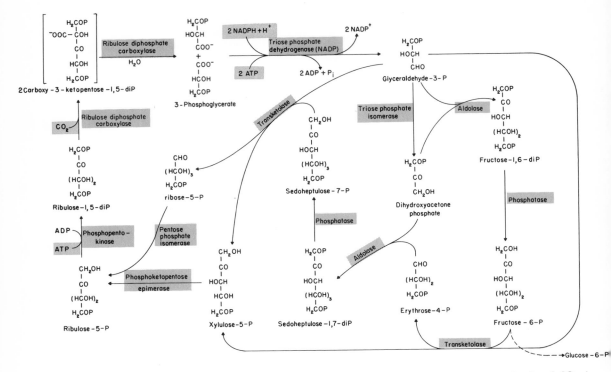

Fig. 15-26. The Calvin cycle. In each turn of this complex sequence of reactions, one molecule of CO_2 is incorporated into carbohydrate and one molecule of ribulose diphosphate acceptor is regenerated.

from PGA requires ATP, the other product of light-dependent reactions in the quantasomes essential for the continued operation of the Calvin cycle.

For every turn of the cycle, one CO_2 is incorporated into the organic matter of the cell, and one molecule of ribulose diphosphate (RuDP) is regenerated. The balance of this complex series of reactions is summarized as follows:

$$3C_5 \xrightarrow{3ATP} 3RuDP$$

$$3RuDP + 3CO_2 \longrightarrow 6PGA$$

$$6PGA \xrightarrow{12H} 6C_3$$

$$2C_3 \xrightarrow{6ATP} C_6$$

$$C_6 + 2C_3 \longrightarrow C_5 + C_7$$

$$C_7 + C_3 \longrightarrow 2C_5$$

Net change: $12H + 3CO_2 \xrightarrow{9ATP} C_3 + 3H_2O$

Therefore, for every molecule of CO_2 entering the cycle, four [H]'s and three ATP's are needed, a stoichiometry that agrees with the observation that, in the light reactions of the quantasomes, four [H]'s and three ATP's are generated for each O_2 evolved, and that an energy input of 8 or more quanta of light is required.

Finally, it is important to appreciate the intimate involvement of the photosynthetic carbon cycle with other aspects of cellular metabolism. The relationships between photosynthetic intermediates and the essential building blocks of plant cells are summarized in Figure 15-27. The key intermediate, 3-phosphoglyceric acid, the first stable product of CO_2 assimilation in photosynthesis, is a precursor common to pathways leading to carbohydrates, fats, and amino acids. All major plant constituents can be derived from PGA.

Alternate Routes of Carbon Fixation

Within the past decade, it has become

apparent that, although the Calvin cycle is the major route of CO_2 into cellular intermediates, there are circumstances where CO_2 entering the photosynthetic machinery is diverted. One alternate route is utilized under conditions of low CO_2 concentration (not really such an unusual circumstance). Here the entry of CO_2 into the photosynthetic machinery appears to follow the Calvin cycle except that CO_2 plus ribulose diphosphate is split into glycolic acid plus triose phosphate. The **glycolic acid pathway**, as its name implies, results in the accumulation of glycolic acid as an early product and is a more direct route to amino acid synthesis than is the Calvin cycle, in which carbohydrates tend to accumulate as excess

products of photosynthesis. The glycolic acid pathway assumes a significant role in photosynthetic carbon metabolism primarily under what could be considered "CO_2 starvation" conditions for green cells. In terms of plant function, this glycolic acid pathway has been implicated in the mechanism of stomatal opening, providing a means for increased entry of CO_2 into the internal space of leaves during periods of maximum photosynthesis (see Chapter 16).

When plants accumulate glycolic acid they have a tendency to oxidize this compound. The resulting **photorespiration** can be detected as an increased rate of oxygen uptake in the light or as a high CO_2 compensation point. Photorespiration occurs in cellular organelles called

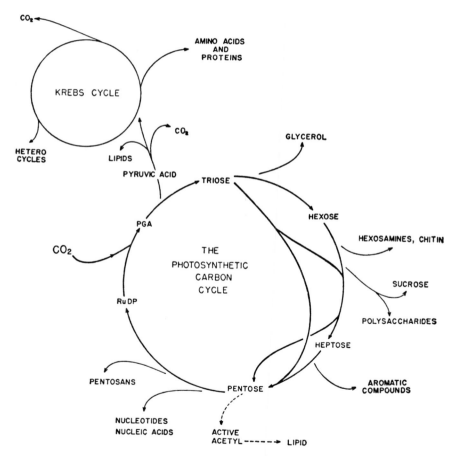

Fig. 15-27. Summary of pathways by which carbon compounds are synthesized from CO_2 by plants. From. J. A. Bassham and M. Calvin: *The Path of Carbon in Photosynthesis.* Prentice-Hall Inc., Englewood Cliffs, N. J., 1957.)

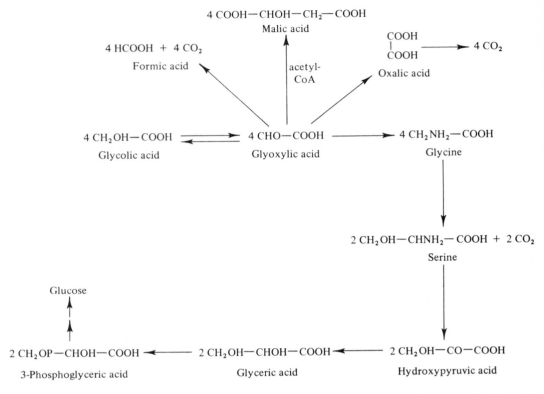

Fig. 15-28. Intermediates of the glycolic acid pathway of carbon reduction and related reactions. Note that the amino acids glycine and serine are early intermediates. After I. Zelitch: *Photosynthesis, Photorespiration, and Plant Productivity*, Academic Press, New York, 1971.)

leaf peroxisomes. Those conventional plants in which CO_2 fixation is via the Calvin cycle and which exhibit photorespiration are referred to as C_3 **plants**.

There are in addition some plants, mostly tropical species, which have an additional system for accumulating CO_2 from the atmosphere by incorporating it into four-carbon compounds, the dicarboxylic oxalacetic, malic, and aspartic acids. Such C_4 **plants** have an unusual pattern of leaf anatomy consisting of concentric cylinders: a central vascular bundle is surrounded by a **bundle sheath** (a dense single layer of dark green cells); a loosely packed layer of mesophyll cells lies outside the bundle sheath. An all too familiar temperate plant exhibiting C_4 anatomy and metabolism is crabgrass.

In the so-called C_4 plants CO_2 is first incorporated into four-carbon dicarboxylic

acids by the leaf mesophyll cells. These compounds later are decarboxylated and the CO_2 released is utilized to form the intermediates of the Calvin cycle by cells of the bundle sheath. Thus the eventual accumulation of new carbon compounds is via a sequence of reactions identical to that in the C_3 plants.

C_4 metabolism appears to be an evolutionary advance that provides a physiological adaptation to intense heat and water stress. In arid environments with intense sunlight and high day temperatures, C_4 plants exhibit higher photosynthetic efficiencies, higher temperature optima, and more efficient utilization of water. Stomatal resistance is high so that water loss from leaf surfaces is retarded. Although the import of CO_2 into the leaf is also slowed, the preliminary storage of CO_2 in dicarboxylic acids provides a mechanism for concentrating this essential substrate in photosynthetic tissues. Efficiencies are high because there is no diversion or loss of the intermediates of carbon fixation in the reactions of photorespiration. Many C_4 plants appear to lack the photorespiratory apparatus.

C_4 metabolism is a relatively recent discovery. Investigations of this new dimension of photosynthesis are being carried out in many parts of the world, especially in the United States and Australia.

SOME CONCLUSIONS ABOUT PHOTOSYNTHESIS

In summary, the over-all balance of raw materials consumed and products generated by the photosynthetic apparatus is as follows:

$$CO_2 + H_2O \xrightarrow[\text{chlorophyll}]{\text{8 quanta of light}} CH_2O + O_2$$

In order for this over-all change to be accomplished, four [H]'s and three ATP's generated within the quantasomes are utilized in the reduction of CO_2 and the regeneration of a new CO_2 acceptor molecule.

In spite of the seemingly endless factual details that have been amassed by investigators of photosynthesis since the time of Priestley and Ingen-Housz, it is wise to remind ourselves that this is still an incompletely understood process. Neither the most skilled chemist nor the most complex chemical factory can dupli-cate what is now produced by green plants in their harvesting of the energy of sunlight. No man-made substitute for the process of photosynthesis is available at this time, nor will it be feasible for chemical factories to take over the work of photosynthesis in the foreseeable future. As long as test tube photosynthesis is still a dream, it is essential that we understand and appreciate the role of green plants (from the microscopic diatoms of the ocean to the giants of our redwood forests) in maintaining our atmosphere and food chains. The contribution of photosynthesis to the maintenance of life on this planet cannot be overemphasized. We cannot exist without it.

USEFUL REFERENCES

Bassham, J. A., and Calvin, M. *The Path of Carbon in Photosynthesis.* Prentice-Hall Inc., Englewood Cliffs, N. J., 1957.

Devlin, R. M., and Barker, A. V. *Photosynthesis.* Van Nostrand-Reinhold Co., New York, 1971.

Levine, R. P. The mechanism of photosynthesis. *Scientfic American, 221(6): 58-70,* 1969.

Olson, J. M., Hind, G., Lyman, H., and Siegelman, H. W. (eds.) Symposia in Biology, No. 19: *Energy Conversion by the Photosynthetic Apparatus.* Clearinghouse for Federal Scientific and Technical Information, Springfield, Va., 1967.

Rabinowitch, E., and Govindjee. *Photosynthesis.* John Wiley & Sons, Inc., New York, 1969.

San Pietro, A., Greer, F. A., and Army, T. J. *Harvesting the Sun: Photosynthesis in Plant Life.* Academic Press, New York, 1967.

Zelitch, I. *Photosynthesis, Photorespiration, and Plant Productivity.* Academic Press, New York, 1971.

Plants and Water; Translocation and Mineral Nutrition

WATER: AN ESSENTIAL INGREDIENT OF ALL LIFE

Life arose in the sea so it is not surprising that all living things, whether plants or animals are dependent on water for survival. Plants share with animals many of the following reasons for needing water: (1) Many constituents of protoplasm, including carbohydrates, proteins, and nucleic acids, are hydrated. Their physical and chemical properties change if water is removed. Ordinarily protoplasm contains up to 95 per cent water. Dehydration usually results in the death of cells. (2) Water molecules are essential components of many of the chemical reactions that occur in protoplasm. Cleavage of polymers such as carbohydrates and proteins by addition of water (**hydrolysis**), adding together molecular subunits by splitting out water (**condensation**), utilization of water as a substrate for photosynthesis, and the formation of water from molecular oxygen, electrons, and protons in the terminal oxidation step of respiration are but a few examples of the involvement of water in cellular metabolism. This unique liquid also serves as the medium within which many chemical reactions occur. (3) Water is a solvent, carrying in dissolved form all of the inorganic salts needed to sustain plants. (4) Water also has a universal role in maintaining the shape and rigidity of plants, particularly structures such as leaves, flowers, and fruits that lack woody supporting tissues. Most plant cells have large central vacuoles filled with water plus some dissolved chemicals. These relatively massive bags of fluid within the thin layers of protoplasm are responsible for maintaining the turgor of cells, tissues, and whole organs. The phenomenon of **wilting** is no more than a loss of turgor due to excessive loss of water from the vacuoles of plant cells. (5) Water is essential not only to the proper internal functioning of plant cells, but also must be present in thin films covering cell surfaces. Water molecules even fill the spaces between cellulose microfibrils in the cell walls. Extracellular water in plants provides a medium for transfer of dissolved substances from the environment into cells (gases from the atmosphere as well as minerals from the soil) and from cell to cell.

THE CHEMICAL AND PHYSICAL UNIQUENESS OF WATER

Water is a wholly remarkable and unique fluid. The properties of water on which life depends not only make the surface of this planet a suitable habitat for life but also permit water to take part in the innermost chemistry of cells.

Water is a liquid at temperatures between 0 and 100°C. Other molecules of comparable size such as methane and ammonia have much lower melting and boiling points and so are gases at planet earth temperatures. Water, in a sense, behaves as though it had a much higher molecular weight, due to the close association of its molecules. Water reaches its maximum density at a temperature of 4°C, *i.e.*, just above its freezing point. A consequence of this unusual property is that as water freezes, it expands and the less dense ice floats on top. If ice formed on the bottoms of bodies of water, lakes in the temperate parts of the world would be at least partially filled up with ice most of the year.

Furthermore, water has a high **specific heat**. To raise the temperature of one gram of liquid water one degree centigrade, the input of one calorie of heat is required. The raising of the temperature of other common fluids requires far less energy. For ethyl alcohol the heat required is 0.456 cal./gm./$^{\circ}$C.; for benzene, 0.428. Because the temperature of water changes slowly, climates near oceans and even large inland lakes tend to be moderate. To evaporate water, *i.e.*, to change it from a liquid to the gaseous state, requires an input of 540 cal./gm., whether over a fire causing the water to become steam or merely in evaporation from any moist surface. Most of the water lost from plants is in the form of water vapor. Because of its large **heat of evaporation**, evaporating water can cause considerable cooling. The loss of water vapor from large bodies of water also moderates the temperature of the remaining fluid. To convert liquid water to ice (its solid state), 80 cal./gm. must be removed. Another way of looking at this biologically important property of water is to remember that water stores heat, so to speak. When the temperature falls to 0°C., water does not immediately crystallize. It gives up 80 cal. of heat per gram before turning to ice. Since ice cyrstals are highly damaging to protoplasm, in fact usually fatal, this property of water affords a temperature cushion for living things. One reason that ice cubes are so effective in cooling liquids, such as in iced tea, is that 80 cal. of heat have to be absorbed from the surrounding liquid in order for a gram of ice to melt, *i.e.*, to be converted from solid to liquid.

The freezing and boiling points of water are changed by the presence of dissolved substances. Freezing point depression and elevation of the boiling point are proportional to the concentrations of dissolved substances, and these changes can be used to estimate the molecular weight of a dissolved compound. One factor in the development of frost hardiness of plants during the fall months (when average temperatures are decreasing) is the increase of sugars and other materials within the vacuoles, with the consequence that temperatures must drop below 0°C. for ice crystals to form.

Two additional properties of water are important for plants. A high surface tension and the tendency of water molecules to stick together tenaciously, combined with their ability to wet and adhere to surfaces means they will move through small cavities in soil or in plant cells by capillary action. A final characteristic of water is its transparency to visible light, which makes it possible for aquatic plants to carry on photosynthesis. Water absorbs both the short wavelength ultraviolet and the long wavelength infrared portions of the spectrum. Without a protecting shield of water, life could never have arisen, because many cellular constituents, including the nucleic acids, are destroyed by ultraviolet radiation. The absorption of infrared wavelengths makes water an effective heat shield.

THE CELLULAR BASIS FOR WATER MOVEMENT

Plant cells are enclosed in cellulose walls, very porous boxes that offer practically no resistance to the movement of water molecules or solutes. The cellulose walls of plant cells provide a physical limit to the expansion of the protoplasm and vacuole. Control over what substances enter and leave plant cells rests in the membranes on the outer and inner surfaces of the protoplasm, the **plasma membrane** and **vacuolar membrane**. These membranes are **selectively permeable**; that is, they permit practically unhindered passage of some molecules such as water, are impassable to others such as proteins, and permit intermediate degrees of movement of other molecules such as dissolved sugars and salts. The advantages of selective permeability are obvious. If a cell must import essential substrates from its environment, it must provide for the entry of these materials. On the other hand, a cell must prevent the loss of its essential structural components and nutrients.

Several kinds of mechanisms are involved in the entry, movement, and loss of water and solutes such as sugar and minerals in plants. Both the uptake of water by the root and the loss of water vapor by evaporation from the leaves follow the principles of **diffusion**. Diffusion is nothing more than a random movement of molecules from a region of higher to one of lower concentration. This is probably the only physical mechanism involved in the movement of water molecules from the moist intercellular

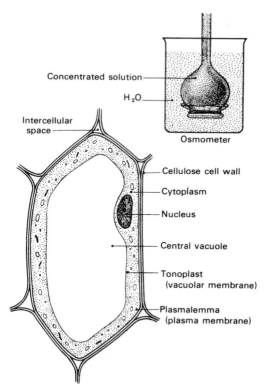

Concentrated solution

H_2O

Osmometer

Intercellular space

Cellulose cell wall

Cytoplasm

Nucleus

Central vacuole

Tonoplast (vacuolar membrane)

Plasmalemma (plasma membrane)

Fig. 16-1. The semipermeable membranes of plant cells, the plasma membrane and the vacuolar membrane, control the movement of substances into and out of cells. These membranes obey the same physical laws as a simple thistle tube osmometer.

solute molecules, there is a net tendency for solvent molecules to diffuse into the concentrated solution. The extent to which solvent molecules move into the solution is proportional to the concentration of solute and to the absolute temperature. The **osmotic pressure** of the internal solution can be expressed either in terms of concentration (molarity of solute) or as pressure (atmospheres). It turns out that the potential osmotic pressure of a solution inside a semipermeable membrane follows the same physical laws as do gases. In other words, the osmotic potential of a solution is dependent on the total concentration of dissolved molecules and ions. A 1-molar solution of sugar (sucrose) at $0°C.$ has an osmotic pressure of 22.4 atmospheres. A 1-molar solution of salt (NaCl) under the same conditions would have twice this osmotic pressure value because of its complete dissociation into Na^+ and Cl^- ions.

Each plant cell is, in effect, a tiny osmometer. When placed in pure water, cells take up water molecules due to the osmotic potential of substances dissolved in their vacuoles and protoplasm. Uptake is accompanied by a swelling of cell contents until they press against the cell walls, *i.e.*, become **turgid**. In a fully turgid cell the counterpressure exerted by the cell wall on its contents just balances the tendency for water molecules to enter the cell in response to the osmotic pressure of the vacuole and cytoplasm. If the external environment of the cell has a higher osmotic potential than its contents, water molecules diffuse outward at a rate faster than they enter. Consequently the contents of the cell decrease in volume. If these circumstances persist, the vacuole and protoplasm contract away from the cell wall. The cell becomes **flaccid** and finally **plasmolyzed**.

The phenomenon of plasmolysis provides one of several possible means of determining the osmotic concentration of the contents of cells. Individual cells or thin strips of tissue placed in a series of solutions of differing molarities can be observed under the microscope. That concentration in which cells are at the state of incipient plasmolysis is the one having an osmotic concentration equivalent to that of the cell contents. An alternate method involves the use of uniform pieces of tissue (potato tubers are favored materials for student

spaces of stems and leaves outward through openings in the surfaces of these structures, the lenticels and the stomates.

The structural components of a plant's vascular, *i.e.*, conducting, tissues, the **xylem** and **phloem**, are long and tubular. The xylem elements especially, with their cylindrical vessel cells joined end to end, provide a means for movement of water and solutes from the roots to the top of the shoot. Xylem elements are structurally and functionally analogous to the plumbing system in our homes. Within both systems there is a mass flow of liquid under hydrostatic pressure gradients.

Another phenomenon involved in the transfer of water and solutes is **osmosis**. When a concentrated solution is separated from pure solvent by a semipermeable membrane, *i.e.*, one that is permeable to the solvent but not to the

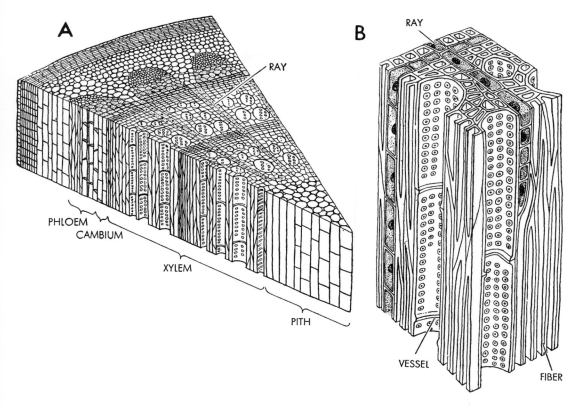

Fig. 16-2. Conduits for movement of dissolved materials in plants. A, arrangement of vascular tissues in a woody stem. Xylem, composed mostly of dead cells, is the tissue through which water and minerals absorbed by the roots are transported upwards to all parts of the shoot. Phloem, located just outside the cambium, is the route of transport for a variety of dissolved substances (including carbohydrates and minerals) to all parts of a plant. B, enlarged section of xylem showing arrangement of vessels, fibers, and ray parenchyma cells. (From S. Biddulph and O. Biddulph: The circulatory system of plants. Scientific American, *200:* 44-49, 1959.)

laboratories because they contain a single type of thin-walled parenchyma cell). Weighed slices of tissue are placed in a series of solutions of different concentrations and are left to equilibrate for several hours. They are then blotted dry and reweighed. The solution causing neither gain nor loss of water (and weight) is equivalent in osmotic concentration to that of the cell contents in the experimental tissue.

TRANSPIRATION: THE MASS FLOW OF WATER THROUGH PLANTS

The discovery of blood circulation in animals by William Harvey in 1628 provided an impetus in the search for a comparable circulatory system in plants. The first major publication dealing with the possible circulation of sap and related aspects of plant metabolism was the book **Vegetable Staticks**, published in 1726 by Stephen Hales, who concluded that there is no circulation of sap in plants comparable to the circulation of blood in animals. Instead the movement of fluids in plants has been found to be like the action of a wick: massive amounts of water are drawn from the soil solution into the roots, pulled upward through the stems, and finally move out through the stomates of leaves and the lenticels of the stems as water vapor. This unidirectional mass flow of water in plants with a continual loss from aerial parts is called **transpiration.**

Transpirational water loss can be measured very simply by using methods first introduced

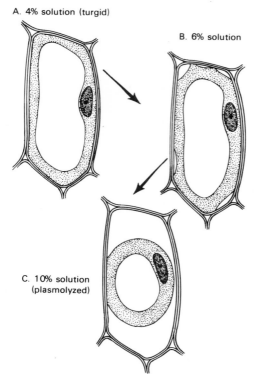

A. 4% solution (turgid)

B. 6% solution

C. 10% solution (plasmolyzed)

Fig. 16-3. Plant cells respond to the osmotic concentration of their environment. A, in water or a hypotonic solution of sugar a plant cell is turgid. B, in a solution of osmotic concentration equal to that of its contents a cell is in the state of incipient plasmolysis. C, in a hypertonic solution there is a net loss of water from the cell, the contents contract away from the cell wall, and the cell becomes plasmolyzed.

by Stephen Hales. All one has to do is follow the loss of weight of an entire potted plant or of a shoot or even a single leaf by weighing at regular intervals. The only necessary precaution in the case of a potted plant is to cover the pot so that only the water loss from the shoot is measured. When working with shoots or single leaves, the cut end of a shoot or the petiole of a leaf is immersed in a reservoir of water so that the excised piece of plant does not dry out. In the latter case it is the water loss from the closed reservoir that is measured as water is drawn from it to replace that lost in transpiration.

A more recently invented device called a **potometer** can be used if one wishes to measure the volume of water lost in transpiration. With this simple instrument, transpired water is replaced from a reservoir connected to a calibrated capillary tube. A small air bubble is introduced into the tube and the position of the bubble is recorded at intervals as it moves along the tube. (A 0.1-ml. volume calibrated pipette is frequently used in home-made potometers.)

As soon as scientists began to measure transpiration rates, they discovered that rates vary not only with the species of plant used in the experiment but also with a number of environmental conditions and even the time of day. Obviously, there must be some mechanism by which the rate of water loss from the leaves may be regulated. If we look at the surfaces of leaves, we find that most of the epidermal cells look much like pieces of a jigsaw puzzle, forming a tightly interlocking covering. Dispersed in the layer of epidermal cells are pores called **stomates**, openings between two kidney-shaped **guard cells**. The entire complex of opening plus surrounding cells is called the **stomatal apparatus.** When the guard cells are turgid, their inner surfaces pull apart and the pore or **stoma** is open. Conversely, when the guard cells become flaccid, the surfaces forming the edge of the stoma collapse and the stoma is closed. Stomates regulate the movement of gases between the intercellular spaces of the leaf and the atmosphere. Closed, they drastically reduce the exchange of gases and loss of water vapor from leaves. With intermediate degrees of opening, the rate of gas exchange is proportional to the extent of opening. There is a diurnal pattern of stomatal opening and closing that corresponds closely to diurnal changes in rates of transpiration, strongly suggesting a cause and effect relationship. Assuming that the opening and closing of stomates is the means for regulating rates of transpiration, it appears that the crucial question is: What mechanism regulates the opening and closing of the stomates?

Mechanism of Stomatal Opening and Closing

Environmental factors such as high humidity, light, and low CO_2 tension promote opening while dry air, darkness, and high CO_2 tension

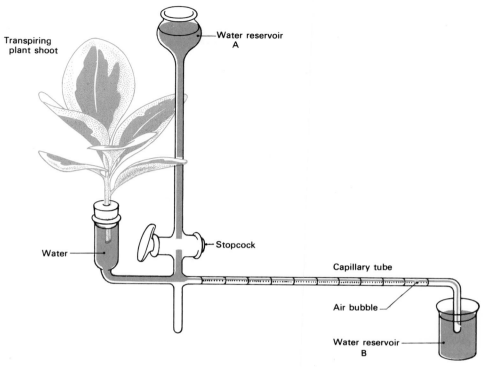

Fig. 16.4. A potometer is a simple device used to measure the volume of water lost in transpiration by following the movement of a small air bubble along a calibrated capillary tube. The bubble is introduced by lifting the end of the capillary tube out of water reservoir B. The air bubble can be moved back to the right end of the capillary tube by opening the stopcock and permitting water to flow from reservoir A.

promote stomatal closing. Some studies have indicated that the action spectrum for light-induced stomatal opening corresponds closely to that of photosynthesis with maxima in the blue and the red portions of the spectrum. It was therefore concluded that photosynthesis is somehow involved, although the fact that this light-dependent response saturates at very low light intensities was quite mystifying. A further observation, that light-dependent stomatal opening requires oxygen, added to the mystery.

For many years it seemed that light-dependent stomatal opening could be explained very simply on the basis of the photosynthetic capacity of the guard cells. These kidney-shaped cells that change in turgor, and thus make the space between them appear or disappear, differ from all other epidermal cells by having chloroplasts. Thus, it appeared that opening of the stomates in the light could be the result of increased osmotic pressure from the photosynthetic production of soluble carbohydrates. Dark closure is, according to this hypothesis, due to condensation of soluble carbohydrates into insoluble starch. The difficulty with this hypothesis is that starch-sugar conversions are sensitive to the water content of cells rather than to light. When humidity is high there is little relationship between the relative starch and sugar contents of the guard cells and the degree of stomatal opening. When humidity is low, guard cells can open but still contain starch.

Recently, Israel Zelitch, working at the Connecticut Agricultural Experiment Station in New Haven, proposed a mechanism for light-dependent stomatal opening that is consistent with most of the facts. It will be recalled from the chapter on photosynthesis (Chapter 15) that when CO_2 concentration is very low there is a change in the products of CO_2 fixation from carbohydrates to glycolic acid. Zelitch

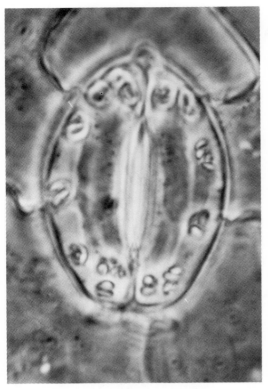

Fig. 16-5. Stomatal apparatus of a sunflower (*Helianthus annuus*) leaf. Photomicrograph of living cells of the lower epidermis as seen under phase contrast optics, × 1,750.

observed that the opening of stomates in the light is inhibited by hydroxy sulfonates, chemicals that are known to be competitive inhibitors of the enzyme glycolic acid oxidase. This inhibition can be reversed by providing an excess of the substrate, glycolic acid. In order for stomates to open in the light, glycolic acid must be oxidized to glyoxalic acid in a reaction requiring oxygen. Glycolic acid is formed in photosynthesis via a pathway for CO_2 reduction that operates only when CO_2 concentration is unusually low. When normal atmospheric or higher concentrations of CO_2 are present, the glycolic acid pathway becomes inoperative and is replaced by a pathway leading to carbohydrates. Realizing this, we can understand why high concentrations of CO_2 can inhibit light-dependent stomatal opening (because glycolic acid is not formed), and why

CO_2 inhibition, like that caused by the hydroxy sulfonates, can be reversed by glycolic acid. It appears that a lowering of CO_2 concentration in the intercellular spaces within the leaf is a key factor in the opening of stomates.

Stomatal opening and closing provides a mechanism not only for the regulation of water loss from the leaves but also the regulation of gas exchange between the internal tissues of the leaf and the atmosphere. A high rate of photosynthesis leads to rapid depletion of CO_2 in the air spaces of the leaf. The change from a pathway leading from CO_2 to carbohydrate synthesis to one in which glycolic acid is formed insures that stomates will stay open to provide maximum access of this essential substrate to photosynthesizing cells.

Very recently some new observations have reopened the question of what mechanism regulates the light-dependent changes in permeability of guard cells. It has been shown in several laboratories that the increased turgor of guard cells in the light is accompanied by a rapid uptake of potassium ions. Other studies by one of the authors of this book with an albino mutant sunflower have revealed that light-dependent stomatal opening can occur even in nonphotosynthetic albino leaves. A low intensity opening of stomates in the sunflower mutant is mediated by phytochrome (the light-sensitive pigment involved in many aspects of the photophysiology of plants from seed germination to flowering). Far red light promotes, while red light inhibits, opening of the mutant sunflower stomates. Wild type sunflower stomates respond in the same way as those of the mutant to far red light, but in addition exhibit a high intensity opening which is most responsive to blue light. It appears likely that the low intensity, phytochrome-mediated system may be responsible for the permeability changes resulting in the flux of potassium ions into the guard cells.

There is a striking parallel between light-dependent stomatal opening and the sleep movements of leaves in leguminous plants such as **Mimosa pudica** (the sensitive plant) and **Albizzia**. Leaflet orientation is controlled by the turgor of motor cells located in the upper and lower sides of the swelling (**pulvinus**) at the base of the petiole. Dark closing (or folding

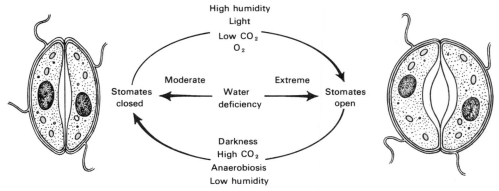

Fig. 16-6. Summary of stomatal responses to environmental factors.

together) of leaflets is accompanied by a movement of potassium ions out of the ventral motor cells and into the dorsal motor cells. Pre-irradiation with red light is necessary for this movement of potassium. Opening of the leaflets is promoted by blue light (and also by far red light) and is accompanied by a reverse flow of potassium. Ruth Satter and Arthur Galston of Yale University, the scientists who have studied these leaflet movements, have suggested that "potassium flux is a general prerequisite for turgor changes in all species with moving leaves or leaflets." The same condition appears to apply to turgor changes in the guard cells.

Adaptations That Affect Transpiration Rates

Plants exhibit a variety of adaptations that influence transpiration rates. Many such adaptations have to do with the spacing and placement of stomates or with the shape and the surface to volume ratio of leaves. Land plants are able to survive in a wide array of habitats ranging from very wet to very dry. Often the amount of available water varies with the season. One way that plants manage to survive during parts of the year when water supplies are limited is to shed their leaves and then grow a new set of these appendages when conditions are again favorable. Winter months are usually a time of water deficit for terrestrial plants, not because of any lack of precipitation but rather because water in its solid state cannot be absorbed. Even in areas having very cold

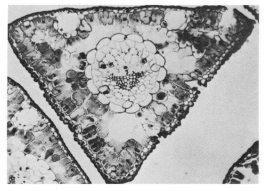

Fig. 16-7. White pine leaf cross section as seen under a microscope. Note reduced surface relative to volume, sunken stomates, and cutinized epidermal cells. (From E. J. Fisk and W. F. Millington: *Atlas of Plant Morphology: Portfolio I—Photomicrographs of Root, Stem and Leaf.* Burgess Publishing Co., Minneapolis, 1959.)

winters, there are plant species that retain their leaves, including most of the conifers, the rhododendrons, and hollies. If we look at the structure of the leaves of these plants we can understand why they can be retained. In the conifers needle-like leaves are much reduced in surface area and the stomates are located not on the outer surface but rather in cavities below the surface. Leaves of the conifers and of the evergreen rhododendrons and hollies have very thick waxy cuticles covering the epidermal cells. Also, there are fewer stomates per unit of surface area than in species that lose their leaves at the end of the warm growing season. Thus, there are three aspects of leaf anatomy that affect transpiration rates: surface area, the number and location of stomates, and thickness of the cuticle.

The water problems of many desert species have been resolved by drastic reduction of leaves and development of fleshy, water-storing stems. Plants adapted to very moist or aquatic habitats do not have the problems of water conservation faced by species growing in deserts or even in areas of moderate supplies of soil moisture. Although the stomates of aquatic species are not particularly important for the regulation of water loss, they still serve an important function in facilitating the exchange of gaseous substrates and products of photosynthesis and respiration. Therefore, it is not surprising to find that the floating leaves of water lilies have stomates only on their upper surfaces. Many aquatic plants have internal tissues made up mostly of air spaces (the so-called **aerenchyma** tissues) that provide for gaseous reserves and often help to assure buoyancy so that leaves will float.

WATER MOVEMENT FROM THE ROOTS TO THE SHOOTS

Vascular tissues, both xylem and phloem, provide a continuous conducting system from the root, through the stem and into the leaves where branching strands, the **veins**, extend to all parts of the leaf blade. There are several kinds of evidence that the xylem (the woody center of tree trunks) is involved in the transport of water. First, of the two kinds of vascular tissue in plants, xylem and phloem (see Chapter 9), only the xylem has sufficient volume to account for the massive amounts of water that must be moved. Second, it is possible to remove the bark of a tree (and thereby the phloem, a procedure usually referred to as **girdling**) without interfering appreciably with the upward transport of water. On the other hand, destruction of the xylem results in immediate cessation of upward movement of water accompanied by obvious wilting of the leaves. Third, water can be labeled, either by means of hydrogen or oxygen isotopes or by means of soluble dyes. The label can be seen to rise in the xylem but not in the phloem. (This can be neatly demonstrated in a stalk of

celery.) Finally, the contents of xylem cells are dilute and watery compared to phloem cells. Thus, the major function of xylem is the transport of the massive amount of water needed to replace that lost in transpiration.

Phloem, on the other hand, is involved in the transport of solutes; its major function is the distribution to all parts of the plant of the excess products of photosynthesis. Evidence of phloem transport will be presented later when the translocation of sugars and other metabolites is discussed.

Water enters plants through the **root hairs**. These extensions of individual epidermal cells of the root provide a large surface in close contact with the films of water that surround soil particles. Water molecules entering the root through the root hairs and other surface cells are transferred through the peripheral cortex to the endodermis. Walls of most cells in the endodermis are **suberized**, *i.e.,* they contain a waterproof corky material and present a barrier to the movement of water. Some cells of the endodermis, called passage cells, remain thin-walled and are thought to provide a pathway for free movement of water and solutes to the central core of xylem tissue.

Once in the xylem, water is carried upward through non-living cells, **vessels** and **tracheids**. These tubular cells are arranged one on top of another. The end walls of vessel elements usually disintegrate at maturity so that there is no barrier to water movement. Although tracheid end walls remain intact, they have tapered ends with many thinner areas called **pits**. Pits, also found on the side walls of tracheids and vessels, facilitate the passage of water from cell to cell.

Root Pressure and Guttation

The crucial question which we must now attempt to answer is: What forces propel water upward through the rather remarkable "plumbing" system just described? We could sum up the state of our ignorance by simply saying that movement of water is a consequence of **root pressure** and **transportational pull** exerted by the leaves.

After their shoots are cut off, the stumps of plants often exude liquid from their cut stem surfaces. The amount of such exudates can be measured and is often surprisingly large: in 1936 Crafts collected 551.7 ml. (well over ½ quart) of exudate from a 450-gm. (about one pound) squash root in 24 hours. If a tube is connected to the cut stem, liquid will rise in the tube to amazing heights. In 1726, Stephen Hales reported in *Vegetable Staticks* that liquid rose to a height of 21 feet in a tube connected to the stem of a cut vine. Furthermore, Hales observed that the liquid rose more rapidly during the day than during the night.

Root pressure is related to the metabolic activity of roots and is correlated with an accumulation of dissolved substances in the xylem cells. Root pressure decreases when temperatures are lowered, when any of the

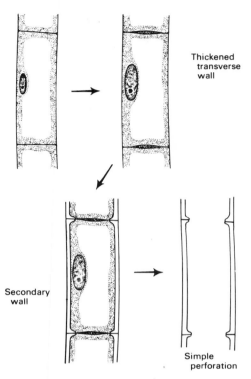

Fig. 16-9. Development of xylem vessel elements. End walls disintegrate and protoplasts disappear at maturity. As in pieces of pipe joined end to end, there is no barrier to water movement. (Redrawn from K. Esau: *Anatomy of Seed Plants*. John Wiley & Sons, Inc., New York, 1960.)

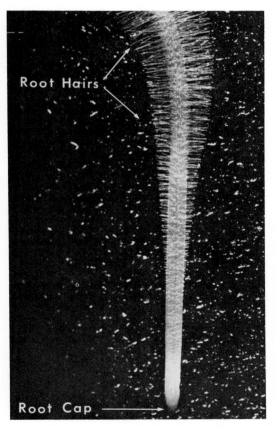

Fig. 16-8. Root hairs are the site of entry of water and dissolved minerals from the soil into a plant. (From T. P. O'Brien and M. E. McCully: *Plant Structure and Development*. The Macmillan Co., Collier-Macmillan, Ltd., London, 1969.)

essential inorganic nutrients are lacking, or if certain metabolic poisons are present. This would not be true of a purely physical process. It is probably correct to conclude that root pressure is a consequence of the active accumulation of nutrients from the soil solution at the expense of energy made available through respiration. A positive pressure on contents of the xylem can occur not only in severed stems but also in intact plants; it is responsible for the phenomenon of **guttation**, or the exudation of liquid water from leaves. Water is expelled either through the stomates or from special structures called **hydathodes** at the edges of leaves. Guttation usually occurs at night when water loss through transpiration is reduced. It can readily be seen in the laboratory or classroom when seedlings of corn or oats are placed under a bell jar.

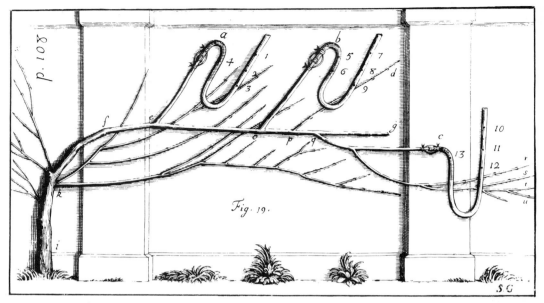

Fig. 16-10. The first U-tube pressure gauges were made by Stephen Hales to measure sap pressure. (Photographed from Hales' original publication, courtesy of the Library of Congress.)

Fig. 16-11. Strawberry leaf with drops of water exuded by guttation. (Photograph by J. A. Herrick, from V. A. Greulach and J. E. Adams: *Plants: An Introduction to Modern Botany.* John Wiley & Sons, Inc., New York, 1967.)

Although root pressure is a probable contributing factor in the ascent of sap in plants and certainly is involved in the uptake and translocation of dissolved mineral nutrients, it cannot explain how water gets to the tops of trees. At best, roots can exert a hydrostatic pressure of only a few atmospheres. One atmosphere can push a column of water to a height of 10.4 meters. Pressures of 10 atmospheres or more would be required to move water to the heights of over 100 meters attained by the tallest trees.

Cohesion-Tension Theory of Sap Ascent

It is reassuring in a time when new scientific knowledge is accumulating at an almost explosive rate to find that some theories are durable. The best available explanation for the ascent of sap, the **cohesion-tension theory**, was proposed in 1894 by a plant physiologist, H. H. Dixon, and a physicist, J. Joly. According to this theory, the evaporation of water from leaf cells creates a water deficit in them. To replace water lost in transpiration from the leaves, water is withdrawn from the ends of nearby conducting tissues. This results in a tension which is transmitted downward on the continuous columns of water in the vessels and tracheids of the xylem. This tension, or "pull" on the water within the xylem, is a result of the cohesive forces holding the water molecules together and draws water molecules into the roots. The cohesiveness of water molecules is a consequence of their tendency to form hydrogen bonds with other water molecules. A mass flow of water thus moves up the plant to replace water lost in transpiration. This theory

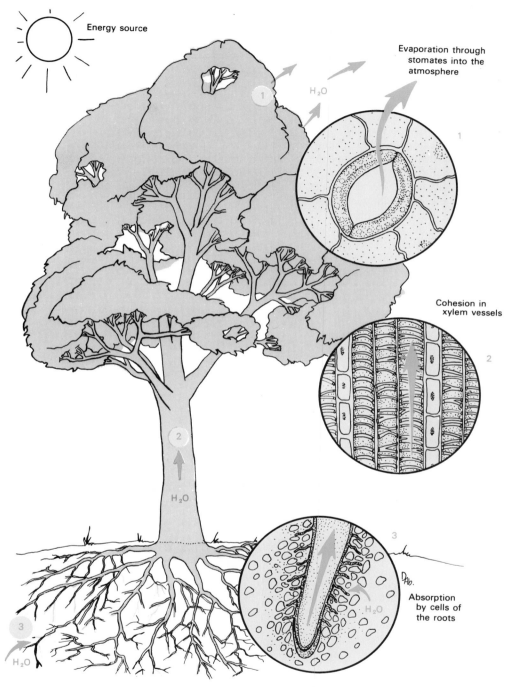

Fig. 16-12. The cohension-tension theory. Evaporation of water from leaves (1) creates a water deficit. To replace water lost in transpiration, water is withdrawn from the conducting tissues. This results in a tension which is transmitted downward through the xylem (2). The "pull" on water in the xylem is a result of cohensive forces holding water molecules together and draws water into the roots (3). A mass flow of water thus moves up the plant to replace water lost in transpiration.

predicts that tensions on the xylem ought to cause a decrease in stem diameter during times of maximum transpiration. It is relatively easy to demonstrate that tree trunks contract slightly during the day and are of measurably greater diameter at night when rates of transpiration are low.

Further support for the cohension-tension theory comes from a physical model constructed by Askenasy in 1895. A porous clay cup filled with water is connected by a long glass tube to a reservoir of mercury. Cohesive forces exist between water molecules in the glass tube and there are adhesive forces between water and mercury at their interface. As water molecules evaporate from the surface of the clay cup, capillary forces draw water into the pores of the cup. In such a model system, the column of mercury can be drawn upward far in excess of the height it can be raised by atmospheric pressure. Comparable results are obtained if the porous clay pot is replaced by a transpiring shoot with leaves.

INORGANIC NUTRIENTS—PASSENGERS IN THE TRANSPIRATIONAL STREAM

Compared to animals, plants are amazingly modest in their nutritional requirements. Using water from the soil, CO_2 from the air, and energy from the sun, photosynthesis provides the carbon skeletons of all compounds needed as substrates for respiration or as structural components for plant and, indirectly, for animal cells. In addition, only a dozen or so elements absorbed from the soil as soluble inorganic ions are needed to sustain healthy plant growth and development. Thus, only about 16 out of the 100 or more elements found in the earth's crust are needed to make a plant. Of these, carbon and oxygen are obtained only from the atmosphere. Hydrogen, combined with oxygen in water, is obtained from the soil solution. From these three elements a plant can synthesize carbohydrates and lipids. With just one more element, nitrogen, most amino acids can be manufactured. Nitrogen is also a component of nucleic acids and porphyrins. Although nitrogen is absorbed from the soil in the form of NO_3^- (nitrate) or NH_4^+ (ammonium) ions, the ultimate source of this element is the gaseous

nitrogen making up about 80 per cent of our atmosphere.

In recent years the great importance of a fifth element (phosphorus) has become clear. For the release of energy from reserve nutrients, plants need water from the soil and O_2 from the atmosphere. In order for carbohydrates to be metabolized, however, they must be phosphorylated. The element phosphorus is also a component of all nucleic acids, the important energy carriers adenosine triphosphate and adenosine diphosphate, the pyridine nucleotides nicotinamide adenine dinucleotide and nicotinamide adenine dinucleotide phos-

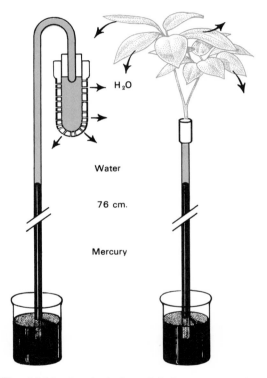

Fig. 16-13. A physical model demonstrates that evoporation can raise water to greater heights than atmospheric pressure alone. As water evaporates from a porous clay cup (left) or a plant shoot (right), water is pulled up the glass tube. Adhesion of water to mercury causes mercury to be pulled up from the reservoir. Atmospheric pressure alone can push a column of mercury to a height of 76 cm. In such a model mercury can be pulled up to 100 cm. or more. (After V. A. Greulach and J. E. Adams: *Plants: An Introduction to Modern Botany*. John Wiley & Sons, Inc., New York, 1967.)

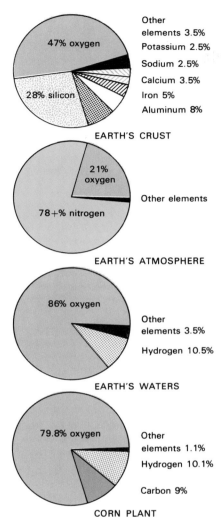

EARTH'S CRUST

47% oxygen

28% silicon

Other elements 3.5%
Potassium 2.5%
Sodium 2.5%
Calcium 3.5%
Iron 5%
Aluminum 8%

EARTH'S ATMOSPHERE

21% oxygen

78+% nitrogen

Other elements

EARTH'S WATERS

86% oxygen

Other elements 3.5%

Hydrogen 10.5%

CORN PLANT

79.8% oxygen

Other elements 1.1%

Hydrogen 10.1%

Carbon 9%

Fig. 16-14. Diagrammatic summary of the distribution of elements (per cent by weight) in a corn plant and in the earth's waters, atmosphere, and crust. (Redrawn after V. A. Greulach and J. E. Adams: *Plants: An Introduction to Modern Botany.* John Wiley & Sons, Inc., New York, 1967.)

phate, and of the phospholipids essential in the structure of membranes.

If we think further about the details of cell chemistry discussed in Chapter 3, we realize the need for other essential elements. Calcium is found in the calcium pectates of the middle lamella, the cement that holds together plant cells. Sulfur is found in the amino acids cystine, cysteine, and methionine. The sulfhydryl bonds

formed between sulfur-containing amino acids have an important role in maintaining the special configuration of protein molecules. We have also encountered sulfur in the bond that holds acetate groups to the coenzyme A molecule in acetyl-CoA, the initial substrate for the Krebs cycle.

Magnesium is essential to plants because of its presence at the center of the tetrapyrrole nucleus of the chlorophyll molecule. Magnesium ions are also needed for the proper functioning of a number of enzymes. Mg^{2+} ions are involved in the aggregation of structural subunits of the ribosomes and thereby play a role in protein synthesis. If we add only one more element, potassium, that is essential for the activity of many enzymes and the maintenance of membrane permeability, we find that we have already listed the six essential mineral nutrients needed in largest amounts by plants, the so-called **major elements**: nitrogen, phosphorus, calcium, sulfur, magnesium and potassium.

Another half-dozen or so elements are also essential but are needed in only trace amounts and therefore are usually referred to as the **minor elements**. The minor elements required by all plants are iron, manganese, zinc, copper, molybdenum, boron, and chlorine. Once again we recognize from considerations of the chemistry of cells or of special processes such as photosynthesis that the trace elements are components of molecules essential to the maintenance of cell metabolism. Iron is a component of the cytochromes and of ferredoxin and is present in the prosthetic groups of a variety of enzymes including catalase and cytochrome oxidase. Manganese is necessary for the activity of a number of enzymes including one involved in photosynthetic oxygen evolution. Zinc is needed for an enzyme in the biochemical pathway from the amino acid tryptophan to the plant hormone indoleacetic acid (auxin). Copper is found not only in plastocyanin in the photosynthetic apparatus but also in a variety of oxidizing enzymes including ascorbic acid oxidase, polyphenol oxidase and cytochrome oxidase (an enzyme containing both iron and copper). The roles of molybdenum, boron, and chlorine are somewhat more obscure. Molybdenum is involved in nitrogen metabolism; boron in the translocation of carbohydrates; and chloride ions are essential for activity of chloroplasts.

In addition to the rather short list of elements universally needed, a few inorganic ions are needed by plant cells having rather special requirements. For example, silicon is essential for the cell walls of diatoms and also accumulates in the stems of horsetails. The symbiotic nitrogen-fixing bacteria all require cobalt, a peculiarity of the enzymes associated with the steps in nitrogen fixation. It should be noted that except for the symbiotic and free-living nitrogen-fixing bacteria, plants are different from animals in not requiring cobalt. For animals, this element is an essential component of cyanocobalamin, or vitamin B_{12}. Although vitamin B_{12} is not synthesized by higher plants, it is manufactured by many soil microorganisms (bacteria and fungi) and by intestinal bacteria. In animals, vitamin B_{12} is bound to polypeptides. Animal proteins are a good source of this vitamin for human diets while plant proteins are not. Sodium is another element essential to animals that is ordinarily nonessential for plants. Only plants adapted to soils with a high salt content (the **halophytes**) absorb significant amounts of sodium. The nonessential Na^+ ions help to raise the internal osmotic pressure of plant cells to counteract that of their salty surroundings.

It is not unusual to find a variety of nonessential elements in plant tissues. Some sixty different elements have been detected in plants including gold, silver, lead, mercury, arsenic, and selenium. Some of these nonessential elements have no effect on plants but can be extremely toxic to herbivorous animals grazing on them. Certain plants, notably about two dozen species in the genus *Astragalus*, a legume somewhat resembling alfalfa, have the unique ability to extract selenium from the soil. These accumulator species have been shown to contain from 100 to 10,000 times higher levels of selenium than other plants native to the western plains of the United States. Because of the effects of the high selenium levels of accumulator species on animals grazing on them, they have been called "loco" weeds (from the Spanish *loco*, crazy). Cows and horses eating such selenium-containing plants develop alkali poisoning or "blind staggers" that can be fatal. Why the selenium analogs of

sulfur compounds formed by accumulator plants are harmful only to the animals grazing on them is one of the mysterious differences between plants and animals. There is some evidence that selenium is a micronutrient essential for the accumulator species. They certainly grow more rapidly in the presence than in the absence of selenium. Perhaps plants can tolerate the selenium analogs because of their ability to make more of the usable sulfur compounds while animals must survive on what they consume.

Fig. 16-15. A "loco" weed (*Oxytropus lambertii*). This species is closely related and very similar to *Astragalus*. After feeding on "loco" weed, range animals become nervous, jerky in their movements, and generally wild. (Courtesy U. S. Department of Agriculture.)

Mineral Requirements of Plants

An interest in the conditions needed for healthy plant growth is as old as the science of agriculture. Controlled experimentation on the effects of fertilizers and a search for the elements essential for plants go back to the beginnings of our knowledge of chemistry. One of the earliest attempts to discover what plants need in order to grow was described by van Helmont in the early 1600's. He placed a 5-pound willow sapling in a container of dried soil weighing 200 pounds. Nothing except rainwater was added to the container. After five years, van Helmont carefully removed the willow tree and dried and reweighed the soil. By this time the tree weighed 169 pounds. After 5 years the soil weighed 199 lb. 14 oz., and van Helmont attributed this apparent weight loss to experimental error. In spite of the simplicity and logical nature of van Helmont's experiment, he arrived at a wrong conclusion because, on the basis of his experimental data, he believed that plants need only water and air.

Hydroponics: Culture of Plants in Solutions of Inorganic Salts

By the 1800's, studies of the effects of fertilizers on agricultural yields had provided some evidence that the elements calcium, potassium, and phosphorus are taken up from the soil by the roots of plants. However, it was not until the late 1800's that a straightforward methodology was developed for determining experimentally what elements are essential for plant growth. The German plant physiologists Sachs and Knop, using techniques similar to present-day water culture of plants (**hydroponics**), grew plants with roots immersed in nutrient solutions consisting of dissolved inorganic salts. Their objective was to determine the minimum number of elements needed to support plant growth. Knop found that a rather simple mixture of salts could support plant growth indefinitely. Knop's solution (which still is used with only minor modifications) contained the following salts: $Ca(NO_3)_2$, KNO_3, KH_2PO_4, $MgSO_4$, and $FePO_4$.

It is obvious to us that Knop did not add to his nutrient solution any of the trace elements except for iron. The relatively impure salts available in his day assured ample contamination with all of the other essential minor elements. This is a fine example of the necessity

for long-term suspended judgment and a proper lack of dogmatism in interpreting scientific results, no matter how carefully obtained.

Table 16-1 shows the amounts of various component salts needed to prepare a liter of balanced nutrient solution in which a variety of common plants will grow well.

One of the practical advantages of nutrient solutions is that they enable us to delete just one element at a time to determine how a deficiency in one essential nutrient affects plants. The other important fact about the nutrition of plants that can be demonstrated by hydroponics is that for healthy growth, plants need only water, CO_2, sunlight, and a rather modest assortment of inorganic salts. Does this conflict with the present enthusiasm for organic gardening? Not really, because when organic materials, whether decomposing leaves, coffee grounds, etc., are used as fertilizer, the gradual decomposition of these substances by the microorganisms of decay results in a release of inorganic nutrients and a final breakdown of organic compounds to CO_2 and H_2O. The decomposing organic materials contribute inorganic chemicals to the nutrition of plants and at the same time modify the consistency of soils, making clay soils light and more porous, sandy soils better able to absorb and retain water.

Symptoms of Mineral Deficiency

When grown in nutrient solutions or soils that are deficient in one or more essential elements, plants exhibit deficiency symptoms. Their growth is stunted and abnormal just as surely as the growth of humans is stunted and

Table 16-1
Components of a balanced nutrient solution for plants

Salt	Weight per Liter
	gm.
$Ca(NO_3)_2$	0.8205
KNO_3	0.50055
$MgSO_4$	0.24076
KH_2PO_4	0.13609
$FeCl_3$	0.0145
H_3BO_3	0.00286
$MnCl_2 \cdot 4H_2O$	0.00181
$ZnCl_2$	0.00011
$CuCl_2 \cdot 2H_2O$	0.00005
$Na_2MoO_4 \cdot 2H_2O$	0.000025

abnormal when their diets lack vitamins or contain the wrong kinds or insufficient amounts of protein.

The symptoms of specific mineral deficiencies are most readily demonstrated when plants are grown in nutrient solutions lacking one essential element but containing balanced and adequate amounts of all others. Some symptoms, such as the **chlorosis** (yellowing due to the absence of chlorophyll) that develops when supplies of iron are inadequate, are so striking that they can be recognized even under the far more complex circumstances of home gardens or in agricultural crops. Most deficiencies are difficult to analyze under field conditions because soils are rarely poor in only one element. The deficiency symptoms characteristic of each of the essential mineral nutrients are listed in Table 16-2 along with information about the forms in which each is absorbed and its functions in plant metabolism.

Inorganic Nutrients and Crop Yields

Soil deficiencies can be alleviated and elements removed from soils in the form of harvested crops can be replaced by the application of appropriate amounts and kinds of fertilizers. One almost universal problem is how to determine what fertilizers to add. Because of its complexity, an analysis of the soil rarely provides an adequate answer to the question of what or how much should be added. Additionally, deficiency symptoms in the field offer difficulties in interpretation. The pH of the soil and the relative amounts of other elements can influence the availability and uptake of specific essential nutrients. For instance, calcium ions, because of their role in maintaining the integrity and functioning of cell membranes, can influence the uptake of other elements. Calcium deficiency sometimes results in the uptake of excess amounts of other nutrients and it is not uncommon when calcium is limiting to observe symptoms of magnesium toxicity rather than of calcium deficiency. It turns out that the best indicator of what should be added to the soil to increase plant growth is the plant itself. Analyses of leaves have shown that for any crop there is a predictable relationship between mineral content and yield. Once this relationship has been established, the best way to determine whether more of a specific element should be added is to analyze a random sample of leaves for the element in question.

Commercially available fertilizers come in a variety of forms with a terminology mysterious to the uninitiated. The numbers used to describe them indicate nothing more complicated than the relative amounts of the three major elements present. For instance, in a fertilizer labeled 5-10-5, the first figure stands for the percentage of nitrogen in the total mixture, the second for available phosphate, and the third for potassium.

Generally, the yields of a field of corn or tomatoes in the home garden are dependent on a host of conditions including climatic factors such as temperature and rainfall, the presence or absence of predators and parasites, and the adequacy of minerals in the soil. Yield is influenced by the relative sufficiencies of each factor essential for growth, including each of the essential mineral elements. An inadequate supply of any one, even a trace element, can severely reduce yields. This generalization, also called the **law of the minimum**, is true for all living organisms. It was first enunciated by Liebig (1846) at the time when fertilizers were being introduced into agricultural practice.

If agricultural yields can be increased drastically by applying the elements essential for plant growth, to what extent can we utilize our knowledge of mineral nutrition to increase world food supplies? Unfortunately, the application of fertilizers will not result in miracles because of a second universal law, the **law of diminishing returns**, stated by Mitscherlich at the beginning of this century. Once adequate amounts of a given element are available, further increase will have no effect on growth or yield. In fact, overabundance of certain elements can even have undesirable effects on yield. Excess nitrogen leads to lush vegetative growth, a condition that is desirable in lawns or pastures, but which suppresses flowering and thus is not desirable when fruits or seeds rather than leaves are to be harvested. Many a home gardener has learned this lesson when his lush tomato plants failed to set fruit.

Accumulation of Minerals From the Soil

Some aspects of the selective permeability of cell membranes have already been noted in

Table 16-2
Elements essential for plant growth

Elements	Form in Which Absorbed	Biochemical Function	Deficiency Symptoms
Major Nitrogen	NO_3^- or NH_4^+	Present in all amino acids, nucleic acids, and porphyrins	Chlorosis, especially in lower leaves, with loss as deficiency becomes more severe. Some species develop purple or red veins in leaves due to increased anthocyanin synthesis. Excess nitrogen produces lush, dark green foliage with weak stems and abundant vegetative growth.
Potassium	K^+	Maintenance of membrane permeability	Chlorosis, sometimes with mottling of leaves and development of dead areas in their tips and margins; stems often weak.
Calcium	Ca^{2+}	Synthesis of middle lamella; enzyme activity	Rapid deterioration of terminal growing regions of shoots and roots; malformation of youngest leaves.
Phosphorus	$H_2PO_4^-$	Present in phospholipids, nucleic acids, high-energy phosphate compounds, and coenzymes	Leaves dark green in color, tendency for red or purple anthocyanin accumulation; development of areas of dead tissues on leaves, petioles, or fruits, often resulting in leaf or fruit drop.
Magnesium	Mg^{2+}	Present in chlorophyll; needed for enzyme activity	Chlorosis of leaves developing upward from the base of the plant often accompanied by death of portions or entire leaves.
Sulfur	SO_4^{2-}	Present in some amino acids and vitamins; involved in secondary and tertiary structure of proteins	Chlorosis, especially of younger leaves at tops of plants. Even older leaves become pale green if deficiency is severe. Root system may be larger than normal.
Minor Iron	Fe^{2+}	Present in cytochromes, catalase, and other porphyrins; needed for chlorophyll synthesis	Chlorosis developing first in areas between veins of youngest leaves. Chlorotic leaves persist and often remain on the plant for long periods; except for color, leaves appear healthy.
Copper	Cu^{2+}	Present in plastocyanin, ascorbic acid oxidase, cytochrome oxidase, and polyphenol oxidase	Withering of tips of young leaves; plants often appear wilted even when ample supplies of water are available.
Manganese	Mn^{2+}	Enzyme activity; needed for photosynthetic oxygen evolution	Young leaves become progressively paler and develop brown or gray dead spots. Chlorosis and dead areas develop first in areas between veins of leaves and leaves are soon lost.

Table 16-2–*Continued*

Elements	Form in Which Absorbed	Biochemical Function	Deficiency Symptoms
Zinc	Zn^{2+}	Needed for auxin synthesis	Chlorosis of lower leaves at tips and margins. Leaves remain clustered on short branches; "little leaf" disease of citrus crops.
Molybdenum	Mo^{3+} or Mo^{6+}	Nitrate reductase	Chlorosis of leaves; early symptoms resemble those of nitrogen deficiency; with prolonged deficiency leaves become mottled, chlorotic areas are puffed in appearance, leaves become twisted and distorted.
Boron	$BO_3{}^{3-}$ or B_4O^{2-}	Involved in translocation of carbohydrates	Death of growing regions, no new leaves formed. Deficiency often called "top sickness." Fleshy organs such as fruits disintegrate with browning of internal tissues.
Chlorine	Cl^-	Needed for photosynthesis	Reduced shoot and root growth. Early foliar symptoms resemble those of manganese deficiency; later, depressions form in areas between veins.
Essential to some plants but not to all			
Cobalt	Co^{2+}	Needed by nitrogen-fixing organisms	
Silicon	$H_2SiO_4{}^{2-}$	Present in cell walls of diatoms and horsetails	
Sodium	Na^+	Taken up by halophytes	
Vanadium	$VO_3{}^-$	Needed by certain algae and fungi	

connection with the water relations of plants. To some extent, dissolved minerals are carried along in the transpirational stream, but a look at the relative concentrations of elements within plant cells and in the environment of roots indicates that for many essential elements there is an active uptake from the environment and accumulation against concentration gradients. The mechanism of uptake and the ways in which ions are able to pass through selectively permeable membranes is poorly understood in spite of extensive research. Obviously if ions are accumulated inside cells, work has to be done in order to move them against concentration gradients. It has been shown that factors inhibiting aerobic respiration (lack of oxygen or specific poisons) also inhibit the rates of ion accumulation. Therefore, the energy expended in the process of active uptake must come from the respiratory cycle. There is some evidence that there are specific absorption sites on cell membranes for particular ions. Enzymes at the absorption sites may participate in the transfer of ions across the membrane.

All minerals are taken up as positively or negatively charged ions. Sometimes the root cells secrete ions in exchange for ions absorbed. H^+ ions are exchanged for positively charged ions such as K^+, while the uptake of negatively charged ions such as $NO_3{}^-$ is neutralized by secreting bicarbonate or the anions of organic acids such as malate ions. Secretion of large numbers of hydrogen ions will make the surrounding medium more acid. In this way the metabolism of plant cells can have yet another kind of effect on their environment.

Nitrogen Fixation

A great pool of nitrogen exists in the earth's atmosphere (which is 80 per cent nitrogen). In spite of the apparently inexhaustible supply of this essential element, plants often exhibit symptoms of nitrogen deficiency because they cannot use gaseous nitrogen. Most plants can only absorb this element through their roots in the form of nitrate or ammonium ions. A relatively small number of organisms have the ability to convert atmospheric nitrogen into a form that plants can absorb. The very few kinds of organisms, all prokaryotic, capable of doing this very essential conversion include: certain of the blue-green algae, notably members of the genus *Nostoc*, some of the photosynthetic bacteria such as *Rhodospirillum*, a few free living bacteria existing on organic matter in the soil, including the aerobic *Azotobacter* and the anaerobic *Clostridium*, and finally species of such genera as *Rhizobium* living symbiotically in the root nodules of certain higher plants, especially legumes such as peas and beans. Although there is great diversity in the mode of nutrition among the nitrogen-fixing organisms, from saprophytic to photoautotrophic to symbiotic, they all share the distinct advantage of being able to utilize atmospheric nitrogen and being independent of the soil for supplies of this essential element. Because the symbiotic nitrogen-fixing bacteria add soluble nitrogen compounds to the soil, it is common agricultural practice to rotate crops, including a leguminous crop in every cycle of rotation.

In the nitrogen fixation process, atmospheric nitrogen is transformed to ammonia, which is either utilized directly or oxidized to nitrate by other microorganisms. Within plant cells, nitrates must be reduced back to ammonia through the action of the enzyme nitrate reductase. This enzyme contains molybdenum and is the only known component of plant cells for which this trace element is required. It has been shown that plants can grow perfectly normally in the absence of molybdenum if nitrogen is provided in the form of ammonia rather than nitrate. Plant cells utilize ammonia to make amino acids (which are then incorporated into proteins) and a host of other nitrogenous compounds including the nucleic acids.

The nodules formed in plant roots invaded by nitrogen-fixing bacteria are interesting both structurally and biochemically. The initial invasion probably takes place through the root hairs. An infection "thread" spreads to other root cells and causes a curious proliferation of root cells. The interior of the resulting nodule contains large host cells, many of them filled with bacteria. The nodules of legumes usually are pink because of the presence of **leghemoglobin**, a red pigment chemically similar to animal hemoglobin, that appears to be necessary for nitrogen-fixing activity.

Fig. 16-16. Root nodules formed by a strain of *Rhizobium leguminosarum* on the roots of a young pea plant (*Pisum sativum*). (From R. Dubos: *The Unseen World.* The Rockefeller Institute Press, New York, 1962.)

The Nitrogen Cycle

Through the work of the nitrogen-fixing microorganisms, vast amounts of atmospheric nitrogen are continually being transformed into soluble ammonia and nitrates in the soil. These fixed forms of nitrogen are absorbed from the soil solution by plant roots and are utilized by plants to form amino acids and ultimately proteins. The leaves, fruits, and seeds of plants are ingested by animals and transformed into animal proteins and nitrogenous wastes such as urea. The primary consumers of plants (herbivores) are not infrequently consumed in turn by other animals (carnivores) with the result that a given atom of nitrogen may pass through a complex and lengthy food chain. Ultimately all plants and animals die; their tissues are decomposed into simple breakdown products by microorganisms in the soil. Nitrogenous wastes may be absorbed by plants and recycled again through living systems, or free nitrogen may be released into the atmosphere through the action of soil denitrifying bacteria on nitrates.

TRANSLOCATION: TRANSPORT OF THE PRODUCTS OF PLANT METABOLISM

One further aspect of the transport of materials in plants remains to be considered. Thus far we have concentrated on the movement of raw materials, water and dissolved minerals, from their point of entry, the roots, to the tops of plants. In the shoot, these raw materials are utilized. The leaves of the shoot have in their chloroplasts the factories where raw materials from the soil and the atmosphere are converted to usable organic compounds. The shoots of plants import simple inorganic materials which are converted at the expense of absorbed solar energy into the more complex substrates for respiration, protein synthesis, and the complex structural components of all plant cells. The products of the leaves are "exported" to the rest of the plant. In the absence of a

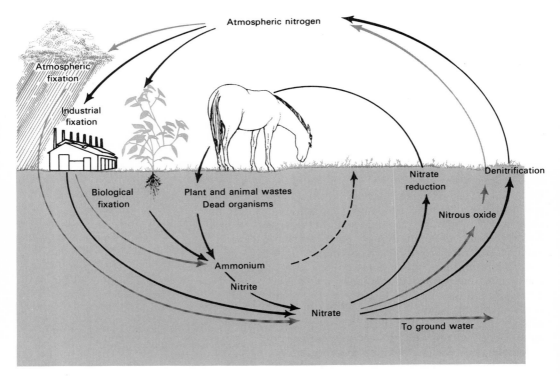

Fig. 16-17. The nitrogen cycle. The supply of atmospheric nitrogen is virtually inexhaustible but it must be converted to NH_4^+ or NO_3^- by nitrogen-fixing microorganisms before it can be absorbed by plants.

closed circulatory system and with a "pumping" system that appears to be unidirectional, how is this accomplished?

The oldest evidence of a movement of materials from the leaves was obtained in the late 1600's by Malpighi, the Italian who also discovered blood capillaries and many other items of microscopic anatomy. He removed rings of bark from the stems of various kinds of trees. Such **girdling** often resulted in a swelling of the tissues above the ring because the movement of dissolved substances through the phloem was blocked. Malpighi further noted that such swelling above a girdle occurred only during those seasons of the year when the trees had leaves. It was not until 1837 that Hartig discovered the sieve cells of the phloem and only in 1860 did he demonstrate that sap flowed from the cut bark of trees. Hartig's evidence that phloem cells in the inner layers of the bark are involved in the downward conduction of substances exported from the leaves was refuted or ignored until the 1920's—mostly because it was contrary to the diffusion theories of translocation in vogue in his day.

Phloem, a complex tissue made up of a number of different cell types including sieve elements, companion cells, phloem parenchyma, sclereids and albuminous cells, is located in the inner portion of the bark region. It is a difficult tissue to study because as stems increase in diameter during secondary growth, the layers of phloem just outside the cambium get pushed outward. Consequently, older parts of the phloem are crushed and distorted. **Sieve elements**, also called **sieve tubes**, are the conducting cells of the phloem. At maturity, the nucleus of the sieve tube disintegrates; pores develop in the end plates of these cells and their cytoplasm becomes disorganized. Recently there have been reports describing strands of protoplasm, present in mature sieve elements. These **transcellular strands** which appear in some materials to be tubular and bound by a membrane while in others to consist of slime or fibrils, connect to strands in adjacent cells through pores in the end plates. It remains to be seen whether the transcellular strands of sieve elements are indeed functional structures or only artifacts resulting from the methods used to prepare phloem for observation in the light or electron microscope.

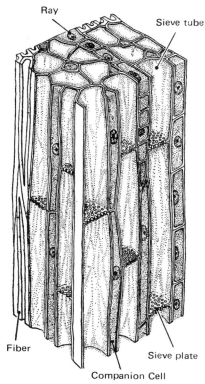

Fig. 16-18. Conduits involved in transport of products of plant metabolism. Enlarged section of phloem showing arrangement of sieve tubes, companion cells, ray parenchyma, and phloem fiber cells. See Figure 16-2 for location of phloem in woody stem. (From S. Biddulph and O. Biddulph: The circulatory system of plants. Scientific American, *200:* 44-49, 1959.)

The contents of phloem cells are rich in carbohydrates. The concentrations of sugars in phloem exudates are highest during the day, when photosynthetic rates are greatest, and are lowest at night. It has recently become possible to sample and analyze the contents of single sieve tubes by taking advantage of the feeding habits of aphids and other insect parasites. After they have inserted the tips of their minute tubular stylets into the phloem, the aphids are anesthetized with a stream of CO_2. The insects are cut away with a fine splinter of glass, leaving the severed stumps of stylets still inserted. Sap continues to exude from the severed stylets for several hours and can be collected with a capillary tube.

Another approach to studies of what is

Fig. 16-19. A, an aphid feeding on a willow stem. (Photograph by J. S. Redhead, University of Durham, England.) B, photomicrograph showing stylet of an aphid inserted into a phloem sieve cell. (Photograph by R. Kollmann, Botanisches Institut, Universität Bonn.) (From M. Richardson: *Translocation in Plants*. St. Martin's Press, New York, 1968.)

translocated in plants and by what route involves labeling the products of photosynthesis by introducing radioactive substrates. Not only the identity of the components translocated but also their location and rate of movement can be determined by tracer techniques. The first application of radioisotopes to a biological problem was in a study of the uptake of lead isotopes by bean seedlings done by the Danish investigator Hevesey in the early 1900's. All tracer studies are based on the assumption that living cells do not discriminate between radioactive and nonradioactive isotopes of an element in their environment. Thus, the uptake of any isotope is in proportion to its relative abundance. Using radioactive forms of several essential elements including phosphorus and calcium, Orin and Susan Biddulph and their associates at Washington State University followed the uptake and distribution of these elements in intact bean plants. By harvesting plants at intervals after exposure to the tracer, they were able to follow the movement of elements absorbed through the roots to all parts of the plant and to follow the redistribution of tracer following its incorporation into the products of metabolism.

Plants grown in an atmosphere containing ^{14}C-labeled CO_2 form radioactively labeled products of photosynthesis. In many plants it

has been shown that radioactively labeled compounds are translocated both upward and downward from leaves. The sugar present in greatest amount is the simple disaccharide sucrose. If a piece of stem is killed by steam or hot wax, radioactivity no longer moves past the dead tissue. If thin sections cut from a stem translocating labeled sugars are placed next to pieces of photographic film, the film becomes exposed only in the area of the phloem cells.

Tracer techniques have also been used to estimate the amounts of sugars and inorganic nutrients translocated per unit time and their velocity of travel. The movements of amino acids, plant hormones, insecticides, herbicides, and viruses have been studied.

The facts of translocation are firmly established. The redistribution of essential nutrients within plants is well documented. A demonstration of the uptake of radioactive phosphorus by the roots, its movement first to mature leaves, and somewhat later accumulation in the apical meristems, has become a routine exercise for beginning students of plant physiology. But we have not yet answered the question of how translocation, frequently in a direction counter to the transpirational stream, actually works.

Of the several available theories of phloem transport, none is completely satisfactory. It has been suggested that sugars move through sieve tubes by a process of simple diffusion. Unfortunately, diffusion is much too slow to account for measured rates of solute movement. A second theory suggests that substances moving in the phloem are carried by actively streaming protoplasm. Even though an end-to-end surging of the contents of young sieve tubes has been observed, it is difficult to imagine how such a mechanism can account for the rapid movement of dissolved substances in the phloem sap.

Mass Flow Hypothesis

The best available hypothesis of phloem transport was proposed in 1930 by Münch. According to the mechanism suggested by Münch (which differed very little from Hartig's view in 1860), photosynthesizing leaves act as a "source" of translocated material and the roots as a "sink." Water drawn upward through the xylem by the transpirational pull of the leaves is drawn into the leaf cells which contain high concentrations of carbohydrates. This uptake

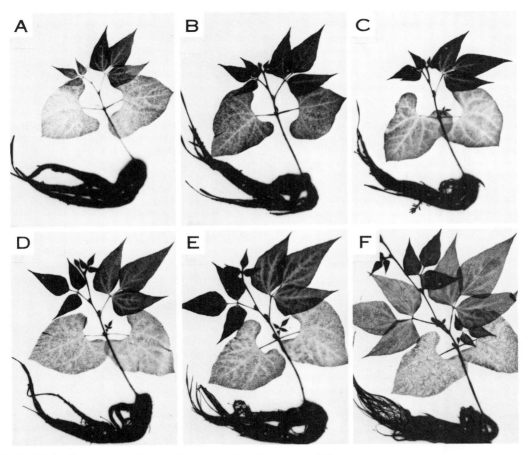

Fig 16-20. Radioautograms showing the progressive distribution of ^{32}P administered by a one-hour exposure of bean roots to a nutrient solution containing 88 μc. of ^{32}P per liter. Bean plants were harvested at the following times after exposure to ^{32}P: A, 0 hr.; B, 6 hr; C, 12 hr.; D, 24 hr.; E, 48 hr.; F, 96 hr. Harvested plants were dried and then left in contact with X-ray film for 17 days prior to development of the film. Note continuing high concentration of labeled phosphorus in young leaves and stem apices indicating great mobility of this element. (Courtesy O. Biddulph, from O. Biddulph, S. Biddulph. R. Cory, and H. Koontz: Circulation patterns for phosphorus, sulfur and calcium in the bean plant. Plant Physiology, *33:* 293-300, 1958.)

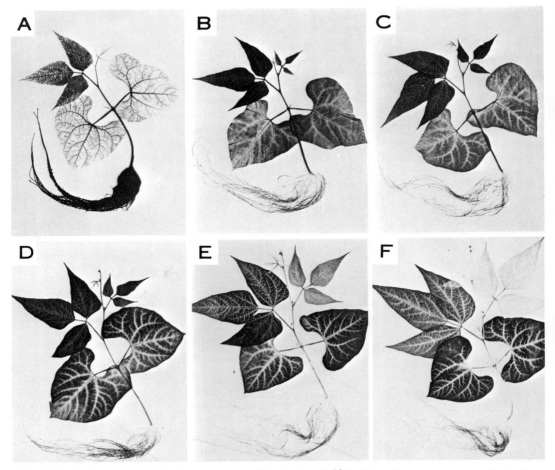

Fig. 16-21. Radioautograms showing progressive distribution of ^{45}Ca. Procedures were comparable to those described in legend to Figure 16-20. Calcium is relatively immobile following import via the transpirational stream. Tissues developing after initial uptake of radioactive calcium contained little label. (Courtesy O. Biddulph.)

of water due to osmotic forces causes increased hydrostatic or turgor pressure in the leaf cells. At the same time the roots are utilizing the products of photosynthesis for respiration, growth, or storage, and because the osmotic pressure thus tends to be lowered in the root cells, these "sinks" for the products of photosynthesis tend to have a reduced hydrostatic or turgor pressure. According to this picture, there is a mass or pressure flow of solutes from their source to sites of utilization.

The **mass flow** concept can be illustrated most easily in terms of a physical model:

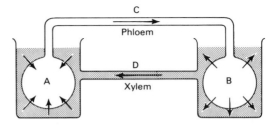

Two cells, A and B, permeable only to water, are connected by tube C. Cell A, representing the photosynthetic leaf cells, has a high osmotic concentration while B, representing the root

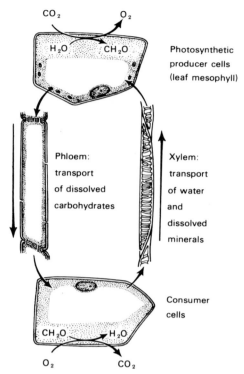

Fig. 16-22. The mass flow hypothesis. Water is drawn up through the xylem and into leaf cells by the transpirational pull of leaves. Hydrostatic or turgor pressure of leaf cells forces a flow of solutes (mostly carbohydrates) through the phloem to the roots. Photosynthetic cells of the leaf are producers of carbohydrates that are translocated to the consumer cells of the roots which utilize the products of photosynthesis for respiration, growth, or storage.

cells, has a low osmotic pressure. If such cells are placed in containers of distilled water, water will enter cell A, creating a hydrostatic pressure in A that forces fluid through connecting tube C into cell B. Water will be forced out of cell B and the flow just described will continue until the osmotic concentrations of cells A and B are equalized. If we add to our model the circumstances existing in plant cells, of continual production of soluble carbohydrates in A and their continual utilization in B, then we see how this process can persist so long as the sources and sinks of translocated material remain operative. In the veins of the leaf are thin strands of vascular tissue made up of

xylem and phloem. Through these strands there are simultaneous import of water and dissolved minerals through the xylem and export of the products of photosynthesis through the phloem.

One of the requirements of this hypothesis for phloem transport is that a gradient of turgor pressure must exist between shoot and root cells. Although it is difficult to demonstrate gradients in turgor pressure, it is relatively simple to show gradients in concentration of dissolved substances. When leaves are present and photosynthesis is taking place, concentrations of sugar in the contents of phloem cells are highest near the top of a stem and progressively decrease toward the base. When leaves are removed experimentally, or in trees during the late fall and winter (after leaves have been dropped), such gradients no longer exist. In early spring there is usually a massive transport of reserve carbohydrates upward. Advantage is taken of this upward movement when sap is collected from maple trees and evaporated to make maple syrup and sugar.

DIFFERENCES BETWEEN CIRCULATION IN PLANTS AND ANIMALS

In considering the means that plants have evolved for solving the problems of transporting water and nutrients, we have noted the absence of a pump and a system of circulation that is open rather than closed as it is in animals. Viewed at the level of individual cells, we find practically no difference between plants and animals: the same laws of diffusion and osmosis plus active transport prevail and control the movement of molecules through selectively permeable membranes. The problems of transport become increasingly complex for multicellular forms primarily because greater distances separate the site of entry or synthesis of a given ion or molecule and the location of utilization. In plants, translocation takes place through the specialized cells of the vascular tissues, the xylem and phloem. Transfer of molecules from one part to another is an intracellular as well as intercellular phenomenon.

Although crude outlines of the mechanisms

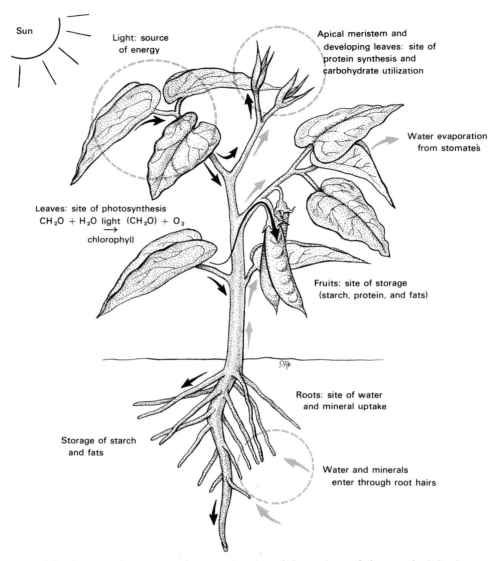

Sun

Light: source
of energy

Apical meristem and
developing leaves: site of
protein synthesis and
carbohydrate utilization

Water evaporation
from stomates

Leaves: site of photosynthesis
$CH_2O + H_2O$ light $(CH_2O) + O_2$
$\xrightarrow{}$
chlorophyll

Fruits: site of storage
(starch, protein, and fats)

Roots: site of water
and mineral uptake

Storage of starch
and fats

Water and minerals
enter through root hairs

Fig. 16-23. Patterns of movement of water, minerals, and the products of photosynthesis in plants.

of transport in vascular plants have emerged, they are still incompletely understood. The hypotheses discussed in this chapter are not completely adequate to account for what can be observed in nature, and many problems remain to be solved even though the circulation of water, minerals, and metabolites was one of the first aspects of plant physiology to be attacked experimentally. The new technologies of radioactive tracers and electron microscopy have stimulated renewed efforts to answer questions first posed centuries ago.

USEFUL REFERENCES

Biddulph, S., and Biddulph, O. The circulatory system of plants. Scientific American, *200:* 44-49, 1959.
Crafts, A. S. *Translocation in Plants.* Holt, Rinehart and Winston, New York, 1961.
Galston, A. W. *The Life of the Green Plant.* Prentice-Hall Inc., Englewood Cliffs, N. J., 1964.

Gauch, H. G. *Inorganic Plant Nutrition.* Dowden, Hutchinson and Ross, Inc., Stroudsburg, Pa., 1972.

Meidner, H., and Mansfield, T. A. *Physiology of Stomata.* McGraw-Hill Book Co., New York, 1968.

Richardson, M. *Translocation in Plants.* St. Martin's Press, New York, 1968.

Sprague, H. B. (ed.) *Hunger Signs in Crops,* 3rd ed. David McKay Co., New York, 1964.

Steward, F. C. (ed.) *Plant Physiology: A Treatise,* vol. II: Plants in Relation to Water and Solutes. Academic Press, New York, 1959.

Steward, F. C. (ed) *Plant Physiology: A Treatise,* vol. III: Inorganic Nutrition of Plants. Academic Press, New York, 1963.

Sutcliffe, J. *Plants and Water.* St. Martin's Press, New York, 1968.

Hormonal Regulation in Plants

Before recorded history some farmer probably wondered how seedlings achieve the proper orientation with shoots growing up and roots down, regardless of whether the seeds are planted right side up or upside down. About a century ago Charles Darwin asked the same question and began a long investigation into the effects of light on the direction in which plants grow. This work on plant **tropisms**, as such movements are called, led directly to a series of investigations which are continuing today, for it led to the discovery of plant hormones. The concept of specific chemical communication between different parts of plants was proposed by the founder of experimental plant physiology, Julius Sachs, in the 1880's; however, the idea of plant hormones was generally rejected until the isolation of auxin in 1928 by Frits Went, a Dutch army draftee who worked at night in his father's laboratory in Utrecht.

It is now clear that growth and development in both plants and animals are controlled by chemical regulators, the hormones. The synthesis of a hormone usually occurs at a site removed from the target organ on which it has its effect. Characteristically, very small amounts of these specific chemicals produce amazingly large effects. Consequently, it is not surprising that natural and synthetic ⋅ growth-regulating hormones have been used for many purposes ranging from the control of shape and appearance of ornamental plants and the chemical weeding of crops and lawns to the defoliation of jungles as a tactic of warfare. Whether used to promote rooting and thus speed the propagation of horticulturally useful species or in the eradication of unwanted weeds, the chemical control of plant growth provides a powerful

tool for agriculture. In many ways the era of "made-to-order" plants is at hand.

NATURALLY OCCURRING PLANT HORMONES

At the present time there are three major classes of hormones known to be naturally occurring and essential for normal development of plants. These are the **auxins**, the **cytokinins**, and the **gibberellins**. Recent investigations indicate that two additional classes of compounds fall into the category of naturally occurring regulators of plant growth: **abscisic acid** and **ethylene**. Still other kinds of plant hormones have long been postulated, such as the flowering hormone, **florigen**, and **wound hormones**. It is probably safe to predict that our list will expand in the years ahead. In addition to the naturally occurring chemical regulators of plant growth, a number of synthetic herbicides and defoliants have been developed. These products of the chemical industry have enabled us to grow weed-free lawns or crops and have made the harvesting of certain plant products much easier. The widespread use of plant poisons and defoliants by military forces has been the subject of much controversy and concern over long-term ecological effects. The subject of plant hormones is of general human concern and has raised serious ethical questions extending beyond the realm of pure science, but which have to be dealt with very soon.

Experimental Approaches to the Nature of Hormone Action

One of the first steps in studying any hormone is to prove that a particular tissue

Fig. 17-1. Julius Sachs, founder of experimental plant physiology, who in the 1880's proposed the idea of plant hormones. (Courtesy Hunt Botanical Library.)

Fig. 17-2. Frits Went, the first plant physiologist to isolate auxin and to devise an assay for this plant hormone. (Courtesy Hunt Botanical Library.)

extract or chemical can function as a plant (or animal) hormone. This is done by applying criteria similar to those proposed by Robert Koch in establishing the relationship between a disease and its causative agent: (1) removing the source of the substance must change the pattern of growth or metabolism; (2) applying the substance exogenously must restore normalcy; and (3) effects of the substance must be specific.

To answer questions about the effects of a hormone and how these results are achieved requires information about what happens to the gross structure of a plant and also what happens on the cellular and biochemical levels. Effects on the whole plant can readily be seen in most cases. In recent years evidence has been obtained that at the molecular level many (if not all) hormones act as derepressors of genetic information. Thus, in the dogma of molecular biology, a hormone is an effector molecule that turns on genes. Other mechanisms of hormonal action have been proposed in addition to the currently fashionable hypothesis that they are involved in the transcription and translation of genetic information. Hormones may be involved in changing the configuration of proteins or in the activation of enzymes. There is considerable evidence that they affect cell membranes and thereby regulate the entry of essential substrates into cells.

AUXINS: THE MOLECULAR BASIS FOR TROPISMS

Tropisms are growth movements of plants in response to environmental factors such as light or gravity. **Phototropism** is a growth response to unilateral light. Ordinarily this is a positive response in plants, although certain fungi exhibit a negative phototropism. **Geotropism** may be either positive or negative: shoots grow upward (negative response), while roots grow downward (positive response).

Tropisms are of vital importance during the early stages in plant development. Nutrient reserves available to the embryo in a germinating seed are limited. Before these reserves are exhausted, the roots must be oriented so that they can absorb water and inorganic nutrients from the soil; the shoots must emerge from the soil so that developing leaves will be exposed to the sun. Unless the appropriate orientation is achieved within a few days, the seedling will die. Older plants also respond to light and gravity, although their response may be slower. Because of their sensitivity and the rapidity of their growth responses, most studies of the response of plants to light and gravity have been performed on seedlings.

The first studies of tropisms were reported in 1881 by Charles Darwin in a book called *The Power of Movement in Plants*. Experimenting with young seedlings of canary grass, Darwin observed that "when seedlings are freely exposed to a lateral light, some influence is transmitted to the lower part causing it to bend." Such a bending response could be prevented by either cutting off the tip of the seedling or by shading it. The bending response actually occurs some distance below the tip. Clearly there must be some means for transmitting the signal from the tip, where light is perceived to the lower portion of the stem that responds. Additional insights into such bending responses accumulated very slowly. In 1913, Boysen-Jensen demonstrated that cut tips of seedlings could be pasted back in place with a layer of gelatin without interfering with transmission of the stimulus from the tip. Although the stimulus passed readily through gelatin, Paal in 1919 found that mica or foil would prevent its transmission. Paal also showed that severed tips replaced to one side caused curvatures of seedlings resembling those in response to unilateral light. Then in 1928, Frits Went, in his father's laboratory at the University of Utrecht, succeeded in isolating the chemical signal first postulated by Darwin. The technique of isolating the active substance turned out to be amazingly simple: it diffused out of plant tissues into a small block of agar. This technique provided not only a method for separating auxin from plants but also the basis for a sensitive assay technique for auxins used to this day. The assay developed by Went is based on the bending response of decapitated oat (*Avena sativa*) seedlings in response to auxin in a tiny block of agar applied to one side of the cut tip. The bending of seedlings in response to known amounts of auxin is measured to obtain a standard curve for this bioassay. By comparing the response to extracts of plant tissues with the response in the

standard curve, an investigator can determine the concentrations of extractable auxin.

The availability of an assay for auxin led to further insights into the nature of auxin effects on plant growth. Attempts to isolate and chemically identify auxin obtained from plant materials led to the realization that this thermostable substance that is soluble in water, ether, and alcohol is a rather simple molecule: indole-3-acetic acid. The relationship between auxin and tropisms was clarified in the decade that followed.

Blaauw had demonstrated that phototropism in *Avena* seedlings is due to a difference in rate of growth between the illuminated and dark sides of the seedling. Went cut off the tips of illuminated oat seedlings and placed them on agar blocks separated by a razor blade so that the auxin from the illuminated and shaded sides diffused into two separate blocks. When the auxin activity of these blocks was tested in the *Avena* assay, blocks from the shaded sides caused greater curvature. Thus the curvature of intact seedlings appeared to be the result of higher auxin concentration in the shaded side.

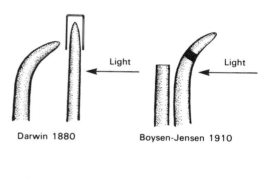

Darwin 1880   Boysen-Jensen 1910

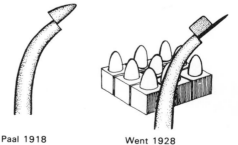

Paal 1918   Went 1928

Fig. 17-3. Early studies of plant hormones. (Redrawn from K. V. Thimann: Growth and growth hormones in plants. American Journal of Botany, *44:* 49-55, 1957.)

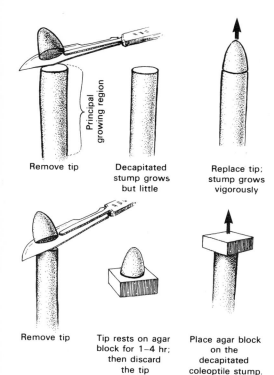

Remove tip   Decapitated stump grows but little   Replace tip; stump grows vigorously

Remove tip   Tip rests on agar block for 1–4 hr; then discard the tip   Place agar block on the decapitated coleoptile stump. Coleoptile grows

Fig. 17-4. Coleoptile tips of oat (*Avena sativa*) seedlings have growth-promoting properties.

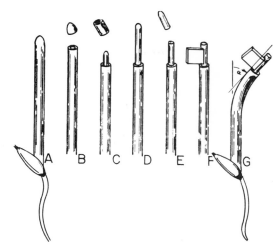

Fig. 17-5. *Avena* assay for auxin. A, 3-day-old dark-grown oat seedling ready for use in auxin assay. Seedlings are exposed to a few hours of red light during the second day of germination to retard elongation. B, apical section of 1 mm. is removed to eliminate seedling's source of auxin. C, second decapitation 3 hours after first to remove regenerated auxin-synthesizing capacity. D, primary leaf (inside coleoptile) is pulled up to sever its connection to apical meristem of seedling. E, tip of primary leaf is removed. F, agar block containing known amount of auxin or extract to be tested for auxin activity is placed on shelf provided by cut end of coleoptile. G, after 90 minutes curvature is measured. Constant temperature (25°C.) and high humidity (85 to 90%) must be maintained during growth of seedlings and assay. (From A. C. Leopold: *Auxins and Plant Growth*. University of California Press, Berkeley, 1955.)

At about the same time, Dolk, another Dutch investigator, reported parallel experiments demonstrating that an asymmetric distribution of auxin is also involved in geotropism. Of the total auxin diffusing from the tip of an oat seedling placed on its side, 62.5 per cent was in the lower half. When similar assays were performed on roots, a comparable redistribution of auxin was found. But roots turn downward while shoots grow up. It was soon realized that increased auxin concentrations may promote growth under some circumstances but inhibit under others.

Other Effects of Auxin on Plant Growth and Development

In 1933, Thimann and Skoog proposed that auxin produced in the plant apex can inhibit the development of lateral buds. Thus began the accumulation of a long list of ways in which auxin is involved in the morphology of plants.

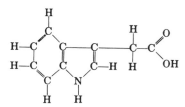

Fig. 17-6. Structure of auxin, indole-3-acetic acid.

Apical dominance, a growth pattern in which lateral bud development is inhibited, can easily be shown to be the consequence of high concentrations of auxins exported from the apical regions. If the terminal portions of a plant are removed, lateral buds develop; if an

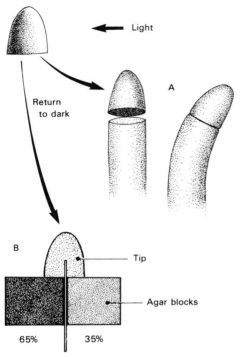

Fig. 17-7. Experiments of Went demonstrating asymmetric distribution of auxin caused by light. After unilateral illumination, an oat coleoptile tip placed on a decapitated seedling in the dark can cause a bending response (A). When placed on agar blocks separated by a sheet of mica (B), more auxin is collected from the shaded than from the illuminated side of the coleoptile tip.

external supply of auxin is applied to the cut ends, development of lateral buds will remain suppressed. The extent of lateral bud development establishes the branching pattern of a plant. The upright conical form of trees such as the spruces results from suppression of lateral growth near the apex of the plant. In other trees, terminal buds often form flowers; further shoot growth is due to development of lateral buds and a much branched pattern results. The relationship between the source of auxin production and growth form is the basis of an old horticultural practice, **pruning,** or cutting back the terminal branches of plants to promote a fuller, bushier growth.

It was also learned in the 1930's that externally supplied auxin stimulates the formation of adventitious roots. This knowledge has been used to promote the rooting of cuttings, the method used to propagate many horticultural species. For example, hollies are notoriously hard to propagate, yet they form roots more readily when auxin-containing dusts or pastes are applied to the ends of cuttings.

In 1936, La Rue showed that auxins play a role in **leaf abscission,** the process by which leaves drop off in the fall of the year. Later it was demonstrated that in the loss of any plant organ (whether leaves, flowers, or fruits), the formation of an abscission layer is associated with a decrease in the auxin content of the organ. At about the same time Gustafson showed that auxin applied to immature flowers could stimulate the development of ovaries prior to pollination. Parthenocarpic fruits, developing from ovaries without fertilization, have been grown in this way, and the normal development of fruit appears to be stimulated by auxin.

In 1937, Thimann proposed a generalized scheme for the effects of auxin on the growth of various plant parts. Auxin is effective not only at amazingly low concentrations but also over a very wide range of concentrations, encompassing 10 orders of magnitude from 10^{-11} to 10^{-1} M. Distinctly different concentrations of auxin are needed to stimulate the growth of stems, buds and roots. For example, a concentration that promotes stem elongation can inhibit the development of buds or roots. From this generalized picture we can more readily understand why the increased concentration of auxin on the lower side of a seedling has opposite effects on the shoot and the root.

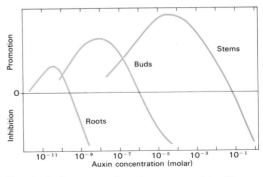

Fig. 17-8. Summary of auxin concentration effects on growth of roots, buds, and stems.

The higher auxin content stimulates shoot elongation but inhibits root growth.

Recent Studies on the Nature of Auxin Action

What seemed like a very neat and comprehensive picture of auxin action by the end of the 1930's left unanswered some crucial questions that could not be approached experimentally until the development of modern tracer techniques. Using ^{14}C-labeled indoleacetic acid, a number of modern workers tried to repeat the classic experiments on tropisms of Went and Dolk. Radioactively labeled auxin was applied to young oat seedlings that were then illuminated from one side or placed horizontally. To everyone's dismay and confusion, no differences in radioactivity could be detected between the upper and lower or lighted and shaded sides of the seedlings. Were the earlier experiments that so clearly showed asymmetric distributions of auxin in error and the classic theory of auxin involvement in tropisms invalid? Even if the old theory were still true, it did not explain how more auxin could be extracted from the shaded or lower side of a plant.

The key to ending the confusion over experiments with ^{14}C-labeled auxin was the realization by Thimann and his associates that the method of labeling is critical. Indoleacetic acid (IAA) is oxidized by an enzyme in plant cells. Uniformly labeled IAA, *i.e.,* having ^{14}C in both the carboxyl and indole portions of the molecule, when oxidized by this oxidase-peroxidase enzyme, loses its carboxyl group and is converted to a derivative of oxindole that has no auxin activity. Yet most of the initial radioactivity remains in the inactive breakdown product. Under these conditions there is no relationship between the presence of radioactivity and the location of active hormone. If, however, the ^{14}C label is introduced only in the carboxyl group of indoleacetic acid, then the radioactivity and biological activity of the auxin remain together. Recent tracer experiments indicate that there is a lateral movement of auxin in both geotropism and phototropism. In the response to light, neither photodestruction nor a light influence on longitudinal transport are detected.

Fig. 17-9. K. V. Thimann, whose investigations have contributed much to our understanding of auxin action. (From Annual Review of Plant Physiology, Vol. 14, 1963.)

Receptors for Tropic Responses

It has now been firmly established that the effector molecule in both geotropism and phototropism is indoleacetic acid and the asymmetric distribution of this effector molecule is a consequence of its lateral transport. Two further basic questions remain: What are the receptors for these responses to the external environment and what organelles are responsible for detecting gravity and light? For gravity to be detected, there must be something within plant cells that falls in response to this force. To detect light, there must be a colored photoreceptor that absorbs light energy.

At various times it has been proposed that starch grains or plastids, structures which can become oriented along the bottoms of cells, might be involved in the geotropic responses of plants. However, the response to gravity occurs just as well in the absence as in the presence of these structures. The best available hypothesis at this time is that gravity is detected by the movement of cytoplasm and vacuoles. Pressures against membranes could affect the rates of auxin transport.

Although there is still some uncertainty about the receptor for phototropism, its action spectrum has been measured with great precision by Thimann, Curry, and others. The most effective wavelengths are in the blue region of the spectrum with peaks at 445 and 475 nm. and a shoulder at 425 nm. There is an additional region of effective wavelengths in the ultraviolet between 360 and 380 nm. While the visible portion of the action spectrum for phototropism closely resembles the absorption spectra of certain of the carotenoid pigments, the additional peak in the ultraviolet suggests that a flavin type compound may be involved. Unfortunately, no single known compound has an absorption spectrum identical to the measured action spectrum. Not only is there uncertainty about the chemical nature of the photoreceptor, but also about where it is located. Recently Thimann reported observing rectangular or rhomboidal bodies, smaller in size than mitochondria, in *Avena* and also in the photosensitive parts of certain fungi. These structures are bounded by a single membrane and are filled with crystalline material made up at least in part of protein. They are found between the cell wall and the protoplasm. In time it may be shown that these particles contain a carotenoid-protein complex that serves as the photoreceptor for phototropism. The question will then arise: How does the absorption of light energy by these particles control the lateral transport of auxin?

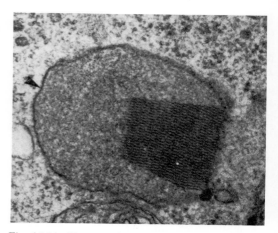

Fig. 17-11. Electron photomicrograph of rhomboidal particle which may be the receptor for tropic responses. (Courtesy K. V. Thimann, from K. V. Thimann: Tropisms in plants. Embryologia, *10:* 89-113, 1967.)

Chemical Specificity of Auxins

Since the identification of auxin as indoleacetic acid, a large number of related chemicals have been shown to possess hormonal activity. It is now known that molecules having an unsaturated ring and an appropriately attached acid side chain will have auxin-like effects on plant growth. The unsaturated ring may be as small as a benzene ring or as large as an anthracene ring. Substitution of other groups for hydrogens in the unsaturated ring and the position of the substitution can have large effects on activity. For example, introducing two chlorine atoms in the *meta* and *para* positions of phenoxyacetic acid vastly increases its growth-promoting activity to the point of making 2,4-dichlorophenoxyacetic acid (2,4-D) one of our most potent weed killers. The nature of the side chain is also important: an even number of carbons must be present because side chains with an odd number of carbons are readily broken down. A knowledge of how molecular structure affects auxin activity has made it possible to manufacture made-to-order herbicides and growth regulators. The 2,4-D mentioned above is one of the oldest herbicides and still the most widely used. By 1966, 2,4-D production in the United States exceeded 100

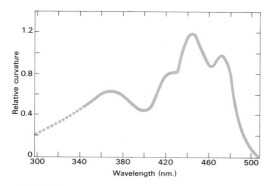

Fig. 17-10. Action spectrum of phototropism.

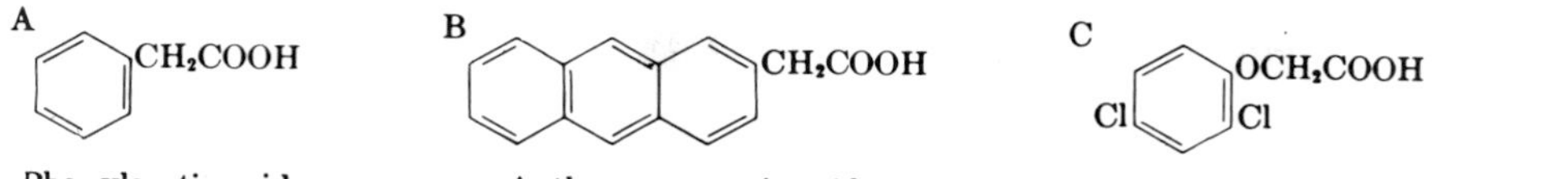

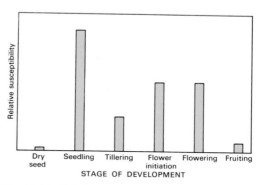

Fig. 17-12. A, phenylacetic acid; B, anthracene acetic acid, C, 2,4-dichlorophenoxyacetic acid, a widely used herbicide that is more toxic to dicots than to monocots.

million pounds per year. This weed killer appears to work by promoting growth to such an extent that the plant outgrows its resources. An interesting aspect of the way 2,4-D affects plants is that it is a **selective herbicide**, toxic to dicots (broad leaved plants) at concentrations that have no noticeable effects on monocots (parallel veined grasses and their relatives).

Factors in Auxin Herbicide Susceptibility

For auxin-like chemicals to be effective as selective herbicides, they must not only be absorbed by and translocated throughout the target plants but also produce toxic reactions. There is much well-documented information indicating that a plant's susceptibility to auxin herbicides changes during development. In the monocots, ungerminated seeds are relatively insensitive, while young seedlings are far more sensitive than later stages of vegetative growth. Susceptibility increases during flower initiation, remains high during flowering, and falls during fruit set. The dicots are similarly most sensitive during the early seedling stages and tend to be less susceptible later. There appears to be a close correspondence between susceptibility and growth rate.

The fact that dicots are generally far more susceptible to the auxin herbicides than are monocots is usually explained by differences in the morphology of these two groups of seed plants. In the dicots, meristems (or the actively growing regions) of the shoot are relatively more exposed, located terminally rather than near the base of the plant as in grasses. Furthermore, their broad leaves provide more surface for the absorption of herbicides applied as sprays. In addition, factors such as differences in the wettability of leaf surfaces (which would affect the rates of absorption) and in rates of translocation following absorption would influence susceptibility. Because of the

Fig. 17-13. Relative susceptibility of grains to auxin herbicides applied at various stages of development.

great differences between species in the effects of particular compounds, the auxin herbicides are highly selective. Their use as weed killers has another decided advantage. The herbicidal action of these compounds is rapidly lost in the soil; that is, they are biodegradable and leave no toxic residues. They are effective in extremely low concentrations and are therefore inexpensive. Most important of all, they are relatively nontoxic to animals and humans.

THE GIBBERELLINS: PROMOTERS OF ELONGATION

Although the **gibberellins** were first discovered at about the same time as auxin, an awareness of the existence of this kind of plant hormone and research on its role in the growth of plants lagged for a quarter century. Scientists in Europe and the United States remained preoccupied with auxin research while ignorant of the existence of the gibberellins because the latter were discovered by Japanese scientists who published their findings in Japanese journals. These investigators were trying to solve a problem of great economic importance. Early in this century Japan suffered serious losses from a fungus disease of rice, the *bakanae* ("foolish seedling") disease. Infected plants were unusually tall, weak stemmed, pale green, and at times failed to set fruit. With crop losses as high as 40 per cent, the Japanese had good reason for wanting to learn the cause of this disease and how to control it. It was soon discovered that the cause is a *Fusarium* type fungus called *Gibberella fujikuroi*. In 1926, Kurosawa demonstrated that a cell-free extract of the fungus applied to rice plants could produce the symptoms of excessive growth characteristic of the *bakanae* disease. It was not until 1935 that another Japanese investigator, Yabuta, crystallized that active ingredient in the fungus extract and called it **gibberellin**. Gibberellin research continued in Japan, unnoticed by Western scientists until after the end of World War II. In the early 1950's, the gibberellins were suddenly "discovered" by the rest of the world, and intensive studies began in the United States, England, and elsewhere. A number of gibberellins have been characterized chemically and they have been extracted not only from fungi but from a number of higher plants as well. Like the auxins, the gibberellins are considered to be naturally occurring plant hormones.

Chemically, the gibberellins are far more complex than auxin. They are diterpenoids having five rings. Biological assays for gibberellin activity are based on the fact that these compounds are able to reverse genetic dwarfism. Thus the promotion of stem growth in dwarf peas and the elongation of dwarf corn have been used to determine the amounts of gibberellin in extracts of plant materials. The changes in growth patterns resulting from application of gibberellins provide a ready explanation for dwarf and rosette patterns that are seen in plants like lettuce and cabbage. A head of cabbage is essentially a tall plant with many leaves at intervals (nodes) along the stem, but in which the growth of the internodes has failed. The result is that successive new leaves are produced immediately above the older ones. Application of gibberellins to such species results in extensive stem growth or internode elongation and 4 to 5 meters tall cabbages have been produced. The normal growth patterns of cabbage, lettuce, and dwarf corn or peas are undoubtedly a result of the low levels of gibberellins normally produced in these species.

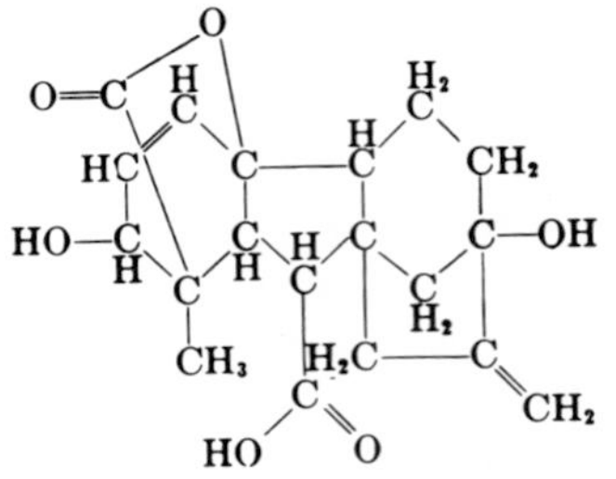

Fig. 17-14. Structure of gibberellin.

In addition to their obvious role in promoting stem elongation, gibberellins have far-reaching effects on seeds. Their application can break dormancy, an indication that the natural termination of dormancy is accompanied by an increase in the gibberellin content of seeds. In certain seeds such as lettuce, where light is needed for germination, gibberellins can substitute for light in promoting germination. The synthesis of enzymes needed to digest food reserves during the germination of seeds is triggered by gibberellin produced in the embryo. This last example of how gibberellins affect seeds provides an amazing story of how a plant hormone can act on the molecular level.

Gibberellins and the Derepression of Genes

Although by 1940 the Japanese had established that gibberellin promotes the germination of barley and rice seeds, it was not until 1960 that a Japanese (Yomo) and an Australian (Paleg) simultaneously reached the conclusion that gibberellin is the chemical which activates the genes for production of digestive enzymes. The seeds of cereals consist mainly of an embryo and the endosperm, or food storage tissue. The latter is surrounded by cells of the **aleurone** layer. During germination, the starch stored in the endosperm is hydrolyzed. As long ago as 1890, Haberlandt had recognized that in order for digestion of the endosperm to occur, the cells of the aleurone layer had to secrete an enzyme (α-amylase). This enzyme is formed only if the embryo is present. If barley seeds are cut in half, starch in the embryoless halves does not liquify. But, as Yomo and Paleg showed, if gibberellin in concentrations as low as 2×10^{-11} M is applied to the embryoless halves, the digestion of starch can be observed

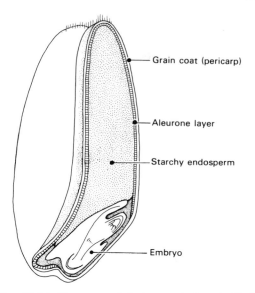

Fig. 17-16. Structure of barley seed. Longitudinal section showing position of embryo, endosperm, aleurone layer, and seed coat.

within 48 hours. More recently, Varner has proved that the α-amylase secreted by the aleurone layer is formed by *de novo* synthesis of the enzyme. Here we have a clear-cut case of a hormone, gibberellin, acting as an effector molecule to derepress the genetic information needed for the synthesis of a specific enzyme protein. Without the aleurone layer, gibberellins have no effect on the endosperm tissue. Gibberellin synthesized in the embryo triggers the production of many other enzymes during germination. In addition, it activates enzymes involved in the breakdown of cell walls, an effect that not only speeds the digestion of food reserves but also weakens the seed coats and facilitates the emergence of the root of the developing embryo.

Other Effects of Gibberellins on Development

Gibberellins have been shown to affect many other aspects of plant development. They can induce flowering in certain species requiring specific photoperiods or temperature regimes. The bypassing of photoperiodic requirements for flowering or the light requirements for germination by gibberellin is not particularly

Fig. 17-15. Cabbage plants treated with gibberellic acid grow to heights of several meters. Untreated plants are shown at bottom left. Cabbage is a biennial that "bolts" and forms flowers only during its second year of growth. Gibberellins cause flowering during the first year. (Courtesy S. H. Wittwer, Michigan Agricultural Experiment Station, Michigan State University, East Lansing, Mich.)

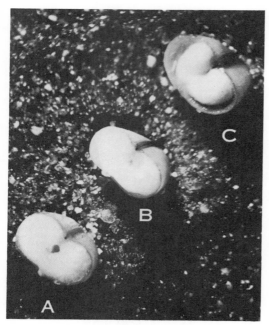

Fig. 17-17. The cut surfaces of three barley half seeds (embryo-containing halves removed) have been treated as follows: A, water control; B, gibberellin solution, 1 part per billion; C, gibberellin solution, 100 parts per billion. Forty-eight hours after treatment, digestion of endosperm tissue is very evident in C, less so in B, while there is no evidence of digestion in the control (A). (Photograph by J. E. Varner, from J. van Overbeek: The control of plant growth. Scientific American, *219:* 75-81, 1968.)

mysterious. It has been shown that the induction of flowering as well as the start of germination is accompanied by increases in gibberellin content. Thus light may promote germination by stimulating the synthesis of gibberellin in the embryo. Developing fruits and seeds are good sources of gibberellins, and it appears likely that the gibberellins are involved in the development of these structures. One poorly understood effect of gibberellins is their influence on the sex of flowers. Perhaps this effect is in some ways parallel to the hormonal control of secondary sex characteristics in animals. Gibberellins promote the formation of **staminate** (*i.e.*, male) flowers in cucumbers and squashes. In these species auxin has the opposite effect: promoting the formation of **pistillate** (*i.e.*, female) flowers.

Synthetic Antigibberellins

In recent years the manufacturers of agricultural chemicals have produced compounds which have effects on the growth patterns of plants just opposite to those produced by gibberellins. AMO-1618 (an ammonium salt of pyridine carboxylate), CCC (a chlorinated choline) and Phosphon (a chlorinated phosphonium salt) act as inhibitors of stem elongation. Application of these compounds to plants grown in greenhouses during the winter results in shorter, stockier plants with larger, darker green foliage. The effects of these plant growth regulators can be counteracted by gibberellins. Recently it has been shown that growth retardants such as AMO-1618 and CCC act by suppressing gibberellin synthesis. They are now being used as selective inhibitors to investigate the physiological and molecular action of gibberellins.

THE CYTOKININS: CELL DIVISION FACTORS THAT ACT SYNERGISTICALLY WITH AUXIN

Although the cytokinins are a class of naturally occurring plant hormones that have been recognized rather recently, the idea of chemical control of cell division is an old one.

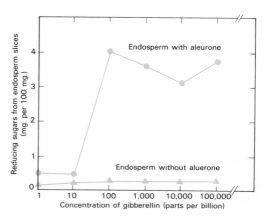

Fig. 17-18. Aleurone layer cells must be present for gibberellin-induced α-amylase synthesis. There is no synthesis of enzyme and no digestion of starchy endosperm (no reducing sugars formed) when gibberellin is applied to endosperm in the absence of aleurone. (Redrawn from J. van Overbeek: Plant hormones and regulators. Science, *152:* 3723, 1966.)

Early in this century Haberlandt discovered that phloem exudates could induce cell division. In potato tuber cells, crushed cells promote cell division near a wound, while wound healing does not occur as readily if the contents of injured cells are rinsed away. After plant tissue culture techniques had been developed in the 1940's and 1950's, investigators, including Steward at Cornell and Skogg at the University of Wisconsin, attempted to devise completely defined media capable of sustaining indefinite growth of plant tissues *in vitro*. Both investigators found that a factor in coconut milk was necessary for sustained growth of excised plant tissues and the existence of yet unknown plant hormones was suspected.

In 1955, C. O. Miller separated a compound from yeast DNA that stimulated cell division. This compound was found to be a purine derivative, 6-furfuryl amino purine. It was called kinetin, and the generic name kinin was proposed because it promoted **cytokinesis** (cell division). The name kinin was not retained, however, because at about the same time, zoologists had proposed that this term be used for a different class of biological products—polypeptides that act on smooth muscles and nerve endings as do insect and snake venoms. The plant hormones having the same effects as the initially isolated purine derivative are now referred to as the **cytokinins**.

The cell-division-promoting compounds isolated originally from such strange sources as yeast and herring sperm DNA must now be viewed as artifacts. However, the biological activity of these breakdown products of nucleic acids was very real. By 1964, Leathem had isolated and characterized an active factor from young kernels of sweet corn. This substance was the first naturally occurring cytokinin identified and has been called **zeatin**. Chemically, it is 6-(4-hydroxy-3-methyl-trans-2-butenylamino) purine, not surprising because DNA contains much purine. Zeatin has since been obtained from peas and spinach. Furthermore, zeatin analogs and closely related chemicals have been extracted from a variety of sources. One such close relative, 6-(γ,γ-dimethylallylamino)-purine, has been obtained from hydrolysates of the transfer RNA's for serine and tyrosine from a number of animal and plant sources.

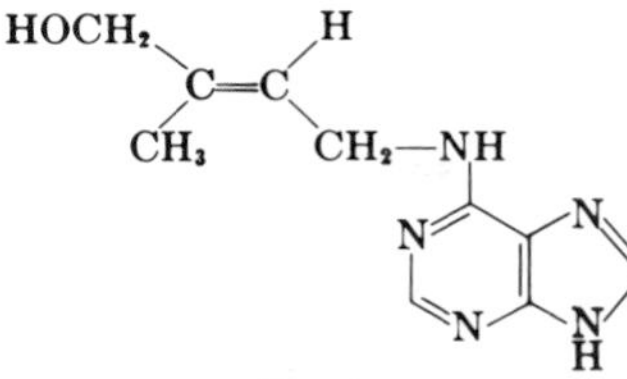

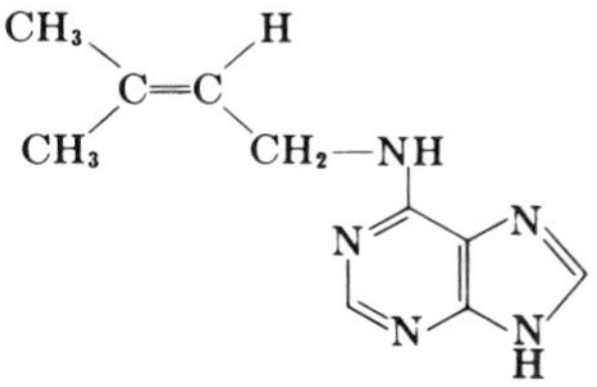

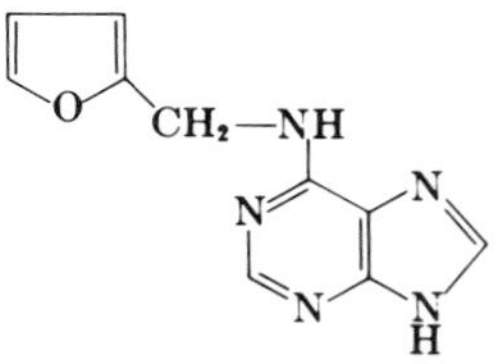

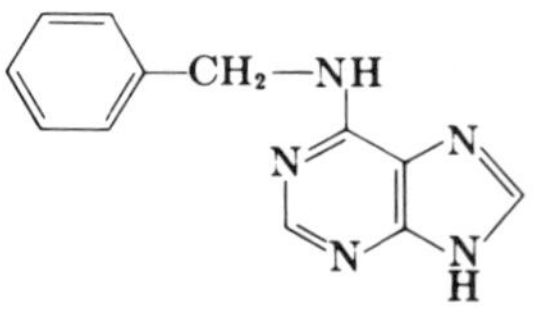

Fig. 17-19. Structures of four cytokinins. Zeatin and 6-(γ, γ-dimethylallylamino)-purine occur naturally in plants. Kinetin and 6-benzyladenine are synthetic cytokinins.

At first, the only known effect of the cytokinins was their ability to promote cell division in cultures of plant tissues such as tobacco callus (tissue of thin-walled cells developed on wound surfaces) where growth is dependent on added cytokinins. A very strange aspect of this promotion of cell division is that cytokinins alone have relatively little effect, yet they have a remarkable capacity to act synergistically with other hormones. This is illustrated by the absence of growth in tobacco callus in the presence of kinetin alone and the great promoting effects of kinetin in the presence of auxin. The range of effective kinetin concen-

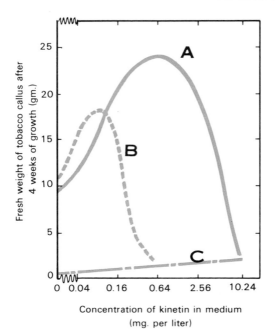

Fig. 17-20. Kinetin acts synergistically with auxin in promoting growth of tobacco callus cultures. A, kinetin at concentrations indicated plus 26 parts per million auxin; B, kinetin plus 4 parts per million auxin; C, kinetin alone. Note that kinetin alone has practically no growth-stimulating effect. (Redrawn from J. van Overbeek: Plant hormones and regulators. Science, *152:* 3723, 1966.)

trations is dependent on the level of auxin present. In addition to promoting cell division, cytokinins were soon found to have profound effects on the differentiation of cultured plant tissues. The ratio of auxin to cytokinin in a medium is a controlling factor in determining the kind of differentiation that will take place: high auxin to cytokinin ratios favor root formation while low auxin to cytokinin ratios lead to shoot initiation. This kind of relationship was not too surprising once the effect of auxin/cytokinin ratio had been established because in nature new shoots form on the stumps of trees and adventitious roots are produced on the base of a cutting or in the region just above a girdle on a stem. A plant's main source of auxin is its shoot system with major sites of synthesis in the apical meristems and leaves. A girdle interferes with the transport of auxin down the stem and auxin accumulates above the girdle. The high auxin cytokinin ratio above the girdle promotes formation of adventitious roots. Cutting off the supply of auxin by girdling or by chopping down a tree results in a low auxin/cytokinin ratio below the girdle or at the top of the stump. The low ratio promotes the formation of buds and development of new shoots.

It is also evident that auxin/cytokinin ratios are important in the control of apical domi-

Fig. 17-21. By controlling the relative amounts of cytokinin and auxin in culture media, Skoog and his collaborators at the University of Wisconsin were able to grow roots and shoots from cultures of undifferentiated tobacco stem cells. Flasks (left to right) contained the following concentrations of 6-(b,γ-dimethylallylamino)-purine: 0, 0.04, 0.2, 1.0, 5.0, and 25.0 micromoles per liter plus standard amounts of minerals, sucrose, vitamins, and auxin. Cultures had grown for 6 weeks. (From J. van Overbeek: The control of plant growth. Scientific American, *219:* 75-81, 1968.)

nance, a phenomenon we have already discussed in terms of auxin concentration. As you will recall, lateral buds develop when apical meristems (and the source of auxin) are removed. The development of lateral buds can be inhibited by applying auxin to replace that formerly provided by the apical meristem. The action of applied auxin can be overcome by simultaneously applying cytokinins.

Cytokinins and Delay of Senescence

An additional and perhaps very different effect of the cytokinins that is of obvious commercial application is their generally inhibiting action on changes associated with senescence. The synthetic cytokinins, kinetin and 6-benzylaminopurine, have been used in the postharvest preservation of vegetable crops such as asparagus, broccoli, and celery. Not only is the fresh appearance of such crops maintained, but also their storage life is extended and their nutritional value is preserved. In part, these antisenescence effects of the cytokinins are the result of reduced respiratory rates. The breakdown of chlorophylls as well as hydrolysis of proteins is retarded. When cytokinins are applied to the leaves of intact plants, a transport of nutrients toward the treated areas can be observed.

Initially, the only available bioassay method for the cytokinins was their promoting effect on cultures of plant tissues *in vitro*, a slow and laborious technique. More recently, rapid assays based on the expansion of young leaves employing discs cut from radish or etiolated bean leaves have been developed.

INTERACTIONS OF AUXINS, GIBBERELLINS, AND CYTOKININS

Some specific interactions of the naturally occurring plant hormones have already been noted. Expecially intriguing is the fact that in regulating differentiation, the cytokinins and auxins interact. It is the ratio of their concentrations rather than the absolute concentration of either that determines what pattern of growth will occur. Studies with plant tissue cultures have clearly shown that not only auxin and cytokinin but also gibberellins must be present for maximum growth. With completely defined media it is now possible to obtain at will practically any desired pattern or rate of growth in isolated plant tissues.

In the development of plants from seeds, the plant hormones tend to act in sequence. In the earliest phases of germination, the production of enzymes needed for hydrolysis of food reserves is triggered by the gibberellins. As cell division begins in the developing embryo, the cytokinins are essential for the promotion of nucleic acid and protein synthesis. Later, when the seedling has emerged and the proper orientation is essential, the auxins play a major role in tropic responses to light and gravity. The levels of hormones in developing plants are continually changing. Because of the ways in which the auxins, gibberellins, and cytokinins interact, a change in the concentration of one can affect the action of constant levels of the others. Table 17-1 provides a summary of some of the facts concerning the chemistry, sites of synthesis and the action of auxins, gibberellins, and cytokinins at the subcellular, cellular, and morphological levels.

ETHYLENE—THE GASEOUS HORMONE INVOLVED IN RIPENING FRUITS AND LEAF ABSCISSION

Ethylene, $CH_2{=}CH_2$, is a simple hydrocarbon occurring in natural gas. Between 1860 and 1870 there were reports of defoliation of trees and many varieties of herbaceous plants after accidental exposure to illuminating gas. It was not until the early 1900's, however, that ethylene was identified as the biologically active component of this fuel. For many years the striking effects of ethylene on plants, including the promotion of **leaf abscission**, induction of **epinasty** (the downward bending of petioles), and a hastening of the ripening of fruit were regarded as interesting curiosities.

With the introduction of the technique of gas chromatography, a sensitive means for measuring the concentration of ethylene in mixtures of gases became available. It is now a proven fact that ethylene is a product of plant metabolism. Its production is prevented by temperature extremes (*i.e.*, near $0°$ or above

Table 17-1

Some comparisons of the auxins, gibberellins, and cytokinins[a]

Property	Auxins	Gibberellins	Cytokinins
Chemical nature of the naturally occurring hormone	Indole-3-acetic acid	Gibberellic acid or gibberellin A_3 (5-ringed diterpenoids)	Zeatin; 6-(4-hydroxy-3-methyl-2-butenylamino)-purine
Structural formula			
Major sites of synthesis	Most active synthesis in apical meristems and enlarging tissues, leaves	Roots, embryos in germinating seeds	Root apex, developing fruits, germinating seeds
Subcellular and molecular effects	Promotion of protoplasmic streaming	Triggering the *de novo* synthesis of enzymes in the aleurone layer of cereal grains	Prevention of chlorophyll and protein degradation, attraction and retention of metabolites and water; maintenance of nucleic acid and protein synthesis; reduction of respiration rates
Effects at the cellular level			
Division	Growth of tissue cultures; cell division in cambium and fruits; inhibition of growth of lateral buds	Cell division in subapical regions of shoot meristems; cell division in cambium and fruits; termination of dormancy in buds and seeds	Cell division in tissue cultures when auxin is present; promotion of growth of lateral buds and fruits
Enlargement	Elongation of stems; growth of flowers and fruits	Elongation of stems, growth of flowers and fruits; termination of dormancy in buds and seeds	Growth of stems and expansion of leaves; termination of dormancy in seeds
Differentiation	Adventitious root formation in cuttings, cambial differentiation; formation of pistillate (female) flowers	Flower initiation in photoperiod-dependent species; inhibition of root formation; formation of staminate (male) flowers	Roots form when auxin/cytokinin ratio is high; buds form when auxin/cytokinin ratio is low
Morphological effects	Tropisms are caused by asymmetric distribution of auxins; involvement in apical dominance and fruit development	Excessive stem growth; reversal of genetic dwarfism	Reduction of shoot growth relative to root development

[a] Adapted in part from A. Lang: Intercellular regulation in plants. In *Major Problems in Developmental Biology*, ed. by M. Locke. Academic Press, New York, 1966, p. 257.

40°C.) and low oxygen tensions. In the natural ripening of fruits there is a sharp increase in the rate of ethylene formation. This so-called **climacteric,** when respiration rates increase sharply, can be induced by adding ethylene to the atmosphere where harvested fruits are stored, or it can be delayed by rapid air circulation which prevents the accumulation of naturally produced ethylene.

The production of ethylene generally occurs

in those parts of plants where auxin content is high. In fact, the production of ethylene can be increased very rapidly (within 15 to 30 minutes) by applying auxin. The capacity to produce ethylene in response to exogenous auxin is related to the age of the tissue and falls off rapidly during senescence. In promoting leaf abscission, ethylene has its greatest effects on old leaves and leaf drop can be prevented by applying auxin.

Leaf Abscission

A century ago von Mohl had shown that two kinds of changes are involved in leaf fall: (1) There is a differentiation of cells at the base of the petiole with the formation of an **abscission zone.** (2) The cells in this zone separate from each other with their walls intact. The work of Abeles and his associates at Fort Detrick, Maryland, has contributed much to the understanding of how ethylene is involved at the molecular level in the process of leaf abscission. According to the aging-ethylene hypothesis proposed by Abeles, the essential role of ethylene is to serve as an effector molecule triggering the synthesis of enzymes responsible for cell separation in the abscission zone. As would be predicted from this hypothesis, the stimulation of mRNA and protein synthesis (including increases in cellulase) by ethylene

have been detected in the abscission zone. Inhibitors of DNA-dependent RNA synthesis (actinomycin D) and protein synthesis (cycloheximide) retard abscission and block the usual ethylene-triggered changes in the separation layer.

The realization that ethylene should be classified as a naturally occurring plant hormone has greatly simplified the interpretation of many diverse observations concerning the effects of specific chemicals on leaf abscission. It has now been established that the many seemingly unrelated chemicals that are effective as defoliants all have the common effect of

Fig. 17-22. Epinasty of tomato petioles caused by ethylene. Plant at right was exposed to ethylene; plant at left is a control. (Courtesy F. B. Abeles, U. S. Department of Agriculture, Beltsville, Md.)

Fig. 17-23. Promotion of flowering in *Billbergia pyramidalis* by ethylene. Plant at left was exposed to 1 part per million ethylene for 24 hours. This photograph of treated and control plants was taken 6 weeks later. (Courtesy F. B. Abeles, U. S. Department of Agriculture, Beltsville, Md.)

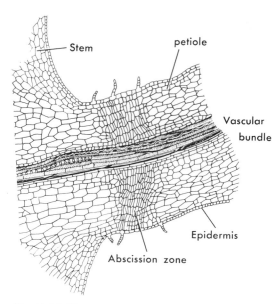

Fig. 17-24. The abscission zone differentiates in tissues at the base of the petiole. As a leaf becomes senescent, the cells in this zone separate. Ethylene promotes these changes by inducing the synthesis of cellulases. (From J. G. Torrey: *Development in Flowering Plants.* The Macmillan Co., New York, 1967.)

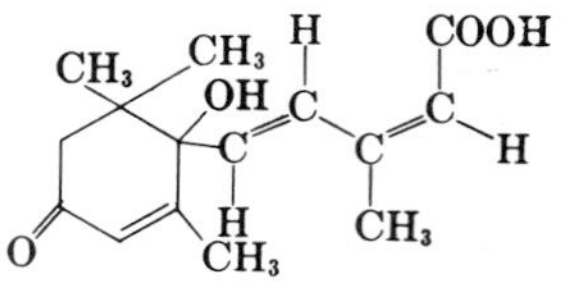

Fig. 17-25. Structure of abscisic acid. This naturally occurring growth retardant is also called abscisin II and dormin.

stimulating ethylene production and therefore have a single physiological basis for their action. Although the impetus for research on the chemical control of leaf abscission has come from military needs, peaceful applications of the results of such chemical warfare research are numerous. The machine harvesting of some crops such as cotton is much simpler if leaves are absent. If some easy means were available for controlling the time when leaves are shed from trees in the fall, homeowners would be everlastingly grateful for the simplification of their autumn chore of raking leaves.

ABSCISIC ACID—A NATURALLY OCCURRING INHIBITOR OF DEVELOPMENT

Abscisic acid (also called **abscisin II** or **dormin** in the literature of plant growth-regulating chemicals) is a carboxylic acid with a structure related to that of vitamin A. It was originally found in the leaves of woody plants

grown under conditions of short daily photoperiod and in mature cotton fruit. Present evidence indicates that the occurrence of this hormone is practically universal in higher plants. Abscisic acid induces dormancy in buds, inhibits germination, retards plant growth (especially elongation) and hastens senescence. As might be expected, it interacts in complex ways with other plant hormones. There is a direct antagonism between abscisic acid and the gibberellins: abscisic acid inhibits and the gibberellins promote elongation of stems, germination of seeds, and the sprouting of buds. The action of one can be counteracted by applying the other. Abscisic acid and the cytokinins have antagonistic effects in lettuce seed germination, leaf senescence, and the over-all growth of plants.

The mode of action of abscisic acid has not been established with certainty, but its antagonistic action toward other hormones, particularly the gibberellins, strongly suggests that it acts as a repressor that prevents the synthesis of messenger RNA specific for enzymes essential in the developmental process. Its effectiveness in causing dormancy of buds and seeds is of obvious advantage to plants during the winter months. The absence of dormancy in the buds and seeds of temperate plants would mean sure death for both the individual plant and its progeny.

MORPHACTINS: RECENTLY SYNTHESIZED CHEMICALS AFFECTING PLANT GROWTH AND DEVELOPMENT

During the past several decades, chemical companies have begun a systematic testing of newly synthesized products for activity as plant growth regulators. Among the most interesting of these are fluorene-9-carboxylic acid, and its derivatives, collectively referred to as the **morphactins** (from the term *morph*ogenetically *acti*ve substances). The over-all effect of these compounds is a general retardation of growth. When dilute solutions are sprayed on vegetation, they have little effect on the parts of plants already formed, while further growth is inhibited. The advantages of such compounds for those of us who appreciate green lawns but are too lazy to mow them are obvious. The most intriguing effects of the morphactins for those who are interested in how plants grow are their effects on phototropism and geotropism. Seeds germinated in the presence of these

Fig. 17-26. Effect of morphactin (IT 3233, 6×10^{-5} M) on the phototropic response of oat seedlings. Treated seedlings on left, controls on right. (Courtesy A. A. Kahn, from A. A. Kahn: Physiology of morphactins: Effect on gravi-and photo-response. Physiologia Plantarum, *20:* 306-313, 1967.)

Fig. 17-27. Effect of morphactin (IT 3233, 6×10^{-5} M) on geotropism in seedlings of timothy (*Phleum praetense*) after 96 hours of growth. Left, treated; right, control. (Courtesy A. A. Kahn, from A. A. Kahn: Physiology of morphactins: effect on gravi- and photo-response. Physiologia Plantarum, *20:* 306-313, 1967.

compounds show absolutely no response to unilateral illumination or gravity.

THE FUTURE—MADE-TO-ORDER PLANTS?

With currently available herbicides, natural and synthetic hormones, and plant growth regulators and inhibitors, it is possible to eliminate all but a single wanted species from a farmer's fields, to grow plants of practically any desired shape and dimensions, and to induce abscission of plant parts such as leaves that interfere with the mechanical harvesting of other structures; in short, to grow plants to order. Although this approach to agriculture is still in its early trial and error stages, the future application of our expanding knowledge of the chemical control of plant growth and development has tremendous potential. Unlike the insecticides that tend to persist and accumulate in plants and later in herbivores and carnivores in the food chains, many of the herbicides and chemical regulators of plant growth tend to be degraded rapidly in the soil and are essentially nontoxic to animals. Some exceptions to this generalization have been found, however, and the indiscriminate use of weed killers is to be avoided. The destruction of undesirable plants often has profound effects on host-parasite and predator-prey relationships. Furthermore, as herbicides are more thoroughly tested for possible harmful effects on animals, we are beginning to learn that some (for example, the auxin relative 2,4,5-T) can influence some stages of animal development and can produce **teratogenic effects** (*i.e.*, developmental abnormalities). With growing concern about the environment, there is a healthy increase in caution about the introduction of any new chemical into the biosphere. No matter how desirable the short-term effects might appear to be, we must thoroughly test all new agricultural chemicals for their possible long-term consequences.

USEFUL REFERENCES

Galston, A. W., and Davies, P. J. *Control Mechanisms in Plant Development*. Prentice-Hall Inc., Englewood Cliffs, N. J., 1970.

Galston, A. W., and Davies, P. J. Hormonal regulation in higher plants. Science, *163:* 1288-1297, 1969.

Gould, R. F. (ed.) *Gibberellins: Advances in Chemistry Series No. 28.* American Chemical Society, Washington, D. C., 1961.

Hegelson, J. P. The cytokinins. Science, *161:* 974-981, 1968.

Leopold, A. C. *Auxins and Plant Growth.* University of California Press, Berkeley and Los Angeles, 1955.

Letham, D. L. Chemistry and physiology of kinetin-like compounds. Annual Review of Plant Physiology, *18:* 349-364, 1967.

Phillips, I. D. J. *The Biochemistry and Physiology of Plant Growth Hormones.* McGraw-Hill Book Co., New York, 1971.

Steward, F. C. (ed) *Plant Physiology: A Treatise,* Vol. VIB: Physiology of Development: The Hormones. Academic Press, New York, 1972.

Thimann, K. V. Tropisms in plants. Embryologia, *10:* 89-113, 1967.

van Overbeek, J. The control of plant growth. Scientific American *212:* 75-81, 1968.

Reproduction in Plants

Most plants (and a great many animals) have the capability of forming new individuals both asexually, that is, by vegetative propagation (either normally or with human assistance), and by sexual reproduction. Within each mode of procreation, there are countless variations. Yet, despite such diversity, all asexual means of reproduction share the advantage of producing offspring identical to the original organism in genetic endowment, while sex, whether in plants or animals, provides for the recombination of inherited characteristics, the basis for evolutionary change in populations.

Consider the case of the common potato, which is a highly heterozygous organism. If potatoes are grown from seed, the resulting plants show enormous variability. Some even produce bright red or bright purple potatoes, most of them very small in size or number and highly unsatisfactory for farmer and consumer alike. If new plants are propagated asexually by cutting out and planting the buds ("eyes") from the tubers, then all the offspring produce potatoes like the parent, whether good or bad.

Clearly, if the objective of a horticulturalist is to obtain new plants from a desirable but heterozygous strain, to plant a new orchard of apple or grapefruit trees, or to propagate a newly discovered variety of flowering shrub, seeds from that genetically mixed strain cannot be used. The techniques of vegetative propagation are necessary to obtain new plants with precisely the characteristics of the original. On the other hand, a plant breeder wishing to obtain new types of plants will use sexual reproduction. To obtain a new strain of disease-resistant wheat, for example, thousands of plants are grown from seed and tested for disease resistance. To obtain a new kind of corn combining uniform height for easy mechanical harvesting with high yield, the plant breeder would begin by crossing plants having each of the desired traits (see Chapter 7).

Genetic manipulation is the most modern development in a long history of agriculture beginning with the first domestication of plants in Neolithic times. Over thousands of years, desirable crop varieties have been selected. The exploration of new continents has led to the importation of new species of agriculturally useful plants. The recent development of new and better varieties of crop plants is undoubtedly the most promising means available for increasing agricultural productivity.

ASEXUAL REPRODUCTION IN PLANTS

The Adaptive Value of Asexual Propagation

One of the strangest generalizations that can be made about plants is that unstable or changing environments (whether shifting sand dunes, a strip of land cleared for a new road, or a puddle of rainwater) tend to be inhabited by species capable of reproducing asexually. This is borne out by common experience. Everyone has noticed that in places where water drains slowly after a rain, "blooms" of algae develop rapidly, only to disappear when the surface of the soil again dries. The shifting sand dunes along our coasts are inhabited by grasses and other seed plants that spread by means of **stolons** or **rhizomes**, modified stems that spread

Fig. 18-1. Beach grass on the dunes of Cape Hatteras, North Carolina. These plants spread by means of modified stems growing just below the surface of the sand. By preventing wind and wave erosion, they help to stabilize the environment.

horizontally at, or just below, the surface of the soil. Vegetative propagation of these plants permits them to spread out around established individuals. In doing so, they help to stabilize the environment and make possible the invasion of other species having more specialized requirements for the germination of seeds. Thus, the asexually reproducing plants play a major role in ecological succession, a topic that will be discussed in detail in Chapter 36.

Some Examples of Asexual Reproduction

Plants in Which Cell Division Equals Reproduction

When unicellular organisms divide, two new individuals are formed. Here the replication of the nuclear DNA and division of the cytoplasm give rise to daughter cells genetically identical to the parent cell. Procreation of the pro-

karyotic bacteria and blue-green algae, of the unicellular eukaryotic algae and fungi among plants, or of the protozoa in the animal kingdom can be just this simple: in single-celled organisms cell division is a means of asexual reproduction.

In the bacteria, this kind of multiplication is commonly termed **binary fission**, a division into two equal parts. In many yeasts new cells are produced by **budding**, the process by which mature cells produce one or more daughter cells. The small daughter cells, or buds, often remain attached to the parent cell for a time, sometimes forming clumps or chains of cells. Some species of yeasts multiply by fission as do the bacteria.

Spore Formation

In many of the unicellular plants (both prokaryotes and eukaryotes), cells respond to unfavorable environmental conditions, by forming spores (see Chapters 4 and 10). In the process there is usually a condensation of the protoplasm with a reduction in the water content and the formation of a new cell wall. Such changes make it possible for spore-forming species to survive conditions that would probably be lethal for physiologically active cells. Spores can usually withstand high temperatures, ultraviolet light, and many chemical agents, precisely the factors normally used in maintaining asepsis. Although asexual spore formation provides a mechanism for surviving unfavorable conditions, spores are not necessarily formed only in response to hostile environments; they can form under favorable conditions as well. In fact, the mechanism of asexual spore formation is a poorly understood phenomenon. The advantages of spores are obvious, however. When conditions are again favorable, spores absorb water, germinate, and are transformed into physiologically active cells. In a strict sense, when a single cell forms a single spore, reproduction has not occurred. However, such a spore may be carried to a new location and there found a new colony. It is important to remember that these spores are very different from the spores formed in groups of four as a result of meiosis that are part of sexual reproduction.

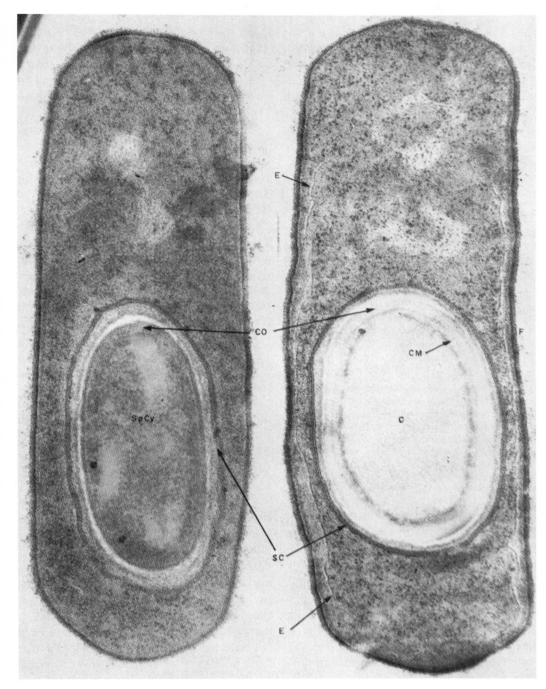

Fig. 18-2. Developing spores of *Clostridium bifermentens* (sectioned preparation seen under the electron microscope). Labeled structures: E, exospore membrane; CO, cortex; SC, spore coat; CM, cortical membrane; C and SpCy, spore cytoplasm and nuclear structures. (From P. D. Walker, R. O. Thomason and A. Baillie: Journal of Applied Bacteriology, *30:* 444, 1967.)

Asexual Reproduction in Multicellular Plants

With the association of cells to form filaments, colonies, or more complex plants, cell division results in an increase in the size of the organism rather than producing new individuals. Asexual means of reproduction were not abandoned by multicellular organisms, however. They only became more varied.

Filaments or colonies of the kinds found in the many groups of algae can form new individuals by fragmentation. In the fungi we find a variety of forms of vegetative reproduction. Pieces of mycelium, when removed, will continue to grow and form new hyphae. Asexual spores are borne on special hyphae. In most of the fungi, asexual means of reproduction predominate and sexual reproduction is a rare occurrence—in some cases unknown (see Chapter 10).

The more complex plants have also evolved specialized means for vegetative propagation. In the simplest of terrestrial forms, the mosses and liverworts, new individuals are produced by continually dying behind and growing ahead. New and separate plants are formed when disintegration of the posterior end reaches a branch point. There are, in addition, more elaborate means of asexual reproduction in these simple plants. Multicellular structures called **gemmae** can form on the surfaces of the vegetative thalli. Once detached from the parent plant, they germinate and form new plants (Chapter 11). The mosses and liverworts share with their more complex evolutionary descendents the ability to regenerate entire new plants from practically any piece, be it leaf, stem, or rhizoid.

Adaptations for Vegetative Reproduction in Higher Plants

The spread of populations of many of the seed plants ranging from beach grasses to strawberries occurs in nature by rather minor modifications of ordinary plant parts. Strawberry plants have special stems, called **stolons** or **runners,** that grow above the soil surface. If they touch the soil they develop roots and new shoots that can form a new plant. The standard method for propagating strawberry varieties in agriculture is by this vegetative means. Many

Fig. 18-3. Raspberries and blackberries can be propagated by tip layering in which roots form adventitiously on branches covered with soil.

other plants can form roots on branches covered with soil. Blackberries, raspberries, and many other plants do this in nature even though they do not form specialized stems like the stolons of strawberries. The horticultural practice of **layering,** where a branch is purposefully buried so that roots will develop, is based on the capacity of stems to form such adventitious roots when brought into contact with the soil.

Other plants, including the irises and a great many grasses, have specialized underground stems (**rhizomes**), that send up aerial shoots. Many varieties of lawn grass have been selected precisely for this tendency to spread rapidly by means of rhizomes. Some trees, including aspens, and many weeds, such as goldenrod, morning glory, and crabgrass, proliferate by this means.

Tubers are another kind of specialized underground stem formed when the ends of rhizomes enlarge as a result of cell proliferation and the accumulation of food reserves. The best known example of a tuber is the familiar potato. This bulbous underground stem, like all stems, has nodes and internodes but its leaves are reduced to small scales. Buds develop in the axils of these nonfunctional leaves. When planting potatoes in the spring, farmers cut the tubers into pieces having one or more such buds or "eyes." In the soil, the buds sprout to form new shoots and roots. Functionally, tubers serve two purposes: food storage and reproduction.

Two further types of modified stems that are important structures of vegetative propagation in the higher plants are **bulbs** and **corms. A bulb** is really nothing more than a bud. There is a short stem at the base with many fleshy, scale-like leaves growing from its upper surface.

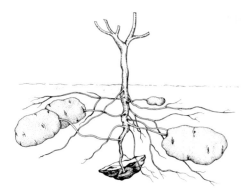

Fig. 18-4. Potato tubers are modified stems with nodes and internodes. Leaves are reduced to small scales. Pieces of potato with one or more buds are used to propagate this plant. (Redrawn from V. A. Greulach and J. E. Adams: *Plants: An Introduction to Modern Botany.* John Wiley & Sons, Inc., New York, 1962.)

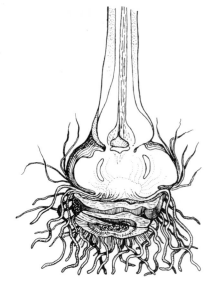

Fig. 18-6. A corm consists mainly of stem tissue. Its outer surface is covered with scaly leaves.

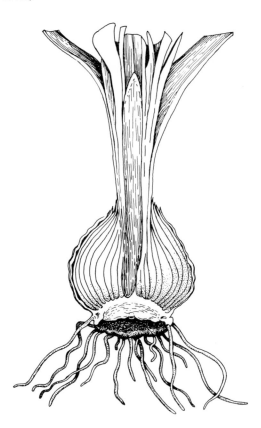

Fig. 18-5. A tulip bulb consists mostly of fleshy storage leaves growing from the very short stem at its base.

Adventitious roots develop from the base of a bulb. The bulk of any bulb is comprised of the fleshy storage leaves. Buds developing in the axils of these fleshy leaves can be separated from the parent bulb and used to propagate the species. Many familiar spring flowers including hyacinths, tulips, and daffodils are grown from bulbs. The readily available food reserves stored in the over-wintering bulbs are a major factor in the early flowering of many of these species. Animals, too, sometimes take advantage of such food reserves: the onion is a bulb which appears regularly on our dinner tables. **Corms** are sometimes referred to as "bulbs" by the uninitiated, but differ from bulbs in consisting primarily of stem tissue. The outer surfaces of corms are covered with leaves that are usually thin and scaly. Adventitious roots form at the base as they do in bulbs. The crocuses and gladioli are familiar garden flowers propagated by means of corms.

In still other common garden varieties, reproductive structures are formed on the shoot. The tiger lilies form aerial **bulbils** about the size of peas in their leaf axils. These, too, are nothing more than modified stems surrounded by fleshy leaves. When they abscise and fall from the parent plant, they sprout to form new individuals.

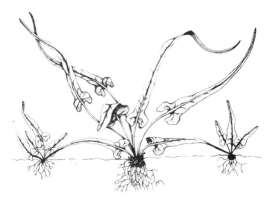

Fig. 18-7. In the walking fern new plants form when the tips of leaves touch the soil. (Redrawn from V. A. Greulach and J. E. Adams: *Plants: An Introduction to Modern Botany.* John Wiley & Sons, Inc., New York, 1962.)

Even leaves may be the means for vegetative propagation. The leaves of African violets and begonias readily form adventitious buds and roots when removed from the plant and placed in contact with the soil. In the walking fern, new plants are formed when the tips of the leaves touch the soil. In bryophyllum tiny plantlets develop in notches along the edges of the fleshy leaves. When shed from the parent plant, they form roots and develop into new plants. In some species long photoperiods promote plantlet formation while short days promote flowering. The tiny species of *Lemna* (duckweed), flowering plants found floating on the surfaces of ponds and streams, usually reproduce asexually by forming new individuals from leaves.

Thus in nature all of the vegetative organs of higher plants (roots, stems and leaves) have been modified by one or another species as a means for asexual reproduction. Vegetative propagation is not restricted to the vegetative organs of plants by any means. Some types of aerial bulbs develop from flower buds. In species such as the common dandelion, seeds can be produced parthenogenetically (*i.e.,* without fertilization). Recent tissue culture experiments indicate that embryo-like development is possible from almost any diploid cell, if appropriate nutrients and hormones are provided. Even some haploid cells such as developing pollen grains can give rise to morphologically normal plants (see Chapter 9).

Rooting and Grafting—Horticulturally Useful Methods for Asexual Propagation

Two common horticultural means of vegetative propagation—stem cuttings and grafting—are feasible because of the ability of plants to replace missing parts and to repair wounds. The many commercially important plants that are propagated by means of stem cuttings include chrysanthemums, roses, sugar cane, grapes, many kinds of fruit trees, and most ornamental shrubs. Although cuttings from many of these plants readily form adventitious roots, a knowledge of the effectiveness of plant hormones in promoting root development (see Chapter 17) has made it possible to propagate even varieties that do not readily form adventitious roots. Adequate moisture is absolutely necessary during rooting. The bases of cuttings are usually stuck into moist sand, vermiculite (modified) mica), or peat. The tops are kept moist by means of sprays, mists, or plastic covers. Often the formation of roots is promoted if slightly higher temperatures are maintained around the bases of cuttings.

Grafting is no more than an artificial means of providing roots for a cutting. The cutting, which may be a piece of stem or only a bud, is called the scion. The plant or the root system onto which it is grafted is called the stock. Many kinds of woody and herbaceous dicots from apple trees to sunflowers have been grafted. The probability of success is greatest when scions and stocks are the same or different varieties of a single species. However, both interspecific and intergeneric grafts are possible. Such mixed grafts are not uncommon in fruit trees where flowers and fruits of numerous varieties and species can be produced on a single composite tree. Grafting is important in horticulture for two purposes. The first is to make possible the asexual propagation of a highly desirable variety, *e.g.,* a fruit tree, which is heterozygous and so will not breed true. Many varieties of apples, pears, grapes, and other kinds of fruit-bearing plants are propagated in this way from a single fortunate mutation. The second purpose is to make it possible to use a more vigorous or disease-free stock to propagate a scion that is difficult to grow on its own root system. Ordinarily ornamental roses are grafted onto wild rose stocks; many fruit varieties including most grapes are grown on less productive but more disease-resistant root systems.

SEXUAL REPRODUCTION IN PLANTS

Advantages of Sexual Reproduction

Vegetative propagation, whether by means of modified stems or roots, or even leaves that are

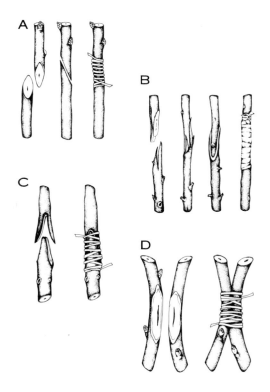

Fig. 18-8. Some widely used grafting techniques. A, "splice" graft in which obliquely cut stock and scion are fitted together to match cambium layers. B, "whip and tongue" graft. Tongue cleft aids in holding stock and scion together. C, "saddle" graft. Obliquely cut stock forms a blunt point or saddle. Scion is cut to fit over tapered end of stock. D, "approach" graft. Similar shields are cut from the two components. Whatever the type of graft, stock and scion are securely tied together with waxed string, nursery tape, etc. Often waterproof coverings such as wax or plastic film are used to retard water loss.

shed and then sprout to form new plants, provides a means of making new individuals with genotypes that are well adapted to a given environment. However, any change, be it unseasonable weather, the arrival of a new insect pest or disease-causing microorganism, or slight changes in soil or in average annual rainfall, threaten the survival of asexually reproducing species. A fact of evolution (and of life) is that change within a population is the only effective way to combat change in the environment. Survival in the evolutionary sense of adapting to change and of exploiting new environments is dependent upon sex because only in sexual reproduction is variability assured.

Meiosis and fertilization are the two mechanisms that assure the variability of progeny. During meiosis there is an independent segregation of genes on separate chromosomes. Fertilization not only multiplies the amount of variation but makes possible the combination of favorable mutations which have appeared in different strains, something impossible in a-sexual reproduction.

It is important to remember that not all variability is genetic. Variability remains even in plants that have been inbred for a number of generations and have become homozygous. A classic demonstration of such variability which is dependent on environment can be found in the work of the Danish biologist Johannsen. Early in this century Johannsen inbred the progeny of a single bean plant for several generations. He gathered seeds varying in size and weight from this population of plants grown under the most uniform conditions possible. He then selected the largest and smallest seeds and grew another generation of plants from them. When the seeds from these two sets of plants were compared, they had the same average size as the batch of seeds from which large and small ones had been selected for planting. This demonstrated that selection for a particular characteristic (large seeds) is effective only when there is genetic variability for this characteristic in the population. In plants, where development is profoundly influenced by environmental factors, it is more obvious than in animals that it is not final form but rather the potential for development that is inherited.

Factors Which Promote Sexual Reproduction in Plants

The annual succession of blooms—the crocuses and violets of the spring to the chrysanthemums and asters of the fall—has long stirred human interest. Moreover, an understanding of this phenomenon is of great practical importance to florists the world over. As a result of modern investigations by plant physiologists, important new information is available to answer some of the ancient puzzles about why higher plants blossom when they do and even whey algae and fungi go sexual.

To the long recognized but only vaguely understood factors like temperature and nutrition, and·the mysterious "ripeness-to-flower," modern work has added various chemical agents. Equally important is the discovery of the role of photoperiod. We have already mentioned how plants such as bryophyllum reproduce vegetatively under conditions of long days and short nights but form flowers and thus reproduce sexually when the days are short and the nights long.

The development of the structures involved in sexual reproduction is usually triggered by environmental factors such as temperature and light that change with the seasons. Other internal and external conditions, including chemical growth regulators and mineral nutrition, also influence the onset of flowering. Most plants, as they become older, have an increased tendency for the changes associated with sexual reproduction to be initiated. For the flowering plants, this tendency has been referred to as **"ripeness-to-flower"** and some species will form flowers regardless of environmental conditions once an appropriate stage of maturity has been reached.

Environmental changes serve the useful function of synchronizing reproduction, thus promoting cross-fertilization and variability. A further advantage of seasonal flowering is that it permits a plant to utilize all of its available resources for vegetative growth during part of the growing season. This enables it to compete more effectively with other species.

Temperature

Both diurnal and seasonal fluctuations in temperature can affect plant development. In Chapter 9 we noted that some plants require a period of cold following a time of vegetative development in order for sexual reproduction to occur. The cold treatment of germinating seeds (a process called **vernalization**) is an important agricultural practice (particularly in Russia) that makes it possible to plant and harvest certain cereal grains in the same growing season. Without such treatment, these varieties of rye and wheat would have to be planted in the fall and then harvested the following summer, and they could not be grown at all in climates too cold for fall plantings to survive the winter.

Seasonal fluctuations in temperature are also important in the breaking of dormancy, whether in buds or in underground storage organs such as the bulbs, corms, and tubers that form in response to decreasing day length. A period of days to months at temperatures below 10°C. is usually required for the breaking of dormancy in buds. In certain bulbs and corms, the formation of flower primordia requires specific temperatures. In tulips, for

Fig. 18-9. Day length and chemicals are used to control form and flowering of ornamental plants. United States Department of Agriculture plant scientist H. M. Cathey has used short days to keep the petunia plant at left short and vegetative. Tall and flowering plant at center was grown under long days. Plant at right was sprayed with a growth retardant (which inhibited elongation) and grown under long days to promote early flowering. (Courtesy U. S. Dept. of Agriculture.)

example, bulbs must be maintained near 20°C. for flower primordia to form. Their further development requires a drop in temperature to 9°C. (about 48°F.) or less for about 14 weeks. Such thermoperiodic requirements correspond to seasonal fluctuations of temperature in regions where these plants are native. Because of these temperature requirements, tulips cannot be grown successfully in warm climates, even though bulbs imported from colder regions may bloom during the first year.

Diurnal thermoperiodicity can influence not only the rate of vegetative growth of plants, but also certain aspects of their reproductive development. For unknown reasons, most plants grow best when temperatures at night are lower than during the day. In certain species such as squash and cucumbers, where both male (staminate) and female (pistillate) flowers are formed on the same plant, the relative numbers of male and female flowers are influenced both by photoperiod and by diurnal fluctuations in temperature. Long days and warm nights promote the formation of male flowers, while cool nights decrease and warm nights with short-days increase the production of female flowers.

Chemical Growth Regulators

Certain of the naturally occurring plant hormones and chemical plant growth regulators can substitute for the effects of light and temperature on dormancy and flowering. Gibberellins, 2-chloroethanol, and thiourea are effective in breaking dormancy of potato tubers; while naphthaleneacetic acid (a synthetic auxin) and maleic hydrazide have been used to prolong dormancy and prevent the sprouting of potato tubers during storage. The flowering of many cold-requiring species is promoted in the absence of cold by gibberellins. Thus, this plant hormone is potentially useful as an alternative to vernalization. It is particularly effective in those species whose flowering is accompanied by the shift from a rosette growth pattern to one with greater internode elongation.

Many examples of the chemical control of sexuality in plants have been discovered. In the more primitive plant groups such as the algae and the fungi there is evidence that the production of gametangia, the maturation of the sex organs, the release of gametes, and the attraction of sperms to eggs are influenced by specific chemicals produced by these organisms at particular stages of the life cycle. The investigation of specific chemical action in the production of gametangia has been studied in considerable detail in the ferns. The haploid, gametophyte prothallus of the fern can be cultured in the laboratory on synthetic media. Grown singly in test tubes, the prothallia develop normally except for the fact that they do not form male gametangia (**antheridia**). If grown in the presence of other gametophyte plants, or if some medium from cultures of **old** gametophyte plants is added, then antheridia are formed. When sufficient quantities of antheridium-inducing factor are added to the culture medium, practically every cell of the prothallus differentiates and forms an antheridium. This factor appears to be produced in old prothalli but only younger gametophyte plants can respond to it (hence an isolated prothallus can form no antheridia). Female gametangia (**archegonia**) can be formed on old gametophyte fern plants even though no antheridia have been formed. The production of male gametangia in neighboring plants through the production of a specific chemical stimulus insures that sperms will be available to fertilize the eggs formed in the archegonia. The antheridium factor appears to be specific to each family of ferns.

Mineral Nutrition

We must not forget that mineral nutrition can influence the vegetative and reproductive development of plants. As you will recall from Chapter 16, an excess of the essential element nitrogen promotes vegetative development. We take advantage of this effect of excess nitrogen to grow lush, dark green lawns. However, a comparable zeal in applying fertilizer to tomatoes or other species grown for their fruits has disappointing consequences. Overfed plants respond in a way that is somewhat comparable to the situation in overfed animals. Instead of getting fat, however, their excess nutrients are utilized for a superabundance of vegetative growth, while there is very little flowering and fruit set. Under circumstances of nitrogen deficiency, plants respond in the opposite way and flower after very limited vegetative growth. Obviously this quickened reproduction of a starving plant can have important survival value for the species.

Photoperiodic Control of Flowering— The Reason for Giant Tobacco and September Soybeans

One of the amazing aspects of the photoperiodic control of flowering is that such a widespread and, one would think, obvious phenomenon was not discovered by plant scientists until the present century. Not until the classic publications of W. W. Garner and H. A. Allard appeared in the early 1920's was it realized that relative lengths of day and night are controlling factors in the onset of flowering. These two workers and their associates at the United States Department of Agriculture Plant Industry Station in Beltsville, Maryland, were puzzled by the strange behavior of two kinds of plants.

One plant, a mutant variety of tobacco, (*Nicotiana tabacum*, var. *Maryland mammouth*), when grown in the field during the summer, attained a height of over 3 meters but persistently remained vegetative. This was frustrating to those who wanted to obtain seeds for commercial use of the mutant or wished to use

Fig. 18-10. H. A. Allard (right) collaborated with W. W. Garner in pioneering studies of the photoperiodic control of flowering in plants. In this photograph taken in 1960, Dr. Allard is discussing more recent aspects of light-controlled plant growth with Dr. H. A. Borthwick. (Courtesy U. S. Dept. of Agriculture.)

it in plant breeding experiments. When cuttings of this plant were grown in the greenhouse during the winter months, they flowered profusely and set seed after only limited vegetative growth. Plants grown from seeds produced by winter-grown plants again grew to unusually large size and failed to flower in the field during the following summer. Another plant that had a puzzling behavior was the Biloxi variety of soybean (*Glycine max*). No matter when these soybeans were planted, whether in May, June, or July, they all flowered in September. This meant that those planted early in the season grew vegetatively for 4 months while those planted in July flowered when much smaller after only 2 months' growth. It appeared that all of the plantings responded in unison to the same environmental signal.

After considering a number of possible factors such as temperature, light intensity, and mineral nutrition, Garner and Allard realized that relative length of day and night was the condition controlling the flowering response in both the Maryland mammoth tobacco and the Biloxi soybean. Garner and Allard were able to induce or prevent flowering in tobacco and soybean experimentally by shortening or lengthening the day. Shortening was accomplished by placing plants in dark chambers, lengthening by turning on artificial light at night. This knowledge has proved to be of great practical importance, especially to florists who want plants to blossom at particular times.

Classification of Plants According to Their Photoperiodic Requirement for Flowering

Garner and Allard found that flowering plants can be classified according to their responses to photoperiod into the following three categories: (1) **short-day** plants, (2) **long-day** plants, and (3) **day-neutral** plants. Short-day plants flower only when subjected to day lengths shorter than a particular number of hours during each 24-hour cycle of light and dark. The Maryland mammoth variety of tobacco, the Biloxi soybean, and the plantlet-forming bryophyllum belong in this category. Other examples of short-day plants are the

Fig. 18-11. S. B. Hendricks, United States Department of Agriculture scientist whose studies of the action spectra of phytochrome-mediated plant processes led to the prediction of many of the properties of this important photoreceptor. He is shown here measuring soybeans grown under controlled photoperiod. (Courtesy U. S. Dept. of Agriculture.)

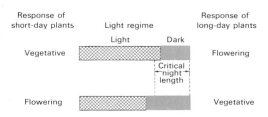

Fig. 18-12. Summary of responses of short-day and long-day plants grown under controlled photoperiods.

particular species. For example, many kinds of plants flower in response to a 14-hour photoperiod; *i.e.*, 14 hours of light and 10 hours of darkness. This happens to be shorter than the critical photoperiod necessary for the short-day cocklebur and longer than the critical number of hours of light needed by the long-day black henbane.

Patterns of Seasonal Changes in Day Lengths and Their Effects

Following Garner and Allard's discovery of photoperiodism in plants, zoologists found comparable control of reproductive cycles in animals such as aphids and birds. In addition, there are in animals other seasonal responses related to photoperiod such as changes in fur color and the migration of birds. Seasonal change in day length thus appears to be a signal used universally as a cue for changing environmental conditions. The amount of annual change in day length increases with distance from the equator. In the tropics, there is relatively little fluctuation in day length or temperature. Seasonal changes in these two climatic factors increase with increasing latitude. Representative patterns of the annual changes in day length at various locations differing in distance from the equator are shown in Figure 18-13. In 1957 Ferguson demonstrated an important aspect of climate by plotting mean monthly temperatures against day lengths. The climatic cycle for any location appears as an orbit with a range that increases at higher latitudes. Obviously such climatic cycles have great influence on the distribution of plants and animals. Tolerance of temperature extremes and capacity to respond to the lengthening or shortening photoperiods that

cocklebur, chrysanthemum, aster, ragweed, and poinsettia. In **long-day** plants, flowering is initiated under conditions where day length exceeds a critical number of hours. Examples of long-day plants include spinach, black henbane, and some varieties of barley. **Day-neutral** plants are indifferent to photoperiodic conditions in that after reaching a critical age or size, they flower regardless of day length. Many common plants, including tomatoes, peas, and sunflowers, are day-neutral.

One strange thing about the terms "long day" and "short day" is that they do not necessarily imply anything about the absolute length of day needed to induce flowering in a

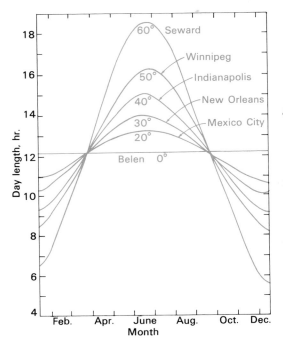

Fig. 18-13. Seasonal changes in day length increase with distance from the equator. Curves indicate the number of hours from sunrise to sunset as a function of time of year for six cities at the latitudes indicated. (Redrawn from A. C. Leopold: *Plant Growth and Development.* McGraw-Hill Book Co., New York, 1964.)

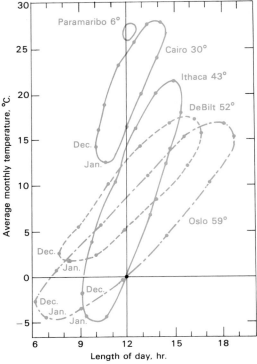

Fig. 18-14. Graphic representation of annual climatic cycles. The climates of five cities are shown as plots of monthly means for day length against average monthly temperature. (Redrawn from A. C. Leopold: *Plant Growth and Development.* McGraw-Hill Book Co., New York, 1964.)

signal climatic changes determine whether a given species is able to survive in a particular climatic cycle. It should come as no surprise that plants native to the tropics, where seasonal changes in photoperiod and temperature are small, tend to be day-neutral. On the other hand, in temperate climates, where the need to respond to seasonal changes in temperature is greater, short-day and long-day plants are common.

Other aspects of plant development besides the onset of sexual reproduction are influenced by photoperiod. Long days promote onion bulb development and are favorable for the vegetative growth of most plants. In some species growth patterns are influenced by photoperiod. Internodes remain very short during vegetative growth in spinach and mullein; *i.e.*, in these plants the leaves form a **rosette** close to the

ground when photoperiods are short. However, when their critical day lengths are exceeded, they "bolt" as the onset of flowering is accompanied by rapid internode elongation. Short days generally promote tuber formation, leaf abscission, and the dormancy of buds. All of these modifications facilitate survival during the cold winter period that follows.

The Mechanism of Photoperiodic Control of Flowering

For a time after the discovery of the photoperiodic control of flowering, the length of the light period was thought to be the single critical factor. However, it was soon discovered that in short-day plants, flowering could be inhibited not only by artificially prolonging day length, but also by interrupting the dark period. Such interruptions can be surprisingly short—usually light of only a few minutes' to an hour's duration is sufficient provided that the interruption comes at about the middle of the dark period. This discovery indicated that slow biochemical reactions taking place in the dark are required for flowering to occur in short-day plants and that they perhaps should have been termed "long-night" plants. In long-day plants such dark reactions probably proceed more rapidly and they could be referred to as "short-night" plants.

A clue about the mechanism of photoperiodism came from asking a logical question about light interruption of the dark period: "What wavelength is effective?" It turned out that red light is as effective as white light. Furthermore, the effects of red light can be canceled out by subsequent exposure to far red light, provided that the exposure to far red light follows within about a half hour. The effectiveness of red light and the antagonistic effect of far red light are characteristic of responses mediated by **phytochrome.** We encounter many such light-regulated responses in plant development, including light-dependent seed germination, the straightening of seedlings as they emerge from the soil, leaf expansion, and synthesis of the

anthocyanin pigments (see Chapter 9). All of these responses have the same action spectrum and are triggered by light-dependent changes in phytochrome. You will recall that phytochrome can exist in either of two interconvertible forms. P_R, the blue form with an absorption maximum at about 660 nm., is converted on absorption of red light to P_{FR}, having an absorption maximum in the far red at about 730 nm. P_{FR} is thought to be the enzymatically active form of phytochrome. In the flowering response, it appears that several events must occur during the critical dark period. First, P_{FR} formed during the preceding illumination must be converted back to the red absorbing form. This pigment conversion is thought to be only a preliminary to further dark reactions which are necessary to produce a chemical which can be translocated to the apical meristems where the changes associated with the induction of flowering occur.

There is an abundance of experimental evidence supporting the hypothesis that the diffusible chemical stimulus which is responsible for flower induction is formed in the leaves. It is then translocated to bud primordia where it causes a shift from development of leaf buds to development of flower buds. In some species, such as the cocklebur, it is necessary to expose only a single leaf or branch to the appropriate photoperiod to induce flowering in the entire plant. When photoperiodically induced branches are grafted onto stocks maintained in light regimes that inhibit flowering, the uninduced stocks will flower. Neither the nature of this diffusible flowering hormone (sometimes called **florigen**) nor the sequence of metabolic changes that it induces in the bud primordia are known. Nevertheless, there have been many practical applications of our knowledge of the photoperiodic control of flowering. Plants grown in greenhouses are commonly brought into flower at times when the market demand is high. During winter months, day length can be extended by means of ordinary incandescent lamps. During the summer, days can be shortened by covering plants with special kinds of black cloth manufactured specifically for use by florists. These practices are now so widespread that we have come to expect that poinsettias will be blooming at

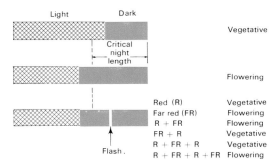

Fig. 18-15. Short-day plants grown under conditions of short days and long nights can be kept vegetative by a short flash of light during the dark period provided that the interrupting light is of the appropriate color. A red flash keeps plants vegetative, while a far red flash has the same effect as a continuous dark period. When red and far red flashes are given in sequence, the response is determined by the color of the last flash. (After A. W. Galston: *The Life of the Green Plant,* 2nd ed. Prentice-Hall Inc., Englewood Cliffs, N. J., 1961.)

Christmas time and that chrysanthemums will be available at any time from midsummer until late fall. There has even been limited use of photoperiodic controls for field crops. It is possible to obtain greater yields from sugar cane when flowering is prevented by brief exposure during the night to flood lights or flares.

ANNUALS, BIENNIALS, AND PERENNIALS

Plants may be classified according to their flowering behavior as annuals, biennials, or perennials. Annual plants usually grow from seeds, flower, and die within one growing season. Biennials grow vegetatively during the first year and flower and die during the following year. Perennials persist for many years and usually flower repeatedly. Although such classifications are consistent with what occurs in nature, many species fall into one group or the other depending on the conditions that they encounter. After all, a biennial is probably nothing more than an annual that needs a period of cold before flowers can develop. Many annuals turn out to be perennials when planted in a different climate. This is particularly true for a number of plants that are annuals in temperate climates but continue to flourish as perennials when grown in the tropics. For example, common geraniums grow to almost tree-like proportions after several years in California.

It is perhaps more meaningful to classify plants according to the number of times they flower regardless of environmental factors. **Monocarpic** plants, or those flowering only once followed by death, include such species as the common garden pea (*Pisum sativum*, a true annual in every sense), the century plant (*Agave*) that grows for 5 to 20 years before flowering, and certain bamboos that develop vegetatively for 2 to 50 years. True perennials, **polycarpic** plants, on the other hand, flower and fruit repeatedly over a period of many years and include apples, pears, and orange trees.

FLOWER DEVELOPMENT

The onset of sexual reproduction, be it flowering in the seed plants or formation of gametangia in the gametophytes of more primitive plants, often means drastic changes in the nature and products of the meristems, or growing regions. In the seed plants, instead of new leaves and internodes, the apical meristem forms grossly modified shoots called flowers. Flowering signals an end to the juvenile stage of plant development; it is also frequently associated with the onset of senescence and the ultimate death of the individual.

Once the internal and external stimuli necessary for the induction of floral primordia have been received by a plant, buds are formed that are destined to develop into the specialized shoots that we recognize as flowers. We have already examined the role of flowers as the sexual phase of angiosperm life cycles in Chapter 14. It is necessary here to review only the major features of what happens. The working parts of a flower are the male **stamens** and the female **pistils**. The showy, often very colorful **petals** and the green leaf-like **sepals** are not essential, except to attract insects or birds

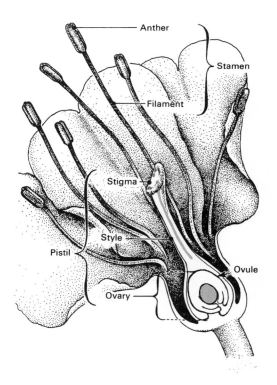

Fig. 18-16. Structure of flower parts essential for reproduction, the stamens and pistils.

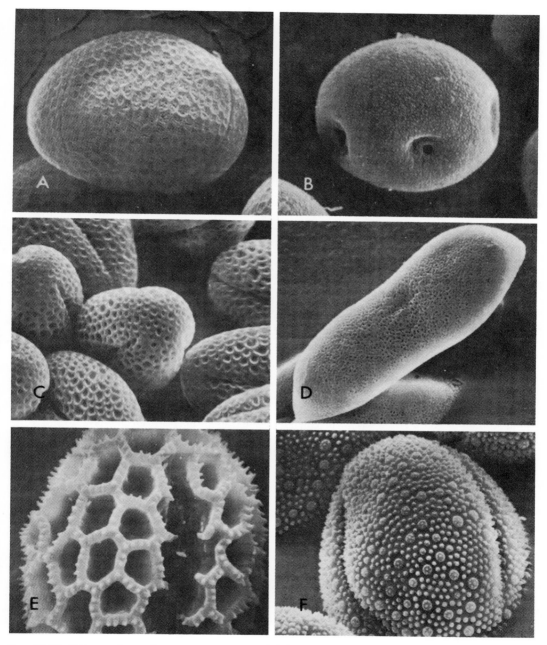

Fig. 18-17. Pollen grain surface structures as revealed by scanning electron microscopy. A, touch-me-not (*Impatiens grandiflora*), × 2,800; B, hornbeam (*Carpinus*), × 1,750; C, wallflower (*Cheiranthus*), × 1,850; D, blue-eyed grass (*Sisyrinchium bermudiana*), × 1,100; F, flax (*Linum austriacum*), × 1,000. (From P. Echlin. Pollen. Scientific American, *218* (no. 4): 80-90, 1968.)

that serve the useful purpose of disseminating pollen.

Stamens usually have an enlarged part called the **anther**, borne on a stalk called the **filament**. Within the anther, specialized cells called **microsporocytes** undergo meiosis to form haploid **microspores**. The nuclei of these haploid cells usually divide before they are released from the anther. The resulting cells are called **pollen grains**. These cells often have elaborately sculptured cell walls. Their very beautiful surface features are a factor in their distribution by animal vectors because they contribute a certain "stickiness."

Pistils are frequently vase-shaped with an enlarged base (**ovary**) connected by a tubular **style** to the **stigma**. Within the ovary there are one to many ovules containing the **megasporocytes**. These cells undergo meiosis to form **megaspores**. Usually only one of the four haploid megaspores formed from each megasporocyte develops further. Through successive divisions of the megaspore nucleus, an **embryo sac** containing eight or more haploid nuclei develops.

A unique feature of sexual reproduction in the seed plants is the double fertilization that occurs. After landing on the sticky surface of the stigma, pollen grains germinate. By this time a second nuclear division has occurred and three haploid nuclei, one **tube nucleus** and two **sperm nuclei**, enter the developing pollen tube. Extensive growth of the pollen tube (up to many centimeters) is necessary before the two sperm nuclei can be released into the embryo sac. Growth of the pollen tube appears to be guided by the tube nucleus which disintegrates when this job is completed; nutrients to sustain such phenomenal growth appear to come from the tissues of the style and ovary. Once released into the embryo sac, one sperm nucleus fuses with the egg nucleus to form the diploid zygote; the second fuses with the polar nuclei to form the **primary endosperm nucleus**. This second product of fertilization gives rise to the endosperm tissue, which is polyploid in many species. Fertilization stimulates further development of the ovule which at maturity forms the **seed**. The surrounding tissues of the ovary can enlarge. Mature ovaries, with their one to many embryo-containing seeds, are called **fruits**.

Flowers may be bisexual (containing both stamens and pistils) or unisexual (**staminate** or **pistillate**), depending on the plant species. Those plants with unisexual flowers may have both on the same plant and such plants (for example, corn) are called **monoecious**. An alternate arrangement is for male and female flowers to be borne on separate plants, and such species as hollies and the ginkgo are termed **dioecious**. With separate staminate and pistillate plants, cross-pollination and the resulting recombination of genetic traits and variability of progeny are assured. Cross-fertilization is also guaranteed in many monoecious plants (even those with bisexual flowers) by mechanisms such as self-sterility or the maturation of stamens and pistils at different times. Still other species have a flower structure that guarantees self-pollination. One classic example of such a naturally inbreeding plant is the common garden pea, *Pisum sativum,* the object of Gregor Mendel's well-known early studies of genetics.

Agents of Pollen Dispersal

Some of the most impressive examples of interaction between plants and animals are found in the adaptations of many flowers that facilitate pollen dispersal by specific animal carriers. Various plant species have flowers with colors, structures, odors, or nectar that attract particular pollinating agents, including bees, moths, butterflies, flies, beetles, birds, and bats. Often the shapes of flowers are especially adapted for pollination by a particular animal species. The elaborate structures of orchids provide fascinating examples of modifications appropriate for insect pollination that were first described over a century ago by Charles Darwin. One such modification not recognized by Darwin is a very clever trick whereby a male insect mistakes the pattern on an orchid flower for the female of his species and tries to copulate with it. Examples of this phenomenon of **pseudocopulation** are found in an Australian orchid species pollinated by a fly and in two European and North African genera pollinated by bees. The specificity of such a device for pollination insures that pollen will not be distributed indiscriminately to other species of flowers. There are examples of plants restricted

Fig. 18-18. Flower of the bee orchid (*Ophrys apifera*). Note resemblance of central part of the flower to a bee. (Photograph by R. H. Noailles, from P. Jaeger: *The Wonderful Life of Flowers*. E. P. Dutton and Co., Inc., New York, 1961.)

in their geographical distribution to the areas where their pollen-carrying agents can survive. For example, the natural range of monkshood in the Northern Hemisphere corresponds to that of the bumble bee. Alfalfa tends to be infertile in California unless appropriate kinds of bees are imported to insure pollination. Wind-pollinated species predominate at high altitudes or extreme latitudes where few insects and birds can survive. Further specific examples of plants pollinated by particular agents and their structural modifications are summarized in Table 18-1.

Recent studies have indicated that some insect-pollinated flowers have patterns on their petals that humans cannot see but that are visible to insects with eyes sensitive to ultraviolet light. Ultraviolet-absorbing compounds in these flower petals are distributed in patterns that can be seen in photographs taken with ultraviolet sensitive films or with television cameras equipped with ultraviolet transmitting lenses.

There are many examples of flower structures specifically adapted for pollination by a particular agent. Bee-pollinated flowers often have petals that serve as landing platforms where bees alight. Such flowers secrete nectar from glands at the base of a tube of petals. As a bee reaches for the nectar with its long, slender tongue, its body hairs pick up pollen from the flower's stamens. Bees tend to feed on one species of flower at a time and thus tend to distribute pollen to other flowers of the same species where it will do the most good. In bee-pollinated flowers, stamens and pistils are grouped together so that a bee simultaneously picks up pollen and delivers enough pollen from other flowers to fertilize many ovules.

PLANT DISPERSAL

Whether reproduction is by asexual or sexual means, some mechanism for the dispersal of progeny is essential to the survival of plants. After all, a plant is destined to spend its entire life in the spot where the spore or seed from which it develops happens to germinate. Unlike animals that can migrate as the seasons change or search for favorable habitats, plants flourish, exist, or perish in response to the environment that they encounter by chance. The immobile plants are dependent on factors such as wind, water, and the freely moving animals for the dispersal of their progeny. Animals frequently are rewarded for this service just as they are for their important role in pollination. Many kinds of fruits are consumed, seeds and all. Because seeds often have tough, indigestible protective layers, they are excreted in the feces, usually at considerable distances from the spot where they were eaten. Some seeds contain abundant food reserves in the form of starch or oil which are as useful to an animal consumer as they are for the plant embryo. Such seeds are frequently transported over long distances by ants or squirrels and tend to be deposited in locations where they can germinate. Other kinds of seeds are marvelously adapted for being attached to the fur of passing animals. This amazing variety of hooks, spines, and bristles that facilitate this

Table 18-1
Plant adaptations to agents of pollen dispersal

Pollinating Agent	Kinds of Flowers Pollinated	Special Features of Flowers and Their Distribution
Bees	Orchids Verbena Violets Blue columbine Larkspur Monkshood Bleeding heart Mints	Nectar Showy, bright colored petals, often blue or yellow (bees are color blind to red); often closed at night when bees do not fly Stamens and pistils grouped together
Moths	Morning glory Tobacco Jimson weed Yucca Phlox Evening primroses Orchids	Nectar Heavy fragrance Open late in the afternoon and night when moths fly Most plentiful in the tropics
Butterflies	Red columbine Fuchsia Marigolds Butterfly bush	Often red or orange Open in the daytime
Flies	Black arum Dutchman's pipe Some lilies	Dull colors Rank odors resembling the carrion, dung, humus, etc., on which flies feed Common in the Arctic, at high altitudes and in shady woods
Beetles	Magnolias Pond lilies California poppy Wild roses Dogwood Elder Spirea Buckthorn	Fruity, spicy or sweet odors Most abundant in the tropics Buried ovules
Birds	Red columbine Fuchsia Passion flower Eucalyptus Hibiscus	Colored but odorless Often red or yellow Most common in the tropics and warm temperate climates Petals often fused into a nectar-containing tube
Bats	Calabash Sausage tree Candle tree Trumpet vines	Usually open only at night Large, often white Fermenting or fruit-like odors
Wind	Oaks Grasses Grains Pines	Freely exposed stamens and pistils Often lack petals Bright colors, odors, nectar, etc. are lacking Feathery brushy or fleshy stigmas Separate male and female flowers Plants often dioecious Common in the arctic and cold temperate regions

kind of hitch-hiking is known to all who have walked through fields and woods in the fall of the year.

The wind, which is the agent of pollination

Fig. 18-19. Nectar-collecting honeybee on alfalfa blossom also picks up pollen, which is the white mass packed on hind leg. (Courtesy U.S. Dept. of Agriculture.)

of many seed plants, also serves to transport seeds. Many kinds of devices enable seeds to be carried great distances by air currents. We have all seen the parachute-like structures of dandelion and milkweed seeds carrying their tiny packets of DNA on the wind. Those of some trees such as the maples, elms, and lindens have wing-like structures that enable these relatively heavy seeds to glide with the wind. These rather large seeds are exceptions to the general rule that structures dispersed by air currents must be small and light. The microscopic spores formed by bacteria, fungi, and other primitive plant groups are natural components of the dust carried everywhere by air currents.

There are some rather remarkable ways of catapulting seeds and spores from the structures in which they are formed. As some fruits dry, their walls split open, forcibly ejecting the seeds. In the mosses and liverworts there are many examples of structures that expand and contract in response to changes in atmospheric moisture content and therby expel spores at times when the conditions for their germination are optimal. *Pilobilus,* an unusual fungus that can easily be cultured on fresh horse dung, expels its spores with considerable force,

Fig. 18-20. The marsh marigold (*Caltha palustris*) is one of many flowers that have an ultraviolet reflection pattern visible to insects but not seen by human eyes. A, flower as we see it appears evenly yellow; B, ultraviolet reflection pattern seen by insects is observable with a television camera that has an ultraviolet transmitting lens. (Courtesy T. Eisner.)

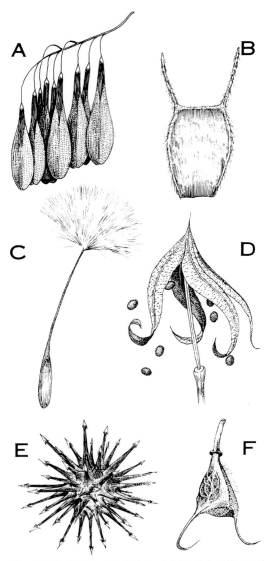

Fig. 18-21. Seeds and fruits specialized for dispersal. A, winged fruit of ash (*Fraxinus*), ×0.6; B, barbed fruit of bur marigold (*Bidens*), ×7; C, airborne fruit of cat's ear (*Hypochaeris*), ×3; D, mechanically dispersed seed of touch-me-not (*Impatiens*), ×4; E, barbed fruit of *Acaena* with grapnel-like spine tips, ×7.5; F, fruit of *Proboscidea,* ×0.3.

sending them over distances as great as six feet and always aimed, with amazing accuracy, at a source of light.

No consideration of plant dispersal would be complete without acknowledging the involvement of mankind. Since the beginnings of agriculture, humans have gathered and preserved the seeds of plant varieties having unusually good yields, desirable flavors, useful fibers, and the like. With the movement of peoples and especially since the exploration of the Western Hemisphere beginning in the 15th century, mankind has transported innumerable plant species from their places of origin to new areas. Many tales can be told about the ways in which newly introduced plant species have changed the course of history. The potato, a native of the Western Hemisphere, was probably introduced in Europe by the Spaniards. Sir Walter Raleigh is credited with the first cultivation of the potato for food in Britain in 1586. For the next 250 years, both potatoes and people prospered; in some areas such as Ireland, potato tubers became the mainstay of human diets. The widespread failure of potato crops in the mid-1800's caused by a fungus disease brought about famines and was the precipitating cause of a large migration to the New World.

The Western Hemisphere was the original home of many other plant species that have been transported by man to Europe, Africa, and Asia to be perpetuated and modified for agriculture. Tobacco, corn (maize), tomatoes, the sugar beet, and peanuts are good examples of such New World plants that have become widespread in their distribution. In fact, the agriculture of this century is truly cosmopolitan in that crops of very diverse origins are grown in most of the temperate regions. The production of soybeans, native to China, is now rapidly increasing because of their extensive use as a raw material for industries manufacturing oils, soaps, and meat substitutes. Among crops that were already cultivated by ancient civilizations, barley and coffee originated in Ethiopia, oats and wheat in Iran and Afganistan, oranges and peaches in China, rubber along the Amazon, and cotton in India. Our knowledge about the regions where plants originated comes from the detective work of plant geographers and is based on the premise that the greatest numbers of varieties of a given plant are found in its original home. With increasing distance from this center, fewer kinds are found. Information about the early use of a plant species in agriculture comes from the remains of early civilizations (excavated tombs, villages and cities), and even from ancient paintings and sculpture.

With the eras of exploration and settlement of the earth now at an end, the worldwide dispersal of edible plant species is fairly complete. Plant products are exported from those regions where they are most easily grown to wherever a market demand exists. For the future we can look to the botanical sciences for improvement of many plant species to produce higher yields and enhanced nutritional value. As remote areas become more accessible, and there is increased contact with formerly isolated peoples, new knowledge may become available of medicinally useful plants that are now restricted in their distribution. Perhaps we will see in the next century a worldwide dispersal of pharmacologically valuable plant species comparable in extent to the dissemination of edible ones in the past.

USEFUL REFERENCES

Baker, H. G. *Plants and Civilization.* Wadsworth Publishing Co., Inc., Belmont, Calif., 1970.

Briggs, D., and Walters, S. M. *Plant Variation and Evolution.* McGraw-Hill Book Co. (World University Library), New York and Toronto, 1969.

Cook, S. A. *Reproduction, Heredity and Sexuality* (Fundamentals of Botany Series). Wadsworth Publishing Co., Inc., Belmont, Calif., 1964.

Garner, W. W., and Allard, H. A. Effect of the relative length of day and night and other factors of the environment on growth and reproduction in plants. Reprinted in *Papers on Plant Growth and Development,* ed. by W. M. Laetsch and R. E. Cleland. Little, Brown and Co., Boston, 1967.

Hillman, W. S. *The Physiology of Flowering.* Holt, Rinehart and Winston, New York, 1962.

Lawrence, W. J. C. *Plant Breeding.* St Martin's Press, New York, 1968.

Mahlstede, J. P., and Haber, E. S. *Plant Propagation.* John Wiley & Sons, Inc., New York, 1957.

Schery, R. W. *Plants for Man,* 2nd ed. Prentice-Hall, Inc., Englewood Cliffs, N.J., 1972.

7

THE WORLD OF LIFE: PROTOSTOME PATTERNS

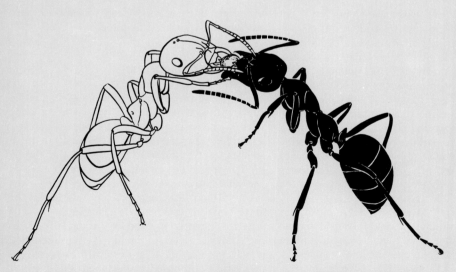

The exchange of liquids among the members of the same colony plays a key role in the social organization of most social insects—termites, wasps, bees, and ants.

(By permission from Wilson, E. O. *The Insect Societies*, The Belknap Press of Harvard University Press, Cambridge, MA., 1971.)

Nematodes and Related Animals

NEMATODES

No other group of animals so little known to the general public has anything like the ecological, economic, medical, or scientific importance of the nematodes, sometimes called roundworms. They are second only to insects in the damage they inflict on agriculture and they cause numerous debilitating and fatal diseases of man and animals. Nematodes live in the widest possible variety of environments: around the roots of grass and within the human eye, in the depths of the ocean and in glacial pools in the high Alps, and in all types of soils from swamps to sandy deserts. No species of vertebrate is known that does not harbor parasitic nematodes nor are crustaceans, mollusks, insects, or centipedes free from them. The U. S. Department of Agriculture estimates that annual nematode damage exceeds $2 billion. The crops seriously affected include beets, citrus fruits, cotton, mushrooms, peanuts, potatoes, rice, tobacco, tomatoes, and wheat. The human suffering due to nematodes, especially in underdeveloped parts of the world, cannot be measured. Nematodes are one of the classic animals for the study of chromosome behavior and have very recently become popular among biochemists working on basic problems in the development of the nervous system.

Over 15,000 species have been named, and conservative estimates place the total in excess of half a million—far more than all other animals combined except for the arthropods. The abundance of individuals is also enormous. Some 300 million nematodes of a single kind occur per acre in sugar beet fields in Utah and Idaho. At least 3,000,000 people in the United States alone suffer from infections of the nematode *Ascaris* and even more from hookworm.

The phylum **Aschelminthes** (*ascus*, sac, + *helminthos*, worm), to which the nematodes belong, also contains several small groups of relatively unimportant creatures. The best known are rotifers, semi-microscopic-forms long the delight of microscopists and students because of their jewel-like forms and lively activities. About 1,500 species are known.

Nematodes, like other members of the aschelminthes, are characterized by a cluster of traits which sharply separates them from the flatworms and, in certain respects, from all other animals. Nothing certain is known about the origin or evolutionary relationships of these tremendously successful animals.

Nematode Organization

The body of a nematode is a **non-segmented tube within a tube,** *i.e.,* a body wall, a body cavity, and a gut extending from an anterior mouth to a posterior anus located, in most species, just anterior to the tip of the tail. A highly protective **cuticle** covers the body and prevents growth so that Aschelminthes must **molt** as their size increases. As in insects, the stage between each molt is called an **instar.** The anterior end seen head-on is radially symmetrical. The body cavity is a **pseudocoel,** which means that it is not lined by a special membrane composed of flattened cells as in the true coelom of the earthworms and vertebrates.

Nematodes are easily identified as such, even though to identify species of nematodes is often difficult. Unlike either flatworms or earthworms, nematodes are encased in a flexible but inelastic cuticle so that they cannot change either their length or their diameter to any appreciable degree. As a result they move like animated wires forming undulating S's and C's.

Nematodes range in size from microscopic to over two feet in length. Most soil and aquatic species are semi-microscopic, many parasitic species grow to be an inch or so long, and some, like the common *Ascaris,* which inhabits the intestine of man and many of his domestic animals, grow to be a foot long. The sexes are separate except in rare instances. The gonads are the only organs lying in the body cavity.

The Anatomy of Success

Functional Systems

Judged by their success, nematodes must possess one of the most efficient body organizations known. The skeletal, muscular, excretory, and, to a large extent, the nervous systems are all built into the body wall. The **cuticle** is virtually an exoskeleton providing both protection for the worm and support for the muscles of locomotion. It is a noncellular layer composed of a horny material, **keratin**, that is extremely resistant to solvents and digestive enzymes. In parasitic species the cuticle is organized in five or six specialized layers. In many nematodes the cuticle bears various external projections and markings. Immediately beneath the cuticle is the epidermis (sometimes called the hypodermis or subcuticle), which secretes the cuticle.

The **muscular system** of a nematode is very different from that found in any other phylum. The muscle cells lie immediately under the epidermis and are divided into four quadrants by a mid-dorsal and a mid-ventral nerve cord and a right and left excretory duct. All the muscle fibers are longitudinal; there is no antagonistic set of circular muscles as in all other kinds of worms. The muscle cells are giant cells of a fixed and relatively small number, usually about 24 per quadrant, although the large parasitic genera like *Ascaris* may have a total of 150 muscle cells per quadrant. The contractile fibers lie in the periphery of the cytoplasm of these big muscle cells, always on the side adjacent to the epidermis and in many species extending well up on the sides of the cells, making a U-shaped structure in cross section.

The common relationship between muscles

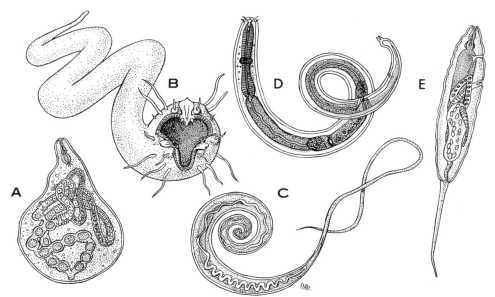

Fig. 19-1. A variety of nematodes. A, gourd-shaped adult female of the rootknot nematode; B, a carnivorous species; C, the intestinal whipworm; D, an earth-dwelling species; E, the tailed nematode. (**Highly magnified.**)

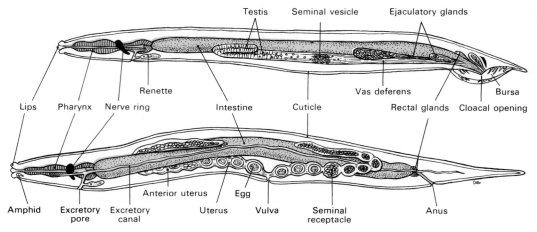

Fig. 19-2. Structure of a generalized nematode: male (above) and female (below).

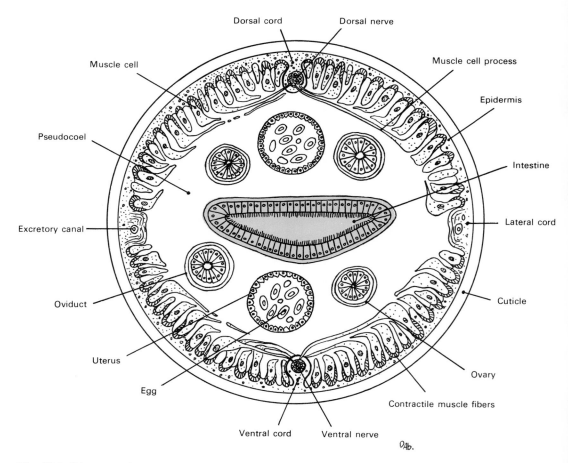

Fig. 19-3. Diagrammatic cross section of a nematode. Note that in marked contrast to the rest of the animal kingdom, muscle cells send extensions to the nerve cords.

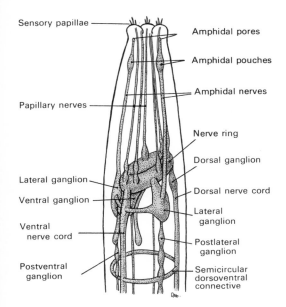

Fig. 19-4. Nervous system of a nematode head. More than 50 per cent of the cells of the entire body are nerve cells.

Labels on figure:
Sensory papillae — Amphidal pores — Amphidal pouches — Amphidal nerves — Papillary nerves — Nerve ring — Dorsal ganglion — Lateral ganglion — Dorsal nerve cord — Ventral ganglion — Lateral ganglion — Ventral nerve cord — Postlateral ganglion — Postventral ganglion — Semicircular dorsoventral connective

and nerves is reversed in the nematodes. Instead of nerves extending from a brain or nerve cord to the muscles, a long protoplasmic fiber extends out from each muscle cell to the nerve cord. The processes of the muscle cell in the two dorsal quadrants extend to the dorsal nerve cord; those of the two ventral quadrants extend to the ventral nerve cord. The dorsal and ventral muscle cells act as a pair of antagonists. There is evidently no neural mechanism making it possible for the two right or the two left quadrants to contract together. Consequently, nematodes can bend and lash only in a dorsoventral plane, although as soon as they begin to do so it causes them to lie on their sides unless they are in a fairly thick medium. There is, however, some mechanism enabling anterior muscle cells to contract separately from posterior ones. Consequently, an undulatory motion is possible and can easily be seen in nematodes.

The **nervous system** of a nematode is complex and evidently plays an important role in the nematode way of life, for more than 50 per cent of the cells of the entire animal are nerve cells. In *Turbatrix,* the common vinegar eel, 251 cells are nerve cells out of a grand total

of 432 cells in the entire body. The only part of the complex nematode nervous system that can be seen in a living worm is the **circumpharyngeal nerve ring.** This encircles the pharynx or esophagus and looks like a doughnut worn at a jaunty angle.

The **sense organs** include the amphids, situated like ears on either side of the head, various sensory papillae and bristles, a pair of eyespots in a few species, and possibly the phasmids, a pair of strange cells, one on each side of the body near the tail. Cilia, otherwise absent in nematodes, have been seen by the electron microscope in some of these sensory structures.

The **excretory systems** of nematodes are extremely simple, at least anatomically. In the mid-ventral region, at about the level of the nerve ring, there is a single giant excretory cell called the **renette.** It opens to the exterior via a mid-ventral pore. In the larger parasitic species there is, in addition, a pair of excretory tubules running the length of the worm, one in each of the lateral cords and connected to the excretory pore by a cross tube. Very little is known about how these structures function.

The **digestive system** follows a similar pattern in all nematodes. The mouth, however, shows considerable variation. As with insects and mammals, this is correlated with the diet. Many are carnivorous, eating protozoans, small annelids, rotifers, tardigrades, and even other nematodes. Others are herbivorous, living especially on roots; some live on decaying material; and many are parasitic. Thus, some species have relatively smooth lips and merely suck in their food while others have three or more strong chitinous jaws; still others have sharp spears with muscles to push them out of the mouth and into plant or animal tissues. The first part of the gut is a muscular sucking region called by some workers the **pharynx,** by others the **esophagus.** It may have one or two muscular bulbs which are sometimes provided with teeth. The intestine is usually a straight tube without bends or attached glands. There are no circulatory or respiratory systems.

Reproduction and Development

The reproductive system of nematodes is one of the most efficient known. A female *Ascaris* in the human intestine, for example, lays eggs at the rate of about 200,000 per day. The gonads are ideally structured for the study of chromosome behavior during meiosis, gamete formation, fertilization, and early development. The testis and vas deferens form a continuous tube to the cloaca through which sperms are ejaculated. The ovary, oviduct, and vagina likewise form a continuous duct. At the innermost end in each case are primordial germ cells which proliferate by mitosis. Farther down the tube are primary spermatocytes or oocytes in the first meiotic division, followed by secondary spermatocytes or oocytes in the second meiotic division. Then come sperms which, in marked contrast to other animals, are amoeboid rather than flagellated, or eggs. In females all the stages of fertilization, cleavage, and embryonic development can be easily observed. Thus all these events can be seen in sequence like reading a line of print.

In many species the eggs contain small worms when laid. Very commonly the larvae or juveniles pass through five instars or stages separated by four molts or ecdyses before becoming adult, but there is little change in body form. In parasitic species like the hookworm or the frog lung nematode, the larval forms are usually the ones which first enter the host.

BRYOZOANS

The **bryozoans** are utterly unfamiliar to most people, yet they are extremely common in both marine and freshwater environments. Hence, they must play a significant role in the ecology of their habitats. As their name implies (*bryon*, moss), these animals bear some resemblance—extremely superficial, of course—to the mosses. They are found covering rocks and shells along seacoasts and covering submerged logs and twigs in clear, quiet freshwater ponds. Some of the freshwater colonies secrete large masses of jelly a foot or more in diameter within which the colony lives. Other species grow in a branching, almost vine-like manner. They are found in highly polluted water but are rare when the dissolved oxygen falls below 30 per cent saturation. Most species avoid bright sunlight, and many can grow in shaded spots such as under ledges or in water pipes. The

marine species form either a tightly packed and very fine mosaic over the rocks or a very finely branched growth resembling seaweed. They are of almost no importance to man, although in some places the large jelly masses produced by freshwater species have clogged the intake grates of hydroelectric plants or created problems in water pipes.

Under a dissecting microscope the most conspicuous feature of bryozoans is their **crown of ciliated tentacles**. In most marine species this is a single circle; in freshwater species it is usually a double row arranged like a horseshoe. The whole crown can be rapidly jerked into the protective tube, or cecum, within which the animal lives. The cilia of the tentacles sweep algae, protozoans, small crustaceans, and the like into the mouth at their base. The alimentary canal is U-shaped or Y-shaped,

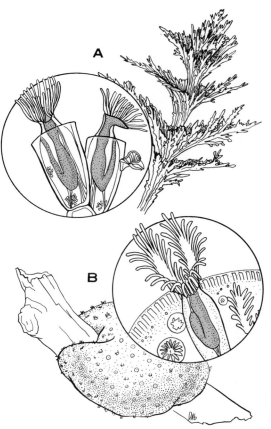

Fig. 19-5. Bryozoans, life-sized and highly magnified. A, a marine species; B, a freshwater species.

with the anus outside the crown of tentacles. There is a true coelom, one or more retractor muscles, and an ovary and testes, since most species are hermaphroditic.

ROTIFERS (TROCHELMINTHES)

The **Rotifers** or Trochelminthes (*trochos,* wheel, + *helmins,* worm) are semi-microscopic animals with a structure somewhat resembling a short, fat nematode bearing a pair of ciliated disks on its head. The motion of the cilia gives the illusion of revolving wheels so convincingly that the pioneer microscopist, Leeuwenhoek, thought that they actually revolved. Their lively behavior under a microscope have made them favorites of students and investigators, but the ecological, economic, medical, or other importance of rotifers is negligible. They have a puzzling life history involving many generations of parthenogenetic females, followed by the appearance of males and sexual females, apparently when environmental conditions be-

come unfavorable. Rotifers can be used as indicators of the pH of the water. Some species prefer acid, some neutral water.

BRACHIOPODS (PHYLUM BRACHIOPODA)

The **brachiopods** are notable because they are so abundant as fossils in extremely ancient rocks. In fact, they are widely useful as indicators in typing rocks and identify the period to which they belong. They are abundant in rocks of all ages, from the earliest Cambrian to the present. In each age there is a characteristic brachiopod fauna. From age to age some species became extinct, others appeared, others that were rare became abundant, and *vice versa.* A remarkable fact is that the two very different present-day forms have existed almost unchanged during the entire course of evolution from the beginning of the Paleozoic era.

Brachiopods look superficially like bivalve

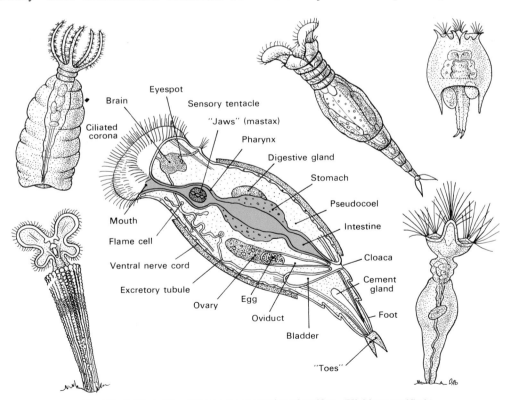

Fig. 19–6. Diversity of form among species of rotifers. (Highly magnified.)

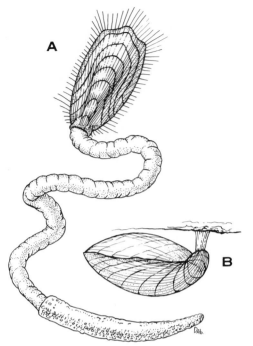

Fig. 19-7. The two basic forms of brachiopods. A, *Lingula* with horny shell and long tube, a sand liver; B, *Terebratulina*, the "lamp of learning" brachiopod attached by its "wick" to the underside of a rock. (Approximately life size.)

mollusks, but on opening the shells, instead of the muscular foot and fleshy visceral mass of a clam, the interior of the shell seems empty except for a pair of coiled, ciliated arms called

lophophores, which resemble watch springs when seen from above. Furthermore, the pair of **shells** of a brachiopod are dorsal and ventral, while in clams one shell is on the right and one on the left. In other words, if you cut a clam in the plane passing between the two shells, the clam will be cut into a right and left half which are mirror images of each other. To cut a brachiopod into two symmetrical halves, it is necessary to cut each shell in half and cut through the attachment stalk as well. Lining the shell of a brachiopod is a **mantle** which secretes the shell. The structure of the shell is different from that found in mollusks, and in one group of brachiopods it is largely chitinous. Protruding from the edge of the mantle are golden chitinous **setae** like those of the polychaetous annelids.

USEFUL REFERENCES

Bird, A. F. *The Structure of Nematodes*. Academic Press, New York, 1971.
Goodey, T. *Soil and Freshwater Nematodes*. John Wiley & Sons, Inc., New York, 1951.
Hyman, L. H. *The Invertebrates*, Part III: Acanthocephala, Aschelminthes, and Entoprocta. McGraw-Hill Book Co., New York, 1951.
Lee, D. L. *The Physiology of Nematodes*. W. H. Freeman and Co., San Francisco, 1965.
Pennak, R. W. *Fresh-water Invertebrates of the United States*. The Ronald Press Co., New York, 1953.
Sasser, N. J., and Jenkins, W. R. (eds.) *Nematology*. University of North Carolina Press, Chapel Hill, 1960.

Annelids

The **earthworms** and other annelids constitute a phylum of only modest size (about 8,000 species) but are of considerable ecological and scientific importance. As Charles Darwin pointed out long ago, earthworms mold the surface of the ground and profoundly affect local soil conditions. By weighing the castings of earthworms found after rain within a square yard, Darwin was able to calculate that they bring 2 to 18 tons of earth to the surface of each acre per year. Actual amounts depend on soil type, moisture, and other climatic factors such as temperature. More recent investigations show that in certain tropical regions as much as 100 tons per acre per year may be brought to the surface. Whenever present, earthworms are a major factor in soil building.

The marine annelids, called **polychaetes**, are provided with gills and are often brightly colored. Their eggs have long been favorite material at seaside laboratories for research on fertilization and early development. Earthworms also have been much used in studies on regeneration. Both terrestrial and marine annelids have served as the experimental material for investigations ranging from the transmission of a nerve impulse to sexual reproduction. Earthworms are carriers of a tapeworm of chickens and a lung worm of pigs, but no annelids are known to transmit or cause a human disease.

All annelids are segmented with a **body plan** consisting of a tube within a tube. The gut runs straight from mouth to anus and the cavity between it and the dermomuscular body wall is lined with an epithelial membrane, the peritoneum, as in vertebrates. Such a cavity is known as a coelom. As in the crustaceans and insects, there is a **ventral ganglionated nerve cord** extending the length of the animal and connected by a pair of nerves which encircle the pharynx to the brain which lies dorsally. The cleavage of the egg is spiral, producing a beautiful jewel-like embryo. The aquatic larva, when present, is a minute, top-like ciliated creature called a **trochophore**.

ANNELID DIVERSITY

The approximately 8,000 species of annelids fall into several very small groups of uncertain status plus three large classes: the **Polychaeta** which are almost exclusively marine, the **Oligochaeta** which are freshwater or terrestrial and include the earthworms, and the leeches or **Hirudinea**.

Polychaetes can usually be identified as such because each segment bears a pair of leg-like lateral appendages, **parapodia**, bearing more or less conspicuous chitinous bristles, **setae**. Many possess large colorful **gills** and tentacles. The sexes are separate. Of the two orders of polychaetes, members of the first, the **wandering species** or errantiformes, swim or crawl freely in mud burrows on oceanic bottoms, in coral reefs, or wharf pilings. One segment is much like another except for the head, which commonly has palps, chitinous jaws, and, in some species, large well-developed eyes. Familiar examples include the clam worm, *Nereis,* and the famous palolo worm, *Eunice,* which swarms out of the coral reefs of the South Pacific during the fall breeding season in such numbers that the natives celebrate with a

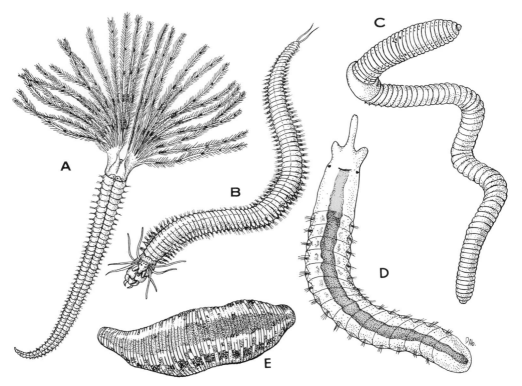

Fig. 20-1. Variety of form among annelids (not to scale): A, *Sabella,* a tube-living polychaete with fan of gills; B, a sand-dwelling polychaete, the clam worm, *Nereis (Neanthes);* C, *Lumbricus,* a common earthworm and oligochaete; D, *Stylaria* (greatly enlarged), a transparent freshwater oligochaete; E, *Hirudo,* the medicinal leech, a hirudinean.

feast at which the featured dish is palolo worm. Several small freshwater forms have been reported from the rivers and lakes in North America and Russia.

Members of the second order, the **sedentary species** or sedentariformes, spend their adult lives in tubes from which their gills or tentacles may be extended. Jaws and palps are lacking but minute eyes may be present on the gills. Familiar examples include the fan worm, *Sabella,* sometimes seen as a showpiece in marine aquaria, and *Chaetopterus,* a brilliantly luminescent worm with elaborate parapodia that lives within a U-shaped parchment tube that, except for its tips, is buried in the mud.

The **oligochaetes,** such as the earthworms and their more or less transparent and much smaller freshwater relatives, lack parapodia, possess only a **few setae** (hence the name oligochaete), are **hermaphroditic,** and bear,

when in the breeding condition, a swollen, ring-shaped glandular region, the **clitellum,** which secretes a proteinous capsule over the fertilized eggs.

There are many species of earthworms or angleworms but only two are easily identified. *Lumbricus terrestris,* the big night crawler or dew worm, can be recognized because the tail end of the worm becomes noticeably flattened when the worm crawls backwards and because the creased edges of the prostomium (a kind of proboscis) extend back to touch the second segment. *Eisenia foetida*, called the "brandling" by Izaak Walton, is also known as the tiger earthworm because of its transverse cinnamon brown stripes, one to a segment. It is common around barnyards but rare in gardens. The tiny transparent freshwater species, like *Nais,* are common in ponds. *Tubifex,* which is often a sign of some kind of organic pollution, is intermediate in size, blood red in color, and lives with its head buried in soft mud and its tail waving in the water. They are sometimes so numerous that they form red mats on the bottom of ponds and can often be bought as food for tropical aquarium fishes.

Leeches form two groups: those with and those without teeth. The former are often found on freshwater snails. The toothed leeches attack vertebrates and include the large medicinal leech, *Hirudo,* once used by physicians for bloodletting and now sometimes as a source of the anticoagulant **hirudin.** In many ponds such large species can be found attacking turtles, fishes, and mammals.

The very small aberrant groups of annelids include *Bonellia,* which is famous for what is probably the most extreme form of **sexual dimorphism** known. The female is about the size of a walnut, but the male is a tiny, ciliated creature about the size of a paramecium which lives most of its life within the female genital tract. *Bonellia* larvae that develop in isolation become females; those developing near a female become males.

ANNELID ORGANIZATION

The skin of the larger annelids is covered by a cuticle of collagen, a fibrous protein. Its changing structure during the growth of the worm is now under investigation with the electron microscope, in part because changes in collagen are believed to play an important role in aging and in certain diseases in man. Beneath the epithelial cells, which secrete the collaginous cuticle, is a layer of circular **muscles** and, under them, a layer of longitudinal muscles. Here is seen the familiar pattern of muscular organization from coelenterates to man, namely pairs and sets of antagonists. The circular muscles make the worm long and thin; the longitudinals make it short and stout. The **coelomic fluid** provides what rigidity annelids possess when the muscles contract. Thus, it acts as a skeleton. When the muscles relax, the pressure is released from the fluid and the worm is flaccid.

In all but a few species there is a closed **vascular system** consisting of a pulsating dorsal blood vessel just above the intestine and several pairs of sausage-shaped hearts in the anterior

segments which pump blood down into a ventral blood vessel below the gut. Blood flows toward the head in the dorsal vessel. The rhythmic contractions of the blood vessels and hearts are affected by the same drugs that affect the human heart and must therefore be under nervous control—a good illustration of the underlying unity of life.

The exchange of O_2 and CO_2 between blood and tissues and between blood and the water or moist soil in which the worms live occurs through **capillaries.** In most species there is red hemoglobin consisting, as in vertebrates, of an iron-containing porphyrin complexed with a protein. Like vertebrate **hemoglobin,** it can be made incapable of carrying O_2 by poisoning with carbon monoxide. In two or three polychaetes such as *Glycera,* often used for fish bait, the hemoglobin is confined within red blood cells. Several species contain green respiratory pigments called chlorocruorins,

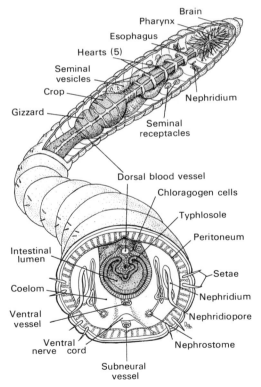

Fig. 20-2. Typical annelid organization—actually a group of mutually supporting organ systems packaged in a dermomuscular tube.

which are iron porphyrins of high molecular weight. Their evolutionary or functional meaning is unknown.

The **diet** of annelids is vegetarian. Earthworms ingest leaves which they often pull into their burrows, thus aiding the soil-building process. The marine polychaetes are either filter feeders, straining out food particles on their gills, or vegetarians eating larger algae. The **digestive tube** usually begins with a muscular pharynx which enables the worm to ingest earth and food. An esophagus carries the food from the pharynx past the hearts and (in earthworms past the reproductive organs) to a crop. This is immediately followed by a muscular gizzard which opens into the intestine. The earthworn esophagus bears numerous glands that secrete calcium carbonate granules. A long series of investigators from the day of Darwin up to the present have studied these glands without finding a convincing reason for their existence. Do they neutralize some acidity? Do they help grind up leaves? When the answer ultimately comes, it may be trivial or it may open up important new insights into animal physiology. Lying along the intestine are masses of yellowish-brown cells, called **chlorogogue tissue**, that probably correspond to the liver of vertebrates. There is evidence that chlorogogue cells store glycogen and lipids and give off ammonia and urea.

The **excretory system** of annelids has been the subject of much interest, in part because an annelid kidney or **nephridium** bears a striking resemblance to a nephron, the structural unit of the vertebrate kidney. There is usually a pair of nephridia in each segment, although in some tropical forms there are many in each segment. Each nephridium begins with a ciliated funnel which collects coelomic fluid from one segment and passes it into a triple-coiled tubule in the postjacent segment as shown in Figure 20-3. Note the opportunity for the well-known principle of countercurrent exchange to operate as tubules fold back on themselves on the way to the storage bladder. The functional significance of this in the earthworm is still unclear. The tubules are highly vascularized as is true in the vertebrate nephron.

The Nervous System

The annelid nervous system is the prototype of the system found in all the mollusks and arthropods, *i.e.*, in most animals. A dorsal **brain**, the suprapharyngeal ganglion, connects by a pair of **circumpharyngeal nerves** to a **ventral chain of ganglia** running the entire length of the animal. Peripheral nerves reach muscles, skin, tentacles, and sense organs. There is also a **visceral nervous system.** As with the vertebrate autonomic nervous system which serves the viscera, this subsidiary system is directly connected to the central nervous system. It should be noted that an earthworm can be taught to learn a T maze and will retain this information after the removal of the brain. However, after a new brain has been regenerated, the previously learned behavior is lost. Evidently the new brain cells "know" nothing about T mazes and impose their "ignorance" on the rest of the nervous system.

Three notable features of the annelid nervous system are the **neurosecretory cells**, the giant nerve fibers, and the sense organs. The neurosecretory cells stain in the same way as those of man, and apparently have a similar function in regulating various visceral responses. The **giant nerve fibers** (three in earthworms, more in some polychaetes) extend the entire length of the worm in the ventral nerve cord. These conduct impulses with great speed, permitting the entire worm to contract simultaneously and thus jerk back into its tube or burrow, the chief defensive response against predators. Each giant fiber is composed of units one segment in length. Where two units abut there is a giant,

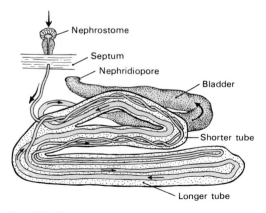

Nephrostome

Septum

Nephridiopore

Bladder

Shorter tube

Longer tube

Fig. 20-3. Earthworm nephridium, as seen under high power of a light microscope.

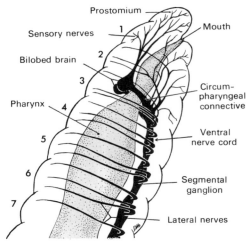

Prostomium

Sensory nerves — 1

Mouth

Bilobed brain — 2

3

Circum-pharyngeal connective

Pharynx — 4

Ventral nerve cord

5

6

Segmental ganglion

7

Lateral nerves

Fig. 20-4. Earthworm nervous system.

flat synapse remarkable for its size, the speed with which it transmits, and for its ability to transmit in either direction. Thus the worm can escape the robin at its head or the mole at its tail.

Some polychaetes which swim freely as adults, have large well-developed **eyes** complete with lens, direct retina, and pigment cup. A few annelids possess statocysts consisting of deep, flask-shaped inpocketings of the skin containing sand grains. Thus both light and gravity detectors are present.

REPRODUCTION AND DEVELOPMENT

Probably the most widely known and in-structure reproductive behavior of annelids is seen in the breeding swarms of such species as the palolo worm of the South Pacific. Most of the year these worms crawl about in the crevices of the coral as asexual forms called **atokes**. They continue to grow in length by the addition of new segments in a zone of cell division immediately anterior to the anal segment. After a time a new head with large eyes forms roughly in the middle of the worm. Behind it the sturdy crawling parapodia of the atoke become flattened swimming organs; in some species, eyes even develop on each parapodium, and in all species gametes develop. The posterior sexual half of the worm, called an **epitoke**, now breaks free and swims off close to the moonlit surface of the ocean to reproduce

or perhaps to provide food for an annual Polynesian feast. The pedestrian atoke crawls back into the reef to grow another epitoke.

The time of the reproductive swarming is synchronized by an **annual rhythm** limiting it to one, or in some species two, months of the year, a lunar rhythm which places the activity within a certain day of the month, and a diurnal rhythm placing it within a certain time after sunset. It has been claimed that approximately 90 per cent of the palolo worms which swarm do so within a two-hour period once a year. Whether or not these periodicities are internal self-sustaining or due purely to the cumulative effects of environmental influences is still under investigation.

The control of such rhythms is very poorly understood. It is known that the development of the sexual epitoke is somehow controlled by a hormone formed in the brain. If the brain is removed from an atoke, epitokous character-istics appear. Likewise, if the brain of a young atoke is implanted into a mature one, epitokous traits are inhibited. Control of reproduction by cells in or close to the brain will be found in many other places in the animal kingdom, including man.

In **polychaetes** the gametes, either eggs or sperms, are merely shed into the sea water. Hence, synchrony is of great importance be-cause the ocean is so immense and the chance of gametes meeting is so small. In most polychaetes there are no permanent gonads. Sex cells develop from the peritoneum lining the coelom. Gametes may be shed through the nephridia, via special reproductive tubules, or the body wall may simply disintegrate. The fertilized eggs develop into tiny ciliated, top-shaped trochophore larvae which meta-morphose into segmented adults.

In earthworms there are fixed gonads, a pair of **ovaries** attached to the anterior wall of the 13th segment, and two pairs of **testes** on the anterior wall of the 10th and 11th segments in most species. Copulation in earthworms varies slightly from species to species, but in all the worms lie side by side with their heads in opposite directions while sperms are exchanged between the male gonopores and the sperm receptacles of the partner worm. The worms separate and for many weeks thereafter each worm can fertilize its eggs with sperms obtained

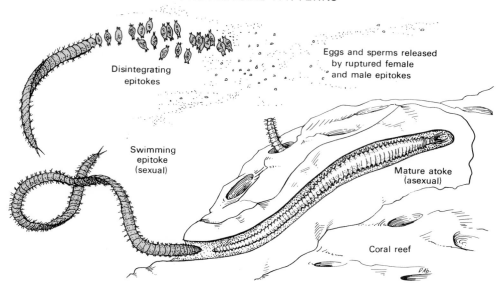

Fig. 20.5. Palolo worms during the annual breeding swarm of this polychaete from a coral reef. (Life size.)

during copulation. A slime tube is secreted around the anterior portion of the worm and the **clitellum** secretes a tough, proteinaceous encircling band. The eggs and sperm are extruded beneath the slime tube and come to be enclosed in this band. The band slips off the worm and constricts at each end, forming a **cocoon** about the size and shape of a grain of wheat in which anywhere from one or two to eight or ten wormlings will develop. Leeches are also hermaphroditic and possess a clitellum which secretes a cocoon in much the same way as is done by earthworms.

The small freshwater oligochaetes rarely reproduce sexually. Instead, a single segment somewhere near the middle of the worm experiences an explosion of cell divisions until it becomes jammed with nuclei. The posterior portion of that **fission segment** then forms the four or more head segments of a new worm, while the anterior portion of the segment becomes many small tail segments. The original

worm then breaks into two: one with a new tail and old head, and one with a new head but an old tail. Almost nothing is known about the neural, hormonal, physiological, or biochemical control of these amazing phenomena.

USEFUL REFERENCES

Barnes, R. D. *Invertebrate Zoology.* W. B. Saunders Co., Philadelphia, 1963.

Brinkhurst, R. O., and Jamieson, B. G. M. *Aquatic Oligochaeta of the World.* University of Toronto Press, Toronto, 1971.

Buschsbaum, R. *Animals Without Backbones,* rev. ed. The University of Chicago Press, Chicago, 1948.

Buschsbaum, R., and Milne, L. J. *The Lower Animals: Living Invertebrates of the World.* Doubleday and Co., Garden City, N. Y., 1960.

Dales, R. P. *Annelids.* Hillary House, New York, 1963.

Edmondson, W. T. (ed.) *Ward and Whipple's Fresh Water Biology,* 2nd ed. John Wiley & Sons, Inc., New York, 1959.

Laverack, M. S. *The Physiology of Earthworms.* The Macmillan Co., New York, 1963.

Mollusks

Mollusks have been important as food since well before the beginning of written history. This is known by the enormous piles of shells found making up the kitchen middens or refuse piles of many Stone Age peoples in Europe. The American Indians, who were living in a Stone Age culture before the arrival of Europeans, also left huge piles of clam and oyster shells along the coast of North America. Along San Francisco Bay there are enormous Indian shell mounds estimated to represent an accumulation extending back over 3,000 years. In North Carolina some mounds are so large that they serve as a source of road-building material. The royal purple of the ancient Greeks and Romans was obtained from a gland in a Mediterranean snail, *Murex*, which secretes a yellow pigment that turns intense purple upon exposure to light. Huge piles of these shells are still to be seen on the shores of Syria and adjacent regions. Today the *United Nations Economic Yearbook* records the total world harvest of mollusks in millions of tons annually.

Mollusks form a well-defined phylum of some 70,000 living and 25,000 fossil species including animals as different as a clam and an octopus. They are all bilaterally symmetrical with rudimentary or no segmentation and soft bodies (*molluscus*, soft) often enclosed in a calcareous shell. The body is typically composed of four parts. A **visceral mass** contains the digestive and reproductive organs. There is a ventral **muscular foot**, a **head** (absent in some groups), and a **mantle** of skin which secretes a shell in most species. The flexible muscular tongue called the **radula** bears long parallel rows of minute teeth (in an octopus the radula is within a beak like that of a parrot).

Modern studies indicate that mollusks are related to annelids notably because the most ancient forms are segmented and possess a well-developed coelom which is greatly reduced in other mollusks. The larval form of mollusks, when present, is a **trochophore**, as in annelids. The trochophore may metamorphose directly into the adult or, especially in marine species, become a free swimming **veliger** with two prominent ciliated lobes on its head.

There are now six known living classes: the very ancient Monoplacophora, the rock-dwelling chitons (Amphineura), the elephant tusks (Scaphopoda), the snails (Gastropoda), the bivalves (Pelecypoda), and the squids and octopuses (Cephalopoda).

THE MONOPLACOPHORA

To the amazement of zoologists, a population of living monoplacophorans (*mono*, single, + *plakos*, flat surface, + *phora*, bearing), was discovered by a Danish oceanographic vessel in water 3,590 meters (roughly 2 miles) deep off the west coast of Mexico. These organisms are more ancient than the coelacanth fish found two decades ago in deep water north of Madagascar off the east coast of Africa. The coelacanths were supposed to have been extinct since the end of the Mesozoic period (the Age of Reptiles) 70 million years ago. The most recent fossils of the primitive monoplacophoran mollusks date from the Silurian period, over 300 million years in the past. No wonder there is no popular name for them.

As the word monoplacophoran suggests, these ancient animals bear a single, broad, rounded shell somewhat like the shield of a

Fig. 21.1. The octopus, the timid terror of the seas which fills the ecological niche in the ocean that the tiger does on land, a loner which hunts by stealth. (Courtesy D. P. Wilson.)

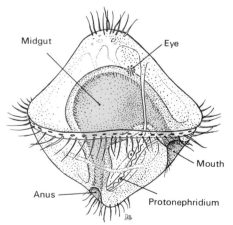

Fig. 21-2. The trochophore larva of a typical mollusk as seen under high power of a light microscope.

classical Greek warrior. Under this shell the animal is bilaterally symmetrical with an anterior mouth, a strong radula in the pharynx, and a posterior median anus. On each side of the flat, rounded, snail-like foot is a row of gills, forming five pairs in all. Internally there are five pairs of nephridia and five pairs of gill hearts. There are five pairs of strong retractor muscles for the foot plus three more pairs in the "head." The body is clearly in five segments and the head evidently represents three more segments. The coelom is well developed and the sexes are separate, as in polychaetes. No copulatory organs have been found; presumably gametes are shed directly into the water. The specimens collected are in the Zoological Museum of the University of Copenhagen.

THE CHITONS (AMPHINEURA)

Members of the class Amphineura (*amphi*, double, + *neuron*, nerve) are usually found in shallow water along the seacoast where they eat the algae that grow over rocks. The adults vary, according to species, from about 1 to 30 centimeters in length. Their oval body is protected dorsally by eight transverse shell plates which give the chitons a **segmented appearance**.

The body plan of a chiton may be taken as typical of mollusks. When seen in longitudinal section, the entire body is covered by a mantle, which covers the upper surface of the chiton and lies immediately under the shell plates it has secreted. As its name suggests, the mantle hangs down like a cloak on either side of the body. Extending under the lower surface of the chiton is a thick muscular foot. The space

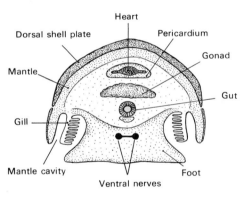

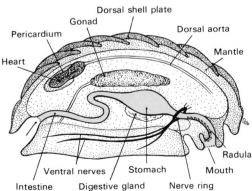

Fig. 21-3. A simple rock-dwelling mollusk, the chiton, shown in diagrammatic cross- and longisections.

between the mantle and the visceral mass and foot, known as the mantle cavity, opens freely to the sea water. At the anterior end of the animal is a short proboscis with a mouth containing the characteristic radula. The ribbon of fine teeth on the radula acts like a flexible file to rasp off algae growing on rocks and other objects. Within the body above the gut is the **heart**, lying in a small cavity, the pericardium. The **gonad** lies above the gut and opens into the mantle cavity via a pair of ducts that end above the edge of the foot near the posterior end of the body. The sexes are separate. The larvae are free swimming trochophores.

The **nervous system** consists of a circumesophageal ring from which a pair of ventral nerve cords extend back through the foot, and a pair of lateral nerves extend back on either side of the body. The **gills** can be seen by turning the animal upside down or by examining a cross section. They hang down in the mantle cavity between the mantle and the visceral mass and foot. This is also the normal position of the gills in clams and other mollusks. Chitons have from 6 to 80 pairs of gills.

The muscular foot on the large species of chitons found in the West Indies is eaten as "sea beef."

THE ELEPHANT TUSK SHELLS (SCAPHOPODA)

Members of the class Scaphopoda (*skaphe*, boat, + *podos*, foot) are the least important group of mollusks. In the past, their shells were extensively used as money and ornaments among American Indians of the west coast. Along the east coast of the United States scaphopods are known to occur only in deep quiet water on the bottom of a few Maine coves.

The edges of the scaphopod mantle are fused together along the mid-ventral line, thus enclosing the animal in a tube open at both ends. The mantle tube is slightly curved and just small enough at the posterior end to make the shell it secretes resemble a miniature ivory tusk. A somewhat knob-like foot protrudes from the anterior end along with several sensory ciliated tentacles which conduct food to the mouth and radula. Gametes reach the exterior through the right nephridial duct.

THE SNAILS AND OTHER GASTROPODA

The snails and their relatives, members of the class Gastropoda (*gaster*, belly, + *podos*, foot), are an enormous and important group of animals. Anyone who wishes to make a complete collection of snail shells from all over the world will have to accumulate some 50,000 species. In Asia and Europe snails are regarded as edible delicacies. In semi-tropical regions several species can become destructive to citrus fruits and other crops. In oyster beds a small carnivorous snail, *Urosalpinx*, the oyster drill, can cause extensive and costly damage.

The structure of a typical gastropod is essentially like that of a chiton (described above), except that the main body mass, the so-called visceral mass, containing the intestine and gonads along with the heart and nephridia, has been pulled up into a great hump, usually coiled. Within the mantle cavity is a gill-like organ of taste and smell, called the **osphradium.**

In the more primitive snails and their bizarre relatives, the intestine is more or less straight and the anus and gills are posterior in position. However, in the overwhelming majority of gastropods, the larva undergoes a curious asymmetrical growth known as **torsion.** By this process the left side of the visceral mass grows more than the right, causing the body to rotate counterclockwise with respect to the head and foot through 180° when viewed from above. In the end, the visceral mass with the intestine is twisted around a half-circle so that the anus along with the gills is brought forward on the right side. The nerve cords are thrown into a crossed loop. The evolutionary significance of this curious maneuver may be the advantage derived from bringing the gills forward in the shell.

The coiling of the visceral mass, as seen in the familiar snail shell, is a separate process which primarily occurs later in development. Instead of growing higher and higher into a long top-heavy point, the visceral mass merely winds around an axis at right angles to the axis of torsion.

Freshwater and terrestrial snails and slugs are called **pulmonates** because they breathe by means of a lung, consisting of the mantle cavity which in other mollusks is widely open to the water. Although cross-fertilization is the rule, pulmonates are hermaphrodites. Most are herbivorous. Marine snails possess gills, are mostly carnivorous, and have separate sexes. The eggs are enclosed in leathery capsules often attached in strings. The larger species are widely used in chowder and as fish bait.

There are also two very strange groups of gastropods which continue to baffle zoologists. The sea butterflies or **pteropods** are blue or pink creatures only two or three centimeters long—but they swim through the polar seas in such enormous swarms that they color the ocean for miles and constitute one of the major foods of the large whales. The other group, called **nudibranchs**, crawl over seaweed and eat hydroids. Often they are brilliantly colored. Most are small, although the "Florida hare" grows to be the size and shape of a football.

THE BIVALVES (PELECYPODA)

The pelecypods (*pelekys*, hatchet, + *podos*, foot) include the clams, oysters, mussels, shipworms, and scallops, which are anatomi-

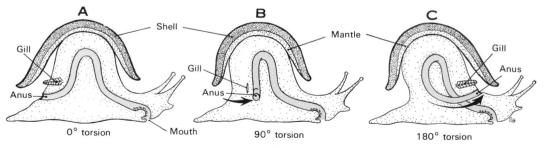

Fig. 21-4. Diagrams showing the process of torsion in a snail.

Fig. 21-5. Photo of two living nudibranchs feeding on marine hydroids. (From E. F. Ricketts and J. Calvin: *Between Pacific Tides,* 3rd ed., rev. by J. W. Hedgpeth, Stanford University Press, Stanford, 1968.)

Short rounded columns of muscle, the anterior and posterior **adductor muscles**, along with the hinge ligament, connect the right and left shells. A remarkable fact about them is that each muscle is really double. The crescentic region of each, located toward the ends of the shell, is the **locking muscle**. This muscle can close the shells only very, very slowly, but it has great strength and endurance and can hold the shell closed indefinitely against the pull of the hinge ligament which tends to open the shells and the pull of enemies. The main part of the adductor muscle, usually pink in color, is the **closing muscle**. It can snap the shells shut, but rapidly fatigues. If the locking muscle is cut, the closing muscle soon tires and the shells open. It is the adductor muscle of scallops that is eaten. Another remarkable feature of the muscles is that they gradually change their position. As the shell grows longer, the muscles not only become larger but also gradually move apart, stationing themselves near the anterior and posterior ends of the shell.

In adult oysters there is only one adductor muscle. It moves down the shell from the hinge as the oyster grows, and very often it is possible to see the scar of the old muscle attachment marking a path leading from the muscle back toward the hinge. The growth mechanism involved is little understood, but one is reminded of the way the interior of the long bones in children is eroded away as the marrow cavity enlarges, while at the same time new bone is deposited on the outside as the bone grows larger.

In pelecypods the head and radula are completely missing, the mouth merely a hole into the intestine. The soft visceral mass hangs down in the mantle cavity between the two valves of the shell. At its lower edge is the more or less hatchet-shaped muscular foot. Small muscles called the anterior and posterior **retractors** pull in the foot. In many species there is also a **protractor** muscle which helps extend the foot. The foot itself is muscular and is protruded in part when blood is forced into it.

The foot functions in digging the clam into the mud and sand and in "walking." Oysters, which have no foot, are completely stationary and grow stuck to a rock or to another oyster by the left shell. Scallops, which also lack a foot, swim vigorously by allowing the mantle

cally highly specialized and economically the most important of the mollusks.

The bivalve mantle, and consequently the shell, is divided into a right and left half. A most interesting problem presented by the mantle is how it secretes the inner calcium carbonate and the outer horny layers of the shell. Knowledge of the enzymatic mechanisms involved should shed light both on bone and tooth formation.

In most clams the right and left shells are closely similar. The easiest way to tell right from left is by the position of the **siphons** (popularly called the "neck"), which are posterior, and of the hinge, which is dorsal. Raw oysters are usually served on their left shells because the left shell tends to be more cup-shaped than the right.

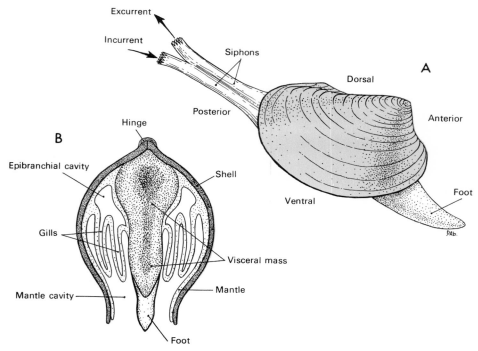

Fig. 21-6. A clam. A, as it lies mostly buried in the sand with only the ends of the siphons exposed; B, diagrammatic cross section. The excurrent siphon is continuous with the epibranchial cavity.

cavity to fill with water and then clapping their shells together to squirt the water out dorsally.

On breaking open a clam a large cavity is revealed within which lies the soft visceral mass ending with the tough ventral and pointed foot. Lying over the visceral mass can be seen the gills, and pressed against the shell is a thin membrane, the mantle. The cavity between the mantle and the visceral mass and foot is the mantle cavity. The greatly reduced coelom lies within the visceral mass.

The **gills** of clams and pelecypods generally are their key specialization and the respiratory, digestive, circulatory, and, in many species, reproductive systems are all tied in with them. The current of water pulled in through the **incurrent siphon** by the action of the cilia on the gills carries both dissolved oxygen and food particles. The food is entangled in cilia and mucus, passes down the gill filaments to the lower edge of the gill, and then forward in a continual stream to the mouth and into the stomach and intestine. This intestine makes a loop or two in the visceral mass above the foot

before ending near the **excurrent** or **dorsal siphon**. Near the stomach is a digestive gland, the liver, and a long blind tube opening into the stomach. This blind tube secretes a glistening, translucent rod of protein called the **crystalline style** which is slowly pushed into the stomach, where the protein breaks up, releasing a starch-digesting enzyme. An anatomical peculiarity of the clam's intestine is that it passes directly through the ventricle of the heart.

The slow beating of the heart may be readily observed after one shell and the mantle have been removed. The blood of most mollusks contains a blue copper-containing protein. This oxygen-carrying respiratory pigment is called **hemocyanin**. The red in the muscles of the radula is due to a **hemoglobin**.

The reproductive system of clams and their relatives is both simple and effective. A single oyster may produce 60 million eggs a year. The gonad lies in the visceral mass among the coils of the intestine. A short genital duct empties the gametes—eggs or sperms—into the mantle cavity. In many freshwater clams and in some marine species—the European oyster, for example—the gills are used as brood sacs for the developing embryos.

Most marine pelecypods are **protandrous**; *i.e.*, they are males when small, and become females when they attain full size. The larva is a trochophore. In contrast, in the large freshwater clams, the sexes are separate and the larva is known as a **glochidium**, which is a temporary but obligatory parasite on fishes. The edge of each valve of the shell of the glochidium is provided with a sharp hook or tooth that fastens the animal to the gills of a fish. Later the glochidium drops off and matures in the stream bed. The fingernail clams of quiet ponds follow still another pattern. They are hermaphroditic and brood their young in their gills until the adult form is attained.

The pelecypod **nervous system** consists of a pair of head ganglia encircling the esophagus near the anterior adductor muscle and nerve cords extending from it to the rest of the body. The best known and certainly the most beautiful sense organs of these mollusks are the eyes that adorn the edge of the scallop's mantle like a row of luminous pearls.

Economically the most important clams are freshwater forms like *Unio* and *Anodonta*, which are sought as food and for mother-of-pearl, and the marine hard clam or littleneck clam, *Venus mercenaria*. The latter was much used by east coast American Indians as food and for the manufacture of purple wampum. "Quahog," the common name applied to this species, is from the Algonquian language. Various other species are nearly as important, especially the soft shell, long neck, or steamer clam, *Mya arenaria*. In the rivers of the Mississippi drainage there are 15 or 20 species of freshwater clams which form the basis of a profitable pearl button industry. True pearls can be found in freshwater clams as well as in oysters. These are formed when a minute foreign particle, perhaps a fine grain of sand, becomes surrounded by a covering of mantle epithelium which in turn secretes the pearl.

Perhaps the only destructive pelecypod is *Teredo*, the shipworm. This elongate clam uses its modified bivalve shell to bore into wood. Wharfs and ships that are unprotected by tar soon become completely riddled by these animals.

THE SQUIDS AND OCTOPUSES (CEPHALOPODA)

The class Cephalopoda (*kephale*, head, + *podos*, foot) includes the squids and octopuses. This class now contains only about 400 living species, though there are some 10,000 known fossil forms. Cephalopods are the only invertebrate group which can boast large and potentially dangerous animals competing with the vertebrates in size. The giant squid of the cold waters of the northernmost Atlantic attains a body length of 6 meters (20 feet), and has a pair of grasping tentacles over 9 meters (30 feet) long and eight shorter ones about 3 meters (10 feet) in length. Maximum diameter is probably about 5 meters (15 feet).

The body plan of a cephalopod can be easily understood by imagining a snail in which the mouth with the radula has been moved back into the center of the foot, and the foot itself drawn out around the mouth into eight or more tentacles. The visceral mass in the squid has a stiff internal shell, while in an octopus the body is a mere bag. Note that the siphons of both the clam and squid are posterior.

The cephalopods are undoubtedly mollusks. Everything about their adult anatomy is molluscan: the mantle cavity with its contained gills, the siphon formed from the mantle edge, the radula, the visceral mass, even the ink sac. However, in flagrant disregard of the evolutionary theory that related animals have similar developments, the development of cephalopods is very different from that of other mollusks. There is no trochophore or other larval stage, and the young develop directly into the adult form.

Living cephalopods fall into two major groups. The nautilus has many fossil genera but only one living genus, *Nautilus pompilius*, the pearly nautilus of song and story. It is edible and is found chiefly in the Indian and Pacific oceans. The second group includes the octopus and the squid. Both can be dangerous because the mouth is always armed with a formidable parrot-like beak, and the tongue is a radula. The saliva of these animals is rich in serotonin, an amine normally present in minute amounts in vertebrate brains and other tissues. In large amounts serotonin is a potent poison. Adult men have died a few hours after a bite by a

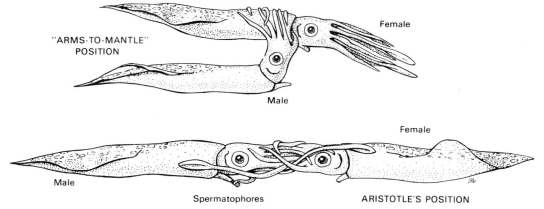

Fig. 21-7. Squids mating in which the male introduces a long package of sperms (spermatophores) into the mantle cavity or beside the mouth of the female from which the eggs will be fertilized immediately after they are laid.

small (45-cm. diameter) octopus. Both squids and octopuses vary in size from species a few inches long up to true giants.

In the course of their evolution, mollusks have undergone a successful **adaptive radiation** into many diverse environments. With arthropods and vertebrates they share the conquest of the land; snails inhabit the land everywhere except in the driest and coldest regions. On the rain-swept islands of the South Pacific nearly every valley has its unique variety of land snail. In fact, several zoologists have speculated as to why no octopus has become terrestrial. Perhaps this is something evolution has in store for the future. In any case, inasmuch as crabs are a favorite food of the octopus, there is no obvious reason why an octopus, which can move over land breathing with moist gills within the mantle cavity in much the same way land crabs breathe, should not chase beach and land crabs, especially on small islands where

such crabs swarm and octopuses lurk along the shore. Anyone who has ever watched the deadly efficiency of a "wolf pack" of squids shooting back and forth and catching the fish attracted to a strong light held over a ship's side must have wondered why squids were not the predominant large animals of the sea.

USEFUL REFERENCES

Abbott, R. T. *American Seashells.* D. Van Nostrand Co., Princeton, N. J., 1954.

Morris, P. A. *A Field Guide to the Shells of Our Atlantic and Gulf Coasts.* rev. ed. Houghton Mifflin Co., Boston, 1951.

Morris, P. A. *A Field Guide to the Shells of the Pacific Coast and Hawaii.* Houghton Mifflin Company. Boston, 1952.

Pennak, R. W. *Fresh-Water Invertebrates of the United States.* The Ronald Press Co., New York, 1953.

Wells, M. J. *Brain and Behavior in Cephalopods.* Stanford University Press, Stanford, 1962.

Wilbur, K. M., and Yonge, C. M. *Physiology of Mollusca.* Academic Press, New York, 1964.

22

Crustaceans, Insects, and other Arthropods

Arthropods touch the life of mankind in an endless variety of ways, both good and bad, although the bad often predominates to a disastrous extent. Many tens of thousands of people around the world suffer from mosquito-transmitted malaria; armies have been brought low by lice-borne fevers; famines and death have resulted from the ravages of crop-eating insects. The problem of DDT and its persistent poisoning as it passes along food chains would not exist were it not for the serious menace of insects.

Probably the most important species of arthropods from the strictly human point of view are the small crustaceans, **copepods**, which often swarm in the oceans where they constitute the chief link in the food chains between the photosynthetic algae and small fishes and other animals. **Insects** play a similar role on land. In addition, many arthropods play a vital role as agents of pollen dispersal in the sexual reproduction of land plants. Certain arthropods (butterflies and social insects, for example) are objects of endless delight to watch or to study. Lobsters offer delights equally real, although quite different. Certainly from a medical or agricultural point of view, arthropods are the most important animals to know something about, with the exception of the vertebrates and perhaps the protozoa. Approximately 1,150,000 species of arthropods have been described and named, of which about 1,000,000 are insects. The total number of described species for the entire remainder of the animal kingdom from protozoa to vertebrates is about 225,000. Thus arthropods vastly exceed all other phyla in number of species. They often exceed other groups in number of individuals and are clearly the most successful of all animals.

Arthropods can be defined with much justification as annelids which knew what to do with segmentation. They are constructed on the annelid body plan—a straight gut running from one end to the other of a segmented animal with a nervous system like that in an annelid. The body cavity or coelom between gut and body wall is reduced to almost nothing. The thin exterior cuticle is strengthened in arthropods by calcium and chitin into a rigid or semi-rigid exoskeleton. Appendages comparable to the parapodia found in polychaetous annelids develop into six or eight to dozens of pairs of legs in the arthropods.

All arthropods are characterized by the five following traits:

1. **Jointed legs**, as their name indicates (*arthron*, joint, + *pod*, leg).

2. **Serial segmentation** in the annelid manner.

3. A **chitinous exoskeleton**. A result of this feature is the necessity for periodic molting (ecdysis) during growth.

4. A type of nervous system consisting of a **dorsal brain** or pair of cerebral ganglia, a pair of **circumpharyngeal connectives** around the anterior part of the gut, and a **ventral nerve cord** consisting of a chain of more or less fused ganglia.

5. A greatly **reduced coelom** and an "open" **vascular system**; that is, the arteries do not empty into capillaries but into intercellular

cavities or **hemocoels**. No trochophore larva is ever present.

SPIDERS AND THEIR RELATIVES (CHELICERATES)

The spiders, scorpions, ticks, mites (the arachnoids), and *Limulus,* the horseshoe "crab," plus a number of less well-known forms constitute the chelicerates. They **lack jaws** and obtain their food by sucking juices. Adjacent to the mouth is a pair of segmented appendages, the **chelicerae**, which end in either claws or fangs which, in the case of spiders, are hollow and conduct poison into whatever is bitten. Furthermore, chelicerates always **lack the antennae** which are such a conspicuous characteristic of the insects, crustaceans and other members of the jawed or mandibulate arthropods. Lastly, chelicerates have only **six pairs of appendages**: four pairs of legs, a pair of palps (usually called pedipalps), and the pair of chelicerae. Look closely at any tarantula or at *Limulus.*

CRUSTACEANS AND INSECTS—THE MANDIBULATE ARTHROPODS

As their name implies, these animals possess **mandibles** or jaws which do not end in a claw. Even in a mosquito, where the mouth parts are highly modified for piercing and sucking, jaws are present. Mandibulate arthropods also possess antennae, two pairs in crustaceans, one pair in centipedes, millipedes, and insects. Instead of the constant six pairs of appendages characteristic of the chelicerate arthropods, mandibulates often have many.

The Crustaceans

Crustaceans, class Crustacea (*crusta,* shell), are gill-breathing mandibulate arthropods with two pairs of antennae, and, generally, **biramous** (*bi,* two, + *ramus,* branch) or **two-branched appendages**. A larval form, when present, is a minute "three-legged" creature called a **nauplius**, a familiar sight at certain seasons in both fresh and salt water.

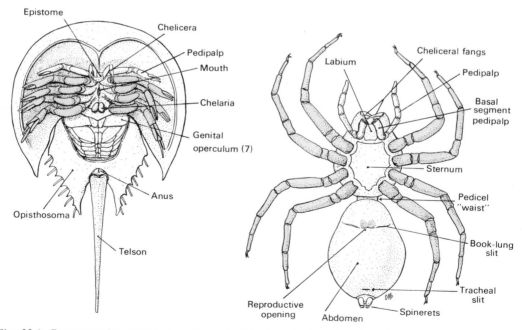

Fig. 22-1. Representative chelicerate arthropods. *Limulus,* the horseshoe "crab" and an orb-weaving spider. Note that, like the disease-carrying ticks, they both have four pairs of legs and two pairs of mouth parts.

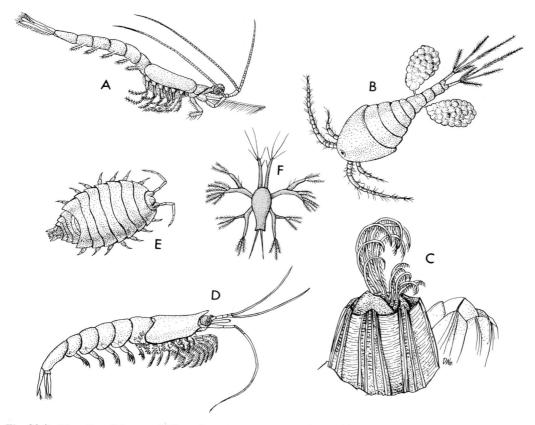

Fig. 22-2. Diversity of form and life style among crustaceans. A, mysid, sometimes called the kangaroo shrimp because the female (shown here) carries her brood in a ventral sac. Mysids are abundant marine filter feeders which strain out minute organisms. B, copepod with characteristic pair of egg sacs. Copepods are a chief link between primary producers, *i.e.,* algae, and secondary consumers, *i.e.,* larger animals, in both fresh and salt water. C, barnacle. Living fixed to some object, it uses its curly legs to kick food into its mouth. D, euphausid. These occur in such enormous numbers that they constitute a chief food of the larger whales. E, common pill "bug" found under boards and in damp leaf piles. F, nauplius, the larval form of all marine crustaceans. (Not to scale.)

Importance

Most people think of crustaceans as important chiefly as a source of food—shrimp, crabs, and lobsters. However, their ecological importance is vastly greater. In all oceans and in all bodies of fresh water, crustaceans, mostly tiny copepods, form the link between the primary producers in the food chain, the algae, and all the larger animals.

Structure

The structure of crustaceans can be best understood by regarding a crustacean as a modified polychaete worm. In the primitive crustaceans the paired appendages, one pair to a segment, are closely alike from head to tail. Each appendage, like the parapodia of a polychaete, is biramous, *i.e.,* two-branched with a dorsal and a ventral portion. Such a situation can be seen in the appendages of the beautiful little fairy shrimp found in woodland ponds in early spring. Close to the head and at the end of the tail the appendages are specialized and, therefore, different in structure.

In the highly evolved forms like the lobster or crayfish, the original biramous nature of each appendage is still more or less evident, but these appendages have been highly modified in

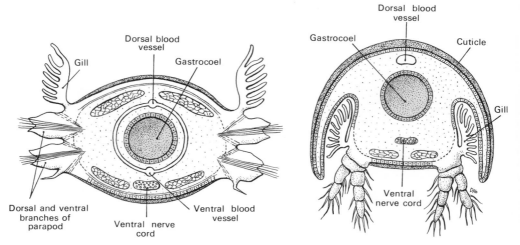

Fig. 22-3. Comparison of a polychaete worm (left) and a crustacean (right) seen in cross section.

the course of evolution into structures useful in feeling out the environment, manipulating food, fighting, breathing, walking, reproducing, swimming, or escaping enemies. Of the 19 pairs on an adult male lobster, only three of the abdominal appendages are alike, even though in a very young lobster the appendages are very similar and clearly show the ancestral two-branched condition. These facts were used by T. H. Huxley in his classic exposition and defense of Charles Darwin's theory of evolution.

The Crustacean Body

The **muscles** of crustaceans, like those of other animals, are arranged in antagonistic sets of flexors and extensors. The chief muscle mass of the body is composed of the flexors of the abdomen, which account for most of its bulk. These are the muscles which produce the vigorous escape flips of the "tail." Dorsal to them is a much thinner layer of extensors which straighten out the abdomen. These are the chief muscles eaten in boiled lobster.

The **digestive system** of a lobster consists of a short esophagus, a two-part stomach containing teeth, the so-called gastric mill, and a straight intestine. Near the stomach is the yellowish-green hepatopancreas, also a gourmet delicacy, which serves much the same functions of secreting digestive fluids, storage of

food, and processing nitrogenous wastes that the liver of vertebrates does.

The **circulatory system** consists of a dorsal heart from which arteries lead to the various parts of the body. There are no capillaries conducting blood to veins, but the arteries merely open into tissue spaces from whence blood works its way back to the heart.

The **excretory system** consists of a pair of "green glands" located in the head, one at the base of each of the larger antennae. They appear to be a highly modified version of the earthworm nephridium with a thin-walled sac which empties via a nephrostome into a glandular labyrinth of interconnecting glandular passages. The gonads, whether ovaries or testes, are Y-shaped and lie below the heart with the arms of the Y on either side of the intestine. Many species of crustaceans carry the developing eggs with them for long periods of time. In many marine crustaceans there are several larval stages beginning with the nauplius.

The **nervous system** follows the annelid pattern. There is a brain or a pair of cerebral ganglia between the eyes and above the esophagus, a pair of connectives, and a ventral chain of ganglia, clearly one ganglion per segment in the abdomen, but with more or less fusion in the thorax. The chief sense organs are the tactile, and probably also olfactory, first and second **antennae**; the **eyes**, which are compound and which are composed of many visual units, ommatidia. In daylight each ommatidium is enclosed in a sleeve of pigment which is rolled up at night (Fig. 22-6).

Molting

Increase in the size of a lobster, or any other crustacean, is possible only immediately after a molt. Interest in molting has been stimulated both by the problems of the crabbing industry and of medicine. Soft crabs—that is, crabs which have recently molted—are prized as delicacies and bring the highest prices. Soft crabs are quiescent, rarely come to bait, and are therefore seldom caught. Consequently, it is necessary to keep large quantities of crabs waiting in enclosures to molt. The mortality and consequent loss is very high. A practical method of inducing molting is obviously desirable. Enough is now known to make such control possible in the foreseeable future. Medical researchers are also interested in this problem because of its relationship to the absorption and deposition of bone, especially in aging men and women.

In the actual process of molting the calcium is absorbed from the exoskeleton, which becomes soft, and is deposited instead in the large **hepato-pancreas** (liver) and, at least in crayfish, in the **gastroliths** (*gaster,* stomach, + *lithos,* stone) secreted by the gastric epithelium of the anterior wall of the stomach between the epithelium and the chitinous lining of the stomach. After molting, the two gastroliths are exposed to the gastric juice, dissolve, and the absorbed lime is made available for redisposition in the new exoskeleton. When about to molt, the animal seeks a sheltered corner, arches itself, and pulls its thorax and head and then its abdomen out of the old shell.

The period or stage between each molt, *i.e.,* each ecdysis, is called an **instar**. The first instar in crustaceans is a minute creature with one median eye and three pairs of legs, the nauplius.

Adult crayfish usually molt twice a year, once in the spring and once in the fall. If the eyestalks of a non-molting individual are removed, calcium begins to be withdrawn from the exoskeleton, the gastroliths begin to enlarge, and in about 15 days molting occurs. The mortality is heavy, but those crayfish that survive molt again within about 15 days. Molting can be inhibited by implanting eyestalks from non-molting donors into animals without eyestalks. The responsible organ in the eyestalk is the **sinus gland,** an inconspicuous body close to the optic nerve, and an associated structure, the **X organ. Neurosecretory cells** in the X organ and in the brain form a molt-preventing hormone which is stored in the sinus gland. The sinus gland itself is mostly a mass of nerve fiber endings. Similar results have been attained in other species of crustaceans.

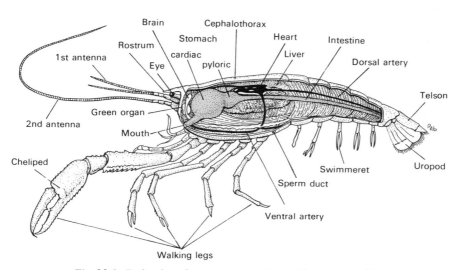

Fig. 22-4. Body plan of a crustacean such as a lobster or crayfish.

Fig. 22-5. Eyes of the nocturnal beach ghost crab, *Ocypode.* (Courtesy W. H. Amos.)

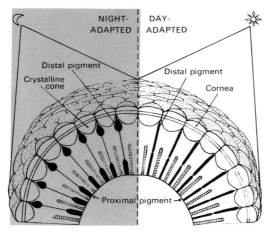

Fig. 22-6. Diagram of a section through a crustacean compound eye, dark-adapted ommatidia on the left, light-adapted on the right.

A second pair of glands, the **Y organs**, are located in the anterior part of the thorax, which in most crustaceans is near the base of the jaw muscles. The Y organs secrete a

hormone, probably closely similar to ecdysone of insects, which induces molting. During most of the life of the animal the hormone of the Y organ is held in check by the sinus gland hormone. When the brain and X organ fail to produce the sinus gland hormone, molting occurs under the action of the Y organ.

It is worth noting that extracts of mammalian pituitary, when injected into crustaceans, produce results that are similar to those of eyestalk extracts.

Ecology of Crustaceans

Most crustaceans are light-shy if not actually nocturnal. Probably the most abundant and ecologically important crustaceans of the ocean after the **copepods** are certain shrimp, **Euphausids,** which form much of the "krill" eaten by the larger species of whales. These shrimp migrate up close to the surface at night and then go down into deeper water every day. Fiddler crabs often swarm on muddy tidal flats

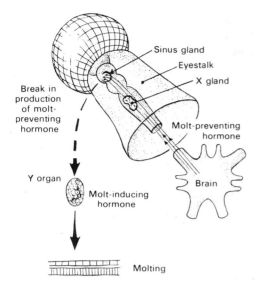

Fig. 22-7. Diagram of relationships of hormone-producing structures, which control molting and metamorphosis, in adult crayfish eyestalk.

at low tide and high noon, but they are exceptions.

An analysis of population dynamics has been begun in relatively few species. The abundance of small freshwater and marine crustaceans is a determining factor and possibly the key factor in controlling the growth rate and size of population among fishes. Marine populations are notoriously difficult to estimate and virtually impossible to count. This is especially true of crustaceans, which molt and are thus unsatisfactory animals to tag.

Some of the larger crustaceans are known to **migrate**. *Callinectes,* the blue crab of Chesapeake Bay, moves down the bay toward the Virginia Capes into saltier water every year to breed, and returns several hundred miles up the bay to feed during the rest of the year. Land crabs migrate to the sea to breed. The spiny lobster, *Panulirus,* forgets its fear of light and can be seen crawling over the bottom in broad daylight on its way to deeper water at breeding season.

Barnacles

No discussion of crustaceans would be complete without some mention of barnacles.

As a young man Charles Darwin established his reputation as a qualified scientist by his monumental work on barnacles. It occupied him for eight years. Their strange life history with its profound metamorphosis from a swimming nauplius to the encrusted fixed adult helped to support the theory of evolution.

Barnacles, or cirripedes (*cirrus,* curl, + *pedis,* foot) are found on rocks in most oceans and are a problem because they settle and grow on the bottoms of ships. They have been known from remote antiquity, but until a surprisingly recent date everyone supposed them to be mollusks because of their heavy calcareous shells. A little over a century ago an English army surgeon and amateur naturalist took them out of the Mollusca, where they had been placed by Linnaeus and Cuvier, and showed that they are truly arthropods, a classification that can be verified by anyone who observes a barnacle in an aquarium or who dissects one. Their most conspicuous feature is their biramous jointed and curly legs, with which the barnacle kicks food particles into its mouth. Further proof that barnacles are indeed crustaceans is to be seen in the typical nauplius larvae that develop from their eggs. The nauplius swims freely for a short time, then affixes itself to a rock, ship bottom, or some other object and undergoes a metamorphosis into the adult.

The Insects

The often fantastic and beautiful forms of insects, their bites and stings, strange habits, highly structured societies, and amazing life histories have attracted amateur collectors and thinkers as diverse as Darwin the scientist, Maeterlink the poet, and Rothschild the financier. Insects also have a direct and often tragic impact on the lives of people whether or not they even know what an insect is.

The damage inflicted by insects to crops, to natural vegetation, to buildings, and to an endless variety of other things is well known and costs billions of dollars annually. Fewer people are aware of the catastophic effects of insect-transmitted diseases and that, as long as the diseases exist, the insects responsible for their transmission will remain a constant threat to mankind. In spite of all the control measures, malaria, which is transmitted by a particular kind of mosquito, remains the world's most common human disease. Regions where flea-transmitted bubonic (black) plague is endemic among rodents, and hence a continual danger to man, exist in many parts of the world—Southeast Asia, Morocco, Central Africa, around the Caspian Sea, parts of Argentina and Brazil, and in our own Southwest.

To realize what such diseases mean in human terms one needs only to look briefly at history. Malaria was a serious problem in the South Pacific in World War II as in many other wars. It produced disastrous effects in medieval Italy and in classical Greece. Bubonic plague threw the Roman world of Justinian (565 A.D.) into confusion. Whole towns and countrysides were deserted, crops rotted in the fields, and it is estimated that at least half the population of the western world died after suffering so intense that people jumped from housetops to escape their agonies.

In the pandemic of bubonic plague (Black Death) in the 14th century two out of every three students at Oxford died. All over Europe there were depopulated towns and farms. The resulting economic and social crisis led directly to the political and religious turmoil of the 14th and 15th centuries.

Clearly no contribution of science to mankind has been greater than the knowledge that insects can carry human diseases. This fact, combined with the vast importance of insects to agriculture, places **entomology**, the study of insects, high in the ranks of biological sciences. Table 22-1 lists the more important insect-borne diseases. The story of how some of these momentous discoveries were made should be part of the cultural heritage of every civilized man and will be found in the chapter on parasitism.

With over a million species and vast numbers of individuals, it cannot be denied that except for the development of intelligence, insects are the dominant animals on this planet. Why are insects dominant? No certain answer is at hand. The best that can be done is to find the features which characterize insects and which in combination appear to make them so successful and so dangerous to man.

The Machinery of Success: Insect Structure and Function

One of the best ways to gain insight into the nature of the efficient machines for living that insects truly are is to compare a relatively

TABLE 22-1
Some important diseases transmitted by insects

Disease	Causative Agent	Carrier	Discoverers of Carrier
Malaria	Protozoa *Plasmodium vivax*	*Anopheles,* mosquito	Ross, 1897; Grassi, 1898; Manson and Sambon
African sleeping sickness	*Trypanosoma gambiense*	*Glossina palpalis,* tsetse fly	Bruce, 1895; Castellani, 1903
Chagas fever	*Trypanosoma cruzi*	*Rhodnius,* bloodsucking bug	Chagas, 1909
Bubonic plague (Black Death)	Bacteria *Bacillus pestis*	*Pulex irritans,* common flea *Xenopsylla cheopsis,* rat flea	British Indian Plague Commission at Bombay—best proof, 1907
Typhoid	*Bacillus typhosus*	*Musca domestica,* housefly	Vaughan and Veeder, 1898
Yellow fever	Viruses Yellow fever virus	*Aedes aegypti (Stegomyia),* mosquito	Reed, Carroll, Lazear, Finlay, and others, 1900
Typhus fever	Typhus virus (Rickettsia bodies)	*Pediculus,* human louse; *Polyplax,* rat louse	Nicolle, Comte, and Conseil, 1909
Elephantiasis	Worms *Filaria bancrofti*	*Culex, Aedes,* and *Anopheles,* mosquitos	Patrick Manson, 1878

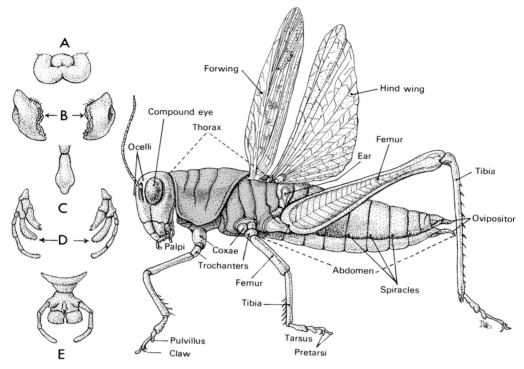

Fig. 22-8. External anatomy of a grasshopper. Mouth parts from anterior to posterior: A, labrum; B, mandibles; C, tongue; D, maxillae; E, labium with labial palps.

generalized insect like a grasshopper with a specialized species like a honeybee. This knowledge is also basic to a full understanding of means to control insects.

External Anatomy. Both the grasshopper and the honeybee show the typical tripartite division into head, thorax, and abdomen. The **head** carries the chief sense organs and the mouth. The **compound eyes** occupy prominent bilaterally symmetrical positions. Three simple eyes, or **ocelli**, form a triangle with one corner near the top of each compound eye and the third in the middle of the "forehead." The **antennae** are organs of touch, smell, and chemoreception. The **mouth parts** consist of seven elements.

The **thorax** is divided into prothorax, mesothorax, and metathorax. The **prothorax** bears legs; the **mesothorax** and **metathorax**, both legs and wings. The six segments of a leg are a short **coxa** and **trochanter**, followed by a long **femur**, a long **tibia**, and a shorter **tarsus** and **pretarsus**, the latter equipped with a pair of claws to hold onto rough surfaces and between them a flap-like **pulvillus** to adhere to flat surfaces.

Insects' legs show remarkable adaptations. The prothoracic legs of bees are provided with a semicircular brush and a hinged scraper which work together as an antennae cleaner. The mesothoracic leg has a spur used to pick off the scales of wax that are secreted between the abdominal segments. The metathoracic legs bear pollen baskets composed of the concave broad side of the leg, plus surrounding long bristles. The metathoracic legs of grasshoppers are highly modified for jumping.

The segments of the abdomen are covered by plates of exoskeleton, the dorsal **terga** and the ventral **sterna**. Along each side is a row of **spiracles**, small holes one pair to a segment, which open into the **tracheae** through which the insect breathes. Both alimentary canal and reproductive organs have a common opening at the posterior end of the abdomen. In males there are usually clasping organs, while in females the external genitalia take the form of a pair of dorsal and a pair of ventral blade-like structures forming an **ovipositor**. In some species the ovipositor is longer than the rest of

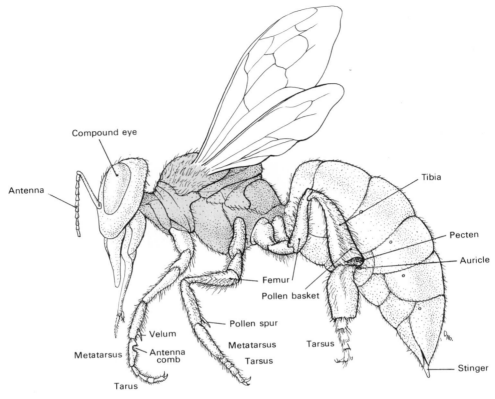

Fig. 22-9. *Apis,* the honeybee.

the body, and in many of the hymenopterous species the ovipositor is a stinger. Thus, drones are unarmed.

One trait all insects possess is an **exoskeleton** composed largely of **chitin**. Many entomologists have extolled the advantages of this protective armor. Chitin has the properties of an ideal plastic. It is strong, wear-resistant, and waterproof. It does not soften with heat or become brittle at freezing temperatures. It will stand up against alcohols and other organic solvents. It is not injured by any acids or alkalis found in the natural environment. In fact, boiling in strong acids is one of the few treatments that will destroy the chemical integrity of this remarkable substance. The chitin molecule is a polysaccharide containing nitrogen and consists of a long chain of glucose molecules with added acetyl groups (from acetic acid) and amino, *i.e.,* $-NH_2$, groups.

It must be remembered that some other organisms can also form chitin. The cell walls of some molds, the jaws and setae of polychaete annelids, and the "pen" of a squid are all formed of chitin. Apparently none of these other groups has used chitin so well and in such a happy combination with other advantageous traits. Over the chitin in many insects is a waxy layer of protein and lipid material giving a double protection.

A second trait common to all insects is **small size**. The very largest insects scarcely attain the size of the smallest mammals. Small size is no doubt connected with the limitations imposed on growth by an exoskeleton which must be molted for increase in size to occur. This is not only physically more difficult in large than in small organisms, but a really big animal by vertebrate standards would, while soft, be crushed by its own weight after molting and die a helpless mass suffocated from crushed lungs or tracheae, just as a stranded whale suffocates because its short ribs cannot support its lung cavity.

The small size of insects has sometimes been regarded as a limitation and a disadvantage, but smallness may well be one clue to the success of the insects. On the scale of things on this planet, small size greatly increases the availability of cracks, crannies, and protective microenvironments. A level field for a fox or a rabbit is a veritable jungle for an ant or a beetle.

A second important effect of smallness is that it greatly increases the strength of an insect in relation to objects in its environment. A man begins to have serious trouble in lifting and moving objects only. slightly heavier than himself. An ant or a wasp can move an object 10 or more times its own weight. This is not because insect muscle is stronger than vertebrate muscle, but because of the geometry of size. The weight of a muscle increases with the cube of its dimensions, while its strength increases with the cross-sectional area. In other words, for any increase in size the weight of a muscle increases in proportion to the cube of its dimensions, while the strength increases only in proportion to the square.

A third and perhaps the most important trait of insects is the **ability to fly**. Insects have been enjoying the advantages of living in the air age for at least 250 million years. In many ways flight much more than compensates for lack of size. It enables insects to escape enemies, overcome obstacles, and disperse species into new and distant environments. A rabbit may run and hop two or three miles in a day, although all the evidence indicates that rabbits cover that much ground only under the most unusual conditions of crowding. However, a winged ant, once it attains some height above the ground, may be carried 10 miles by the wind.

True, flight has been attained independently in the course of evolution by several groups of animals, the prehistoric reptiles, the birds, a group of mammals (the bats), and at last artificially by the chief of the primates, man himself. But none surpasses the easy mastery of the air seen in dragonflies, honeybees, mosquitos, or a thousand other insects; and none possesses the other traits which, combined with flight, seem to be an explanation for the dominance of insects.

A fourth feature of insects which seems to have been important in their success is the **character of their nervous system** and the type of **behavior** that goes with it. An important fact about insects is that they have a complex nervous system but almost no intelligence. Many other invertebrates lack intelligence—jellyfish and flatworms, for example—but they have simple nervous systems. Vertebrates, especially the higher mammals, have complex nervous systems but of a type that permits, and indeed necessitates, a highly flexible type of behavior. Learning plays a dominant role in their lives. By contrast, insects are like spiders: their behavior is predominantly instinctive; that is, based primarily on more or less complex, stereotyped, unlearned reaction patterns which are characteristic of species and are adaptive. Learning and experience play only a minor role.

In the insects, behavior patterns are almost as diagnostic of the order, family, or even species as are anatomical characters. This is true of the kind of plants on which a given species of beetle will lay its eggs, the shape of the mud nest built by a wasp, the shape and location of the cocoon spun by a moth caterpillar, the way a dragonfly mates and then lays its eggs under water, or the way a female cricket approaches a chirping male. In fact, it is possible to call a mature female cricket to the telephone although the male may be in another city. Male mosquitos will fly toward a tuning fork vibrating at the same frequency as the wings of females of their own species. The list of such acts could fill many volumes.

Some of these instinctive responses are relatively simple, and knowledge of them can be of economic importance. For example, males of many species of moths will travel upwind for as much as a mile to reach a female. The large feathery olfactory antennae of the males of these species reflect this behavior. The females secrete a special sex attractant in the form of a highly volatile chemical. These odoriferous substances have been isolated and analyzed in several cases. As cited in Chapter 1, they are long-chain molecules of simple construction. The formula of the sex attractant of the silk moth, as worked out by Butenandt, is shown here.

$$\underset{\displaystyle \text{CH}_2-\text{CH}_2-\text{CH}_2-\text{CH}_2-\text{CH}_2-\text{CH}_2-\text{CH}_2\text{OH}}{\overset{\displaystyle \overset{\text{H H H H}}{\underset{|\ \ |\ \ |\ \ |}{}}}{\text{H}_3\text{C}-\text{CH}_2-\text{CH}_2-\text{C}=\text{C}-\text{C}=\text{C}-\text{CH}_2-\text{CH}_2-\text{CH}_2-}}$$

A similar molecule has been found as the sex attractant of the female cockroach and the female gypsy moth. The attractant of the latter, manufactured and sold under the name Gyplure, is used to entice males into death traps.

The advantage of this instinctive type of behavior is that it furnishes its possessor with prefabricated answers to meet the expected events of life. The insect does not waste much time learning and making mistakes. A male mosquito responds to the wing vibration of a female mosquito. If for any reason either in his hereditary make-up or in his environment, he fails to respond, he also fails to leave descendants and the unresponsive nervous system fails to be perpetuated. In the vast majority of insects the parents die shortly after laying eggs, so that there is no possibility of imitation or other forms of learning while protected by parents as in birds and mammals. For most insects life is short, and it is evidently an advantage—a very great advantage—to be provided with a ready-made built-in response mechanism.

The rich endowment with instincts does not mean that insects learn nothing or that vertebrates posses no instincts. A wasp learns the location of its nest. A rooster raised in a soundproof incubator, deafened at hatching by surgical removal of both cochleas (the organs of hearing), and then raised in isolation will crow while making appropriate postures and wing flappings when mature. The difference is one of degree, but it is enormous.

A fifth and almost universal feature of insect life is **metamorphosis**. What is its adaptive meaning? Primarily it is a device to provide a single organism with two bodies: one, the **larva**, is specialized for eating and growing; a second, the **imago** or adult, is specialized for reproduction and dissemination of the species. One investigator calls the adult "a flying machine for reproduction." Very often the two bodies are adapted for entirely different environments. The larva may be aquatic, live inside a tree trunk, or even reside in the nasal passages of a sheep, while the adult flies freely in the air.

The study of insect metamorphosis has been motivated by the basic human desire to understand mysteries, especially spectacular mysteries. There are also more mundane motives. Ecdysone, the insect molting hormone, has been chemically identified as a peculiar

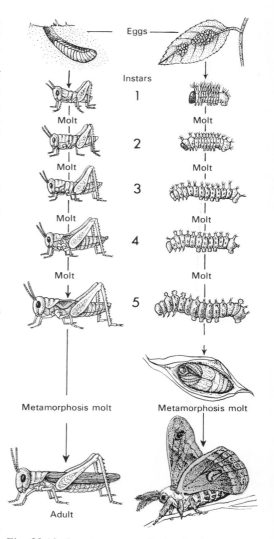

Fig. 22-10. Insect metamorphosis. Gradual or hemimetabolous (left) and sudden or holometabolous (right).

lipid found in many plants. Analogues are now being manufactured as potent insecticides which prevent metamorphosis. They are highly specific against particular kinds of insects but harmless to other animals.

There are two major types of insect metamorphosis. One is the gradual type like that of the grasshoppers, where the wings develop as little external paddle-like flaps on the "shoulders." The other or sudden type of metamorphosis is familiar in the butterflies and bees.

The egg hatches into a caterpillar or maggot. Wing development occurs within the body. Neither type should be regarded as being more advanced than the other. They are two somewhat different methods of achieving the same ends.

The anatomical facts in both types of metamorphosis are basically similar, but much more drastic changes take place at the two final molts in the case of the sudden or complete type. In this type the next to the final molt yields the **pupa**. During the quiescent pupal instar, certain organs are broken down, digested by phagocytic white blood cells, and used to build the new adult. This is true of the muscles and salivary glands, which are very different in the caterpillar or the maggot than in the adult. Other structures remain, or merely continue their growth. This is true of parts of the gut, the nervous system, and the gonads, although the nervous system and other organs are extensively modified. Lastly, many structures are formed anew. The wings, legs, eyes, and mouth parts of the imago are formed from clusters of cells called **imaginal disks**, because

they form structures characteristic of the adult or imaginal instar. These disks can be removed from one larva and implanted into the body of another, where they will develop into extra organs. When the pupa molts, the adult emerges.

The **mechanism of metamorphosis** has been known only in recent years and is by no means completely understood. The first clues came from tying ligatures around caterpillars a short time before metamorphosis. If this is done at different body levels, the part of the body in which the brain is located molts to the pupal stage while the part on the other side of the ligature from the brain remains in the caterpillar stage. Figure 22-12 shows a similar experiment using the readily available bluebottle fly maggot. Implanting certain glands into the part of the larva that has been separated from its own brain by a restriction results in that part metamorphosing also. At first it seemed that metamorphosis in each insect might be different, but as has been discovered with respiration and many other matters, the underlying mechanisms turned out to be remarkably alike.

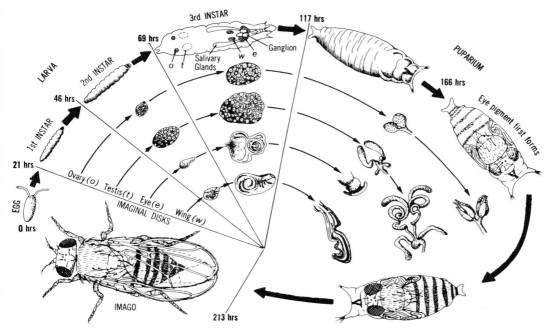

Fig. 22-11. Life cycle and metamorphosis in *Drosophila,* showing the development of ovaries, testes, eyes, and wings from imaginal discs. (Courtesy D. Bodenstein. From M. Demerec: *Biology of Drosophila.* Hafner Publishing Co., Inc., New York, 1950.

Large cells in the brain, **neurosecretory cells,** secrete a growth and differentiation hormone. This hormone functions, as do the hormones of the pituitary gland at the base of the brain in vertebrates, by stimulating a second gland. This is the **prothoracic gland**, in moths and bugs located in the prothorax; in flies it lies close beside the brain. The hormone of the prothoracic glands, **ecdysone**, acts on the skin and other structures, inducing them to grow and molt. Thus ecdysone seems to correspond to the hormone of the Y organ in crustaceans. Chemically ecdysone is a steroid and thus is related to vertebrate sex and adrenal cortical hormones. However, this system is regulated by a hormone from a pair of small glands, the **corpora allata**, located a short distance from the brain. Each corpus allatum (*corpus*, body, + *allatum*, brought to) is connected to a **corpus cardiacum** (*cardiacum*, heart-like), the whole making up a so-called **ring gland**, which in turn is connected to the brain by a pair of nerves. The hormone of the corpora allata is known as the **juvenile hormone** because, as long as it is abundant, larval molts occur but not meta-

morphosis. If the corpora allata are surgically removed, the following molt will result in metamorphosis into the adult form. If extra corpora allata are implanted before the final molt, the molt will not yield the adult but another juvenile or larval form.

If such a hormonal control of metamorphosis exists, then it should be possible to produce dwarf insects by removing the source of the juvenile hormone, the corpora allata, and to produce giant insects by adding such bodies. These feats have been achieved. By removing the corpora allata of a silkworm caterpillar in the next to the last larval instar, a Japanese worker has obtained dwarf adults. By such removal in the second to last larval instar, he has obtained still smaller dwarfs. Similar results have been attained by American workers with cockroaches. Similarly, giant-sized walking stick insects, grasshoppers, and bugs (*Rhodnius*) have all been produced by implanting extra corpora allata.

How are these three hormones regulated in the normal insect? Much remains to be learned here, but a beginning has been made. The nervous system plays an important role, as is shown by the direct connection of the corpora allata with the nervous system and the dependence of both these and the prothoracic glands on the neurosecretory cells in the brain. Work on the American silkworm, *Platysamia (Hyalophora) cecropia,* has shown that the brain must be chilled for a certain time before it will release the hormone that activates the prothoracic gland. This is an important mechanism in the life history of any insect which has but one generation per year, because it prevents an individual from developing through metamorphosis and hatching in the late summer or fall. Since the brains of these moths must be exposed to at least two weeks of cold, a sudden cold spell in the early fall will not set the insect off to a premature start. This situation is similar to the cold requirements for flowering of many biennial plants.

The action of the juvenile hormone from the corpora allata and the growth and differentiation hormone of the prothoracic glands seems to be in an antagonistic balance. Implanting extra prothoracic glands in early larval stages has an effect similar to but less dramatic than removal of the corpora allata. There is also

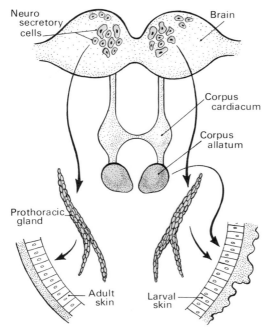

Fig. 22-12. Insect brain and glands of molting and metamorphosis.

evidence that the act of molting in some way, through the nervous system or through the body fluids, influences the endocrine activity of the brain, thus constituting a feedback system. In some insects, like the bug *Rhodnius,* nervous impulses from the stomach after a full meal appear to be an important stimulus to set off the neurosecretory activity of the brain. Recent studies on *Platysamia (Hyalophora) cecropia* have shown that brains which are not secreting a hormone are electrically silent (that is, send out no brain waves) and that cholinesterase, an enzyme essential for neural function, is absent.

Evolutionary Relationships

Insects are an extremely ancient group. Fossil cockroaches are found in Carboniferous strata laid down some 300 million years ago, tens of millions of years before the appearance of dinosaurs. Many fossil insects are beautifully preserved in amber (fossil gums from conifers) and in ancient deposits of volcanic ash and sand from what must have been lake bottoms. Many genera present in this 50-million-year-old material are represented by living species, and some of these ancient specimens are indistinguishable from present-day forms.

The evidence at hand very strongly suggests that insects arose from millipede-like ancestors by neoteny, *i.e.,* the attainment of sexual maturity by a larval stage which thereby becomes the adult. The millipede larva has three pairs of legs, tracheae, Malpighian tubules, and several other traits that make it closely resemble an insect. Millipedes are mandibulate arthropods, the familiar slow-moving vegetarians found under boards and in woodlands.

Kinds of Insects: An Overview

The classification of insects brings a manageable system to this stupendous array of animals and reveals their evolutionary kinship. Four major characteristics provide the basis for insect taxonomy.

1. The first criterion for insect classification is whether or not wings are present. Several small orders of very primitive, somewhat crustacean-like, insects constitute the subclass **Apterygota** (*a,* not, + *pteryon,* wing). All other insects belong to the subclass *Pterygota,* in which the character of the wings forms an important and easily usable taxonomic feature.

2. Among the winged or **pterygotous** insects, the type of **metamorphosis** is a major taxonomic trait. Again there are two major groups. In the first group (**Exopterygota**), wings develop on the outside of the body. Here are found the cockroaches, grasshoppers, and termites, among others. The egg hatches into a form resembling a miniature adult with small wing pads. These larval forms are called **nymphs** and undergo a series of molts, gradually becoming more and more like the adult until the adult or imago is reached. Because the differences between each stage or instar are small, this type of metamorphosis is also called **gradual, direct, incomplete,** and **hemimetabolous** as well as exopterygotous. The nymphs in most species have large compound eyes. In the second group (**Endopterygota**) are found the butterflies, beetles, flies, ants, bees, and many others. The **larval** form is usually a worm-like grub or caterpillar with tiny eyes or none. The next to the final molt commonly produces a quiescent stage, a **pupa,** from which the adult will emerge. This type of metamorphosis is termed **sudden, indirect, complete,** and **holometabolous.**

3. A third important basis for insect classification is the nature of the **mouth parts,** structures that are easily studied under a dissecting microscope. They reveal not only the evolutionary relationships of the various species but also furnish important clues about the food of the insect. The basic primitive pattern of mouth parts can be seen in a cockroach or grasshopper, but even in a mosquito all the parts present in a grasshopper can be identified, although highly modified as a stiletto for withdrawing blood.

4. A fourth commonly useful taxonomic criterion is the nature of the **external genitalia.** In fact, differences of this kind seem to have been important in evolution by making the exchange of genes impossible between sympatric species, *i.e.,* species occupying the same general area.

Many other traits are used in taxonomy to identify and distinguish between the multitudes of insect species. Those concerned with external structures such as the eyes, antennae, legs, and surface bristles are especially useful.

Such is the nature of evolution's most successful product on this planet. Whether a similar form or something radically different from anything known on earth has been successful on some other planet, only the future can tell.

USEFUL REFERENCES

Bates, M. *The Natural History of Mosquitoes.* The Macmillan Co., New York, 1949.

Borror, D. J., and Delong, D. M. *An Introduction to the Study of Insects,* 3rd ed. Holt, Rinehart & Winston, New York, 1971.

Brues, C. T., Melander, A. L., and Carpenter, F. M. *Classification of Insects, Rev. ed.* Museum of Comparative Zoology, Cambridge, Mass., 1954.

Burch, J. B. *How to Know the Eastern Hand Snails.* Wm. C. Brown, Co., Dubuque, Iowa, 1962.

Cloudsley-Thompson, J. L. *Spiders, Scorpions, Centipedes, and Mites.* Pergamon Press, New York, 1958.

Erlich, P. R. *How to Know the Butterflies,* 2nd ed. Wm. C. Brown Co., Dubque, Iowa, 1961.

Farb, P. *The Insects.* Time-Life Nature Library, New York, 1962.

Green, J. *A Biology of Crustacea.* Quadrangle Books. Chicago, 1961.

Hiffler, J. R. *How to Know the Grasshoppers.* Wm. C. Brown Co., Dubuque, Iowa, 1963.

Jaques, H. E. *How to Know the Beetles.* Wm. C. Brown Co., Dubuque. Iowa, 1951.

Jaques, H. E. *How to Know the Insects.* W. C. Brown Company. Dubuque, Iowa, 1951.

Kaston, B. J. and Keston, E. *How to Know Spiders.* Wm. C. Brown Co., Dubuque, Iowa, 1953.

Klots, A. B. *A Field Guide to the Butterflies.* Houghton Mifflin Co., Boston, 1951.

Lutz, F. E. *Field Book of Insects.* 3rd ed. G. P. Putnam's Sons, New York, 1948.

Snodgrass. R. E. *A Textbook of Arthropod Anatomy.* Cornell University Press, Ithaca, N. Y., 1952.

Waterman, T. H. (ed.) *The Physiology of Crustacea: I, Metabolism and Growth: II, Sense Organs. Integration, and Behavior.* Academic Press, New York,

Wilson, E. O. *The Insect Societies.* The Belknap Press of Harvard University Press, Cambridge, Mass., 1971.

THE WORLD
OF LIFE:
CHORDATE
PATTERNS

Vertebrate Beginnings: Primitive Chordates and Echinoderm Relatives

The chordates will always be of special importance and interest to mankind because we ourselves are chordates. The primitive ones like the tunicates and amphioxus stand close to the beginning of the long evolutionary trail which led to mankind, and their traits remain with us to this day. Our pituitary gland, for example, although it is located in the head, controls the activity of the gonads; its fore-runner in the tunicates is also located close to the brain, where its position is readily explained. The chordates number about 55,000 species, mostly vertebrates like ourselves. They were established as a phylum in 1874 by the great German evolutionary theoretician, Ernst Haeckel, largely on the epoch-making discoveries of Alexander Kowalevsky on amphioxus and the tunicates.

CHORDATE CHARACTERISTICS

All chordates, from the simplest worm-like marine forms to the most sophisticated of the vertebrates, possess at some stage in their life histories the same basic set of anatomical features.

The long axis of the body of every chordate is strengthened by a stiff but flexible rod, the **notochord**. This notochord confers on the chordates a swimming ability that enables them to completely outclass other organisms of comparable size and shape. The difference in locomotion between a chordate and a segmented worm is like that between two men each standing in the stern of a boat and trying to propel it forward, one by moving a rigid oar back and forth, the other by wiggling a rope.

Both methods work but the rigid oar is vastly superior. The vertebrates improve on the notochord by converting it during embryonic development into a series of bony disks, the **vertebrae**.

A second feature common to all chordates and not found in any other phylum is a set of pharyngeal **gill slits**. These are paired openings on either side of the pharynx or throat, so placed that water taken in at the mouth can pass to the exterior again through these slits. In all nonvertebrate chordates the gill slits remain functional throughout life and serve as organs of both respiration and food catching. Among the vertebrates the gill slits remain throughout life only in fish, but in mammals and birds they are easily seen in embryonic stages. Gill slits are obvious in a human embryo about 30 days old. In land animals, of course, the gill slits grow closed. The first pair forms the Eustachian tubes of the ears and others help form some of the endocrine glands.

A third common feature of chordates is a **dorsal tubular nervous system**. This tube, made up of nerve cells and their extensions, runs the length of the animal just dorsal to the notochord. In vertebrates it develops into the central nervous system, composed of the brain and spinal cord. This dorsal neural tube presents a marked contrast to the ventral chain of ganglia characteristic of nonchordates like the insects and annelids, or to the nerve nets of the coelenterates.

A number of other important traits characterize chordates but are not diagnostic because chordates share them with other groups. Chordates are **bilaterally symmetrical**, with right and left, anterior and posterior, and dorsal and ventral sides. Like annelids and arthropods, chordates at some stage in their lives are **serially segmented**. Even in man this segmentation is clearly evident in the backbone, ribs, nervous system, and body musculature. Like annelids, most chordates have a spacious **coelom** or body cavity between the body wall and the alimentary canal.

THE TUNICATES (UROCHORDATA) AND THE ORIGIN OF THE PITUITARY GLAND

The most easily found and studied of the primitive chordates are the tunicates, so named by Lamarck because the body is covered by a tough protective "tunic" or coat. They are found in all oceans, some species living attached to rocks or wharves, others swimming in the high seas far from shore. Their practical importance arises because they are what navi-

gators call **fouling organisms** that form undesirable growths on ships' bottoms. They were the first animals in which alternation of generations was discovered. This was the achievement of a German poet and naturalist, Chamisso, and led directly to a very important practical application, an understanding of the life history of tapeworms. This was followed by the discovery that the malaria parasite likewise has a double life cycle with sexual reproduction in the mosquito and asexual reproduction in man.

The familiar tunicates vary greatly in appearance. Some look like black velvety bananas, others like gnarled brown potatoes or scarlet thumbs attached to rocks. The fertilized eggs develop into minute **tadpoles** with a basic structure similar to frog tadpoles. These larvae are equipped with a notochord, segmented body musculature, and a long dorsal nerve cord innervating the muscles. Thus, they are well

Fig. 23-1. From the strange creatures in the foreground began the long journey to mankind.

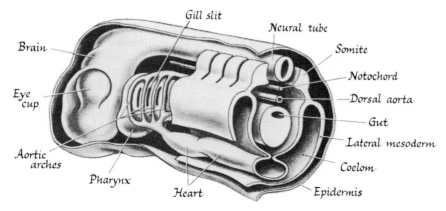

Fig. 23-2. Basic chordate structure as seen in a vertebrate embryo in the gill slit stage.

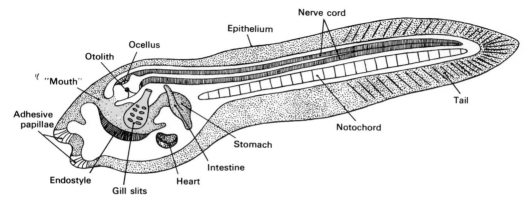

Fig. 23-3. A tunicate tadpole. Although the basic structure is closely similar to a frog tadpole, the mouth usually does not actually form.

equipped for their role as site-seekers to disseminate the species. Once settled on a rock or other object, the tadpole metamorphoses into the adult.

In several ways tunicates are among the strangest creatures known. Although they are animals and closely related to the vertebrates, they secrete cellulose, the material of which their tunics are formed. Their long tubular hearts, which can be readily seen when the tunic is removed, pulsate first in one direction and then in the other, reversing their beat every few minutes. Their blood cells do not contain hemoglobin but in many species are bright green, yellow, red, blue, or purple from various oxidation states of the vanadium they contain. It has recently been discovered that, when

injured, these **vanadocytes** cause the blood to clot.

A Typical Tunicate

Ciona may be regarded as typical of tunicate structure, since it is relatively simple and is considered by specialists probably to be representative of the most primitive and generalized type. The adult sedentary *Ciona* is cylindrical, with two funnels or siphons at the free end. The one at the tip is the mouth, or incurrent siphon, while the one just beside it but lower down is the excurrent atrial siphon.

Like clams, tunicates are **filter feeders.** The cilia of the gills produce a current of water which sweeps food particles into the pharynx where they are caught on the gills in mucus secreted by a long groove, the **endostyle**, running along the mid-ventral line of the pharynx. The mucus is swept upwards and then posteriorly into the stomach and intestine, wich curves upwards in a "U" ending near the excurrent siphon.

The brain lies close to the surface between the two siphons. Just below and in close contact with the brain and with a ciliated funnel opening into the dorsal part of the pharynx is the **subneural gland**, believed to be the forerunner of our pituitary. When gametes from another tunicate enter the feeding current and are drawn into the ciliated funnel, the subneural gland signals the gonads to release sperms or eggs and fertilization is assured. Thus the strange position of the pituitary, which controls the gonads but which is located at the base of the vertebrate brain, can be explained. Not only is the subneural gland in the same position, at the base of the brain, as is the pituitary gland, but extracts of it will stimulate the growth of ovaries in adolescent mice and initiate the contraction of uterine muscles.

AMPHIOXUS (CEPHALOCHORDA) AND THE ORIGIN OF VERTEBRATES

Amphioxus has long been regarded as the prototype of all the chordates, including those with backbones. The lancelet, as amphioxus is sometimes called, looks something like a flattened worm or a very thin little eel with a pointed head, no eyes and no paired fins. It is only about 5 to 8 centimeters (2 to 3 inches) long, translucent, and a very rapid swimmer. Its habitat is sandy bottoms near shore but below low tides southward from the Virginia Capes on the Atlantic and from San Diego on the Pacific. On the coasts of China it is abundant enough to serve as human food.

According to present evolutionary theory, vertebrates arose when the tadpole larvae of sedentary forms like the tunicates took to swimming up rivers against the current and living on the rich food supply being washed down from the newly occupied land. Amphioxus apparently is one of those pioneers which omitted the sessile adult form and became sexually mature while remaining in its juvenile body form, *i.e.*, became **neotenous**. Amphioxus

has remained in river estuaries and has spread out along the seacoasts. It has also remained a filter feeder like the tunicates. Its life is spent more or less buried in the sand. Water currents set up by the cilia on the gills enter through the mouth, pass through the gill slits and out into an atrial cavity, and thence to the exterior through a pore which corresponds to the excurrent siphon of a tunicate.

As in the tunicates there is a deep, ciliated, and mucus-secreting **hypobranchial** (below the gills) **groove**, the **endostyle**, extending along the ventral side of the pharynx between the gill slits. This groove on the floor of the throat is believed to correspond to the vertebrate **thyroid gland** for three reasons. It lies in the same position in the pharynx as the embryonic thyroid of any vertebrate. Furthermore, when the larval form of the lamprey eel, which is a vertebrate and which closely resembles amphioxus (including the possession of an endostyle), undergoes metamorphosis into the adult, the

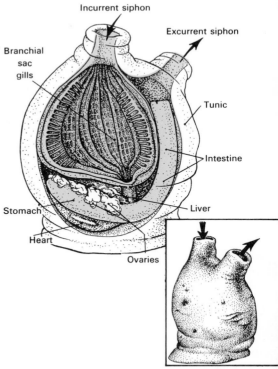

Fig. 23-4. A sessile adult tunicate. The connection between the esophagus at the base of the enormous gill sac and the stomach is hidden by the liver (× 1/3).

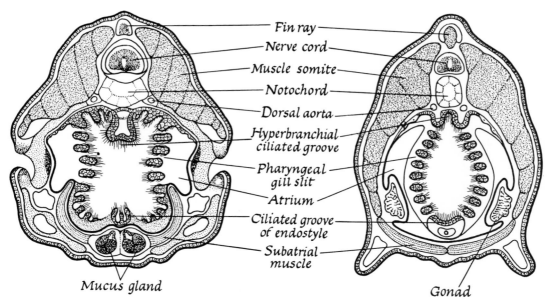

Fin ray
Nerve cord
Muscle somite
Notochord
Dorsal aorta
Hyperbranchial ciliated groove
Pharyngeal gill slit
Atrium
Ciliated groove of endostyle
Subatrial muscle
Mucus gland
Gonad

Fig. 23-5. Comparison of amphioxus (right) with the larva of a very primitive vertebrate, *i.e.*, with the ammocoetes larva of a lamprey.

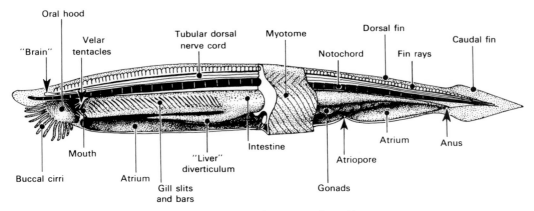

Oral hood
Velar tentacles
Tubular dorsal nerve cord
Myotome
Dorsal fin
Caudal fin
"Brain"
Notochord
Fin rays
Mouth
"Liver" diverticulum
Intestine
Atrium
Anus
Buccal cirri
Atrium
Atriopore
Gill slits and bars
Gonads

Fig. 23-6. Diagram of structure of amphioxus.

endostyle develops into the thyroid. Further evidence for this theory is now at hand. It is well known that the thyroids of frogs and mammals concentrate iodine in their cells, where it is incorporated into the molecules of hormone secreted by the gland. Recent studies using radioactive iodine show that the endostyle of amphioxus does indeed concentrate iodine. Even more convincing is the additional discovery that anti-thyroid drugs like uracil likewise inhibit the accumulation of iodine by the endostyle of amphioxus.

The origin of all the primitive chordates is very obscure. The first to appear may have been sessile filter feeders like *Rhabdopleura*, living on fine detritus drifting in the sea, or perhaps the worm-like Enteropneusta (hemichordates), neither of which are regarded by some experts as chordates, although they show similarities to both them and the echinoderms.

ECHINODERMS

Starfish, sea urchins, and other echinoderms are chiefly of interest to children and biologists. *Arbacia*, the most common sea urchin along the western shores of the Atlantic from Cape Cod to Yucatan and throughout the West Indies, has become a standard laboratory animal for the cell physiologist. The eggs of sea urchins and starfish are favorite experimental material for embryologists. The facts of fertilization and early development of the egg were first clearly worked out on echinoderm eggs at the famous Naples biological station. The remarkable powers of regeneration of these animals have long challenged biologists for an explanation. A starfish can be torn into pieces and each fragment give rise to a complete animal. Sea cucumbers, when attacked, can eviscerate themselves and then later regenerate a complete set of replacement internal organs. In the Orient and in the Mediterranean, sea cucumbers are eaten, as *trepang* in Singapore, *bêche-de-mer* in Marseilles. Sea urchin eggs are also edible.

Starfish eat oysters and clams. The infamous crown-of-thorns starfish has suddenly become very destructive to coral reefs in widely separated parts of the South Pacific and the Red Sea for quite unknown reasons. Although factors as different as DDT and radiation fallout have been blamed, there were similar outbreaks many decades ago.

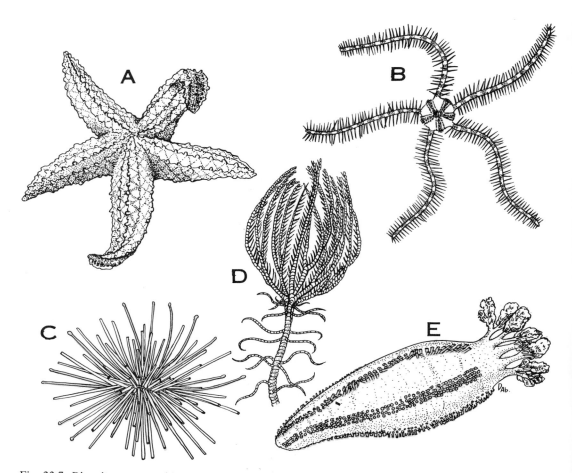

Fig. 23-7. Diversity among echinoderms. A, starfish; B, brittlestar; C, sea urchin; D, crinoid or sea lily; E, sea cucumber. The tube feet are colored green. (Reduced ×½.)

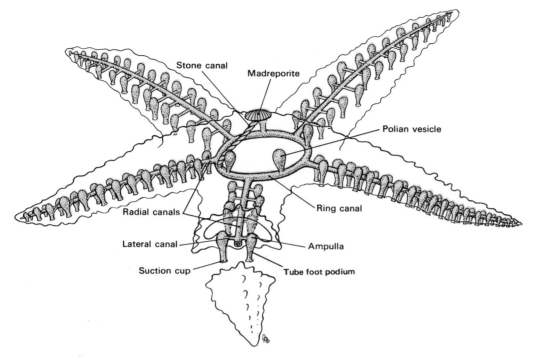

Fig. 23-8. Water vascular system of a starfish.

Echinoderms as a Phylum

Echinoderms have been known from ancient times. Aristotle described the beautifully precise and complex jaw apparatus of sea urchins over 2,000 years ago. About 1800 Lamarck brought together in one group most of the animals we know as echinoderms. It is now clear that they constitute a well-defined phylum of about 6,000 living and over three times as many fossil species. Exclusively marine, they occur from tide lines to great depths.

It would be hard to imagine animals more different from the vertebrates than starfish and their relatives, yet echinoderms are clearly related to us as vertebrates, and far more closely than are the social insects, for example. The characteristic echinoderm features include:

1. The possession of a calcareous endoskeleton, commonly with spines. It is this trait that gives them their name (*echinos*, hedgehog [spiny] , + *derma,* skin). Even the longest spines of a sea urchin, those which appear like exoskeletal structures, are in fact actually

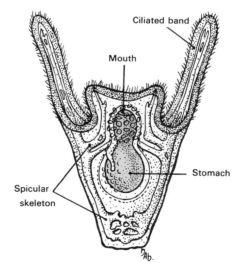

Fig. 23-9. Pluteus larva of a sea urchin. The gut is colored green.

covered by a thin layer of living cells which have secreted the bony spine.

2. A conspicuous radial symmetry as adults. Usually it is a five-rayed symmetry, although

there are exceptions. From the St. Lawrence River north there is found a common species of starfish which normally has six arms, and from Puget Sound north there is a very different species which also possesses six arms. *Solaster* may have 20 or more rays.

3. A water-vascular system, which is a unique feature of the echinoderms. This is a locomotor system consisting of a circular duct, the **ring canal** around the esophagus, and five long water-filled ducts, the **radial canals**, that extend out from it, one along each ray. There is also a calcified canal that extends from the ring canal up to the aboral surface, *i.e.*, the surface away from the mouth. On the aboral surface this **stone canal** ends in a sieve-like disk, the **madreporite**, often called the "eye." It is thought to serve as an intake valve for the water-vascular system. Arranged in pairs along the five radial canals are the **tube feet**. These are hollow muscular tubes. On the inner end of each is an expanded bulb and on the protruding end, a suction disk. By contracting the muscles of the bulb and simultaneously relaxing those of the protruding tube, the foot can be greatly extended, and by contractions of the musculature on one or another side of the tube, the foot can be bent in any direction.

Like the vertebrates, echinoderms not only have a calcareous endoskeleton but a capacious coelom and a nervous system which, in part, lies as a ribbon exposed to the surface like a vertebrate embryo's neural plate. This can be most easily seen along the ventral midline of the arm of a starfish. The development of echinoderms shows basic similarities to vertebrates. The larva is bilateral, and in a sea urchin is called a **pluteus**. The somewhat simpler larva of a starfish, called a **bipennaria**, closely resembles the larvae of certain very primitive chordates. The adult echinoderm is formed by a profound metamorphosis of the larva.

USEFUL REFERENCES

Berrill, N. J. *The Origin of Vertebrates.* Clarendon Press, Oxford, 1955.

Bigelow, H. B., and Farante, I. P. Lancelets. In *Fishes of the Western North Atlantic.* Yale University Press, New Haven, 1948.

Boolootian, R. A. (ed.) *Physiology of the Echinodermata.* John Wiley & Sons, Inc., New York, 1966.

Harvey, E. B. *The American Arbacia and Other Sea Urchins.* Princeton University Press, Princeton, 1956.

Hyman, L. H. *The Invertebrates*, Part IV: Echinodermata. McGraw-Hill Book Co., New York, 1955.

Nichols, D. *Echinoderms.* Hillary House Publishers, New York, 1962.

Fish and Amphibians

24

All vertebrates possess the basic chordate traits, *i.e.*, gill slits in the pharynx (at least as embryos), a dorsal tubular nervous system, and a notochord extending along the length of the animal immediately ventral, *i.e.*, below the tubular nerve cord. The feature distinguishing the vertebrates from other chordates is that during embryonic life, a backbone develops around the notochord, which itself disappears during subsequent growth. This stiffening of the long axis of the body is the central advance in passing from the prochordates to the vertebrates. It represents a true breakthrough in locomotion conferring a new and higher order of speed and agility.

Fish and amphibians are placed in the same taxonomic group because they possess many structural similarities having to do with adaptation to aquatic or semi-aquatic environments. They lack the key adaptation needed for terrestrial living, an amnion. The vertebrate amnion is a fluid-filled sac surrounding the embryos of reptiles, birds, and mammals, permitting the embryo to develop within an egg laid on land or retained within a uterus, in the case of the higher mammals. Hence, fish and amphibians are grouped together as **anamniotes**.

FISH

Fish fall into two basic subdivisions: the **Jawless fish**, including the cyclostomes, *i.e.*, lampreys, and hagfish which are eel-like species lacking paired fins and parasitic on other fish; and the **jawed fish**, which possess two pairs of fins, an anterior and a posterior pair, and include the sharks and all the various bony fish.

The Jawless Fish (Agnatha)

The cyclostomes (*cyclos*, circle, + *stoma*, mouth), or lampreys and hagfish, differ radically from all other vertebrates in their complete lack of jaws. Their mouths are merely round sucking holes. In the most common group, the lampreys, the mouth has a rasping tongue and is surrounded by a fleshy hood, so the anterior end of the animal resembles a plumber's suction cup. Cyclostomes also differ from all other vertebrates in a complete absence of paired limbs, either fins or legs, and in having a larva, called **ammocoetes**, which closely resembles the primitive chordate amphioxus. There is no doubt that cyclostomes are the most primitive of all vertebrates and closely related to various very early fossil forms which also lacked jaws.

Lampreys and hagfish resemble ordinary eels, but eels are merely bony fish essentially like trout with jaws and paired fins but with bodies drawn out into a snake-like form. There are many additional internal differences. To breed, lampreys migrate from the sea up rivers where their amphioxus-like larvae live as filter feeders while eels migrate from fresh water into the mid-Atlantic to breed. Landlocked lampreys in the Great Lakes migrate up rivers and streams to breed. Hagfish are entirely marine.

All present-day cyclostomes parasitize other fish. The lampreys, which grow to be almost a meter long, attach themselves to the side of a fish, rasp a hole in its side and suck out its blood. Hagfish burrow into the interior of fish.

The importance of lampreys is two-fold. They are regularly sold for food and are considered a great delicacy by gourmets. Unfortunately, the destruction of other fish which they have wrought in the Great Lakes in recent years has far overbalanced the value of lampreys as food. In 1829 the Welland Canal around Niagara Falls opened the Great Lakes to the sea lamprey. The lampreys did not thrive in Lake Erie (perhaps because it is shallow), but once they had passed through it into the deep lakes beyond, serious trouble began. Lake trout and other commercial fish were found in increasing numbers with lamprey scars. The number of fish that were caught fell precipitately. In Lake Huron, for example, 1,743,000 pounds of lake trout were caught in 1935, against 940,000 in 1940, only 173,000 in 1945, and a negligible 1,000 pounds by 1950. Since then control measures have been at least partially successful. Since these lamprey live as larvae in streams, it has been possible to attack them at this vulnerable stage with specific poisons which apparently have only a neglible effect on other organisms in the stream. Still more recently salmon have been introduced into Lake Michigan with dramatic success. How they escape lampreys is uncertain.

Jawed Fish (Ichthyagnatha)

Sharks, Skates, and Rays

The elasmobranchs (*elasmos*, plate + *branchia*, gill), so-called from the plate-like character of their gills, which are well known to anyone who has ever slit a shark's throat, are also called Chondrichthyes (*chrondros*, cartilage, + *ichthyes*, fish). Their skeletons are composed not of bone but of cartilage. Unlike the cyclostomes but like all other vertebrates, the elasmobranchs have jaws. Externally their gill openings are visible as five vertical slits on either side of the pharynx. In the other groups of fish, only one slit is visible externally because the gill slits are covered by a protective flap-like operculum. The body is covered with a characteristic type of scales (**placoid**) resembling teeth in structure and showing transitional stages from typical scales to typical teeth at the edges of the mouth. Sharks are commonly regarded as the wolves of the sea. Recent studies by Eugenie Clark have shown that they are not only dangerous but also intelligent.

Fig. 24-1. Lampreys attacking trout. (Kindness of R. E. Lennon and the U. S. Fish and Wildlife Service.)

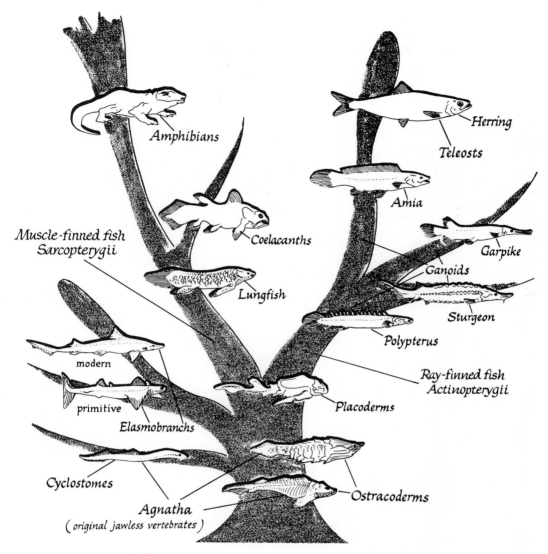

Fig. 24-2. Diversity and evolutionary relationships among fish.

Ray-Finned Fish (Actinopterygii)

The familiar ray-finned fish represent the first great deployment of vertebrates. In contrast to the scant 3,000 species of elasmobranchs, there are a bewildering 20,000 species of the ray-finned. As their name implies, they have merely **flexible dermal rays** to support their fins with an occasional one hardened into a spine. Thus they stand in sharp contrast to the elasmobranchs which possess a cartilaginous skeleton extending out into their paired fins and the lungfish which have both a bony skeleton and muscles supporting and moving their paired fins. Most of the species in this group are teleosts, easily identified by their relatively thin flexible scales overlapping like the shingles on a roof.

The social and economic importance of fish is great everywhere except in desert regions, and is especially so for maritime countries like Norway, England, and Japan. One of the great complications in the fishing industry, and one felt quite keenly in small countries like Norway and Japan where fisheries represent a large proportion of the total national income, is the erratic nature of fish populations. One year fish may be superabundant and there is a glut; perhaps the next year there will be very few. Despite prolonged investigations, the migrations and the rise and fall of fish populations are very poorly understood and remain one of the important challenges of oceanography for the future.

Fig. 24-3. Flying fish. (Kindness of H. E. Edgerton.)

Lungfish (Sarcopterygii)

The most notable fact about the muscle-finned fish is that they quite clearly gave rise to all the terrestrial vertebrates—amphibians, reptiles, birds, and mammals. Their scientific name, Sarcopterygii, refers to the flesh or muscles on the fins (*sarcos*, flesh, + *pteron*, wing or feather). The muscles on the paired fins are supported almost to the tip by a jointed bony skeleton. These fish are also sometimes called Choanichthyes (*choane*, funnel, + *ichthyes*, fish) to refer to the nostrils which only fish of this group possess. The nostrils of elasmobranchs and of teleosts and other ray-finned fish are merely little blind sacs lined with an olfactory epithelium. The nostrils of the muscle-finned sarcopterygians are like those of air-breathing terrestrial vertebrates and connect the opening on the snout or nose with the inside of the throat. These nostrils are indeed funnels. Present-day lungfish occur in Australia, Africa, and Brazil.

AMPHIBIANS

Frogs, toads, salamanders, and other living members of the class Amphibia are sharply separated from the fish on one side and from the reptiles, birds, and mammals on the other. They differ from the fish by having pentadactyl, *i.e.*, five-toed, legs instead of paired fins. Living species differ from the scaly reptiles by a lack of external scales, and by laying eggs which have to develop in water or at least in a very damp place.

The egg, embryo, and larva of the amphibians are almost indistinguishable from those of the lungfish. Prominent among the fish-like traits of the larval or tadpole stage characteristic of amphibians are the functional gills and gill slits. The lateral line of sense organs along each side of the body, and a tail fin that usually extends well forward along the back. The fin never has supporting rays as in fish, but is important in swimming. Tadpoles excrete nitrogenous wastes as simple ammonia as fish do, instead of converting them to urea like frogs and mammals.

Relatively abrupt metamorphosis from the aquatic larva into a gill-less adult is characteristic of amphibians. This has long been

Fig. 24-4. End of the road; spring peepers. Modern amphibians are an evolutionary blind alley. (Kindness American Museum of Natural History.)

known to be controlled by the thyroid gland, but only recently has it been discovered that the hypothalamus of the brain plays an essential though incompletely understood role.

The frogs, and to a lesser extent the salamanders, have proved to be extremely useful animals for biological teaching and research. They are clean, cheap, readily available, and certainly far outrank the guinea pigs as laboratory animals. The economic importance of the amphibians is negligible.

Evolutionary Relationships

The ancestors of the amphibians were clearly muscle-finned fish (*i.e.*, Sarcopterygii), probably closely related to our present-day lungfish which, when first hatched, so resemble larval salamanders that only an expert can distinguish them. The amphibians represent a striking case of pre-adaptation because, as first pointed out by Alfred Romer at Harvard's Agassiz Museum, their three major adaptations for terrestrial life—legs, lungs, and eggs that can be laid on land—were all adaptations developed by aquatic ancestors which helped them continue to survive in the water. Legs enabled primitive salamanders to clamber from a drying water hole to a wetter place. Lungs made air breathing possible. The eggs that could be laid out on the bank meant that they could be placed out of reach of egg-eating predators in the water. These important adaptations opened the way for terrestrial life as a byproduct. It was a classical case of what is called **pre-adaptation**, *i.e.*, the development of an adaptation in one environment which incidentally adapts an organism to live in a different environment.

A rather confusing aspect of modern living amphibians is that they are all highly modified from their primitive ancestors. The fossil record shows that the first amphibians were similar to scaly lizards and laid reptile-like eggs with protective shells. This type of egg is still found in some tropical amphibians. Frogs in particular are highly specialized for jumping and very different from any generalized vertebrate.

There are only three living groups of amphibians. The salamanders are characterized by the possession of tails throughout life. The frogs and toads possess tails only as larvae in the tadpole stage. In the tropics there is a small group of worm-like amphibia which lack legs.

USEFUL REFERENCES

Fish

Bigelow, H. B. *Fishes of the Western North Atlantic.* Yale University Press, New Haven, 1948.

Brown, M. E. (ed.) *The Physiology of Fishes*, Part I: Metabolism; Part II: Behavior. Academic Press, New York, 1957.

Carlander, K. D., *Handbook of Fresh-water Fishery Biology*. W. C. Brown Co., Dubuque, Iowa, 1955.

Hardisty, M. W. and Potter, I. C. *The Biology of Lampreys.* 2 vols. Academic Press, New York, 1972.

Lagler, K. F., Bardach, J. E., and Miller, R. R. *Ichthyology.* John Wiley & Sons, Inc., New York, 1962.

Amphibians

Bishop, S. C. *Handbook of Salamanders: The Salamanders of the United States, of Canada, and of Lower California.* Comstock Publishing Co., Ithaca. N. Y., 1968.

Conant, R., *A Field Guide to Reptiles and Amphibians.* Houghton Mifflin Co., Boston, 1958.

Moore, J. A. (ed.) *Physiology of the Amphibia.* Academic Press. New York, 1964.

Oliver, J. A., *The Natural History of North American Amphibians and Reptiles.* D. Van Nostrand Co., Princeton, 1955.

Wright, A. A. and Wright, A. H. *Handbook of Frogs and Toads: The Frogs and Toads of the United States and Canada.* Comstock Publishing Co., Ithaca, N. Y., 1933.

Reptiles, Birds, and Mammals

25

Birds and mammals are so much like some of the reptiles in basic structure, such as brain, skeleton, muscles, heart, kidneys, lungs, reproductive system, and embryonic development, that some comparative anatomists and paleontologists have suggested that both are really special kinds of reptiles. Both are warm-blooded, *i.e.,* **homoiothermic**, which is a key trait conferring on them far greater freedom from their environment than any reptile ever had. Birds maintain a constant body temperature by insulating their bodies with feathers and mammals with hair, both of which evolved from the scales of their reptilian ancestors.

Reptiles, birds, and mammals all possess a special membrane, the **amnion**, which grows completely around the embryo, enclosing it in its private pond and thus enabling the embryo to develop within an egg laid on the land or within the uterus of a mammal. Hence these three groups are known as **amniotes**.

REPTILES

Reptiles may be defined either as scaly tetrapods or as "cold-blooded" *i.e.,* **poikilothermic, amniotes**. In the fossil record they can be distinguished from the amphibians with great difficulty if at all.

Importance to Mankind

The reptiles are of little economic, medical, or even esthetic importance to man. Compared with the fish, birds, mollusks, and many other groups, their human importance is negligible. The most valuable of all reptiles is easily the friendly green turtle, *Chelonia mydas*, of the warmer parts of the ocean. Both its eggs and its flesh find a ready market. The average adult green turtle ranges between 300 and 500 pounds in weight, but individuals up to 850 pounds are known. The diamondback terrapin of brackish bays along the seacoast is the species most prized for soup. The pugnacious hawksbill turtle, *Eretmochelys imbricata*, like the green turtle, inhabits the warmer seas and is esteemed for its eggs and flesh. It is the species which provides natural tortoise shell.

BIRDS

Birds are feathered bipeds, but this definition tells little about the class Aves, except by implication. Birds, like mammals, are homoiothermic amniotes, *i.e.*, they maintain a constant and high body temperature, and the embryo grows a protective amnion enclosing itself in a sac of fluid during development. Both the anatomy and the physiology of birds are profoundly modified by their dominant adaptation, the ability to fly.

Importance to Mankind

From before the dawn of history birds have touched human life at many points. Stone Age men snared and shot them both for food and for their feathers. Ever since, the plumage, the songs, the migrations, and the behavior of birds have fascinated a large segment of mankind. The Egyptians, the Greeks, and other peoples of antiquity developed an extensive bird lore, part observation, part myth. Well over

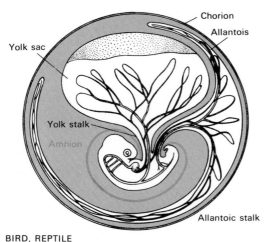

BIRD, REPTILE

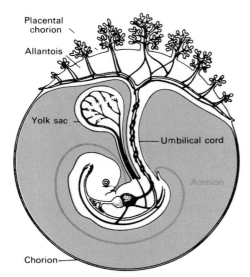

MAMMAL

Fig. 25-1. Diagram showing the relationships of the fetal membranes in reptiles, birds, and mammals. Amnions in green.

100 different kinds were named—not only the familiar eagles, owls, swans, ravens, but various kinds of wrens, swallows, flamingos, herons, ostriches, shearwaters, and even the Indian talking mynah bird. The famous birds of Diomedes, whose eerie wailing cries and nocturnal flights have shivered the spines of many an ancient mariner, were shearwaters. These are the birds that Saint Augustine, a native of the Mediterranean, mentions in his great book, *The*

City of God. Caged songbirds were as common in the marketplace of Athens in the age of Pericles as they are today in the little market under the shadow of Notre Dame of Paris.

Since World War II there has been a great resurgence of interest in bird behavior among zoologists. A series of challenging new theories has led to a host of new observations and the discovery of new facts.

The most important bird economically is the domestic chicken. No one knows precisely how or when it was domesticated. It is not mentioned in Homer or the Old Testament, but by the great age of Greek civilization it was commonly known as the Persian fowl. According to Chinese legend, the chicken was introduced into that culture about 1400 B.C. from the West. Today jungle fowl closely similar to old-fashioned black and brown barnyard chickens roam forests and thickets from eastern India to the Malay Peninsula.

Evolutionary Relationships

Birds evolved from reptiles. Both the structure of living birds and the fossil record are very clear about this. The skeleton, the muscles, the heart, the character of the scales, and the manner of development all point in that direction. Beautifully preserved fossil birds have been found dating from the Mesozoic (the Age of Reptiles), some 150 million years ago. *Archaeopteryx* and *Archaeornis* both possessed long reptilian tails with feathers along each side, three prominent clawed fingers on each wing, solid bones, and a complete set of teeth. Their skeletons bear a very close resemblance to those of many of the smaller dinosaurs, with long hind legs adapted for running and jumping, shorter front legs, and a generally light body build. Other fossil birds have been found in later rocks. Some of these are transitional between the long-tailed *Archaeopteryx* and the stub-tailed modern birds.

Like reptiles and mammals, the birds have undergone **adaptive radiation** which fits them into many diverse environments and types of life. Almost 30 orders exist of which our familiar song or passerine (literally "sparrow-like") birds are only one.

Birds and Light-Darkness

The flight of birds has been envied and studied since the days of Leonardo da Vinci, who dissected birds to learn the secret of flight. In our own day the major excitement in the study of birds are the discoveries being made about their behavior (discussed in Chapter 34) and other discoveries concerning how their reproduction and migration are controlled by photoperiod, *i.e.*, the number of hours of light and darkness birds experience within every 24-hour period.

Reproductive Activity

Some birds, like the common herring gull, *Larus argentatus*, are determinate layers. Once such birds begin to lay, they lay a fixed number of eggs in a clutch—three in the case of this gull—regardless of any known external stimuli. Other birds are indeterminate layers and tend to continue to lay eggs until a given number is present in the nest. The eastern flicker, *Colaptes auratus*, for example, normally lays five to nine eggs in a clutch. If, early in the series, an egg is carefully removed every time a

Fig. 25-2. Rattlesnakes in combat. (Kindness of the Zoological Society of San Diego)

Fig. 25-3. Gannet braking its flight before landing. Note the small black bastard wing or alula on the leading edge of each wing. These alula correspond to the thumb or big toe of other vertebrates. (Kindness of G. Harper Hall.)

clutch of eggs is laid, broodiness sets in, but it can be brought to a premature end by preventing the bird from setting. During incubation a special **incubation patch** develops on the ventral surface of the bird's body. The down feathers over this region are molted and the skin becomes highly vascular. The result is a hot spot against which the eggs are pressed.

The generally more primitive birds are called **precocial**, because their chicks hatch completely covered with down and are able to follow their parents and feed themselves almost immediately after hatching. Precocial species, such as ducks, chickens, or quail, usually lay large numbers of eggs. Infant mortality is high. The passerine birds, like thrushes and warblers, hatch helpless and nearly naked. Such **altricial** (*altrix*, nurse) nestlings require prolonged parental care. Altricial birds typically build well-constructed nests and lay few eggs. Infant mortality is much lower, but the ecological equilibrium maintains the numbers of both types about constant.

new one is laid, a flicker will continue to lay up to 60 or 70 eggs. Domestic fowl are indeterminate layers. Different breeds vary, but commonly a hen begins a series by laying an egg soon after dawn. Then each succeeding day an egg is laid, approximately an hour later than on the day before. This continues until the laying time falls late in the afternoon, when a day is skipped and the egg is laid early the following morning. Domestic chickens lay their maximum number of eggs the first year. Then, under comparable conditions of food and care, the number of eggs a hen lays in any year is roughly 75 per cent of the number laid the previous year.

The primary condition controlling bird reproduction, by controlling size of the gonads, is the number of hours of daylight. It is common practice to give domestic hens added hours of light during winter in order to increase egg production.

Incubating the eggs may be done by the female alone, as in chickens, or by both parents, as in the majority of birds, or in rare cases by the male alone. **Broodiness** is a definite psycho-physiological condition dependent on the internal hormonal state. As long as a hen is laying eggs she will not incubate them. When a

Migration

Annual migrations are crucial in the lives of many species of birds. Not only large birds like storks, but many of moderate and even small size migrate long distances and cross the equator twice every year. Many species of plovers, terns, swallows, and warblers do so. Many other species migrate only relatively short distances: bluejays, juncos, finches, and robins, for instance. In many species of birds males come north earlier than females. A large flock composed exclusively of males of the showy red-winged blackbird, makes a never-to-be-forgotten sight. In some species young and old birds migrate together, but in others first-year birds migrate alone—a startling fact. Many species migrate at night. Such flights are studied quantitatively by patient observers watching the full moon through telescopes.

What is the **adaptive purpose** of migration? Very clearly migration takes birds from the crowded tropics or subtropics into otherwise largely unoccupied regions to breed where food is abundant and the days are longer than they are near the equator.

MAMMALS

Mammals are vertebrates with hair and subcutaneous mammary glands. Except for a few primitive egg-laying species like the Australian duckbilled platypus, all are viviparous. As in birds, the mammalian heart is completely divided into four chambers: a right atrium and ventricle and a left atrium and ventricle.

Because man himself is a mammal, the study of this group complies with the Socratic imperative, "Know thyself." Knowledge of the mammals illuminates human psychology and social anthropology; it is indispensable to the medical sciences; and because most of our domestic animals are mammals, their study is important to agriculture.

Mammals fall into three subclasses: the monotremes, the marsupials, and the placentals.

Monotremes (Prototheria)

The **monotremes** are a very small group of egg-laying mammals limited to Australia, New Guinea, and adjacent islands. They are believed to have diverged from the main mammalian stock at extremely remote times. The body is covered by hair and is provided with subcutaneous mammary glands. There are no nipples, and the milk merely exudes through many tiny pores on the ventral side of the body. The skeleton is extremely reptilian, and so is the urogenital system, which opens to the exterior in common with the alimentary canal through a cloaca.

Overlying these very primitive traits are some specializations. The Australian duckbilled platypus is adapted for a semi-aquatic, semi-burrowing life along the edges of streams. In addition to the duck-like bill, the clawed toes are webbed. Also in this subclass are spiny anteaters found in Australia and adjacent regions.

Marsupials (Metatheria)

The **marsupials** are characterized by a pouch or **marsupium**, which is usually supported by a pair of **epipubic bones**. The young crawl into the marsupium immediately after birth, which takes place in a very much earlier stage of embryonic development than in placental mam-

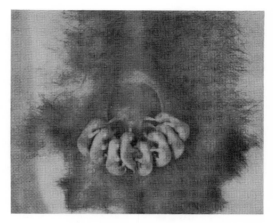

Fig. 25-4. Young marsupials attached to their mother's nipples in the pouch. (Kindness General Biological Supply House.)

mals. A litter of 20 newborn opossums scarcely fills a teaspoon. The front legs are precociously developed to facilitate crawling into the pouch. Intrauterine life in a placental mammal like a cat lasts nearly 60 days; in an opossum of comparable size it lasts scarcely two weeks, but the young spend about 50 days in the pouch. The sides of the baby oppossum's lips grow together around the teat, which extends far down the throat of the young animal. Milk is squeezed into the infant by special muscles of the mother. All of this apparatus—marsupium, epipubic bones, special muscles to express milk into fetuses too young to suckle, precociously developed forelegs, and lips which grow together around the teats—is clearly a highly specialized and coordinated group of adaptations.

The brain is relatively smaller in marsupials than in modern placental mammals and, most surprisingly, there is no corpus callosum, the chief right-left connection between the cerebral hemispheres. The teeth are unspecialized and the urogenital system is reptile-like in that both alimentary canal and urogenital system emerge together through a cloaca.

Placentals (Eutheria)

Mankind is a **placental mammal**, a fact which conditions much of human life. Placental mammals have a long gestation period during which a **placenta**, formed from both embryonic and more or less maternal tissue (depending on the species), makes possible an exchange of food and wastes between mother and offspring. The overriding trait which sets placental mammals apart from all other animals is their intelligence, which influences all their relationships with each other and with their environment. Their highly plastic behavior, subject to profound modification by experience, stands in marked contrast to the stereotyped innate responses characteristic of insects and even of reptiles and birds.

Correlated with the development of intelligence are three key anatomical structures. Without a highly complex brain, especially a **cerebral cortex** or its equivalent, not much intelligence would be possible in any case. Without a **placenta** vivipary would be impossible and vivipary guarantees that the mother will be present when the young emerge into the world.

The concomitant development of the **mammary glands** and the resulting dependence of the young mammal on its mother's milk binds young and adult together. All this insures the young mammal of parental protection during the early crucial stages of learning the ways of the world in which it will live. The young insect or spider has little need for parental protection because it hatches into its world with a reasonably complete set of innate responses. These responses are adequate for most of the situations it will face, and they can be modified only to a slight degree in any case. The obligatory association of the young and the adult mammals also provides an opportunity for the direct imitation of the adult by the young, a form of learning which is, of course, important in many species. Thus, for launching into the world an animal that depends on intelligence, viviparity sets the stage, the mammary glands enforce the association of the immature and the experienced, while the cerebral cortex capitalizes on the opportunity.

Evolutionary Relationships

The first mammals appear in the fossil record in rocks dating from the Jurassic period during the Age of Reptiles (the Mesozoic era), some 150 million years ago. These early mammals were all small and generalized, somewhat like a very large shrew or a small mongrel dog. They can be distinguished from the closely related mammal-like reptiles by their skulls, teeth, ear bones, and pelvis. The skull has a relatively larger brain case, and the lower jaw is composed of a single bone on each side instead of three or more. The side teeth of mammals are differentiated into complex molars and premolars, while those of reptiles form a row of mere conical pegs. The middle ear of all mammals contains a chain of three little bones, the stapes, incus, and malleus, between the tympanum (eardrum) and the cochlea, whereas the mammal-like reptiles have but one and retain the other two as small bones at the angle of the jaw.

The fossil record reveals virtually nothing of the internal structures of primitive mammals, nor of the evolution of hair from scales. It should be remembered that some mammals still possess reptilian epidermal scales, covering the tail in rodents and most of the body in one group of anteaters.

At the beginning of the Age of Mammals, the Cenozoic era, some 70 million years ago, all the known mammals were still no larger than medium-sized dogs. Most of the major lines of mammalian evolution were at least foreshadowed, although there were no highly specialized forms. During the past 70 million years the mammals have undergone an explosive evolution, radiating outward in many adaptive directions, much as did the reptiles before them. And in line after line—horses, elephants, cats, deer, etc.—the mammals have followed Cope's law that in the course of evolution, over millions of years, a race of animals tends to become larger in body size. In this also they resemble the reptiles. Note, however, that in both groups there are lines, like the rodents, where there has been little if any tendency to increase in size.

Primates

Of the 15 or more major groups of placental mammals, one, the **primates**, is of special human interest. The primates include man and all the various apes, monkeys, lemurs, and similar forms. It is a very large, heterogenous

Fig. 25-5. Four primates. None is typical because there is no such thing as a typical primate. A, *Presbytis*, the languar. A 20-year-old female. Native of woodlands in India. B, *Cynopithecus*, the Celebes black ape, a fruit lover of rainforests in the East Indies. From lowlands up to over 2,030 meters (6,500 ft.). C, *Nasalis*, the proboscis monkey of Borneo. Mother with infant. D, *Cercocebus*, the Mangabey of tropical Africa. (Kindness of J. R. Napier and P. H. Napier. From J. R. Napier and P. H. Napier: *Handbook of Living Primates*. Academic Press, New York, 1967.)

495

group which has been virtually impossible for taxonomic specialists to define. The technical distinction between primates and other placental mammals is based on apparently trivial features. The primitive number of incisor teeth between the canines in placental mammals is six—as in dogs and cats. Primates have only four incisors.

The key fact about primates is that they evolved as an **arboreal group** and are still arboreal, with a few exceptions, such as man and the baboons. In an arboreal environment, where one has to jump from branch to branch, natural selection strongly favors the development of grasping forelimbs to catch onto the next bough. This requirement led to the **opposable thumb** and great **manipulative ability** in the hand. Once such a structure has evolved, natural selection would then favor mutations increasing brain size with a resulting increase in the ingenuity with which the versatile hand is used.

Life in the trees also places a high premium on the ability to judge distances correctly, and on spying things out by **sight** rather than tracking them by smell. Hence natural selection would press strongly in favor of the **visual centers** in the brain at the expense of olfactory centers, and in favor of mutations advantageous to **binocular vision**. There is a strong tendency among the primates for the eyes to face forward and for the snout to be reduced.

Except for some remarkable investigations by a handful of workers on three or four species, very little has ever been learned about man's closest relatives. The primates as a group are surely a rich field for future research.

Our most primitive relatives among the primates are the lemurs and the strange little creature, *Tarsius*. True lemurs are found only on Madagascar and nearby islands, but closely related animals, the loris and the bush baby (*Galago*), are found from the East Indies across southern Asia and into Africa. They are all small nocturnal animals, arboreal and omnivorous, subsisting on insects, worms, birds' eggs, fruits, nuts, etc. The tail is long and bushy but never prehensile, *i.e.*, used as a grasping organ. Some species appear strangely human; others resemble cats or squirrels.

Tarsius is the only living representative of its suborder. Limited to the East Indies, it resembles a lemur but has a better developed brain, enormous anteriorly directed eyes resembling those of owls, and, like owls, it looks in different directions by turning its neck a full half-circle if need be. Its retina contains only rods and consequently the eyes are most useful at night.

The anthropoids constitute a suborder of the primates which includes man. It is customarily divided into two major groups. The **ceboid** or platyrrhine (flat-nosed) monkeys live in tropical America. The nasal septum is broad and the nostrils face somewhat laterally. Unlike any of the Old World monkeys of Africa and Asia, the ceboids have **prehensile tails**. In this they are like several other and unrelated South American animals (opossums, for instance). Two families are recognized. The family Hapalidae, or marmosets, are very small monkeys. The family Cebidae includes, among others, the organ-grinders' monkey, the spider monkey, and the howler monkey. The latter lives in large "clans," and its social behavior has been extensively recorded.

The **cercopithecoid** or catarrhine (downward-nosed) anthropoids include the Old World monkeys, apes, and man characteristic of southern Asia and Africa. The nasal septum is relatively narrow. Three familes are recognized: the Cercopithecidae (*cercopithecus,* longtailed monkey), the Simiidae (*simia*, ape), and the Hominidae (*homo*, man).

The family Cercopithecidae includes the baboons, mandrills, proboscis monkeys, the physiologists' macaque *i.e., Rhesus,* and others. They are highly socialized animals often living in large colonies.

The family Simiidae includes the four great apes: the gibbon, the orangutan, the chimpanzee, and *Gorilla*.

The family Hominidae includes but one living species, *Homo sapiens*. The fossil record so far has revealed very little about the relationship of this family to other cercopithecoids.

USEFUL REFERENCES

Reptiles

Bellairs, A. *The Life of Reptiles.* Universe Press, New York, 1970.
Carr, A. *Handbook of Turtles.* Comstock Publishing Co., Ithaca, N. Y., 1952.

Conant, R. *A Field Guide to Reptiles and Amphibians.* Houghton Mifflin Co., Boston, 1958.

Birds

Bent, A. C. *Life Histories of North American Birds.* Dover Publications, New York, 1961 (paperback).

Griffin, D. R. *Bird Migration.* Doubleday & Co., New York, 1965.

Peterson, R. T. *A Field Guide to the Birds*, rev. ed. Houghton Mifflin Co., Boston, 1947 (paper original 1968).

Robbins, C. S. Bruun, B., and Zim, H. S. *A Guide of Field Identification of Birds of North America.* Golden Press, New York, 1966.

Walcott, C. Bird navigation. Natural History, *81:* 32-43, 1972.

Mammals

Buettner-Janusch, J. (ed.) *Evolution and Genetic Biology of Primates.* Academic Press, New York, 1965.

Burt, W. H., and Grossenheider, R. P. *A Field Guide to the Mammals.* Houghton Mifflin Co., Boston, 1952.

Green, E. L. (ed.) *Biology of the Laboratory Mouse,* 2nd ed. McGraw-Hill Book Co., New York, 1966.

Hall, E. R., and Kelson, K. R. *Mammals of North America.* Ronald Press, New York, 1959.

Napier, J. R., and Napier, P. H. *Handbook of Living Primates.* Academic Press, New York, 1967.

Norris, K. S. *Whales, Dolphins, and Porpoises.* University of California Press, Berkeley, 1965.

Schultz, A. H. *Life of Primates.* Universe Books, New York, 1969.

9
ANIMAL FORM AND FUNCTION

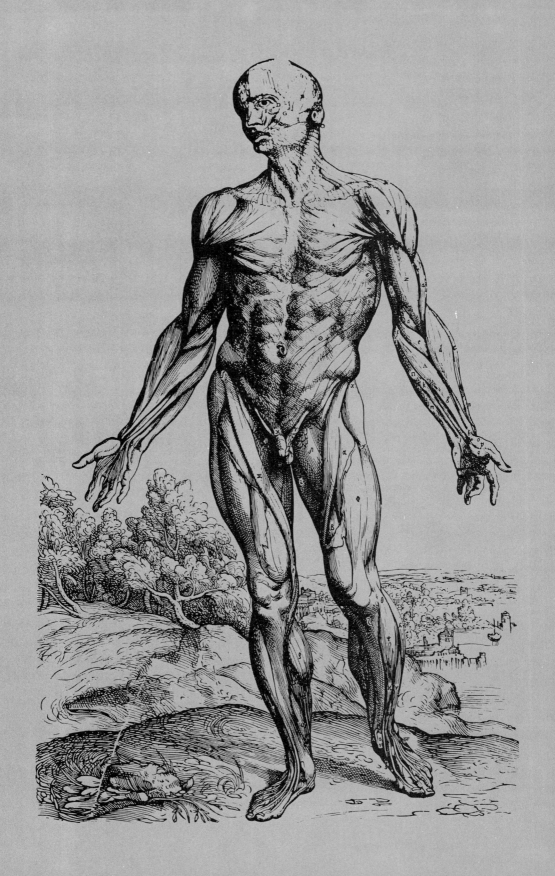

Skeletal and Muscular Systems

26

For the life of man and the other vertebrates, the importance of a skeleton can hardly be exaggerated. The same holds true for the insects and other arthropods. On a planet having the gravitational pull of the earth, we would all be as hopelessly doomed without our skeletons as astronauts projected into orbit without space-suits, although for different reasons. Without skeletal support, terrestrial animals of any appreciable size would rapidly suffocate, for there would be no rib cage to support the lungs of vertebrates or framework for the trachea of arthropods. In fact, under the pull of gravity, we would all slump down into more or less formless blobs.

Some knowledge of the skeleton is a pre-requisite for any complete understanding of how most animals function. The skeleton comes first in the book which marks the beginning of the modern study of human anatomy, Andreas Vesalius' *De Humani Corporis Fabrica* (1543), and it remains first in many studies of anatomy.

MAJOR TYPES AND FUNCTIONS OF SKELETONS

Two major types of support have appeared in the course of evolution: **exoskeletons** and **endoskeletons**. Exoskeletons, covering the entire surface of an animal, have been enor-mously successful. The vast majority of species and of individual animals are enclosed in an exoskeleton, since this is a characteristic of all the insects, crustaceans, and other arthropods. Except for the sponges there are only two important groups of animals with endo-skeletons—the starfish and other enchinoderms,

and the vertebrates. Both possess calcareous support which is always covered by living tissue.

The Skeleton's Function

In addition to providing necessary support, a major function of the animal skeleton is to make possible **locomotion** of a more rapid and powerful sort than is found in any of the worms and all but a few of the mollusks. The notochord, running along the length of all chordates, and the backbone which develops around it in the vertebrates, is what gave our primitive ancestors a decisive advantage over the worms. The paired fins of fish and the paired limbs, whether legs or wings, of other vertebrates (though all are built on the same pattern), show wide diversity of adaptations for different modes of locomotion. The same is true of the legs of insects and of crustaceans. In both groups parts of the skeleton act as levers for the pull of muscles; in vertebrates the muscles are outside the skeleton, whereas in arthropods they are within.

In vertebrates the formation of red blood cells (and some white ones also) takes place in the marrow of the long bones of the limbs and in the sternum between the tips of the ribs.

In arthropods a major function of the skeleton is **protection**. The entire body is encased in armor, chitin, which is strengthened, in the case of lobsters and other crustaceans, by calcium. In vertebrates, this protective function is largely limited to the skull, which protects the brain, and the neural arch of the backbone, which encases the nerve cord down the back.

The advantages of an exoskeleton are clearly

501

very great, as is shown by the enormous success of arthropods. An exoskeleton, however, has one great disadvantage. The obvious difficulty with living inside an exoskeleton is that growth in size is impossible without molting. This process, termed **ecdysis**, is a highly dangerous and physically difficult feat. A soft crab, for example, is at the mercy of its enemies in the brief period immediately after shedding its old skeleton but before its new exoskeleton has been secreted and hardened. Why are there no really big terrestrial arthropods? For any animal that must shed its skeleton to grow, there is an ecological reason imposed by the relationship of size (weight) to the pull of gravity on our planet. To be without any skeletal support is bad enough even temporarily for an aquatic organism, but for a terrestrial one it would be disastrous, because lungs or tracheal sacs and cavities in which oxygen and carbon dioxide exchanges occur would collapse.

Structure and Composition of Exoskeletons

The structural material of exoskeletons is a nitrogen-containing polysaccharide called **chitin**. It clearly rivals the best plastics in many respects and surpasses most because fortunately it is biodegradable. Were it not, the world would be a vast trash pile of discarded insect and arthropod shells. Chitin is not too difficult to analyze. On hydrolysis in acid, it breaks down into glucosamine, *i.e.,* the sugar glucose with an $-NH_2$ group added close to the keto-end of the molecule. Glucosamine is found in numerous components of the living world, including the antibiotic streptomycin and the blood anticoagulant, heparin. Chitin itself is found in a wide variety of places: the bristles of annelids, the teeth and beaks of the octopus, and the cell walls of many fungi.

The glucosamines which make the nitrogen-containing polysaccharide, chitin, are linked in a long chain much as in cellulose. Furthermore, chitin is often combined with other components—sometimes pigments that confer the familiar browns, blacks, reds, and yellows to the bodies of insects and their relatives, and sometimes with calcium or some other strengthening material..

Fig. 26-1. An arthropod emerging from its old exoskeleton. The dragonfly in the middle figure has experienced one of the sometimes fatal accidents of life with an exoskeleton: its head became caught during ecdysis. (From K. von Frisch: *Biology*. Harper & Row, Publishers, New York, 1964.)

The properties of chitin vary enormously, depending on the other materials incorporated with it. It may be glass-clear as in the cornea of the eyes of insects and crustaceans, more or less permeable to water and other solvents. It may be thin and very flexible or else quite rigid. It is indigestible by most animals.

The **exoskeletons** of insects and crustaceans are very similar in structure. There is a noncellular basement membrane separating the exoskeleton from the rest of the body. Over this is a single layer of secretory cells, the epidermis. Outside the epidermis lies a relatively thick chitinous cuticle covered in turn by a thin epicuticle. The latter is nonchitinous and contains waxy substances, fatty acids, and cholesterol. No wonder insects are so resistant to poisonous sprays.

The Vertebrate Endoskeleton

The endoskeleton of vertebrates and, more particularly, of mammals consists of two basic divisions. The **axial skeleton** consists of the skull and the backbone plus the ribs, sternum,

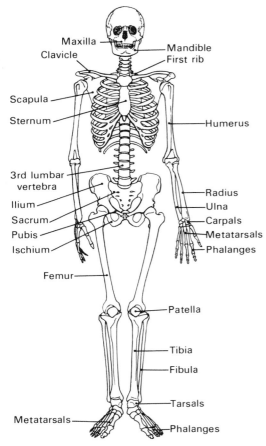

Maxilla
Clavicle
Mandible
First rib
Scapula
Sternum
Humerus
3rd lumbar vertebra
Ilium
Sacrum
Pubis
Ischium
Radius
Ulna
Carpals
Metatarsals
Phalanges
Femur
Patella
Tibia
Fibula
Tarsals
Metatarsals
Phalanges

Fig. 26-2. Human skeleton.

and the remnants of the gill arches seen in the throat. The **appendicular skeleton** is composed of the pectoral and pelvic limb girdles and the two pairs of limbs.

The **skull**—especially the human skull—has been an object of study for many centuries. Classical Greeks, medieval Arabians, the men of the Renaissance, the poet Goethe, and many others have contributed to our knowledge, but it was T. H. Huxley, Charles Darwin's champion, who established the modern view of the nature of the mammalian skull. The skulls of all mammals, including man, are built on the same plan, have the same number of bones, and bone for bone are homologous throughout. Only the relative sizes and shapes of the individual skull bones are different.

A comparison of the skull of a man and a cat will demonstrate this fact. Notice (Fig. 26-3) that the cartilages and bones of the throat, the hyoid bone at the base of the tongue, the thyroid cartilage (Adam's apple), and the other derivatives of the seven embryonic gill arches are basically the same. The chief difference between the two skulls lies in the much greater growth of the bones which protect the human brain, *viz.,* the frontal, temporal, parietal, and occipital.

In mammals, the **backbone** is divided into five regions—cervical, thoracic, lumbar, sacral, and caudal. With the exception of only two or three species, there are precisely seven **cervical vertebrae** in the necks of all mammals—men, giraffes, whales, or mice. The differences in length in these animals are due entirely to differences in lengths of the seven cervical vertebrae. Why this is so is unknown and is all the more puzzling because birds with long necks, like swans and flamingos, possess many more cervical vertebrae than do birds with short necks, like owls. The **thoracic vertebrae** bear the **ribs**, one pair per vertebra. There are usually 12 to 14 ribs, depending on the species. There are 12 in both sexes of *Homo sapiens.*

The **appendicular skeleton,** consisting of **limb girdles** and **limbs,** has been highly modified in many of the mammals, but the basic pattern is unchanged from the ancestral reptiles. Each limb is supported by a tripod of three bones. The bases of the tripods press against the axial skeleton, *i.e.,* against the backbone. At the apex of the tripod, the three bones meet the

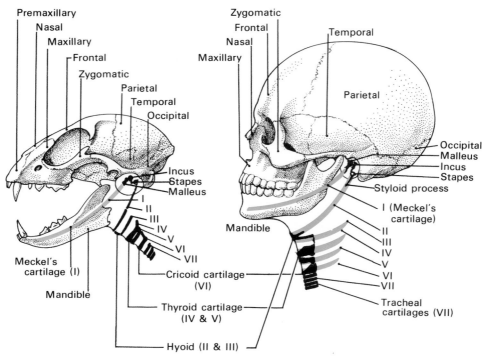

Fig. 26-3. Human and cat skulls showing their basic homology bone for bone. The Roman numerals indicate the gill arches, which are represented with solid black where ossified. (From G. B. Moment: *General Zoology*, 2nd ed. Houghton Mifflin Co., Boston, 1967.)

first—*i.e.,* the most proximal—bone of the leg. This is the humerus in the front leg or arm, and the femur in the hind leg. The pair of tripods that support the front legs or the arms constitute the **pectoral girdle**; the pair that support the hind legs, the **pelvic girdle**.

It is often asked why the pelvic girdle of mammals, through which the young must be born, should be fused in such a rigid manner. The pectoral girdle, in marked contrast, is very loosely connected with the axial skeleton via shoulder muscles and ligaments. It seems probable that the very different construction of the two limb girdles and their contrasting connection with the backbone is the result of adaptation for running and jumping. The rigid pelvis transmits the main push of the hind legs to the backbone and thus to the whole skeletal frame simultaneously. However, after making a leap, mammals from horses to rabbits land on their front legs. The flexible muscles of the pectoral girdle can help absorb the shock of landing.

The skeletons of front legs (or wings or arms)

and hind legs are homologous bone for bone throughout the entire range of mammalian adaptations. In the forelimb are a single **humerus**, then an **ulna** and **radius** side by side, with the ulna making a hinge joint at the elbow, a set of **carpals** in the wrist, five **metacarpals**, at least in the embryo, and five **digits** also, at least in the embryo. In the hind limb are a single **femur**, a **tibia**, and slim **fibula** side by side in the shank, a set of **tarsals** in the ankle, and the five **metatarsals** and five **digits**, at least in the embryo. Compare the arm of a man, the flipper of a seal, the wing of a bat, and the column or foreleg of an elephant.

Units of the skeleton are held together by **ligaments**, bands of tough connective tissue composed of a protein, largely collagen. Similar bands that attach muscles to bones or to each other are called **tendons**.

Structure and Growth of Bones

On the level of tissues and cells, bones are formed of **Haversian systems**. Each "system"

Canaliculi   Haversian canal   Lacuna   Interstitial lamellae

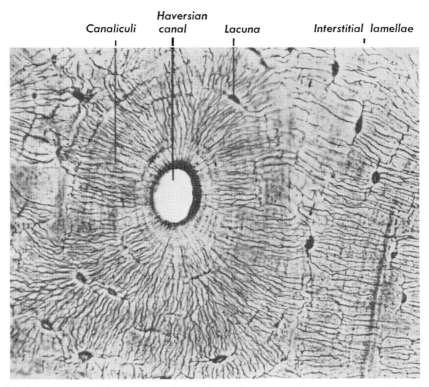

Fig. 26-4. Cellular structure of bone as seen in a thin section under a light microscope. The cell bodies with nuclei lie in the lens-shaped laculae with cytoplasmic extensions running out through minute canaliculi. (From F. R. Bailey: *Textbook of Histology*, 16th ed. The Williams & Wilkins Co., Baltimore, 1971.)

consists of a small canal, carrying blood vessels and nerves, surrounded by concentric layers of bone separated by bone-forming cells lying in microscopic cavities, **lacunae**, and interconnected by minute passageways, **canaliculi**.

The long bones of the limbs are composed of the **shaft** at each end of which there is a disk of bone, the **epiphysis**. Between the shaft and the epiphysis is a growth disk or epiphyseal line. Two long bones move against each other in a joint at the distal surface of the epiphysis. Thus the function of the epiphysis is to protect the zone between it and the shaft where the bone grows in length. Surrounding the bone is a special membrane, the **periosteum**. The inner layer of the periosteum forms new bone, thus adding to the diameter of the shaft. At the same time within the marrow cavity are special cells which eat away the bone, thus enlarging the cavity.

The **growth of bones** is controlled by many factors going back ultimately to the genetic make-up of the individual. There is a regular order in which the epiphyses of the various human bones appear and then later fuse with the shaft. X-ray photographs of the bones of the hand and wrist give an excellent indication of physiological age as contrasted with chronological age. Bone formation is profoundly influenced by vitamin D, by the pituitary, thyroid, and parathyroid glands, and by mechanical forces. Vitamin D, a sterol related structurally to the sex hormones, appears to facilitate the action of an enzyme concerned with the absorption of calcium and phosphorus from the gut and their reabsorption in the kidney. This enzyme is also active in bone, where it breaks down organic phosphorus compounds into the inorganic phosphates which help to form bone.

PRINCIPLES OF MUSCLE ACTION

A muscle can do only one thing: contract. It cannot push. The result of this basic fact can be

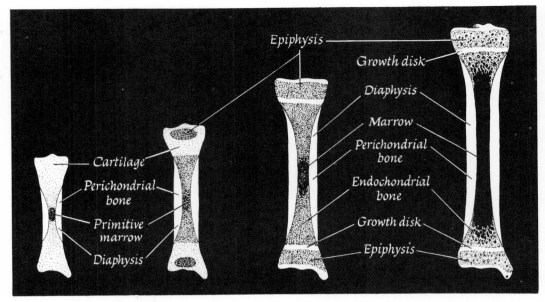

Fig. 26-5. Structure and growth of a long bone shown diagrammatically. (From G. B. Moment: *General Zoology*, 2nd ed. Houghton Mifflin Co., Boston, 1967.)

clearly seen in the way muscles are organized in animals from hydra to man. Almost everywhere muscles are arranged in **antagonistic sets.** This is the meaning of all the groups of **flexors** which bend arms, wings, and legs and of the **extensors** which straighten them. It is why the iris of the eye is provided with both a circular **sphincter** muscle which closes the pupil and a radiating **dilator** muscle which enlarges the pupil. In a simple coelenterate like hydra or in an earthworm there is a set of longitudinal muscles extending the length of the animal. Their contraction makes the body shorter and plumper. The antagonists are a set of circular muscles at right angles to the longitudinal ones. Their contraction makes the animal longer and thinner.

Basic Vertebrate Muscle Pattern

The muscles of a shark, the marine dogfish (*Squalas acanthias*), for example, show the basic vertebrate muscle arrangement with great clarity. The body muscles are divided into a series of more or less zig-zag segments or **myotomes** on each side of the body. The significance of such segmentation is that with a single muscle along each side, the fish could only bend into the shape of the letter C. With a series of muscle segments, each of which is able to contract independently, waves of contraction can pass along the animal, pushing the body and tail against the water so that forward motion results. This muscular segmentation and the paired segmental series of nerves that go with it persist in all vertebrates.

In the course of evolution from fish to man, the segmented muscles of the trunk have undergone relatively little change, but the muscles which move the legs have fanned out over the muscles of the trunk. Muscles on what are merely simple paired fins in a shark have undergone a great increase in size and complexity over the legs, arms, or wings of the higher vertebrates.

Histologically, *i.e.*, as seen under a light microscope, muscles are of two types, striated and smooth. The striations of striated muscle are very fine and incredibly regular cross markings at right angles to the direction of contraction. They can be seen in living animals —in *Daphnia*, the crustacean water flea, for example—and even ground beef shows them. Until recently the meaning of the striations was completely unknown. Striated muscles are often called **skeletal muscles**, because they are usually attached to bones, or **voluntary muscles**, because for the most part they are

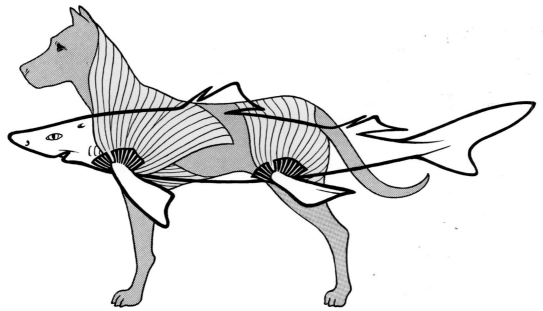

Fig. 26-6. Diagram showing the evolutionary development of the muscles moving the paired limbs from the dogfish (black) to the dog (green).

under the control of the will. These muscles can contract rapidly, but they also fatigue rapidly.

Smooth muscles lack fine cross striations. Smooth muscles are also called **visceral muscles,** because they are found in the walls of the stomach, uterus, and other viscera, or **involuntary muscles,** because they cannot be directly controlled by the will. They contract and fatigue more slowly than striated muscle.

Heart or **cardiac muscle** has special properties. Although striated, cardiac muscle is physiologically involuntary. Its fibers appear to branch, although the evidence from electron microscope studies and from growing heart muscle cells *in vitro* clearly shows that this is not so. The cells are not lined up in parallel like ordinary skeletal muscle; neither do the individual cells fuse into a giant multinucleate network as seems to be the case under a light microscope.

The relationships of muscle types is shown in the following scheme:

Voluntary	Skeletal ⎱	
		Striated
	⎰ Cardiac ⎰	
Involuntary	⎰ Visceral	Smooth

Modern Knowledge of Muscular Contraction

For centuries there has been an impassable gulf between the anatomist who studied form and the physiologist who investigated function. The anatomist knew about the arrangement of extensors and flexors. The physiologist knew about such things as the speed of contraction and relaxation, the characteristics of fatigue, and even some of the chemical prerequisites and end products of muscle action. But no one knew what goes on inside the muscle itself. A muscle was what a physicist would call a "black box,"—impenetrable, perhaps unknowable. Modern biochemistry and the electron microscope have now combined to give us a look inside that "black box."

It has long been known that a muscle can contract in the absence of oxygen but that under such anaerobic conditions the **glycogen,** which represents stored energy in the muscle, disappears and **lactic acid** accumulates. After a relatively short time the muscle becomes fatigued, or at least will no longer respond to further stimulation. If oxygen is supplied, the lactic acid does not accumulate, even though the glycogen is broken down into glucose. As we now know, the glucose passes down the

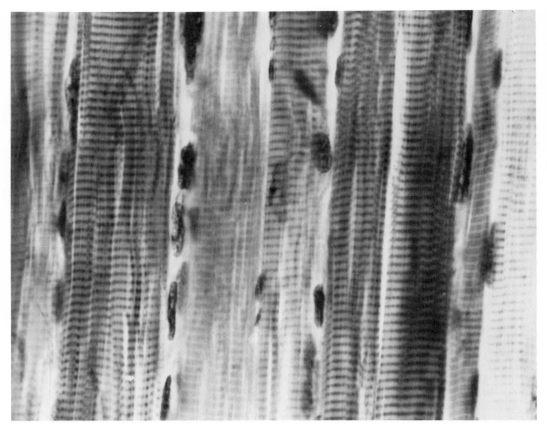

Fig. 26-7. Appearance of striated muscles under high power of a light microscope. (Kindness of Ward's Natural Science Establishment.)

Embden-Meyerhof glycolytic pathway and is finally completely oxidized in the Krebs cycle and electron transport system of the mitochondria. It is hardly surprising that muscles are rich in mitochondria. They require energy in abundance.

When muscle contractions take place faster than the mitochondrial enzymes can provide energy as ATP packets, pyruvic acid accumulates at the end of the glycolytic pathway faster than the Krebs cycle can take it up. This excess pyruvic acid is promptly converted into lactic acid:

$$CH_3-\overset{\overset{\displaystyle O}{\|}}{C}-COOH \underset{\underset{\displaystyle lactate}{\text{dehydrogenase}}}{\overset{\overset{\displaystyle NADH_2 \quad NAD}{\longrightarrow}}{\rightleftharpoons}} CH_3-\overset{\overset{\displaystyle OH}{|}}{\underset{\underset{\displaystyle H}{|}}{C}}-COOH$$

Pyruvic acid   Lactic acid

The lactic acid is later reconverted into pyruvic acid or even glycogen in either the muscle itself or the liver, and oxidized in the mitochondrial Krebs cycle when oxygen becomes available. This is the chemical basis for being out of breath after violent exercise. One is "repaying" an "oxygen debt" as the Krebs cycle catches up with the material supplied to it by glycolysis.

That ATP and the numerous mitochondria play an important role in contraction is certain. But in addition to adenosine triphosphate there is another high-energy phosphate compound in muscles, **phosphocreatine** in the vertebrates and **phosphoarginine** in most invertebrates. These compounds are formed from creatine or arginine at the expense of ATP which is degraded to the less energy-rich ADP (adenosine diphosphate). It is possible that the energy-rich creatine is the compound directly involved with the mechanics of contraction. No one yet knows.

A spectacular breakthrough into the "black box" of muscular contraction was achieved by the Hungarian-American Nobel prize winner, Albert Szent-Györgyi. He succeeded in extracting two kinds of protein from striated muscle, **actin** and **myosin**. When the two are mixed and then squirted out of a fine pipette into very dilute KCl solution, the combined **actomyosin** will contract when ATP is added as an energy source. Neither actin nor myosin alone will contract, no matter how much ATP is added; moreover, ATP is degraded to ADP in the process. These experiments strongly suggest that contraction is due to some kind of relationship between the two proteins. Actomyosin, or some closely similar protein, has now been extracted from the muscles of animals as far removed from man as sea anemones. Apparently the structural basis of muscle action is generally the same throughout the animal world.

A second breakthrough came from analysis with an electron microscope by H. E. Huxley. The electron photomicrographs of striated muscle confirm the presence of dark bands alternating with lighter ones as revealed by the

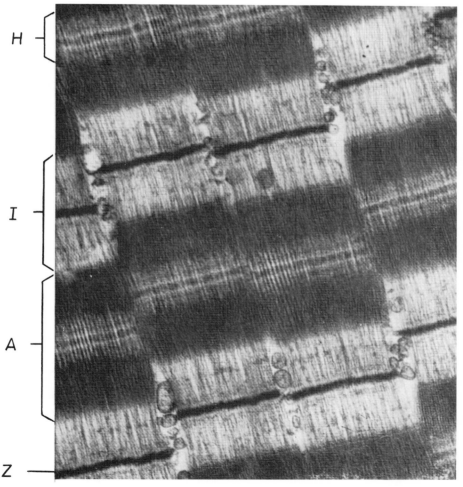

Fig. 26-8. Electron microscope view of striated muscle. A represents a dense anisotropic band transected by the lighter H (Hensen's) band, I an isotropic band transected by the very dense Z line (disk) bounding a sarcomere. (From H. E. Huxley, in *Bailey's Textbook of Histology*: The Williams & Wilkins Co., Baltimore, 1971.)

Fig. 26-9. Cross section of a striated flight muscle of an aphid. Note that each of the thicker myosin filaments is surrounded by six evenly spaced thinner actin filaments. According to present evidence diagrammed in figure 26-10, through which band was this section cut? Approximately × 140,000. (Kindness of David S. Smith.)

light microscope. Each dark band is bisected by a narrow light one, and each lighter band by a very narrow dark line. Terminology has varied a bit but the wide dark bands are usually called **A bands**, A standing for anisotropic, meaning that the molecules here are so completely oriented that they transmit polarized light only in a particular plane. The A bands can be shown by chemical extraction techniques to be mostly myosin. The lighter stripe across the A band is called an **H (Hensen's) band**. The large light bands are usually known as **I bands**, I for isotropic. The very dark narrow lines bisecting the I bands are terminal or **Z lines**. The I bands turn out to be mostly actin. A single unit from Z line to Z line is called a **sarcomere**. Because muscle fibers are elongate cylinders, each line is really a disk.

With the electron microscope H. E. Huxley was able to show—and many others have since

confirmed and extended his work—that one end of each of the thinner actin fibers is attached to a Z line (disk) and the other end extends out into the sarcomere between the thicker myosin fibers. Thus the actin and myosin fibers mesh

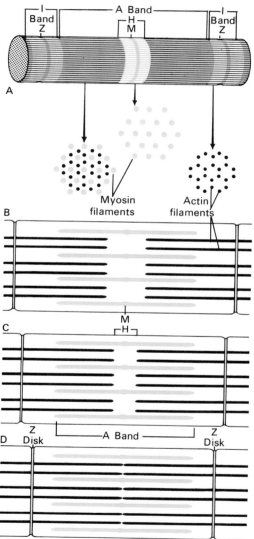

Fig. 26-10. Diagram showing the structure of a muscle fibril as revealed by the electron microscope in a relaxed condition, A and B, and during contraction, C and D. The thicker myosin filaments are shown in green, the thinner actin filaments in black. Note that a section through the H band shows only myosin filaments and through the I band close to the Z disks only actin.

much as the bristles of two brushes would if pushed together. Cross sections viewed with an electron microscope show that when a muscle contracts, the actin fibers move further in between the myosin fibers and thus shorten the sarcomere much as an extension ladder shortens when the two parts slide together. Cross sections through the I bands near the Z lines at the ends of sarcomeres show only thin actin filaments. Sections through the H stripe in the center of the A band show only heavier myosin filaments. Sections through the portion of the A bands adjacent to the I bands show both types of filaments. If this theory is correct, the A bands should not become narrower during contraction but the I bands should. They do.

This sliding filament theory, in contrast to any accordion-like folding theory, is also compatible with what the electron microscope has shown to be true of vertebrate smooth muscle, invertebrate striated and smooth muscle. Thus once again an advance in biochemistry brings not only new knowledge but also a simplification and order on a vast scale. The black box has been opened.

What pulls the actin and myosin filaments along each other? No one can be certain but myosin filaments show regularly spaced, finger-like protrusions which reach out and touch the actin filaments. But how these crossbridges work is not known. Calcium ions and ATP are known to be required for contraction. Phosphocreatine can be presumed to play an essential role. The supply of ATP is provided in large part by the mitochondria which are found between the myofibrils. A fine tubular network or reticulum penetrates muscles, making possible the exchange of energy-bearing compounds and waste products. Muscles contain small amounts of a third protein, **tropomyosin**. This may be in the Z bands at the ends of the finger-like extensions of the myosin. But until much more is known about how the actin and myosin pull themselves together, the gap as old as Aristotle between form and function will not have been completely bridged.

USEFUL REFERENCES

Ballard, W. W. *Comparative Anatomy and Embryology*. The Ronald Press Co., New York, 1964.

Gray, H. *Anatomy of the Human Body*, 25th ed. Lea & Febiger, Philadelphia, 1959.

Hoar, W. S. *General and Comparative Physiology*. Prentice-Hall Inc., Englewood Cliffs, N. J., 1966.

Wilkie, D. R. *Muscles*. St. Martin's Press, New York, 1968 (paperback).

Vascular Systems and Oxygen Transport

27

A vascular system to carry oxygen and food to all parts of the body and to take away carbon dioxide and other wastes is a clear necessity for any animal much larger than a flea. Furthermore, a sound knowledge of the vascular system is a prerequisite for understanding almost any other aspect of animal physiology. It was no accident that no one knew how the human body functions until Harvey demonstrated convincingly that the blood is pumped by the heart via the arteries to the body and returns to the heart via the veins. Before his time everyone thought that the blood merely surges back and forth in the veins and that the arteries contained air.

HARVEY'S DISCOVERY OF CIRCULATION

After graduating from Cambridge University, William Harvey (age 19) left for northern Italy, then the world center of scientific learning. Here he made one of the great landmark discoveries in the growth of human knowledge. Like his contemporary, Newton, and like Darwin he combined the work of his predecessors with original observations of his own to achieve a new synthesis. At Padua Harvey became familiar with the work of Vesalius who had published, a generation earlier, the first accurate account of human anatomy based on actual dissection. He also learned that there are flap-like valves in the veins which had recently been discovered by his own teacher, the famous Hieronymus Fabricius.

Possibly the most crucial contribution to Harvey's thinking was the argument of Michael Servetus that there is a pulmonary circulation in which blood passes from the heart to the lungs via the pulmonary arteries and returns to the heart via the pulmonary veins. Servetus had been burned at the stake before Harvey arrived in Padua for his political and religious beliefs, and all his books consigned to the flames. But it is difficult to burn all copies of a book. Matheus Columbus (a professor of anatomy and no relative of Christopher) stole the ideas of Servetus word for word. Reprehensible as this was, it nevertheless made the concept of pulmonary circulation available to Harvey.

Harvey marshaled his argument along three main lines. The first was anatomical. He noted the correspondence in size between the arteries and veins of any organ. That was not proof, but did constitute a necessary condition for circulation. He showed that the valves in the veins prevented blood from flowing in any direction except toward the heart. He also showed that the valves, at the place where the arteries leave the heart, permit blood to leave but not to enter the heart. As seen in Harvey's own figure, if the veins in the forearm are made to stand out by applying a tourniquet above the elbow, it can be demonstrated by squeezing blood along the vein with a finger that blood will flow in a direction toward the heart but not in the opposite direction.

Harvey's second line of reasoning was quantitative. He counted the number of times the heart of a dog beat per minute; then he killed the animal and measured both the amount of blood the heart forces into the arteries with each beat (by measuring ventricular capacity) and the total volume of blood in the animal. By

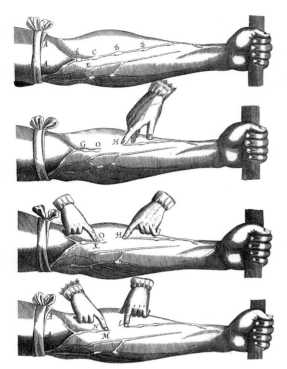

Fig. 27-1. Harvey's experiment to show that blood can be made to flow only toward the body in the veins. (From C. D. Leake: *Harvey's Exercitatio Anatomica: De Motu Cordis et Sanguinis in Animalibus.* Charles C Thomas, Springfield, Ill., 1928.)

simple arithmetic he then showed that in a dog or a sheep the heart pumps out many times this total volume every hour. In man, the heart pumps out three to five liters per minute. In other words, the volume of blood passing through the heart every minute is as great as the total volume of blood in the body. Where can all this blood pumped by the heart come from or go to, if it does not travel within a closed circuit?

Last, Harvey backed up these arguments by a wide array of experiments on dogs, snakes, snails, and even insects and shrimp. If the veins entering the heart are tied off, the heart becomes empty of blood. If the arteries leaving it are tied, the heart remains permanently gorged.

VASCULAR SYSTEMS

Animals have evolved three major patterns of blood vessel arrangement. There is the annelid-arthropod pattern, seen in earthworms and lobsters, which Harvey studied in shrimp to supplement his observations on snakes and dogs; a molluscan system with some very remarkable features; and, third, the vertebrate system.

Annelid-Arthropod System

In the annelid-arthropod system blood is pumped by a pulsating dorsal blood vessel located above the gut. In an earthworm the dorsal vessel squeezes blood toward the hearts in a series of waves of contraction. In the anterior part of the body are five pairs of sausage-shaped hearts which pump blood downward (ventrally) into a ventral blood vessel beneath the gut. From this, smaller arteries carry blood to fine capillaries ramifying throughout the body. These capillaries empty into veins which lead back to the dorsal blood vessel. This is a **closed circulatory system** such as is found in man. In a newly hatched earthworm the scarlet blood can easily been seen under a dissecting microscope as it follows the course just described. In crustaceans and insects the system is similar to the earthworm's, except for the oblong-shaped heart in place of the long pusating dorsal vessel, and the lack of capillaries. Arteries empty blood into tissue spaces, **sinuses**, from which it is carried back to the heart via veins. This is an **open system**.

Molluscan System

The molluscan system is also an open one like the arthropod scheme and includes a centrally located dorsal heart which, in clams, is easy to see pulsating just below the hinge of the shell. In oysters, which are commonly eaten alive, the heart is visible beside the firm, rounded adductor muscle. The octopus and the squid have three hearts: one centrally located which pumps blood to the body, plus two gill hearts, one at the base of each gill, which pump venous blood into the gill artery. It is comparable to having a heart at the base of each lung.

Vertebrate System

All vertebrate vascular systems are modifications of a basic closed system pattern clearly

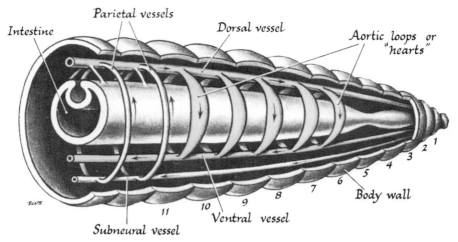

Fig. 27-2. The basic annelid-arthropod vascular system as seen in the anterior end of an earthworm. The arterial portion is colored green. (From G. B. Moment: *General Zoology,* 2nd ed. Houghton Mifflin Co., Boston, 1967.)

recognizable in fish and salamanders and in the embryos of birds and men. The heart is ventral to the pharynx or throat in fish and the embryos of higher vertebrates. In adult reptiles, birds, and mammals it has moved posteriorly during development to occupy the familiar position in the chest. This is why the heart is innervated by the **vagus nerve** which comes directly from the brain, enters the heart while it is under the pharynx, and then is pulled down into the chest with the heart.

From the vertebrate heart a ventral artery runs anteriorly below the pharynx. In fish and in embryos of the higher forms, six pairs of **aortic arches** curve up on either side of the pharynx between the gill slits and join to form the **dorsal aorta** above the pharynx. The dorsal aorta is the main blood vessel of the body and runs to the tail just ventral to the backbone and dorsal to the gut. Branching off are a pair of **carotid arteries** to the head, **subclavian arteries** to the front legs or arms, **renal arteries** to the kidneys, a **coeliac** and other unpaired arteries to the intestine and other viscera, a pair of **iliacs** to the hind legs.

The arteries divide into smaller and smaller vessels and finally into thin-walled **capillaries**. It is only through the capillary walls that exchanges between the blood and the tissue cells take place. Food is absorbed from the intestine and distributed to other parts of the body, oxygen is obtained in the lungs and carbon dioxide given off. As soon as microscopes were

invented people began looking for the capillaries. Leeuwenhoek observed the comb of a young rooster, a rabbit's ear, and even a bat's wing. When he finally used a tadpole's tail he wrote:

A sight presented itself more delightful than any mine eyes had ever beheld; for here I discovered more than 50 circulations in different places . . . . For I saw not only that in many places the blood was conveyed through exceedingly minute vessels, from the middle of the tail toward the edges, but that each of the vessels had a curve or turning, and carried the blood back toward the middle of the tail, in order to be again conveyed to the heart.

Blood is returned to the heart in vertebrates by a system of veins, the **postcaval vein** from the abdomen being the largest. In the lower vertebrates there are two portal systems: one, the **renal portal veins** which take blood from the hind legs or fins to the kidney, and another, the **hepatic portal veins** which carry blood from the intestine to the liver. In mammals like ourselves only the hepatic portal system is present. A portal vein is a vein which carries blood to a structure other than the heart.

VERTEBRATE HEART

The vertebrate heart always begins in the embryo as a straight, pulsating tube squeezing

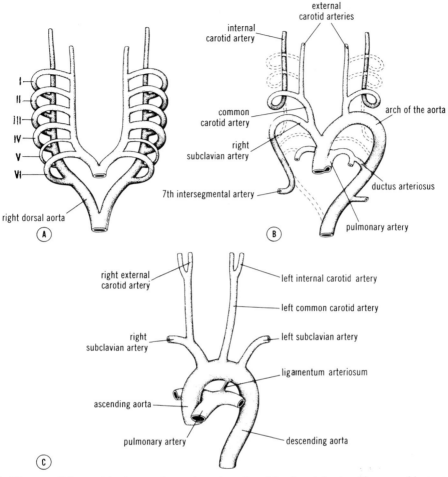

Fig. 27-3. Diagram of the vertebrate vascular system. A, as found in all vertebrate embryos and in some fish and amphibian adults. B, as seen in some amphibians and in mammalian embryos. C, as seen in an adult mammal. By international convention in anatomy, these figures are drawn as though the animal were being dissected lying on its back. Therefore the dorsal aorta is represented as beneath the ventral. (From J. Langman: *Medical Embryology*. The Williams & Wilkins Co., Baltimore, 1969.)

blood anteriorly. In fish it remains a straight tube consisting of a linear series of three single chambers, *viz.,* a thin-walled **sinus venosus** which receives blood from all of the body, an **atrium** which receives the blood from the sinus, and a muscular **ventricle** which takes the blood from the atrium and pumps it out into the **ventral aorta** and up through the pairs of aortic arches into the dorsal aorta.

In mammals the embryonic tubular heart bends in the course of development into a U-shaped structure so that the arteries and veins all leave or enter the heart at the same end. At

the same time the interior of the heart becomes divided into right and left halves, making two atriums and two ventricles. The sinus venosus becomes reduced to a small area called the **pacemaker** on the wall of the right atrium, so called because it initiates the heart beat. Oxygen-poor blood from the body enters the right atrium, passes into the right ventricle, and thence to the lungs via the pulmonary arteries. Blood returning from the lungs, as doomed Michael Servetus noted so long ago, enters the left atrium, goes into the left ventricle, and thence out into the dorsal aorta. The blood

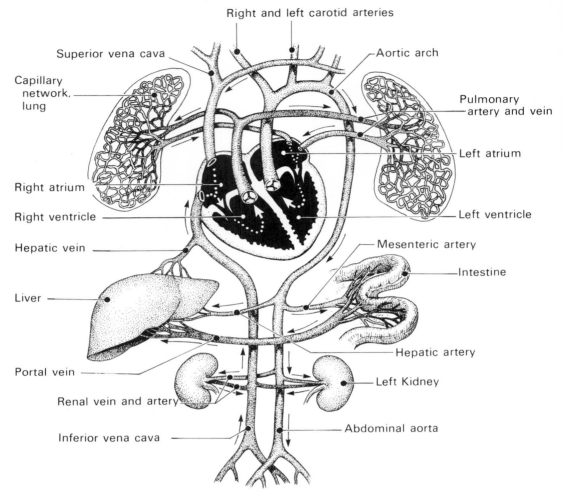

Fig. 27-4. The mammalian vascular system. (From G. B. Moment: *General Zoology*, 2nd ed. Houghton Mifflin Co., Boston, 1967.)

passes through the heart twice to complete one circuit, but the time required is very short. In a rabbit it takes only 7 or 8 seconds; in a human only a bit over ¼ minute.

Because of its great medical importance and intrinsic interest, the study of the heart (cardiology) has produced an enormous body of knowledge. We will mention only the most significant aspects. Heart muscle is striated and composed of single cells like skeletal muscle but the fibers form a network rather than all being parallel. Physiologically, cardiac muscle exhibits four outstanding properties.

First, it is inherently **rhythmic**. Isolated heart cells from an embryo beat when grown *in vitro*

and even bits of adult heart muscle isolated from nerves will beat. Second, after each contraction, heart muscle has an extremely long refractory period during which it cannot be stimulated to contract. This insures that **diastole** (dilatation) and filling of the heart will occur. It is impossible to throw the heart into tetanus, or cramp, which happens easily to skeletal muscle.

A third characteristic is the way the heart automatically adjusts the strength of its contraction to the amount of work it has to do. If it is filled with a large volume of blood, as in violent exercise, its walls are stretched and the contractions become correspondingly more powerful. This is Starling's "law of the heart."

Last, heart muscle requires a proper salt balance to beat rhythmically. Sodium, potassium, and calcium must all be present in the fluid surrounding the organ. The ratio of these necessary ions is closely similar to that in sea water. The molecular mechanisms of the effects of salts are still unclear, but in general, high concentrations of calcium or sodium stop the heart in **systole** (contraction); high concentrations of potassium stop the vertebrate heart in diastole. These discoveries were originally made on the frog heart by Sidney Ringer and led directly to all the physiological salt solutions now in use for intravenous administration and many other procedures.

The rate of heartbeat in adults is under nervous control. The heart can be completely stopped by stimulation of the vagus nerve. The heart also receives accelerator nerves from the sympathetic system. One of the most dramatic indications of neural control of heart rate is the slowing, **brachycardia,** which occurs when a seal dives or even when its nose is held under water. The onset is prompt, so fatigue cannot be involved. A seal's heartbeat falls from 80 to 7 or 8 beats per minute. Similar brachycardia has been found in every diving animal studied, including ducks, crocodiles, porpoises, hippopotami, and men. Try holding your face under water while a friend takes your pulse. The adaptive value of diving brachycardia is obscure, but a brachycardia also occurs during hibernation. In the light of these facts it is not surprising that heart rate can be controlled by conditioned reflexes in dogs and humans.

LYMPHATICS

The fluid directly bathing the cells is not whole blood, which is confined within capillaries, but **lymph**. The composition of lymph is essentially that of blood minus the red and white cells. The lymph vessels resemble capillaries and veins, possessing valves which permit lymph to flow only toward the heart. The flow is one-way and empties via the great veins in the chest into the right side of the heart. Lymph is continually replenished by seepage from the capillaries.

Where the legs join the body, in the armpits and neck, and at several other sites, the lymph passes through glandular **lymph nodes**. The lymph nodes are the site of production, during adult life, of **lymphocytes,** the white blood cells which produce antibodies against foreign proteins, invading bacteria, etc. Lymph nodes also "filter" out foreign material.

BLOOD

Blood consists of cells and a fluid, the **plasma.** In mammals, red cells, or **erythrocytes,** lack nuclei, except during their developmental stages which occur chiefly within the marrow of the long bones and sternum. The white cells are called **leukocytes**. About 75 per cent have a granular cytoplasm and a very irregular, or polymorphic, nucleus. These are the **polymorphonuclear leukocytes,** popularly called polymorphs. About 70 per cent of all leukocytes are polymorph **neutrophils,** so called because their cytoplasm stains in most types of dyes. Polymorph **acidophils** (or eosinophils), which stain with acid dyes, constitute about 4 per cent, and the polymorph **basophils** about 1 per cent or less. The polymorphs ingest bacteria and cell debris. The remaining approximately 25 per cent of the leukocytes are mostly either small **lymphocytes** or large **monocytes.** Monocytes easily become phagocytic, engulfing foreign particles. They probably are the same cells, known since the time of Metchnikoff about a century ago as macrophages, which play an important role in the first steps of immune reactions. The lymphocytes (plasma cells) manufacture the antibodies. The absolute and relative abundance of various leukocytes is a useful diagnostic tool in medicine.

OXYGEN-CARBON DIOXIDE TRANSPORT

Water will carry some dissolved oxygen. When exposed to ordinary air, which is approximately 21 per cent oxygen, water holds only about 0.6 ml. of oxygen per 100 ml. of water at 30°C. The red hemoglobin of some annelids will make it possible for 100 ml. of blood to hold about 10 ml. of oxygen, the hemoglobins of fish about 15 ml., and of birds and mammals up to 20 ml., or even 30 ml. of oxygen in the case of porpoises, which are adapted for lengthy dives. The significance of hemoglobin is that it combines easily with oxygen when there is a high oxygen pressure, becoming scarlet oxyhemoglobin, and gives it up as readily when

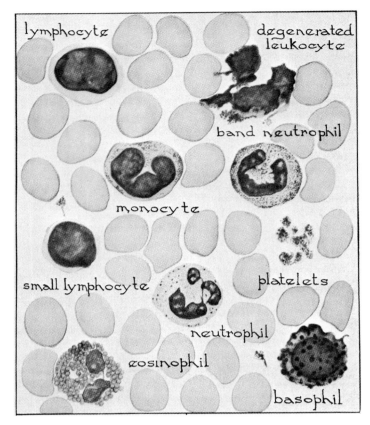

Fig. 27-5. The erythrocytes and leukocytes of normal blood (Hastings' stain). (Reproduced by permission from A. W. Ham: *Histology,* 6th ed. J. B. Lippincott, Philadelphia, 1969.)

the oxygen pressure (concentration) falls as it does in the tissues of the body some distance from gills, lungs, or moist skin. Hemoglobin is thus a very special kind of compound.

Deep within the body, carbon dioxide (more accurately carbonic acid) produces acid conditions which facilitate the breakdown of oxyhemoglobin into oxygen and hemoglobin. The H_2CO_3, which is more acidic than hemoglobin, takes a potassium ion away from the hemoglobin and becomes potassium bicarbonate, $KHCO_3$. Once back in the lungs, the hemoglobin gains oxygen, becoming more acidic and therefore able to recapture the potassium. This means that potassium bicarbonate is reconverted to carbonic acid which, under the influence of the enzyme carbonic anhydrase in the red cells, breaks down into H_2O and CO_2 which is lost to the surrounding atmosphere or water.

$$CO_2 + H_2O \underset{\text{carbonic}}{\overset{\text{anhydrase}}{\rightleftharpoons}} H_2CO_3$$

Oxygen-carrying pigments present many intriguing puzzles. **Hemoglobin** is found not only in vertebrates but in earthworms, many marine annelids, water fleas among crustaceans, some pond snails, *Ascaris* (a common intestinal parasitic worm), and even certain ciliates. What does such a wide distribution indicate if not that hemoglobin is a very ancient pigment which must have appeared early in evolution or that it is derived from some similar and widespread molecule? Chemically hemoglobin is a double molecule consisting of heme united with a protein, globin. Heme is a close chemical relative of chlorophyll, of cytochrome, and of vitamin B_{12} (cyanocobalamin). It is a typical porphyrin having iron in the center where chlorophyll has a magnesium atom. The protein differs somewhat from species to species. Vertebrate hemoglobin from fish to man is composed of heme plus four polypeptide chains, each with a molecular weight of 17,000.

The other most widespread respiratory pigment is **hemocyanin,** which is a copper-containing compound lacking porphyrin. It is a light French blue when oxygenated, colorless when reduced. This is the pigment in the blood of lobsters, *Limulus,* some spiders, the octopus, and various other animals.

COAGULATION

The coagulation of the blood appears simple, yet careful investigation has shown it to be one of the most complex of all known biological processes. At least 30 factors may play some role. From the point of view of the organism, the problem is that unless the blood clots promptly, death by hemorrhage might follow even a minor scratch. On the other hand, if the clot blocking a cut should extend into the arteries or veins and cause even a small fraction of the blood within the body to clot, death would also result.

From the welter of observations and theories the following facts stand out. The jelly-like clot or **thrombus** is composed of a fibrous protein, **fibrin,** which is present in normal plasma in an unpolymerized form called **fibrinogen.** In order for fibrinogen to polymerize into fibrin, calcium ions must be present. This is why citric acid, which combines with calcium to form calcium citrate, can prevent clotting. What produces the conversion of fibrinogen to fibrin? This is done by another blood protein, the enzyme **thrombase** sometimes called simply **thrombin.**

USEFUL REFERENCES

Adolph, E. F. The heart's pacemaker. Scientific American, March 1967.

Guyton, A. C. *Circulatory Physiology: Cardiac Output and Its Regulation.* W. B. Saunders Co., Philadelphia, 1963.

Ingraham, V. M. *The Hemoglobins in Genetics and Evolution.* Columbia University Press, New York, 1963.

Spain, D. M. Atherosclerosis. Scientific American, August 1966.

Vander, A. J., Sherman, J. H., and Luciano, D. S. *Human Physiology.* McGraw-Hill Book Co., New York, 1970.

Vroman, L. *Blood.* The Natural History Press, Garden City, N.Y., 1968 (Paperback).

Nutrition and Respiration

Nutrition is simultaneously a personal and an international problem of prime importance. Not enough of the right kinds of proteins very early in life may result in permanent mental retardation. Too little of the right vitamins at any time of life will bring serious trouble. Too much fat, especially cholesterol, in middle age greatly increases the chances of heart disease. Little understood nutritional factors absorbed from intestinal bacteria seem to play an important role in health. Even intelligent people have been led to believe that calories in the form of carbohydrates are worthless because they are "empty," by which it seems to be meant that such things as polished rice or white bread are without nutritional value because they are not a complete and adequate diet in themselves. There are many proteins that by themselves cannot support human life because they lack certain essential amino acids. Nor can man live by vitamins alone.

There is also a philosophical interest in nutrition. If one stops to think, it is an amazing thing that what you eat turns into you, while the same food eaten by someone else may turn into a beautiful young girl, a crabby old man, or even a parrot.

Everywhere the problems of adequate nutrition are more than strictly scientific questions because political and economic factors are deeply involved. This is especially true in parts of the world where the population is larger than the resources of the country can support.

The problems of nutrition and respiration are basically inseparable because the purpose of eating is two-fold, to build and maintain the structure of the body and also to furnish the energy that makes it function.

ANATOMY OF NUTRITION

For most vertebrates nutrition begins with the use of teeth. So clearly do teeth reflect the whole life of their possessors that a great comparative anatomist, Sir Richard Owen, paraphrased Archimedes' famous remark on discovering the principle of the lever, "Give me where to stand, and I will move the earth," by asserting, "Give me a tooth, and I will reconstruct the animal." This is an exaggeration, of course, but one that contains much truth. Carnivores, like the cat and seal, have large canines and pointed molars. Herbivores, like the deer and horse, have reduced canines or none, with molars that are grinders with flat, corrugated tops. Omnivores, like man and other primates, have generalized teeth.

The **pharynx** or throat follows immediately behind the buccal or mouth cavity which contains that versatile organ, the tongue. It is in the pharynx that the pathway of the food crosses the route of air to and from the lungs in reptiles, birds, and mammals.

The **esophagus** is merely a muscular tube conducting food from the pharynx past the heart and lungs into the stomach, which lies just below the diaphragm, a transverse muscular partition separating the coelom around the lungs from that surrounding the stomach, liver, and intestine. The structure of the wall of the esophagus is essentially the same as that of the stomach or intestine. The cavity or lumen of all three is lined by an epithelium composed of squamous, *i.e.,* thin and flat, cells in the esophagus and tall, columnar ones in the stomach and intestine. This epithelium is supported by a thin connective tissue sheet, or lamina, and a very thin muscular sheet. These constitute the **mucosa**, although this word is sometimes used to refer solely to the epithelium. The submucosa encircling the mucosa consists of loose, fibrous connective tissue with numerous blood vessels and, in some parts of the digestive tract, glands. Outside of the submucosa is a layer of circular muscles, then a layer of longitudinal muscles, and finally, on the very outside, an epithelial covering. All these muscles are of the nonstriated type.

The **stomach** is an enlarged and specialized portion of the gut tube. The portion which the esophagus enters is known as the **cardiac region**, since it is nearer the heart. The opposite end is the **pyloric region**. The **pylorus** is the orifice between the stomach and the intestine, kept closed by a circular, *i.e.*, sphincter, muscle.

The first region of the **small intestine** into which the stomach empties is the **duodenum**. The common bile duct, carrying the secretions of both pancreas and liver, enters the duodenum not far from its origin at the pylorus. The remainder of the small intestine, *i.e.*, the **ileum**, is much longer than the duodenum and leads into the shorter large intestine, or **colon**. The colon ends in a short straight section, the **rectum**, which leads to the exterior via the **anus**. Where the small intestine enters the large there is in most vertebrates a longer or shorter blind sac, *i.e.*, a diverticulum, termed the **cecum**, and vermiform appendix in man.

CHEMISTRY OF DIGESTION

Mammals are equipped with an impressive battery of glands associated with the gut. Some are unicellular; others, like the pancreas, are large and complex. The function of all these glands is to secrete digestive enzymes that hydrolyze large molecules, *i.e.*, split them with the addition of water, into smaller molecules that can be absorbed through the wall of the digestive tract. Carbohydrates are broken down into glucose and other **simple sugars**, lipids into **glycerol** and **fatty acids**, and proteins into **amino acids.**

Table 28-1 shows the main sources of mammalian digestive enzymes and their actions. Like other enzymes these are very sensitive to pH. **Ptyalin**, the amylase in saliva, works at a pH close to neutrality; the proteolytic gastric pepsin requires an acid pH; the pancreatic and

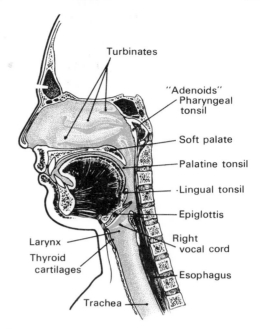

Fig. 28-1. Sagittal section through a human head to show the buccal and pharyngeal cavities where the pathways of air and food cross. (From G. B. Moment: *General Zoology,* 2nd ed. Houghton Mifflin Co., Boston, 1967.)

intestinal enzymes require an alkaline environment for their activity.

Gastric juice is a clear fluid containing about 0.4 per cent hydrochloric acid and two enzymes, **pepsin**, which splits proteins by hydrolysis into short chains of amino acids called peptides, and **rennin**, which clots milk. Recent evidence indicates that a lipase is also normally present in the stomach. Why doesn't the stomach digest itself? Such cannibalism is physiologically possible. The lining of the stomach is normally protected against pepsin by a coating of mucus. The enzyme-secreting cells form pepsinogen which becomes active pepsin only after it comes in contact with the acid gastric juice.

When the acid stomach contents pass through the pylorus into the duodenum, the pancreas pours out **pancreatic juice**. The duodenum secretes a hormone, **secretin**, which is carried by the bloodstream to the pancreas. Digestion in the intestine of mammals is due in part to the pancreatic juice which contains enzymes capable of hydrolyzing all three major classes of foodstuffs. Multicellular glands in the

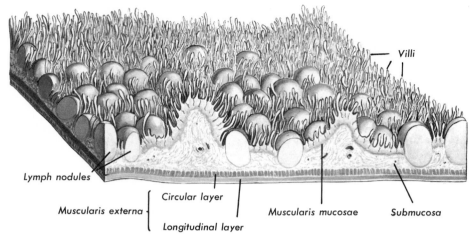

Fig. 28-2. Lining and wall of the human ileum as it would appear under high power of a dissecting microscope. (From F. R. Bailey: *Textbook of Histology*, 16th ed., The Williams & Wilkins Co., Baltimore, 1971.)

Table 28-1
Principle digestive enzymes in mammals
Substances shown in italics are absorbed through the walls of the intestinal villi.

Enzyme	Occurrence	Substrate	Chief Products
Ptyalin	Saliva	Starch	Maltose
Pepsin	Gastric juice	Protein	Polypeptides
Amylases	Pancreatic juice	Starches	Maltose
Lipases		Lipids	*Glycerol* and *fatty acids*
Trypsin		Proteins	Peptides
Chymotrypsin		Polypeptides	Peptides
Peptidase		Peptides	*Amino acids*
Sucrase	Intestinal juice	Cane sugar	*Simple sugars*
Maltase		Maltose	*Simple sugars*
Lactase		Milk sugar	*Simple sugars*
Lipase		Lipids	*Glycerol* and *fatty acids*
Erepsin		Peptides	*Amino acids*
Peptidases		Peptides	*Amino acids*

submucosa of the duodenum secrets mucus and probably more digestive enzymes.

The lining of the intestine is not only covered with ridges or folds, but the surface, including the surface of the folds, is covered with finger-like projections called **villi**. Each villus is covered with columnar epithelium and contains a core of fibrous connective tissue with capillaries and a lymph vessel. In between the villi are numerous "post-holes," called **crypts of Lieberkühn**. The mucosal epithelium not only covers the villi and the general lining of the gut but extends down into the crypts where cell division is frequently seen. It has been estimated that the epithelial cells of the

intestine are completely renewed every one and a half to two days.

ABSORPTION

Most of the digested food is absorbed in the intestine by the villi just described. The glucose and amino acids are passed as such into the capillaries, whence they are carried to the liver in the hepatic portal vein. The absorption of fats is more complex and less understood. The glycerol is taken into the villi as such. The fatty acids unite with bile salts forming water-soluble compounds which enter the cells of the villi. Within the villi the fatty acids reunite with the

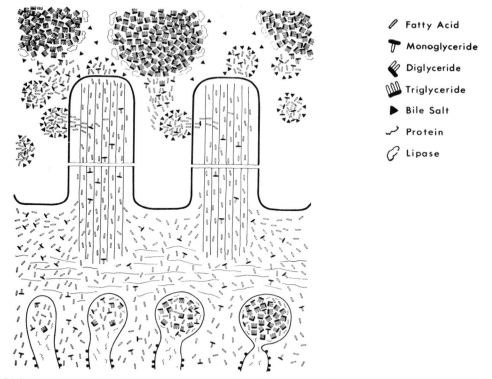

Fatty Acid

Monoglyceride

Diglyceride

Triglyceride

Bile Salt

Protein

Lipase

Fig. 28-3. Diagram of the process of lipid absorption into the epithelial cells lining the intestine. Note how lipase molecules break up oil droplets with the help of bile salts into fatty acids and (mono)glycerides which enter the cells and reassemble as intracellular lipid droplets. (From F. R. Bailey, *Textbook of Histology,* 16th ed. The Williams & Wilkins Co., Baltimore, 1971.)

glycerol, forming lipids again. Some lipids are absorbed directly as glycerides, *i.e.,* neutral fats (esters) formed as usual of a union of glycerol and a fatty acid. Most of this fat passes as a milky emulsion in the lymphatics of the mesentery up to the thoracic duct and through it into the left subclavian vein to the heart.

The absorption of whole protein molecules is of negligible importance for nutrition but is significant in food allergies and in the absorption of bacterial toxins. Most absorption is due to **active transport,** requiring the expenditure of energy. This is evident because very low concentrations of glucose and amino acids can be absorbed from the intestine into the bloodstream where the concentration of these substances is much higher, and because metabolic poisons quickly stop absorption.

ROLES OF THE LIVER

The liver plays a key role in the fate of the digested food carried to it from the intestine by the hepatic portal vein. No organ, not even the brain, has been the object of such intense study by physicians and soothsayers from remote antiquity. Yet it was not until the latter years of the past century that something of its actual function was first learned.

Glucose is converted by the liver into glycogen under the influence of insulin. It is stored until the glucose in the blood falls to a threshold level when the glycogen is reconverted into glucose.

The fate of amino acids in the liver raises important questions and highly controversial views about how much protein is necessary in the diet. Only a small proportion of the

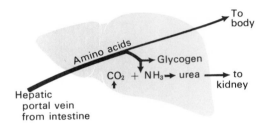

Fig. 28-4. Diagram of mammalian liver showing the course of amino acids from the intestine via the hepatic portal vein and their fate within the liver. Note some continue via the hepatic vein to the body in general.

absorbed amino acids reaching the liver via the hepatic portal vein from the intestine are passed through to be utilized in building body protein. Most amino acids have their amino groups taken from them, a process called deamination, a two-step process in which the amino acids are oxidized by removal of two H's and then hydrolyzed to form a keto acid and ammonia. The keto acids corresponding to the various amino acids are generally metabolized as energy sources in the Krebs cycle.

The synthesis of urea is a major function of the liver. The ammonia, whether from deamination or oxidation of amino acids, is poisonous except in very low concentrations. That "clever chemist," the liver, combines two molecules of ammonia with one of another waste product, CO_2, to form a colorless, odorless, and harmless compound, urea, which is carried by the bloodstream to the kidneys and there excreted. The synthesis of urea occurs when NH_3 and CO_2 unite with ATP to form a carbamyl phosphate which, in turn, unites with a long-chain amino acid called ornithine. The ornithine unites with citrulline and this compound with still another and then with arginine, after which the resulting compound plus water splits into urea, $(H_2N)_2CO$, plus ornithine. Hence urea is said to be formed in the **ornithine cycle.**

How could anyone even suspect that there is such a thing as the ornithine cycle? The

$$2NH_3 + CO_2 \rightarrow H_2N\overset{\overset{\displaystyle O}{\|}}{-}C-NH_2 + H_2O$$

Fig. 28-5. The over-all formula for the synthesis of urea from ammonia and CO_2.

discovery was made by the same Hans Krebs who later worked out the details of the citric acid cycle named after him. Krebs was studying the formation of urea by adding various amino acids to slices of liver fresh enough for the enzymes to be still active. He found that added ornithine or arginine greatly increased the production of urea, especially in the presence of added ammonia. This opened the door.

In addition to the synthesis of urea from the breakdown products of amino acids and the storage of glucose as glycogen, the liver secretes bile, destroys worn-out red blood cells, makes prothrombase, stores vitamins, detoxifies miscellaneous harmful substances, and plays a role in lipid metabolism, as well as synthesizing its own proteins. The liver is indeed a versatile as well as a clever chemist.

NUTRITIONAL REQUIREMENTS

The Calorie Problem—Carbohydrates and Lipids

Carbohydrates are the primary fuel of life for man and most animals. Every major human culture has relied on some carbohydrate-rich food for energy—wheat, potatoes, rice, maize, breadfruit. Lipids are a more concentrated source of energy. If both are lacking, then body proteins will be utilized. It will be recalled that at the "lower" end of the glycolytic pathway, acetyl-coenzyme A serves as a central metabolic crossroads where many kinds of molecules as well as pyruvic acid from the anaerobic breakdown of glucose are fed into the Krebs aerobic cycle as acetyl-CoA. Lipids and amino acids from proteins are converted into acetyl groups, $CO\cdot CH_3$ which, linked to coenzyme A (a derivative of the vitamin pantothenic acid), enters that "meat grinder" the Krebs cycle. Lipids serve as essential constituents of cell membranes, as reservoirs of stored energy, and as carriers of the fat-soluble vitamins.

The **fuel** or **energy value** of any food is measured in heat units called **calories.** One large Calorie (kilocalorie), the unit usually employed, is defined as the amount of heat required to raise the temperature of 1,000 grams of water $1°$ centigrade. The total caloric content of food can be determined by complete combustion in a heat-measuring device called a **bomb calorimeter.** The powdered food is ignited by an electric spark. The heat liberated is measured by the rise in temperature of the water in a jacket that surrounds the combustion chamber. A man who is doing hard physical labor requires from 3,500 to 5,000 calories a day. A person doing sedentary or semi-sedentary work burns about 2,500 calories daily. To maintain life, about 500 calories a day are necessary. Although dietary proteins, unless present in minimal amounts, regularly serve as a source of energy, *i.e.,* calories, the primary role of proteins is quite different, and to that we now turn.

THE PROTEIN PROBLEM

There is no doubt whatsoever that proteins or, more precisely, amino acids, are essential in the human diet. How much protein does a child or adult need for normal growth and health? The answer to this question is a fighting issue for some people and is often clouded by ancient beliefs and modern prejudices. The fact that as much as 90 per cent of the ingested amino acids are quickly converted in the liver into glycogen and used as a source of energy plus urea, which is merely excreted, has cast very serious doubt on the need for large amounts of protein. Not only do prejudices obstruct rational answers, so also does fundamental ignorance of many aspects of the scientific side of the problem. What are some of the important facts?

Severe protein deprivation, especially in young children, leads to a serious disease, **kwashiorkor**, meaning "red boy," so called because of the characteristic dermatitis. It is so prevalent in parts of Africa, Central and South America, and Asia that special United Nations committees have been studying ways to eliminate the disease.

Far less dramatic but probably much more widespread than kwashiokor is brain damage, shown especially by impairment of learning and other intellectual abilities, in young children having had protein-deficient diets. This conclusion has been reached in very careful studies of children in Central American villages by an international group of investigators including Rene Dubos of the Rockefeller University, Craviots, Ramos-Galvan, and others.

Heredity sets only the upper limits of intelligence in man as in any other primate.

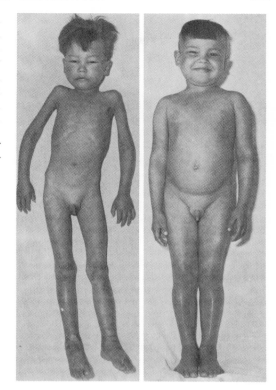

Fig. 28-6. Boy suffering from kwashiorkor before and after treatment with a diet of mixed plant proteins containing all the essential amino acids. (Kindness of M. Behar.)

How fully the hereditary potential is realized obviously depends on many factors. Too little thyroid hormone, for example, will turn an otherwise normally intelligent child into a retardate. In these protein deficiency studies, the effects of heredity were eliminated by the reasonable assumption that children, fed adequate diets and superior in mental ability, had the same genetic potential as their permanently mentally retardate relatives who had started out in life without enough of the right kind of protein.

It turns out that it is not so much the amount of protein that is important but the kind of protein. Each protein is composed of a different array of amino acids in different proportions. Since amino acids are necessary for synthesis of the proteins that make up the cytoplasm and nucleus of cells, as well as enzymes, muscles and other specialized structures, and compounds such as melanin, thyroxin, and adrenaline, an animal requires the right amino acids in the proteins it eats.

The amino acids which an animal must have in its diet for normal growth and health are called **essential amino acids** and a protein which has them all is known as **complete**. This does not mean that no additional amino acids are needed by the animal but only that from the "essential" ones any necessary additional acids can be made by the animal itself. In man, 23 different amino acids have been identified. Ten are essential in the above sense. The remaining 13 can be made from the 10 by our metabolic machinery. In the body there are special enzymes, **transaminases**, which convert one amino acid into another.

In general, proteins from animal sources —meat, fish, eggs, cheese—are complete. Many plant proteins are not. This is the case with beans, peas, or grains from which the germ has been removed. Fortunately some plant proteins complement one another. For example, Indian corn, *i.e.,* maize, is very poor in lysine but contains the other essential amino acids including methionine, which is deficient in beans. Beans, however, possess plenty of lysine. Obviously, when these two protein sources are mixed, their deficiences cancel out. Because there is no significant storage of amino acids in the body, these two proteins must be eaten at the same meal to be effective in supplementing each other. No wonder that the Algonkian Indians did so well on their succotash.

A major breakthrough in very recent times is the development of a lysine-rich corn by a Purdue University team lead by E. T. Mertz. Unfortunately the high lysine mutant known as "opaque-2" is recessive so that breeding the corn for seed presents special problems. It also requires a somewhat different kind of milling and preparation for cooking. But when tested on piglets, those on opaque-2 grew over 300 per cent faster than those on "normal" corn. Spectacular results have now been obtained with malnourished children in Latin America.

An individual is said to be in nitrogen balance when the amount of nitrogen excreted as products of protein breakdown equals that ingested. Rapidly growing children and young animals, pregnant women, individuals recovering from severe illness characteristically show a positive nitrogen balance, excreting less than they ingest. Most authorities claim that somewhere between 21 and 35 grams (about an ounce) of protein per day is enough for an average man; 21 to 27 grams for a woman.

Vitamins

While it has been commonly accepted that vitamins are essential for physical well-being, it is now evident that vitamins, specifically ascorbic acid and niacin, are also essential for mental health. How a knowledge of vitamins was won is an important part of our cultural heritage.

The discovery of vitamins grew out of the needs of explorers. For centuries, **scurvy** had been the nightmare of sailors and explorers as well as the curse of boarding schools and prisons. In his voyage around the Cape of Good Hope in 1498, Vasco da Gama lost almost two-thirds of his crew from scurvy. In the words of Dr. Logan Clendening, a noted medical historian:

> First a sailor's gums would begin to bleed. Then some of his teeth would fall out. The stench from his mouth would be horrible. Then great blotches would appear on the skin. The wrists and ankles would swell. A bloody diarrhea set in. Parts of the flesh would rot out. Finally, exhausted, delirious, loathsome, the poor wretch would pay his debt to nature.

The first recorded successful effort to control this scourge took place during an exploration of the St. Lawrence River in 1536 by Jacques Cartier. After he had lost 26 men, the rest were restored to health by a pine needle concoction prepared by the native Indians. In view of the dramatic nature of the need, it is difficult to understand why nothing came of this incident.

Eighty years later, John Woodhall, a surgeon in the employ of the East India Company which regularly sent ships around the Cape of Good Hope, mentioned in print the value of lemon juice in preventing scurvy, as did James Lind 137 years later. But such is the conservatism, stupidity, or perhaps mere inertia of men that it was not until 1768, over 230 years after Cartier's men were saved by pine needles, that the famous Captain Cook set out in *H.M.S.*

Endeavour and proved the efficacy of lemons on a long voyage in the South Pacific. Even today British sailors are called "limeys" from the lemons and limes that formed a part of their diet. In comparatively recent times the agent in citrus fruits which is responsible for preventing scurvy was named vitamin C and has been identified as **ascorbic acid.**

The actual chemical identification of vitamins came about in a very interesting way. In the 19th century a Russian investigator, Lunin, had found that when animals are fed on highly purified diets of carbohydrates, fats, and proteins, they sicken and and die even though all three major food components are present. Many years later Hopkins, a pioneer biochemist working in a small basement laboratory at Cambridge, reinvestigated this old puzzle of Lunin and found in milk what he called an accessory food factor needed in minute amounts for health. This announcement stimulated work in several laboratories and by 1912 the first such substance was isolated from rice polishings by Casimir Funk. It was the anti-beriberi factor, which turned out to be thiamine (B_1). Funk coined a new word by calling the factor a **vitamin.**

Modern Knowledge of Vitamins

Until rather recently, nothing was known about the biochemical function of vitamins. Then almost simultaneously, physiologists discovered a "yellow enzyme" essential for the oxidation of foodstuffs within the cells and nutritionists found a yellow vitamin, **riboflavin,** essential for normal growth in young birds and mammals and for the prevention of various eye and skin disorders in adults. The "yellow enzyme" and riboflavin were found to be intimately associated, and from this fact came the knowledge that vitamins can function as structural components of enzymes.

Most, and probably all, enzymes consist of two parts, a larger protein portion and a smaller non-protein portion. The protein part of the enzyme molecule is called the **apoenzyme;** the smaller non-protein part, the **coenzyme.** If the coenzyme is very firmly bound to the apoenzyme, it is often called the **prosthetic group.** A coenzyme or prosthetic group is usually regarded as a vitamin for any particular animal

Fig. 28-7. Chickens showing the usual fowl symptoms of beriberi (polyneuritis) due to thiamine deficiency. (Kindness of the University of Wisconsin.)

only if the animal cannot manufacture it for itself.

The knowledge that vitamins form the non-protein part of enzyme molecules is of much more than purely theoretical interest. It makes it possible to form medically useful drugs like sulfanilamide. Such substances are essentially fake vitamins, substances with which the protein part of the enzyme will unite but which will not make a functional enzyme. In such a way sulfanilamide blocks the enzymes of the bacteria.

Up to the present, somewhat more than a dozen vitamins have been discovered. Many are water soluble, like vitamin C and the B vitamins. The rest are fat soluble. Some are heat stable; others are readily destroyed by heat. Table 28-2 gives the chemical name, the letter symbol (when one exists), dietary sources, and the function of most of the vitamins.

Special interest attaches to **niacin,** the pellagra-preventing vitamin, because of its certain relation to a severe mental derangement. Pellagra is usually said to be characterized by three D's: dermatitis (a roughening of the skin), diarrhea, and dementia. It should be four, because death is the end result. Because niacin cures the severe psychosis of pellagra patients, it is now being tried in mental derangements not associated with pellagra. Some real success has been reported. Niacin forms part of two coenzymes essential for the utilization of carbohydrate in muscles, yeast, and presumably all cells. These are nicotinamide adenine dinucleotide (NAD) and nicotinamide adenine dinucleotide phosphate (NADP). In stability,

Table 28-2
Important vitamins

Solubility	Chemical Name	Letter Symbol	Functions	Dietary Sources	Heat Stability
Water	Ascorbic acid	C	Anti-scurvy, mental health	Citrus fruits, tomatoes, leafy vegetables	Labile
Water	Thiamine	B₁	Anti-beriberi, coenzyme in pyruvate metabolism	Milk, meats, leafy vegetables	Fairly stable in acid
Water	Cyanocobalamin	B₁₂	Anti-pernicious anemia, synthesis of purines	Milk, whey, soybeans, cotton seed	
Water	Niacin, nicotinic acid		Pellagra-preventive, NAD and NADP nucleotides, mental health	Peanuts, liver, chicken, fish, whole wheat	Stable
Water	Folic acid		Normal growth, blood formation	Meats, eggs, beans, yeast, leafy vegetables	Fairly stable
Water	*Para*-aminobenzoic acid		Normal growth	Many foods	
Water	Riboflavin	B₂	FAD respiratory coenzyme	Whey, most foods	Stable
Water	Pantothenic acid		Part of coenzyme A	Peanuts, lettuce, eggs	
Water	Pyridoxin	B₆	Anti-dermatitis, anti-acrodynia (in rat)	Egg yolk, wheat germ, yeast	
Water	Inositol		Anti-alopecia (in mouse), anti-fatty liver (in rat)	Rootlets, sprouts, fruit, lean meat, yeast, milk	
Water	Biotin	H	Anti-egg white injury	Egg yolk, kidney, liver, tomatoes, yeast	
Lipid	Retinol	A	Healthy skin and mucous membrane, night vision	Milk, yellow and green vegetables	Stable
Lipid	Calciferol	D	Anti-rickets	Fish liver oils, butter, egg yolk	Stable
Lipid	Menadione	K	Normal blood clotting	Most foods	Stable
Lipid	Tocopherol	E	Herme synthesis, anti-sterility, normal muscle development	Leafy vegetables, meat, yolks	Stable

niacin is at the opposite pole from ascorbic acid. It is stable to heat, light, acids, alkalis, and oxidizing agents. Yeast, lean pork or beef, and liver are rich in niacin. Corn (maize) is very deficient in it, and most cases of pellagra occur in regions where corn is the chief article of diet.

ROLE OF INTESTINAL FLORA

Most vitamins are obtained from food. However, evidence by Richard Barnes and his associates at Cornell and by other investigators in this country and abroad indicate that the intestinal microflora contribute important amounts of vitamins to the host. Gnotobiotic animals, *i.e.*, germ-free rats, chickens, and guinea pigs, show unmistakable symptoms of deficiencies of vitamins in the B group—thiamine, niacin, riboflavin, and folic acid—as well as ascorbic acid (vitamin C).

Antibiotics are regularly fed to many domestic animals to improve growth. The mechanism is poorly understood; the problem is complicated by evidence that some intestinal bacteria secrete toxic products. It is significant that antibiotics do not exert either a vitamin-sparing or a growth-promoting effect on germ-free animals.

In general, antibiotics like penicillin, which are easily absorbed and reach only the stomach and duodenum, increase vitamin production of the intestinal microflora. Antimetabolites like the sulfonamides, which are poorly absorbed by the small intestine and hence reach the large intestine and cecum, depress vitamin production.

There is still much to be learned about this whole problem. It was discovered in 1908 that germ-free ducks and chicks benefited from inoculation with intestinal bacteria, although this work was noticed by few. It has been known for many years that cows and other ruminants receive not only B vitamins but also proteins from their symbiotic stomach micro-organisms. Recently it has been found that adding urea to the diet of a cow increases the protein available to her. Apparently her gastro-intestinal bacteria and ciliates utilize the urea to form amino acids which somehow become available to the cow. So the problem is not new. What may lead to significant break-throughs are the new methods for using germ-free animals and antibiotics. Who knows? There may be a factual basis for those wild tales of the superabundant health of certain remote peoples who eat yogurt, a cheese-like food rich in *Lactobacillus bulgaricus.*

RESPIRATION

From the earliest times respiration has seemed close to the essence of life itself. In fact, the word "spirit" is derived from the Latin *spirare* (to breathe). Hence there has always been an insistent philosophical interest in the meaning of respiration. In modern times urgent practical problems involving respiration have arisen in mining, in submarine and diving operations, in aviation, in the use of anesthetics, and in the diagnosis and treatment of metabolic diseases.

Knowledge of respiration has advanced in four major steps which correspond, in a very general way, to the four centuries during which research has pushed ahead. The facts and principles established at each stage are permanently valid and will be presented in order, since they are essential to a firm understanding of this vital topic.

Early Research

In the 17th century Robert Boyle and his young friends, all in their early 20's, put to the test the old idea that life is like a flame. They obtained definite proof that a mouse cannot live long in air in which a candle has burned itself out, nor can a candle burn in air in which

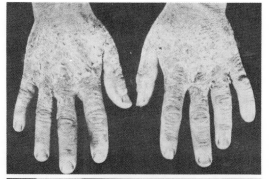

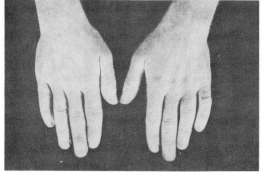

Fig. 28-8. The hands of a person with pellagra before and after treatment with niacin. (Kindness Southern Medical Association.)

a mouse has suffocated. Clearly there is some basic connection between what a flame and what breathing does to air.

The Classic Laws of Respiration

The over-all laws of respiration were established in the 18th century largely by the work of Lavoisier, one of the discoverers of oxygen, and Simon Laplace, mathematician and astronomer. These men placed animals in ingenious calorimeters (heat meters) designed to measure the amount of heat produced. They also made extensive chemical tests of air before and after it had been inhaled by men and animals. This and subsequent work proved that:

1. In respiration, animals obey the **law of conservation of energy**. Both a breathing guinea pig and burning charcoal give off the same amount of heat energy when the same amount of oxygen is used up.

2. In respiration, animals obey the **law of conservation of matter**. During respiration the

Fig. 28-9. Lavoisier in his laboratory experimenting on breathing. Note Madame Lavoisier on the right taking notes and an assistant feeling the pulse of the subject of the experiment. The expired air is being collected in a glass bell jar inverted over water. From a contemporary print. (Kindness of the Johns Hopkins University Institute for the Study of the History of Medicine.)

amount of O_2 burned is equivalent to the amount of CO_2 produced. In a closed system there is no change in total weight.

The 19th century refined these fundamental observations and thereby added the third and fourth laws of respiration.

3. The **respiratory quotient**, that is, the volume of CO_2 given off divided by the O_2 consumed, is an indication of the type of food being burned. As can be seen from inspection of the equation:

$$C_6H_{12}O_6 \text{ (sugar)} + 6O_2 \rightarrow 6H_2O + 6CO_2$$

when sugar is burned the respiratory quotient is one. Fat, however, has far less oxygen in proportion to carbon and hydrogen than sugar does; hence fat requires proportionally more oxygen to oxidize it completely into water and carbon dioxide. For example, the formula for beef fat is $C_{57}H_{110}O_6$. Simple arithmetic will show that 81.5 volumes of oxygen will be required for every 57 volumes of carbon dioxide produced. Hence the respiratory quotient (RQ) will be less than one, in this case 0.7.

The RQ of proteins is intermediate between those of fats and carbohydrates. The amount of protein being metabolized is usually determined, however, by measuring the amount of nitrogen-containing compounds in the urine.

4. In mammals the **basal metabolic rate** (BMR) depends on size. The basal rate is the rate at which oxygen is used when the animal is at complete rest. The smaller the animal, the higher the BMR. The reason for this inverse relationship is clear. Mammals maintain a constant body temperature above their surroundings and hence lose heat to it. Any solid object can lose heat only from its surface. By the facts of solid geometry, the smaller an object is the more surface it has *in proportion* to its volume or mass. Therefore it follows that to maintain a given temperature a small animal must burn more glucose per pound of flesh than a large one.

Inspection of Figure 28-10, where the weight of several homoiotherms (warm-blooded animals) is plotted on semi-log paper against their respiratory rate, will reveal that as size goes down, metabolic rate skyrockets. Because

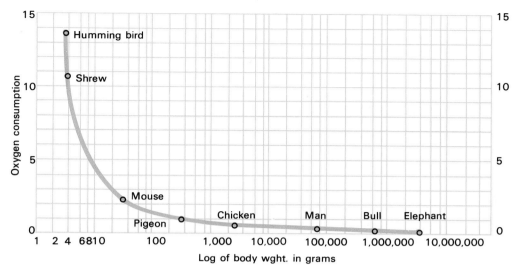

Fig. 28-10. Respiration and body weight. The oxygen consumption is given in terms of milliliters of O_2 consumed per gram of body weight.

of their high rate of oxidation, very small mammals, like shrews, which weigh only 3 or 4 grams, must eat almost continuously and have voracious appetites. They consume approximately their own weight every 24 hours or starve to death. It is the equivalent of a 150-pound men eating three 50-pound meals a day! The graph also makes it clear that about 2.5 grams is the lower limit for size of a warm-blooded animal. Below that, food intake presents an impossible problem.

The 19th century also turned its attention to the actual mechanics and to the important physiological controls of respiration as well as the transport of O_2 and CO_2 in the blood.

Mechanics of Breathing

There are three general mechanisms for getting oxygen into cells and carbon dioxide away. Feathery **gills**, which circulate blood in close proximity to water, are found both in vertebrates and invertebrates. **Tracheas** are found in insects and a few other invertebrates. They almost completely permeate the body with a tree-like network of interconnected tubules. Vertebrates, except for most fish, possess internal and highly vascular sacs, the lungs. In the higher reptiles, birds, and mammals air is drawn into the lungs when a partial vacuum is created by the enlargement of

the chest cavity. This is accomplished by the contraction of the muscles of ribs and diaphragm. The intercostal muscles between the ribs raise them and thereby enlarge the chest. When the **diaphragm**, a dome-shaped sheet of muscle separating chest from abdominal cavity, contracts it flattens the dome and thus enlarges the cavity of the thorax and the volume of the lungs. During expiration the size of the chest cavity is decreased.

Air travels from the pharynx down the trachea and either the right or left **bronchus** into the corresponding **lung**. Each lung is attached only in the region of its bronchus. The rest of the lung hangs freely in its own pleural cavity. If a stab wound is made into a pleural cavity, air will be sucked in, the vacuum will be destroyed, and the lung will collapse, making breathing impossible with that lung.

Regulation of Breathing

The control of breathing is under both chemical and nervous auspices and constitutes a beautiful self-regulating or 'feedback mechanism. Increasing concentrations of carbon dioxide in the blood stimulate a breathing center in the medulla, part of the brain stem. This center is actually paired (as are the eating centers in the brainstem), one on the right and one on the left of the midline. Each of these centers is double (again like the eating centers) with an expiratory and an inspiratory sub-center. There are important chemoreceptors in the aorta and the **carotid bodies** on the carotid arteries in the neck which sense the pH and probably O_2 and CO_2 concentration in the blood and inform the parts of the brain where breathing is controlled.

Impulses pass down the **phrenic nerves** from the brain to the diaphragm, cause it to contract, and thus produce inspiration. If the two phrenic nerves are cut, the diaphragm no longer contracts. Expansion of the lungs in turn stimulates sensory nerve endings, which send inhibitory impulses back to the brain via the two vagus nerves. Inhibition of the respiratory center in the medulla permits the diaphragm to relax, producing expiration. Impulses leading to exhalation also arise from the carotid body, and from the aortic body. Exercise, involving movements of the joints of arms, hands, legs, and feet, produce stimuli that accelerate the rate of breathing.

The role of carbon dioxide in stimulating breathing is of paramount importance to both anesthetists and high-altitude flyers. If a man breathes into a closed system in which his expired carbon dioxide is allowed to accumulate, his rate of breathing will increase markedly. If the expired carbon dioxide is absorbed, for instance by KOH, his breathing will increase only slightly. Note that in a closed system the decline in oxygen concentration will be the same in both cases.

The effects of lack of oxygen depend on the rapidity with which it occurs. If an aviator suddenly loses his oxygen supply or a miner walks into a pocket of methane or other gas, he is likely to "black out" very suddenly and completely. If the loss of oxygen is gradual, the results are quite different, but more or less the same whether the loss is due to carbon monoxide poisoning, alcoholism, or ascent into high altitudes. At first there is commonly a sense of well-being and competence. As the oxygen lack persists, there comes a period of loss of judgment and unstable emotions, commonly accompanied by muscular inco-ordination, faulty vision, and poor memory. Fixed and irrelevant ideas are frequent. Finally, a feeling of sublime indifference and extreme

weakness may end the series. In the case of continued deprivation of oxygen, these symptoms are followed by extreme nausea, convulsions, and finally, death.

Too much oxygen is also dangerous. Various symptoms of toxicity begin to appear when the oxygen pressure (concentration) in the gas breathed by most mammals begins to exceed 0.8 atmospheres (*i.e.,* about four times the normal level of O_2 in air). Excess oxygen is used in medicine to alleviate the effects of pneumonia, heart, and other diseases. Patients in oxygen tents are usually given air containing slightly over 50 per cent oxygen. An atmosphere of pure oxygen can be tolerated by man and in some conditions may be beneficial, but it may lead to extreme hypoventilation, resulting in coma.

The greatest hazard to human lungs, even in a smog-filled city, is tobacco smoke from cigarettes which, in some poorly understood way, leads to lung cancer. The evidence amassed in this country and in Europe is extensive and detailed. It would easily convince everyone if it were not for the understandable fact that no one likes to think that one of his established habits is potentially dangerous. The evidence is statistical, but it must be remembered that all evidence, in that famous final analysis, is statistical. Statistical reasons are the only ones for believing Koch's classic postulates about the germ theory of disease.

The Grand Synthesis

The present century has built with amazing success upon the discoveries begun in the 17th century by Boyle and Lavoisier about respiration and by certain navigators about fatal diseases that can be cured by diet. Consequently, for the first time in human history, we know how the food we eat and the air we breathe are united to provide energy—at the end of an electron transport system in our mitochondria. We even know how food is split apart chemically in digestion and the resulting units incorporated into our own characteristic proteins, *i.e.,* into us, by DNA-RNA-ribosomal-enzymatic machinery. This modern insight into the secrets of what our predecessors called "the flame of life," not only represents a tremendous achievement, it is also a great simplification because it sweeps into a single conceptual scheme all animals from the lowest to the highest. The double task of the future is to learn what is the best diet to provide abundant health, both physical and mental, and how to make this diet available to all people.

USEFUL REFERENCES

Frandson, R. D. *Anatomy and Physiology of Farm Animals*. Lea & Febiger, Philadelphia, 1968.

Guthrie, H. A. *Introductory Nutrition*. C. V. Mosby Co., St. Louis, 1967.

Hötzel, D., and Barnes, R. H. Contributions of the Intestional Microflora to the Nutrition of the Host. In *Vitamins and Hormones, Advances in Research and Application*, Vol. 24, ed. by R. S. Harris *et al.* Academic Press, New York, 1966.

Keys, M., and Keys, A. *Eat Well and Stay Well*. Doubleday and Co., Garden City, N. Y., 1963.

Pantelouris, E. M. *Introduction to Animal Physiology and Physiological Genetics*. Pergamon Press, Oxford, 1967.

Staub, N. C. Respiration. In *Annual Review of Physiology*, Vol. 31, ed. by V. C. Hall *et al.* Annual Review, Inc., Palo Alto, 1969.

Internal Controls: Excretion, Homoiothermy, Immunity

Homeostasis is the name given to the relatively stable internal conditions that living organisms must maintain for survival. The maintenance of homeostasis is the purpose of eating and breathing, and hence of all the innumerable adaptations in teeth and claws and in muscles and enzymes wonderfully fitted for carrying on the business of life. All the escape and protective devices of animals and plants, from the thick blubber beneath the skin which protects a whale against the frigid polar seas to the waxy cuticle which protects a desert plant from water loss, are ultimately involved in maintaining homeostasis.

Within animals there is a wide diversity of control mechanisms upon which their lives depend. One of the most important and best known mechanisms is the excretory system which, by removing waste and water from the blood, maintains a constant internal environment. In birds and mammals, a high and constant body temperature is of great importance, giving them a freedom from the environment and conferring a superiority over cold-blooded or poikilothermous animals like insects and reptiles, which slow down when the temperature falls. Other homeostatic mechanisms are concerned with eating and drinking.

Many constant conditions in vertebrates are maintained by beautifully balanced pairs of centers within the hypothalamus in the lower part of the brain. The endocrine system is in large part a homeostatic system of many feedback circuits which regulate the concentration of hormones in the bloodstream. The ability of animals to produce antibodies maintains the integrity of the organism against bacteria and other foreign invaders. This chapter will be devoted to three of the major regulating systems: those concerned with excretion, with the control of body temperature, and with the production of immunity.

EXCRETION AND A CONSTANT INTERNAL ENVIRONMENT

Life arose in the sea. Consequently, protoplasm has an osmotic concentration like that of the ocean from which it arose. As a result, any organism living in fresh water, or descended from one which did, has a problem. Because the osmotic concentration of fresh water is very low, cells tend to take up water through the cytoplasmic membrane and will burst unless this excess water is excreted. From amoeba to man, the primary function of excretory organs is the elimination of excess water. In marine invertebrates there is no such problem because the osmotic concentration of the salt water environment is much higher. Thus excretory organs, although present, are much less important.

The distinction between secretion and excretion is a fuzzy one. **Excretion** is usually defined as the concentration and elimination of metabolic waste materials, including water, which have been within cells. **Secretion** is defined as production by a cell, gland, or tissue of a substance synthesized or at least concentrated within cells and released through their membranes. Thus urine can properly be called an excretion, although wastes are secreted into it.

Vertebrate Excretion

In higher vertebrates the three chief excretory organs are the lungs, the liver, and the kidneys. The lungs eliminate waste carbon dioxide and other volatile substances. Ether can be smelled on the breath of postoperative patients for many hours after anesthesia. The breath of long untreated diabetics smells of acetone, a volatile product of acetoacetic acid, which in this disease accumulates in the blood faster than it is converted into acetyl-CoA and fed into the Krebs cycle.

The liver synthesizes urea from CO_2 and NH_3, which is later excreted by the liver. It is also responsible for the chemical processing of a multitude of substances in the blood and the changing of many into forms which are then eliminated by the kidneys.

The first function of the kidneys is to hold constant the **osmotic pressure of the blood** at a physiologically appropriate level. Were the blood to become too concentrated, the cells would shrivel; if too dilute, they would swell up and burst. Happily, you cannot dilute your blood no matter how much water you drink. With great precision, the kidneys will excrete exactly the correct amount of the additional water. When water intake is restricted or water loss by evaporation increased (a man walking on a desert at 110°F. loses a quart of water per hour from lungs and skin), then the kidneys excrete a smaller volume of more concentrated urine.

A second important activity of the kidneys is to regulate the **pH of the blood plasma**. If the plasma becomes acidic, *i.e.,* if the pH falls, the kidneys excrete more hydrogen ions. If the blood becomes more alkaline, the kidneys excrete more bicarbonate. Since the concentration of sodium, calcium, and potassium bicarbonates in the plasma influences the osmotic pressure of the blood, it is obvious that the regulation of blood pH is closely related to the regulation of its osmotic pressure.

The elimination of **waste nitrogen** is a third major function of the kidneys. Most of the protein in the human diet is deaminated (has its nitrogen removed) as its constituent amino acids pass through the liver. This excess nitrogen is in the form of ammonia, which is combined with CO_2 and converted into **urea** by the liver. Mammals also excrete nitrogen in the form of **creatinine** in a small but constant daily amount which is independent of the amount of protein ingested but depends on the total

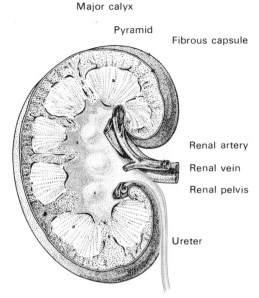

Major calyx

Pyramid

Fibrous capsule

Renal artery

Renal vein

Renal pelvis

Ureter

Papillae of pyramid

Medullary substance

Cortical substance

Fig. 29-1. A human kidney seen in longisection. The function units, called nephrons, lie primarily in the cortex.

muscle mass of the individual. This is not too surprising since creatinine is a constituent of muscles.

Most mammals also secrete small amounts of nitrogen as **uric acid**. This compound may be built up from ammonia or other simple nitrogen compounds or it may be derived from the breakdown of the purines in nucleic acids. Uric acid is itself a purine not too different from adenine and guanine.

The kidney excretes a very wide variety of additional substances, mostly **waste products** of liver metabolism or injurious substances processed by the liver. Included is **urochrome**, a yellow pigment which gives urine its characteristic color. Urochrome is a degradation product of bile pigments ultimately derived from worn-out red blood cells. Among the virtually endless number of substances removed from the blood by the kidneys are glucose (if its concentration exceeds a threshold level), ketones, and other products of deranged lipid and carbohydrate metabolism, breakdown products of coffee and many other substances, including vitamins, hormones, and even a few amino acids.

The Anatomy of Vertebrate Excretion

The gross anatomy of the excretory system is simple and essentially the same from fish to man. A pair of kidneys lie on either side of the backbone, in mammals a short distance posterior to (below, in man) the diaphragm. Each kidney is served by a short **renal artery**, coming directly from the dorsal aorta, and by a renal vein. From each kidney a **ureter** carries urine to the **urinary bladder** in the pelvis. From the

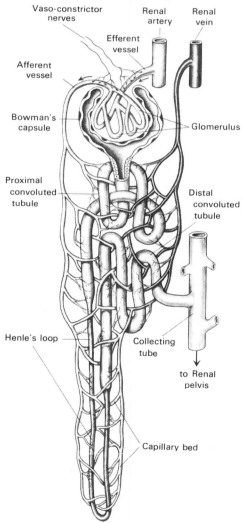

Fig. 29-2. Diagram of a single nephron. The glomerulus plus Bowman's capsule together constitute a renal corpuscle. (From G. B. Moment, *General Zoology,* 2nd ed. Houghton Mifflin Co., Boston, 1967.)

bladder the urine is conducted to the exterior by the **urethra.**

The microscopic anatomy of the kidney was first explored by Malpighi. He noticed small round structures near the edges of the kidney. They can be seen with a good hand lens without dissection close to the thin edges of a frog's kidney. These little bodies are termed Malpighian or **renal corpuscles.** Each renal corpuscle is at the beginning of a long **tubule.** Together they form the functional unit of the kidney, the **nephron.**

Each renal corpuscle consists of a tuft of capillaries called the **glomerulus** and its surrounding double-layered capsule called **Bowman's capsule.** The cavity between the capsule layers opens into the beginning of the uriniferous tubule which produces urine. Immediately after its formation at Bowman's capsule the tubule becomes convoluted, and then forms a long straight loop about 3 cm. long known as **Henle's loop.** When a loop gets back to the renal corpuscle it forms a second or distal convolution. The distal convolutions empty into collecting tubes which in turn empty into the pelvis of the kidney where the ureter originates.

How the Kidney Works

How does the kidney do all of these remarkable things? This question is not only of evolutionary interest but also of the greatest medical importance. Over a century ago, William Bowman discovered the capsule that bears his name and immediately proposed a theory of kidney action: **glomerular filtration** followed by **tubular secretion** of wastes. Carl Ludwig, the leader of the great Leipzig group of physiologists, immediately proposed a different theory, glomerular filtration followed by **reabsorption** of useful molecules. Thus arose a classic controversy which stimulated research the world over for decades and finally led to our present knowledge.

Filtration takes place from the glomerulus as long as the hydrostatic pressure of the blood exceeds its osmotic pressure. A hydrostatic pressure of about 75 mm. of mercury is imparted to the blood by the beating of the heart and tends to force its fluid constituents through the thin membranes of the capillaries

The labels on the figure read:
Vaso-constrictor nerves, Renal artery, Renal vein, Efferent vessel, Afferent vessel, Bowman's capsule, Glomerulus, Proximal convoluted tubule, Distal convoluted tubule, Henle's loop, Collecting tube, to Renal pelvis, Capillary bed

of the glomerulus into Bowman's capsule. On the other hand, blood proteins, urea, and other osmotically active blood components all tend to pull liquid back into the blood. This osmotic pressure is about 30 mm. of mercury, roughly equivalent to a 0.1% solution of sucrose.

This theory can be proved in several ways. If the hydrostatic pressure of the blood falls to a point where it is no greater than the osmotic pressure, due to a heart weakness, bleeding, or drugs, the formation of urine will stop. If back pressure is applied up the ureter into the kidney, the formation of urine will stop when the applied pressure plus the osmotic pressure equals the hydrostatic pressure. Finally, direct proof can be obtained by collecting fluid from Bowman's capsule. This fluid on analysis shows the same concentrations of salts, glucose, amino acids, urea, and other substances of small molecular size as does the blood plasma. It lacks the blood cells and the blood proteins which are filtered out.

What about **reabsorption**? This obviously occurs because if the glucose concentration in the blood does not exceed the threshold value of about 160 mg./ml., none of it appears in the urine. Water, also, is reabsorbed, for the urine may be more concentrated than the fluid in Bowman's capsule. Water absorption is under the control of the antidiuretic hormone (ADH) of the posterior lobe of the pituitary gland.

What about **secretion** of materials into the tubules? The first convincing evidence that tubular secretion as well as reabsorption is a fact came with the discovery that salt water teleost fish have either no glomeruli whatever or have reduced ones. The ugly little toadfish, *Opsanus tau*, of north Atlantic coastal waters must form its urine by secretion, for it completely lacks renal corpuscles, *i.e.*, glomeruli and Bowman's capsules.

The first clear evidence for **active transport**, *i.e.*, active secretion, in the mammalian kidney was gained by the use of an obvious method

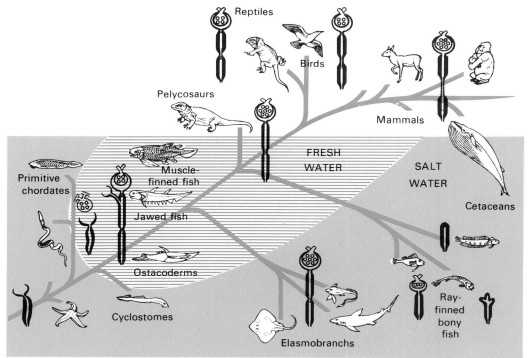

Fig. 29-3. Evolution of the vertebrate kidney. Note that the bony fish shown on the right which returned to salt water have reduced or even lack glomeruli where water is filtered from the blood. This is an adaptation to the high osmotic pressure of sea water which prevents fish blood from absorbing large amounts of water. Only the sharks retain well developed glomeruli but they counter the osmotic pressure of the sea by retaining urea in their blood. (Redrawn from Scientific American.)

(obvious, once you think of it). The dye phenol red is readily excreted in the urine after injection into the bloodstream. The dye could be in the urine entirely due to filtration and then concentration in the tubules by its failure to be reabsorbed with the watery part of the filtrate. The dye could be in the urine because of active secretion into the tubules. Or its presence might be the result of both processes. To test whether or not active secretion was taking place, the blood pressure in an experimental animal was lowered until filtration stopped. Phenol red was injected and later the tubules were observed to see if they were concentrating the dye. They were, especially in the proximal convolutions. In mammals, at least, there appears to be an active secretion of urea (in addition to that present due to filtration) and other substances into the tubule.

Filtration into Bowman's capsule from the glomerulus is a passive process as far as the cells there are concerned. The necessary energy is provided by the heartbeat. In contrast, reabsorption and secretion against a concentration gradient are both active processes which require energy at the site, *i.e.,* in the tubule cells. Numerous mitochondria and abundant ATP would, therefore, be expected in tubular cells engaged in these activities. This expectation is fulfilled by electron microscope studies. Indeed the kidney has been found to require more energy per hour per gram of tissue than does an active muscle like the heart.

BODY TEMPERATURE

Animals are easily divided into those which have little or no control over their body temperature, the so-called cold-blooded animals, or **poikilotherms**, and those which have an internal control which maintains their body temperature at a constant level, the warm-blooded birds and mammals or **homoiotherms**.

BEHAVIORAL, ANATOMICAL, AND PHYSIOLOGICAL ADAPTATIONS TO ENVIRONMENTAL TEMPERATURES

Both poikilotherms and homoiotherms have behavioral ways of controlling their body temperature. The animals of a sand dune are active mostly at night and burrow in the sand to avoid the searing heat of mid-day. By night the white ghost crabs search the beach for scraps of food but only their tracks are visible by day. Desert insects, toads, snakes, and rats keep out of the sun.

Most homoiothermous animals possess anatomical adaptations to preserve heat. Fur, wool, and feathers are all excellent insulators, as is a thick layer of subcutaneous fat common to seals, walruses, porpoises, and whales.

Perhaps the most general adaptation against cold is the result of a fact of solid geometry. The larger an object, whether sphere, square, or oblong, the smaller is the surface compared with the volume. Because any object can lose heat to its surroundings only through its surface, the larger an animal is, the smaller its surface relative to volume and the more easily it maintains its body temperature. The present relevance of this fact is expressed in **Bergman's rule**: the farther north the members of a species live, the greater their size. Thus the largest Virginia deer are not in Virginia but in Canada. The largest tigers are not in India but in Siberia. Allen's corollary states that the colder the climate the shorter the ears and other extremities.

From the medical standpoint and for the life of any bird or mammal, it is the physiological thermostat in the brain that is of paramount importance. The temperature control center, like so many other centers regulating bodily activities, is located in the **hypothalamus.** It has been known since the beginning of this century that a brain center controls temperature because brain injuries sometimes result in high fevers with few if any other symptoms. The precise location of the control site had to wait for modern instrumentation. The use of delicate probes like those used to locate the eating, drinking, and "pleasure" centers showed that body temperature was indeed regulated by cells in the hypothalamus, located at the base of the brain above the optic chiasma where the nerves from the eyes cross.

But we all know that temperature changes are keenly felt by sensory nerve endings in the skin. Stimulation of these endings leads to shivering, ruffling of feathers or panting and other cold- or heat-combatting actions. What kind of thermoregulating center is in the hypothalamus? Does it receive and coordinate

Fig. 29-4. Allen's corollary of Bergman's rule. The colder the climate, the smaller the ears and other extremities. A, Arizona jackrabbit. B, Oregon jackrabbit. C, varying hare from Minnesota. D, Arctic hare from northern Canada. (From G. B. Moment, *General Zoology,* 2nd ed. Houghton Mifflin Co., Boston, 1967.)

reports coming to it from the skin or is it a thermostat which has its own way of knowing what the temperature is before sending out the appropriate neural messages?

The best available evidence indicates the brain center takes its own temperature as well as sending out appropriate messages. If the blood supply to this part of the brain is somewhat raised, skin capillaries dilate and the temperature of the body as a whole will fall. If cooled, skin capillaries contract, shivering begins, and the body temperature rises. What does this leave for the cold and heat sensors in the skin to do? Apparently impulses from the skin lead animals into overt behavior which helps control temperature. For example, a dog too near a hot stove moves away.

Modern interest in hibernation, in which there is a drastic fall in body temperature, has been motivated in part by the thought that when enough is learned about how an animal's thermostat can be turned down (and turned up again) it would be very useful information for people on space trips. A hummingbird, for example, hibernates every night with an important saving of energy. It will be recalled that because of its very small size, the metabolic rate and the resultant consumption of food is extremely high in these tiny birds. Some species of bats also enter at least a pseudo-hibernation every day. Much has been learned about hibernation in many species on a purely descriptive level. The stimulus of cold is usually essential, but by itself is not enough to induce hibernation. The animal must be in a state of readiness. Many animals store fat prior to hibernation.

IMMUNITY

The phenomena of immunity include not only our ability to resist diseases but also the problems of skin and organ transplants, of blood groups, and of a host of allergies. If cancers are due to some action of viruses, as now seems likely, then it may be possible to develop artificial immunity to them as it has been for poliomyelitis.

Human knowledge about immunity grew out of attempts to overcome a loathsome and often fatal disease, smallpox. In the 18th century Lady Mary Montagu brought back to England from Turkey the practice of acquiring immunity against smallpox by inoculation with material from a mild case. If the inoculated individual came down with a mild case, he would be permanently immune to this dread disease. However, much too frequently severe disfigurement and even death was the result, so that this sort of inoculation was a kind of Russian roulette. About 50 years later a teen-age medical apprentice, Edward Jenner, wondered about combining the Turkish method of inoculation using actual smallpox material and the folklore that anyone was immune to smallpox who had ever had cowpox (a very mild bovine ailment resembling smallpox). In May of 1796 Jenner, by then a doctor, began a historic test. In so doing he faced the same kind of acute ethical problems that heart transplants and genetic counseling involve. He inoculated a healthy boy with cowpox material. After the boy had well recovered from the cowpox, Jenner injected him with material from a case of smallpox. The boy failed to contract smallpox and Jenner became a hero throughout Europe. But what if the theory had been wrong and the boy had developed the disease and died?

Today, as a result of the work of Louis Pasteur, Paul Ehrlich, and a host of others, a vast body of knowledge about immunity has been gained although the gaps in this knowledge are enormous. Immunity is of two sorts. **Natural immunity** is that which is characteristic of a species. Humans do not get bird malaria or distemper, which is highly contagious and often fatal for dogs. **Acquired immunity** involves some kind of response by an individual to a foreign invasion. Three different mechanisms of immune response are known. **Cellular immunity** is due to leukocytes which phagocytize, *i.e.*, engulf, foreign bacteria. **Humeral immunity** is due to proteins, specifically gamma globulins, circulating in the blood and lymph. These globulins recognize foreign proteins and somehow unite with them, causing agglutination and precipitation. A third immune agent, **interferon**, has recently been brought to light and may be as important as the other two. When a virus enters a cell, the cell may give off a protein, interferon, about the size of a hemoglobin molecule, which interferes with the synthesis of new viral protein, i.e., with the translation of viral RNA.

Any substance which elicits an immune response is an **antigen**. Most antigens, but not all, are proteins. The proteins, specifically the **gamma globulins**, which are produced in response to the antigen and which react with it are called **antibodies**. There is some evidence that the major part of all antibody molecules is the same, namely globulin, and that a small part is custom-made to recognize and react with the homologous, *i.e.*, corresponding, antigen.

Rejection of grafts may be a very special case. Autografts of skin or other structures from one part of an individual to another part of the same individual present no problem. But homografts between two individuals of the

same species or heterografts between individuals of different species are rejected after a period of weeks. It is clear that this rejection is an immunological phenomenon because if, after the rejection of the first graft, two new grafts are made one from the same donor and the other from a new donor, the graft from the original donor is rejected much more quickly than it was the first time, while the graft from the new donor takes as long as the first did to be rejected the first time. This experiment shows that there is an **immunological memory** and that the antibodies responsible for these rejections are highly specific. Because identical twins can accept each other's skin and other grafts permanently it seems certain that genetic factors control tissue compatibility. Studies with inbred strains of mice and analysis of human data show that mammals carry at least a dozen of these **histocompatibility genes**. They seem similar to the various sets of genes which produce the different blood groups.

There are several ways of **suppressing immunity**, which is important in organ and skin transplant operations. If an animal is injected with cells of another at or around the time of birth, when adult it will not be able to recognize tissues of the donor as foreign and will consequently accept grafts from that individual. X-irradiation or chemicals which destroy the lymphoid cells which produce antibodies will result is a loss of ability to synthesize antibodies. Various antimetabolites which interfere with protein synthesis also inhibit antibody production. Finally, the removal of the thymus gland very early in life results in a lack of the various sorts of lymphocytes and, hence, no antibodies. Obviously all these methods, except the first, have serious side effects, not the least of which is a high susceptibility to infections.

There are thousands, and perhaps millions, of different proteins against which an individual can make specific antibodies. How is this feat possible? How, indeed, do the cells know not to make antibodies against the other cells composing the body of which they themselves are a part? Many unanswered questions remain, but if the thymus gland is removed at birth, the lymph glands remain very small—apparently because they have not received via the bloodstream the seed cells which produce lymphocytes. At the same time, ability to produce antibody is reduced or absent. By various labeling techniques is it possible to show that globulins are produced by small lymphocytes called **plasma cells** and that large white blood cells called **macrophages** play some role.

There are two theories to account for how all the thousands of kinds of antibodies can be formed. According to the **instructional theory** of Pauling and Haurowitz, foreign protein somehow acts as a template or model which instructs the lymphocytes to form the specific antibody globulin to match the foreign protein. The discovery that foreign protein complexes with host RNA supports such a view. Possibly, the normal protein-synthesizing machinery can be adapted so that a protein can make another protein which is complementary to it, as RNA is complementary to DNA.

The other theory, which is now in general favor, is the **clonal selection theory** proposed by the Australian investigator, Burnet. According to this view, during very early development approximately 10,000 different kinds of lymphocyte stem cells are produced by mutation. (Highly mutable genes are well known in corn and other organisms.) Each stem cell and all its descendants are able to produce some specific antibody. Presumably, any cells which have mutated in such a way that they form antibody against their own organism die or are killed. This may help explain how permanent tolerance can be obtained by early exposure to a foreign protein. In any case, according to the selection theory, when a foreign protein enters the bloodstream it comes into contact with macrophages and small lymphocytes. The lymphocytes which are of the right sort are somehow stimulated to proliferate, which may account for the delay in antibody response, and then to synthesize the appropriate gamma globulins.

It has recently been discovered, in what is a beautiful example of one of the ways biological science advances, that there are two types of lymphocytes involved. Graduate student Timothy Chang, in Ohio, attempted to demonstrate for a class the development of immunity in chickens after they are inoculated with salmonella bacteria. He obtained his chickens from fellow student Bruce Glick but all failed to develop immunity! It turned out that Glick had previously removed the bursa of Fabricius, a finger-shaped gland attached to the chicken hind gut, in an effort to discover its function. Soon after, Professor Robert Good in Minnesota, learning of these facts, organized a team to remove the thymus from some newly hatched chicks and the bursa from others.

As adults the chickens developed some immunity but those without a thymus could not reject foreign skin or organ grafts and those without a bursa did not make gamma globulin. Both types of lymphocytes are present in humans, the T or thymus-dependent cells and the B or bone marrow derived cells. (A bursa of Fabricius is not present in mammals.) The whole field abounds in important and challenging problems.

USEFUL REFERENCES

Grollman, S. *The Human Body, Its Structure and Physiology*. The Macmillan Company, New York, 1964.

Hoar, W. S. *General and Comparative Physiology*. Prentice-Hall Inc., Englewood Cliffs, N. J., 1966.

Ingram, V. M. *The Hemoglobins in Genetics and Evolution*. Columbia University Press, New York, 1963.

Krogh, A. *Comparative Physiology of Respiratory Mechanisms*. University of Pennsylvania Press, Philadelphia, 1959.

Smith, H. W. *The Kidney: Structure and Function in Health and Disease*. Oxford University Press, New York, 1951.

Snively, W. D. *Sea Within Us: The Story of Our Body Fluids*. J. B. Lippincott Co., Philadelphia, 1960.

Guyton, A. C. *Basic Human Physiology*. W. B. Saunders Co., Philadelphia, 1971.

Vroman, L. *Blood*. The Natural History Press, Garden City, N. Y. 1968.

Endocrines

The most important fact about hormones is that knowledge of them can confer enormous power over the lives of men, animals, and plants. Too little thyroid hormone and you will become a virtual vegetable; too much and you will become so hyperexcitable that life is almost unbearable. Without the pituitary growth hormone, we would all be Tom Thumb midgets; with too much, either circus giants or victims of acromegaly with enormous bony hands and feet, heavy elongated faces, and short lives. The Pill, one of the most widespread agents of birth control, is a hormone product.

Insect pests that annually cause damage to crops worth billions of dollars will probably be controlled in the future not by DDT, a long-lasting and indiscriminate killer, but by insect hormones which either force the bugs to molt too often or prevent their attainment of maturity. Such hormones are both selective and biodegradable.

Behavior in animals including man is profoundly influenced by hormones. A female canary will never sing unless a pellet of male sex hormone is implanted under her skin. She then sings so well she can be sold as a male but will continue to sing only as long as the pellet lasts, six to eight weeks. Hormone deprivation in humans can result in marked psychological and personality changes.

Newer knowledge about hormones is advancing in three principal areas. First, it is becoming clear that in both vertebrates and invertebrates endocrine glands are under the control of the nervous system at some point. Second, the mechanisms by which hormones produce their results are being discovered and, third, knowledge of the effects of hormones on behavior is being greatly expanded.

Animal glands fall into two main classes: exocrine and endocrine. The digestive glands, for example, are known as **exocrine** glands because they possess ducts which carry the chemical agents from the gland of origin and empty it into some body cavity. The glands which produce hormones are called **endocrine** because they lack such ducts. Their products are taken up by the bloodstream and are carried to one or more "target organs," perhaps the comb and feathers of a rooster or the lining of the uterus of some mammal. It is on the target organ that hormones have their effects.

Among vertebrates, most endocrine glands secrete under the stimulus of specific hormones produced by the pituitary gland (itself an endocrine gland) on the underside of the brain. If the pituitary is removed, the thyroid, gonads, and adrenal glands stop secreting and may even degenerate. On the other hand, these glands exert an inhibitory influence on the pituitary by way of the hypothalamic brain centers which control the pituitary. In other words, the pituitary turns these glands on and they reach back, so to speak, and turn the pituitary off with their own hormones, although indirectly. The result is a self-regulatory **feedback mechanism** which may result in a periodic rise and fall in hormone level. This is the explanation of the familiar female reproductive cycles in mammals. The same feedback mechanism may also maintain a steady-state level of hormones. Over these various cycles and responses the nervous system imposes a measure of control, sometimes slight, sometimes complete.

FOUNDATIONS OF ENDOCRINOLOGY

The effects of castration on the growth and behavior of man and other vertebrates have

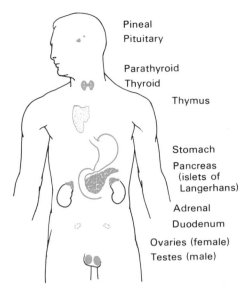

Pineal
Pituitary

Parathyroid
Thyroid

Thymus

Stomach

Pancreas
(islets of
Langerhans)

Adrenal

Duodenum

Ovaries (female)

Testes (male)

Fig. 30-1. Position of the human endocrine glands.

been known from remote antiquity. Eunuchs were commonplace in ancient civilizations and often held important posts in government and the military services. Castrati were used as professional singers in Europe until about a century ago. The practice of removing the gonads from horses, bulls, and chickens is also very ancient.

It has always been known that there is a great deal of variation in the results of castration. Some of this variation may be due to the activities of the adrenal glands, which are known to produce masculinizing hormones under certain conditions. The bearded lady of the circus is usually a victim of misbehaving adrenal glands. Psychological factors also play an important role.

Scientific knowledge of hormones dates only from the mid-19th century. Johannes Müller and Jacob Henle, the most eminent physiologist and anatomist, respectively, of the times, asserted that the ductless glands had little or no effect on animals. Shortly after this "authoritative" pronouncement, A. A. Berthold, a man otherwise unknown to history, who lived near a small provincial German university, showed by castration and by the transplantation of the testes that the gonads of a rooster are essential for comb growth and normal rooster behavior. At the same time Thomas Addison, in London's famous Guy's Hospital, described a strange fatal disease in which the victim turns a metallic

purple-gray color and grows progressively weaker and more emaciated until death ensues. Autopsy always shows a diseased adrenal gland. This was the first accurate account of a disease associated with an endocrine disorder. It is only within the present decade that Addison's disease has been brought under even partial control through the use of adrenocorticotrophic hormone (ACTH) and synthetic adrenal steroids. The discovery of J. F. von Mering and Oscar Minkowski in 1889 that diabetes develops in dogs after removal of the pancreas is also an important landmark in the history of endocrinology.

The first clear and convincing proof that a substance secreted by the cells of one organ could be carried by the bloodstream and produce a specific effect on a distant target organ was provided by the work of Bayliss and Starling in 1902. They discovered that as food enters the intestine from the stomach, the intestine liberates a chemical, **secretin,** which signals the pancreas to secrete digestive enzymes. With the help of a professor of Greek, they coined the word hormone and the science of endocrinology was truly launched.

Methods of Study

A more or less standard procedure for the identification and study of endocrine glands has been developed. The first step is usually examination of the suspected cells with an ordinary light microscope. The cells of endocrine glands—pituitary, gonad, adrenal, and the rest—suggest their secretory function by their structure. The cells tend to be cuboidal and filled with granules. The gland is well vascularized. With an electron microscope an extensive endoplasmic reticulum and Golgi complex are visible.

The orthodox test for actual endocrine activity is a double one. The gland is removed surgically. A sharp lookout is kept for accessory glands in unexpected places. Then the results are observed. The gland is implanted in another animal, or better yet, an extract is made and injected. Both aqueous and fat-soluble extracts must be tried. Often the results have been spectacular. Sometimes, however, the effects of glands have been missed completely because no one knew what effects to look for or else removed the glands at the wrong time in the life of the animal. This happened over and over again in the case of the adrenal, the thymus, and the pineal glands.

The third phase in the analysis of a hormone is chemical identification. To accumulate enough hormone sometimes requires almost superhuman efforts. To get enough of the insect growth hormone, for example, the glands of several barrels of insects have to be dissected out, homogenized, and the extract concentrated. Once the active material has been identified, the fourth step is the laboratory synthesis of the hormone molecule. With the identification of any hormone goes a study of its role in the life of the animal and of how its secretion is controlled.

Chemical Nature of Hormones

Chemically, hormones fall into two major groups. Some are steroids; others are polypeptides or simple amino acids.

The sex hormones and the hormones of the adrenal cortex are **steroids**; so also is ecdysone, the molting hormone of insects. This means they are fat-soluble compounds closely related to vitamin D, cholesterol, and bile salts. Other steroids include potent cancer-producing agents, embryonic inductors, and the very useful heart stimulant, digitalis.

All steroids are built around the same chemical nucleus consisting of four joined carbon rings. Three are six-carbon rings and one is a five-carbon ring. These rings are given conventional letters and their carbon atoms are given numbers for identification. A keto group (which is merely a carbon atom with two bonds to an atom of oxygen and two single bonds to other atoms),

$$\begin{array}{c} O \\ \| \\ -C- \end{array}$$

in position 17, for example, makes a 17-ketosteroid, a substance excreted by men. All 19 carbons in the basic steroid configuration are derived via acetyl-CoA, mostly from acetic acid. Once again the pivotal position of acetyl-CoA becomes apparent. The various hormones are produced by adding keto, hydroxy, methyl, or other groups to one of the 19 carbons.

The hormones of the pituitary, thyroid, and pancreatic glands are proteins, or, in the case of the thyroid, simply an amino acid. These

hormones are water soluble. The hormone insulin from the pancreas, for example, consists of two chains, one of 21 and one of 30 amino acids, held together at two fixed points by disulfide bonds, *i.e.*, the —S—S—, between two cystines. It will be recalled that cystine is an amino acid containing sulfur. The whole insulin molecule has a molecular weight of about 5,700. Glucagon, the other pancreatic hormone, which has an effect on blood sugar opposite to that of insulin, consists of a single chain of 29 amino acids and has a molecular weight of 3,485.

How Do Hormones Work?

There are three important theories of hormone action. The most recent and convincing is that at least some hormones work at the **gene level**. There is also good evidence that hormones work by changing the **permeability of cell membranes** and thus change metabolism by accelerating or blocking the entrance of certain substrates into the cell. Since various substances—lactose, for example—are known to pass into cells by an active process requiring

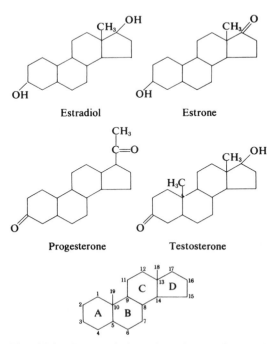

Fig. 30-2. Structural formulas of several steroid hormones and of the steroid nucleus. Estradiol, estrogen, and progesterone are female vertebrate hormones; testosterone is a male hormone.

enzymes, appropriately called permeases, it is possible that hormones which affect cell permeability may work by changing the rate at which genes synthesize permeases, but this is not certain. A third theory holds that hormones act on **enzymes**, perhaps by acting as coenzymes. In all cases there appear to be specific receptor sites on the surface membranes of the target organs which bind the hormone molecules to which the target organ responds.

The action at gene level theory originated with work in Germany on the insect hormone **ecdysone**. Several investigators showed that ecdysone when injected activated specific genetic loci on the chromosomes of the salivary glands of certain insect larvae and that this activation revealed itself by puffing of specific regions (Fig. 8-17). Other workers investigating bacteria demonstrated that the antibiotic actinomycin blocks the formation of new RNA. Here, then, was the idea, *i.e.,* hormones turn genes on, and the tools to test it.

In several laboratories in the United States and abroad investigators working with different hormones on different animals have obtained convincing evidence that hormones initiate DNA-dependent RNA synthesis, presumably by combining with genetic repressors. New mRNA, of course, means new enzymes. New enzymes mean changes in metabolism and a wide range of possible new products.

Clear evidence comes from studies of female sex hormone. If the ovaries of a rat are removed, the endometrium lining the uterus remains permanently in the reduced, anestrous condition. If estrogen, female sex hormone, is then injected into the rat, an increase in uterine mRNA can be detected within half an hour. After three or four hours an increase in protein can be measured and the uterus begins to grow into the estrous condition. These facts in themselves support the theory. After hormone injection, new mRNA appears first, followed later by new protein. Confirming evidence has been obtained by administration of actinomycin D, an inhibitor of DNA-dependent RNA synthesis. If actinomycin is injected before the sex hormone, no new RNA appears and there is no subsequent increase in protein. The uterus remains anestrous.

The more interesting of the two older theories holds that the hormone insulin lowers the concentration of blood glucose by facilitating glucose use. It will be recalled that all carbohydrates must be converted to glucose before they can be metabolized, and that the first step in glucose utilization is its phos-

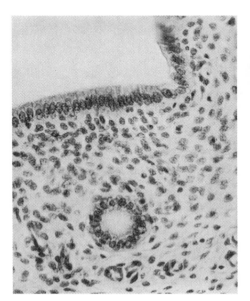
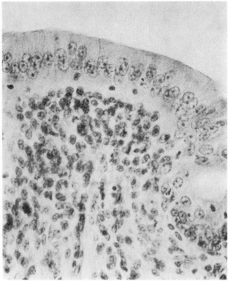

Fig. 30-3. Rat uterus. On left, appearance after castration. Note the thin line of epithelial cells (along top edge). On the right, the appearance after administration of estrogen. Note much thicker endometrial epithelium while the underlying mesodermal cells remain much the same. (Kindness of Sheldon Segal.)

phorylation by ATP and the enzyme hexokinase:

$$glucose + ATP \xrightarrow[\text{kinase}]{\text{hexo-}} glucose\text{-}6\text{-}phosphate + ADP$$

Insulin by itself has no effect on the action of hexokinase but hexokinase is inhibited by some of the pituitary and adrenocortical hormones. Insulin blocks this inhibition and thereby increases the utilization of glucose.

THE THYROID

A mammal's thyroid gland forms two lobes pressed against either side of the trachea and connected with each other by a narrow band of thyroid tissue, so that the entire gland resembles a pair of saddle bags. The thyroid glands of all vertebrates are found in the same general region of the throat.

Microscopic examination reveals that the gland is composed of thousands of more or less spherical follicles made up of secretory cells enclosing a colloidal material. The gland is so highly vascularized that more blood flows through it, in proportion to its size, than through any other organ except perhaps the adrenal glands.

Marked underfunction, **hypothyroidism**, in small children produces **cretins**. These pitiful individuals are greatly stunted, woefully feebleminded, and have characteristically bloated faces, bodies with loose, wrinkled skin, and coarse, sparse hair. Basal metabolism, body temperature, and heart rate all are abnormally low. If thyroid deficiency occurs in adult life, the thyroid may enlarge greatly and form a swelling in the throat called a **goiter**. Sometimes goiter is accompanied by many symptoms of cretinism; in other cases it is not, perhaps depending on whether or not the enlarged gland supplies minimal needs.

Cretins and goiterous persons used to be common in localities far from the sea, such as isolated valleys in the Alps and Pyrenees and in the interior of continents, where there is a marked deficiency of iodine in the diet. Dramatic recoveries can be accomplished by feeding thyroid gland, iodine, or the synthetically made hormone thyroxine to hypo-

thyroid patients. The blank face gains expression, the bloated body assumes a normal shape, and the mind brightens.

Hyperthyroidism, oversecretion of the thyroid, sends the basal metabolism to abnormal heights and produces a hyperactive and unpleasantly irritable animal or person with protruding eyeballs (exophthalmia), a rapid heartbeat, high metabolic rate, and some thyroid enlargement, *i.e.,* goiter.

The hormone of the thyroid gland is an amino acid, **thyroxin**. It is an iodine-carrying modification of another amino acid, tyrosine. That iodine is concentrated by the thyroid gland can be readily shown by injecting

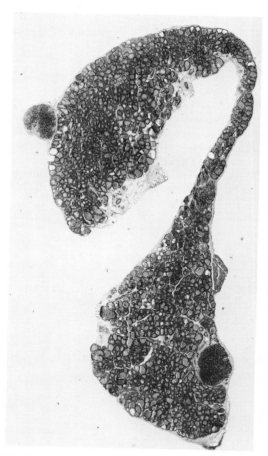

Fig. 30-4. Section through a rat thyroid gland showing the two lobes connected by the isthmus. The dark parathyroid glands can be seen attached superficially to one lobe and almost completely imbedded in the other. (× 20).

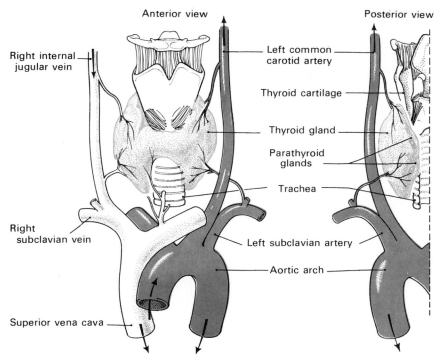

Anterior view

Posterior view

Right internal jugular vein

Left common carotid artery

Thyroid cartilage

Thyroid gland

Parathyroid glands

Trachea

Right subclavian vein

Left subclavian artery

Aortic arch

Superior vena cava

Fig. 30-5. Human thyroid gland and associate structures. Note the arteries and veins serving the gland and also the pair of parathyroid glands on the posterior surface of the left thyroid lobe.

radioactive iodine into a mammal. Within an hour the iodine shows up in the cells constituting the walls of the thyroid follicles. Later the radioactive material appears within the follicles.

Thyroxin causes mitochondria to swell and increase in number but whether this is the cause or a result of the increase in rate of metabolism is uncertain.

The activity of the thyroid gland is dependent on a thyrotropic hormone from the anterior lobe of the pituitary. Thyroxin itself inhibits the formation of thyrotropin either by acting directly on the anterior pituitary or through inhibiting cells in the hypothalamus, the action of which is required for the formation and release of thyrotropin by the pituitary.

THE PARATHYROIDS

The parathyroids are four small glands either on or embedded in the thyroid, two on each side. In some way they control calcium and potassium metabolism. The target organs are the bones and the kidneys. In dogs and carnivores generally, parathyroidectomy leads to a serious lowering of blood calcium levels, muscular spasms increasing in severity into total tetany, and ultimate death. The symptoms are usually much less severe in omnivorous and herbivorous animals.

CALCITONIN

It has recently been discovered that both the thyroid and the parathyroids secrete a hormone essential for health, named calcitonin. The effect of calcitonin is just the opposite of that of the parathyroid hormone. It depresses the concentration of calcium ions in the blood, causing them to be deposited in the bones.

THE PANCREAS

The pancreas is an elongated gland located beside the intestine close to the stomach. It is sometimes sold, along with the thymus gland, as sweetbreads. It secretes a digestive pancreatic juice via the pancreatic duct into the intestine, and also two hormones, insulin and glucagon, elababorated in nests of cells called the **islets of Langerhans**. These islets usually lie scattered throughout the body of the gland.

Lack of **insulin** causes diabetes mellitus, a disease described by the ancient Greeks as the ailment in which "the flesh melts away into urine." Nearly 2,000 years later a 17th-century English anatomist and physician noticed that the urine of diabetics is sweet, so to the classic symptoms of excessive thirst, excessive urination, and great emanciation, a fourth was added, excretion of sugar.

About 200 years later still, in typical 19th-century experiments, von Mering and Minkowski made a surprising discovery. Removal of the pancreas of a dog causes diabetes! It could not be from lack of the digestive juice, because if the pancreatic duct was brought to the surface of the body so that all of the pacreatic juice merely dropped to the ground, the dogs did not develop diabetes. In 1900 Eugene Opie, a member of the first class of the Johns Hopkins Medical School, discovered that in diabetic patients the islet cells were atrophied.

The two kinds of cells, arbitrarily called alpha and beta cells, make up most of the islets of Langerhans. The **alpha cells** are somewhat larger than the beta cells, lie near the periphery of the islets, and have a granular cytoplasm which stains differently from the granules of the smaller, more centrally placed beta cells. It is the **beta cells** that atrophy in diabetes, and they are the ones that secrete insulin. Moreover, a synthetic pyrimidine called **alloxan** causes diabetes when administered to animals. The beta cells degenerate. The way alloxan knocks out the beta cells is unknown at this time, but it is interesting to note that pyrimidines form part of nucleic acids and to recall that thiouracil, an inhibitor of the thyroid gland, is also a pyrimidine.

The alpha cells secrete a hormone, **glucagon**, which has the opposite effect on blood glucose from insulin. Instead of decreasing the level of glucose, glucagon increases it. The existence of such a hormone came to light after it was found that patients from whom the entire pancreas had been removed (because of cancer) required less insulin to keep the blood sugar down to

normal levels than patients with naturally occurring diabetes. This indicates that the pancreas produces something which tends to increase the concentration of glucose.

THE ADRENALS

The adrenal glands of mammals are a pair of more or less rounded, highly vascularized structures situated either close to or against the kidneys—hence the name **adrenal**. The central part of each gland is called the **medulla** and secretes two similar hormones which are modifications of the amino acid tyrosine and are often grouped together under the term **adrenaline**. The outer part of the gland, the **cortex**, secretes several steroids of which the best known is **cortisone**.

The hormones synthesized by the medulla are **epinephrine** and **norepinephrine**. They differ only in that epinephrine has a methyl group, *i.e.,* a CH_3 radical which norepinephrine lacks.

The effects of lack of cortical hormones have been known since Thomas Addison described the strange and fatal disease mentioned earlier. Oversecretion or deranged secretion of the adrenal cortex is also a serious affliction. Some of the cortical hormones have a masculinizing

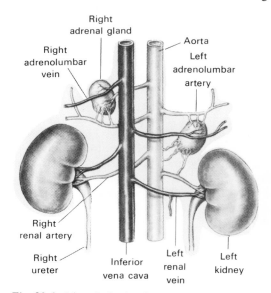

Fig. 30-6. Adrenal glands of a cat. Note the abundant blood supply. (From G. B. Moment: *General Zoology*, 2nd ed. Houghton Mifflin Co., Boston, 1967.)

effect. If oversecretion begins in childhood, the voice deepens, and facial, axillary, and pubic hair develop in a typical masculine way even in genetic girls. Muscular development may produce an adult-like dwarf of herculean conformation. In genetic boys, although the penis may attain adult size, the testes remain infantile or even abnormally underdeveloped. Under normal conditions, cortisone promotes healing, plays an important role in muscle contraction, promotes fat deposition (often in the face), and affects excitation thresholds in the nervous system and the proportions of the different types of leucocytes. It has various other poorly understood functions such as counteracting shock.

In populations of mammals such as woodchucks and mice, enlargement of the adrenal glands and decrease in reproductive competence accompany great crowding, *i.e.*, stress. This response may be consistent and may act as a brake on population growth.

The secretion of the adrenal cortex is under the usual feedback control by the anterior pituitary, which secretes a polypeptide ACTH, adrenocorticotrophic hormone. Without it the cortex fails to secrete.

THE GONADS

It seems highly probable that gonadal hormones produce their effects on the growth and pigmentation of their various target organs by initiating the transcription of new mRNA. This certainly seems to be the case with female sex hormone and the lining of the uterus. The effects which sex hormones have on the behavior of vertebrates appear to be due to direct action of the hormone on specific cells in the brain, specifically in the hypothalamus.

It should be noted that although the primary **estrogen** (female sex hormone) produced by the Graafian follices of the ovary is **estradiol**, this hormone is often found in a slightly modified form called **estrone** in many tissues including the ovary itself, the placenta, the adrenal cortex, and even the testis. Paradoxical as it seems, the testes and urine of stallions are among the richest sources of estrogen. Of course they carry even more male sex hormone, **androgen**. There has never been any satisfactory explanation for this fact. One of the commercial sources of estrogen is the Mexican yam or barbasco root. Why this plant should produce such a hormone is also a mystery.

Progesterone is secreted by the corpora lutea of the ovary after fertilization; it is necessary for the continuance of the pregnancy and has a molecular structure similar to estradiol. So also do the androgens, **testosterone**, the form in which male sex hormone is secreted by the interstitial cells in the testis, and **androsterone**, the slightly modified form in which it is excreted in the urine.

All of these gonadal hormones are under the regulation of the anterior lobe of the pituitary, which secretes gonadotropic hormones essential for the formation of sex hormones. They in turn inhibit the secretion of their respective gonadotropins.

THE PITUITARY

The pituitary gland, or **hypophysis**, is commonly called the "master gland," for it exerts a regulatory control over most of the other glands of internal secretion. It is beginning to appear, however, that the pituitary is not so much a master as an executive officer carrying out the instructions of the **hypothalamus**. In adults, the pituitary is located on the floor of the skull just behind the optic chiasma where the optic nerves cross as they enter the underside of the brain—a very awkward place to reach surgically. This is an inconvenience we apparently owe to our remote tunicate-like ancestors.

All the hormones of the pituitary are polypeptides. Indeed some of them are molecules large enough to rank as proteins. The **anterior lobe** of the pituitary secretes somatotropin (the growth hormone), gonadotropins, thyrotropin, adenocorticotropin, a corpus luteum-stimulating, or luteinizing, hormone (LH), and a lactogenic hormone. In general these **anabolic hormones**, especially somatotropin, promote protein synthesis and a general buildup of body structures. There is occasional talk that athletes are injected with these hormones to increase their strength. How effective such measures are depends in part on how close an individual is to his maximum size. ACTH stimulates the synthesis of cortical hormones and inhibits the synthesis of fat in adipose tissue.

Table 30-1
Chemical Regulators

Source	Hormone	Chemical Nature	Action
Anterior pituitary (adenohypophysis)	Anterior pituitary growth and other hormones, GH, FSH, LH, TSH	Protein	Stimulates thyroid, gonads, and mammary glands; bone growth; protein synthesis and anabolic metabolism generally
	Anterior pituitary, ACTH	Protein	Stimulates cortisone synthesis and fat breakdown
Posterior pituitary (neurohypophysis)	Antidiuretic hormone, ADH and oxytocin	Protein	Stimulates renal reabsorption of water; contraction of uterine and mammary musculature
Intermediate lobe of pituitary	Intermedin	Polypeptide	Darkens skin by causing melanophore expansion
Thyroid	Thyroxin	Tyrosine derivative	Stimulates metabolism; promotes protein synthesis; metamorphosis in amphibia
	Calcitonin		Lowers blood calcium
Parathyroid	Parathyrone		Maintains blood calcium
Islets of pancreas	Insulin	Protein	Controls carbohydrate utilization; depresses blood sugar level; promotes protein synthesis
	Glucagon		Increases blood sugar level
Pineal gland	Melatonin	Tryptophan derivative	Slows ovarian development and estrous cycle
Adrenal cortex	Cortisone	Steroid	Promotes healing, fat deposition, muscular development; influences neural excitation thresholds, white blood cells
	Aldosterone	Steroid	Activates production of enzymes increasing Na reabsorption in kidney and elsewhere
Ovary	Estrogen (estradiol)	Steroid	Female secondary sexual traits; in adult estrous cycle; growth of uterus and mammary glands
Corpora lutea of ovary	Progesterone	Steroid	Stimulates growth of uterine lining and milk-secreting cells
Testis	Androgen (testosterone)	Steroid	Masculinizes the brainstem reproductive rhythm center; embryonic development of male reproductive structures; male secondary sexual traits in adult
Insect prothoracic glands	Ecdysone	Steroid	Induces molting hormone
Insect corpora allata	Juvenile hormone	Steroid(?)	Prevents attainment of adulthood
Crustacean Y organ	Ecdysone or similar hormone		Induces molting
Crustacean X organ and brain	Juvenile or similar hormone	Steroid(?)	Prevents molting

Neurohormones			
Neurosecretory cells of brain	See under Pituitary (both anterior and posterior lobes and the pituitary portal vein)		
Adrenal medulla	epinephrine (adrenaline)	Tyrosine derivative	Raises blood pressure; increases heart rate; liberates liver glycogen as glucose
Adrenal medulla	Norepinephrine	Tyrosine derivative	Similar to epinephrine
Nerve cell endings	Acetylcholine		Transmits information across synapses
Brain, octopus salivary gland, plant nettles	Serotonin	Cyclic organic compound	Similar to adrenaline but more violent; neural functions uncertain

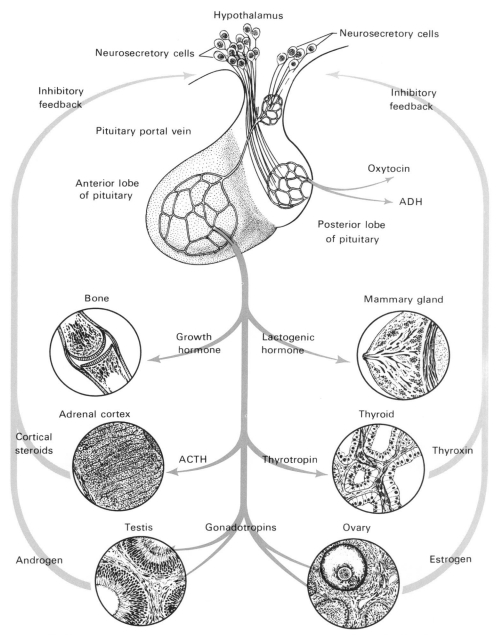

Fig. 30-7. Feedback control between the anterior pituitary and other endocrine glands.

The **posterior lobe** stores and discharges two hormones, the antidiuretic hormone (ADH) and oxytocin. These are produced in neurosecretory cells located in the hypothalamus above the optic chiasma. The secretion moves down into the posterior lobe via the axons of these neurosecretory cells. The anti-diuretic hormone controls the formation of urine by increasing water reabsorption in the kidney tubules. Failure of the posterior pituitary results in diabetes insipidus, which is unrelated to a lack of insulin. Large amounts of sugar-free urine are produced. **Oxytocin,** also known as **vasopressin** because in large doses it causes arteries to constrict, is a hormone which stimulates smooth muscles. It is important in the release of milk on suckling and almost certainly in stimulating uterine contractions during childbirth.

The **intermediate lobe** of the pituitary produces a melanophore-dispersing hormone, **intermedin,** in lower vertebrates and probably in mammals as well.

The cells of the **anterior lobe** are of three major types according to staining and other characteristics. There are **acidophils,** so called because their cytoplasmic granules stain with acid dyes like eosin or orange G. These cells are often called eosinophils or alpha cells. There are **basophils,** often called beta cells, in which the cytoplasm stains with basic dyes like methylene blue. Third, there are cells which resist staining and are therefore called **chromophobes.** They constitute roughly half of the cells present.

Many attempts have been made to identify which cells secrete which hormone. Some, if not all, of the acidophils secrete growth hormone. Several genetic strains of dwarf mice show a hereditary lack of alpha cells and no pituitary growth hormone. Human pituitary (Tom Thumb) midgets are also deficient in alpha cells. Some of the beta cells or basophils evidently secrete a hormone related to the gonads. After removal of either testes or ovaries, many of the basophils become enlarged as though engorged with secretion.

The importance of the anatomical relationships of the pituitary gland can scarcely be exaggerated. Not only does it lie adjacent to the underside of the hypothalamus but each lobe is functionally tied to the hypothalamus in a very special way. The posterior lobe is directly innervated by nerve fibers coming down the stalk of the gland from the hypothalamus. These are called **neurosecretory cells** because secretion granules can be observed to form in the cell bodies which lie in two groups (each called a "nucleus") in the brain. These granules then move down the axons into the posterior lobe of the pituitary.

The anterior lobe of the pituitary is also connected to its own set of neurosecretory cells in the hypothalamus, but via the bloodstream. These nerve cells send their axons down into the stalk of the pituitary where they end among the meshes of a small capillary network. The capillaries pick up the secretion from this group of neurosecretory cells and carry it in a **pituitary portal system** which almost immediately breaks up into a second set of capillaries within the anterior lobe. The blood supply of the posterior lobe is quite separate from the system of the anterior lobe.

The evidence for a regulatory **feedback control** between the pituitary and the hormones produced by ovaries, testes, thyroid, and adrenal cortex is unassailable. If the anterior pituitary is removed, the thyroid gland atrophies, the gonads fail to produce sex cells or sex hormones, reproductive cycles stop, and the adrenal cortex fails. If one of these glands, a target organ for the pituitary, is removed, the level of the corresponding pituitary trophic hormone increases greatly. In the case of removal of either male or female gonads, characteristic "castration cells" appear in the anterior lobe. Injection of sex hormones or thyroxine, for example, depresses the amount of gonadotropin or thyrotropin secreted by the pituitary.

It is not yet clear whether the inhibitory effect of the various hormones is directly on the cells of the pituitary or on the neurosecretory cells in the hypothalamus, or on both. Here is some of the evidence. When part of the thyroid of a rat is removed, the pituitary increases its release of thyrotropin. The remaining piece of the thyroid gland then increases in size and ultimately the level of thyroxine in the blood is restored to normal levels. At the same

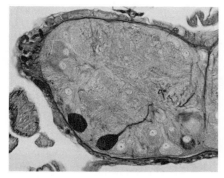

Fig. 30-8. Neurosecretory cells in the central nervous system of an earthworm stained dark purple by the Scharrer-Gomori aldehyde-fuchsin method. Such neuroendocrine cells appear identical with those seen in vertebrate brains. Note the axon carrying the darkly stained secretion.

time the level of thyrotropin returns to normal. So far there is no hint of a role for the hypothalamus and its neurosecretory cells. However, this sequence of events can be completely blocked by electrically destroying the cells in a very small region of the hypothalamus. If the thyroid is then removed, the pituitary behaves as though it knew nothing about it.

THE PINEAL BODY

The pineal body is a small rounded structure on the dorsal side of the thalamic portion of the brain opposite the pituitary on the underside of the thalamus. In the adult frog the pineal body underlies the skull, but in mammals it is overgrown by the enormous cerebral hemispheres and the cerebellum. In many lizards the pineal body is a third eye; in frogs it contains **cone cells** closely similar under an electron microscope to the cones of the visual retina.

Over the years many investigators have attempted to discover what the pineal gland does. It has long been known that tumors of this structure in children are commonly associated with precocious sexual maturity, or paradoxically, sometimes with delayed sexual development. The answer was found only recently. Precocious sexual maturity is associated with tumors of the supporting tissue around the pineal gland which apparently choke the structure into inactivity while overgrowth of the pineal itself is associated with delayed puberty. These facts suggest that the pineal exerts an inhibitory effect on gonadal development.

A real breakthrough came with the discovery by Virginia Fiske at Wellesley that continuous illumination results in a decrease in the weight of the rat pineal body and an increase in the size of the ovaries. Other investigators have shown that exposure of male hamsters to cycles of one hour of light and 23 hours of darkness will cause atrophy of the testes but that this effect can be prevented by removal of the pineal gland. It was then discovered that there is a circadian (roughly 24-hour) cycle in the formation of a pineal enzyme which produces **melatonin.** Melatonin is a hormone derived from the amino acid tryptophan. It is worth noting as a typical example of how laborious the identification of a hormone can be that extracts of over 200,000 cattle pituitaries were prepared and purified in this work by A. B. Lerner and his co-workers. This hormone slows down the estrous cycle and inhibits ovarian development; it is produced only in the pineal body.

THE THYMUS

The thymus, like the pineal body, has been an enigmatic gland, often thought to play some role in retarding sexual development because it undergoes marked involution with the onset of sexual maturity. Many experimenters have removed the gland, which lies in the upper chest beneath the breastbone. Extracts have also been injected, but no significant results have followed either procedure. It has now been found that if the thymus is removed promptly at birth, antibody-forming lymphocytes (plasma cells) will be lacking. Apparently the lymphocyte stem cells in the various lymph glands, which produce these blood cells in later life, come from cells which originate in the thymus. There is also some evidence that the thymus secretes some hormone-like material essential for proper lymphocyte development.

USEFUL REFERENCES

Barrington, E. J. W. *Hormones and Evolution.* D. Van Nostrand Co., Princeton, 1964.
Guyton, A. C. *Function of the Human Body.* W. B. Saunders Co., Philadelphia, 1969.
Kruskemper, H. L. *Anabolic Steroids.* Academic Press, New York, 1968.
Scharrer, E., and Scharrer, B. *Neuroendocrinology.* Columbia University Press, New York, 1963.
Turner, C. D. and Bagnara, J. T. *General Endocrinology,* 5th ed. W. B. Saunders Co., Philadelphia, 1971.

Reproduction

3 1

BIOLOGICAL MEANING OF REPRODUCTION

Reproduction involves two conflicting requirements if life is to continue and if evolution is to take place. First is the basic requirement of self-duplication. Second is the production of novelty.

Without the first, obviously, life would not continue to exist. On the molecular level reproduction means replication of DNA. For cells it means mitosis. On the level of multicellular organisms reproduction means either some kind of asexual division in which fairly large replicates can be produced directly, or some kind of sexual reproduction in which the organism is reduced to the level of single cells. Why this second method?

The answer is the production of novelty, or variation. In a very real sense this is the opposite of reproduction. At the basic molecular level novelty is the result of two factors: errors in the replication of deoxyribonucleic acid, *i.e.,* gene mutations, and the inherent properties of the DNA, for the kinds of errors that can occur reflect the chemical nature of nucleic acids.

The basic novelty produced on the molecular level is compounded on the level of cells by sexuality, that is, by the making of new combinations and permutations of chromosomes in the cycle of meiosis and fertilization. The variation produced by meiosis alone may be very great. With 23 pairs of chromosomes, each human may produce over 8,000,000 kinds of eggs or sperms. Fertilization multiplies these two possibilities by each other in the production of the zygote. 8×10^6 multiplied by

8×10^6 equals 64×10^{12}, the number of possible kinds of zygotes from any marriage, and what a very large number it is—64,000,000,000,000! Equally important, sexual reproduction makes it possible to bring together in a single individual and its descendents beneficial mutations which arose in different individuals. Clearly then, sexual reproduction is a highly important part of the machinery of evolution. This is, in fact, what sex is all about from the biological point of view. From the evolutionary point of view, asexual reproduction can advance only at a snail's pace.

This does not mean that asexual reproduction is unimportant among living things. Once a well adapted animal or plant has been produced as a result of mutation, sexual combinations, and natural selection, asexual reproduction can produce standardized duplicates in enormous numbers. Along side of sexual reproduction, asexual reproduction is frequent among protozoans, coelenterates, the parasitic flatworms, tapeworms, some annelids, and even in certain insects.

ADAPTATIONS TO INSURE THE MEETING OF GAMETES

Temperature and Light

If eggs and sperms are to meet, the spawning of animals which merely discharge their gametes into the sea water must be synchronous. In animals which mate, there must be mechanisms which bring both sexes into a state of sexual readiness and together at the same time of the year. Two of the earliest and most widespread of such adaptive mechanisms are responsiveness to temperature or light.

Temperature is said to be the chief factor in triggering the shedding of gametes into the water for oysters and many other aquatic organisms. However, some animals such as the hydroids among the coelenterates have been very clearly shown to shed eggs and sperms at very definite times after the stimulus of light.

Throughout most of the vertebrates, and perhaps in all, photoperiod, namely the relative number of hours of light and dark in every 24, sets the timing for annual breeding periods. This discovery was made by William Rowan in Edmonton in northwestern Canada. Like many people before him, he was impressed by the astonishingly precise timing of migration and breeding of several species of birds. Unlike anyone preceding him in human history, he discerned the explanation. The temperature, food supply, and weather conditions varied greatly from year to year. The one factor that was precisely constant was the increasing number of hours of daylight beginning after the winter solstice on December 22. By artificially increasing the length of daylight each day, Rowan was able to prove that his theory that gradually increasing day length is the stimulus which leads to the annual breeding and migration. Juncos in outdoor cages in the bitter cold of the Alberta winter responded to increasing periods of daylight by a dramatic increase in the size of their gonads. When released, such birds flew northward in what resembled a migratory flight. This phenomenon resembles the control of flowering which Garner and Allard of the U. S. Department of Agriculture had demonstrated. Many plants blossom not under the stimulus of nutritional or thermal conditions but when the photoperiod is right for them.

Since Rowan's epoch-making discovery, photoperiod has been shown to affect reproduction in many animals, especially birds and insects. In the crayfish, the ovarian cycle and molting can be manipulated by controlling the photoperiod. On the practical side, in all industralized nations, hens are stimulated to lay more eggs by artificially increasing day length.

The mechanisms of photoperiodic control of reproduction are very imperfectly understood. Apparently light perceived with the eyes stimu-lates the neurosecretory cells of the hypo-thalamus to send their secretion down into the anterior hypophysis via the pituitary portal system described in the previous chapter. The role of light, acting through the pineal gland, has been discovered in the case of rats and a few other mammals. It is interesting to note that there are long-day and short-day plants (see the chapter on plant reproduction). The sheep and goat are short-day animals which come into estrus or "heat" with the decreasing daylight hours of the fall. The raccoon and horse are long-day animals which come into estrus with the lengthening days of spring. In all cases, the reproductive activity of the individuals of the species is coordinated. Moreover, it is coordinated at such a time that the young will hatch or be born at a favorable time of the year.

Sounds, Sights, and Smells

The mating calls of animals are familiar, although many of them serve equally as a warning to rival males. This is especially true of bird song which is produced almost exclusively by males and is an invitation to a female and a warning to other males.

Visual stimuli which attract one sex to the other are common. The flashing of fireflies is a conspicuous case. The two common species in the eastern United States illustrate this nicely. The low-flying *Photinus pyralis* emits a yellow-ish-orange flash while flying upward an inch or two. Individual flashes may be several minutes apart. The higher flying *Photuris pennsylvanica* emits a yellowish-green light in a series of three or four closely-spaced flashes, giving a twinkling effect. Females answer the flashing of males by flashing themselves about a second after the flash of the male.

Teleost fishes, among the animals with color vision, recognize the sex of members of their species by sight. Among the sticklebacks, males have a scarlet belly in the breeding season. In birds, of course, color and pattern play a predominant role in sex identification and attraction. The fantastically gorgeous and exotic plumage of male birds of paradise, peacocks, and turkey gobblers illustrate how important visual stimuli are.

The overwhelming majority of mammals are

color-blind. Bulls can no more see red than they can see blue or yellow. Typically, mammals find members of the opposite sex by odor. An odoriferous material secreted by one animal which affects a second is called a **pheromone**, a class of substances discussed in Chapter 1. The most familiar, though undetectable by the human olfactory epithelium, is the pheromone secreted by a female dog in estrus. Similar substances are produced by mares, cows, does, and other mammals. Primates like ourselves are an exception. In this order of mammals, vision, not olfaction, is the dominant sense, for primates are primarily arboreal. In most primates the distinction between the sexes rests on differences in bodily form and behavior, although in male baboons secondary sexual traits include bright blue buttocks and fierce bright blue and red faces. Pheromones which are sex attractants play vital roles in the lives of many insects. Knowledge of them has proven useful in agriculture, making it possible to attract males to poisons or into traps.

Courtship and Mating

Animals present a wide range of courtship and mating behavior ranging from some of the flatworms at one extreme where no known equivalent of courtship exists and sperms are merely injected into the body of the receiving individual through the epidermis at almost any point. In most species of flatworms, however, the penis is inserted into a special passageway, the vagina.

The spiders represent an opposite extreme. The sex of a spider is extremely difficult to discover until after the final molt, when a male spider is conspicuous because of a complex bulbous apparatus on the end of its palps. Any spider that looks as though it were wearing boxing gloves is a mature male. Before mating, the male spins a small flat web on which he deposits a drop of semen. He then carefully fills the bulbs on the ends of his palps with this sperm-laden fluid. Thus equipped, he begins to seek a female. Male spiders have a number of signs which act as stimuli for the female, conveying the message that "here is a potential mate, not a meal." Some species signal the female by fantastic posturings with the legs. Other males begin by striking the female with

their legs, while in still other species the males commence by tentatively and gingerly touching the tip of one of the female's legs. Many orb weavers begin by tweaking the web of the female. Males are very likely to be eaten by the females, if not before mating, then afterwards. In the act of copulation, if in spiders it can be called that, the male places the seminal bulb at the end of his palp against the genital opening of the female. Here the transfer of sperms takes place.

In the vertebrates courtship behavior often involves elaborate and highly ritualized displays. These species-specific patterns have been carefully studied in birds, particularly in different kinds of wild ducks. The males of each species show stereotyped hereditary sequences of head bobbing, wing spreading, tail waggings, preening, and swimming in circles. By contrast, in members of the parrot family known as "love birds," initial courtship seems to consist of merely looking at a member of the opposite sex. This happens well before the onset of sexual maturity in either member of the pair. After such mutual inspection for a period as short as four minutes, a pair of love birds are firmly mated for the rest of their lives. The prolonged and irreversible nature of this response suggests another example of very early and long-lasting learning, the attachment of offspring to parent known as **imprinting** which occurs in many birds and probably in some mammals.

The chief point about reproductive behavior in birds is its highly stereotyped nature within any one species, where there is little variability, and the enormous range of differences found

Fig. 31-1. One of the much studied courtship rituals of a male duck. These extremely rigid patterns are characteristic of each species, and females do not respond to off-beat behavior. (From I. Eibl-Eibesfeldt: *Ethology: The Biology of Behavior.* Holt, Rinehart and Winston, New York, 1970.)

between different species. In some birds, chickens and quail, for example, and in many of the larger hoofed mammals, there is little in the way of courtship but there is combat between males while the females remain passive onlookers.

PARENTAL CARE

Care of the young is an essential feature of reproduction in birds and mammals, in marked contrast to the rest of the animal kingdom, where parental care is extremely rare. Aristotle claimed that the male catfish guards the eggs and newly hatched fry. So unusual is such action that European zoologists regarded this tale as just one of the myths Artistotle accepted until Louis Agassiz in modern times discovered that male catfish in North America behave as Aristotle had described 2,000 years before! Female wolf spiders can sometimes be seen carrying spiderlings on their backs.

The necessity for parental care arises partly because of the evolution of intelligence, *i.e.*, the ability to learn. If an animal is born with a full set of instincts and innate responses, *i.e.*, prefabricated answers to the problems of life, the young are equipped to fend for themselves very early. If learning is to occur, then some protected period is required during which learning can take place. This is not the only factor in the evolution of parental care but it is an important one.

The kind of parental care and whether or not the male plays a part is adaptively correlated with the number of offspring, infant mortality, and, in birds, with the number of eggs laid and the stage of development of the young at birth. In penguins, one of the most intensively studied of birds, both sexes cooperate in incubating the single egg and feeding the young. Among ducks and chickens the male has a minor role after the eggs are fertilized. In such species the female lays a large number of eggs and the so-called **precocial** chicks are able to move about and feed themselves within a few hours after hatching. Infant mortality is high. Among song birds both sexes are deeply involved in incubating the eggs and in the care and feeding of the young. Such species lay only a small number of eggs, the helpless young, known as **altricial**

(*altrix*, nurse), receive intensive parental care, and infant mortality is much lower.

In mammals the requirement for milk enforces at least a measure of maternal care which is usually both intensive and extensive. Paternal care varies widely. Male baboons show a great interest in very young baboons and solicitude for them. The males of the larger hoofed mammals commonly ignore their offspring except in defending them as members of the herd. Male foxes, wolves, and lions cooperate with their females not only in the care and education of the young but in the business of living generally. Male tigers live alone and are thought to eat newborn kits if the opportunity arises. How all these very different patterns of reproductive behavior are coordinated with all the other anatomical, physiological, hormonal, and behavioral adaptations in the lives of these animals is complex and very imperfectly understood—a challenge for future study.

THE ANATOMY OF REPRODUCTION

The Male Reproductive System

The male gonads or testes vary greatly in shape and number throughout the animal kingdom. Meiosis is, however, always the same. In nematode worms the testis has been much used to study meiosis because it is a continuous tube with primordial diploid germ cells at the distal end and mature haploid sperm at the proximal end. In between, the various stages of meiosis can be read like a printed sentence from left to right.

The testes of arthropods and of lower vertebrates, fish, and salamanders are much alike. Each testis is elongate and composed of a series of squarish compartments. Within each compartment all the cells are in approximately the same stage. At one end of the testis are primordial germ cells, and at the other are mature sperms, so that the observation of meiosis is not too difficult. Sex chromosomes were first discovered in the testis of the squash bug when it was noticed that in females there are two complete matching sets of chromosomes while in males there is one pair which does not match.

The testes of reptiles, birds, and mammals are all very similar. Each testis is composed of a

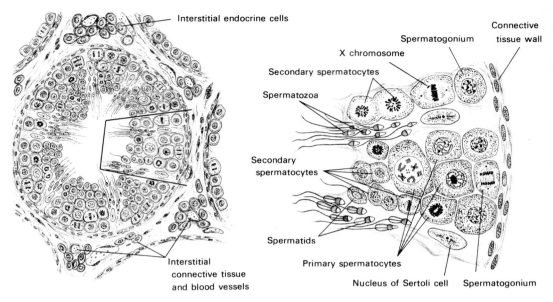

Fig. 31-2. Portion form cross section of a human testis, typical of all higher vertebrates. (After L. B. Arey: *Developmental Anatomy*. W. B. Saunders Co., Philadelphia, 1965.)

mass of **seminiferous tubules** in which meiosis and sperm development take place. The diploid **spermatogonia**, which multiply by mitosis, lie around the periphery of the tubule; the **primary and secondary spermatocytes** lie closer to the lumen of the tubule; and finally the haploid **spermatids** lie closest to the center. As in other animals, the spermatids develop into mature sperms while embedded in the cytoplasm of nurse cells, usually called **Sertoli's cells.**

Between the seminiferous tubules are masses of interstitial or **Leydig's cells** which secrete **testosterone**, the male sex hormone.

Mature sperms pass through the lumens of the seminiferous tubules into the **epididymis**, a clump of coiled tubules lying against the testes, and finally into the **vas deferens.**

The **accessory sex organs** in different mammals are much the same. Most species possess a **seminal vesicle**, probably more accurately called a seminal gland because it is known to pour a secretion into the vas deferens, but whether it stores sperms is uncertain. Surrounding the junction of the vas deferens with the urethra from the urinary bladder is the **prostate gland**. Its secretion forms a major part of the **seminal fluid**. The position is unfortunate because in older males it has a tendency to enlarge and

obstruct the flow of urine. A very small pair of glands, **Cowper's glands**, open into the urethra just below the prostate. Their secretion has a lubricating and acid-neutralizing function.

The seminal fluid is ejaculated into the vagina of the female from the **penis**, which is brought into a state of erection when specialized cavernous tissue within it becomes engorged with blood. In some mammals, notably the carnivores such as dogs and seals, the penis is further strengthened by a special bone.

The puzzling fact about the male reproductive system in mammals is the existence of the **scrotum**. This is essentially a muscular sac which acts as a thermoregulator, keeping the temperature of the testes several degrees below body temperature. If the temperature of the scrotum of an experimental animal is kept at body temperature either by insulated wrapping or by surgically placing it within the abdomen, spermatogenesis ceases, and the seminiferous tubules degenerate. This is probably why **cryptorchid** individuals, whose testes lie within the abdomen, are sterile. Yet the curious fact remains that in elephants, seals, and porpoises, the testes are normally and continuously abdominal. They are, also, in birds where the body temperature averages about 40 to 43°C.

(104 to 108°F.), a dangerous fever for a mammal! In all females the ovaries are abdominal.

Secondary Sexual Characteristics

The **secondary sexual characteristics** of male mammals differ greatly among the different orders. The lion's mane and the moose's antlers are familiar examples. In porpoises and rodents secondary sexual characteristics are virtually nonexistent, except in behavior. In mammals, secondary sexual characteristics develop under control of gonadal hormones; in some they are independent of sex hormones; and in some they are due to a combination of hormones and more local genetic action. In the common English house sparrow, for example, the color of the bill is a very sensitive indicator of the concentration of sex hormone in the blood,

while the color pattern of feathers is independent of sex hormones and can even be seen in birds castrated while still in their nestling down. In other species of birds—chickens, for example—feather type and pigmentation are controlled by sex hormones.

The Female Reproductive System

The ovaries of animals vary considerably from one group to another. In flatworms and echinoderms, they are irregular in shape. In the much studied nematode worms, each of the two ovaries consists of a long tube in which meiosis occurs in a neat order as one looks along the ovary, continuous with the oviduct and vagina. In insects the ovary is usually a cluster of tubules, each with a single row of developing egg cells.

In lower vertebrates, fish and amphibians,

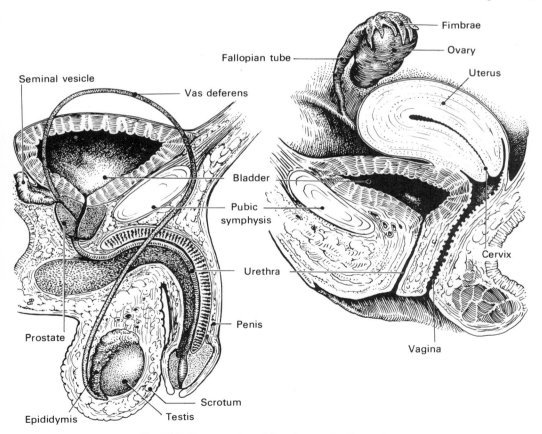

Fig. 31-3. Human male and female reproductive systems.

the ovary is a hollow sac, although, as anyone who likes to eat fish roe knows, in the breeding season each of the two ovaries becomes solid with eggs with no central cavity apparent. In the higher vertebrates, the ovaries are solid and covered with a continuation of the peritoneal epithelium which lines the coelom. A peculiarity of birds is that only the left ovary and left oviduct develop.

The peritoneal epithelium over the surface of the ovary is known as the **germinal epithelium.** From it groups of cells containing descendants of the primordial germ cells migrate into the interior of the ovary. Here the germ cells begin meiosis. Each egg cell grows enormously in size and is surrounded by layers of cells between which a cavity forms. This cavity enlarges and becomes filled with **follicular fluid** rich in **estrogen,** the female sex hormone secreted by the cells surrounding the cavity. Outside these cells is a thin layer of smooth muscle fibers. The entire capsule is known as a **Graafian follicle.** The number of ova which mature at any one time depends on the species: normally one in women (the right and left ovaries are thought to alternate), three or four in vixen, seven or eight in mice. Some very delicate hormonal balance apparently controls this number.

In mammals, the **vagina,** or birth canal, located between the openings of the urethra and the alimentary tract, is merely a tube lined with mucoid epithelium which leads into the **uterus** or womb. At the opening of the uterus into the vagina is a thick muscular region, the **cervix.** In most mammals the uterus is more or less double. The inner end of each horn of the uterus is continued as an **oviduct,** known in human anatomy as a **Fallopian tube.** The oviduct on each side ends in a thin, funnel-shaped membrane that more or less encloses the ovary.

Fertilization takes place at the upper end of the oviduct. The stimulus of mating induces the release of a hormone, **oxytocin,** which causes uterine and oviductal contractions that carry the sperms up the oviduct in remarkably quick time; in some species only a few minutes are required to reach the upper end. The function of the swimming of sperms is chiefly to prevent them from becoming stuck on the inner walls of the reproductive tract and perhaps to assist in the actual process of fertilization.

If pregnancy occurs, the developing egg and its surrounding cells signal the ovary by hormones. The result is that the cells of the ruptured follicles do not degenerate but instead proliferate and differentiate into a new structure, called from its yellowish color a **corpus luteum.** There are as many corpora lutea as there were ruptured follicles. They secrete **progesterone,** a hormone essential to maintain pregnancy.

REPRODUCTIVE HORMONES

It will be recalled that both male and female sex hormones are **steroids,** differing only slightly in the chemical groups that are attached to the basic sterol nucleus. Male sex hormones are termed **androgens,** and female sex hormones **estrogens. Prolactin,** a pituitary hormone, stimulates the secretion of milk. **Oxytocin,** another proteinaceous pituitary hormone, causes the flow of milk by producing contraction of smooth mammary muscles. Oxytocin also stimulates the contractions of the female reproductive tract which carry sperms up the uterus and oviducts.

Estrous Cycles

Ovulation in mammals occurs at more or less regular intervals: once a year for deer, every

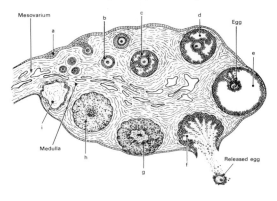

Fig. 31-4. Section of a mammalian ovary. The logical arrangement of the developing follicles in a clockwise direction is an artistic convention for the sake of clarity. Normally the stages are completely mixed. e, Graafian follicle; h, corpus luteum.

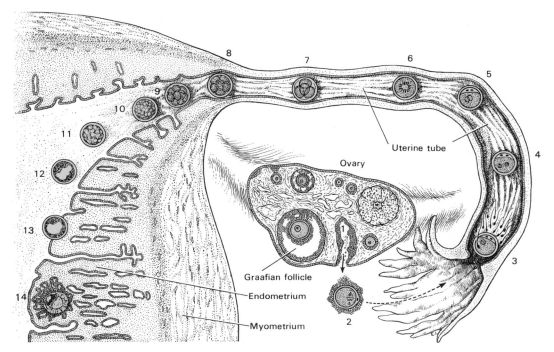

Fig. 31-5. Human fertilization, cleavage, and implantation. All stages in the development of the ovum are shown in green. (From G. B. Moment: *General Zoology,* 2nd ed. Houghton Mifflin Co., Boston, 1967.)

four to six months for dogs and cats, approximately every 28 days for women, every 21 days for mares and cows, and every 4 or 5 days for rodents. The period of ovulation, or **estrus**, in many species, notably cats and cows, is accompanied by a period of restlessness and sexual receptivity known as "heat." In the higher primates the lining (**endometrium**) of the uterus undergoes growth, which prepares it for implantation of the blastocyst, culminating at about the time of ovulation. If no implantation and pregnancy take place, the hormones which have stimulated the growth of the endometrium recede, and the lining of the uterus is shed along with a certain amount of blood. This phenomenon is called **menstruation** from *mensis,* meaning month. Menstruation will be discussed later in this chapter. A species with but one estrous period a year is termed monestrous; a species with several, polyestrous.

The timing of estrus is under complex control. As mentioned above, in a very few cases—the rabbit and cat, for example—although the estrus condition is due to other factors, the actual shedding of the eggs from the ovaries is due to the stimulus of copulation. Psychic or at least neural factors are important in the mouse. A group of female mice showing a great deal of asynchrony and variation in their estrous cycles will become regulated by the presence of a male, even though he is separated from the females by a fine mesh cage. Since merely the used bedding from a male will produce this effect, it is clear that the actual stimulus is odor. The chacma baboon has an extremely regular sexual cycle of 32 days, but this will be interrupted if the female is permitted to watch a fight between two other baboons. (This would result in an automatic limitation on population growth when numbers became great enough to make fighting frequent.)

Photoperiod, that is, the number of hours of daylight and of darkness, also has a profound influence on reproduction. In the rabbit light stimulates the retinal cells and sends impulses back to the brain. Artificially increasing the hours of daylight will bring wild rabbits into estrus even in the dead of winter. It has recently been shown that continuous 24-hour

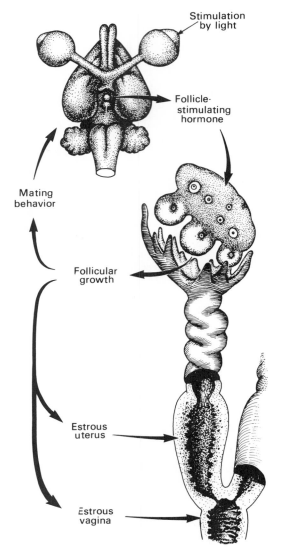

Fig. 31-6. Light and the neuro-endocrine control of reproduction in the rabbit. Other mammals are similar, although some, probably including man, do not respond to the number of hours of light and dark. (After G. B. Moment: *General Zoology,* 2nd ed. Houghton Mifflin Co., Boston, 1967.)

illumination results in an increase in the weight of the rat ovary and a decrease in the pineal gland, as discussed in the previous chapter.

The anterior lobe of the pituitary secretes **gonadotropin**, the **FSH** (follicle-stimulating hormone) mentioned earlier, which stimulates the Graafian follicles to secrete estrone, the **estro-**

gen produced by the ovary. It has five main effects. (1) It stimulates the growth of the lining of the vagina and uterus. (2) It stimulates the growth of the milk-secreting ducts of the mammary glands and stimulates the development of female secondary sex characteristics in general. (3) It activates female reproductive behavior. (4) It stimulates the anterior pituitary to secrete a luteinizing hormone, LH, which triggers the release of eggs by the follicles and stimulates the growth of the corpora lutea. (5) It inhibits the secretion of the follicle-stimulating hormone of the pituitary. Thus it saws off the limb that supports it, for once the follicle-stimulating hormone of the pituitary is no longer present in the bloodstream, the follicles in the ovary regress, and the concentration of estrogen falls drastically. At this point the pituitary is again free to form FSH, the follicle-stimulating hormone, and so the cycle is repeated. There is no better example of biological feedback.

From these facts it might be concluded that the anterior pituitary and ovary constitute a self-sustaining oscillating system. Perhaps they do, and the role of the eyes and the pineal gland is to turn the oscillator on or off in such a way that it will be correlated either with the seasons or the time of day.

Uterine Cycles

In most mammals the **endometrium,** that is, the epithelial lining of the uterus, undergoes rhythmic cycles along with ovulatory and estrous cycles. The effect of estrogen on the endometrium of an ovariectomized rat can be seen in Figure 30-3.

As now understood, the full sequence of events in the menstrual cycle runs as follows: Estogen, which stimulates the proliferation of the lining of the uterus, also stimulates the pituitary to secrete the luteinizing hormone (LH). The LH not only triggers ovulation, but also stimulates the growth of the corpus luteum, which as already described secretes progesterone which further stimulates the final stage of growth of the uterine lining, making it ready for implantation and the nourishment of an embryo. If pregnancy does not take place, the corpus luteum degenerates, the concentration of progesterone consequently falls, the

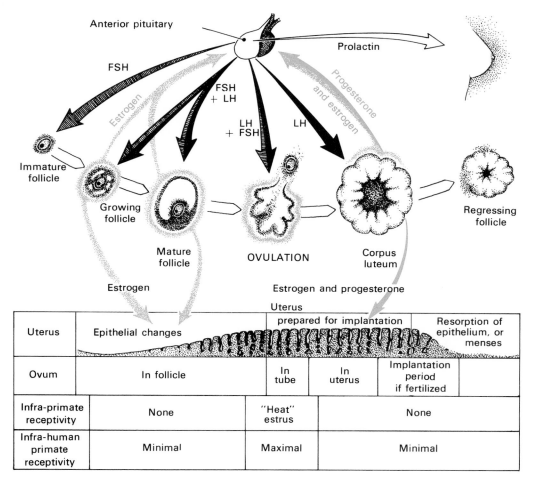

Fig. 31-7. Pituitary-ovary-uterus relationship. The two lower lines indicate behavioral correlates in infra-primate and in non-human primate species. (After G. B. Moment: *General Zoology,* 2nd ed. Houghton Mifflin Co., Boston, 1967.)

lining of the uterus loses its hormonal support and sloughs off in the characteristic bleeding. Menstruation can therefore be inhibited by suitable injections of progesterone. Ovulation usually occurs on or about the 14th day after the onset of menstrual bleeding, when the uterine lining is prepared to receive the early embryo, but again there is considerable variation.

If pregnancy occurs, the placenta itself acts as an endocrine gland, for it secretes both estrogen and the luteinizing hormone. This additional supply of LH prevents the corpus luteum from degenerating. Consequently the supply of progesterone is maintained and the lining of the uterus is not sloughed off.

HORMONAL CONTROL OF REPRODUCTION

One of the most important events of modern history was the discovery that hormones can be used to increase or decrease fertility. Barren women who desire children can sometimes be helped by precisely the correct dosage of pituitary gonadotropin to stimulate the development of a Graafian follicle and ovulation.

Of an even greater human import is the

discovery that for reasons as yet imperfectly understood, oral administration of various combinations of estrogen and progesterone inhibit either ovulation or the implantation and development of the embryo. Some hormone preparations cause a marked thinning of the uterine lining. Others are thought to block the response of the ovary to gonadotropin from the pituitary. There is much talk about *the* Pill, but actually there are many different pills available, each with somewhat different hormones and hormone combinations and of different degrees of purity. It is not surprising that there are sometimes unwanted side effects.

There is much concern about the number of users of the Pill who develop blood clots (thromboses) in the veins or lungs. Present data show that this complication is almost exclusively limited to patients belonging to blood group A. Why this should be is a puzzle. In any specific case the patient and physician must decide on the wisest course of action, weighing the dangers of side effects against the dangers of excessive pregnancies. Recent tests suggest that it may be possible to develop long-lasting agents. A single dose of the progesterone derivative depo-medroxyprogesterone acetate, seems to be effective for six months, but unfortunately it has highly undesirable side effects, including frequent uterine bleeding.

Various contraceptive intrauterine devices, IUD's, have been tested, with considerable though not complete success. These plastic rings or coils may be expelled or cause bleeding or cramps. In a study group of over 10,000 women in the United States and Puerto Rico, the frequency of pregnancy was only 3 per cent but 20 per cent experienced some kind of difficulty. Immunological techniques have been tried but offer little promise of success because of the great difficulty in producing antibodies specifically against one particular tissue.

Anti-androgens are also available. Most are derivatives of progesterone. They have been used to control unwanted facial hair in women with deranged adrenal glands and to control prostatic cancer in men.

It should be remembered in connection with the problem of conception control that simple surgical procedures have long been available. Either the vas deferens in the male or the oviducts in the female can be tied off. The fact that these surgical procedures result in permanent sterility, although leaving the individual's sexuality unaffected, may be a large part of the reason why they have never been widely accepted.

The biological as well as sociological basis for the importance of human birth control rests primarily on the great advances in modern medicine which began with the establishment of the germ theory of disease. These advances have not extended the potential life span—the traditional "four score and ten" with occasionally a little more—but they have greatly increased life expectancy. In the 17th century the mean life expectancy at birth was about 25 years. Today it is about 67 years for men and 74 for women in Western industrialized countries.

Over the long course of history, human birth rates have more than compensated for high mortality rates. Save for times of pestilence, famine, and war with their accompanying toll of human lives, there has been a persistent trend of population growth. So long as vast areas of the earth's surface were still sparsely occupied, these ever-increasing numbers of humans could readily be accommodated. We have now reached the point where population

Table 31-1
Gestation and litter data

Animal	Gestation Period	Litter Size	Life Span
Elephant	21 to 22 months	1	65 years
Whale	10 to 12 months	1	75 years
Horse	11 months	1 or 2	21 years
Cow	9 months	1 or 2	13 years
Human	9 months	1 to 5	70 years
Deer	8 months	1 or 2	12 years
Bear, black	7 months	1 to 3	22 years
Sheep	5 months	1 or 2	12 years
Pig	4 months	8 to 25	
Lion	108 days	2 to 4	20 years
Guinea pig	68 days		
Dog	63 days	3 to 12	20 years
Cat	56 days	3 to 12	20 years
Rabbit	32 days	4 to 7	8 years
Rat	20 to 21 days	4 to 10	3 years
Mouse	20 to 21 days	5 to 15	3 years
Opossum	13 days in utero, 50 in pouch, and 30 free nursing		

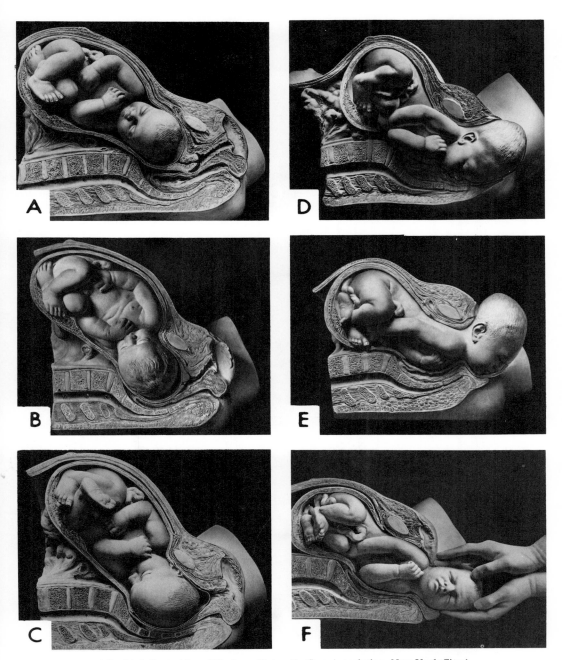

Fig. 31-8. Parturition. (Kindness Maternity Care Association, New York City.)

densities in many areas exceed what can be adequately supported by available natural resources. Clearly birth rates must ultimately be adjusted so that the frightful consequences of overpopulating this planet with our own species can be avoided. A valid analogy can be made with the reproductive situation in birds. Mankind is passing from a condition comparable to

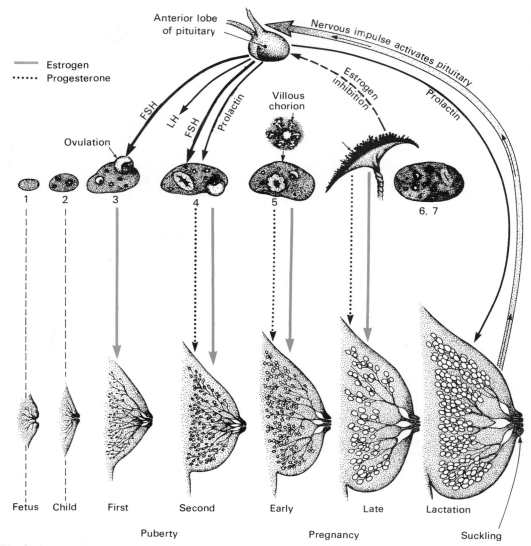

Fig. 31-9. Hormonal and neural control of lactation. (From G. B. Moment: *General Zoology,* 2nd ed. Houghton Mifflin Co., Boston, 1967.)

the precocial species, where there are many offspring and a high mortality, to a state more like that of the songbirds, where there are very few offspring but where the parents give their young prolonged care with consequent lower mortality.

GESTATION AND BIRTH

Gestation

The period of growth of the embryo in the uterus is known as the period of **gestation** or pregnancy. Ever since the time of Aristotle, zoologists have tried to find some firm correlations between the size of the animal, the number of young in a litter, and the length of the gestation period. In a general way, the larger the animal, the longer the period of gestation (see Table 31-1).

Parturition

The factor which triggers the uterine contractions which expel the infant are still unknown.

The pain involved in this process varies greatly, depending on many factors both physical and psychological. The first stage of labor is the period during which the os or mouth of the uterus becomes dilated. The second stage is the descent and expulsion of the infant and the third stage is the period following the birth of the infant and before the expulsion of the placenta and membranes. The later stages of parturition are always times of potential danger, but many highly civilized people from Mme. Curie to an increasing number of present-day women have found no anesthetic necessary or desirable. Apparently in the final phases there is hypnotic numbness. Like other branches of medicine, obstetrics is a blend of art and science requiring skilled judgment.

LACTATION

The process of lactation, or milk production and secretion, is under endocrine control. The rudiments of the mammary glands are laid down in embryos of both sexes. In females at puberty the estrogen from the ovarian follicle cells stimulates the growth of the milk ducts and hence the general enlargement of the mammary glands. During maturity, and especially during pregnancy, progesterone from the corpus luteum stimulates the growth of alveoli, which are rounded glands at the tips of the branches of the milk ducts. After the birth, the maternal pituitary secretes a new hormone, **prolactin**. This lactogenic hormone, first discovered in pigeons, causes the actual secretion of milk. Its production seems to be due to the sudden fall in level of estrogen caused by the loss of the placenta, which has been secreting large quantities of both estrogen and progesterone. Estrogen has been used clinically to inhibit milk secretion. Mammals are not all alike in this respect, for small amounts of estrone increase milk production in cows and ewes.

Many secondary factors also govern lactation. Lack of the adrenal cortex or of thyroxin or insulin inhibits the mammary glands. Lack of an abundant supply of water in the diet also greatly diminishes milk secretion. Last, the mechanical stimulus of sucking stimulates the hypothalamic region of the brain which, in turn, results in greater production of oxytocin by the pituitary. This hormone stimulates contraction of muscles within the mammary gland which squeeze milk to the outside. This fact can be effectively demonstrated by injecting oxytocin into an anesthetized lactating animal. Hormones from the adrenals, thyroid, and posterior lobe of the pituitary also influence this long familiar but still incompletely understood process.

USEFUL REFERENCES

Arey, L. B. *Developmental Anatomy,* 7th ed. W. B. Saunders Co., Philadelphia, 1965.

Karmel, M. *Thank You, Dr. Lamaze: A Mother's Experience in Painless Childbirth.* J. B. Lippincott Co., Philadelphia, 1959.

Millen, J. W. *Nutritional Basis of Reproduction.* Charles C Thomas, Publisher, Springfield, Ill., 1962.

Nalbandov, A. V. *Reproductive Physiology; Comparative Reproductive Physiology of Domestic Animals, Laboratory Animals, and Man.* W. H. Freeman and Co., San Francisco, 1965.

Pincus, G. *The Control of Fertility.* Academic Press, New York, 1965.

The Nervous System

32

The nervous system presents problems of a depth and complexity not found elsewhere in the living world. Of all the objects known to science, none even remotely approaches the brain in intricacy of structure; and within it lies that ultimate mystery, consciousness. Long after the problem of how the genetic code in DNA transforms an egg cell into an organism is solved, the problems presented by the nervous system will remain to challenge scientist and philosopher alike. Anyone eager to explore areas beyond the fringe of knowledge—for example, "psychokinesis," the direct action of mind on matter—should remember that whether or not thought can directly influence the way dice fall, conscious purposes and physical actions are united within the central nervous system in some utterly unknown way. Answers to these deepset questions are for future centuries. Meanwhile, we can look at the nervous systems found on this planet and learn something about the ways in which they work. Much actually has been learned already. A beginning has been made even on the biochemical basis of memory.

NEURONS, THE BASIC UNITS

Although the nervous systems found in higher animals—an octopus, a man, or a honey bee—are very different in gross organization, the unit in every case is the same: the single nerve cell or **neuron**. Neurons occur in a great variety of forms but all are characterized by thread-like cytoplasmic extensions, often called **processes**. Some are elaborately branched, others are unbranched; some are microscopic, others are over a meter long. These are the units

which compose both the simplest and the most intricate wiring patterns of animal nervous systems from jellyfish to man. The telephone and computer analogies to the nervous system are useful and valid, as far as they go. What must be remembered is that neurons are more complex than the component parts of a computer and that the versatility and range of abilities of a complex nervous system not only exceed those of a computer but transcend them into new dimensions. The arguments in favor of the continued presence of astronauts in the space program are based on this greater range of capabilities and adaptability of the human nervous system compared with computerized packages of instruments.

In this day of team research, it is worth taking time to note that the **neuron theory**, which was the basic scientific breakthrough establishing the foundation of modern knowledge of neural function, was the work of three loners. Wilhelm His in Switzerland and Santiago Ramon y Cajal in Spain worked in relative isolation from the bustling laboratories of Europe at the turn of the century. Both came to the conclusion that all nervous systems, including brains, are actually built up during embryonic stages by individual cells which remain the functional units. Both worked with preserved, sectioned, and stained tissues. This is still the method used today to investigate the location of control centers within the brain, even though modern staining methods will reveal neural hormones and enzymes, completely unknown to these pioneers. Few accepted the ideas of His and Ramon y Cajal until another loner, a young American named Ross Harrison, demonstrated that individual nerve

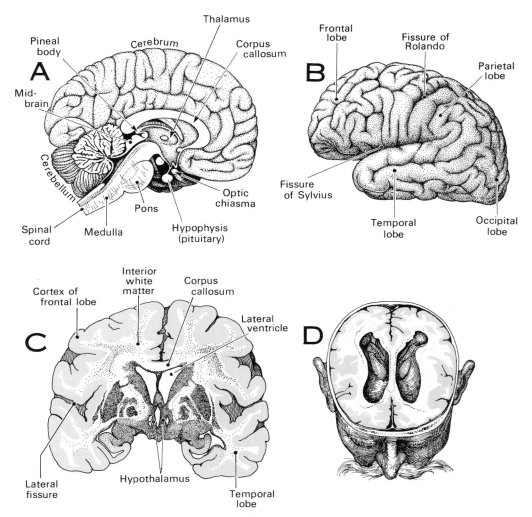

Fig. 32-1. The human brain. A, mid-sagittal section; B, view of left side; C, cross section; D, coronal section. The gray matter composed of the bodies of nerve cells is colored green.

fibers can be observed growing from single living cells cultured in a glass dish. This famous observation was the beginning of animal tissue culture *in vitro.*

Each neuron consists of a cell body, the nucleus with its surrounding cytoplasm, a region often called the **perikaryon**, and from one to numerous filamentous extensions. Within the cytoplasm are numerous elongate granules called **Nissl bodies**, which are now known to contain RNA. The precise function of this RNA is uncertain. The only function ever suggested for RNA beyond protein synthesis is some role in memory, but this is far from proven and would seem unlikely, although not impossible, in the case of motor neurons. It

has been noted that when poliomyelitis virus attacks the nervous system, it characteristically causes disappearance of RNA in the large motor (ventral horn) cells of the spinal cord. It is the destruction of some or all of these cells which results in paralysis and muscular atrophy.

The internal structure of nerve cells as revealed by recent studies with the electron microscope is remarkably similar in organisms from coelenterates like hydra and the jellyfish to vertebrates such as man. In addition to the abundant RNA in the cytoplasm surrounding the nucleus, there are numerous mitochondria and several areas of endoplasmic reticulum with ribosomes. In other words, both an energy source, in the mitochondria, and the machinery

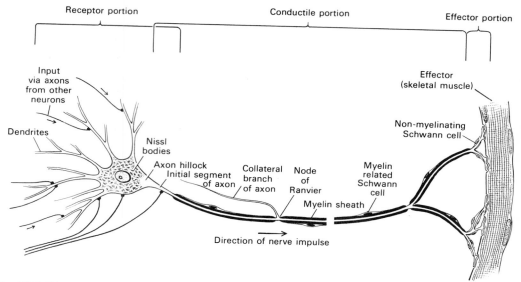

Fig. 32-2. Diagram of a single motor neuron. The cell body with the nucleus is at the left. (From *Bailey's Textbook of Histology*, 16th ed., ed. by W. M. Copenhauer *et al.* The Williams & Wilkins Co., Baltimore, 1971.)

for making new protein, in the Nissl bodies and the ribosomes, are available. In addition, there are usually various granules, sometimes containing glycogen, and vesicles. The structure which carries the actual neural disturbance or impulse is the cytoplasmic membrane that surrounds both the cell body and the longer or shorter dendrites and axons.

A nerve fiber carrying impulses toward the cell body of a neuron is a **dendrite**; a fiber conducting impulses away from the cell body is an **axon**. Each neuron usually has several short dendrites, often extensively branched, and one axon which may or may not be branched.

Despite their almost bewildering variety of form, nerve cells fall into three great classes. **Motor** or **efferent neurons** conduct impulses outward, away from the central nervous system to effector end-organs, muscles or glands. (Note that both "efferent" and "exit" begin with *e*.) **Sensory** or **afferent neurons** conduct impulses toward the central nervous system from free sensory nerve ends or from nerve endings in complex sense organs. **Association, internuncial,** or **interneurons** (all three terms are in common use), conduct impulses from one neuron to others within the central nervous system or some of its ganglia.

The visible nerves, readily seen on dissection looking like silvery-white cords, are essentially cables composed of a few dozen to hundreds or even thousands of individual nerve fibers, *i.e.*, axons or dendrites. In the case of nerves to the hands and feet, the axons of the motor neurons and the dendrites of the sensory neurons are very long because the cell bodies are located in or close to the spinal cord. The receptor or effector ends may be two meters away in an adult giraffe.

Most of the cranial nerves to and from the brain and the spinal nerves coming off the spinal cord are myelinated. That is, each nerve fiber is covered with an insulating **myelin sheath** made of a series of **Schwann cells** that have wrapped themselves around the nerve fiber. It is the myelin, composed of fatty material, which gives nerves their characteristic glistening appearance. The outermost membrane of the myelin sheath is conspicuous under a light microscope and is known as the **neurolemma**. At intervals the myelin sheath is interrupted by exposed zones called the **nodes of Ranvier**. Myelinated nerve fibers have been found to carry impulses faster than unmyelinated ones. Among the nerve cells, both around and between cell bodies, axons, and dendrites, are numerous supporting cells called **glia cells**. The interneurons or association cells

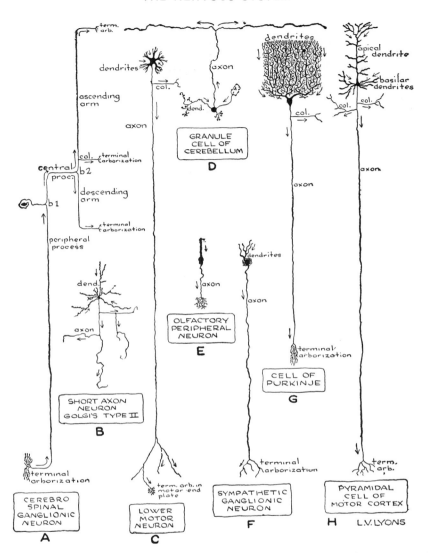

Fig. 32-3. Types of nerve cells. A, sensory or afferent neuron; C and H, motor or efferent neurons. The Purkinje cells are found in the cerebellum. (From *Bailey's Textbook of Histology,* 16th ed., ed. by W. M. Copenhauer *et al.* The Williams & Wilkins Co., Baltimore, 1971.)

occur in many forms. Compare, for example, a pyramidal cell from the cerebral cortex with a Purkinje cell from the cerebellum.

The Synapse

A **synapse** is the spot where the terminations of an axon transmit an impulse to another neuron. Synapses may be located at the tips of the dendrites of the receiving cell or directly on the cell body. The transmitting ends of the efferent fibers swell into tiny synaptic knobs or **end bulbs** containing several small mitochondria and several dozen small synaptic vesicles which seem to contain the transmitter chemicals. These end bulbs do not actually come in contact with the receiving neuron, but are separated from it by a cleft visible with a dissecting microscope. A very important fact about large motor nerve cells is that they

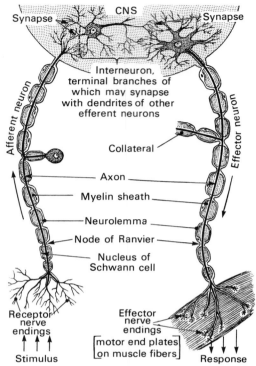

Fig. 32-4. Diagram of a theoretical reflex arc with motor and sensory nerves showing myelination surrounding the axon and absence of the myelin sheath at the nodes of Ranvier.

receive impulses from hundreds of end bulbs of many different sensory and internuncial nerve fibers which form synapses directly against their cell bodies. This anatomical arrangement has two functional results. Impulses arriving from many different sources may cause the same motor nerve cell to fire off a message over what is clearly the **final common pathway** to a given muscle. A second result is that if several weak stimuli arrive from several different sources, none of sufficient strength by itself to cause the motor nerve to discharge, their **summation** may attain the necessary threshold for a motor nerve response.

THE VERTEBRATE NERVOUS SYSTEM

The adult vertebrate system consists of two main subsystems. The **central nervous system** (CNS) is made up of the **brain** and the **spinal cord**. The famous **gray matter** which is located

as a surface layer of the brain is composed of cell bodies of neurons; the **white matter** within the brain is made up of myelinated nerve fibers. In the spinal cord the fiber tracts are peripheral, and the gray matter is centrally placed so that it resembles a butterfly (roughly) when the spinal cord is cut transversely.

The **peripheral nervous system** consists of the nerves which extend from the CNS to the various parts of the body. It should always be remembered that the central and peripheral nervous systems form one single, closely integrated system. It is convenient in analyzing structure and function to regard them separately and so this primarily verbal and methodological distinction has come into use. The peripheral nervous system of vertebrates is itself separable into two divisions: the **somatic nervous system** which innervates the skeletal muscles, the skin, the major sense organs, etc., and is generally under the control of the will; and the **autonomic nervous system** which innervates the smooth muscles, especially of the viscera, the heart and blood vessels, etc., and is not under conscious control. (If it be objected that the concept of consciousness is not a scientific concept, the authors will not argue. One can avoid using the term, but at some cost in convenience and perhaps honesty.) The autonomic system in turn is composed of two divisions: the **sympathetic** and the **parasympathetic**.

The Somatic Peripheral Nervous System

The peripheral nerves leave the central nervous system in pairs, one member of each pair going to the right and one to the left side. Nerves leaving the brain are known as **cranial nerves**; those leaving the spinal cord as **spinal nerves**.

In mammals there are 12 pairs of cranial nerves numbered consecutively starting from the most anterior and continuing back to the beginning of the spinal cord. They also bear names sometimes indicative of their function and sometimes of their anatomical relationships.

The pairs of spinal nerves vary in number: there are several dozen in some fish, 10 in a frog, and 31 in a man. In mammals, they are

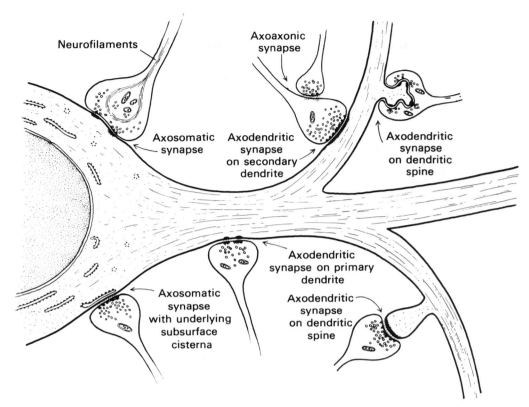

Fig. 32-5. Diagram showing several positions and types of synapses. Note that most are between an axon and a dendrite but that some of the end bulbs impinge directly on the receiving cell body. Both mitochondria and secretory granules are visible in the end bulbs. (From *Bailey's Textbook of Histology,* 16th ed., ed. by W. M. Copenhauer *et al.* The Williams & Wilkins Co., Baltimore, 1971.)

commonly grouped as **cervical, thoracic, lumbar, pelvic** or **sacral,** and **caudal** spinal nerves.

Regular segmentation of the nerves is a conspicuous feature of the human as well as of the shark's nervous system. Each nerve has a double root by which it connects to the spinal cord. There is a dorsal (posterior in an erect man) sensory root with a ganglion and a ventral (or anterior) unganglionated motor root. The nuclei and cell bodies of the sensory nerves constitute these ganglia. The nuclei and cell bodies of the motor nerves lie within the spinal cord, specifically within the lower wing of the "butterfly," often called in medicine the **anterior horn.** The spinal nerves innervate the muscles of the trunk, limbs, and skin.

The separate functions of the dorsal and ventral roots were discovered in the early years of the last century by Charles Bell in England

and Francois Magendie in France. Actually, Galen, the physician of the Roman Emperor, Marcus Aurelius, taught that when nerves to muscles are injured, movement is abolished, and that cutting nerves to the skin eliminates feeling. How he knew this is problematical, perhaps from observing the results of sword wounds. It was not until a millenium and a half later that Bell sought to determine by experiment whether there are two kinds of nerves, as Galen had indicated. Bell severed the ventral roots of spinal nerves in cats and observed paralysis, which demonstrated that the ventral roots are motor or efferent. A decade later Magendie showed that cutting the dorsal roots abolished sensitivity and so established what has been termed ever since the Bell-Magendie law: dorsal roots are sensory; ventral roots are motor.

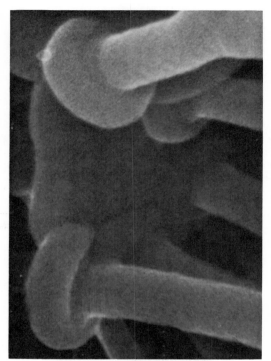

Fig. 32-6. Synaptic end bulbs, ×60,000, as seen with a scanning electron microscope. These are from a California marine snail but are closely similar to those in vertebrates. (Kindness of E. R. Lewis.)

The Autonomic Peripheral Nervous System

This system presides over organs and responses which are not under the direct control of the will—tear glands, heart, lungs, genitalia, gastrointestinal tract, blushing, sweating, and the like. It consists of two divisions, sympathetic and parasympathetic. The two are antagonists and each end-organ is innervated by both. Where one excites, the other inhibits.

Both parasympathetic and sympathetic nerves consist of two sets of neurons, **preganglionic** ones that have their cell bodies within the central nervous system and **postganglionic** ones that have their cell bodies in ganglia outside the central nervous system.

The **sympathetic division** is composed of nerves having preganglionic neurons with cell bodies within the gray matter of the thoracic and lumbar regions of the spinal cord, and

axons which pass out of the cord in the ventral motor roots of spinal nerves and then reach one of the chains of **sympathetic ganglia** by what is called the white ramus or branch of a spinal nerve. In a ganglion, the preganglionic fiber may (1) synapse with a postganglionic fiber; (2) continue up or down the chain of sympathetic ganglia; or (3) continue out into the solar plexus or some other visceral ganglion and there synapse with a postganglionic fiber. The postganglionic fibers then extend for some distance, often considerable, to reach their end-organ.

Preganglionic sympathetic fibers release a chemical, acetylcholine, and are hence called **cholinergic.** Acetylcholine stimulates the appropriate postganglionic neurons. Postganglionic sympathetic fibers do not release acetylcholine at their ends but release epinephrine (adrenaline) and therefore are called **adrenergic.** Their stimulation may produce dilation of the pupil, acceleration of the heartbeat, inhibition of gastrointestinal mobility, constriction of the pyloric, ileocolic, and anal sphincters, relaxation of the urinary bladder, secretion of sweat, and erection of hair. Acetylcholine is very widely distributed throughout the animal kingdom. So far epinephrine appears to be confined to vertebrates and earthworms.

The **parasympathetic division** consists of nerves having preganglionic neurons with cell bodies, some located within the brain and others within the pelvic region of the spinal cord. The preganglionic fibers travel all the way to the end-organ—heart, stomach, colon, etc. Within their end-organ they synapse with very short postganglionic fibers which terminate in the end-organ. The parasympathetic nerves to the eye and salivary glands are exceptions, for they do enter ganglia and there synapse with postganglionic fibers which travel some distance to their end-organs. Both pre- and postganglionic, parasympathetic nerves, like preganglionic sympathetic fibers, release acetylcholine and are termed **cholinergic.** The effects of the parasympathetic nerves are opposite to those of the sympathetic nerves.

Ample evidence from many sources attests to the effectiveness of the relationship of the autonomic system and the highest centers in the brain. For example, the response of blushing can be induced by rather sophisticated causes of embarrassment.

Twelve pairs of nerves arise directly from the undersurface of the Brain to supply Head and Neck and most of the viscera.

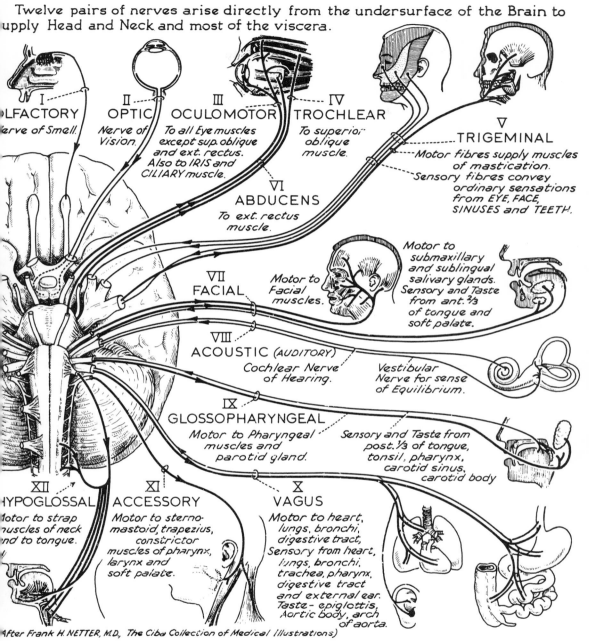

I OLFACTORY
Nerve of Smell.

II OPTIC
Nerve of Vision.

III OCULOMOTOR
To all Eye muscles except sup. oblique and ext. rectus. Also to IRIS and CILIARY muscle.

IV TROCHLEAR
To superior oblique muscle.

V TRIGEMINAL
Motor fibres supply muscles of mastication. Sensory fibres convey ordinary sensations from EYE, FACE, SINUSES and TEETH.

VI ABDUCENS
To ext. rectus muscle.

VII FACIAL
Motor to Facial muscles.
Motor to submaxillary and sublingual salivary glands. Sensory and Taste from ant. 2/3 of tongue and soft palate.

VIII ACOUSTIC (AUDITORY)
Cochlear Nerve of Hearing.
Vestibular Nerve for sense of Equilibrium.

IX GLOSSOPHARYNGEAL
Motor to Pharyngeal muscles and parotid gland.
Sensory and Taste from post. 1/3 of tongue, tonsil, pharynx, carotid sinus, carotid body

XII HYPOGLOSSAL
Motor to strap muscles of neck and to tongue.

XI ACCESSORY
Motor to sterno-mastoid, trapezius, constrictor muscles of pharynx, larynx and soft palate.

X VAGUS
Motor to heart, lungs, bronchi, digestive tract, Sensory from heart, lungs, bronchi, trachea, pharynx, digestive tract and external ear. Taste - epiglottis, Aortic body, arch of aorta.

After Frank H. NETTER, M.D., The Ciba Collection of Medical Illustrations)

Fig. 32-7. Cranial nerves and the structures they innervate. The brain is viewed from the underside; cerebral hemispheres are colored green. (From A. B. McNaught and R. Callander: *Illustrated Physiology,* 2nd ed. The Williams & Wilkins Co., Baltimore, 1970.)

THE NERVOUS IMPULSE

A question of great human interest has always been: "What is a nervous impulse?"

Only a little over a century ago many people believed that the nature of the impulse which carries messages along a nerve could never be known. Whenever it is asserted that a particular

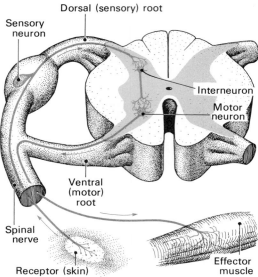

Fig. 32-8. Cross section of a spinal cord, showing a pair of spinal nerves each with its dorsal and ventral roots. The gray matter where the cell nuclei are located is colored green.

problem can never be solved, the case of the nervous impulse is worth remembering. The most eminent physiologist of the first half of the 19th century, Johannes Müller, maintained that even the speed of a nervous impulse was unknowable. Yet the print was hardly dry on Müller's assertion when one of his own students, Hermann Helmholtz, showed that this could indeed be discovered and measured.

Helmholtz's method was simplicity itself. He used the frog which Müller had reintroduced into physiology (Bell and Magendie had used mammals) and exposed the great sciatic nerve which innervates the gastrocnemius muscle in the calf of the leg. He stimulated the nerve at two points, one fairly close to the muscle, the second at a carefully measured distance along the nerve from the first. By very accurately measuring the time elapsing between application of the stimulus and contraction of the muscle, he discovered that the greater the distance of nerve over which the impulse had to travel, the longer the time between the stimulus and the contraction. Simple arithmetic gives the speed of the nervous impulse. It varies somewhat from species to species and from nerve to nerve. Nerves lacking the myelin sheath con-

duct impulses very slowly. In mammalian nerves at body temperature, the speed is about 100 meters (325 feet) per second, faster than a man can run but far slower than sound (about 300 meters per second).

New facts and new theories about the action of nerves have been accumulating ever since. These fall logically into three groups: what happens to a nerve cell at the time of stimulation; transmission of the nervous impulse along the nerve, and the means by which a nerve delivers its message either across a synapse or to an end-organ such as a muscle.

Irritability is a common and presumably universal attribute of protoplasm. In nerves, this property has become specialized and heightened. A nervous impulse is a wave of **electronegativity** easily detected with a student-type galvanometer and a pair of electrodes. This wave of negativity is apparently not a mere accompaniment of the nervous impulse but is itself part of the driving mechanism. With the wave of negativity goes a sudden and profound **change in permeability** of the cell membrane. A nervous impulse is an "all-or-none" affair. Either a neuron fires off or it doesn't. The only requirement is that the exciting stimulus—whether mechanical, chemical, or electrical—must attain a certain intensity, usually called the threshold. The wave of negativity neither increases nor decreases as it moves. The action at every point along the fiber is a local action.

The old analogy of the powder train is a good one here. In fact a spectacular demonstration can be made by pouring a very narrow path of smokeless gunpowder along the top of a laboratory table and then carefully igniting one end. The gunpowder train analogy illustrates the fact that the impulse does not vary with the strength of the initial stimulus, provided only that the stimulus was sufficient to ignite the reaction. If one or more branches are made in the powder path, the initial spark may result in two, four, or many burning powder trains. This illustrates the **principle of amplification.** Because nerve action is local, total response depends on branching neural connections, not incident energy at a sense organ.

Modern ideas about the nature of this self-propagating event, the nervous impulse, are based on the theory of Julius Bernstein at the turn of the century. He pointed out that the wave of negativity could be explained on the following basis. If the outside of the nerve cell and its fiber are positive and the inside negative, and if the cell membrane is semipermeable, keeping the more negatively charged ions inside and the positively charged ones outside, then a nervous impulse could be a wave of breakdown in selective permeability permitting a relatively free passage of ions through the membrane so the membrane potential disappears.

To test the theory that the outside of a nerve fiber is positive and the inside negative, a fiber large enough to place one electrode actually inside the fiber and the other outside is required. The best place to find such giant nerve fibers is in the squid, where they're on the inner side of the mantle, a muscular body wall loosely surrounding the animal. No dissection is necessary, as the fibers lie on the surface.

The diameter of a single fiber is as large as an entire vertebrate nerve made up of hundreds of nerve fibers. Using these very large fibers, A. L. Hodgkin and A. Huxley at the Marine Biological Laboratory in Plymouth, England, were able to show, 40 years after Bernstein had proposed the electrical depolarization theory of the nervous impulse, that it was in fact correct. The

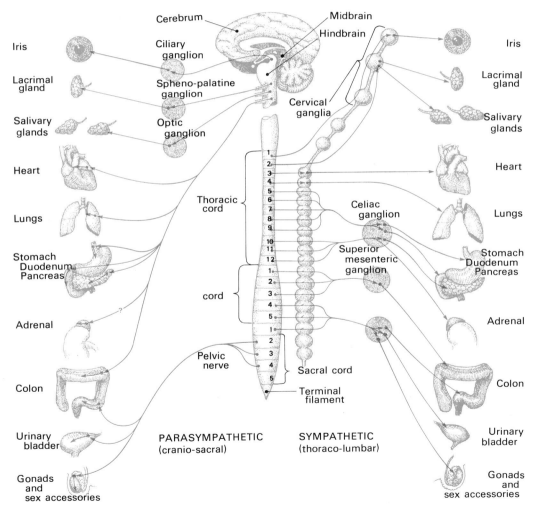

Fig. 32-9. The human autonomic nervous system showing the parasympathetic or cranio-sacral division on the left and the sympathetic or thoraco-lumbar on the right.

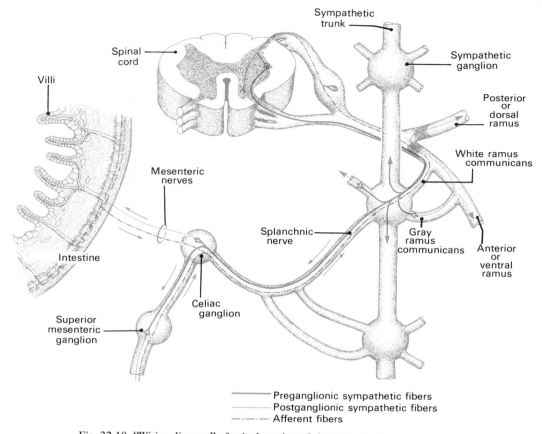

Fig. 32-10. "Wiring diagram" of spinal portion of the sympathetic nervous system.

electrical potential between inside and outside falls from about +70 millivolts outside, to zero and then to a reversed polarity, with the inside actually positive sometimes by as much as 50 millivolts before the normal resting condition is restored. The characteristics of the electrical change as an impulse passes along a nerve fiber can be handsomely observed if electrodes on a nerve are connected to an amplifier and then to an oscilloscope (an instrument similar to a television tube.)

The explanation of how this wave of negativity is propagated has not been easy to obtain, and all the answers are not in yet. In a resting cell there is about 20 times as high a concentration of potassium ions on the inside as on the outside. The concentrations of sodium and chloride ions are far higher outside than inside. The wave of negativity induces a transient change in permeability which permits positive sodium ions to enter. This further depresses the potential so that the wave passes like fire along a fuse. There is some evidence that acetylcholine is important in this process. Metabolic energy is required to restore the resting potential and the marked differences in ion concentration. This can be shown by the use of metabolic poisons. The mechanism which maintains the higher external concentration of sodium by the expenditure of metabolic energy is referred to as the **sodium pump** (or sodium-potassium pump).

SYNAPTIC TRANSMISSION OF NERVOUS IMPULSES

The properties of synapses, where an impulse passes from one neuron to another, and of myoneural junctions, where a nervous impulse passes from a nerve fiber to a muscle, are of the utmost importance. Some of the most powerful poisons known exert their deadly effects at these points. Among these poisons is curare, which South American Indians use on the tips of their arrowheads, the modern anticholinesterase type of nerve war gas, and some of the most dangerous anti-insect sprays. Drugs of various sorts exert their effects here also. It is even thought by some that changes in synaptic connections are at the basis of learning. In the everyday functioning of the nervous system, synapses act as valves which transmit in one direction only, and of course the nervous system could not function without its thousands of billions of connecting points.

What happens when an impulse reaches a synapse at the end of its run? Axons terminate in one or a cluster of small **synaptic knobs** (or end bulbs) which press within about 20 nanometers (10^{-9} m., the kind of gap only an electron microscope can see) of a dendrite or nerve cell body. Within the knob, the vesicles evidently discharge a transmitter substance, either acetylcholine or epinephrine, when the impulse arrives. If the nerve fiber is a stimulating fiber, the transmitter substance lowers the membrane potential of the receiving or postsynaptic cell on the far side of the gap. The lowered potential of that cell is called by neurophysiologists and some psychologists the **excitatory postsynaptic potential** or **EPSP**. In some instances the discharge of a single end knob can lower the postsynaptic potential to the approximately 50 millivolts necessary to trigger a wave of depolarization, *i.e.*, a wave of electronegativity commonly called a nervous impulse. If the discharge of a single end knob is not sufficient to produce an EPSP as low as 50 millivolts, then there will be no impulse shot off unless stimuli from several other nerve fibers reach the cell membrane. If enough end knobs discharge transmitter substance within a short enough interval of time, the excitatory postsynaptic potential will be low enough to initiate a nervous impulse. When several nerve fibers act together, it is called **spatial summation**. If the EPSP is produced by repeated discharge of a single end knob, it is called **temporal summation**. These types may occur together.

After the discharge of acetylcholine at a synapse, a special enzyme, acetylcholinesterase, destroys the acetylcholine and permits the receiving nerve to restore its membrane potential and thus be ready to respond to another stimulus, *i.e.*, another dose of acetylcholine.

If the nerve carrying the initial impulse is an inhibitory nerve, like the vagus branch to the

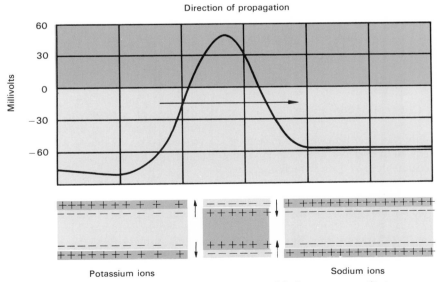

Fig. 32-11. Diagram of the electrical charges which accompany and in large part constitute a nervous impulse. The electropositive regions are shown in darker green than the electronegative.

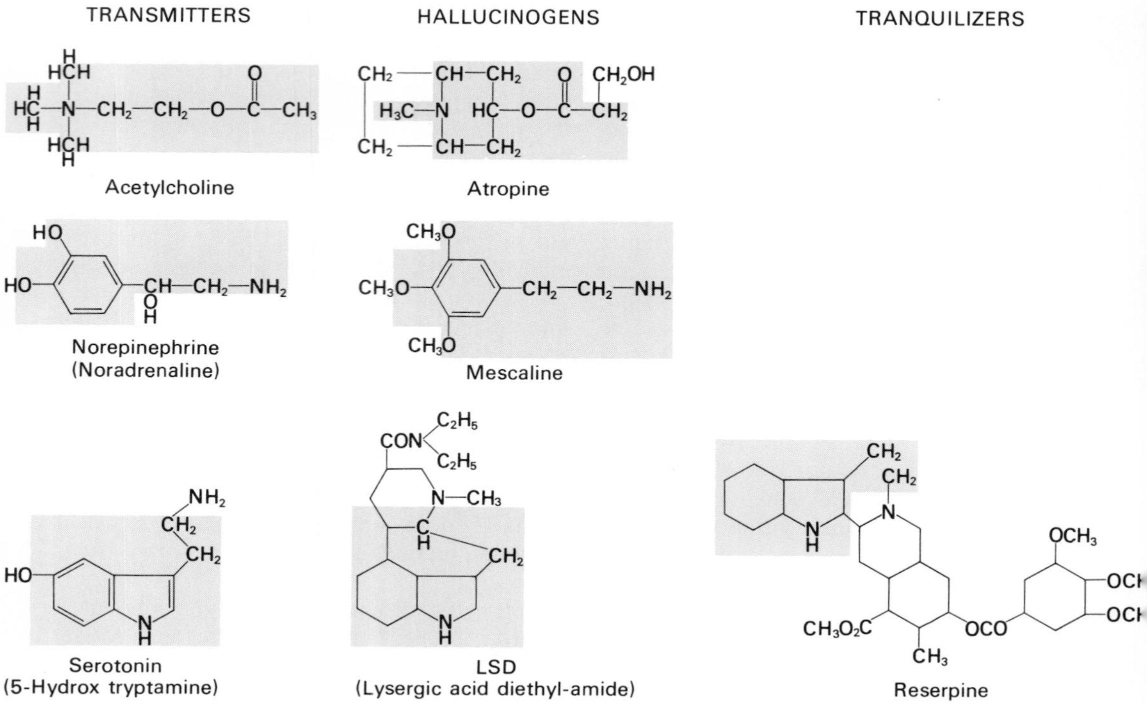

TRANSMITTERS   HALLUCINOGENS   TRANQUILIZERS

Fig. 32-12. Part of the evidence that hallucinogens and tranquilizers act at the synapses. Similarities of molecular structure are shown in green. (Read from left to right.)

heart, the acetylcholine does not lower the membrane potential of the receiving neuron to any appreciable extent. It may even raise it, causing the membrane to be hyperpolarized. In some way an inhibitory message over a synapse stabilizes the membrane potential and either prevents or makes more difficult the triggering of an impulse. This new potential is called the **inhibitory postsynaptic potential** or **IPSP**.

Chemical Control of Synapses

Much study has been devoted to effects of biochemically active agents on synaptic transmission. Presumably, any agent which interferes with acetylcholine, epinephrine, norepinephrine or any other naturally occurring transmitter substance would interfere with the functioning of the nervous system. Such agents might facilitate or block transmission by combining with a natural transmitter or acting on enzymes necessary for its removal or by affecting the receiving membrane. Lysergic acid diethylamide (LSD), for example, mimics serotonin (5-

hydroxytrypamine), which is a derivative of the amino acid tryptophan. Serotonin in very small amounts is found in neural tissue of vertebrates, mollusks, and arthropods. It is the poison in the deadly saliva of an octopus, in the sting of a hornet, and in the venom of some spiders. Mental diseases, especially schizophrenia, have been attributed to disturbances of serotonin metabolism. Anticholinesterases are so effective that a few drops in the eye can kill a man. Knowledge of such potent and often psychopharmacologically active chemicals can furnish a two-edged sword able to wreak havoc or bring healing into human lives.

REFLEXES

Modern ideas about the action of the nervous system began with the French mathematician and philosopher René Descartes in the 17th century. He proposed that the unit of action of the nervous system is the **nervous reflex**. By this he meant that a stimulus, such as light striking the eyes, would set off an impulse to the brain which would be reflected back from the brain through nerves to muscles in the arm or some other organs.

One of the first men to find experimental confirmation for such a view was a versatile 18th-century clergyman and amateur biologist, Stephen Hales, who also discovered blood pressure. Briefly, Hales found that one can destroy a frog's brain, by decapitation if necessary, and still not destroy responses to stimuli. A frog so treated is known as a **spinal animal**. If a small square of paper soaked in vinegar is placed on the back of such a frog, one of the hind legs will flick it off with great accuracy. Of course there is little flexibility in the reaction, for all "originality" is lost with the brain. If the spinal cord is destroyed by pushing a wire down inside the backbone, these spinal reflexes cease permanently. Here then is clear proof that such responses can be mediated solely by the spinal cord, without any assistance from the brain.

Spinal men have occasionally been produced by war or other accidents. In the case of one army lieutenant, a piece of shrapnel completely severed his spinal cord between the fifth and sixth ribs. He retained control of his arms, but his legs were completely without sensation, nor was he able to move them at will. However, if a toe was pinched, his leg would bend at the knee and pull the foot back.

Since Hales' day much has been learned about reflexes. A number of reflexes concerned with breathing and with the heartbeat have been discussed in previous chapters. Salivation at the sight or odor of food is a famous reflex in dogs studied by Ivan Pavlov, but common in man as well. An easily observed reflex is the pupillary light reflex, although the neural pathways involved have turned out to be more complex than in a spinal reflex. Stand before a window and cover one eye completely for about two minutes. Then look in a mirror as your readmit light to the eye. You will observe that the pupil becomes smaller. This is due to a contraction of the sphincter muscle in the iris.

Although reflex action is simple to demonstrate in the pupil of your eye or with a spinal frog, the neural mechanisms underlying reflexes are complex. Think for a moment of the intricate interconnections and delicate adjustments necessary to place the toes of a frog's foot on a precise spot anywhere on the frog's back! Moreover, in an intact frog, the beautifully coordinated behavior of those long legs is not only directed by reflexes within the spinal cord but it is also under the overriding control of the brain. The simple picture of a reflex as an arc composed of a cutaneous sense organ, a sensory nerve passing into the spinal cord via a dorsal afferent root, making synaptic connections with a single interneuron within the cord which in turn stimulates a motor nerve leaving by a ventral efferent root and ultimately causing a muscle to contract, is a greatly over simplified picture; in fact, a fairy tale for children.

Understanding the reflex basis of action within the nervous system is well begun by reading Sir Charles Sherrington's classic *Integrative Action of the Nervous System*. Although originally published in the early part of this century, it has often been reprinted because this work brought knowledge of reflex action to a culmination which has not yet been appreciably altered, even though much more has been learned.

There are a number of rules, simple to state, which govern reflex action. To produce a reflex response, a stimulus must first attain a certain intensity; unless this **threshold** is achieved, nothing happens visibly. The threshold can be reached by a variety of means. The stimulus, whether a beam of light or a pinch on the toe, can be made more intense. A series of subthreshold stimuli may be applied until the animal responds. This phenomenon is called **temporal summation. Spatial summation** is also possible when an animal is given several stimuli, each subthreshold by itself, in different parts of the body. A most important point to remember is that the relationship between the intensity of the stimulus and the size of the response is a very indirect one.

Locomotion and Anti-Gravity Reflexes

Basic to locomotion, whether of legs or wings, hands or jaws, is the requirement that extensor muscles must relax when flexor muscles contract. This is just as true of a man as of an earthworm. Were the sets of antagonistic muscles not to act in such a coordinated manner, a man would try to stand up and sit

Dorsal sensory root

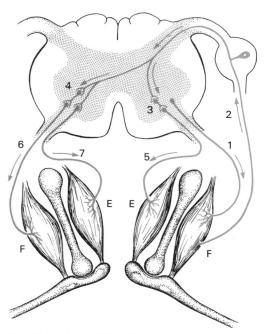

Fig. 32-13. A simplified diagram of neural pathways in a simple leg reflex. E denotes an extensor and F a flexor muscle. A motor (efferent) impulse via nerve 1 causes the flexor in the right leg to contract. This contraction stimulates a sensory "spindle" in the muscle to send an afferent impulse up sensory nerve 2 to the spinal cord, where it branches and stimulates very short interneurons (Renshaw neurons). These, in turn, inhibit nerves 5 and 6 to the right extensor and the left flexor, preventing them from contracting. This is known as reciprocal inhibition. Note also that the nerve which stimulates the Renshaw neurons may also stimulate the left extensor to contract.

down simultaneously. All this is rather obvious, including the need for smoothly coordinated contractions and relaxations to prevent animal actions from being more jerky than those figures in an old silent film.

What is not so evident is the way in which the nervous system is organized to produce the smooth perfection of the most commonplace acts: walking, swimming, flying, or merely moving the eyes around. The neurological process which underlies these motions and thus makes animal life possible is based on **reciprocal inhibition**, not infrequently called Sherrington's

inhibition. Briefly, reciprocal inhibition means that when a motor nerve sends an impulse to a muscle, causing it to contract, another impulse is sent to the antagonistic muscle inhibiting its contraction.

Closely integrated with these reflexes are the anti-gravity reflexes which maintain our posture. Sensory inputs coming from these reflexes and from various sensory endings in tendons and joints are important components of the proprioceptive system which keeps us continually informed about the position of the various parts of our bodies. It is an important system not only on land, but also in space travel.

THE BRAIN

The human brain has been called many things: "that great ravelled knot"; the "computer in the skull." And justifiably, because of all the structures known to man, the brain is the most complex, by several orders of magnitude. In fact, no computer remotely approaches the brain in complexity, much less in versatility. The number of neurons in the human brain has been calculated at about 12 billion. Vast numbers of them each receive branches from the axons of 10,000 other nerve cells. The number of possible combinations and permutations staggers the mind. Perhaps anyone familiar with Gödel's proof (that in mathematics there must always be an unproved axiom) may think it unfair to ask the brain to understand itself.

Brain Size and Intelligence

A great deal has been made of the relation of brain size to intelligence. In general, the larger the brain, the greater the potential intelligence. A large absolute size appears necessary to provide for complex associative activities, including feedback mechanisms of self-adjustment. However, absolute size must be corrected for total body size. An elephant or a whale has a larger brain than a man in absolute terms, but not in proportion to body weight. The great size of an elephant's brain is presumably due to the large number of neurons essential to the sending and receiving of signals to the enormous body mass, and does not represent an increase in the higher associative centers. At the other end of the size scale, the proportionality rule again breaks down. Insectivores like moles, with extremely modest intellectual abilities, possess brains larger than man's in proportion to total body size.

Actual brain size varies widely in the higher primates. In chimpanzees and gorillas the brain ranges from 325 to 650 cubic centimeters in volume. In fossil man-apes it ranged from 450 to 700 cc.; in fossil Java man 750 to 900 cc.; in Peking man, 800 to 1,200 cc.; and in Neanderthal man, 1,100 to 1,550 cc. Although his brow was low and his features decidedly simian, brain size in Neanderthal man falls easily within the normal range of variation that is found in contemporary human populations.

Many attempts have been made to discover correlations between brain size or convolutions, on the one hand, and degree of intelligence and achievement in man, on the other. Except for obvious deformities such as those in microcephalic idiots, these efforts have not succeeded. Most men of outstanding intellectual ability have had brains in the upper half of the frequency distribution curve for size, but there have been a surprising number with brains well below average. It is not difficult to believe that along with size there may well be other factors—a slightly different twist of metabolism, for instance—that determine potential intelligence. And actual achievement is very clearly the result of many factors in addition to intelligence.

General Brain Structure

The basic structure of the brain is the same in all vertebrates whether fish, frog, bird, or man, because it always develops as five swellings at the anterior end of the embryonic neural tube. The names of these five major parts are the same in all vertebrates, and so are the basic structures developed from each. The difference is in the degree to which different parts are developed in different vertebrates. The inner central fluid-filled cavity connects all parts of the brain and continues down the spinal cord. This is why a bubble of air carefully injected into the central canal of the spinal cord of a patient who is sitting up will move up and into the cavities, **ventricles** as they are called, of the

brain. Followed by X-rays such air bubbles can be used in the diagnosis and location of growths and other obstructions in the central nervous system.

The Cerebrum

The most anterior part of the brain develops two cerebral hemispheres, right and left. The roof or pallium of these hemispheres constitutes the cerebrum and is especially large in the mammals. In birds, the roof of the cerebral hemisphere is almost paper-thin. How then can birds learn so much? We now know that the characteristic song of most birds is somehow encoded in their brains. So also are the instructions about where and what kind of a nest to build, together with the appropriate construction methods. Perhaps even more remarkable, a small white-crowned sparrow carries within its skull navigational programming which will enable it to reach its own nesting grounds in southern Alaska even after being transported in a closed cage from its winter home in California to Maryland. How a bird's brain can be programmed for such complex behavior is an unanswered question.

Discovering the answers to questions like these, not just for birds but for all animals, is motivating much of the research concerning brain structure. Remember that it is in its organization, its cytoarchitecture, that the characteristic features lie which confer on it such remarkable powers. The difference between the brain of a fish and that of a fisherman, it should be recalled, does not lie primarily on the cellular level. The neurons of a jellyfish, a shark, and a dog are much alike. Far less does the difference lie on the molecular or atomic level. It is the design of the brain that is of utmost importance.

In primitive mammals like the insectivores (shrews, for example) and rodents, the surface or **cortex** of the brain is smooth, but in many others including the primates, the cortex is highly convoluted. In man a deep, more or less horizontal fissure, the **fissure of Sylvius**, separates the **temporal** lobe below from the **frontal** and **parietal** lobes above. The frontal lobe is separated from the parietal lobe by the vertical **fissure of Rolando** which runs upward from the fissure of Sylvius to the top of the

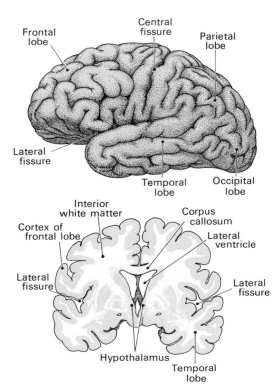

Fig. 32-14. Human brain in the side view and in cross section. The central fissure is also called the fissure of Rolando and the lateral fissure or sulcus the fissure of Sylvius.

cerebrum. The region at the posterior and lower end of each hemisphere is known as the **occipital lobe**.

If the cerebral hemispheres are cut in half from left to right, it can be seen that the cortex, or superficial zone, is gray, whereas the inner portions are white with rounded gray masses here and there. The grayness of the gray matter is due to the presence of several billion cell bodies of neurons. The white of the white matter is due to the myelin sheaths of the axons of neurons. It is a mistake to regard intellectual ability as resident solely in the gray matter. The intercommunication systems represented by the white matter are equally essential. The rounded gray masses within the lower part of the brain are relay and shunting centers of neurons and are called **basal ganglia** or **basal nuclei**. (This, of course, is to use the word "nucleus" in its primary meaning of a "kernel" or central region.) The four largest of the basal

ganglia make up the **corpus striatum**, which develops from the floor of the most anterior part of the brain, the telencephalon, and which constitutes the main bulk of the forebrain in birds.

The cerebral cortex of a higher mammal has six layers of cells each itself many cell layers thick. Most of the incoming fibers from other parts of the nervous system end in the fourth layer after branching extensively and making a wide variety of synaptic connections. It is now believed that a basic unit of cortical organization is a simple feedback loop. The evidence is both visual from stained sections and physiological from electrical studies. The function of all these reverberating circuits is unknown.

The white matter immediately below the cortex contains innumerable fibers. Some of these axons pass from the cortex down into the basal ganglia or directly into the spinal cord. Others connect the two cerebral hemispheres. It is these right-left communicating fibers which make up the **corpus callosum**. Still other bands of fibers run at right angles to the right-left fibers and connect anterior and posterior cortical areas.

The corpus callosum long stood as one of the outstanding puzzles of brain organization. It is easily the most massive fiber tract of the brain. Yet when it was completely cut in human patients in the course of surgery for tumors, no observable defects of any kind, subjective or

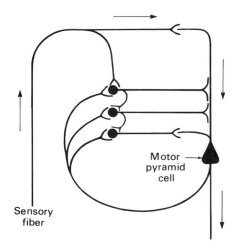

Fig. 32-15. A simple feedback circuit between cells of the cerebral cortex.

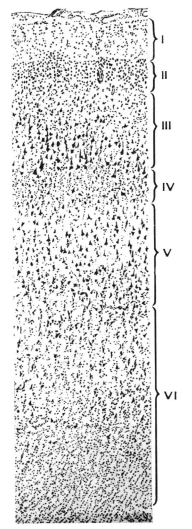

Fig. 32-16. The six layers of cells in the human cerebral cortex. (From *Bailey's Textbook of Histology,* 16th ed., ed. by W. M. Copenhauer *et al.* The Williams & Wilkins Co., Baltimore, 1971.)

objective, resulted. Cats or monkeys which have had the corpus callosum cut are virtually indistinguishable from intact animals in normal daily activities. But if the optic chiasma is cut as well as the corpus callosum, so that the right eye is connected only to the right side of the brain and the left eye only to the left side, the answer to the old puzzle can be found. If a cat or monkey so prepared is taught to make some discrimination using only its right eye—for

instance, press a round but not a square button to get food—it will be found when the right eye is covered and the left eye tested that the left side of the brain knows nothing about choosing the round button. If only the optic chiasma is cut while the callosum is left intact, then tricks learned with one eye are easily performed using the other. So it was learned that the function of the corpus callosum is to keep each side of the brain informed about what the other experiences. If something is learned in one hemisphere, the corpus callosum enables a duplicate **engram** or memory trace to be established in the other hemisphere.

By means of electrical stimulation, surgical removal, and the recording of small electrical changes on the brain surface that accompany various activities, it has been possible to learn a fair amount about the general functions of different regions of the cerebrum. Injuries in the occipital lobe region result in impaired vision. There are important speech areas along the temporal lobe and on the lower part of the frontal lobe. Motor and sensory centers for the skeletal muscles are lined up along the fissure of Rolando, the motor centers on the anterior ridge or gyrus, the sensory centers along the posterior gyrus. Curiously enough, the body is represented upside down. If a point close to the top of the brain just anterior to the fissure of Rolando is stimulated electrically, the foot will twitch. If points lower down are stimulated, muscles in the legs, trunk, arms, neck, eyelids, and other parts of the head will move, in that order. Most of the frontal lobes represent a "silent" area of indefinite function, concerned with the higher associations.

Other Brain Areas

The structure of the second region of the brain, the **diencephalon**, is much the same in all vertebrates. Ventrally, there is the optic chiasma where the optic nerves enter the brain. Just posterior to the chiasma is a slight swelling, the **infundibulum**, to which is attached the pituitary gland. The cavity within the diencephalon is the third ventricle. It is connected anteriorly with the first and second ventricles that lie, respectively, in the two cerebral hemispheres, and posteriorly to the fourth ventricle, which is mostly in the medulla

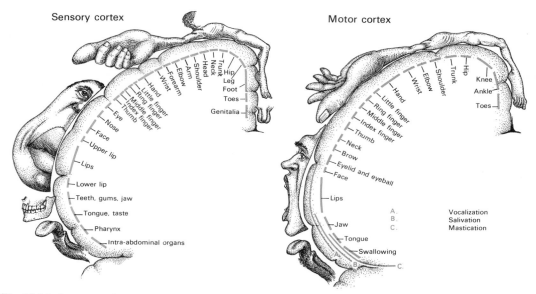

Fig. 32-17. Sensory and motor homunculi showing the position of neural centers along the ridges on either side or the central fissure. (Adapted from W. Penfield and T. Rasmussen: *Cerebral Cortex of Man.* The Macmillan Co., New York, 1950; and various other sources.)

oblongata. Dorally there is a small pineal body, the function of which is only beginning to be known. The greatly thickened sides of the diencephalon are termed the thalami: the **epithalamus** above, the **thalamus** to the sides of the third ventricle, and the **hypothalamus** below. Most of the thalamus in mammals is in fact a neothalamus not present in reptiles. All the connections between thalamus and cortex—and they are legion—are reciprocal. We see here another feedback or reverberating system.

In the hypothalamus are centers for many of the emotional and subrational aspects of life. A center regulating sleep is here. Appropriate electrical stimulation of the hypothalamus in lightly anesthetized cats will provoke ferocious spasms of generalized rage, as though the cat were suddenly faced with a barking dog. Tumors of this region in man have produced long periods of inconsolable grief or continuous and equally inexplicable gaiety. A very narrowly delimited center in the hypothalamus of the rat controls appetite.

From the hypothalamus several fiber tracts run down into the posterior lobe of the pituitary gland. These neurons secrete a characteristic neurohormone which passes down the axons into the gland. It has been demonstrated

that in this way the anterior pituitary may be brought into action by the nervous system.

The **mesencephalon** of mammals still retains, in its roof, paired lobes containing visual functions but it is not the dominant visual center.

The fourth division of the brain consists dorsally of the **cerebellum,** which contains centers controlling the subleties of muscular action. The **pons,** ventrally located, contains many fibers connecting the right and left sides of the brain. The cerebellum forms part of an important feedback system governing voluntary motion. Impulses from the cerebral cortex sweep down through the pons and up into the cortex of the cerebellum. Thence they pass back to a basal ganglion and then via the thalamus back to the cortex of the cerebrum. From the cerebral cortex the impulses may pass directly or indirectly into the spinal cord and thence to muscles, or again to the pons and cerebellum.

The fifth and last division of the vertebrate brain forms the **medulla oblongata** in mammals, as in all other vertebrates. It consists mainly of fiber tracts passing to and from the rest of the brain and the spinal cord. Many of the fibers cross from one side of the brainstem to the other. Most of the spinal nerves (VI through XII) connect with the medulla. Reflexes concerned with breathing and the heartbeat are centered here. Hence injury to the medulla can be quickly fatal.

BRAIN AND BEHAVIOR

In mammals a remarkable series of experiments has revealed the presence of narrowly localized areas within the hypothalamus (the part of the brain associated with the pituitary gland) which govern appetite, thirst, sleep, sex drives, anger, and even generalized pleasure.

In rats, and probably in all mammals including man, a pair of **nuclei** consisting of a very small number of cells in the hypothalamus controls satiety and another pair controls eating. The ventromedial nucleus contains cells that tell an animal that it has had enough to eat and should stop. If this **satiety center** or nucleus on both sides of the hypothalamus is carefully destroyed by an electric needle inserted into the brain, the rat becomes a ravenous eater and as a result becomes enormously obese. Such animals pass through three phases. As soon as they recover from the anesthetic—in fact, while they still may be groggy from it—they attack food and eat voraciously. After about a day, they settle down to a steady program of eating until they have more than doubled their weight. It is as though a 150-pound man came to weigh 300 or even 350 pounds. The third phase is a static one in which the animal maintains its abnormal weight but cuts down on its eating. However, if it is starved for a day or two it can be induced to gorge itself again.

The question now arises as to how the satiety center "knows" when the animal has had enough to eat. It is common knowledge that the level of glucose in the blood rises after eating and falls during starvation. There is now good evidence that it is the action of glucose directly on the satiety centers in the ventromedial region of the hypothalamus that gives the stop signal. If animals are fed glucose to which a toxic atom of gold has been added by a sulfur link (*i.e.,* gold thioglucose), the cells in the satiety center are selectively killed. Evidently these cells have a special affinity for glucose which is far greater than that of any other cells of the body. They take in the glucose and along with it the poisonous gold. As a result of their destruction, the satiety center is put out of order and the rat becomes obese.

Lateral and a bit ventral to each satiety center is a small group of nerve cells. These make up an **eating center**. If both these ventrolateral nuclei are destroyed, a rat will stop eating more or less permanently.

Most remarkable of all is the discovery of a localized **pleasure center** in the hypothalamus. Like so many important scientific discoveries, from X-rays to penicillin, this one was made accidentally by an astute observer. In rats it is possible to insert a fine metal electrode into a specific part of the brain and hold it there permanently by a plastic holder screwed into the skull. Such holders heal in place and apparently are unnoticed by the rat. The experimenter attempted to place the permanent electrode in the sleep center, but missed. The rat concerned was given a small electric shock via the electrodes whenever it approached a certain corner of its cage. Instead of going to sleep, the rat kept running over to the corner where it received the minute electric shock in its hypothalamus. The next step was to place the animal in a Skinner box where it could administer a shock to its own brain by pressing a lever. The rat quickly learned to press the lever to obtain a shock of 0.0004 or fewer amperes lasting less than a second. Such rats pressed the lever anywhere from 500 to 5,000 times an hour! Control rats with the electrode in other parts of their brains pressed the bar only a dozen or so times an hour during random exploratory movements like those of any normal rat.

Micro-electrodes implanted in the correct parts of the hypothalamus of goats, monkeys, and other mammals have confirmed the early discoveries on rats and mice. In monkeys, for example, it is possible to connect such micro-electrodes to small transistor radios strapped to the animal. If one set of electrodes is positioned in the center for angry aggression and another in the center for friendly indifference, then the behavior of the monkey can be dramatically controlled from a distance by radio signals. Dr. Jose Delgado has done the same with bulls used for bullfights. By remote control, a charging bull can be made to "forget" his anger and turn aside peacefully. If and when chemicals are found which have the same effect, entire populations could be pacified by introducing such an agent into the public water supply—a possibility of enormous human import.

Equally spectacular results have been attained in birds. If micro-electrodes are inserted into the correct region of the brain of a hen turkey, she will immediately perform typical

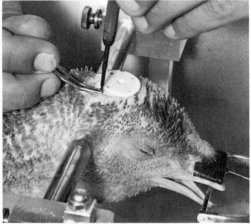

Fig. 32-18. Stages in the insertion of microelectrodes into precisely located points within the brain of a turkey. Note the rigid stereotaxic instrument which makes such an achievement possible. (Kindness R. S. Reese, Pennsylvania Agriculture Experimental Station.)

male courting behavior, spreading and rotating her tail feathers, strutting, etc., when shown a model of a female turkey's head, *provided* a 0.1-milliampere current is applied in one of the electrodes.

Looking at the hypothalamus and the cerebral cortex in a larger perspective, students of animal behavior often divide behavior into two rather general types: appetitive and consummatory. Appetitive is more or less random seeking behavior like the apparently aimless circling of the hungry hawk or the hurried walk of a mature caterpillar. The consummatory act is the dive of the hawk and the grasping of the unfortunate rabbit, or the spinning of a cocoon. The hypothalamus of the vertebrate motivates the appetitive phase.

Thus it now seems certain that centers for appetitive behavior, drives, and general motivation lie in the hypothalamus, its associated region of the brain, and the temporal lobes. The cerebral cortex controls the consummatory acts, that is, the specific manner in which the motivation are expressed.

THE BIOCHEMISTRY OF LEARNING

For many decades men have sought the "memory trace" or "engram" in the brain, but completely without success. Now at last very promising clues are at hand as to the form in which learning is recorded and, in some animals, even where. In both flatworms and rats, several investigators seem to have transferred learning by transferring RNA from the brains of trained to untrained animals. Some of the experiments on flatworms could not be repeated by other workers in other laboratories, but some have been so confirmed. There are those who challenge the results on rats. However, there are enough confirmations to give many people the feeling—it is no more than that—that there is some real connection between RNA and learning. Furthermore, it has been shown that antimetabolites like azoguanine, which interferes with RNA synthesis, and puromycin, which is known to block protein synthesis, also block memory in goldfish and some other animals. Substances which enhance RNA synthesis are said to enhance memory also. Proteins are made according to the specifications of RNA, so it seems highly probable that all these experiments are dealing with the same mechanism.

The most convincing evidence for the importance of proteins in memory, and for a true difference between short- and long-term memory, are some beautifully simple experiments by the Flexners. They trained mice to enter the left arm of a Y maze to obtain a reward. They then waited three weeks and reversed the maze so that the right arm now led to the reward. Within a day after the mice had learned the new correct route, one half of them were injected with puromycin and one half with physiological salt solution. Three days later the mice who had had only saline remembered the new lesson and chose the right arm while those who had received puromycin reverted to choosing the left.

If RNA and proteins are the agents which record experience, it does not mean that there is an RNA code "word" saying, for example, "When you see a red light, turn left in the maze to get the food." On the basis of his long series of experiments on learning in the octopus, the London anatomist, J. Z. Young, formulated the role of RNA in terms of the familiar circuitry of the nervous system. An octopus can be trained to come forward on seeing a crab and grab it, or to retreat when it sees an equally delicious looking crab accompanied by a white square. In both cases the visual image will pass up the same optic nerves to the visual center within the brain. But somewhere within the central nervous system there must be a fork. If the impulses take one branch, the muscles that make the animal advance will be activated. If the impulses take the other path, the muscles causing the animal to retreat will be activated. It may be that the passage of an impulse over the one pathway somehow generates a block to the other. Another possibility, of course, is that instead of certain pathways becoming blocked, certain routes become easier for the passage of neural impulses.

Here it is necessary to recall an often overlooked fact about the central nervous system, namely, the presence of numerous relatively small cells impinging on the longer circuits but having no recognized function. If the neural impulses that resulted from the presentation of crab-plus-white-square and that stimulated the muscles of retreat also stimulated some of those smaller cells to throw up some kind of RNA-protein roadblock on the way to the nerves of the muscles of advance and attack, then the basis for a permanent neural record would exist. The record would be in the permanence of that roadblock. If every time the white square was shown, the retreat muscles were stimulated and the RNA roadblock formed, it would be reasonable to suppose that a habitual or conditioned response had become established. This concept receives very strong support indeed from the well-known and long-established phenomenon of reciprocal innervation and reciprocal inhibition as seen, for example, in the motions of the arms and legs.

How would such a theory explain the transfer of leaning from one animal to another by cross-feeding RNA? It would be necessary to assume that the RNA and the proteins of every neural tract and fiber were uniquely different. This is no difficult assumption. There is much good evidence from embryology that each developing nerve fiber is somehow different from every other.

THE SENSE ORGANS

Sense organs are transducers which translate various forms of energy—light, heat, sound, and certain molecular configurations—into the uniform language of the nervous system, that wave of negativity, the nervous impulse.

More than that, our sense organs, in conjunction with the nervous system of which they are a part, determine in the most fundamental way our conception of what our universe and we ourselves are like. If the genes necessary for the production of the pigments in the retina, essential to see color, are not present, the very concept of color remains completely outside of the experience of the beholder. The redness of red can by no means be described to the color-blind.

Smell

The ability to discriminate odors is the dominant sensory ability of most vertebrates and many other animals. If the olfactory antennae of an ant are severed, it can no longer recognize the members of its own colony and will attack friend and foe alike. Olfaction is a dominant sense in all but a few mammals and is correlated with the fact that most are color-blind creatures of the evening and predawn. Pheromones, which are of great importance in insects and mammals, are mostly odors which attract males to females, are discussed elsewhere (see Chapters 1 and 22).

Many theories have been proposed; none is completely proven. There is much cogent evidence, however, for a theory which was suggested in primitive form by Lucretius in the 1st century B.C. The sensation recognized as a particular odor is due to the size and shape of volatile molecules which allow or disallow them to fit into like-shaped receptor sites on the nerve endings of the olfactory epithelium. Studies on men and frogs indicate that there are about seven primary odors: camphoraceous, musky, floral, peppermint, ethereal, pungent, and putrid. Seven primary odors would require seven differently shaped receptor sites, by no means an impossible demand. Elaborate investigations of the shapes of odorous molecules show that small roundish molecules, regardless of chemical constitution, give a camphor-like odor. All the molecules giving a floral odor are shaped like a key to a Yale lock, elongate with one large rounded end. Narrow elongate molecules give an ethereal odor, very small molecules with positive charges like formic acid are pungent, and so on for all the primary odors. By mixing various combinations of the seven presumably primary odors, it has been possible to simulate a large variety of other odors.

Taste

The chemosensory ability called taste is far more restricted than is smell. In man there are only four recognized tastes: sweet, bitter, sour, and salty. Taste buds sensitive to each are restricted to special zones on the tongue—sweet and salty on the tip, sour along the sides, and bitter at the base. In fish what appear to be taste buds are distributed widely over the body. In butterflies the taste cells are located in the front feet; if the feet are placed in a sweet solution, the coiled tongue will be extended.

From the zoological point of view, one of the most interesting discoveries about taste is that animals far apart in the scale of life have closely similar tasting abilities, or perhaps "preferences." If a man and a blowfly are each asked to rank several sugars in order of sweetness, both will arrange them in the same order. Sucrose is sweeter than cellobiose which is sweeter than lactose which is sweeter than glucose. How do you ask a fly such a question? Dip its feet into a series of solutions of increasing concentrations of a sugar. When the fly reaches the first solution that tastes "sweet," it will lower its proboscis and suck up some of it. By comparing the minimal concentrations of different sugars that will elicit a response, information can be obtained about their relative "sweetness" to the fly.

Hearing

Sound plays a very important role in the lives of many animals. The sonar of bats, whales, and dolphins is the envy of the navy. A blinded bat can locate the source of echoes of its own high-pitched squeaks so accurately that it can

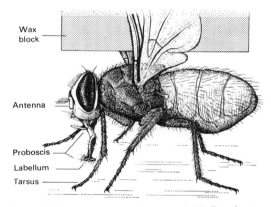

Fig. 32-19. A blowfly with wings held firm in wax tasting sugar water with its feet. (After V. G. Dethier, in G. B. Moment: *General Zooloary,* 2nd ed. Houghton Mifflin Co., Boston, 1967.)

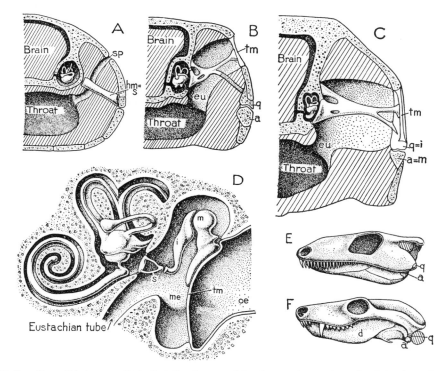

Fig. 32-20. Ears from fish to man. Note that the three semicircular canals, one in each of the three dimensions of space, remain almost the same from start to end. The cochlea, however, undergoes considerable relative enlargement. Note that hm, the hyomandibular bone of a shark, becomes s, the stapes of a mammal. The **quadrate bone, q,** becomes the incus, i, and the articular bone, a, becomes the malleus, m. me, middle ear; eu, Eustachian tube; tm, tympanic membrane or ear drum; oe, outer ear cavity; sp, spiracle of shark. (From A. S. Romer: *The Vertebrate Body,* 2nd ed. W. B. Saunders Co., Philadelphia, 1955.)

avoid piano wires stretched across a room. The function of the powdery scales on night-flying moths is to blur the echo of the bats' squeaks. The abilities of porpoises are equally amazing. A porpoise with opaque rubber cups over its eyes has no difficulty in locating and catching a fish and can even pick out the largest of a group.

The ears of mammals consist of three parts, the outer ear or **pinna** (absent in aquatic forms), the middle ear, and the inner ear. The **middle ear** developes from the first gill slit of the embryo and begins with the **eardrum** and extends down via the **Eustachian tube into the** pharynx. From the eardrum a chain of three

tiny bones, the malleus, incus, and stapes, conducts the vibrations picked up by the eardrum and carries them across the upper end of the eustachian tube to the inner ear.

The **inner ear** consists of the three **semicircular** canals, the organs of equilibration, plus the cochlea, the organ of hearing. The **cochlea** resembles a snail shell made of a coiled tube which is itself divided along its length into three parts. A ventral **scala tympani** and a dorsal **scala vestibuli** enclose between them a **cochlear canal**, roughly triangular in cross section. Extending the length of the cochlear canal and resting on its basement membrane is the **organ of Corti.** It consists essentially of several

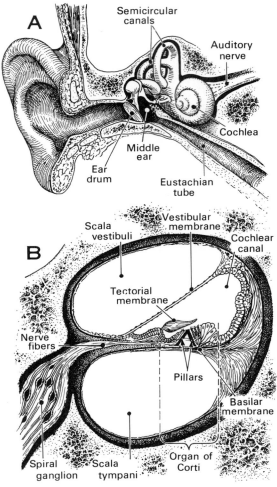

Fig. 32.21. Human ear. Top, general topography. Note chain of three bones from ear drum across middle ear to the cochlea. Bottom, cross section through one of the coils of the cochlea. Note so-called sensory hairs in the organ of Corti.

parallel rows of **sensory hair cells** overhung by a **tectorial membrane.** Different sounds are thought to cause the basement membrane to vibrate to different extents in different regions along its course. This stimulates the sensory hairs of the organ of Corti. The resulting nervous impulses pass along the eighth nerve into the brain where they are interpreted as sounds.

Proprioceptive and Minor Sense Organs

There is a great variety of small sensory nerve endings in the skin and throughout the body, especially in the muscles and tendons. There are special pain endings in the form of irregularly branching naked nerve fibers. There are Meissner's corpuscles for touch, Krause's end bulbs for cold, Ruffini's corpuscles for heat. In fact, histologists have found more specialized nerve endings than psychologists can find sensations. Among the most important are the **proprioceptive endings** in muscles and tendons. The sensory nerves from these endings tell us about our own position, degree of muscle contraction, and motions.

Vision

Merely to summarize what is known about the organs and organelles of vision would require several large volumes. Some light-sensitive structures are as simple as the eyespot of *Euglena.* Others are as complex as the eyes of the higher insects or the vertebrates. In the course of evolution, two major types of eyes have been developed. The differences and similarities of these two major types of photoreceptive organs conform to the requirements for vision imposed by the laws of physics. This fact makes it possible to offer some educated guesses as to what eyes might be like in animals evolving on another plant in another solar system.

The most primitive condition, which can be seen in coelenterates, annelids, arthropods, mollusks, chordates, and even echinoderms, is a flat **plate of photosensitive cells.** From this beginning one line of evolution capitalized on the advantages of invaginating the flat disc into a **cup** of photosensitive cells which can then be called a retina. From the cup it is only a series of slight changes to produce a vesicle which is a cup closed by a transparent lens. This **camera-type eye** exists in certain dangerously poisonous pelagic jellyfish and in all the other groups listed above, including vertebrates.

The second line of evolution, instead of invaginating the disc of light-sensitive cells into a concavity, exploited the advantages of bowing

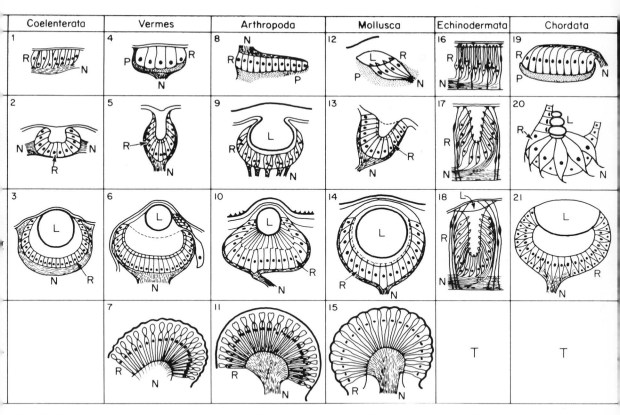

Coelenterata	Vermes	Arthropoda	Mollusca	Echinodermata	Chordata

Fig. 32-22. The types of visual organs found on this planet. I, flat eyes; II, cup-shaped eyes; III, vesicular eyes; IV, compound (convex) eyes. Note that these are not only found among arthropods but also in annelids and mollusks. L, lens; N, optic nerve; P, pigment; R, retina. (From W. S. Hoar: *General and Comparative Physiology.* Prentice-Hall, Inc., Englewood Cliffs, N.J., 1966.)

the disc out into a **convex knob** which becomes a **compound eye**. This type of eye is highly developed in the insects, crustaceans, and other arthropods, but occurs also among polychaete annelids and certain mollusks. No better evidence of the effectiveness of the environment to produce convergent evolution in unrelated animals adapting to the same physical laws can be found that the comparison between the cup-type eyes of a man and an octopus and the compound eyes of an arthropod and certain mollusks.

The vertebrate retina consists of light-sensitive cells, the rods and cones. Color vision is a property of the cones. Areas of the retina where rods predominate, as they do around the periphery, are highly sensitive in dim light but lack color vision.

What happens when light falls upon the retina that sends a pattern of nervous impulses racing up the optic nerves to the brain? If a frog or other vertebrate is killed in the dark and its retina exposed and then observed, it will be seen that the retina is purplish but rapidly bleaches to yellowish-gray in the light. The original color is due to **rhodopsin** (*rhodon,* rose), also called visual purple, located in the **rods**. Rhodopsin itself is a compound molecule, a protein, **opsin**, plus a derivative of vitamin A, a carotenoid called **retinene**. It is for this reason that people whose diets are deficient in vitamin A have trouble seeing in dim light. That

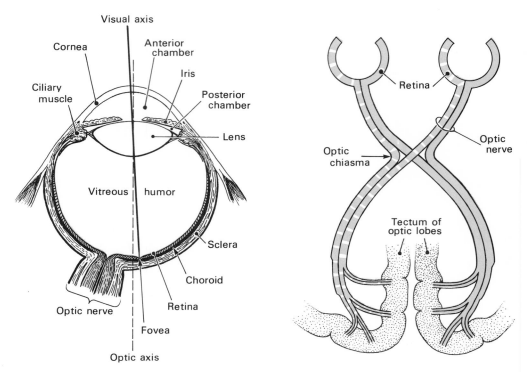

Fig. 32-23. Diagrams of a section through the human eye and of the course of the optic nerves through the optic chiasma into the brain and to the tectum (literally "roof") of the optic lobes of the brain.

rhodopsin is in fact the pigment responsible for vision in dim light has been proven in a very precise and elegant way. It is known from a law of physics (Grotthuss-Draper law) that to be effective, to produce some change, radiation must be absorbed. Otherwise it merely passes through and nothing happens. It is also well known that different substances absorb light of different wavelengths. Rhodopsin best absorbs light of a wavelength of 500 nm. At either side of 500 nm., absorbance falls off rapidly in a characteristic curve. What is the curve of visual sensitivity in dim light of different wavelengths? Exactly the same as the absorption curve for rhodopsin.

The biochemistry of vision has been worked out with great care. When quanta of light hit the rhodopsin in the rods, the molecule is split into the protein opsin and the carotenoid retinene. The results are the generation of a nervous impulse and the loss of the purple

color. The reconstitution of rhodopsin requires a number of steps and may occur over more than one pathway. Either a photochemical reaction or an oxygen-dependent dark reaction can do the trick. Vitamin A from the bloodstream is converted into the immediate precursor of retinene by NAD, nicotinamide adenine dinucleotide.

Until very recently nothing whatever was known about how the rods and cones transduced, i.e., translated, the incident energy of light quanta into a nervous impulse. A beginning has now been made with the electron microscope. Rods are a long series of tightly packed membranes. These appear like a pile of microscopic dinner plates, one on top of the other. In the cytoplasm of the cell bearing the rod, between the rod and the nucleus, are some mitochondria, but, far more interesting, the electron microscope reveals that the rod is connected into the cell body by what is

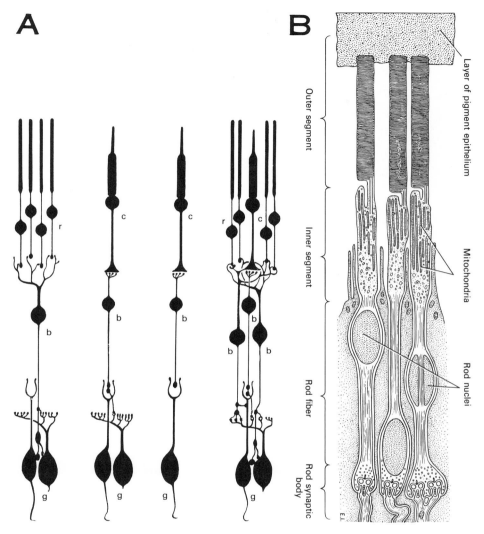

A

B

Layer of pigment epithelium

Outer segment

Inner segment

Mitochondria

Rod nuclei

Rod fiber

Rod synaptic body

Fig. 32-24. Diagram at left shows some of the neural interconnections within the vertebrate retina. b, bipolar cells; c, cone; g, ganglion cell; r, rod. The diagram at right shows retinal rods in the guinea pig eye at high magnification.

unmistakably the basal structure of a cilium. Embryologists have long known that retinal cells have a tendency to produce cilia but no one knew the reason. Light-sensitive structures in many kinds of animals—jellyfish, starfish, scallops, and primitive flatworms, as well as vertebrates—have been found, under an electron microscope, to be composed of piles of more or less twisted masses of membranes. This is true also of the retinular cells in the rhabdomes of

the compound eyes of arthropods where the membranes take the form of masses of micro-tubules rather than a pile of micro-coins as in a vertebrate. Presumably visual pigment is spread in or on these membranes. Since a nervous impulse is a change in electromotive potential across the nerve membrane of the cell, it would not be surprising to find that when light splits rhodopsin into opsin and retinene, a change in membrane potential is produced.

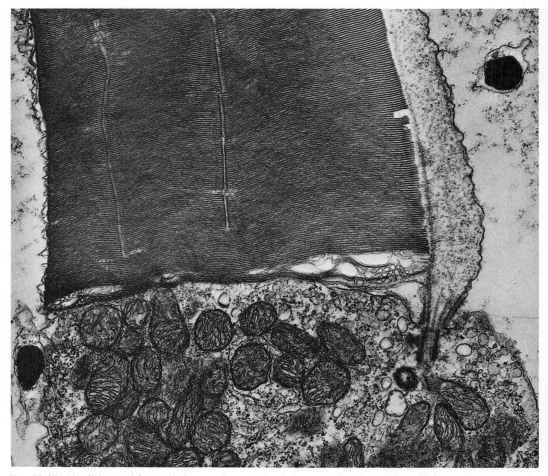

Fig. 32-25. The light-sensitive end of a vertebrate rod cell from the Pacific tree frog, *Hyla regilla*, as seen under an electron miscroscope. Note the resemblance of the stack membrane disks to the structure of a chloroplast. On the right the stack of disks joins the rest of the cell by a stalk consisting of a kinetosome such as is at the base of every cilium. The presence of numerous mitochondria is also evident. (Kindness R. M. Eakin.)

USEFUL REFERENCES

Bullock, T. H. and Horridge, G. A. *Structure and Function in the Nervous System of Invertebrates.* W. H. Freeman and Co., San Francisco, 1969.

Cannon, D. F. *Explorer of the Human Brain: Santiago Ramon y Cajal.* H. Schuman, New York, 1949.

Delgado, J. M. R. *Physical Control of the Mind: Towards a Psychocivilized Society.* Harper Colophon Books, New York, 1971.

Dethier, V. G. *How to Know a Fly.* Holden-Day, San Francisco, 1962.

Gaerluce, G. *Biological Rhythms in Human and Animal Physiology.* Dover Publications, Inc., New York, 1971.

Gazzaniga, M. S. *The Bisected Brain.* Appleton-Century-Crofts, New York, 1969.

Gellhorn (ed.) *Biological Foundations of Emotion.* Scott, Foresman & Co., Glen View and Palo Alto, 1968 (paperback).

MacNichol, E. F. Three-pigment color vision. Scientific American, Dec. 1964.

Sherrington, C. S. *The Integrative Action of the Nervous System.* Cambridge University Press, Cambridge, 1947.

Unger, G. (ed.) *Molecular Mechanisms in Memory and Learning.* Plenum Press, New York, 1969.

Wald, G. Molecular basis of visual excitation (Nobel Prize Lecture) Science *162:* 230, 1968.

Young, J. Z. *A Model of the Brain.* Clarendon Press, Oxford, 1964.

10

THE WEB
OF LIFE

Symbiosis and Parasitism

33

The human importance of parasitism has long been recognized. The Black Death which killed two out of every three students at Oxford during the 14th century was due to the bubonic plague bacillus and is still a serious threat. Some historians believe that massive chronic infection of the population with the malarial parasite was an important factor in the decay of classical Greece and Rome. Even today, a wide variety of bacteria, viruses, and worms infect man and his domestic plants and animals. The annual damage inflicted on wheat, corn, and many other crops by rusts, smuts, and other fungi totals hundreds of millions of dollars.

In contrast, the importance of symbiosis, where two different kinds of organisms live in close association with mutual benefits, has been largely overlooked. Even to biologists the word often suggests a lichen, a double type of organism where a fungus and an alga live together, or some exotic crab which places sea anemones on its shell where they protect the crab and the crab, a messy eater, provides crumbs of food for the anemone. In fact most of our forest trees live in symbiosis with root fungi. There is good evidence that mitochondria and chloroplasts are the descendants of symbiotic bacteria and unicellar green algae, respectively. Thus symbiosis may be basic to all the higher forms of life.

DEFINITIONS

The terminology applied to all the various types and degrees of association between individuals of two species varies somewhat from author to author. In many cases it is practically impossible to determine whether one species is benefited or harmed. We will follow the usage appearing in Pennak's *Collegiate Dictionary of Zoology*.

A **parasite** is an organism which lives in or on individuals of another species from which it derives nutriment—an animal or a plant which has a host that does not benefit from the association. Mistletoe growing on the branch of an oak tree and a copepod living inside the eye of a swordfish are both clearly parasites. Parasites may be **ectoparasites** like lice or **endoparasites** like tapeworms. It should be remembered that it is often rather arbitrary whether a relationship is termed parasitic. Is a mosquito a parasite? Is an aphid which sucks the juices of a plant or a caterpillar that eats the leaves also a parasite?

Commensalism is usually defined as the association of individuals of two species where one derives some benefit and the other is neither helped nor harmed. Some biologists claim that such a relationship never exists because the host always is either injured or benefited, though perhaps very slightly. Firm proof is hard to obtain. Barnacles on whales are commonly cited commensals.

Mutualism is the term for a close relationship where both species are benefited. In addition to lichens and the crab-anemone partnership, one thinks of the hydra with green algae living within its endodermal cells and of the root nodule bacteria of legumes like clover.

Symbiosis, literally "life-together," is sometimes used to designate all of the classes of close association just discussed. Parasitism is antagonistic symbiosis.

Symbiosis is clearly the safest term to use for all these relationships except in the most obvious cases of parasitism.

POSITIVE SYMBIOSIS: MUTUALISM

Lichens

The most undeniable instances of symbiosis in the sense of mutualism are found in plants, notably the lichens. Many problems of very general theoretical importance lie hidden here. How is the growth of two such different organisms coordinated in both vegetative and sexual reproduction? What determines the form, so different in different species of lichens? Is the species concept truly applicable to a situation like this?

The component organisms of a lichen, an alga and a fungus, are members of diverse groups of the plant kingdom. The algal component may be either a prokaryotic blue-green or a eukaryotic green alga, while the fungus may be a Basidiomycete, an Ascomycete, or one of the fungi imperfecti. Both the algal and the fungal components of many lichens have been cultured separately in the laboratory. Thus one cannot argue that their association in nature is a matter of absolute necessity. Undoubtedly, when growing on rocks, the photosynthetic products of the algal cells are utilized by the fungus while the fungus probably supplies water and inorganic nutrients to the alga.

Mitochondria and Chloroplasts

The most surprising recent theory about symbiosis holds that mitochondria are really symbiotic bacteria. Certainly mitochondria and bacteria have much in common. They are about the same in size and shape, and both are self-duplicating by elongation and transverse fission. Mitochondria and chloroplasts contain their own DNA. Both contain phosphorylating respiratory assemblies of molecules in their complex membranes. If it turns out to be true that mitochondria are in fact symbiotic bacteria, the origin of this symbiosis was one of the

Fig. 33-1. Lichens of several kinds on a boulder near the Skyline Drive, Virginia. These dual organisms are composed of an alga and a fungus living together in symbiosis.

major events in the history of life. It is mitochondria that make aerobic respiration possible. They are found in the cells of all organisms that have been examined, higher animals and plants, fungi, algae, and protozoans, with the significant exceptions of the very simple cells of prokaryotic bacteria and blue-green algae. The evidence that chloroplasts are also symbionts, essentially unicellular green algae, rests on the presence of DNA in chloroplasts, which are self-duplicating, and on the close resemblance between these organelles and unicellular algae. This concept is made more plausible by the existence of symbiotic green algae in the green paramecium, *P. bursaria*, in the endoderm cells of the green hydra, and in various other organisms.

Protozoa, Termites, and Cows

The most extensively investigated case of symbiosis among animals concerns the flagellates found in the guts of termites and wood roaches. Termites are completely dependent on these protozoans for the digestion of the wood which they eat, and the flagellates in turn are found only in termites and their relatives. Thus, the termite-flagellate relationship is one of true commensalism. If termites are placed in an oxygen tent for 24 hours, subjected to high temperatures (36°C), or starved for 10 days, their intestinal flagellates die while the termites themselves appear to be unharmed. After treatment, the termites eat wood as usual, but because without protozoa they are unable to digest cellulose, it passes unchanged through their digestive tracts.

Probably the most important case of symbiosis from the practical standpoint is that of the ciliates that swarm in the stomachs and intestines of cows and related animals. The number of these ciliates in a cow is truly prodigious, 100,000 to 400,000 per milliliter of intestinal contents. Are they harmful, useful, or even indispensable to the cow, in the way intestinal flagellates are to the termite? A clear case of necessary symbiosis between ruminants and the microorganisms inhabiting their guts involves the production of vitamin B_{12}. This cobalt-containing vitamin (cyanocobalamin) is essential for animals. In cattle and sheep it is manufactured by their intestinal microflora. In order for the vitamin to be synthesized, cobalt has to be provided in the forage consumed by the host organism. Lush pastures can grow in

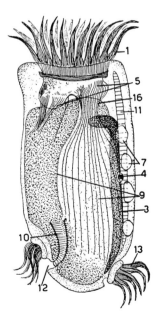

Fig. 33-2. A complex ciliate such as found in the stomachs of horses, cows, sheep, and other herbivores. Note the many organelles including oral cilia (1), macro and micro nuclei (3 and 4), contractile vacuoles (7), a gut-like sac and rectum (9 and 10), and a skeletal rod (11). (From L. H. Hyman: *The Invertebrates*, vol. 1. McGraw-Hill Book Co., Inc., New York, 1940.)

soils deficient in cobalt because this is not an essential element for plants. However, inadequate amounts of cobalt in the diets of cattle causes a nutritional disease in their microorganisms. A consequence of this is a deficiency in vitamin B_{12} in the cattle. In parts of Australia and New Zealand where cobalt deficiency was discovered, cattle and sheep can be provided with continuing supplies of cobalt by lodging "bullets" of cobalt-containing materials in their throats. This is far less expensive than the alternative of supplying B_{12} injections for the ruminants.

Plant Mutualism

An important but often overlooked mutualism exists between most of our familiar trees, pines, oaks, and willows, for example, and several kinds of fungi which grow either on or within their roots. The fungus may be an Ascomycete (the group to which mildews, black and green molds, and *Neurospora* belong) or a Basidiomycete (the group which includes mushrooms and puffballs). Such an intimate association between the root of a higher plant and a fungus is called a **mycorrhiza.** "Fungus roots," as mycorrhizae are sometimes called, tend to be short and swollen. Often only the lateral roots are invaded by the fungus and thus modified. Changes in roots include loss of regions where root hairs are located. The condition of such fungus-infected roots may be regarded as pathological, in which case the fungus would have to be classified as a parasite on the higher plant host. There is evidence, however, that both the fungus and the higher plant benefit from their association. The fungus gains organic nutrients in return for facilitating the uptake of water, minerals, and possibly organically bound nitrogen from the soil. When species in which mycorrhizae occur universally in nature are cultured in the laboratory, seedling growth in sterilized soil is notably slower than in soils inoculated with appropriate species of mycorrhizal fungi.

Mycorrhizae are frequently found in and appear to be beneficial to plants growing in soils of forests or bogs containing much organic matter. The fungi, by digesting organic materials, release minerals that are then available to the higher plants. Thus, they make possible the absorption of organically bound minerals that higher plants ordinarily cannot absorb from the soil.

Higher Animals

Modern biologists share with the old 19th-century naturalists an interest in relatively larger animals which live in some degree of symbiosis with each other. There is a fish which manages to find protection from larger fish by swimming among the tentacles of the Portuguese man-of-war, a fingernail-sized crab which lives within the mantle cavity of oysters, and insects which live in the darkness of ants' nests and beg for food from the ants with their antennae. Perhaps we have now crossed the line and are describing parasites, even though some writers call them commensals.

NEGATIVE SYMBIOSIS: PARASITISM

Traits Characteristic of Parasites

Most parasites, whether plant or animal, share four important traits which adapt them for their way of life. First, there is a **loss of structures**, especially structures having to do with obtaining food. For example, a dodder germinates from a seed and develops like a normal plant with root, stem, and leaves, but as soon as the young plant becomes attached to its host and sends its sap-absorbing holdfasts into the stems of its victims its roots die and the entire plant loses contact with the soil. Leaves are no longer formed except as mere vestiges. The comparatively simple structure of animal parasites is also notorious. Tapeworms and liver

A   B

Fig. 33-3. Effects of mycorrhizal roots on growth of pine trees. A, six-month-old seedlings of *Pinus caribaea,* showing differences in growth between plant inoculated with symbiotic root fungi (left) and uninoculated plant (right), ×¼; B, six-year-old Honduras strain of *Pinus caribaea* trees growing in Puerto Rico. It is not uncommon for this tree to grow 10 feet in one year. (Courtesy Edward Hacskaylo, Forest Physiology Laboratory, U. S. Department of Agriculture.)

flukes are reduced to little more than a smooth muscular sac with a holdfast of some kind on the outside and a gonad on the inside, although rudimentary nervous and excretory systems and, in the case of the flukes, a gut, remain.

Second, many parasites possess **specialized structures** which aid in finding, entering, and remaining within their hosts. The adult has no use for eyes but larval parasites commonly possess them together with special inherited behavior patterns which increase the chances of their finding a suitable host. These behavior patterns may be as simple as crawling up a blade of grass where some host will graze. Other patterns, such as those found in parasitic wasps which seek out special kinds of caterpillars on which to lay their eggs, involve complex sense organs as well as instincts. In such a parasite as an ichneumon wasp there is no loss whatever of the normal complement of organs—sensory, locomotor, digestive, and the like.

Third, most parasites are extremely **prolific**. The probability of any individual finding a suitable host is extremely low. Although greatly exaggerated reproductive powers are commonly thought of as characteristic of parasites, it should be remembered that many animals that are not parasites lay incredibly large numbers of eggs, *e.g.*, oysters, most fish, and nematodes (whether parasitic or free living). Think of the enormous numbers of maple seeds shed year after year. Many parasites are hermaphrodites (*e.g.*, the tapeworms and flukes), but again it is important to remember that many successful parasites, such as *Schistosoma* (a blood fluke) and most nematodes, have separate sexes while many snails and a great many plants are hermaphroditic. The meaning of this difference therefore cannot lie in parasitism as such.

Fourth, although many parasites have only one host and are said to be **monogenetic**, many require two (or more) hosts and are termed **digenetic**. For example, the blood fluke *Schistosoma* lives its adult, *i.e.*, sexually mature, life in humans. The larval forms live in snails. Many disease-causing parasites, including the rickettsias causing Rocky Mountain spotted fever and fungi such as the cedar-apple rust, have alternate hosts. Some parasites are very fastidious about the hosts they infest. The parasite of bird malaria cannot develop in

humans, although it is transmitted and lives part of its life cycle in the common *Culex* mosquitos which must often inoculate people with the sporozoan of bird malaria. Other parasites are rather indiscriminate and can live successfully in many species. The trypanosome causing African sleeping sickness lives successfully in antelope, horses, and camels, as well as humans.

Evolution of Parasites

The evolution of parasitism is most assuredly due to the same combination of mutation, selection, and isolation that underlies the evolution of all other organisms. One view holds that parasites began by being able to live on a single host, evolved to be able to live on several species and then finally became more limited in host specificity. Probably many patterns have occurred in the evolutionary history of parasitism. There are even parasites of parasites, so-called hyperparasites, among insects and the viruses which infect disease-causing bacteria. The well-adapted parasite does not kill its host. Ordinarily the parasite dies of starvation when the host dies. Consequently, it is a selective advantage for any parasite to injure its host as little as possible. Highly and quickly fatal parasites are believed to be recently evolved species.

Host Responses

Host responses to parasites show a wide range. In plants a pathological swelling called a **gall** results from the egg-laying sting of certain insects. Such golf ball sized growths are common on the stems of goldenrod and the twigs of oaks and cedars. Apparently the insects secrete some stimulus to produce these bizarre growth effects. The advantage to the insect which can develop inside such a gall and eat at the plant's expense is obvious. **Hypertrophy** (excessive growth of the host's tissues) also occurs in plant diseases caused by certain fungi such as the "club root" of cabbages and related plants infected with *Plasmodiophora brassicae*. In animals the host responses range from the production of antibodies to overt behavioral

Fig. 33-4. Insect galls on pecan twigs. These growths are caused by small insects (phylloxera) which are closely related to aphids. (Courtesy U. S. Department of Agriculture.)

traits which tend to lessen possible contacts with the parasite or its eggs.

Plant Parasites

Plants, like animals, exhibit all degrees of interaction ranging from competition to states of dependency where a parasite is benefited while the host is harmed and states of interaction where the association of organisms is mutually advantageous. Such interactions occur among many diverse groups of plants ranging from bacteria and viruses to seed plants.

Because of the economic costs and human misery resulting from parasitic infections, subjects such as plant pathology, medical bacteriology, and mycology have been intensively studied. Some examples of parasite-caused diseases of plants are listed in Table 33-1. As can be seen in this brief listing, organisms ranging from the viruses (which lack cellular structure) and the prokaryotic bacteria to flowering plants and a variety of animals can cause diseases of crops with very serious economic and social consequences. Some of the

fungi parasitic on cereal grains produce chemical byproducts that are poisonous to humans or animals eating infected plants. Ergotism, a condition in which contraction of the smaller

Table 33-1
Some examples of widespread disease-causing parasites of plants

Vector	Disease
Prophyta	
Bacteria	Soft rot of vegetables
	Fire blight of pear and apple
	Crown gall
Viruses	Mosaic diseases of tobacco, tomato, cucumber, potato, and bean
	Peach and aster yellows
	Curlytop of sugar beet
Fungi	
Plasmodiophorales	Club root of crucifers (broccoli, cabbage, etc.)
	Powdery scab of potato
Phycomycetes	Downy mildews of grasses, grapes, onions, squashes, and lettuce
	White rust of crucifers
	Late blight of potato and tomato
Fungi imperfecti	Fusarium wilt of tomato
	Early blight of potato and tomato
	White rot of onion and garlic
Ascomycetes	Powdery mildews
	Dutch elm disease—fungus spores are carried by insects; most common carrier is the bark beetle.
	Chestnut blight
	Ergot of grains—alkaloids produced by the fungus can poison humans.
Basidiomycetes	Smuts of corn, oat, barley, and wheat
	Cedar-apple rust
	White pine blister rust
Spermatophyta	
Angiosperms	Mistletoes infecting hardwoods and conifers
	Dodders infecting clover and alfalfa
Aschelminthes	
Nematodes	Root knot
	Nematode diseases of potato, sugar beet, bulbs, and many other crops
Arthropoda	
Insects	Galls caused by wasps, flies, or aphids—parasites cause proliferation of the tissues of the host plant.

in the introduction of parasite-caused plant diseases into new areas. Since 1900, a number of plant pathogens have been introduced into North America by man. The white pine blister rust was first discovered in 1906, chestnut blight in 1904, and Dutch elm disease in 1930. Marked deterioration of many environments

Fig. 33-5. Stem rust of wheat. Pustules caused by the fungus are brick-red in color and elongate in shape. (Courtesy U. S. Department of Agriculture.)

Fig. 33-6. Dutch elm disease, caused by a fungus carried by the bark beetle, is the most destructive shade tree disease in the United States. It will take 80 years to grow another tree the size of this diseased one that was removed in 1966. (Courtesy U. S. Department of Agriculture.)

arteries and smooth muscle fibers can lead to gangrene, has been a serious disorder of epidemic proportions in certain European countries where rye bread is a staple of human diets.

There are fungus-caused plant diseases in which animals participate in the disease cycle as carriers of spores. The Dutch elm disease is spread by the bark beetle. In other cases, man has undoubtedly been an unwitting accomplice and losses amounting to hundreds of millions of dollars have resulted. Strict quarantines have been imposed on the importation into the United States of plants, agricultural produce, and seeds in order to avoid future introductions that, like the chestnut blight and Dutch elm disease, might virtually exterminate or seriously threaten the survival of their host species. Often newly introduced parasites that were not serious pests in the areas where they originated cause havoc when introduced into a new environment where they find more susceptible hosts, a lack of natural enemies, or conditions more favorable for infection or development.

There are some groups of seed plants that exhibit the full range of dependency on other species from a relatively harmless climbing habit to strict parasitism. In the morning glory family (Convolulaceae) are examples of both extremes. *Convolulus,* the common bindweed, twines around any available support (including other plants) but is not dependent on the structure around which it climbs for anything but mechanical support. Its roots absorb water and minerals from the soil; its leaves are large and photosynthetic. In the genus *Cuscuta,* which also belongs to the Convolvulaceae, we find the dodders, obligate parasites that exhibit many of the kinds of structural modifications and adaptations characteristic of parasites in general. The dodders are almost leafless and deficient in chlorophyll (in color they range from pink to pale green or white), adult plants have no contact with the soil, and their

Fig. 33-7. A, damage caused by the elm bark beetle is evident where a section of bark has been removed. Growth of the fungus carried by the beetle blocks conducting tissues of the tree. B, elm bark beetle, ×18. (Courtesy U. S. Department of Agriculture.)

Fig. 33-8. Dodder, a parasitic seed plant, growing on lespedeza. Although their seeds germinate in the soil, dodders absorb nutrients from the plants they parasitize. (Courtesy U. S. Department of Agriculture.)

reproduction can only be described as prodigious. They form many dense clusters of flowers. Seeds of the dodder germinate in the soil and their abundant food reserves are not wasted in the formation of large roots or leaves which later will not be needed. Instead, all available resources are utilized to form long stems that move whip-like in wide arcs until contact is made with another plant. If no contact is made, the seedling dies when its food reserves are exhausted. When a suitable host is found, the root disintegrates and the dodder continues its life as a twining parasite. At points of contact with the host, specialized suckers form.

Dodder seeds are disseminated with crop seeds, hay, movement of farm animals and implements, irrigation and surface drainage and in animal feces. Once farmland becomes infested with dodder it is necessary either to harvest crops before the dodder seeds have matured or to burn all infected plants. Control by means of herbicides or by rotation to non-affected crops is also possible. Among the crop plants affected by these parasites are clovers, alfalfa, sugar beets, onions, and flax. In addition, many ornamental and wild species are affected.

The mistletoes provide further examples of true parasites among flowering plants. The mistletoes are, in addition strict **epiphytes**. Although similar, mistletoes belong to a number of genera, which means they have been evolving for a long time. In Europe a species of *Viscum* is found on hardwoods and conifers. In North America, all true mistletoes belong to the genus *Phoradendron*. They are restricted in their distribution to geographical regions where winters are mild. Their northernmost limits are Oregon, central Colorado, the Ohio River, and southern New Jersey. They are found southward to the West Indies and in central and northern South America. The over 70 different species of *Phoradendron* parasitize many hardwoods (including oak, elm, maple, sycamore, ash, poplar, walnut, and willow) and conifers (including juniper, cedar, and pine).

Mistletoes are **dioecious**. The female (pistillate) plants bear groups of white, yellow, or pink berries with sticky mucilaginous surfaces that easily stick to host plants or to birds that carry them to new hosts. Seeds germinate on

Fig. 33-9. Female plant of western dwarf mistletoe growing on a pine. Note berries on mistletoe and spindle-shaped swelling on pine branch. (Courtesy U. S. Department of Agriculture.)

the host where the developing shoot forms an attachment organ from which protuberances called **haustoria** penetrate the bark through lenticels or buds. Only young shoots are invaded, but once they are penetrated, the haustoria spread through the bark region finally reaching the cambium and ultimately invading the xylem. Once established, these parasitic epiphytes can persist as long as the host tree remains alive. Most mistletoes have chlorophyllous leaves and stems and are dependent on their hosts only for a supply of water and minerals. Because they are photosynthetic, they are found primarily on the uppermost branches of forest trees. The standard technique for harvesting mistletoe for the Christmas market is to shoot it down with a shotgun.

Animal Parasites

Protozoa

The most primitive group of animals, the Protozoa, have developed some of the most devastating of parasites. *Endamoeba histolytica,* the amoeba of **amoebic dysentery**, is only too well known to travelers. Among flagellates, trypanosomes are responsible for many important diseases. Among the best known is *Trypanosoma gambiense,* the causative agent of **African sleeping sickness.** (This is not to be confused with encephalitis lethargica, a form of sleeping sickness due to a virus.) The trypanosome responsible for this disease lives in the bloodstream, then in the lymph nodes, and finally in the central nervous system. In advanced stages the patient sleeps continually, becomes incredibly emaciated, is shaken by convulsions, and finally dies in profound coma. Mortality is high, and large areas of Africa are rendered unfit for human habitation because of its toll. The parasite is transmitted from man to man and from man to the larger mammals, such as cattle, pigs, goats, and antelope, and back to man again by the bite of *Glossina,* the tsetse fly. Many species of trypanosomes parasitize fish, frogs, salamanders, birds, and, of course, mammals. Bloodsucking flies are the usual **vectors,** although one species is known to be transmitted by a leech and another, *Trypanosoma equiperdum,* the cause of a disease in horses and mules, is transmitted from one horse to another during mating.

Most trypanosome diseases can be controlled by injecting compounds of the heavy metals mercury, antimony, or arsenic. Paul Ehrlich, who discovered that heavy metal compounds are effective against the organism of syphilis, also found as long ago at 1907 that heavy metals and some dyes are more poisonous to trypanosomes than to men. Yet much remains to be learned about how these agents work.

Because heavy metal compounds will combine with the $-SH$ (sulfhydryl) group of glutathione, a substance found in most cells, it is believed by some that this is the mode of action of the mercurials and other metal trypanocides. Others hold that they attack the $-SH$ groups on enzymes.

Many of the most serious diseases in animals from earthworms to silkworms, rats, cows, and humans are due to protozoans called Sporozoa. They get their name from the fact that those which are not transmitted from one host to another by a bloodsucking insect usually pass from one victim to another in protective capsules called **spores** which contain either zygotes or juveniles. The most important such disease of domestic animals is **coccidiosis,** which afflicts poultry and cattle. The parasite lives within the cells of the host's intestine. **Malaria,** one of the longest known and most widespread of all human diseases is due to a sporozoan. The discovery of its causative agent and how it is transmitted opened a new era in medicine although today more people still suffer from malaria than from any other single disease.

Malaria. The story of how the cause of malaria was discovered well illustrates the way science advances by the cumulative efforts of many individuals. A good point of departure is the arrival in London in the closing years of the last century of Ronald Ross, a young army surgeon from India. He went to the famous St. Bartholomew's Hospital determined to find out

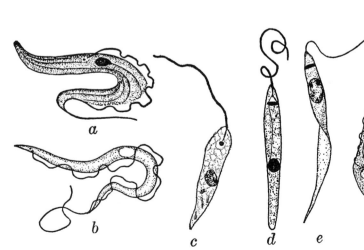

Fig. 33-10. Various trypanosomes taken from (a) a frog, (b) a newt, (c) a man with oriental sore, (d) a house fly, (e) a milkweed plant, and (f) an insect. (From R. W. Hegner: *Invertebrate Zoology.* The Macmillan Co., New York, 1933.)

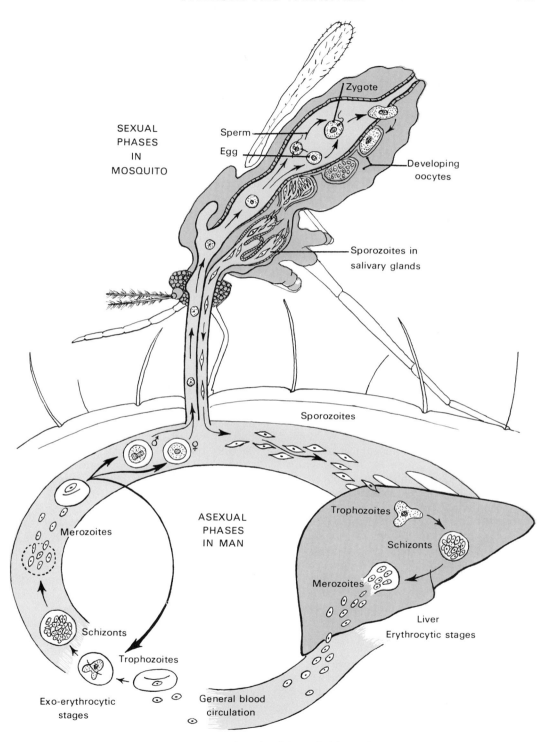

Fig. 33-11. Life cycle of the malarial parasite.

all he could about malaria. There he learned the varied facts that had been uncovered; later he drew them together with new ones of his own into a convincing explanation of the true cause of this great scourge. Following are some of the things he learned.

Some 14 years earlier, Alphonse Laveran, a French Army doctor stationed in North Africa, had found microscopic parasites in the blood of malarial patients. Laveran's discovery had been confirmed by Golgi in Italy who described how the parasites break out of the red blood cells into the blood plasma every time the patient has one of the periodic fever chills characteristic of malaria. Ross learned also of the speculations and observations of Patrick Manson and entered into correspondence with him. Manson practiced medicine along the China coast. He was credited with the discovery that elephantiasis, a disease characterized by great swelling of the legs and caused by a minute worm, was transmitted by the bite of a mosquito. He thought that malaria probably was similarly transmitted.

Several other lines of work contributed to Ross's synthesis. One was the development by Romanowsky of ways to stain blood cells and the malarial parasites in them. Another ingredient that contributed to the final result was the discovery by Danilewsky that many animals such as lizards and birds are subject to malaria.

Ross returned to India and, fortified with all this knowledge, was able to discover how malaria is transmitted from one person to the next. He found that the infection was not carried by bad air but by the bite of a particular kind of mosquito, the *Anopheles,* or, as Ross said, the dapple-winged mosquito. He also discovered the life cycle of the malarial parasite within this species of mosquito. This discovery was the real turning point. Ross then succeeded in transmitting bird malaria from one bird to another by allowing mosquitos to bite first infected birds and subsequently bite healthy ones.

Various workers soon demonstrated the transmission of human malaria by mosquitos. An unpleasant controversy arose as to who really first discovered the cause of malaria. The report of an international commission resulted in a Nobel prize for Ross. But the controversy illustrates the fact that often when a scientific

problem is ready for solution, *i.e.,* when the theoretical and factual backgrounds exist, several people can reach the solution at about the same time. One important aspect of malaria remains to be considered, the social or **public health** aspect. Ross himself made a major contribution with his malaria statistics. While the actual figures he used only hold for the special conditions under which he worked, they illustrate in simple fashion the nature of **epidemiology**, the science of distribution of diseases.

Ross found that only one out of four mosquitos succeeds in biting someone. The chance that a mosquito will live the 21 days required for the sexual phase of the parasite to take place so that sporozoites are in the mosquito's saliva is 1/3, which is another way of saying that within 21 days two out of every three mosquitos are either killed by dragonflies, hit by raindrops, or meet their fate in some other manner. This means that the chance of a given mosquito both biting someone and living three weeks is ¼ × 1/3, or 1/12. This is an application of the **product law** discussed in Chapter 7. From such data the likelihood that a disease will increase or decrease can be predicted. For example, if there is one case of malaria on an island with 1,000 people and one mosquito, the chance that that one mosquito will bite the one malarial patient is ¼ × 1/1,000, or 1 in 4,000. The further chance that this extremely unlucky mosquito will live 21 days and then bite someone is 1/4,000 × 1/3 × ¼, or 1 in 48,000.

The first known therapeutic agent effective against malaria was quinine, an alkaloid derived from the bark of the cinchona tree. Its use was discovered by Peruvian Indians and was introduced into Europe by Jesuit missionaries. It is very effective against the asexual stages in the blood but ineffective against the stages in the fixed cells of liver and spleen. Consequently, patients treated with quinine are usually subject to relapses, sometimes every nine months or so.

The first successful attempt to produce a substitute for quinine was achieved by a team composed of students of the great German pioneer in bacteriology, Robert Koch, and chemists of the I. G. Farben Trust. Other workers in England, the United States, and elsewhere have continued to synthesize possible antimalarial compounds. During World War II over 14,000 compounds were tested in the United States alone. There are now a dozen or more anti-malarials available much as Plasmochin, Atabrine, chloroquine, Daraprim, and others.

Parasitic Flatworms

The trematodes or flukes are parasitic flatworms known from ancient times. Some of the oldest Egyptian mummies contain embalmed trematodes and their eggs, yet our knowledge of their rather complex life histories has been achieved only within the past 75 years. They infect both man and his domestic animals as well as innumerable wild hosts. From the biological as well as medical and veterinary points of view trematodes may be divided into two major groups. One group consists of the **monogenetic flukes** which are external (or semi-external) parasites living on a single host and reproducing only sexually. The most readily found monogenetic trematodes are species which live in the mouths, nostrils, or urinary bladders of frogs. The gills of fish are also often infested.

The other group consists of the **digenetic flukes** which are internal parasites such as liver, lung, intestinal, and blood flukes. They have complex life histories involving both sexual and asexual reproduction and two or more hosts, commonly a vertebrate and an invertebrate. A digenetic trematode, the sheep liver fluke, *Fasciola hepatica,* was the first trematode known, and the first to have its life history worked out. It is prevalent in regions where sheep are raised, but fortunately is rarely found in humans. The Chinese liver fluke, *Clonorchis (Opisthorchis) sinensis,* regularly infests human beings and is widespread in Asia. Infection is contracted by eating uncooked fish. Many frogs that appear in vigorous health harbor lung flukes, (*Haematoloechus medioplexus*), which are conspicuous to the naked eye when the frog's lungs are cut open.

Most digenetic trematodes live in the intestines of their **definitive host,** *i.e.,* the host in which the parasite becomes sexually mature. The most common is the giant intestinal fluke, *Fasciolopsis buski,* found throughout the Orient. The adults grow to be over 7 inches long

and produce an intestinal ulcer at each point of attachment. In addition to physical injury, they produce toxic substances which are absorbed in the bloodstream. Infection, especially in children, is incurred by eating uncooked aquatic plants.

The blood flukes, or schistosomes, have been the scourge of the Nile valley from time out of mind and remain so today in Egypt, over much of Africa, and in large parts of Asia and South America. Schistosomes live chiefly in the small blood vessels of the mesentery of the lower intestine; snails are the intermediate hosts.

Life Cycles of Parasitic Flatworms. The digenetic trematodes living deep within the lungs, intestine, blood, or other tissues of their victims commonly alternate between a vertebrate and an invertebrate host. **Eggs** are laid almost continually and in enormous numbers during adult life in whatever definitive vertebrate host the worm becomes sexually mature. The size and shape of trematode eggs enable the species infesting a man or an animal to be diagnosed. The eggs are passed to the exterior with urine, feces, or sputum. If the egg falls into water, a microscopic ciliated larva called a **miracidium** hatches. To survive, the miracidium must penetrate the tissues of a snail or other small invertebrate within 24 hours. Within the snail the miracidium loses its cilia and grows into a **sporocyst.** This is a small worm-like creature lacking mouth and gut but containing 8 or 10 groups of embryonic cells. Each group of cells forms a worm that may become another sporocyst or a larger worm having at least a rudimentary gut. The worms so formed are called **redias** for their 17th-century discoverer, Francesco Redi (1626-1697), the Italian poet and naturalist known for his pioneer work on the origin of insects from eggs. A redia never leaves the body of the snail but produces, by a process of internal budding, tailed worms called **cercarias.**

A **cercaria** (*kerkos,* tail) superficially looks like a minute tadpole. The body is a small worm with a simple gut and no reproductive structures, but with several adaptations for its life. These include a pair of eyespots, the muscular tail, which is forked in many species, and either some cyst-producing glands or glands which pour out a powerful digestive enzyme that enables the cercaria to penetrate the host's

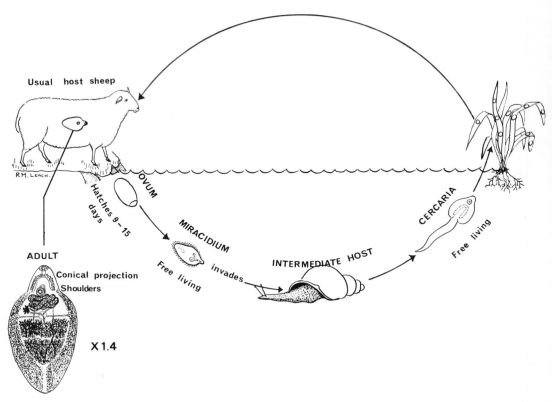

Fig. 33.12. Life cycle of a liver fluke with a snail as intermediate host. Not to scale. (From H. D. Jeffrey and R. M. Leach: *Atlas of Medical Helminthology and Protozoology.* E. & S. Livingstone, Ltd., Edinburgh, 1966.)

skin. The cercarias break out of the redia in which they were formed, pass through the tissues of the snail, and finally swim away in the water.

The cercarias reach the definitive vertebrate hosts in a variety of ways. Those of the schistosomes reach the blood vessels of their victims by directly pentrating the skin of persons who work in irrigated rice fields or walk through puddles. If the proper host of a given kind of cercaria is not a man but a bird or other animal, the cercarias may still penetrate the human skin but are killed on the spot by a local inflammatory reaction called swimmer's itch in freshwater regions and clam-digger's itch around salt water.

In many species such as the sheep liver fluke, *Fasciola,* and the giant intestinal fluke, *Fasciolopsis,* the cercarias become encysted on some plant in or near the water and must wait until eaten by the sheep or man. The cercarias of the

Chinese liver fluke encyst within the muscles of fish and there wait to be eaten. The cercarias of the frog lung fluke encyst in dragonfly larvae and other small aquatic animals until they are lucky enough to be eaten by a frog.

Once within the definitive host the cercaria develops directly into a sexually mature adult. Usually some migration through the body of the host is first necessary to reach the favored spot. The ingested cercarias of the sheep liver fluke burrow across the intestinal wall and into the coelom, whence they reach the surface of the liver. They enter the liver and lodge in the bile ducts. In the case of schistosomes, the cercaria are carried around by the bloodstream before they lodge in the capillaries of the inferior mesenteric and rectal veins.

The human cost of trematodes is great in terms of money and in terms of chronic ill health, or sheer misery, or death. Control measures include eradication of snails as intermediate hosts, care not to eat uncooked food or drink unboiled water in regions where these plagues are endemic, and rigid enforcement of sanitary codes to prevent raw human excreta from contaminating bodies of water.

Tapeworms

Tapeworms (**cestodes**) are flatworms which lack both external ciliation and a digestive system and possess a **scolex**, or head, provided with suckers, hooks, or other adhesive organs. The fully mature adults vary according to the species from small worms 25 mm. or less in length to giants over 10 meters long. The scolex is very small, the elongate, ribbon-like body is usually composed of three or four to three or four thousand segments called **proglottids**. With almost no exceptions, the sexually mature adult lives in the intestine of a vertebrate. No group of vertebrates from fish to man is immune from these debilitating parasites of which there are over 2,000 species (see Fig. 6-17).

The immature bladder worm or **cysticercus** stage is found in the muscles, livers, brains, and other organs of cattle, pigs, rabbits, men, and a wide variety of other animals. Until about a century ago it was believed that tapeworms arose spontaneously from eating too much of the wrong kind of food or from other vague causes. Mankind is indebted to a gynecologist named Kuchenmeister for demonstrating the actual origin of tapeworms. He showed that if you feed bladder worms from raw pork to dogs, tapeworms will appear in the dogs' intestines. He even demonstrated this in human beings by feeding bladder worms to a condemned criminal.

Nematodes

No group of animals that is so little known to the general public has anything like the economic, medical, or scientific importance of the nematodes, or roundworms. Nematodes are second only to insects in the damage they do to agriculture. They also cause numerous debilitating and fatal diseases in man and in both wild and domestic animals. No vertebrate is known that cannot harbor parasitic nematodes, nor are crustaceans, mollusks, insects, or centipedes free from them. The United States

Fig. 33-13. Potatoes grown in nematode-infested soil are too small to be worth harvesting. Nematodes that attack roots deprive plants of essential nutrients. (Courtesy U. S. Department of Agriculture.)

Department of Agriculture estimates the annual nematode damage to our agriculture at about $2 billion. The list of crops seriously injured by them includes beets, carrots, chrysanthemums, cotton, mushrooms, peanuts, potatoes, rice, tobacco, tomatoes, wheat, and many others. Nematology deserves to rank with entomology in agricultural and medical importance.

Nematodes are among the most adaptable animals known, from an ecological standpoint. Not only are there some species that can live in almost any conceivable environment, but many single species live in widely different habitats, from ponds in northern Alaska to jungle lakes in Brazil, in both waters and soils of many types. Most soil and water-living species are semi-microscopic, many parasitic species grow to be 20 or more mm. long, and some, like the common *Ascaris* which inhabits the intestines of humans and many of his domestic animals, grows to be 30 cm. long. In a few cases it has been shown that a single species may have physiologically differentiated races which, although anatomically indistinguishable, nevertheless flourish under different conditions. Perhaps it is a matter of adaptive enzymes; nobody knows. Beyond the simple facts that their reproductive powers are prodigious, and that nematodes occur in enormous numbers, little is known concerning their population dynamics.

In diet many nematodes are carnivorous, eating protozoans, small annelids, rotifers, insect eggs, and even other nematodes. Some are herbivorous, living especially on roots, others live on decaying material, and many are parasitic. With many of the soil-living species the distinction between free living and parasitic becomes very fuzzy. Thus nematodes occupy many riches in the ecology of the soil.

Nematode Diseases of Plants. Among the best known and most destructive of the nematodes injurious to crops is the root knot nematode *Heterodera*. It produces swellings called galls on the roots of potatoes, turnips, tobacco, sugar beets, and some 75 other field and garden crops, vegetables, fruits, and weeds. The juveniles penetrate young roots, which respond by forming a swelling in which the worms mature. The females swell up and when sexually mature become pear-shaped or egg-shaped, though only about a millimeter long. They never leave the gall, but merely spread their eggs in the soil when the root dies and disintegrates. The males resemble normal nematodes and wander about. In hot weather the life cycle requires less than a month. The eggs and juveniles within old galls can withstand drying but fortunately are killed by freezing.

Hookworm. The most widespread and important parasitic nematode is the hookworm, three species of which infest mankind. Adult hookworms live in the human intestine and are attached by their mouths to the villi. They ingest blood and secrete a poisonous anticoagulant that causes anemia and general weakness.

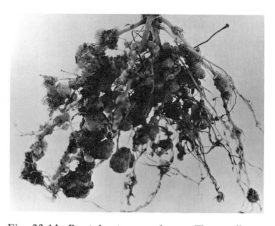

Fig. 33-14. Root knot on soybeans. These galls are caused by nematodes and are not to be confused with the nodules formed by nitrogen fixing bacteria living symbiotically in the roots of legumes. (Courtesy U.S. Department of Agriculture.)

The females are slightly over a half-inch long; the males are appreciably smaller. A mature female lays 5,000 to 10,000 eggs daily and usually lives a year or two.

The eggs already contain embryos when laid, and pass to the exterior or in the host's feces. The eggs hatch in the soil, releasing larvae, which eat, grow, and molt twice in the soil before becoming infective. Optimum conditions are well-aerated topsoil, moist but not wet, absence of direct sunlight, and a temperature of about 90° F. (33°C.). The infective larvae appearing after the second molt are long, thin, so-called **filariform larvae**. They penetrate the skin of the feet, hands, or any other parts of a human body which come in contact with infected soil, and thus cause "ground itch." Within the human body the larvae are carried by the bloodstream to the lungs, where they may cause appreciable damage. From the lungs the growing worms make their way up the windpipe to the throat and thence down into the intestine.

Many people in the United States, mostly in rural areas in the South, are still afflicted with hookworm, *Necator americanus*. The number in Mexico, South America, and Africa also infected is said to run into the millions. Around the shores of the Mediterranean is found a slightly different species, *Ancylostoma duodenale*. In the Orient both species flourish. A United Nations report estimates a world total of over 600,000,000 cases. Control consists of sanitary disposal of human feces, wearing of shoes, and the general avoidance of contact with contaminated soil.

Trichinella. *Trichinella spiralis* is also a very serious threat to human health in the United States and in many other areas. The infection is contracted by eating the insufficiently cooked meat of some omnivorous or carnivorous animal, usually pork. Severe cases, however, have followed eating bear steaks, and it would be a hazard of cannibalism. The larvae lie encysted in muscles or other tissue—brain, for example. When eaten and the cyst digested, the larvae burrow into the host's intestinal mucosa, where they grow and mature. Each female worm deposits about 10,000 larvae which migrate via the bloodstream and encyst in muscles and other sites. It is this encystment that causes most of the damage, especially

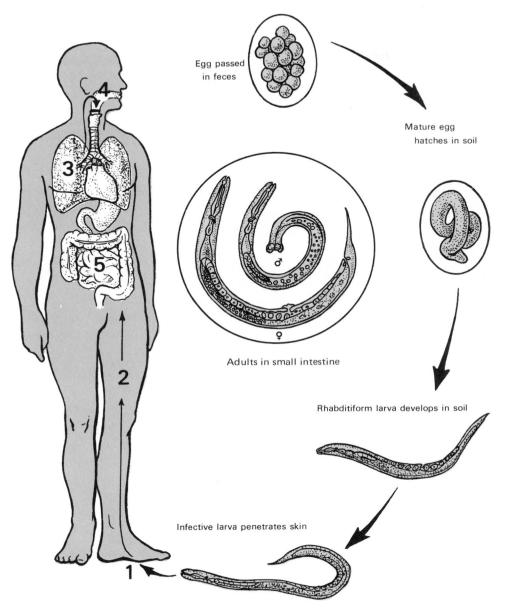

Egg passed
in feces

Mature egg
hatches in soil

Adults in small intestine

Rhabditiform larva develops in soil

Infective larva penetrates skin

Fig. 33-15. Life cycle of a hookworm.

when the site is the heart or brain. Prevention is possible simply by thorough cooking of the meat. This is an ecologically sound practice because it not only protects the human beings concerned but also prevents continuing the cycle by eliminating the possibility of infected garbage which might be eaten by dogs, pigs, rats, or other animals.

Other Animal Parasites

Among insects the ichneumon wasps "parasitize" other insects by laying their eggs on them, usually on the larvae. *Habrobracon* is one which has been much studied and is available for laboratory culture. Other species have been raised to control the numbers of various undesirable insects. Among vertebrates, lampreys, which attack fish and suck their blood, and bloodsucking "vampire" bats are commonly classed as parasites.

USEFUL REFERENCES

Bower, F. O. Parasitism in flowering plants. In *Selected Botanical Papers,* ed. by I. W. Knobloch. Prentice-Hall, Inc., Englewood Cliffs, New Jersey, 1963.

Brown, H. W. *Basic Clinical Parasitology.* Appleton-Century-Crofts, New York, 1969.

Burton, M. *Animal Partnerships.* F. Warne and Co., New York, 1970.

Deverall, B. *Fungal Parasitism.* St. Martin's Press, New York, 1969.

Leclercq, M. *Entymological Parasitology.* Pergamon Press, New York, 1969.

Margulis, L. *Origin of Eukaryotic Cells.* Yale University Press, New Haven, 1970.

Noble, E. R., and Noble, G. A. *Parasitology.* Lea and Febiger, Philadelphia, 1964.

Raff, R. A., and Mahler, H. R. The non-symbiotic origin of mitochondria. Science, *177:* 575-582, 1972.

Read, C. P. *Parasitism and Symbiology.* Ronald Press, New York, 1970.

Read, C. P. *Animal Parasitism.* Prentice-Hall, Inc., Englewood Cliffs, N. J., 1972.

Scott, G. D. *Plant Symbiosis.* St. Martin's Press, New York, 1969.

Animal Behavior

34

The study of animal behavior is one of the major thrusts of contemporary biology. The present renaissance of research in this field grows out of the universal and enduring human interest in what animals do and how they come to do it. This intrinsic interest receives additional force from the theory of evolution. Animals are indeed our brothers in an undeniable biochemical and anatomical sense. Knowledge of animal behavior in general and especially of primate behavior will certainly throw light on the forces underlying human behavior. Gastric ulcers can be produced in monkeys subjected to continual severe worry. The biochemistry of memory is probably the same in a man as in a goldfish, but the overriding fact is that, while man is clearly both a vertebrate and a primate, like other species of primates, he is unique and fully as different from a baboon as a baboon is from either a lemur or a chimpanzee. Anthropomorphism, attributing human personal characteristics to animals, is an ever-present temptation to be avoided as is its opposite, an uncritical extrapolation from animals to man. Therefore, we must come to know ourselves and seek our own behavioral destiny.

THE MODERN REVOLT

The modern explosion of research in animal behavior is the result of a three-fold revolt.

One object of that revolt was an excessive concentration on the study of conditioned reflexes. Under the leadership of the pioneer Russian physiologist, Ivan Pavlov, investigators all over the world studied how a dog will secrete saliva or gastric juice at the sound of a bell or at the sight of a light if such a stimulus is presented several times immediately before the dog is given meat. It was readily admitted that such studies could reveal a great deal about learning and that a wide variety of reflexes in many different kinds of animals from pigs to goldfish can be conditioned. The studies in themselves are valid. The inadequacy lies in the fact that animals do far more than salivate or secrete gastric juice. To study only conditioned responses is to look at only a very small part of animal behavior.

The second object of this revolt was the psychiatric approach to the study of behavior initiated by Sigmond Freud and, in a somewhat changed form, by Carl Jung. From the scientific point of view there are two insurmountable difficulties here. The Freudian school does not provide a scientific methodology based on rigorous evidence. The use of controls is neglected and the statistical treatment of data ignored. Second, the method of collecting the basic data is not only subjective but entirely inapplicable to the study of animals. How can a bird tell you what it dreams about before it starts on a thousand-mile migration to its distant breeding grounds? A spider, soon after hatching, will spin a complicated web in the precise way its parents spun their webs, even though both parents died and their webs were blown away by the winter winds months before the spider hatched. There may well be something here that in some way corresponds to Jung's "racial memory," although Jung died before the importance of DNA was recognized. In any case, neither bird nor spider can be

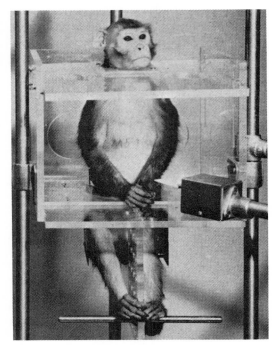

Fig. 34-1. Gastric ulcers in "executive" monkeys. If a monkey in this apparatus is given electric shocks at random over which he has no control, no ulcers develop. If he must interpret signals correctly in order to press the right lever to avoid shocks, then ulcers do appear. (From J. V. Brady: Ulcers in "executive" monkeys. Scientific American, *199:* 95, 1958.)

placed on a psychiatrist's couch and asked to retell its infantile experiences.

The third facet of the revolt was against the popular American school of maze running and puzzle boxes. Here, as with conditioned reflex studies, the objection was not that it is wrong to investigate how animals learn, or fail to learn, to run a maze or escape from a Thorndike puzzle box. The maze today is still a very useful tool. The new widely used Skinner box, which has proved itself a powerful means to investigate such problems as learning, motivation, and concept formation, is a direct development of the old Thorndike puzzle box. The problem is that animals do much more than run mazes or press bars, to escape or receive a reward, when confined in a box.

The conditioned reflex and maze-running approaches came to be fused into extreme environmentalism, a general view of animal

behavior still held by some people. Under the leadership of the late John B. Watson of the Johns Hopkins University, the idea grew that all behavior is due to learning, *i.e.,* conditioning. The concept of the 17th-century philosopher, John Locke, that the mind of every man begins as a clean slate, a *tabula rasa,* seemed to have been validated by modern experimental methods. Many writers, who were really familiar with only two animals, man and the white rat, extended this idea from humans, where it clearly has much truth, to include animals in general. Thus, it came to be widely held that all behavior is the result of a few simple drives and responses plus conditioning based on trial and error or imitation. Instinct especially was anathema, totally inadmissible; innate hereditary factors could safely be ignored because the only meaningful influences are environmental. That any such simplistic view is inadequate and wrong has become evident.

Like most revolutions, the new era in the study of animal behavior had many forerunners. Charles Darwin devoted an entire chapter of his epoch-making *Origin of Species* to the origin and evolution of instincts. Around

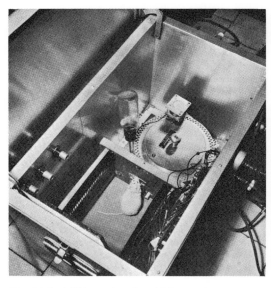

Fig. 34-2. A Skinner box in which an animal presses a bar to obtain a pellet of food and the number and timing of its actions is automatically recorded. (Kindness W. Fleischer, Harvard University News Office.)

the turn of the century, Jean Henri Fabre devoted much of his life to the detailed observation of insects and other arthropods, paying special attention to innate behavior patterns.

Winged ants, for example, emerging from their subterranean nurseries merely climb up a blade of grass, spread their wings, and fly up into the sky. There is no period of learning to fly that remotely resembles that of a child's learning to walk or to roller skate. In the United States, George and Elizabeth Peckham at Cornell made entomological history with their studies of stereotyped, species-characteristic, and complex behavior patterns in wasps. More recently, Oskar Heinroth in Germany published his now classic studies of species-specific courtship and reproductive behavior in different kinds of ducks (See Fig. 31-1).

THE MODERN SYNTHESIS

Four Basics

The present means of attempting to understand animal behavior is an ecumenical one of collecting, testing, and analyzing information obtained by many methods and by investigators of many schools. Four main components of the modern synthetic approach can be distinguished.

First, there is a real concern for observing the whole behavior of animals in their natural surroundings and for attempting to understand the evolutionary and adaptive meaning of that behavior. A scientist who only observed how baboons respond in a Skinner test box might never suspect the complex social organization of a baboon troop. He could not be expected to discover that male baboons, so fierce and powerful that only lions do not fear them, nevertheless make devoted fathers. Furthermore, anyone who studies a wide variety of animals in nature can only be impressed with the complexity and uniformity of their innate behavioral patterns. This emphasis on the full spectrum of behavior as it occurs in nature was the contribution of pioneers like Konrad Lorenz in Germany, Niko Tinbergen in the Netherlands and later in England, and their followers who called their science of the whole behavior of animals **"ethology."**

Fig. 34-3. Winged reproductive ants emerging from their underground home for their first and only flight. (Kindness W. H. Amos.)

Second, and equally important, modern investigators regard controlled laboratory investigations as an essential part of the study of the sum total of behavior. Anyone who observed a troop of baboons as it moved through the edges of the forest could not be expected to discover the extraordinary precision of the 32-day reproductive cycle of the females, nor the way hearing a fight between other baboons, even though unseen, will interrupt that cycle.

Third, it is basic to the modern view that all behavior is gene-dependent. This is because the DNA carries the blueprint according to which the animal develops. Within the DNA are the instructions by which each kind of spider constructs its own kind of web. Clearly there are many steps between the double helix and the web, but the pattern of the web is in the double helix nonetheless. Similarly, the DNA carries the information which results in the kind of brain that endows a dolphin or a person with such remarkable intellectual capacities.

No case more neatly shows the way genes can produce innate differences in behavior between closely related animals than a comparison between two subspecies of the deer mouse, *Peromyscus.* One lives in woods and climbs readily while the other lives on the plains and does not climb even when placed among bushes or trees. If, however, both are raised in cages with no opportunity to climb until fully grown, both climb well when given something to climb. If both are raised in cages with logs to climb, both species begin to climb as soon as their eyes open. The eyes of the plains species open before their leg neuromusculature can provide sufficient support. Consquently, they fall and apparently learn that climbing is a bad thing to do. The eyes of the woodland species do not open until after they can adequately support themselves. They climb successfully from the start and continue to do so.

The question of the role of heredity in behavior usually suggests clear-cut cases of sharply defined bits of behavior which appear to be independent of previous learning. For example, a newly hatched male mosquito does not have to learn what the "song" of a female mosquito is like. He does not begin by exploring many kinds of sounds until he discovers which is a female of the right species. Without previous sexual experience he will fly to a tuning fork emitting the same tone as that of a female mosquito.

Barking styles in dogs follow Mendel's laws. Some breeds of dogs, such as bloodhounds, beagles, and springer spaniels, are known as "open trailers." When following the fresh scent of a rabbit, they "give tongue," a characteristic loud baying and yelling sound. Other breeds, like the airedale, collie, and fox terrier, are "mute trailers," and will follow and even overtake their prey without making any noise. Eight different crosses have been made between these breeds, always with the same result, regardless of which breed was represented by the mother. The first generation always barked when following a "hot" trail, but the bark was always the yapping of the mute trailer rather than the rolling bay of the hound. The collie-hound cross was followed into the second

generation. Random segregation of the factors for hair length and type of bark produced all four possible combinations: short hair and baying, short hair and yapping, long hair and yapping, and finally the double recessive, long hair and baying.

Fourth the modern view is permeated with the realization that biological events, especially behavioral events, occur on many levels of organization, from the molecular and even atomic up to the levels of cells, organs, organisms, and populations. One goose cannot fly in a V formation; only a group of geese can do that. The instructions for geese to fly in that pattern are written in the DNA within the chromosomes of each goose. But between those purine and pyrimidine sequences and that honking, excited, flying V headed to the far north is a long sequence of steps by which each goose, with its characteristic nervous and muscular system, is constructed. The endocrine system must also be properly formed, but above all, the sense organs and the brain must be developed.

Where the Whole Is More Than the Sum of the Parts

All of these steps from the molecular to the anatomical level are important for a complete understanding, but it should never be forgotten that the essential difference between the behavior of animals high and low on the scale of life is not primarily a matter of biochemistry but rather of kind and degree of organization, specifically of cellular pattern. The brain of a man and the brain of a codfish are difficult indeed to distinguish biochemically. The neurons of both contain Nissl substance, which is the ribonucleic acid characteristic of nerve cells. In both cases the motor fibers are coated with lipid-rich myelin. A nerve impulse is a wave of electrical negativity of precisely the same kind in each case, although its speed is a bit slower in the cold-blooded fish. At the synapses where impulses pass from one nerve to another, the chemical compound acetylcholine functions alike in fish and man.

Thus the difference in the behavior of the fish and the fisherman is grounded in the enormous complexity of cellular organization in the man. This applies first and foremost to the nervous system. The cerebral cortex of a fish is a thin sheet; that of a human is a thick, convoluted structure with six layers, each many cells in thickness. While the nervous system is specialized as the determinant of behavior, many other systems play important roles—some more, some less obvious. A fish cannot bait a hook because it lacks hands; a man cannot obtain his oxygen from the water because he lacks gills. Without adrenal glands a man cannot become angry in any complete sense of the word. The most he can do is think about anger. All of these considerations have given new life to the old science of anatomy, especially the study of the gross structure and the cell architecture of the nervous system.

So it becomes evident once again that in passing from one level of organization to another, new properties emerge. When the cells are separated and their contents analyzed, valuable knowledge is obtained, but it throws only an indirect light on the problems of brain function. This is what biologists mean when they say that the whole is greater than the sum of the parts. To resort to a crude analogy, a chemical analysis of the metal sides and the leather head of a drum is inadequate by itself to explain the drum. The leather might as well be a shoe or an apron; the metal nails or hammers. The ultimate problems of the nervous system, which in some way make possible both artistic creation and conscious reason, are certain to be even more subtle and baffling than we can now suppose.

Stimulus and Response

In a nerve-muscle preparation, a stimulus is a relatively simple thing, perhaps a slight electrical charge applied to a nerve, and the response is the contraction of a muscle. The behavior of whole intact animals is built of just such events but in a far different and all-important context. Very commonly the character of the stimulus changes with the total situation. It is the entire pattern which is important, the **gestalt**, as psychologists say.

The classic case is the innate escape reaction of a goose to a hawk flying overhead. If a model shaped as shown in Figure 34-4 is pulled along a wire over the heads of ducks or geese, they pay little attention if it is pulled so that

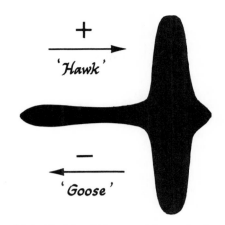

Fig. 34-4. Model which resembles a hawk when moving to the right and goose when moving toward the left.

the long "neck" goes ahead. Perhaps it then resembles a goose. But if the same model is pulled in the opposite direction, ducks and geese scurry for cover. Apparently the long "neck" now appears to be a tail and the short "tail" as a head—and the whole object looks like a hawk.

Of all the potential stimuli impinging simultaneously on an animal (all the sights, sounds, smells, and "feels"), any individual only responds to one or two, presumably those that have some special significance for the animal. When such a stimulus is standard and elicits a predictable response, it is commonly called a sign stimulus. For example, the red breast of an English robin is such a specific sign stimulus that another male robin will attack if it enters his territory. In fact, even a ball of red feathers will be attacked if presented to a male in the breeding condition within his own territory.

Note then three important characteristics of stimulus situations which initiate behavior. (1) The total context makes an essential difference. (2) Not all the physical forces impinging on an animal act as stimuli, but only a very limited fraction of those potential stimuli. (3) The hormonal or physiological condition of the reacting animal makes an important difference.

Sometimes it is difficult to untangle the role of various influences which result in a given response. For example, a billy goat raised in complete isolation from other males will, when he becomes sexually mature, prevent any other billy goat from mating with the she-goats in his flock. The sight of the other male mounting a female precipitates an immediate charge. However, a ram, which will mate with she-goats in the absence of ewes, can copulate with the she-goats without disturbance from the billy. The mating patterns of both goats and sheep are closely similar. However, the odor of the males is very different. Perhaps the proper odor is essential if the visual stimulus is to be effective. Since this whole behavior occurs only in sexually mature males, hormones are clearly involved. The point to remember is that a stimulus-response phenomenon is sometimes a very complex affair. Is the response just described instinctive? Is there a gestalt situation? Is there a sign stimulus? How should the role of odor, or some other difference between rams and billy goats, be regarded?

Appetitive Versus Consummatory Responses

Many responses of animals fall into one of two general types: appetitive or consummatory. Appetitive behavior is more or less random, exploratory behavior. The stimulus is commonly internal, perhaps hunger, which means a lower concentration of blood glucose affecting a nerve center, and perhaps also certain stimuli from the stomach. Consummatory behavior consists of the specific acts that fulfill or consummate the drive or hunger behind the appetitive searching.

Examples abound in widely separated parts of the animal kingdom. A hungry hawk flies hither and yon scanning the ground in the appetitive phase of feeding behavior. The dive and subsequent killing and eating of the rabbit are the consummatory phase. A caterpillar ready to spin its cocoon experiences a compelling wanderlust. When it finds, more or less by chance, a suitable place, the consummatory act of spinning the cocoon is precipitated. In those animals where instinct predominates, consummatory acts are highly stereotyped while in those where intelligence is important, they are learned and highly variable. A hungry spider will always spin a web of a particular type, but who can predict the behavior of a hungry man?

THE WAY OF INSTINCT

Definitions

The word "instinct" has acquired a very bad reputation among psychologists and other critical people, for several reasons. It has been used widely in a very loose sense as a synonym for any habitual or unconscious act, as when an American driving a car in England meets an emergency and suddenly steers over to the right-hand side of the road by "instinct." Worse yet, "instinct" has been used by journalists, literary writers, and others to lend a spurious air of scientific authenticity to ideas which may or may not contain some truth, but for which there is no solid scientific basis. The "work instinct," the "death instinct," and the "leadership instinct" are all examples of this kind of pseudo-scientific nonsense. And to some schools of psychology, instinct is a type of behavior which by definition is unanalyzable.

Charles Darwin, in his *Origin of Species,* began the chapter on instincts by saying: "I will not attempt any definition of instinct." He boldly claimed that everyone understood what was meant and then proceeded to give the example of the cuckoo. This bird lays its eggs in other birds' nests. When the young cuckoo hatches, it ejects the rightful eggs in a particular manner, which is really a gymnastic feat because the rightful eggs are almost as big as the young cuckoo. The North American cowbird also lays its eggs in other birds' nests and, as Lorenz has pointed out, mates when adult, not with the species by which it was raised, but with other cowbirds.

In biology, on the other hand, the word instinct carries a very useful meaning. Konrad Lorenz, one of the leaders in the modern study of instinctive modes of behavior, defines instincts as "unlearned, species-specific motor patterns." An American physiological psychologist, M. A. Wenger, has defined an instinct in essentially the same way as "a pattern of activity that is common to a given species and that occurs without opportunity for learning."

By instincts biologists mean types of behavior which fulfill the following four conditions: They are species-specific *i.e.,* characteristic of a given species, and can be performed without an opportunity for learning although they may change somewhat with repetition. Instincts are more or less stereotyped motor patterns. The more complex ones usually consist of a series of rigid steps which must be followed in a certain unchangeable sequence. Instinctive acts are immediately adaptive, in contrast to the random trial-and-error type of behavior associated with learning. It should be added that although instincts are characteristic of given species, and some are unique to a single species, many are common to several species or even to a whole family, just as anatomical traits are.

The word instinct suggests a fairly complex behavior, like nest building, as compared to a taxis, which is merely the orientation of an organism in a field of force, for example, the caterpillar of the brown-tailed moth exhibits a taxis (a negative geotaxis) by climbing upward when hungry while it spins a particular type of cocoon in a rigid series of steps performed only once in its life with no possibility of learning.

Importance of Instincts

There are several reasons why instincts are important. First, they form a very conspicuous part of animal life, vertebrate as well as invertebrate. The web weaving of spiders, the mating reactions of hens, the migrations of fish, the language of bees, the construction of mud nests by wasps and of hanging nests by Baltimore orioles, the courtship of salamanders—all these and a thousand and one other species-specific, unlearned, and stereotyped motor activities abound in nature.

Because instincts are species-constant they are an aid to zoologists in solving the twin problems of classification and evolution. A particular kind of spider can be identified as surely and far more easily by observing the structure of the web it spins than by looking at its anatomy. Spiders spin webs of four general types. Within each type each species has its own characteristic and uniquely different pattern. Illustrators of detective stories are often rather careless about this but spiders never are. Bird songs are commonly used throughout the world to identify species.

Advantages of Instincts

What is the advantage of instincts over intelligence? Fifty years ago it was the belief of biologists that instinct was a kind of evolutionary precursor of intelligence. This idea has now been almost completely abandoned in favor of the view that the animal kingdom has evolved in two major directions. In one, the arthropod line, instinctive actions predominate; in the other, especially the warm-blooded vertebrates, learning predominates, or at least plays an important role.

The efficiency of instinct is proved by the staggering biological success of insects. The secret is simple: instincts furnish an animal with prefabricated answers to its problems. This is especially important for an animal with a very short life span because no time need be lost in the inevitably wasteful trial and error of learning.

Birds also reveal the advantage of instincts. The type of nest any species builds is determined by instinctive acts. Weaver birds, for example, hand-raised without nests for four generations, will build typical weaverbird nests when given a chance. Try to imagine a pair of robins learning by pure trial and error without any previous experience how to incubate eggs. Should they construct a nest of fine twigs and plant fibers like a mockingbird, or a bulky nest of grass and rootlets like a hermit thrush? Or why not a hanging cup suspended from a forked branch like a vireo? The absurdity of these questions emphasizes the overwhelming importance of instinctive behavior in the lives of birds.

The only known alternative for instinct to enable a bird to build the correct kind of nest would be imprinting, a very early type of learning which will be discussed more fully later. This may play some role in the selection of the nesting site and of the material, although there is no evidence that it does. But as anyone who has tried to put a watch together will testify, knowing what materials to use is a very different thing from knowing how to put them together. In a robin's nest the beautiful bowl of dried mud that lines the rough outer layer is invisible to the nestlings in any case, since it is covered by the inner lining of fine grasses.

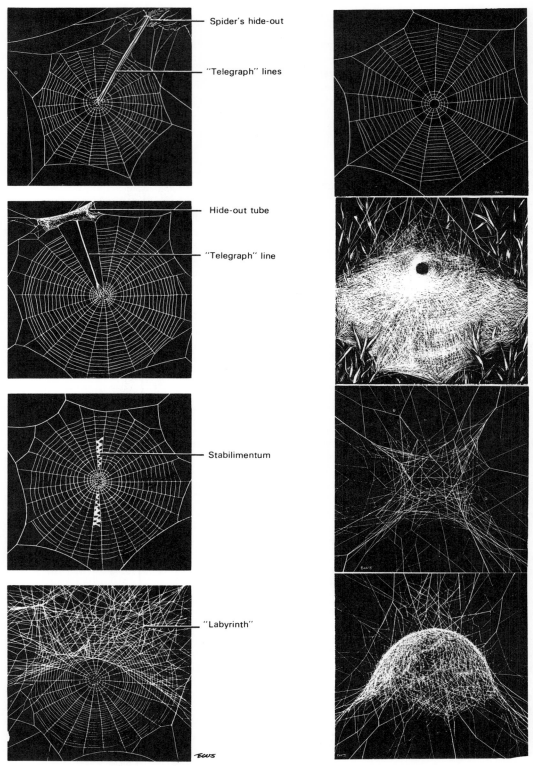

Fig. 34-5. On the right are four basic types of webs. Orb with spokes and spiral, dome, platform and funnel, three-dimensional mesh. On the left are characteristic webs spun by four different species of orb spiders.

In brief, instincts are adaptations in the same way that structures are adaptations. Primarily they are reflections of the structure of the nervous system, although of course the proper organs—wings, silk glands, etc.—must also be present.

Easily Observed Examples of Instincts

Instinctive behavior patterns are easy to see but only if you are at the right place at the right time or can bring certain animals into the laboratory under the proper conditions and have the patience to sit and watch.

Among some of the familiar invertebrates, innate behavior patterns seem so simple that one hesitates to use the word instinct. Certainly the behavior of a jellyfish cannot be compared with that of an arthropod in complexity, yet it is equally species-specific and independent of any known conditioning. Lie on a wharf or on the deck of a boat and watch jellyfish. Unless injured, most species will be doing one of two things. Those that live near shore, like *Aurelia* and *Chrysaora,* the common sea nettle of bays and estuaries, spend their time fishing by first swimming up to, or almost to, the surface of the water, then turning over so the sub-umbrellar surface is uppermost, and slowly drifting to the bottom. When they hit bottom, they turn over and muscular contractions begin which send the medusa upward again in a series of rhythmical contractions. This cycle is repeated hour after hour. Any small animals that become caught in the tentacles are dragged into the mouth and are digested. Jellyfish like the rapidly swimming *Liriope* could not behave in this way since they live where the sea is a mile or more deep. They swim, often in great swarms, but always stay near the surface.

Insects and spiders have long stood as the best examples of animals dominated by instincts. What about instincts in vertebrates, including humans and other mammals? Instincts which are complex, stereotyped, and prefabricated answers to life's problems are really the antithesis of intelligence or learning. Hence, it would not be expected that an animal well equipped with instincts like a spider would be very inventive. Nor would creatures with brains well able to learn be expected to come equipped with a set of highly developed inflexible instincts. In humans, and indeed most other mammals, there seem to be no complex instincts comparable to what can be found in arthropods. Nonetheless, the lives of vertebrates (including mammals) are permeated with innate responses. For example, the behavior of a pseudo-pregnant bitch (to be discussed later), the ways different breeds of dogs bark, and nest building by birds are familiar innate behavior patterns.

THE WAY OF LEARNING

Learning, which is virtually the same thing as intelligence, may be defined as a more or less permanent change in behavior due to past experience recorded in the nervous system. To specify that it is recorded in the nervous system excludes a change in behavior due to a mutilating accident—for example, the loss of a wing. To specify that it is more or less permanent excludes changes which occur after an animal has experienced a full meal or any similar temporary physiological change. A hungry animal and a satiated one usually act very differently, but the hunger will return and with it the appropriate behavior.

It is important to distinguish maturation from learning but it is by no means always easy to do so. Maturation can be defined as changed behavior caused by developmental changes within the nervous system. Before the neurons of a reflex arch have grown into place, clearly the reflex action cannot occur. The classic case is seen in the development of swimming in salamanders. Leonard Carmichael and George Coghill kept developing salamander eggs under complete anesthesia until the control salamanders had passed through all the beginning stages of swimming, from mere twitches to bending like the letter C, to ineffectual wiggles, and finally to swimming actively. The anesthesized salamanders were than placed in pure water and proceeded to swim off as skillfully as the control salamanders without passing through all the preliminary stages which resemble efforts to learn.

Six Major Types of Learning

Habituation

Perhaps the simplest type of learning is called **habituation**. If an animal is repeatedly given a stimulus which is not followed by any reinforcement, *i.e.*, any reward or punishment, the animal ceases to respond after a while. If a shadow is passed over an active barnacle, it will stop kicking its six pairs of double legs and will pull them into its shell. Each time the shadow passes the barnacle responds more slowly and opens out and begins kicking sooner until, after five more passes, it no longer responds. This type of learning is obviously related to habituation to drugs of various sorts. Habituation in this sense is defined as a change which makes necessary an increase in the stimulus, *i.e.*, the dose, to obtain the same result.

Classical Conditioning

A basic type of learning studied with physiological precision is called **classical conditioning**. Its investigation was developed by Ivan Pavlov in Russia and by others in Europe and America. In this type of study the dog, pig, goat, or other animal remains standing in a large box and responds to a stimulus by salivating, raising a foot, or in some other manner. The primary or **unconditioned stimulus** is commonly food or a mild electric shock. If, just before meat is given, the dog sees a red light or hears a bell after a few trials, he will salivate on seeing a red light or hearing a bell even though no meat is given. The light or the bell is then known as the **conditioned stimulus**. Note that the response is a relatively simple reflex type of action.

Instrumental Learning

A third type of learning is **instrumental learning**. The animal is called upon to perform some act, perhaps a rather complicated act, instead of merely exhibiting a relatively simple reflex. The animal must do something, *i.e.*, operate. Hence this is sometimes called **operant conditioning**. Second, the animal is motivated, usually by hunger. The methods are simple. The animal is kept without food or without water

to provide the motivation and is then placed in a maze, a Thorndike puzzle box, or a Skinner serve-yourself box. In the old puzzle boxes the animal had to discover a way to escape. The rat, cat, monkey, or other animal had to push, pull, turn, lift, and otherwise manipulate the various

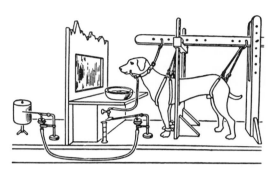

Fig. 34-6. Dog in harness for classical conditioning.

Fig. 34-7. Canary which has learned under operant conditions that the different object in a series, the one tablet among the screws or the one pawn among the bishops, is the object which covers the reward. (Kindness Nicholas Pastore.)

parts of the box, perhaps in a particular order, until he happened on the correct act which opened the door to freedom and food. In the Skinner box the animal learns that by pressing a lever, a simple type of slot machine will serve him a pellet of food or a drink of water. The advantage of this device is that the number of times the lever is pressed can be automatically recorded and thus the rate of learning, the intensity of response, and other factors can be measured objectively. Note that both the Thorndike puzzle box and the Skinner serve-yourself box depend on an initial period of exploratory trial-and-error behavior.

Insight Learning

A fourth type of learning is called **perceptual learning** or, sometimes, **insight learning**. Use of these terms varies and is a reflection of our basic lack of understanding of the process of learning. Insight learning is said to occur when an animal does not go through a period of trial and error, but surveys the problem and then acts correctly on the first try. Insight learning is not regarded as a fourth type of learning by some people, but rather as a variety of trial-and-error learning with the trials and errors made mentally on the basis of previous experience.

Exposure Learning

Exposure learning has been known in a general way for centuries but is still little understood. There are no known rewards or punishments, *i.e.*, no **reinforcements**. A rat will learn a simple maze simply by walking around in it as would a man. Consider the case of the English chaffinch so thoroughly studied by W. H. Thorpe in Cambridge. Only males sing and all have basically the same song but each male has his own particular variation. A male does not begin to sing until a year old (a long time in the life of a bird) yet when he does begin to sing, the song is like the one he heard the father singing when he was a nestling. There was certainly no reward, no "reinforcement" in any ordinary sense of the word. A similar case of the white-crowned sparrow has been investigated by Marler and Konishi. If baby sparrows are raised in soundproof rooms, they

learn whatever "dialect" of white-crowned sparrow song is played to them from a tape recorder. If exposed to both the song of a white-crowned and of a song sparrow, they only learn the white-crown song. If played only the song of a song sparrow, they learn nothing and their vocalizations are mere avian gibberish. Evidently there is something very special about exposure learning in birds.

Imprinting

A sixth and very special type of learning has been discovered in recent years and has been named **imprinting** by the ethologists who discovered it. Typical imprinting takes place only during a brief critical period occurring several hours after hatching or weeks after birth. It is very persistent and, as in exposure learning, there is no obvious reward.

In ducks, geese, and most breeds of chickens, the newly hatched bird becomes imprinted by the first large moving object that it sees within the first hours after hatching. Thereafter, it will follow that object to the exclusion of all others. Normally this first large moving object is the mother. However, if the eggs are hatched in an incubator and the first large moving object is a man, a male duck, or a mechanical floor polisher, the duckling is imprinted by this inappropriate object and will follow it with great and enduring persistence. A remarkable

Fig. 34-8. Gosling following a box on which it has been imprinted. Note that it is ignoring the adult geese in the background. (Kindness A. Ogden Ramsey.)

fact about imprinting, at least in ducklings, is that if the duckling has to work—for example, is forced to scramble over little obstacles—to follow the box or whatever is being used for the imprinting, then the imprinting becomes more firmly established in less time than were the little duck merely walking over a smooth floor. This fact suggests a relationship between imprinting and operant conditioning. Since there is no reward except the act itself, there seems also to be a relation between imprinting and exposure learning.

In puppies it has been found that there is a critical period centering around the sixth and seventh weeks of life when they become most readily and permanently tamed. If puppies have no contact at all with human beings until after they are 14 weeks old, it is virtually impossible ever to make them into friendly pets.

More important results emerged when Harry Harlow and Robert Zimmerman at the University of Wisconsin undertook to develop a completely healthy colony of monkeys free of the infections of various kinds which are passed on from parents to offspring. The obvious method was to rear monkeys from birth in isolation from their mothers. Such monkeys were healthy but surprisingly abnormal in their behavior. They acted frightened and withdrawn. When they became sexually mature, which they did at the normal time, they showed no interest whatsoever in mating and only a few of the females could be bred even by placing them with patient and experienced males.

How widespread imprinting, in either a narrow or a broad sense, may be in the animal kingdom is not yet known. The indications are that the importance of early experience is widespread among vertebrates. In addition to occurring in birds and mammals it very probably takes place among certain fish. The ability of migrating salmon to return to the small tributary in which they were hatched, even though it is but one of many streams in a river system, may be due to imprinting with the characteristic smell of the water.

Zoological Preconditions of Intelligence

What are the zoological preconditions for the development of intelligence and learning? The

Fig. 34-9. Emotionally disturbed young adult monkeys raised in isolation without either contemporaries or parents. (Photo by Sponholz for the University of Wisconsin Primate Laboratory.)

first prerequisite is an appropriately organized central nervous system. Without an efficient retrieval and coordinating center, intelligent action would be impossible.

Furthermore, if an animal is not born or hatched with a set of ready-made behavior patterns to meet all the usual situations of life, it clearly must have a protected period when it can make errors and learn. This means that, in addition to a highly developed brain, some degree of parental care is essential if a high level of intelligence is to be attained. Two key anatomical structures assure mammals of parental care: the placenta which makes viviparity possible, and the mammary glands which both nurture the offspring and make obligatory a prolonged association between mother and offspring, a period during which learning is possible.

The fact that humans and monkeys had tree-living ancestors is now believed to be a primary reason why intelligence could evolve. It will be recalled from our discussion of the primates that an arboreal existence places a high premium on a skillful grasping hand and good binocular vision capable of judging distance since a miscalculated leap could be one's last. Life in the trees also provides a safe environment away from ground predators. Under these circumstances natural selection could favor the more advantageous use of hands and eyes. In contrast, on the ground everything must be sacrificed to quick reactions. If the cat takes too much time to think, the mouse has escaped. If the mouse hesitates to weigh the advantages and disadvantages of alternative courses of action, it gets eaten. But in the trees there is time to think. Natural selection in historical fact did favor the evolution of intelligence in arboreal primates.

HORMONAL BASIS OF BEHAVIOR

In addition to the neuromuscular system, behavior is profoundly influenced by the endocrine system, especially in vertebrates but also in annelids, insects, and mollusks.

The migrations of birds, as well as their songs, are under hormonal control. Female canaries normally never sing. However, if a pellet of testosterone (male sex hormone) is inserted under the skin, a female will sing a typical canary song for several weeks until the pellet is completely absorbed. It is worth noting that the female has the neuromuscular and vocal apparatus for singing and knows what song to sing. What is lacking is the one key substance, testosterone.

The dominating influences of hormones are familiar in mammals and are nowhere more dramatically revealed than in the pseudo-pregnancy which occurs in certain female dogs. Beagles seem especially subject to this affliction. In the absence of mating such bitches show many of the signs of pregnancy after about nine weeks when puppies would be born. Such dogs not only have enlarged mammary glands and secrete milk but will steal a child's doll to defend and "mother." At other times such a dog is no more interested in dolls than in books. Pseudo-pregnancy can be brought to a quick end within two hours or less by an injection of male hormone.

Very convincing evidence shows that sex hormones induce sexual behavior by direct action on centers in the central nervous system. A female cat which has not only been spayed (ovariectomized), but has also had other reproductive organs (uterus and vagina) removed, can be caused to behave like a normal female in estrus (to welcome a male rather than to treat him with angry rejection) by the implantation of minute pellets of a female sex hormone, diethyl stilbestrol, in the right part of the hypothalamus. If female sex hormone is radioactively labeled with carbon-14 and injected into the bloodstream, it will be found that the hormone is selectively localized by special cells in the hypothalamus.

CIRCADIAN RHYTHMS AND BIOLOGICAL CLOCKS

The behavior of humans and many animals, perhaps all, is greatly influenced by a rhythm having a period of approximately 24 hours, hence the name **circadian** (*circa*, approximately, + *dies,* day) rhythms. The common fiddler crab becomes very dark in color during the day and pale at night. Many rodents are active at night and quiescent during the day. Circadian rhythms have been well demonstrated in *Drosophila,* in *Euglena,* and even in the respiration of potatoes. People often have trouble readjusting their rhythm after long east-west flights. Although the existence of rhythms is a fact, what does dumbfound many biologists is that many rhythms will persist for weeks in an unchanging environment. In the pioneer investigations of Frank A. Brown, H. Marguerite Webb, and their collaborators, fiddler crabs were kept in complete darkness in rooms with constant temperature and humidity. Yet rhythms continued. Their persistence under constant conditions has been abundantly confirmed by many workers and in many species.

The real problem is, how do the animals measure time? Some people think that animals carry some kind of internal metabolic clock, but none has ever been identified. Others think that in some way animals tell time by sensing daily changes in some pervasive factor (most likely changes in the earth's magnetic field) which does exhibit circadian rhythms, but no one has been able to prove this theory. It has been demonstrated that both birds and honeybees have an accurate time sense which enables them to compensate for the different positions of the sun in the sky at different times of day. But as little is known about the mechanisms with these animals as with those of fiddler crabs or fruit flies.

Fig. 34-10. Fiddler crabs, day (left) and night (right) adapted. (Kindness H. M. Webb and M. Berlinrood.)

SOCIAL BEHAVIOR

In the sense used here, social behavior is behavior in which two or more animals are responding to each other. Animals may aggregate because each responds in the same way to the same stimulus but, unless there is some inter-animal reaction, we will not regard it as social. Daphnia will collect by the hundreds at the surface of water if the oxygen is depleted, but this is not a social response because there is no interaction between individuals. A single daphnia would move to the surface as would a hundred.

In marked contrast is the behavior of a newly hatched group of fish or squids. Both exhibit a very strong schooling instinct. If a stick is placed in the water and moved through the school, the little fish or squids can be made to separate into two groups as they swim past the stick. But the divided school snaps back into a single group as though drawn together by powerful magnetic forces just as soon as the leading individuals are a few inches beyond the stick. Many birds are gregarious and migrate and nest in flocks. The precise integration of all the members of a flock of pigeons as they maneuver in the air is a striking sight.

Pecking Orders

One of the most obvious social facts is the presence of dominance hierarchies or pecking orders among various kinds of birds and mammals. For example, among a group of hens, one will soon become dominant, able to peck any other hen in the henhouse and drive her

from food. Hen number two feels free to peck any hen except hen number one. The third hen in the hierarchy can peck any hen except numbers one and two. And so the series continues. Similar series have been found in cows, mice, and other animals.

A number of facts should be noted about pecking orders. The first is that they can arise only among animals that can recognize each other as individuals and have an ability to learn. Nothing like a pecking order has been found among worms, ants, or even bees, where life is governed by innate reactions. Even the queen bee does not dominate the colony in the sense of giving directions.

The position of an animal in a pecking order depends on many variables, including his own past experience. Initially it will probably be the largest, most aggressive, and strongest individual which heads the pecking order. Animals low on the list can be raised to a higher status by several methods. Simple isolation from the group for a time may have this result. Perhaps isolation gives animals close to the bottom of the ladder a chance to get enough food and so build up their strength. Mice can be advanced in a pecking order by isolating them from the group and then arranging a series of encounters with other mice in which they will be victors. Apparently this increases their "self-confidence." Doses of sex hormone will cause chickens to rise in the pecking order. Starlings, however, rise in their pecking order after castration.

Knowledge of pecking orders has some practical value. If a cow who is number one in her herd is transferred to another herd where her position is unknown and where the other cows "presume" she is very low on the scale, she will become morose and fail to eat properly, and her milk production will fall. Chickens and other animals very low on their scales may not get enough to eat to develop properly.

On the other hand, for a species as a whole, a pecking order may possess definite advantages because it almost eliminates the wasted motion of continually fighting over food, drink, and space. Pecking orders cannot develop if the population is too large. Probably there are simply too many animals for any one individual to remember who stands where. The rigidity of the pecking order differs greatly among different species. There is much more give and take among canaries and pigeons than among hens.

Territoriality

Most vertebrates exhibit some degree of territoriality; that is, they will defend a given home area against intrusion by others of the same species. Usually this behavior is limited to the breeding season and to males. Territoriality is especially strong in birds where it was first discovered by Eliot Howard who noticed that male birds demark their territories by singing and by driving off other males. Apparently Howard was the first man in history to understand what bird song is all about. Territoriality has also been studied carefully in the little freshwater stickleback and can be readily observed in spring with sunfish or bass which protect "nests" in shallow water near shore.

Territoriality has not been extensively investigated in mammals, although some authors with literary rather than zoological experience regard it as the dominant force in human life. Perhaps the best studied case is the Eskimo dog observed by Tinbergen. The dogs live in packs of 5 to 10 individuals and mark the boundaries of their domain with urine. Puppies have no sense of territory and are forever getting themselves in trouble by trespassing into the territory of other packs. When they become mature they very quickly learn the limits of their own and other territories. Some of the larger carnivores seem to mark off a home range by rubbing odors on trees.

In the primates, the order of mammals to which man belongs, there is much variation among genera and even among species. The much-studied tree-living howler monkey of Central America shows strong group territoriality. Each troop, of anywhere from a dozen to several dozen individuals, is acutely aware of the limits of its territory and defends it by sessions of howling, audible for long distances. Among the Old World monkeys, the rhesus monkeys or macaques are territorial. A group will move around within the same area day after day and fight off intruders with great ferocity, although threats usually suffice.

Baboons are rather different. Each troop has a vaguely defined home range, and there is much overlapping with the home ranges of other troops. Gorillas and chimpanzees, among anthropoid apes, have recently been extensively observed in the wild by Schaller and van Lawick-Goodall, respectively. In neither case is there any territoriality of the strict and intense sort seen in howlers, and nothing is really comparable to the situation in birds. Both gorilla and chimp troops have home ranges of considerable extent. Neither shows any defense of territory. Other groups may pass through and occasionally intermingle with impunity. On the basis of what is now known about primate behavior, there is no justification for the belief that territoriality is the dominant imperative of human life.

The primary biological function of territoriality, according to present knowledge, is to spread the members of a species more or less evenly throughout the available habitat. The result is that bluebirds or howler monkeys will achieve the optimum utilization of the habitat and its food supply. Conflict within the species will be minimized. In both teleost fish and birds, when a neighboring male enters another's territory and is challenged by the owner, the invader retreats. This is probably true of mammals as well. Dogs are certainly much less belligerent when away from home base than when in their own yards.

When two males meet at the boundary between their respective territories, there may be endless skirmishing but there may also be a very interesting phenomenon called **displacement activity**. A stickleback fish may neither attack nor retreat but perform some third and irrelevant activity such as beginning nest building. Apparently when two strong mutually exclusive responses are equally balanced, and the nervous system as a whole is raised to a high state of readiness (perhaps by adrenaline), some third response is set off or disinhibited.

Conflict and Cooperation

Violent conflict and vigorous cooperation are commonplace in the animal kingdom. There is much talk about "aggression" among animals, but scientifically it is necessary to use a neutral term like conflict or **agonistic** behavior. "Aggressive" carries a strong moral overtone of wrongdoing, whereas "**agonistic**" can be freely used to described a combat between two tomcats competing for the same female in a situation where aggressor and innocent victim seem quite irrelevant or an encounter where one animal only attempts escape.

One of the most important facts to remember is the danger, indeed the impossibility, of extrapolation from one species to another, even closely similar, species. Consider the contrasting behavior when a lion and a tiger get into a fight in a circus show involving several lions and tigers. All the lions cooperate and quickly gang up on the tiger. Meanwhile, the other tigers merely sit on their stools and look at the ceiling. As soon as the cooperating lions have killed the first tiger, they move to the second tiger, and then to the third. All this time the surviving tigers continue to remain aloof. Myrmicine ants (which are common under boards and on plants), if deprived of their antennae, attack friend and foe alike. Formicine ants, like the big black carpenter ant, will feed both friend and foe under the same conditions.

Nevertheless, there are a number of worthwhile points which can be made. The stimuli and situations which will elicit conflict are both varied and numerous. Sign stimuli are the most clear-cut. The male stickleback fish will attack almost any red object that appears within its territory (note the gestalt aspect), even though any resemblance to the red belly of another male stickleback is remote. For some carnivores, dogs and wolves for example, the sight of another animal retreating will precipitate an attack. In some circumstances pain or hunger stimulate conflict.

A highly informative experiment on conflict in the rhesus monkey or macaques has recently been carried out in India by Charles Southwick and others. In a group of males, females, and juveniles in a large outdoor enclosure, the daily average of conflict encounters, fights, threats, hostile expressions, and submissions was established. The amount of food was then reduced from a superabundance to a somewhat restricted level. To everyone's surprise there was no general increase in conflict. The food supply was then reduced to a starvation level. The result was a reduction in conflicts. The macaques spent a large amount of time in slow explorations.

Keeping the food level constant, the effect of crowding was studied by reducing the area by half. This did not significantly increase the number of conflicts. What did produce a real pandemonium of aggressive conflict was the introduction of a strange individual, either male or female. Males took the lead in attacking strange males, and females in attacking strange females, but all members of the group joined in the attacks on the newcomers. It is too early to tell whether these results can be extrapolated to other primates.

There are many biological devices which reduce intraspecific conflict. Pecking orders and territoriality have already been mentioned. The ritualization of conflict is commonplace as in the yelling of howler monkeys, the chest beating of the gorilla, and many other stereotyped threat displays. There may be a correlation here with the degree of sociability of the species. Dogs, which are essentially pack animals, have a method of indicating to their opponents that they give in. A vanquished dog or wolf lies down on its back and assumes a more or less puppy-like posture. The cats of various kinds, which are solitary animals, are much more likely to fight until death because they lack a surrender sign.

Positive cooperation is familiar in the coordinated attacks of a pride of lions or a wolf pack, in the defensive circles of bison and cattle, or in the way male baboons work together to protect their troop from marauding leopards and other predators. Among in-

Fig. 34-11. Marching formation of a baboon troop. The central group consists of the clique of dominant males (green) plus mature females either with young or in heat. On the periphery are subordinant males, females, and juveniles. When the troop is threatened, the dominant males move to the periphery on the side of the danger. (From DeVore, I., after I. Eibl-Eibesfeldt: *Ethology*. Holt, Rinehart and Winston, New York, 1970.)

vertebrates the cooperative work of the social insects is well known. The neural and physiological mechanisms underlying either conflict or aggression are little understood. As already noted, there are centers which control aggression in the hypothalamus of vertebrates. Pheromones, chiefly odors, play an important role in activating both conflict and cooperation. Remember the ants minus their antennae or the rams and billy goats.

Natural selection works to produce both conflict and cooperation. Where males have to fight for females, as among deer and lions, or have to protect the females and juveniles against predators as is true for baboons, natural selection selects the best fighters. Where family or group cooperation favors survival as in baboons, bison, and the social insects, natural selection favors the development of cooperative mechanisms. This is called group or kin selection.

Reproductive and Parental Behavior

In both plants and animals, the primary function of sexual reproduction is the production of hereditary variation in the progeny. The basic biological objective of reproductive behavior in animals is to bring gametes together

and, in many species, to nurture the young. With sessile animals which shed their eggs and sperms into the sea, synchronization of spawning is essential. Usually temperature or light is the coordinating stimulus. The famous palolo worms are timed by the moon to emerge from the coral reefs of the South Pacific and to spawn. Pheromones also coordinate spawning. The gametes of oysters, of *Nereis* the clam worm, and of tunicates carry with them some chemical which stimulates other individuals to shed gametes.

Among many teleosts, reptiles, birds, and mammals, the breeding period is controlled by photoperiod. Day length apparently plays an important role in controlling the time of sexual activity in those mammals which have only one breeding period per year, as with deer or sheep. In some animals, psychological factors also exert control.

Courtship is a prominent feature in the sexual behavior of animals as different as spiders, squids, parasitic wasps, birds, and mammals. Not infrequently, the courtship bears at least some resemblance to conflict. In birds of paradise, courtship involves elaborate displays of incredibly fantastic feathers. In some parrots, life-long pairing takes place between two young birds well before either has reached sexual maturity. The immature male merely gazes intently at an immature female and she gazes back. If this mutual exchange continues for more than a few minutes, the pair is mated for life.

In mammals there are the loners like tigers and rats, where males play no part in the care of the offspring and where there is no known permanent male-female bond. There is the dominant male plus harem as in seals, deer, and horses. There are family groups as in lions and wolves. Recent studies in both Siberia and northern Michigan show that the basic social unit of wolves is a family group of four to six individuals. Large packs are conglomerations of several such families. Likewise the amount of parental nurture, above and beyond the basic mammalian requirement of milk for the young, varies greatly, especially in the role of the male. Among elephants and whales one or more female adults, so-called aunts, attend the birth and help protect the newborn against tigers or sharks respectively. Male baboons are very protective of infants; male langurs, a common monkey of Indian woodlands, are very aloof.

Parental Emotions

It is impossible not to believe that parent apes, dogs, cows, dolphins, and other mammals share in some degree parental feelings of love which we experience. Birds likewise appear to share such emotions. However, a caution against anthropomorphism is warranted. The world certainly appears different to a bird. For example, a hen turkey will fight off any animal that approaches her chicks (more properly called poults). Her behavior resembles that of a human parent under similar circumstances, but a bit of study reveals profound differences. A turkey hen will attack a stuffed fox pulled past her nest. If a tape recorder of turkey poult calls is played within the stuffed fox her hostility completely disappears. Moreover, if a hen has never hatched chicks and is deafened, then when her eggs do hatch she will attack and kill her own chicks as though they were rats.

ANIMAL SOCIETIES

As Charles Darwin pointed out a century ago, Wynne-Edwards very recently, and many

thinkers in between, animal societies and the behavior patterns which make them possible form and evolve under the forces of natural selection and (we would add) mutation. Since any society whose members fail to reproduce ceases to exist, it is hardly surprising that the basis of a society is almost invariably a fertilized female or a family. Societies, *i.e.,* complex social systems, have arisen independently many times among different groups of insects and several times among mammals.

Among insects, the termite pattern includes both sexes as workers, soldiers, and reproductives. The tie which binds termite society together is a potent pheromone secreted by the queen. It is licked off her body by the workers and distributed throughout the colony by mutual cross feeding. Not only does it insure "loyalty," it is also an anti-fertility agent for both males and females which prevents them from becoming sexually mature. When termite colonies become extremely large, functional reproductive individuals appear on the distant fringes of the colony because they are so far from the source of the anti-fertility chemical that they escape from its inhibition.

In wasps, ants, and bees, the entire colony consists of the offspring of a single mating of one female. The workers which hatch from her fertilized eggs are under-developed females. Unfertilized eggs develop into haploid male drones. For some obscure reason, societal life works out well for insects, and there are many dozens of very different unrelated social species.

How does natural selection work in cases where the soldiers and workers are sterile and hence leave no offspring? After a swarm of winged functional male and female ants or wasps leaves the nest, each fertilized female finally settles down alone and begins a new colony. If she carries mutant genes which produce superior workers, her colony will be more successful than others and, although the superior workers transmit no genes because they are sterile, the winning genes will be transmitted through the fertile sons and daughters of the original queen.

In birds of many species, flocking occurs only outside of the breeding season. Nesting pairs of most songbirds are antisocial in proportion to the strength of the sexual bond

Fig. 34-12. Scene from the termite underground. Two workers engaged in mutual feeding. (From C. D. Michener and M. H. Michener: *American Social Insects*. D. Van Nostrand Co., Inc., New York, 1951.)

between them. When this wanes, general social cohesiveness increases.

Many kinds of mammals live in social groups. Lions, porpoises, bison, sheep, wolves, deer—the list is a long one. Moreover, in many instances an individual separated from its group is helpless. A baboon separated from the troop is a dead baboon. Primates are especially notable for the variety of their social relationships. There may be only a loose hierarchy, a single dominant male or, as in the case of baboons, a ruling clique of several older males who cooperate with each other to maintain their authority. These dominant males protect the troop against lions.

Several writers have recently claimed that the biological basis of human societies is the endocrinological fact that the human female remains in a potentially reproductive condition all year. It is possible that this characteristic has played a role in maintaining continual social cohesiveness, especially in the beginnings of human societies; but there is no firm evidence for such a theory. Moreover, there is much evidence that a continuous reproductively active state in either sex is not a necessary precondition for permanent social organizations.

Compared with other mammals, different species of primates show little diversity in female reproductive cycles but very great diversity in social organization. One can hardly imagine more striking social contrasts than the retiring, small-familied gorillas, the gregarious and somewhat rambunctious chimpanzees, the solitary gibbon pairs, a baboon troop, or the loosely organized, female-led howler monkeys.

Among many mammals social bonds exist throughtout the entire year, although the females experience estrus only briefly. Wolf packs consist of families and therefore have a sexual basis. However, the pack is a continuing unit for which the widely spaced estrous periods of the females are important but transient incidents. A Merino sheep separated from its flock any day of the year quickly becomes desperate to rejoin its fellows, yet ewes come into estrus only for a few days in early fall. Clearly in both birds and mammals, social cohesiveness transcends reproductive ties.

USEFUL REFERENCES

Altmann, S. A., and Altmann, J. *Baboon Ecology*. University of Chicago Press, Chicago, 1970.

Beroza, M. *Chemicals Controlling Insect Behavior*. Academic Press, New York, 1970.

Dethier, V. G., and Stellar, E. *Animal Behavior: Its Evolutionary and Neurological Basis*, 2nd ed. Prentice-Hall, Inc., Englewood Cliffs, N.J., 1970.

DeVore, I. (ed.) *Primate Behavior. Field Studies of Monkeys and Apes*. Holt, Rinehart and Winston, Inc., New York, 1965.

Frings, H., and Frings, M. *Animal Communication.* Blaisdell Publishing Co., New York, 1964.

Harker, J. E. *Physiology of Diurnal Rhythms.* Harvard University Press, Cambridge, 1964.

Krishna, K., and Weesner, F. M. *Biology of Termites,* vol. 1 (1969); vol. 2 (1970). Academic Press, New York.

Lawick-Goodall, J. van. *In the Shadow of Man.* Houghton Mifflin Co., Boston, 1971.

Lorenz, K. Z. *King Solomon's Ring.* Methuen and Co., London, 1952.

McGaugh, J. L., Weinberger, N. M., and Whalen, R. E. *Psychobiology: The Biological Basis of Behavior.* W. H. Freeman and Co., San Francisco, 1967.

Southwick, C. H. (ed.) *Animal Aggression: Selected Readings.* Van Nostrand-Reinhold, New York, 1970.

Southwick, C. H. (ed.) *Primate Social Behavior.* Van Nostrand-Reinhold, New York, 1963.

Tinbergen, N. *Herring Gull's World.* William Collins Sons and Co., London, 1953.

Wilson, E. O. *Insect Societies.* Harvard University Press, Cambridge, 1971.

Wood, D. L., Silverstein, R. H., and Nakajima, M. *Control of Insect Behavior by Natural Products.* Academic Press, New York, 1970.

Evolution

The theory of the evolution of life from simple molecular beginnings to the phenomenon of man has made a greater impact on the intellectual life of mankind than has any other aspect of biological knowledge. It has profoundly influenced the thinking of many groups of people. The task of this chapter will be to present evolution and its implications from a biological viewpoint, leaving philosophy to the philosophers.

It is now over a century since 1859, when Charles Darwin published his epoch-making book, *The Origin of Species by Means of Natural Selection.* Since that event, new and certain knowledge has been won in many areas.

Darwin's basic concept of evolution has been powerfully supported and, in fact, greatly broadened and deepened. At the same time much of the actual history of life on this planet has been spelled out in detail by new discoveries of fossils and new methods of dating them, especially by the use of such techniques as carbon dating. Even the old question of the origin of life from the non-living material of a planet seems close to solution. The methods of numerical taxonomy and biochemistry have also thrown new light on old problems of evolutionary theory.

THE CLASSIC DARWINIAN THEORY

Charles Darwin was born in England on February 12, 1809, the very day and year of Abraham Lincoln's birth. When about 15, Charles was sent to Edinburgh with an older brother to become a medical student. However, he found the lectures boring and the surgery before anesthetics sickening. At the end of his second year, Darwin transferred to Christ's College, Cambridge, with the intention of studying for the ministry. There he became friends with several undergraduates who were avid beetle collectors and came to know Professor Henslow, a plant hybridizer and man of wide scientific interests, who held weekly open house for undergraduates. Through him Darwin was appointed as naturalist on the naval ship *Beagle* during its five-year voyage around the world (1831-1836).

Darwin formulated his theory that evolution occurs by means of variation and natural selection at the age of 29, soon after his return from the voyage of the *Beagle.* Yet he did not publish the theory until 1859, after he had devoted 22 years to amassing factual evidence and studying all aspects of the problem. Darwin's great achievement, like that of Newton, lay in drawing together into a new synthesis several different lines of scientific advance. Darwin's presentation in his *Origin of Species* made it impossible for anyone who could read to miss the point. At the beginning of each chapter he tells you what he is going to say, and in the body of the chapter he says it clearly and with an overwhelming wealth of factual detail, and at the end of the chapter he tells you again exactly what he has said.

Four Principles

Four primary groups of facts or principles form the basis of Darwin's theory of evolution.

The first is the geological principle of **gradualism**, or **uniformitarianism** as it is also called. This is the theory that the world, with all its mountains, plains, and valleys, was not suddenly created in 4004 B.C. or on any other particular date, but is the result of gradual changes over millions of years. Streams and rivers are constantly wearing down the moun-

Fig. 35-1. Charles Darwin as a young man. (From A. Moorehead: *Darwin and the Beagle*. Harper and Row, Publishers, Inc., New York and Evanston, 1969.)

tains; upthrusts of land continue to produce mountains and earthquakes; sediment still settles on the ocean floor to be pressed into stone. Without this geological background of the long past, a theory of evolution is almost impossible.

The second principle on which Darwin based his theory was one made familiar by Cuvier and a long line of naturalists. It was the realization that many different kinds of plants and animals form a **graded series** from simple to complex, from generalized to specialized. From the time of Aristotle, these so-called ladders of being had been thought of as static, a part of the unchanging order of things. The important point, however, is not that the old naturalists like Linnaeus and Cuvier thought of plant and animal relationships as unchanging, but rather that they saw living things to be very clearly related to each other by their anatomy, and that these relationships permit us to arrange them in logical sequences. The arm of a man, the foreleg of a horse, the foreleg of a cat, the wing of a bat, or the flipper of a seal all show the same basic skeleton. Bone for bone their limbs are **homologous**. Among the insects, the crustaceans, or the coelenterates, everywhere similar basic patterns of structure had become evident. And as explorers and collectors brought in new kinds of animals and plants from the four corners of the world, it became increasingly clear that there are many graded series of types. The German poet and naturalist Goethe was an early proponent of the concept of homology. In his writings he noted and

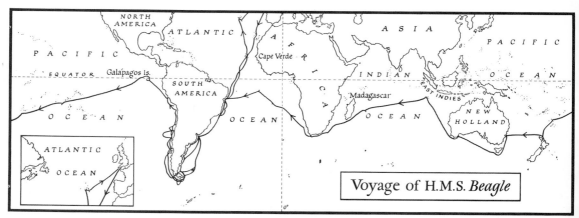

Fig. 35-2. The route of the *Beagle* during its voyage around the world, 1832-1836. (From G. B. Moment: *General Zoology*, 2nd ed. Houghton Mifflin Co., Boston, 1967.)

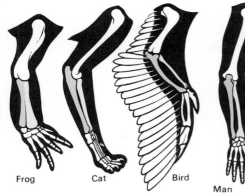
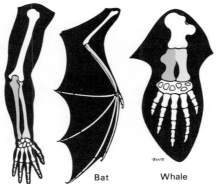

Frog   Cat   Bird   Man   Bat   Whale

Fig. 35-3. Skeletal homology in vertebrate limbs. Radius and ulna of each limb in green. (From G. B. Moment: *General Zoology,* 2nd ed. Houghton Mifflin Co., Boston, 1967.)

illustrated structural similarities in plants as well as animals. Goethe was the first to postulate that flowers are actually modified shoots and that the parts of a flower are morphologically equivalent (homologous) to leaves.

The third body of knowledge that Darwin utilized was from the practice of **plant and animal breeding**. By the early 19th century, plant and animal breeders had produced markedly improved races of wheat, oats, poultry, cattle, and the like by the method of selective breeding, that is, by **artificial selection**. Domestic plants or animals which showed the desired traits were selected by the breeder to be parents of the next generation. Natural selection is the obvious counterpart among wildlife of artificial selection among domesticated plants and animals.

The fourth and perhaps crucial principle that entered into the Darwinian synthesis was the theory of T. R. Malthus on **population growth**. Malthus published his famous *Essay on the Principle of Population* in 1798. The thesis of Malthus is really very simple. He pointed out the rather obvious fact that no population can increase to any significant extent without an increase in its means of subsistence, especially food. He then argued that history shows that whenever there has been an increase in food and living space, the human population has always also increased. His conclusion from all this was gloomy, for he held that only "moral restraint, vice, and misery" such as unwholesome occupations, or extreme poverty, diseases and epidemics, wars, plague, and famine could

hold down the immense power of populations to grow if they are left unchecked.

In any finite environment, whether a test tube or a continent, there are limits to the size a population can attain. For example, one medium-sized female crab will carry over 4 million eggs. If 2 million of her eggs hatch into females, and if each were again to produce 4 million eggs, there would be 2 million times 4 million or 8×10^{12} offspring in the third generation. This prodigious number is the result of the reproductive potential of but three generations from one fertilized crab, yet three generations are as nothing in the life of a species that continues decade after decade, century after century. To illustrate this point Darwin characteristically used the slowest breeding animal known, the elephant, which does not become sexually mature until between the ages of 17 and 20, has a gestation period of 22 months, brings forth but one calf at a birth, and usually stops breeding when about 50 years old. Yet in time, if no forces checked their growth, Africa would be carpeted, coast to coast, with elephants.

The significant fact is that any living population (whether bacteria, elephants, or men) tends to increase geometrically, like compound interest, until checked. This insight of Malthus gave Darwin his clue. In his own words:

I happened to read for amusement Malthus on population, and being well prepared to appreciate the struggle for existence which everywhere goes on from long continued observation of the habits of animals and

Tafel XI
„Eine durchwachſene Roſe von vorn und hinten in meiner Metamorphoſe der Pflanzen
beſchrieben." Original Goethes.

Fig. 35-4. Illustration from Goethe's writings on metamorphosis of plants showing vegetative shoot emerging from a rose blossom. Goethe postulated that a flower is a modified shoot. (From Goethe's original drawing reproduced in W. Troll (ed.): *Goethes Morphologische Schriften*, Eugen Diederichs Verlag, Jena.)

plants, it at once struck me that under these circumstances favorable variations would tend to be preserved, and unfavorable ones destroyed. The result of this would be the formation of new species. Here then I had at last got a theory by which to work.

The theory of **evolution by natural selection** was proposed independently in 1855 in a short essay by Alfred Russel Wallace, a naturalist and explorer. Happily a joint publication led to mutual appreciation rather than bitterness. It is worth recording that Wallace also found his inspiration from reading Malthus.

In formulating Darwin's theory, two facts are basic. The first is the geometric or "compound interest" character of population growth just discussed. The second is the fact that living things vary in many ways and that many of the variations are inherited. These two factors lead to a struggle, which may be quite passive, for existence which results in the survival of the fittest plants and animals, and the ultimate extinction of the less fit. Thus populations tend to change their character. No one in Darwin's time understood variation, although Darwin himself saw its crucial importance for his theory and discussed it. The important distinction between true hereditary variations due to changes in the chromosomes, and nonhereditary variations due to amount of food, exercise, sunlight, or other environmental factors, was not made clear until the work of a Danish geneticist, Johanssen, after the rediscovery of Mendel's laws. For Darwin's argument, this distinction was not crucial. It was enough to know that living things do, in fact, vary widely and that many, if not all, of these variations are inherited. With no true understanding of the physical basis of heredity and mutation, it is little wonder that most of Darwin's immediate successors came to regard natural selection as the virtually self-sufficient cause of evolutionary change.

LAMARCKIANISM

Before beginning a modern critique of Darwin's classic theory, there is a traditional view about evolution which has come down to us from before Darwin's day, and which is still important enough to require brief discussion.

This view, which Darwin accepted and which is even today a hardy perennial among uninformed people in many parts of the world, is Lamarckianism, a belief in the inheritance of the effects of use and disuse. According to this view, the effects of good or poor nutrition, or any other environmentally induced condition in the adult, will somehow produce changes that are transmitted to future generations. This theory was discussed in Chapter 7.

It is a fact that some scientific concepts persist long after they have been shown to be untenable. A notable example of such persistence occurred in Russia during the Stalinist era. Communist dogma at the time asserted that the environment influences heredity. No other views were permitted. Led by Lysenko, Russian geneticists also maintained that hereditary changes could be brought about in plants by grafting or in animals by blood transfusion. Scientists in other countries have been unable to repeat Lysenko's experiments. In fact, Charles Darwin's cousin, Francis Galton, had earlier tried in vain to transmit hereditary traits by blood transfusion between different types of rabbits. It is now generally agreed even in Russia that Lysenkoism is scientifically false and was ultimately counter-productive in terms of agricultural practices. This period of 20th-century scientific history should serve as a

Fig. 35-5. Alfred Russel Wallace, naturalist and explorer, who with Darwin proposed the theory of evolution by natural selection. (From A. R. Wallace: *My Life*. Chapman & Hall, London, 1905.)

Fig. 35-6. Trofirn Lysenko, Russian geneticist whose insistence that acquired characteristics can be inherited is a modern day example of the dangers of mixing science and politics. (From J. Huxley: *Heredity East and West*. Henry Schuman, New York, 1949.)

warning that the use of science to serve political ends can be self-defeating.

THE MODERN SYNTHESIS

The modern, or neo-Darwinian, view of evolution sees four major causes which work together in producing evolutionary change. They are: (1) gene mutation and the subsequent and continual reshuffling of genetic factors through sexual reproduction; (2) natural selection; (3) the effects of chance, often called random genetic drift; and (4) isolation, especially temporal and geographical isolation, and its converse, gene flow between groups.

Mutation and Sexual Reproduction

Mutation is so basic to the subject of evolution that certain relevant aspects merit restatement here. Mutations, whether naturally occurring or produced by radiation or by chemical means, are random in two senses. There is no way of predicting specifically what mutation will occur next or of producing a specific gene change on demand. The ability to produce specific mutations artificially remains for the future, when much more is known about the chemistry of nucleic acids. Mutations are also random in that they bear no necessary

relation to the needs of the organism. In fact, some 99 per cent of all mutations are harmful. This is not at all surprising, since existing animals and plants are the result of millions of years of adaptation. No wonder that the overwhelming probability is that no random change would be for the better. As with a ship or an airplane, making random changes in the blueprint is likely to be disastrous. It should also be remembered (Chapter 7) that whether or not a mutation is advantageous is relative to the environment.

Different genes possess different mutation rates in accordance with variations in their stability. Many cases of back mutations are known. Thus, a gene for normal pigmentation may mutate to a form resulting in albinism, and a gene for albinism may mutate into one for normal pigmentation. The net result of the difference between the mutation rate of a given gene in one direction and its mutation rate in the reverse direction is known as its **mutation pressure**.

Population Genetics and the Hardy-Weinberg Law

Evolution is change with time in a population of genes. Consequently, evolution can be regarded as a problem in population genetics. The basic law of population genetics, formulated in 1908 by an English mathematician, G. H. Hardy, and a German physician, W. R. Weinberg, is known as the **Hardy-Weinberg law**. It is a very simple, common-sense generalization, and also a useful tool in studying the frequencies with which certain genes occur in a population. The Hardy-Weinberg law states that in the absence of mutation, selection, or some other factor causing change, the proportion of genes in any very large population will soon reach an equilibrium and thereafter will remain the same, generation after generation, regardless of what that proportion may be. Whether the genes are dominant or recessive makes no difference.

The Hardy-Weinberg equilibrium forms the basis for the calculations of population genetics. The effects of mutation, selection, and gene flow between populations are all superimposed on this basic equilibrium. If the population is very small, then the Hardy-Weinberg equilibrium cannot be maintained and random genetic drift may occur. If a particular dominant gene is disadvantageous, then the relative number of those genes will decrease through the action of selection. Of course, if a gene confers some advantage, then its numbers will increase in the population, regardless of whether it is dominant or recessive.

Suppose a population is made up of 60 per cent homozygous Rh positive (RhRh) individuals, half males and half females, and 40 per cent homozygous Rh negative individuals (rhrh), half males and half females. If marriages in this population are random with respect to the Rh factor, then it is simple to construct the familiar genetic grid to show what the results will be. This is done by multiplying the gene frequencies in one sex by those of the other. Since 60 per cent of the males are RhRh and 40 per cent are rhrh, the ratio of the two kinds of sperms can be represented as 0.6 Rh:0.4 rh. In a similar way, 0.6 of the eggs will carry an Rh gene, and 0.4 an rh gene. Random fertilization produces the following genotypes in the zygotes:

sperms   0.6 Rh + 0.4 rh
eggs   0.6 Rh + 0.4 rh

zygotes 0.36 RhRh + 2 (0.24 Rhrh) + 0.16 rhrh.

As far as individuals are concerned, only 16 per cent are Rh negative, instead of 40 per cent as in the parent population. However, if we look at the genes in the offspring, we find that their frequencies in the filial population are precisely the same as in the parents. In other words, although 84 per cent of the offspring will be Rh positive and only 16 per cent Rh negative, in 100 individuals the probability is that there will be 120 Rh genes for every 80 rh genes, a ratio of 6:4.

36% RhRh =	72 Rh	
48% Rhrh =	48 Rh +	48 rh
16% rhrh =		32 rh
	120 Rh +	80 rh
	60% Rh +	40% rh

In more general terms, if p = the frequency of one gene and q = the frequency of its allele, and they are the only two alleles at the locus in question, then p + q = 1. In the case just given,

p = 0.6 and q = 0.4, and 0.6 + 0.4 = 1. In a freely interbreeding population, (p + q) sperms $\times$ (p + q) eggs = (p^2 + 2pq + q^2) zygotes. In the case just described, (0.6 + 0.4) $\times$ (0.6 + 0.4) = 0.36 + 0.48 + 0.16.

From this formula it is possible to calculate the actual frequency of any gene in a population if the percentage of homozygous recessives is known. For example, if 0.16 (16 per cent) of a given population are albinos or show any other recessive trait, it is evident from the formula that 0.40 of the alleles at this locus in the population must be recessive. It will be noted that Mendel's laws represent the special case of the Hardy-Weinberg law where p and q are equal. Substituting the values for p and q (in this case p = q = 0.5), then $(p + q)^2$ = (0.5 + 0.5) (0.5 + 0.5) = 0.25 + 0.50 + 0.25, or the familiar 1:2:1 ratio of genotypes found in the progeny of a monohybrid cross. If p represents a dominant gene, the phenotypic ratio here is the familiar 3:1. Indeed, Mendel himself expressed his results in terms of the binomial theorem: $(a + b)^2 = a^2 + 2ab + b^2$.

The Hardy-Weinberg generalization of Mendel's law expresses a stable state and thus represents the absence of evolution. But with such knowledge as a foundation, it is possible to make calculations about how great mutation pressure or selection pressure would have to be to produce a particular amount of evolutionary change. For example, it is possible to determine the number of generations that would be necessary to eliminate an undesirable gene, or to make a desirable gene universal in a population at various selection pressures. Let A be the dominant gene for normal pigmentation and a its recessive allele for albinism. If 100 AA or Aa individuals survive for every 99 aa, the dominant gene is said to be favored by a selection pressure of 0.01. These calculations show that mild selection pressure is not very effective in very large populations, even for dominant genes. For a recessive gene even 100 per cent elimination, whenever it appears in the homozygous state, will result in only a very slow decline in its frequency.

In humans, there are about 5 aa albinos/100,000 persons. Assuming that marriages are at random with respect to this trait, q^2 = 0.00005 aa. Arithmetic will show that there must be 1,420 Aa individuals or carriers/100,000. It is obvious that if either natural or artificial selection were to remove all the 5 aa/100,000 individuals, the total of 1,430 a genes would be reduced to 1,420, a trifling result. At this rate, *e.g.,* by sterilizing all albinos in every generation, it would take 5,000 years to reduce the number of albinos by one-half. On the other hand, it must not be forgotten that drastic selection can be effective much more quickly than this result suggests. The application of DDT to a population of flies, or the introduction of a new variety of wheat that is resistant to a fungus-caused disease, may produce a DDT-resistant strain of flies or a strain of fungi of new virulence simply by the complete elimination of all individuals except the rare resistant or virulent types, respectively.

Natural Selection

There are three different types of selection whether natural or artificial. **Stabilizing selection** tends to eliminate extremes in a population. A sea gull whose wings are too long is at a disadvantage and so also is a gull whose wings are too short. For a bird of a given size with a given mode of life in a given environment, there is an optimum wing length. Natural selection will shorten the lives of individuals which deviate very far in either direction from this optimum. Most selection seems to be of this stabilizing type, which is an anti-evolutionary type of selection pressure since it tends to prevent change.

The second type of selection is **directional**. It favors individuals which vary in one direction and works against whose which vary in the opposite direction. This kind of selection can produce evolutionary change. Speed and maneuverability are advantages for a hawk; their opposites are serious handicaps. Increased ability to survive heat and drought is advantageous to desert animals and plants, their opposites disastrous.

The third type of selection is **disruptive** selection. This is probably the rarest type, but it can also be important in producing evolutionary change. In such selection, the extreme variants at each end of the distribution curve are favored, while those in the intermediate range are at some disadvantage. Such a situation will tend to split a population into two parts,

each of which may finally become a separate species. For example, in a population of tree-dwelling monkeys, the largest and heaviest individuals are able to defend themselves from attack by the others; the smallest and lightest, and therefore the most agile, can escape through the treetops, but the middleweights are neither fast enough to escape nor heavy enough to win against the somewhat larger members of the species.

Sampling Errors or Genetic Drift

In addition to mutation and selection, an important limitation on the Hardy-Weinberg law and almost certainly a very important factor in evolution is the effect of chance in very small populations. This effect is commonly called **genetic drift** or the **founder principle**. When this term is used it must be remembered that no definite "current" is implied, only the purely random sampling that occurs for many genes whenever the population is drastically reduced in size. For such reasons this aspect of evolutionary dynamics is sometimes called **sampling error**, or even the **bottleneck phenomenon**.

As an extreme case, suppose 10 finches happen to get caught in the hold of a grain ship in Argentina and do not escape until they arrive at the coast of China. There they may or may not form a new population, but whether they do or not, such a tiny flock can certainly not be expected to carry a completely representative sample of the genes common to the whole South American population from which they came. Suppose also that 20 per cent of the large South American population carries a dominant gene for an unusually wide white band on the wing feathers. If the flock of 10 were to be representative, two of the finches would have to carry this gene. If only one of the 10 happened to have it, then its frequency would have decreased by one-half. Actually, such a gene could easily be missing altogether in the little flock. Or a gene that was uncommon in the parent population might just have happened to be common among the 10, and so become common in the resulting new population. When all the thousands of genes present in a species are considered, it becomes evident that no very small group of individuals can be truly representative.

Besides the bottleneck phenomenon there are two additional points where pure chance can change the character of a small population. The first of these is in meiosis and fertilization. Many sperms with their genes are lost; many eggs and three times as many polar bodies never form new individuals. In a large population these losses average out, but not so readily in a small population. For example, most of the genes for red hair might get lost in polar bodies. The second point where chance enters is in purely accidental death or survival among the members of a very small population. In large populations the effects of chance, good or bad, tend to cancel.

How small does a population have to be for genetic drift to occur? That depends on the force of natural selection and on mutation pressure, and so may be different for different genes. In putting an infinitely large population of equal numbers of black and white marbles through a bottleneck by drawing a random sample, the overwhelming probability is that a sample of a very few thousand would be representative. In the living world ideal mathematical situations are seldom, if ever, found. Hence, it is impossible to determine exactly how large any particular population of genes has to be in order to avoid the effects of random sampling, *i.e.*, genetic drift.

How important has genetic drift been in the course of evolution? It is thought that in the remote human past, when the primates from which man arose, and the early men, were wandering about in small bands, genetic drift must have been important. Recent careful studies have compared the gene frequencies in a very small population which emigrated from a known part of Germany and settled in Pennsylvania about 200 years ago. Because of their religious beliefs, the Amish have not intermarried appreciably with other people (although over the years some have left the little community and joined the mainstream of American life). If genetic drift is a fact, some of the gene frequencies among this group should be different not only from those of the general population of Pennsylvania but also from those of the large parent population in Germany. This turned out to be true for the ABO and MN

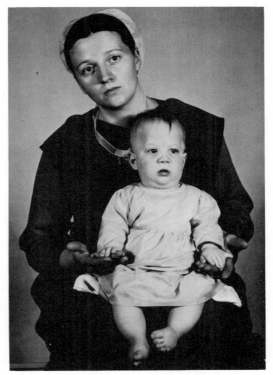

Fig. 35-7. Genetic drift. The Amish (Pennsylvania Dutch) have not intermarried with other people since they emigrated from Germany and settled in Pennsylvania about 200 years ago. Because of genetic drift, their gene frequencies for blood groups and other characteristics such as polydactyly (shown here in the Ellis-van Creveld six-fingered dwarfism) differ from those of the general populations of Pennsylvania and Germany. (Courtesy V. A. McKusick, Johns Hopkins University.)

blood group genes, as well as some other minor hereditary characteristics including polydactyly, the tendency to have extra fingers or toes.

It is also worth noting that natural selection, by drastically reducing a population, may put a species through a bottleneck where genetic drift becomes operative. For example, when an epidemic disease hits a population of wild ducks, there is a rigorous selection for any genes which make for resistance to this disease. All other genes, no matter how important in other situations (genes affecting speed in flight, correct migrating instinct, breeding ability, and success in escaping predators such as foxes, snapping turtles, and hunters) for the time become irrelevant. Only those genes that just happen to be in ducks which also carry the genes for disease resistance will escape through the population bottleneck. But the smaller the bottleneck the more certain that no representative sample of all these other genes will get through. Consequently, when the ducks increase again after the epidemic subsides, the new population will be genetically different.

Isolation

Isolation was regarded by Darwin and his immediate followers as so important that it was held to be a prerequisite without which no evolution was possible. With their pre-Mendelian view that heredity involved a blending of vaguely understood hereditary material ("blood") from the parents, it seemed inevitable. that any deviation would be averaged out as the result of cross-breeding in subsequent generations. The discovery of the Mendelian factors of heredity, the genes, which enter and emerge from crosses unchanged, produced a modification of this view. If a gene or a certain combination of genes confer a selective advantage, it is not necessarily averaged out. On the contrary, it may be preserved and will tend to increase in frequency. In the case of a self-fertile population (such as we find in many plant species) or in a population that is very closely inbred, the proportion of homozygous individuals, whether homozygous for the

Fig. 35-8. Darwin's finches of the Galapagos and Cocos Islands belong to a single subfamily of birds that evolved in isolation. Ground forms presumed to be the most primitive live in arid coastal regions and are mostly seed eaters (9-14). Tree finches live in moist forests and mostly feed on insects (1-8). The warbler finch (7) feeds on small insects in bushes in both arid and humid areas. The woodpecker finch (1) lives on grubs hidden in the bark of trees. Note especially differences in bills. (From D. Lack: Darwin's finches. Scientific American, *188:* 66, 1953.)

dominant or the recessive gene, continually increases. This is because the homozygous individuals mated together can produce only more homozygotes, while a cross between two heterozygous individuals will give rise to only 50 per cent of heterozygous individuals like themselves; the other half of their offspring will be homozygous.

The evidence collected in recent years shows that isolation does play an extremely important role in the origin of new species. In the first place, different but related species are always separated in some way in nature so that they do not interbreed, even though they may be induced to do so under laboratory conditions. Second, wherever there is a barrier between two or more populations, these populations almost invariably show differences.

Allopatric Races

A classic case of the effect of geographic isolation on species formation is that of the snails of the genus *Partula,* which live on the small volcanic islands scattered over the Pacific Ocean. Each island has its own characteristic species, and in addition, every valley on each island has its own peculiar form of the snail. The mountain ridges separating these little tropical valleys are very high and steep, and conditions on them are such as to form a barrier to the snails. Species, subspecies, or races of this kind which occupy different territories are called **allopatric** (*allos,* other, + *patria,* native land). Many such cases are known. *Drosophila* has been studied intensively on a worldwide basis. The fruit flies of the Hawaiian Islands are different from those in North America or in China and Japan.

How isolated do two populations have to be for speciation to occur? This is a hard question to answer because degree of actual isolation is difficult to measure. One of the few cases ever measured is that of the small rodents that live on prehistoric black lava flows in arid parts of the American Southwest. Those flows which are completely isolated by desert sand, or which make contact with rocky regions on less than 10 per cent of their periphery, have endemic races of very dark small rodents. **Endemic races** are races that are native to a particular region. However, when somewhat more than 10 per cent of the periphery of these lava flows is in contact with rocky zones, endemic rodents are not found. Evidently genetic flow between the rodents living on the lava and the general brownish population in rocky areas is too great and natural selection too weak to produce a special endemic race.

A Ring of Races

The effects of geographic isolation in the formation of new species is perhaps most clearly seen in a **rassenkreis** or "ring of races." A dozen or more such **polytypic species,** in which several "typical" forms occur in series, have been investigated. The common harbor gull, *Larus argentatus,* forms such a series in the lands encircling the north pole. Ornithologists believe that the original home of this gull was the eastern edge of Siberia just west of Alaska. In North America are three different races. Extending westward across northern Siberia into Europe and Scandinavia are two additional races, making a circumpolar chain of six races. In the British Isles and coastal Europe, the North American races and the western European races both occur but do not interbreed. However, the European subspecies or races interbreed with the western Siberian ones, these with the eastern Siberian forms, and these in turn with the North American races. If a species is defined, as is customary, as an interbreeding population, then the western European and North American races would appear to be separate species, but the Siberian race would be of the same species as each of the others. It is as though $A = B$ and $B = C$ but A does not equal C!

A similar situation has recently been discovered in the common frog, *Rana pipiens,* a species with a range extending from Canada to Panama. Individuals from the northern part of this range crossed with individuals from New Jersey produce normal embryos. Likewise, New Jersey frogs can be successfully crossed with frogs from Louisiana. But a cross between individuals from the far nothern and far southern parts of the range yield abnormal and nonviable embryos.

From the point of view of a pre-Darwinian taxonomist cataloguing species like a logician, cases of this kind are paradoxical. But in the

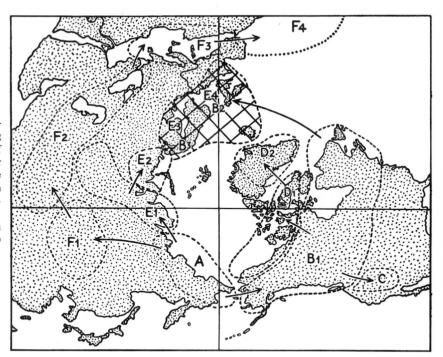

Fig. 35-9. Polar projection map showing the chain of races of *Larus,* the harbor gull, encircling the Arctic ocean. (From G. B. Moment: *General Zoology.* Houghton Mifflin Co., Boston, 1958.)

light of the theory of evolution, they are examples of species in the making. The frogs in the north are adapted for development at low temperatures, those in the south for development at high temperatures. The genes of the two races are just different enough to prevent successful cooperation. Should something happen to remove the middle states group of frogs or the central Siberian population of gulls, the remaining populations at the geographical extremes of the ranges would then quite properly be regarded as distinct species.

Sympatric Species

Species which inhabit the same geographical range are called **sympatric** (*sym* or *syn,* together, + *patria,* native land). Sympatric species are isolated by various mechanisms which prevent interbreeding. Many birds will not mate except after a prolonged courtship of perhaps 10 days or more. This involves a special type of behavior and special responses. If the genetic basis for courtship behavior is lacking in one individual, mating fails to occur. Many closely related species are effectively separated in

nature by differences in breeding season. This is true of the various species of frogs. The wood frog, *Rana sylvatica,* breeds within a few days after the ice is off the ponds. The bull frog, *Rana catesbeiana,* breeds very late in the spring, in May or June. Other species breed at intermediate times.

Reproduction in plants is also seasonal, most commonly controlled by day length (*i.e.,* photoperiod). As cross-pollination can only occur between flowers formed at the same time, there is no possibility in nature that the pollen from a plant blooming in the spring will reach a flower formed during the summer. Some related species of animals and of algae, especially marine species, where the eggs and sperms are cast out into the sea water, are intersterile. Even though mixed together in a small bowl of sea water, the eggs of one species cannot be fertilized by the sperms of the other, or, if fertilized, do not develop into viable embryos.

Temporal Isolation

In the long sweep of evolution, of paramount importance is **temporal isolation**: the simple fact that no group of animals or plants can interbreed and exchange genes with their ancestors. This means that evolution can be progressive under the pressure of natural selection and in the absence of any kind of geographical isolation. The obvious fact that all living populations are continuous with ancestral populations presents a seemingly insoluble logical problem to those who try to apply the species concept to the historical evolution of animals and plants. It is the problem of the ring of races in a polytypic species, extending in **time** instead of in space. In practice, however, there are plenty of gaps in the fossil record so that it is possible to use the same system of nomenclature for fossils and for living forms.

Stratagems and Rules

The modern conclusion is that the four factors discussed above, namely, hereditary variation produced by mutation and sexuality, natural selection, genetic drift, and isolation, are adequate to account for evolution. Within this general framework lie endless devices and stratagems and a number of groups of facts called rules.

Orthogenesis

The tendency of certain organisms to evolve consistently along restricted evolutionary paths over long periods of geologic time is called **orthogenesis,** or straight-line evolution. The evolution of the horse from a dog-sized, four-toed herbivore in the Eocene Epoch 60 or 70 million years ago to the familiar one-toed animal of today is the classic example. But the history of life abounds in examples—the elephant, the camel, certain cephalopod mollusks, and many others. In this sense, orthogenesis is clearly a fact. Once a race begins to specialize in ways which adapt it to a particular mode of life in a particular environment, further modification in the same direction tends to continue for long periods, often until extinction.

There are differences of opinion concerning an explanation for this kind of evolution. The paleontologists of a generation ago, like Henry Fairfield Osborn in the United States and Teilhard de Chardin in France, have seen in the fossil record something more than they believed

mutation and natural selection could explain. There must be, such people feel, some driving force, in the famous phrase of Henri Bergson, an **élan vital**.

Their evidence for inner-directed straight-line evolution is in the broad outlines of the evolutionary history of a particular kind of animal. But as more and more of the fossil record has been uncovered, it has become obvious that the evolution of the horse, to use the classic case, is not a single straight line but a series of branching lines with many blind alleys. Moreover, the study of a fossil record does not show the mutations. These occur between the parents and their offspring, while successive fossils are commonly thousands or tens of thousands of years apart. Even more important, mutations are now known to affect a vast range of characters. Thus, the facts about mutations directly contradict the idea of a directional driving force. The direction comes from the environment, specifically from the relationship between the environment and the adaptations possessed by the species at any particular point in time. An animal already adapted for running and living on the plains or steppes will not be benefited by many mutations that would benefit an animal already fairly well adapted for climbing trees. To repeat, the set of adaptations already possessed by a species determines whether a specific mutation is advantageous and will probably be incorporated into the gene pool of the species or is disadvantageous, as most are, and will be eliminated. If one is seeking inner directedness for evolution, one must look into the potentialities of DNA.

Pre-Adaptation

When a set of adaptations acquired by a group of living things in one environment fits them to live in a new and different environment, they are said to have been **pre-adapted**. A most striking case is the way major adaptations making life on land possible for vertebrates were developed by the muscle-finned fish, which include the lungfish, and were useful to them. Their descendants, the primitive amphibians, were the beneficiaries of this pre-adaptation. Several localities have yielded masses of fossilized remains of these early amphibians

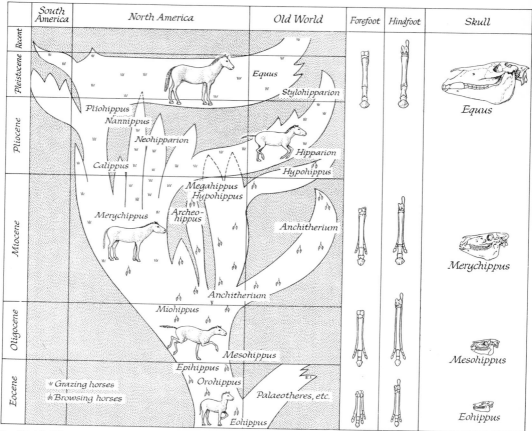

Fig. 35-10. The evolution of the horse. (Adapted from G. G. Simpson: *Horses.* Oxford University Press, London, 1951.)

crowded in what had clearly been a mudhole in some drying lake or river. There is independent geological evidence that the Devonian Period, approximately 300 million years ago, when these little tragedies occurred, was a period of violent climatic changes, of droughts and floods. It is easy to see that lungs would have permitted fish to survive in damp mud during a drought. This is the chief function of the lungs in lungfish today.

Legs, complete with muscles and skeleton, are obviously important for an animal to clamber over the mud in a drying lake or river and to find water. The evidence is that the muscled appendages of the ancestral fish were used not to "invade" the land but to get back into the water after their supply had dried up. Ability to walk over dry land developed slowly later.

The ability of mankind to develop complex civilizations can also be regarded as a pre-adaptation. Our average brain size is certainly no larger now than among the prehistoric Cro-Magnons, nor is there any reason to believe that our intellectual abilities have increased. It may well be that primitive life in very small communities requires greater general competence on the part of all members of the group than does life in a great city.

Another way to view pre-adaptation is to note that once a means for solving a problem is achieved, it tends to be retained and utilized in new environments. Two major breakthroughs in this sense seen in the Plant Kingdom are the development of protective coverings for reproductive parts (first seen in the liverworts and mosses) and specialized conducting cells (first seen in the club mosses and horsetails). Both of these adaptations were essential to the later development of successful land plants and both were retained in all descendant groups.

Neoteny

Neoteny, a phenomenon peculiar to animals, is the appearance of sexual maturity in a larval or juvenile stage. It seems to have been a key event in the evolution of man, insects, the first vertebrates, and other animals. The best-known case of neoteny occurs in the Mexican salamander, *Ambystoma mexicanum.* This lake dweller never loses its gills and tail fin, nor does its skin metamorphose into the adult type of other salamanders. It becomes sexually mature and breeds in the larval condition.

Man has been called a "naked ape." This is merely a manner of speaking, but there is evidence that man is a fetal or neotenous ape. A number of important human characteristics, including the relatively large size of the brain, lack of body hair, and the late time at which the bony sutures fuse, are all fetal traits in other mammals but adult traits in man. From this point of view, man is a modified ape that becomes sexually mature without becoming completely mature anatomically.

Rules of Evolution

There are a number of so-called "laws" or rules which state certain generalizations about the four major causal factors in evolution: mutation, selection, isolation, and genetic drift.

Dollo's rule states that evolution is irreversible. This is generally true, although many reverse mutations for single genes are known. Drosophila that have mutated to eyelessness will mutate back to the eyed condition, but this is very different from having a horse change back into a small five-toed animal the size of a dog. Clearly the web of circumstances in historical events is too complex ever to recur in exactly the same way. The mutations will be different or at least appear in a different sequence, the environment will certainly have changed, and hence natural selection will be different and the effects of chance will be different. Although botanists have said little about it, Dollo's rule apparently applies to plants also. There is no reason to expect that flowering plants will evolve back into ferns.

There is a common tendency for animals to increase in size during the long course of evolution. This is known as **Cope's rule** and is well illustrated in the great prehistoric reptiles, which increased in size until they became extinct. There has also been a very general tendency in many, but by no means all, lines of mammalian evolution for size to increase. Male lions are now so large and clumsy they have to leave the crucial part of hunting to the females while they merely help flush the prey. Meanwhile competition between males seems to be favoring a still greater increase in male size. Perhaps this is an evolutionary trap leading to extinction.

Many **homoiothermal** (warm-blooded) mammals and birds become larger in the colder parts of their range. This is **Bergmann's rule**, which often holds both for individuals of the same species and for related species. The biggest bears are in Alaska, the biggest tigers in Siberia, and the largest Virginia deer not in Virginia but in Canada. The explanation appears to be that a large animal has less surface area in proportion to its volume and so is better able to conserve heat.

A related phenomenon is expressed in **Allen's rule**. The extremities of animals living in colder and colder climates (ears, tails, etc.) are progressively smaller. This is an obvious adaptation against freezing. Bergmann's, Allen's, and other rules of a similar kind (for example, **Golger's rule** that animals living in damp forests tend to be darker than those in dry regions) find a ready explanation on the basis of mutation and natural selection.

Functions of Color and Pattern

The colors and patterns of animals have been molded by mutation and natural selection to serve three functions. The commonest is protective camouflage, which blends the animal into its surroundings. Examples of this abound. A second function is to provide a means of species or sex recognition. This too is widespread and is especially common among mammals, birds, reptiles, fish, and some insects. The cinnamon red of a male bluebird's breast feathers serves as a challenge to other males. A third but less common function is to serve as a warning. Skunks, hornets, and bumblebees are not brilliantly colored and marked for no biological purpose. Skunks present a rather special case because they seem to have the best of two worlds. Close up and in daylight they are extremely conspicuous but at night and from a distance their bold black and white pattern breaks up the outline of the skunk's body and makes it appear like so much light and shadow.

In connection with warning coloration, **mimicry** has developed. There are quite harmless flies that have come, in the course of evolution, to resemble bees. There are famous cases of an edible butterfly (edible for birds,

Fig. 35-11. Allen's rule is illustrated by size of ears in foxes. Ears tend to range from small in the arctic to large in desert foxes. Left, American red fox in Pennsylvania (Photo by Pennsylvania Game Commission); Right, fennec or North African desert fox (Photo by R. Buchsbaum). (From R. Buchsbaum and M. Buchsbaum: *Basic Ecology.* The Boxwod Press, Pittsburgh, 1957.)

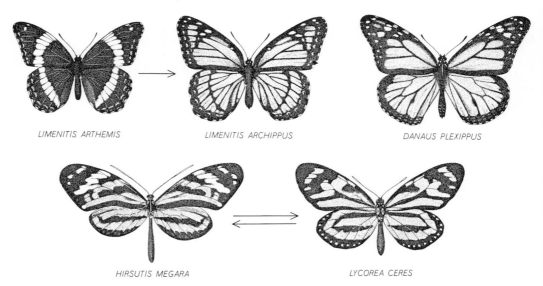

LIMENITIS ARTHEMIS   LIMENITIS ARCHIPPUS   DANAUS PLEXIPPUS

HIRSUTIS MEGARA   LYCOREA CERES

Fig. 35-12. Mimicry. Batesian mimicry (top) where an edible species, *Limenitis archippus,* resembles a bad-tasting one, *Danaus plexippus,* the common milkweed butterfly. Mullerian mimicry (bottom) in which two unrelated species, both unpalatable, gain added protection by resembling each other. (From L. P. Brower: Ecological chemistry. Scientific American, *220(2):* 28, 1969.)

that is) coming to resemble an unrelated but unpleasant-tasting one called the model. Such mimicry, in which one species is noxious and the mimic innocuous, is called **Batesian**. A dramatic verification of the effectiveness of mimicry has recently been obtained by Lincoln and Jane Brower in work with toads. A toad needs to be stung by a bumblebee only once to learn to avoid not only the bumblebees but also the harmless flies which mimic them. In the more common **Mullerian** mimicry, both species are noxious. This probably more than doubles the value of the warning coloration for each species because the predator will find it easier to learn what to avoid.

Color and pattern are important also for the survival of plants. Colors and patterned markings on the petals of flowers serve the essential function of attracting pollen-carrying insects and birds. Especially in those cases where flowers are unisexual, and where male and female inflorescences are borne on separate plants, obligatory dependence between plants and animals has been established. Such plant species can survive only in those areas where the appropriate species of pollinating insect is found.

THE EVIDENCE FOR EVOLUTION

For any science, the question of evidence must always be of central and overriding importance. The evidence that evolution has been a fact of history seems irrefutable to biologists, even though there are intelligent people who profess to find it unconvincing. Each student will have to weigh all the evidence and make up his own mind.

Geographical Distribution

Of the evidence for evolution based on geographical distribution none is more striking than the relationship between the plants and animals on oceanic islands and those on adjacent continents. This kind of evidence forcibly impressed both of the originators of the theory of evolution by natural selection. Charles Darwin, in his voyage as a young naturalist on the *Beagle,* visited the Cape Verde Islands off the West coast of equatorial Africa and the Galápagos Islands off the west coast of

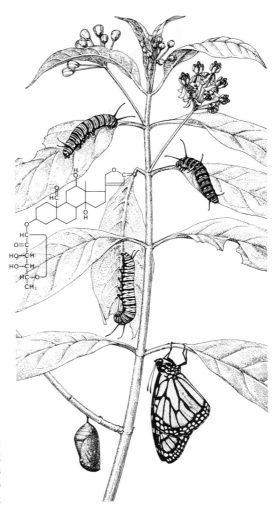

Fig. 35-13. Monarch caterpillars on the milkweed that is their food and from which they derive the bad-tasting substance present in the adult. The formula of this compound, a heart poison, is shown. (From L. P. Brower: Ecological chemistry. Scientific American, *220(2):* 28, 1969.)

equatorial South America, and was profoundly struck by the different flora and fauna existing under similar conditions in geographically different areas. Alfred Russel Wallace studied the life of the long series of islands making up the East Indies.

Each oceanic island has its own native animals and plants. But, and this is the point, in every case they resemble the animals and plants on the nearest continent. The Galápagos Islands are inhabited by turtles, lizards, birds, and other animals that are endemic (native and unique), but are similar to lizards and birds on the adjacent shores of Central and South America. Likewise, the animals and plants of the Cape Verde Islands resemble neither those of South America nor of tropical Asia but those of the adjacent parts of equatorial Africa. These facts indicate that living things came from the nearest continent and then, under the influence of mutation, selection, genetic drift, and isolation, became different from the ancestral populations. There is some geological evidence that the Galápagos Islands were once connected to the South American mainland so that their populations and those on the continent were perhaps continuous in remote ages. Of course, it is also possible for islands to be populated by the descendants of a few land birds blown off course by unusual storm winds, lizards, other animals, the seeds of plants able to survive long sea voyages by clinging to floating logs, and the minute spores of lower plants carried great distances by air currents.

Comparative Anatomy

The facts of comparative anatomy have always been regarded as basic supporting evidence for the fact of evolution, whatever the theory as to its mechanism. This interpretation is itself based on the twin concepts of **homology** and **analogy**. Homology is basic anatomical similarity, regardless of function, both in embryological development and in the adult. The arm of a man, the front leg of a horse, and the wing of a bird are homologous. The wing of an insect, although used for the same function as the wing of a bird, has a radically different basic structure; hence they are merely analogous to each other (see Fig. 35-3).

There are several kinds of homology. **Special homology** involves similar structures, like the forelimb, in two or more different animals while **serial homology** is homology between structures in an anterior-posterior series on the body of a single animal, such as the arm and leg of a man. Both of these homologies were made famous in T. H. Huxley's studies of the appendages of the lobster and the crayfish. The

big claw of a lobster corresponds part for part with that of a crayfish or a crab (special homology) and there is also a basic anatomical correspondence between the big claw and all the other appendages along the lobster's body (serial homology). A basic similarity of structure and development without specific correspondence in the location or in the serial arrangement is known as **general homology**. An example is the anatomical correspondence between all of a shark's scales and its teeth.

The best examples of homologous structures in plants are flowers and vegetative shoots. There can often be seen a transition in a water lily flower, for example, between stamens and petals and between white petals and green leaf-like structures. They, too, stand the test of developmental beginnings, both originating as buds formed in the apical region of an existing shoot (see Fig. 35-4).

The facts of homology do not in themselves prove that evolution has occurred. However, the theory of descent with modification from a common ancestor provides the only known scientific explanation.

Comparative Embryology

Embryology reveals many facts which can be explained easily only on the basis of evolution. How else can the presence of gill slits in the embryos of reptiles, birds, and mammals be explained except by postulating ancestors in which gill slits were functional? The development of the heart and the main arteries from it, and the development of the kidneys also show fish-like stages in the embryos of higher vertebrates.

The modern interpretation of such embryonic structures as gill slits in mammals is that they correspond to the gill slits of an embryo fish rather than of an adult fish. Thus in a sense they are vestigial structures. There are several reasons which taken together seem adequate to explain their persistence. Changes in genes which affect the earliest stages of development will be very likely to produce much more drastic and even lethal results than mutations which affect the later stages when the major organs have been laid down. This in itself will tend to make the early stages of development very conservative. Equally or even more important, natural selection has only a very indirect action on the developmental stages within an egg or a uterus. Its action is largely restricted to such matters as rates of growth and means of obtaining food and eliminating wastes. It is highly significant that it is in this respect that vast changes have taken place; witness the evolution of the placenta and the other fetal membranes. A third factor is often cited as a cause of embryonic conservatism. At least some of the structures in the embryo of a man which resemble the embryo of a fish have a function in the machinery of embryonic development. For example, the primitive kidney or pronephros initiates the development of the pronephric duct, which becomes the vas deferens through which sperms pass to the exterior of the body. Likewise, the embryonic gill slits give rise to some of the endocrine glands (see Fig. 8-14).

Vestigial Organs

Vestigial organs, such as traces of hind legs in certain kinds of snakes, small pelvic bones in whales and porpoises, ear and tail muscles in human beings, and vestiges of toes in horses, can be explained only on the basis of evolution unless one wishes to believe nature is capricious and deceitful. In fact, combined with the fossil record, vestigial organs like these furnish convincing proof of the fact of evolution.

In horses, which walk on one elongated and thickened toe on each foot (the hoof is the toenail), there are vestigial "hand" bones on either side of the main "finger" (the bones horsemen call splint bones). In the recent fossil ancestors of horses the splint bones are much larger and each is tipped with a small hoof, while in the remote ancestral fossils, there are four functional toes on each front foot and three on each hind foot. The geological record shows clearly that all the early primitive mammals were four-legged; therefore, whales and porpoises must be descended from ancestors equipped with hind legs. It is understandable then that whales and some snakes bear signs of hind legs and pelvic bones. The most cogent evidence for evolution is laid down

in the fossil record and to this we now turn after a discussion of how life began.

ORIGIN OF LIFE

All Life from Life?

Until modern times it was commonly believed by ignorant and learned persons alike that life continuously and spontaneously arises from dead matter. Maggots were thought to generate in decaying meat, worms and even frogs from the mud and muck on the bottoms of ponds, clothing moths from a mixture of wool and grease. Only a little over a century ago zoologists believed that tapeworms arose spontaneously in the intestines of their hosts.

The first great blow to the theory of **spontaneous generation** came from the experiments of a 17th-century Florentine physician, Francesco Redi (1626-1698), who showed conclusively that the age-old belief that maggots arise spontaneously in decaying meat was false. Instead, he proved, they develop from eggs laid by flies. His method is a classic example of scientific procedure, for it was clearly conceived to yield a definite answer and made good use of controls. Redi used a wide

35.14. Francesco Redi, the 17th-century Florentine who first demonstrated that flies do not arise spontaneously from decaying meat. (Courtesy Johns Hopkins Institute for the History of Medicine.)

variety of meats, singly and in combinations, for (who knew?) perhaps some meat was unsuitable. Apparently he had access to a good butcher shop and a generous zoo because he records using beef, lamb, venison, chicken, goose, duck, dog, lion, swordfish, and even eel from the river Arno. He placed the meat in two series of jars, one covered with very fine cloth mesh, the other left uncovered. The meat in both covered and uncovered jars became stinking masses, reeking with putrefaction. Yet only in the uncovered jars where flies gained access and laid their eggs did maggots develop. Neither the kind of meat nor its degree of decay made any difference. Increasing knowledge about the life histories and embryological development of insects, frogs, and other macroscopic animals made it clear that animals always develop from previously existing animals, not from non-living material.

As often happens in the history of science, an old question was reopened on a new level by discoveries in later centuries. In fact, the question of spontaneous generation has been reopened three times and is now once again the subject of intense scientific interest. In the 18th century it was the new knowledge of bacteria that raised the question again. Bacteria certainly seem "just to appear" almost everywhere. However, Spallanzani (1729-1799) finally succeeded in showing that if meat or broth is thoroughly boiled and the container sealed while still hot, bacteria never appear and the contents will remain unspoiled indefinitely. This was a discovery of major historical importance, because the canning of food, the control of contagious diseases, and the conduct of surgery are all heavily involved.

Again in the 19th century the question was raised, and given what seemed a final answer by Pasteur (1822-1895). The increasing knowledge of respiration made it seem possible, or even likely, that Spallanzani's heating had driven off the oxygen necessary for life. Moreover, the new chemical theories of fermentation seemed to support the idea of the spontaneous generation of life. Once again the results of Redi were confirmed for bacteria, this time in a long series of investigations by Pasteur culminating in 1864. In brief, Pasteur showed that if beef broth or other foodstuffs were thoroughly boiled in open vessels no bacteria would appear

Fig. 35-15. Lazzaro Spallanzani, the 18th-century abbot who showed that if soups or seeds were boiled half an hour and sealed, no microscopic life appeared in them and hence no spoilage. (Photo of life mask, courtesy Johns Hopkins Institute for the History of Medicine.)

in them so long as the opening to the exterior was a long curved tubular neck bent horizontally so that no dust or dirt particles carrying bacteria could pass up it. From this time on, the phrase, *omne vivum ex vivo,* all life from life, has been a basic tenet of all branches of biology.

The Heterotroph Theory

In recent years the advancing knowledge of biochemistry and genetics, of historical geology, and of the possibilities of natural selection have combined to substantiate the heterotroph theory of life's origin from non-living matter. A **heterotroph** (*hetero,* other, + *trophos,* feeder) is an organism, such as an animal or a fungus, that is dependent on outside sources for complex nutrients, especially on a carbon source more complex than CO_2. An **autotroph** (*auto,* self, + *trophos,* feeder) is not dependent but can

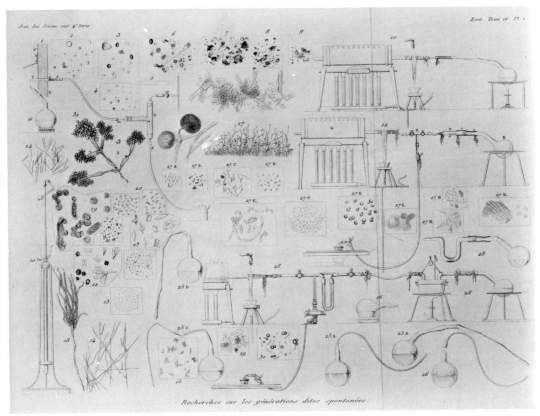

Ann. des Scienc. nat. 4.ᵉ Serie.   *Zool. Tome 16. Pl. 1.*

Recherches sur les générations dites spontanées.

Fig. 35-16. Apparatus used by Pasteur to prove that living organisms are essential for fermentation. (Courtesy U. S. Library of Congress.)

manufacture its own food from very simple inorganic materials, such as water and CO_2. A heterotroph theory of the origin of life has been discussed off and on for a good many years, and in fact was proposed in 1871 by Darwin himself. Much more recently the heterotroph theory has been set forth by J. B. S. Haldene in England and in great detail by A. I. Oparin in Russia. The evidence for this theory does not constitute proof, but it is nonetheless convincing.

The heterotroph theory holds that, before the existence of any life on a planet, complex organic molecules would arise and, once formed, would remain in existence because there would be no bacteria, molds, or protozoa to devour them. The first living thing would be composed of these complex molecules built up into still more complex arrangements. A fundamental characteristic of the proteins and nucleic acids of protoplasm is their ability to duplicate themselves. Whether the first forms of life were "free genes" something like footloose **virus-like particles** or were colloid globules, the so-called **coacervates**, is not certainly known. Coacervation, that is, the formation of discrete colloidal globules separated from their environment by a distinct boundary membrane, is something which occurs readily in non-living material in the laboratory. Such a coacervate composed of self-duplicating nucleic acids and proteins would not be very different from an amoeba.

The Non-Living Basis

Evidence from both astronomy and geology indicates that in the remote past before the appearance of life, the earth's atmosphere lacked appreciable amounts of oxygen but was rich in just those carbon and nitrogen compounds, plus water vapor, that are required to form amino acids, proteins, and nucleic acids. Today the other planets, where it is certainly too cold or too hot for life, have atmospheres of this type. Spectroscopic analysis shows that the "airs" of Jupiter and Saturn are rich in methane (CH_4), ammonia (NH_3), and probably hydrogen (H_2). Space probes have recently confirmed that the atmosphere of Venus contains enormous amounts of carbon dioxide (CO_2); the atmosphere of Mars has both water (H_2O) and carbon dioxide. The earth's seas, after the first dozen million years (it is important to remember that evolution is a matter of some billions of years) would contain salts of many kinds, including phosphates which seem essential for the energy transfers of living material and for building nucleic acids. Recently, two carbon compounds, formaldehyde and formic acid, have been detected in outer space.

The raw materials were certainly present. What were the chances that they would have combined into complex molecules like porphyrins, amino acids, and self-duplicating units like nucleic acids? There is, first, the enormous stretch of time during which the simpler compounds would have had opportunities to come together in all possible ways and combinations. Of course, if carbon, oxygen, nitrogen, and hydrogen could not form amino acids, no amount of time would avail; but these are the elements from which amino acids are made. These are undoubtedly possible reactions.

At least three conditions were present in addition to enormous stretches of time and the essential raw materials that would have greatly facilitated such synthetic reactions. One factor, perhaps the most important, was the intense ultraviolet radiation reaching the earth's surface before there was any appreciable free oxygen in the atmosphere. Free oxygen under the influence of solar radiation forms ozone which today constitutes an ultraviolet absorbing blanket around the upper atmosphere of the earth. A second factor is lightning, which today is still producing nitrogen compounds in the atmosphere that are brought to the earth's surface in rain. Third, in the long course of geologic history, the earth's crust has undergone many upheavals. Mountains have appeared, continents have been submerged, and what is now dry land was once under the sea. In this process arms and bays of the sea were cut off from the ocean and may have dried up completely, leaving salt beds, or may later have been connected with the sea again. The Caspian Sea represents such a cutoff bay. This means that in the past there have been many types of places, from temporary tide pools to inland seas, where evaporation would slowly concentrate salts and organic compounds, thereby increasing the likelihood of new and complex chemical reactions.

Natural Selection at the Chemical Level

As both Darwin and Oparin have emphasized, before any living things existed on our planet, complex compounds which today would be quickly metabolized by microorganisms of various kinds, would have remained. There would have been no decay in the usual sense because the organisms which cause decay did not exist. Under these conditions there would have been a natural selection among the various organic compounds mixing in the primeval seas and pools. Compounds that were less stable or more difficult to form would become less abundant, while the more stable ones would accumulate. Compounds which had the ability to imprint their own organization on other compounds would be favored over those which lacked such an ability. Those well fitted to organize more molecules into configurations like themselves would be favored over those which did so only slowly. The slow ones would be "outbred," so to speak, and might even cease to exist.

Emergence of Life

By the time complex compounds appeared which had the property of organizing, or of catalyzing, the formation of other compounds like themselves, the shadowy borderland between the clearly non-living and the obviously living had already been reached. In considering the probable events during this early period of life's origin, it is important to remember that many of the materials present in the primordial "soup," although relatively simple themselves, are nonetheless the substances out of which the highly complex macromolecules of life are built. Phosphoric acid, H_3PO_4, for instance, is an essential part of the structure of all nucleic acids and as part of ATP is also essential in most energy transfers within living systems. Glycine, composed of a carbon with an amino group and two hydrogens attached to it, plus one —COOH group, is the simplest possible amino acid. Acetic acid, CH_3COOH, is almost the simplest possible organic acid. Yet glycine and acetic acid combine to form porphyrin, a key part of the cytochrome, chlorophyll, and hemoglobin molecules.

The existence of complex, more or less self-duplicating compounds and aggregates of compounds would not prevent natural selection from taking place. If anything, it would be intensified. At first the energy for the synthesis of these compounds would have been supplied primarily by ultraviolet light from the sun and by lightning. But once complex compounds had been formed in some abundance, those that could obtain the energy needed for synthesis by causing the breakdown of other compounds, rather than depending on solar radiation, would clearly have an adaptive advantage. There was probably no free oxygen present, certainly no more than trace amounts, so that these first energy-yielding reactions must have been anaerobic. At precisely what point these self-duplicating, energy-utilizing molecular entities can be said to have been alive is a semantic question.

The view that the first energy-yielding reactions to appear in the history of life were anaerobic is supported by the geological evidence that there was no free oxygen in the atmosphere of our planet in its youth. It is also supported by the fact that the initial stages in the utilization of foodstuffs to provide energy in animals, plants, and bacteria are always anaerobic. In other words, the Embden-Meyerhof glycolytic pathway, which does not depend on free oxygen, precedes the Krebs citric acid cycle and electron transport chain, which require free oxygen. In fact, there are good reasons for believing that the Krebs cycle evolved much later than the glycolytic pathway and that it was tacked on to the end of it, so to speak.

Consequently, biologists no longer believe that the first living things must have been green plants, purple bacteria, or other autotrophs. Obviously, an autotroph must begin with a far more complex set of synthetic enzymes than a heterotroph, which depends on picking up and utilizing already existing compounds of some complexity. To begin with an autotroph is to begin at the wrong end.

Three Primary Life Styles Emerge

Once considerable numbers of primitive self-duplicating units had come into existence as indicated in Chapter 1, they would compete with each other for the dwindling supply of complex nutrients. In such a nutritional crisis, natural selection must have favored two very different lines of evolution. Any unit, organism if you prefer, that required fewer complex molecules in its "diet" would be favored over the unit which required more. For example, an organism which required compounds A, B, C, and D but did not need E because it could make its own E by transforming D would have a selective advantage over the organism which required all five. This line of selection led to the photosynthetic plants, the autotrophs which can build themselves up from simple inorganic salts, CO_2, and water. Water and CO_2 were abundant from the start of life and the fermentations carried out by primitive organisms assured a continuing supply of CO_2. The porphyrins (specifically chlorophyll) provided the light-trapping colored compounds. Once the necessary enzymes were available for converting the trapped solar energy into chemically useful forms (high energy phosphate bonds and complex organic compounds), this line of living organisms and their descendants became nutritionally independent for all time.

Fortunately for us there was another evolutionary answer to the food crisis of those remote times. The second line of selection favored those units which either could move around and so get into a place where the necessary materials for their growth were present or else could engulf other organized units and secure the essential materials that way. This line of selection led to the animals. Once photosynthesis became widespread, it provided a supply of free oxygen. This made aerobic respiration possible for plants but it was of even greater importance for the evolution of oxygen-requiring animals which characteristically depend on vigorous activity, a highly energy-dependent trait. Before this could happen mitochondria had to evolve, either directly in the cells of plants and animals or as symbiotic aerobic bacteria.

Bacteria and other fungi have evolved in a direction which exploits a second nutritional opportunity open to heterotrophs. They obtain their complex carbon compounds from other organisms including dead plants and animals. Without the fungi, decay would not occur and ultimately the earth's supply of CO_2 would become locked in the corpses of plants and animals rather than being recycled. At what point in the history of life on this planet their activities became essential for the continuance of life is unknown.

Laboratory Confirmations

To some of these general arguments, Harold Urey and his student, Stanley Miller, have added laboratory proof. They set up apparatus in which they circulated a sterile atmosphere of water vapor, ammonia, methane, and hydrogen past an electric discharge (to simulate lightning). The water vapor was then condensed, and in this liquid they found after some weeks a variety of complex substances including the amino acids glycine and alanine, two of the commonest in proteins. This experiment clearly ranks with the synthesis of urea by Wohler as a landmark in understanding the chemistry of life. The work of Urey and Miller has been confirmed and extended in other laboratories.

Strong support for the concept that successive mutations increase the synthetic abilities of primitive organisms has been obtained by students of mutations affecting the nutritional requirements of bacteria and fungi, especially the pink bread mold *Neurospora*. Essentially, mutations are occasional errors which occur in the duplication of nucleic acids, and there is no reason to suppose that such events were any rarer in the remote past than they are now.

The Role of Chance

Does all this mean that the origin of life is a matter of chance? This often-asked question can easily lead into an exercise in semantics and scientific double-talk. In the sense that the particular time, place, and manner in which the first self-duplicating unit appeared would be no more and no less predictable than when the number five will turn up in a series of throws of dice, the origin of life was clearly a matter of chance. There is no reason why the first living unit should have been formed from compounds carried down by raindrops after synthesis by lightning rather than from molecules formed on the surface of the sea by ultraviolet radiation. At the same time, the potentialities for combining into amino acids and more complex self-duplicating units had to be present in the original methane, ammonia, and other molecules. If none of the dice used has a side with five dots on it, five will never come up regardless of how many times the dice are thrown. The appearance of life was a matter of chance, but the dice were loaded.

THE HISTORY OF LIFE ON EARTH

Fossil Remains

The vast drama of the evolution of life on this planet is at least partially recorded in the form of fossils in the rocks. Most fossils consist either of the body of an organism which has been gradually mineralized or of an imprint left by an organism. The sandstone of the Connecticut River valley and the Paluxy River in Texas are famous for fossilized footprints left by dinosaurs. Many fossils have not been completely mineralized and still retain traces of organic matter. All of the coal in existence is derived from forests from an age tens of millions of years before the dinosaurs flourished. All of the oil and all of the chalk cliffs are deposits formed from the oil droplets or skeletons of algae or protozoans. By contrast, some remains of extinct and prehistoric life are so recent that they have scarcely been changed enough to be called fossils. In many parts of the United States bones of extinct animals can be found. Preserved in ancient tarpits and in dry caves in the southwest are the remains of saber-toothed tigers, small and large species of elephants, and other animals no longer in existence. In the far north, in both Alaska and

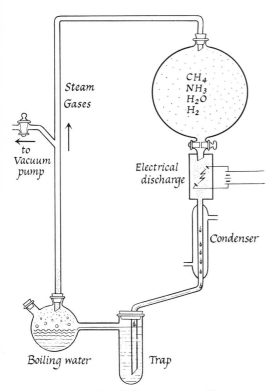

Fig. 35-17. The Urey-Miller apparatus. Water vapor, ammonia, methane, and hydrogen were circulated past an electrical discharge. When condensed liquid was analyzed it was found to contain a variety of complex chemicals including amino acids. (From G. B. Moment: *General Zoology,* 2nd ed. Houghton Mifflin Co., Boston, 1967.)

Siberia, the frozen remains of several woolly elephants known as mammoths have been found. Not only was the fur preserved but even the flesh was still recognizable as meat and was eaten by dogs.

Fossils are found only in sedimentary, that is, stratified, rocks such as slate and are formed in two principal ways. One is the slow accumulation of sand or other sediment on ocean or lake bottoms. The other is the accumulation of material on the land, perhaps in extensive bogs and marshes, which in later ages becomes submerged beneath the ocean by slow changes in the continents. Igneous rock such as granite, which is spewed out of volcanoes or formed the original crust of the

earth, never contains fossils. Obviously older layers underlie more recent ones.

New Methods of Chronology

The new ways of forcing the earth to reveal the age of her rocks and their fossils are continually becoming more precise. At the same time the older methods are still valuable. The rate at which Niagara Falls moves upstream, cutting its gorge as it goes, can be measured. By simple arithmetic it is then possible to estimate how long it has taken Niagara Falls to cut its gorge. Correcting for the probability that more water flowed over the falls in the past when the last glaciers were melting, the answer comes out to somewhat over 10,000 years, an estimate that agrees with estimates made by other methods.

Another method that gives reliable results in determining chronology of the recent past is to count varves in clay. **Varves** are layers of sediment formed where a river runs into a body of quiet water. This is especially true of rivers running down from snow-capped Alps or melting glaciers. Spring flood waters wash down coarse sediment. As the runoff slows down during the summer, the sediment becomes finer and finer. In the winter it is finest of all, to be followed the next spring by another coarse layer. Consequently, by counting the varves in a deposit of clay it is as easy to determine the number of years it was in forming as it is to tell the age of a tree by counting the annual rings revealed by a cross section of the trunk. By comparing the pattern of successive varves in one lake with those in another, it is possible to correlate their dates of origin.

Longer periods are dated by measuring the radioactivity of various embedded materials. Each radioactive element has a specific **half-life,** that is, the time it takes for the level of radioactivity in a sample to decay to one-half the original level. A crystal of uranium-238 will be half lead after 4.5 billion years. By determining the percentage of lead in such a piece of uranium, it is possible to tell its age. The greater the proportion of lead, the older the crystal. Although such crystals form only in molten igneous rock as it cools, it is possible to use them to date strata of sedimentary rock formed by the gradual deposit of silt or other material on the ocean floor. In volcanic convulsions molten igneous rock sometimes breaks through strata of sedimentary rock. Obviously all the strata through which such molten rock breaks must antedate the volcanic action. The molten rock often spreads out on top of the uppermost strata, forming an enormous pancake of igneous rock. Any sedimentary rock above this pancake must of course be younger. Obviously, this method is only useful in measuring long periods.

A recently introduced method is the radio-carbon technique. This provides an accurate clock for periods ranging from a few thousand to about 35,000 years ago, since the half-life of radioactive carbon-14 is $5,568 \pm 30$ years. This carbon is being formed in the earth's atmosphere by the action of cosmic rays on nitrogen and is also continually disintegrating, so that an enduring equilibrium exists age after age. Radioactive carbon built into plants by photosynthesis will not be replenished after they die. Consequently, the longer a piece of wood or part of any animal that ate plants exists, the less radioactive carbon it contains.

Geologic Eras

The Archeozoic: The Origin of Life

The Archeozoic Era, during which life's non-living precursors accumulated and interacted, and in which life finally took form, was an immense stretch of time, probably about two billion years. Geological evidence in rocks of this age indicates a period of intense volcanic activity, mountain building, and extensive erosion with the formation of some sedimentary rocks. Columns of steam are thought to have arisen where molten rock flowed into the sea or violent rains fell on hot rocks. It was presumably the age of the primordial heterotrophs and the time when the animal and plant kingdoms first became differentiated. There are no recognizable fossils in the rocks of this age but there are deposits which may have been made by iron-, sulfur- and carbon-utilizing bacteria.

The Proterozoic: The Origin of Animal Phyla and Plant Divisions

Following the Archeozoic came another era lasting approximately one billion years, the Proterozoic. Fossils occur in rocks of Proterozoic age, but are few in number and extremely difficult to decipher. The time has been too long and the pressures and shearing of the rocks too severe. This also must have been a period of mountain building and erosion with consequent sedimentation. In Glacier National Park there are cliffs of about 10,000 feet of sedimentary rock laid down during this era and yet they represent the deposits of only a small fraction of its duration.

During the Proterozoic Era life was still limited to the sea. Despite the dearth of fossils, it seems quite certain that during this immense stretch of time the major animal phyla were differentiated, because in the earliest strata of the next era, the Paleozoic, representatives of all the major animal phyla are already present. Lost in the tortured rocks of the Proterozoic is the record of how the first metazoans developed. Nor does the record tell how the animal kingdom came to be divided into the two great branches, the deuterostomes (composed of vertebrates, echinoderms, and their relatives) and the protostomes (composed of annelids, arthropods, mollusks, and their kin). The sedimentary rocks of the Proterozoic era do contain some doubtful remains of the simplest plants, the algae and the bacteria.

The Paleozoic: Age of Invertebrates, Fish, and the Earliest Terrestrial Plants

When the Paleozoic began, about 500 million years ago, nearly five-sixths of the entire time during which life originated and evolved had already passed. Nevertheless, stupendous events in the history of life were yet to come—the rise of insects, higher mollusks like the octopus, vertebrates including dinosaurs, birds, and, of course, man. The key event that made possible the rise of higher forms of animal life including man was the invasion of the emerging continents by new forms of plant life, the first of a long line of terrestrial species that eventually forested the land.

The Cambrian Period. The earliest division of the Paleozoic Era is known as the Cambrian Period, because rock strata laid down then were first intensively studied in Wales (Cambria). Cambrian rocks are the oldest in which fossils are abundant. So sharp is the distinction in this respect that all earlier strata are commonly referred to merely as pre-Cambrian. Already present are representatives, small and generalized but unmistakable, of all the major invertebrate phyla of animals: protozoans, sponges, coelenterates, annelids, echinoderms, mollusks, arthropods, and several others. The most abundant larger animals were trilobites, a kind of arthropod. Several kinds of animals living in the Cambrian survive to the present almost unchanged, at least in general form. This is true of some jellyfish and brachiopods. The only major phylum not represented was the chordates. Plant life of the Cambrian Period was primitive. The marine algae served as the basis of food chains on which the diverse forms of animal life were dependent.

The Ordovician Period. The next period of the Paleozoic, the Ordovician, saw two events of paramount importance, the appearance of fresh water and probably land plants and the appearance of the vertebrates. These two events were very possibly connected, since the fossil evidence indicates that the first vertebrates were jawless fish which arose in fresh water. They are supposed to have come from primitive oceanic chordates, probably larval tunicates, that swam up estuaries and rivers where they lived on the rich organic detritus coming down with the current. Until there were land plants, or at least freshwater plants, there could of course be no nourishing debris washed down from the land, so the theory assumes an abundant land vegetation at this time. Botanists generally agree that the first land plants must have appeared during this period although the oldest known remains are of more recent origin. The most abundant animals still seem to have been trilobites; at least, they left the most fossils. Perhaps they had the hardest shells or lived where fossils were most likely to be formed. Notable also were the mollusks and echinoderms.

The Silurian and Devonian Periods. The following two periods of the Paleozoic, the Silurian and Devonian, are memorable for the colonization of the land by both plants and animals. By the end of the Devonian, there were forests of tree ferns, giant club mosses, and giant horsetails covering much of the formerly barren continents. In the sea there were still the marine algae and the invertebrate hordes, but there were now also vertebrates with jaws. These jawed fish evidently swam down the rivers and re-entered the sea. Shark-like fish were numerous as were the first really big vertebrates. More important in the evolution of animal life was the appearance in fresh waters of muscle-finned fish, a group which includes the lungfish and coelacanths. They formed the population from which the amphibians evolved. The way in which the muscle-finned fish became adapted for terrestrial life in the process of becoming adapted for survival in regions subject to drought, and thus gave rise to the amphibians, was discussed as an illustration of pre-adaptation.

The vertebrates were not the only animals to exploit the opportunities of terrestrial life in the Devonian. Several groups of arthropods also became terrestrial. These included centipedes, scorpions, and primitive insects somewhat like cockroaches.

The Mississippian and Pennsylvanian Periods. The next two periods were the earlier Mississippian and the more recent Pennsylvanian. Together they lasted about 50 million years and are often called the *carboniferous periods* because during these times there flourished the great forests of club mosses, horsetails, ferns, and seed ferns which formed our present coal. The climate must have been warm and moist, and much of the land low-lying and swampy to support those lush growths of the same general kind of non-flowering plants that characterized earlier eons. The first primitive seed plants, the gymnosperms, appeared and are thought to be descendants of the seed ferns.

New animal groups also arose. The first small reptiles appeared, scarcely distinguishable from the now common amphibians, and certainly giving no hint of what they would become in a later era. In the Pennsylvanian living things first became able to fly, a remarkable evolutionary achievement 250 million years ago. The first flyers were cockroaches and large dragonflies, very similar in structure to our present-day species. It seems likely that those first wings had the same advantages as wings on the insects of today, namely, to disseminate the species and enable the adults to find new ponds or other suitable places to lay their eggs. The catching of other insects on the wing probably came much later. In the seas, the Pennsylvanian was a great age of echinoderms, especially of sea lilies.

The Permian Period. The Permian Period closed the Paleozoic Era. It was a period of great climatic and geological changes. As the result of movement of the earth's crust, the Appalachian Mountain range along eastern North America was thrust up, carrying with it Paleozoic sediments which had been pressed into fossil-bearing rock strata thousands of feet thick. The climate was dry and cold. In many parts of the world—South America, India, and Australia—there are unmistakable marks of a Permian glaciation apparently longer and more severe than the comparatively recent Pleistocene glaciation from which we are still emerging. All these changes are commonly called the Appalachian Revolution. The lush carboniferous forests of the two preceding periods dried up and the tree ferns all but disappeared. The last of the trilobites, which had been dominant animals since the Cambrian some 350 million years earlier, finally died out. The sea lilies dwindled to a few isolated populations. Generalized reptiles resembling heavily built salamanders and modern-looking insects became the dominant land animals. The most important event from the human point of view was the appearance of the forerunners of the mammals. These were the dog-toothed reptiles, in which the teeth were not merely a row of sharp pegs as in the dinosaurs and reptiles, but were differentiated into incisors, canines, and cheek teeth with several cusps. There were no teeth on the palate as in many reptiles. The best known genus is *Cynognathus* (*kynos,* dog, + *gnathus,* jaw), a lizard-like creature that grew to be about 2 meters long.

The Mesozoic: The Great Age of Reptiles

After the Appalachian Revolution the world entered an era that was more uniformly dry and warm, the Mesozoic. This period saw the rise of the first mammals, probably egg-laying creatures similar to the spiny anteater and the duckbilled platypus of Australia which lay soft-shelled eggs like those of reptiles. During the Cretaceous Period at the end of the Mesozoic Era, Australia became separated from the great land mass of Eurasia and so became a sanctuary for the primitive egg-laying and marsupial mammals. The Mesozoic also saw the origin of the first flowering plants. The gymnosperms continued to flourish and vegetation generally assumed a more modern appearance. The first birds appeared, equipped with two things no living birds possess—teeth and long tails resembling those of a lizard or dog.

The Mesozoic Era is divided into three periods: the **Triassic**, the **Jurassic**, and longest and most recent, the **Cretaceous**. The predominant terrestrial animals during this span of about 130 million years were the dinosaurs. Like other groups of animals before and since, these reptiles experienced one of those evolutionary explosions known as **adaptive radiation**. The reptiles began in the previous era as lizard-like creatures which could lay their eggs on land, relatively safe from the teeming predators of the water. The development of an amnion was the key step here, along with a protective leathery shell for the egg. From this generalized basic population, the reptiles spread out by becoming specialized for many different modes of life in many environments. Biped carnivores like *Tyrannosaurus* preyed on herbivorous dinosaurs like *Triceratops.* In the swamps was the gigantic *Brontosaurus* which attained a length of almost 30 meters. In the sea, marine reptiles competed with sharks and other fish, while in the air were flying pterodactyls.

The Cenozoic: The Age of Mammals and Flowering Plants

The final era of geologic time is the Cenozoic, the age which saw the evolution of the flowering plants and of the mammals including man and the other primates. The climate became progressively more temperate and drier. Hence, the Cenozoic became the age of grass covering extensive prairies, steppes, pampas, and veldts in many parts of the world. The evolution of the horse, camel, and other grass-eating animals was attuned to this spread of grass-covered prairies. It is probable that the thinning of the forests was an important factor in bringing man's precursors down from the trees, though firm proof of this is lacking. The Cenozoic was also the age of birds and of modern insects.

The Cenozoic, which began only about 70 million years ago, is divided into the Tertiary and Quaternary Periods. The Tertiary, in turn, is divided into five epochs: Paleocene, Eocene, Oligocene, Miocene, Pliocene; the Quaternary is divided into the Pleistocene and Recent Epochs.

By the Paleocene Epoch the fossil record shows that the old "ruling reptiles" of the Mesozoic were extinct. Most of the mammals were small, generalized, and dog-like. However, there were a number of species of large and fairly specialized forms. The primitive elephants, for example, were larger than a large hog and probably provided with a well-developed and downward-bending snout. *Hyracotherium (Eohippus)*, the remote ancestor of the horse, was vaguely suggestive of a miniature horse with four toes on its front feet and three on the hind. Its future course of evolution as a running herbivore seems to have been already set.

During the rest of the Cenozoic the mammals underwent the same kind of explosive evolutionary adaptive radiation that the reptiles had experienced in the Mesozoic. Like the reptiles, some of the mammals returned to the sea and became our present whales and porpoises. Later the seals and seacows followed. Others, the bats, became adapted for flying. On land, the mammals diversified into a bewildering variety of species adapted for many kinds of life.

One of the most spectacular and best studied cases is the evolution of the horse, the first fossil record worked out with any degree of completeness. *Hyracotherium*, the 45-cm.-long "horse" of the Eocene, was followed by whole series of populations of descendants. Some were more and some less horse-like, but in the long run natural selection favored those with the most elongated bones in the "hands" and feet, fewer toes, and heavy corrugated grinding molar teeth; in other words, those better adapted for running on the prairie and eating grass. Toward the end of the Eocene, the larger *Orohippus* had not only larger feet, but a third toe larger than the other toes. This genus was followed in the Oligocene by *Mesohippus*, the first recognizable "horse," with a single toe on each foot carrying most of the weight. Its second and fourth toes scarcely touched the ground.

In the later Miocene and the following Pliocene several genera of horses (*Merychippus, Pliohippus*, and others) appeared, again larger in body and more specialized in legs and teeth. The genus *Equus*, which includes as separate species the contemporary horse, ass, and zebra, actually first appeared in the late Pliocene. The members of this genus possess only "splint" bones on either side of the foot as tell-tale evidence of their many-toed ancestry. Horses spread widely in Eurasia, Africa, and North and South America. The charred bones of extinct horses are found in the fireplaces of Stone Age men in California, but long before Europeans arrived on this continent the native horses had become extinct. We do not know why this happened.

More or less fossilized remains of Pleistocene mammals have been found in many locations in North America. Typical species include the Columbian elephant or mammoth, the low-browed mastodon, the saber-toothed tiger, the stag-moose, a giant sloth, and primitive horses. Most exciting, in California there are "fossil fireplaces" containing the charred bones of mastodons. Evidently, early man on this continent killed, cooked, and ate mastodons. For several hundreds of years elephant tusks, "ground ivory," have been found in Siberia. Over 20 frozen, flesh-and-blood remains of mammoths have been found in northern Siberia, and in 1948 the frozen body of a baby mammoth was found during placer mining near Fairbanks, Alaska.

HUMAN ORIGINS

Our ancestry extends back to the origin of life, but our evolution as something very special among mammals began not earlier than sometime in the Paleocene, roughly 60 million years ago. At that time there was no distinction between the shrew-like insectivores and the monkey-like primates. But this primitive insectivore-primate stock possessed two character-

Era and Duration (in Millions of Years)	Millions of years ago (from Start of Period or Epoch)	Period and Epoch and Duration (in Millions of Years)		Characteristic Life (Dominant Organisms)
Cenozoic 70	0.025	Quaternary	1	Herbs and man
		Recent epoch		
	1	Pleistocene epoch		
		Tertiary	69	Angiosperms, mammals, and birds
	10	Pliocene epoch		
	25	Miocene epoch		
	40	Oligocene epoch		
	60	Eocene epoch		
	70	Paleocene epoch		
Mesozoic 130	130	Creataceous	60	Gymnosperms and reptiles
	165	Jurassic	35	
	200	Triassic	35	
Paleozoic 350	230	Permian	30	Lycopods, seed ferns, and amphibians
	250	Pennsylvanian	20	
	280	Mississippian	30	
	325	Devonian	45	Early land plants and fish
	360	Silurian	35	
	450	Ordovician	90	Algae and invertebrates
	550	Cambrian	100	
Pre-Cambrian 2,500 Proterozoic Archeozoic	3,000			Unicellular organisms?

After G. B. Moment: *General Zoology,* 2nd ed. Houghton Mifflin Co., Boston, 1967, and H. J. Fuller and O. Tippo:

Geological Events, Climate, etc.	Advances in Plant Life	Advances in Animal Life
Periodic glaciation.	Increasing dominance of herbs. Extinction of many trees. Increase in number of herbs.	Man; rise of civilization. Extinction of great mammals.
Continued cooling of climate with temperate zones appearing. Cascades, Andes, Coast Ranges formed.	Increasing restriction of plant distribution and of forests. Rise of herbs.	Appearance of man.
Climate greatly changed—cool and semi-arid. Marginal seas. Himalayas, Alps formed.	Restriction of distribution of plants— retreat of polar floras. Forest reduction.	Culmination of mammals.
Climate warm, humid, Pyrenees formed.	World-wide distribution of tropical forests.	Primitive mammals disappear. Rise of higher mammals and birds. First anthropoids.
Climate cool, semi-arid; then warm, humid.	Modernization of flowering plants. Development of extensive forests— to polar regions. Sequoias prominent. Tropical flora in Artic regions.	Modern birds and marine mammals appear.
Mountain glaciers.		
Climate fluctuating. Rocky Mts. and Andes formed. Great Continental seas.	Angiosperms dominant; Gymnosperms dwindling. Modern tropical plant families within Arctic Circle.	Rise of primitive mammals.
Climate fluctuating.	Rapid development of angiosperms—many living genera present.	Extinction of great reptiles.
Climate very warm.	Rise of Angiosperms. Conifers and cycads still dominant.	
Climate warm. Sierras formed. Great continental seas in Western N. America.	First known angiosperms; Caytoniales; conifers and cycads dominant; cordaites disappear.	Primitive birds and flying reptiles (pterodactyls). Dinosaurs abundant. Higher insects.
Climate warm, semi-arid.	Floras not luxuriant; higher gymnosperms increase (Cycadophytes, conifers, ginkgos). Seed ferns disappear.	First mammals. Rise of giant reptiles (dinosaurs).
Climate dry with periodic glaciation. Appalachians, Urals formed. Drainage of seas from continents.	Dwindling of ancient groups, extinction of many. First cycads and conifers.	Rise of land vertebrates.
Period of crustal unrest, alteration of marine and terrestrial conditions.	Dominant lepidodendrons, calamites, ferns, seed ferns, and other primitive gymnosperms (Cordaites). Extensive coal formation in swamp forests.	
Widespread shallow seas on N. America. Acadian Mts. formed.	Dominant lycopods, horsetails, and seed ferns. Early coal deposits.	Rise of primitive reptiles and insects.
Broad shallow seas on N. America during Silurian and Devonian.	Early land plants (Psilophytales-Rhynia, etc.). Primitive lycopods, horsetails, ferns, and seed ferns. First forests.	Rise of amphibians. Fishes dominant.
Taconic Mts. formed.	First known land plants. Algae dominant.	Lungfishes and scorpions (air breathing animals).
Broad shallow seas on the N. American continent.	Rise of land plants (?). Marine algae dominant.	Corals, star fishes, pelecypods, etc. First vertebrates—armored fishes.
Narrow seas within the borders of N. America. Climate warm, uniform over earth.	Algae—especially marine forms.	Many groups of invertebrates—dominance of trilobites.
Grand Canyon, Younger Laurentians formed.		
Rocks, chiefly sedimentary.	Bacteria and Algae.	Worms, crustaceans, brachiopods.
Glaciation.		
Older Laurentians formed.	No fossils found. All organisms probably unicellular or very simple.	
Rocks mostly igneous or metamorphosed; few sedimentary.		

College Botany, rev. ed. Holt, Rinehart and Winston, New York, 1954.

Fig. 35-18. North American mammals characteristic of the Pleistocene epoch. All are drawn to scale with pointer dog (in box) to show relative size. 1, American mastodon; 2, saber-toothed tiger; 3, giant ground sloth; 4, stagmoose; 5, giant beaver; 6, Colombian elephant; 7, Texas horse. (From G. B. Moment: *General Zoology*, 2nd ed. Houghton Mifflin Co., Boston, 1967.)

istics which have left their marks on us to this day. These small mammals were generalized and tree-living.

When did the primates separate from the insectivores? The fossil record is missing, and even today it is impossible to make a clean-cut distinction between these two orders which span the enormous gulf between a man and a mole. Modern insectivores include not only moles but also shrews, which are common almost everywhere though seldom seen because of their small size and nocturnal habits. The primates include man and the anthropoid apes, the Old World monkeys, the New World monkeys (which alone possess prehensile tails),

and the lemurs, now largely restricted to Madagascar and adjacent lands.

The most ancient monkey-like fossils so far discovered were found in lower, *i.e.*, earlier, Oligocene strata in Egypt. *Parapithecus*, as this creature was named, was perhaps close to the base population from which both Old and New World monkeys diverged; but no one can be certain. Our broad shoulders and well-developed collar bones are clearly adaptations for what anatomists call **brachiation**, *i.e.*, swinging from branch to branch. But exactly when our ancestors forsook the trees for the ground is unknown. However, it is practically certain that they were fairly large, since the habit of swinging from branch to branch in contrast to climbing and jumping like the little lemurs is limited to the larger primates.

It does not seem too surprising that our immediate subhuman predecessors left few fossils. Living either in forests or grasslands opportunities for fossilization were scanty, except in occasional swamps, river quicksands, and tar pits. Evidently primitive man was too smart or too agile to get caught in these. However, anatomists have made some informed and plausible guesses about our early history.

Human evolution took place during the end of the Cenozoic. This was a period when over large areas of the earth forests were giving way to grasslands as the climate became drier. With the succulent leaves and buds of forest plants and the diverse animal life of forests vanishing, many animals were faced with a harsh choice: to eat grass or to eat other animals that eat grass.

It is reasonable to suppose that these profound environmental changes affected the predecessors of modern man. The spread of the grasslands would necessitate either a changed way of life, with emphasis on hunting the animals (horses, deer, and mammoths) which ate the grass and shrubs, or a migration southward to keep within the dense forests. Perhaps both happened. Hunting of this kind would put an evolutionary premium on running and communication. Natural selection would then favor the further development of the upright posture and long straight legs, a good brain and ability to communicate, and use weapons and other tools. It would also favor the kind of disposition that makes for mutual cooperation. The breeding units were certainly small. Populations of many millions are found in insects, fish, and marine invertebrates, but surely did not exist in primates until human civilization became industrialized.

Under these changing conditions the primary evolutionary forces of mutation and natural selection produced as much change in the primates as they did in the horses. This must have been the period when genetic forces building on the juvenile form with its proportionately much larger brain and prolonged period of willingness to learn must have come into play. This is the neoteny discussed earlier. In very small populations such as existed in the past, random genetic drift would come into play and produce changes regardless of adaptive significance. Perhaps the nonadaptive differences, if there really are any, among different races of men are to be explained in this way.

The most ancient remains of man-apes, or ape-men, or primitive man have been found in the Olduvai Gorge east of Lake Victoria at the southern end of the Rift Valley in Tanzania. The anthropologist Louis S. B. Leakey and his co-workers have unearthed an abundant deposit of bones, primitive stone tools, and other evidence of a primitive culture. The skeletons in this East African site resemble others found in South Africa and in Australia but are different enough to be given a new name, *Zinjanthropus*. The character of the stone tools dates this settlement about 500,000 to 600,000 years ago. However, the potassium-argon dating method places the remains at the Olduvai site even earlier, about 1,750,000 years ago. The related *Australopithecus,* from both Taung in Bechuanaland in Africa and from Australia may be as ancient as *Zinjanthropus* or even more so. No one really knows.

Dating back into the Pleistocene some 500,000 years ago are the remains of another extremely primitive man or man-like creature called variously the Java man or *Pithecanthropus* or *Homo erectus,* and also the Peking man or *Sinanthropus pekinensis.* In the years since they were discovered, more remains of both have been found with intermediate characteristics so that it seems the Java and Peking groups represent only slightly different forms of the same population. These Java men averaged well under 6 feet in height with a low forehead, heavy bony ridge over the eyes, and heavy chinless jaw with large teeth.

Was *Homo (Pithecanthropus) erectus* truly human? If man is defined as a tool- and fire-using primate, the answer is yes. Along with the humanoid remains are many stones, apparently chopping stones, which seem shaped to fit into the right hand. In the Chinese cave there are also circular charred areas, some deeply burned, which indicate the use of man's greatest tool, fire. There are many bones of extinct deer and other animals split open so as to expose the marrow, and also far too many human skulls in proportion to the scarcity of other human remains. These skulls have all been

Fig. 35-19. Louis B. Leakey (1903-1972) explored the Olduvai Gorge and the surrounding area of Tanzania for more than 40 years. Shown here with his wife working under the shade of an umbrella to protect against the East African sun. (From H. L. Shapiro: Louis B. Leakey, 1903-1972. Saturday Review of Science. *55(44):* 70, 1972.)

broken open. This rather disquieting fact is also evident in the caves of prehistoric man in Europe. The implication of cannibalism is strong.

Neanderthal and Cro-Magnon Man

In marked contrast to all of these early men, who are known from a few scattered bones, are the Neanderthals and Cro-Magnons. The latter people have left numerous remains widely distributed over Europe and extending east into Israel and the Crimea.

The Neanderthal remains were first discovered in the Neander Valley near Düsseldorf, Germany, in 1856. At the time some experts thought that these bones merely represented a pathological maldevelopment of the skeleton of a modern man. But as similar bones were discovered deep in Pleistocene deposits in caves located in many different regions, it became clear that *Homo neanderthalensis* had indeed existed as a widespread population. The skull capacity and hence brain size (1300 to 1700 cc.) fall within the same general size range as

that of modern man, but the proportions of the brain were slightly different. With the bones of Neanderthals have been found numerous hearths, chipped stone axes, flint spear tips, and other possible tools. There is evidence that Neanderthal man buried his dead with stores of food and implements. These men flourished for over 100,000 years during the third interglacial period and into the fourth and most recent glaciation, the Würm-Wisconsin. This is a long time indeed, compared with the trivial 5,000 to 7,000 years of written history. How did they end? Perhaps they were exterminated by their successors, the Cro-Magnons, or perhaps they merged with them.

The Cro-Magnon people appeared in Europe and the Near East as the Neanderthals disappeared 25,000 to 50,000 years ago. Whether the Cro-Magnons had been pushed out of other regions by the increasing cold or actually evolved on location under the stress of the Ice Age, they were physically a superb race, with a brain capacity on the average equal to or greater than our own. They were tall and straight of limb and have left truly magnificent works of art on the walls of caves, so that we flatter ourselves in identifying them as *Homo sapiens*.

The culture of the Cro-Magnons was a Paleolithic, or Old Stone Age, culture in which they chipped but did not grind their stone axes, spear points, knives, lamps, and other implements. In many places they made such extensive use of reindeer bones and antlers that they are sometimes referrred to as the reindeer men. From the bones around their fireplaces and deep in their caves, it is clear that they also ate horses, waterfowl, fish, shellfish, and probably, at least in some localities, other men. The most remarkable fact about their culture is the great number of magnificent wall paintings they have left illustrating reindeer, extinct mammoths, extinct bisons, and extinct types of horses. These people drew pictures of animals now limited to the far north, like the reindeer, on caves in Spain and Israel. They also left engravings and carvings in bone, soapstone, and ivory. They appear to have invented the bow and arrow, perhaps as the development of a stick with a leather thong attached to one end and used to throw a small javelin.

Exactly when man first arrived in North America is unknown, but it appears to have been during or even before the last glaciation. There is abundant evidence of a flourishing Stone Age culture in the southwestern United States on the shores of prehistoric lakes that have not held water since glacial times. On the coast of California there are prehistoric fireplaces containing charred bones of extinct elephants, camels, and horses. Only future investigation can tell who the firebuilders were and what was their relationship to Stone Age peoples in Europe and Asia.

Fig. 35-20. Restoration of Cro-Magnon man, a prehistoric caveman who left magnificent wall drawings in caves in Europe and the Near East. (Courtesy American Museum of Natural History.)

Table 35-2
Pleistocene glaciations and forms of man

Glaciation	Nonglacial Period and Duration	Man Population
	Post-glacial 10,000 to 15,000 years	Contemporary man
Fourth, or Würm-Wisconsin		Cro-Magnon man Neanderthal man
	Third interglacial 120,000 years	
Third, or Riss-Illinoian		
	Second interglacial 300,000 (?) years	
Second, or Mindel-Kansan		
	First interglacial 100,000 years	*Homo erectus* (Java and Peking populations)
First, or Gunz-Nebraskan		
		Australopithecus; *Zinjanthropus*

Fig. 35-21. Drawings of a long extinct elephant (mammoth) were made on cave walls by Cro-Magnon man. (Courtesy Field Museum of Natural History, Chicago.)

PERSISTENT QUESTIONS ABOUT EVOLUTION

Recurring questions about evolution are of two sorts. The first are strictly scientific questions. What are the driving forces of evolution? What is the role of mutation, isolation, sexual reproduction, natural selection, random sampling? What has been the actual history of life on this planet? How adequate is the evidence for what biologists believe to be true? What are the thousand and one evolutionary stratagems which have produced and continue to maintain the wondrous forms of life—orchids, bumblebees, and heliozoans, the octopus and the columbine? These problems we have discussed. Two questions of very general importance remain and to these we now turn before considering the second type of questions which are philosophical.

The Problem of Extinction

It is an article of faith with most ecologists that all species and generous samples of all types of environments should be preserved for the enjoyment, study, and general benefit of present and future generations of mankind. Certainly our planet would be much the poorer without them and at the worst would become uninhabitable if pollutants kill most kinds of vegetation. Clearly the present ecological squeeze threatens many species and entire plant and animal communities with extinction. At the other extreme, there are species that North America at least could well do without. The European gypsy moth, *Porthetria dispar,* was introduced into Massachusetts a century ago by an amateur entomologist who "didn't know it was loaded." By the 1970's it is defoliating whole oak forests on the mountains of eastern Pennsylvania. The vicious Brazilian wild bee, which attacks and kills cattle and even people and is now rapidly spreading northward, is not an ecologically necessary member of the North American ecosystem. Consequently the search for the factors which produce extinction or even the drastic and permanent reduction in population size is now urgent both to preserve threatened species and to reduce species which threaten mankind.

Extinction is an ancient and common event in the history of life. Probably the most spectacular of all the many cases of extinction is that of the great reptiles of the Mesozoic. For over 150 million years the dinosaurs had flourished, experiencing an adaptive radiation which made them dominant on the land, in the swamps, in the sea, and in the sky with forms as diverse as the monstrous brontosaurs and the delicate little hop-skip-and-jump precursors of birds. What happened? One possibility is that they could not survive the widespread change in climate which historical geologists find took place at the close of the Mesozoic. Although the cooling and drying of the regions outside the tropics may account for the lack of the great reptiles in those areas, it is difficult to see why, on this basis, many of them could not still flourish in the tropical regions of Central and South America, Africa, and South Asia or in the ocean.

A second possibility is competition with the primitive mammals. At the close of the Mesozoic the only mammals that had appeared were small and generalized, much like small mongrel dogs or opossums. They are supposed to have contributed to the downfall of the dinosaurs and other reptiles by eating reptile eggs, but such primitive mammals would have had to be ubiquitous and extraordinarily effective to eat all the eggs in so many environments.

A third possibility is that some epidemic viral, bacterial, or even protozoan disease wiped out the great reptile populations. This possibility is as hard to prove as to disprove. It is known that the tsetse fly, which transmits the trypanosome of African sleeping sickness, has made it impossible to keep domestic cattle in large areas of Africa, but (and this is equally important) the native relatives of cattle, like the various gazelles and antelopes, have enough immunity to flourish in these same regions and serve as "reservoirs" of the disease.

One possible explanation is a modern version of the old theory of racial senescence. No modern zoologists believe that time, simply as duration, produces some inevitable racial aging. Too many different kinds of animals have existed unchanged, or with only the most trivial changes, for hundreds of millions of years (sharks, lungfish, some echinoderms, and brachiopods), but the natural course of evolutionary change resulting from mutation and natural selection does tend to make living things more and more narrowly specialized. And excessive specialization can act as a trap. When an organism becomes so highly specialized that it loses the versatility needed to meet changing conditions, in either the living or the non-living environment, then the highly specialized group succumbs to more generalized ones. For example, increase in size is as obvious in many of the mammalian lines of evolution as it was in the reptiles. What evolutionary forces lead to increased size? One obvious factor is competition, especially between males of the same species. Perhaps this leads to a disastrous spiral, in which only the largest males can win females but in which individuals become too large to function efficiently in food-getting and other ways, a situation which has happened to the lion. This disaster may have actually overtaken the big herbivorous dinosaurs such as the three-horned *Triceratops* and the 15-meter-long carnivore, *Tyrannosaurus,* which preyed upon it.

One, all, or none of these four factors may have been important. All we can be certain of is that the evolutionary deployment of the reptiles was remarkably like that of the mammals at a later time. Both began with a small, generalized form, and both gave rise to hundreds of extremely diverse types—the mammals to tigers, rabbits, giraffes, bats, and baboons. The reptiles produced a corresponding diversity but, except for a few descendants, they have

Fig. 35-22. Flight of passenger pigeons shown in an 1885 print. Before their extinction, enormous flocks of these birds darkened the skies of North America. (From G. B. Moment: *General Zoology.* Houghton Mifflin Co., Boston, 1958.)

	10	20	30	40

```
                              9 0 1 2 3 4 5 6 7 8 9 0 1 2 3 4 5 6 7 8 9 0 1 2 3 4 5 6 7 8 9 0 1 2 3 4 5 6 7 8 9 0 1 2 3 4
HUMAN                         G D V E K G K K I F I M K C S Q C H T V E K G G K H K T G P N L H G L F
RHESUS MONKEY                 G D V E K G K K I F I M K C S Q C H T V E K G G K H K T G P N L H G L F
HORSE                         G D V E K G K K I F V Q K C A Q C H T V E K G G K H K T G P N L H G L F
PIG, BOVINE, SHEEP            G D V E K G K K I F V Q K C A Q C H T V E K G G K H K T G P N L H G L F
DOG                           G D V E K G K K I F V Q K C A Q C H T V E K G G K H K T G P N L H G L F
GRAY WHALE                    G D V E K G K K I F V Q K C A Q C H T V E K G G K H K T G P N L H G L F
RABBIT                        G D V E K G K K I F V Q K C A Q C H T V E K G G K H K T G P N L H G L F
KANGAROO                      G D V E K G K K I F V Q K C A Q C H T V E K G G K H K T G P N L N G I F
CHICKEN, TURKEY               G D I E K G K K I F V Q K C S Q C H T V E K G G K H K T G P N L H G L F
PENGUIN                       G D I E K G K K I F V Q K C S Q C H T V E K G G K H K T G P N L H G I F
PEKIN DUCK                    G D V E K G K K I F V Q K C S Q C H T V E K G G K H K T G P N L H G L F
SNAPPING TURTLE               G D V E K G K K I F V Q K C A Q C H T V E K G G K H K T G P N L N G L I
BULLFROG                      G D V E K G K K I F V Q K C A Q C H T C E K G G K H K V G P N L Y G L I
TUNA                          G D V A K G K K T F V Q K C A Q C H T V E N G G K H K V G P N L W G L F
SCREWWORM FLY                 G D V E K G K K I F V Q R C A Q C H T V E A G G K H K V G P N L H G L F
SILKWORM MOTH                 G N A E N G K K I F V Q R C A Q C H T V E A G G K H K V G P N L H G F Y
WHEAT                         G N P D A G A K I F K T K C A Q C H T V D A G A G H K Q G P N L H G L F
FUNGUS (NEUROSPORA)           G D S K K G A N L F K T R C A E C H G E G G N L T Q K I G P A L H G L F
FUNGUS (BAKER'S YEAST)        G S A K K G A T L F K T R C E L C H T V E K G G P H K V G P N L H G I F
FUNGUS (CANDIDA)              G S A K K G A T L F K T R C A E C H T I E A G G P H K V G P N L H G I F
BACTERIUM (RHODOSPIRILLUM)    G D A A A G E K V S K – K C L A C H T F D Q G G A N K V G P N L F G V F
```

A ALANINE	I ISOLEUCINE	R ARGININE
C CYSTEINE	K LYSINE	S SERINE
D ASPARTIC ACID	L LEUCINE	T THREONINE
E GLUTAMIC ACID	M METHIONINE	V VALINE
F PHENYLALANINE	N ASPARAGINE	W TRYPTOPHAN
G GLYCINE	P PROLINE	Y TYROSINE
H HISTIDINE	Q GLUTAMINE	

Fig. 35-23. Amino acid sequences in cytochrome *c* proteins from 20 organisms. Numbering is according to wheat sequence which has 112 amino acids. At 20 positions (green overlay) the same amino acid is found in all. From the raw data of amino acid sequences, a computer has determined the ancestral amino acid sequences that can best account for the observed primary protein structures. These ancestral sequences establish the locations at which the branches of the phylogenetic tree diverged. (After M. O. Dayhoff and R. V. Eck: *Atlas of Protein Sequence and Structure 1967-68.* National Biomedical Research Foundation, Silver Spring, Md., 1968.)

become extinct. Does a similar fate await mammals?

The study of present day populations affords what seem to be more useful insights into extinction. Natural populations of animals and some plants continually undergo fluctuations depending on many factors—climatic and seasonal weather variations, predators and parasites, food supply, and the like. A prolonged cold rainy spell at the time when fawns are born can drastically reduce a year class of deer. When a population is large, the normal fluctuations in numbers do not threaten it with extinction. When a population becomes very small, there is always the danger that in the course of these fluctuations, one will dip to the zero point. Thus, it is not necessary for any

single factor to destroy every breeding pair, only for it to lower the population to the danger level. The extinction of the enormous flocks of passenger pigeons which once darkened the skies of the U. S. Middle West is sometimes attributed solely to over-hunting but it is to be noted that this occurred after the conversion of the extensive oak forests in which they nested into farm lands. This drastically reduced the population. Furthermore, the passenger pigeon was a social nester and, as is true of various sea birds which nest in dense colonies, very small colonies are commonly unsuccessful in rearing their young. The case of the national animal of New Zealand, the strange *Sphenodon* lizard, is instructive. All efforts to stop its decline in numbers failed until sheep

```
            50              60              70              80              90
5 6 7 8 9 0 1 2 3 4 5 6 7 8 9 0 1 2 3 4 5 6 7 8 9 0 1 2 3 4 5 6 7 8 9 0 1 2 3 4 5 6 7 8 9 0 1 2 3 4 5
G R K T G Q A P G Y S Y T A A N K N K G I I W G E D T L M E Y L E N P K K Y I P G T K M I  F V G I  K K
G R K T G Q A P G Y S Y T A A N K N K G I T W G E D T L M E Y L E N P K K Y I P G T K M I  F V G I  K K
G R K T G Q A P G F T Y T D A N K N K G I T W K E E T L M E Y L E N P K K Y I P G T K M I  F A G I  K K
G R K T G Q A P G F S Y T D A N K N K G I T W G E E T L M E Y L E N P K K Y I P G T K M I  F A G I  K K
G R K T G Q A P G F S Y T D A N K N K G I T W G E E T L M E Y L E N P K K Y I P G T K M I  F A G I  K K
G R K T G Q A V G F S Y T D A N K N K G I T W G E E T L M E Y L E N P K K Y I P G T K M I  F A G I  K K
G R K T G Q A V G F S Y T D A N K N K G I T W G E D T L M E Y L E N P K K Y I P G T K M I  F A G I  K K
G R K T G Q A P G F T Y T D A N K N K G I I W G E D T L M E Y L E N P K K Y I P G T K M I  F A G I  K K
G R K T G Q A E G F S Y T D A N K N K G I T W G E D T L M E Y L E N P K K Y I P G T K M I  F A G I  K K
G R K T G Q A E G F S Y T D A N K N K G I T W G E D T L M E Y L E N P K K Y I P G T K M I  F A G I  K K
G R K T G Q A E G F S Y T E A N K N K G I T W G E E T L M E Y L E N P K K Y I P G T K M I  F A G I  K K
G R K T G Q A A G F S Y T D A N K N K G I T W G E D T L M E Y L E N P K K Y I P G T K M I  F A G I  K K
G R K T G Q A E G Y S Y T D A N K S K G I V W N N D T L M E Y L E N P K K Y I P G T K M I  F A G I  K K
G R K T G Q A A G F A Y T N A N K A K G I T W Q D D T L F E Y L E N P K K Y I P G T K M I  F A G I  K K
G R K T G Q A P G F S Y S N A N K A K G I T W G D D T L F E Y L E N P K K Y I P G T K M V  F A G L  K K
G R Q S G T T A G Y S Y S A A N K N K A V E W E E N T L Y D Y L L N P K K Y I P G T K M V  F P G L  K K
G R K T G S V D G Y A Y T D A N K Q K G I T W D E N T L F E Y L E N P K K Y I P G T K M A  F G G L  K K
G R H S G Q A Q G Y S Y T D A N I K K N V L W D E N N M S E Y L T N P K K Y I P G T K M A  F G G L  K K
S R H S G Q A Q G Y S Y T D A N K R A G V E W A E P T M S D Y L E N P K K Y I P G T K M A  F G G L  K K
E N T A A H K D N Y A Y S E S Y K A K G L T W T E A N L A A Y V K N P K A F V L E S K M T F K -  L T K

                        T E M                                          K S G D P K A K
```

were removed from the little islands on which it still existed. Without the sheep, the low shrubbery among which it lived returned and with the restoration of its appropriate environment, *Sphenodon* has increased in numbers. Without a proper environment no restriction on harvesting a plant or animal will prevent its permanent disappearance.

The Problem of Biological Relationships

Although scientists have been classifying living organisms for over 300 years and the theory of organic evolution has been well authenticated and generally accepted the world over for more than a century, both plants and animals are still classified in much the same way as in the day of George Washington; that is, in terms of genera and species grouped into families, orders, classes, and phyla or divisions. No one knows whether an order of placental mammals corresponds to an order of insects or of flowering plants or perhaps to a class or even a family. The theory of evolution indicates that all mammals are more closely related to each other than any one is to a bird or a starfish, and that oaks are more closely related to grass than to pine trees.

However, all these relationships are known only in vague general terms. Man is more closely related to the great apes than to the baboons or the monkeys, but how much more? What is the actual degree of evolutionary difference between any of the primates and a mushroom or a pine tree, or between the great apes themselves? A few years ago questions like these seemed too far-out to be taken seriously. Now quantitative answers are becoming available. Ultimately their answers will be in terms of DNA, but the people working along that line have run into difficulties.

Meanwhile a true breakthrough has been achieved by the comparative study of amino acid sequences in one of the most ancient and universal proteins known, cytochrome *c*. It will be recalled that cytochrome *c* is a respiratory protein located in the mitochondria of all aerobic cells. There are about 110 amino acids in the cytochromes of all organisms tested from man, horse, whale, and chicken to turtle, fly, wheat, fungus, and yeast. The principle of comparison is basically simple. Each change in an amino acid at any point on the sequence represents a single mutation. Therefore, the greater the number of different amino acids in the cytochromes of two organisms, the further apart they are, or, to state this in a more logical

way, the distance between any two or more organisms can be defined in terms of the number of locations in the cytochrome *c* sequences where the amino acids are different. The accompanying table shows the sequences in 21 organisms. It is of interest to observe that at 20 positions the same amino acid is present in all organisms from man to mold. By comparing the sequences of two organisms it is possible to arrive at a hypothetical sequence for a division point or "node" which represents the common ancestor from which the two organisms being compared have diverged. These nodes are computed on the assumption that the pattern requiring the minimum number of mutations is the correct one. Animal and plant relationships derived by this method are shown in Figure 35-24. Much remains to be done, but after three centuries of effort it is at long last possible to assign numbers to the relationships of living things.

Additional questions of a basic nature come to mind. Why, of all the thousands of possible

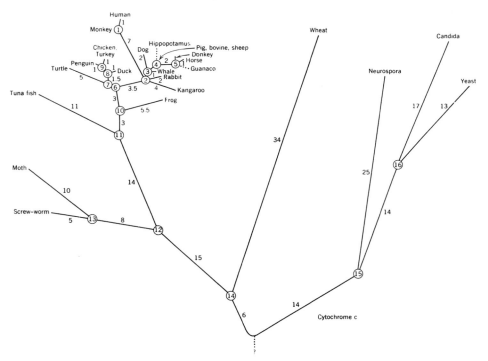

Fig. 35-24. Plant and animal phylogenetic tree based on amino acid analyses of cytochrome *c* proteins. (After M. O. Dayhoff and R. V. Eck: *Atlas of Protein Sequence and Structure 1967-68*. National Biomedical Research Foundation, Silver Spring, Md., 1968.)

amino acids, are only two dozen found in living organisms on this planet? How did DNA come to be the basis of life? Would not any one of a dozen other macromolecules have done as well? What is the explanation for the relation between a given nucleotide sequence and a specific amino acid?

Philosophical Questions

The philosophical questions most frequently asked about evolution concern its meaning and the place of man in nature. These problems have been extensively explored by T. H. Huxley, Teilhard de Chardin, G. G. Simpson, and a host of other biologists. To the age-old question as to whether the appearance and evolution of life were due to chance, we have already given the answers which seem to us to be scientifically justifiable. The specific source of the organic molecules involved, for instance, whether they were the products of reactions energized by ultraviolet radiation or by lightning, and the particular spot on the planet where these compounds became united in such a way that self-reproducing units were formed—all of this was a matter of chance. However, given the inherent properties of the elements of this universe, both chemical and physical, plus energy sources, plus reproduction with variation occurring throughout eons of time, then the emergence and evolution of life are so highly probable that statistically speaking they are as close as an asymptote to the inevitable. And underlying all of these questions is the ultimate mystery of the emergence of conscious intelligence out of the vastness of space, time, matter, and energy.

USEFUL REFERENCES

Briggs, D., and Walters, S. M. *Plant Variation and Evolution.* World University Library, McGraw-Hill Book Co., New York, 1969.

Darwin, C. *On the Origin of Species by Means of Natural Selection* (1859). Doubleday & Co., Inc., New York, 1960.

Dobzhansky, T., Hecht, M. K., and Steere, W. C. (eds.) *Evolutionary Biology,* vol. 6. Appleton-Century-Crofts, New York, 1972.

Hanson, E. D. *Animal Diversity,* 3rd ed. Prentice-Hall, Inc., Englewood Cliffs, N.J., 1972.

Lack, D. *Darwin's Finches: An Essay on the General Biological Theory of Evolution.* Harper Torchbooks, New York, 1961.

Mayr, E. *Animal Species and Evolution.* Harvard University Press, Cambridge, 1963.

Moorehead, A. *Darwin and the Beagle.* Harper and Row, publishers, Inc., New York and Evanston, 1969.

Ponnamperuma, C. (ed.) *Exobiology.* Elsevier Publishing Co., New York, 1972.

Ponnamperuma, C. *The Origins of Life.* E. P. Dutton & Co., New York, 1972.

Romer, A. S. *The Procession of Life.* Doubleday-Anchor, New York, 1972.

Simpson, G. G. *The Meaning of Evolution.* Yale University Press, New Haven, 1960.

Solomon, M. E. *Population Dynamics.* St. Martin's Press, New York, 1969.

Stebbins, G. L. *Process of Organic Evolution,* 2nd ed. Prentice-Hall, Inc., Englewood Cliffs, N.J. 1971.

Volpe, E. P. *Understanding Evolution.* Wm.: C. Brown Co., Dubuque, Iowa, 1967.

Wynne-Edwards, V. C. *Animal Dispersion in Relation to Social Behavior.* Oliver and Boyd, Edinburgh, 1962.

Ecology

Our planet Earth is covered by a seamless garment, the biosphere, within which all plants and animals live. This biosphere is but a thin film of earth, water, and air scarcely two miles thick. Outside of this very special layer no living thing can exist unless encased in an airtight and temperature-controlled capsule. Within this film all living things are in a dynamic relationship with each other and with their non-living environment.

Ecology is the science which deals with all of these complex interrelationships. Because the essential life-supporting qualities of the biosphere within which mankind must live are now threatened by the activities of man himself, no branch of science is more timely or important than ecology. The term ecology is derived from the Greek *oikos*, meaning house or household, and involves the study of the total economy of living organisms in the broadest sense of the word.

Mankind is unavoidably a part of the world **ecosystem**. Thus if we are to maintain our world environment as a desirable or even possible place for human life, it is an urgent necessity that we discover the laws of ecology and live within them, just as an airplane designer must work within the laws of physics. No organism, whether man or earthworm, can leave its environment untouched. Plants also affect their environment and do so in equally profound if sometimes less obvious ways. Thus all the children of DNA, whether bacteria, plants, animals, or people, are bound together in a common destiny which is now threatened by an impending environmental crisis. Extinction has been a commonplace phenomenon in the long course of evolution and it may be that some of us, will become extinct. If such should be the case, it certainly would not be the bacteria that would disappear first.

ECOLOGY AND CURRENT WORLD PROBLEMS

Before taking an intensive look at the science of ecology itself, another important fact must be considered. The problems of the environment which we face are largely the multiple results of industrialization and population growth. Consequently, most of these problems, rather than being strictly biological, are primarily political and economic problems although with important scientific dimensions. No great scientific breakthrough is needed to know that radiation is extremely harmful to all living things nor is any special scientific insight required to realize that mercury is poisonous. Thus, it is of the greatest importance to be certain that all the facts are considered in relation to the sum total of their impact on human life.

The actual situation can be dramatically illustrated by the case of a new nuclear energy power plant on the Chesapeake Bay. The governor of Maryland appointed a Task Force of physicists, engineers, biologists, and other concerned citizens to study the possible effects of such a power plant and to make recommendations for appropriate action. The members of the task force learned that if the plant were built the additional radiation discharged, even though very small, would make it a statistical certainty that over the next decade

Fig. 36-1. Astronaut Edwin Aldrin walking near the lunar module during Apollo 11 extravehicular activity. Note the barrenness of a land without life. (Courtesy National Aeronautics and Space Administration.)

there would be 8 to 10 additional cases of leukemia. Should any such task force accept responsibility for those human tragedies? What criteria should be used to decide? In the Baltimore and Washington, D. C. area most of the electric power comes from coal mined in West Virginia. Only about a year before this terrible question was put to the Task Force, 78 men were killed in a single accident in a West Virginia mine. Fatal and maiming accidents are common in coal mines. Pneumoconiosis, or "black lung," a debilitating and sometimes fatal disease, is also widespread. So the true state of affairs is this: as long as people insist on more electric power and until safer methods of coal mining are developed, if the Chesapeake Bay area is to be free of the very small amount of radiation a nuclear power plant would discharge, then that freedom will be paid for by the lives of coal miners.

We will consider DDT later in this chapter, but it should be pointed out here that the World Health Organization has been the chief proponent of its use. Especially in the under-developed countries of the world, DDT has saved millions of people from death and suffering due to malaria and other insect-borne diseases. When you and your children are dying from disease and from starvation, you could not care less that the American eagle is laying soft-shelled eggs or that traces of DDT can be found in penguins at the South Pole. Although DDT has been banned in the United States, a stark question still confronts us all: what should be done until safer methods of insect control are available everywhere? Birth control pills raise similar questions. Such hormonal preparations sometimes produce harmful and even fatal side effects. Is it enough to say that pregnancy also can produce harmful and even fatal side effects or that overpopulation results in disease and death?

The role of the ecologist is to provide the knowledge on which an intelligent and humane environmental management program can be based. In ecology, as in its sister science, medicine, knowledge can never be complete. Hippocrates, the father of medicine, emphasized long ago that "life is short and the occasion fleeting". In applied ecology, action must often be taken, as it is in medicine, on the basis of the best available information which will, in many cases, fall far short of a complete understanding of all the factors involved or a full foreknowledge of all the implications.

ENVIRONMENTS

One of the first tasks of ecology is to bring some logical order out of the diverse kinds of environments in which organisms live and to discover how plants and animals are interrelated with their environments. The **range** of a species is the entire geographical area over which the animal or plant may be found. For example, the range of *Rana pipiens*, the meadow frog commonly used in laboratories, is from north-

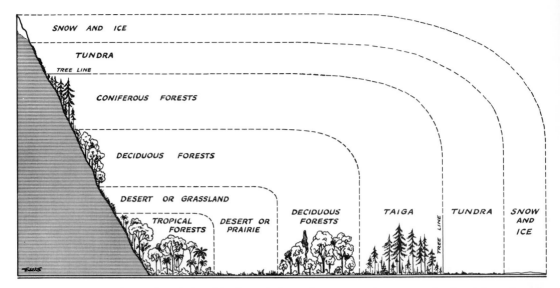

Fig. 36-2. Correspondence of vertical and horizontal biomes from mountaintop to sea level and from the pole towards the equator. (After G. B. Moment: *General Zoology,* 2nd ed. Houghton Mifflin Co., Boston, 1967.)

ern Canada to Panama, and from the Atlantic coast almost to the Pacific. The **habitat** of a species is the particular kind of environment in which it lives. The habitat of *R. pipiens* is damp meadows and along the edges of small streams and ponds. The ecological **niche** occupied by a species is usually defined as its ecological role in its environment, but it is also said that the niche occupied reflects the requirements for living of that organism. The ecological niche of *R. pipiens* is that of a small carnivore in moist locations. As commonly used in ecology, niche is an abstract, functional term.

Terrestrial environments change in a regular and similar way in passing from north to south and from mountain tops to valleys. These changes are sometimes inconspicuous when one group of plants and animals slowly replaces another. Elsewhere the changes may occur in a series of fairly abrupt steps separating well-defined life zones. Each zone supports a characteristic living community called a **biome.** Thus there is a deciduous forest biome, a prairie biome, and so on. The line of contact between two biomes, large or small, is termed an **ecotone.** Theoretically, this is a zone of tension between two communities. Ecotones are often characterized by special animals and plants which thrive only in such transitional regions.

The quail (bobwhite) is such a species which for successful breeding must nest close to the edge of a wooded area, preferably where forest, meadow, and shrub meet.

The Seven Major Land Biomes

Between the North Pole and the equator there are seven major biomes. The north and south polar regions, together with snow-capped peaks, constitute **Arctic zones,** which are similar in many ways though not ,identical. Except in a more or less sporadic form the Arctic zones are largely without life.

Tundra

South or below the Arctic zone is the **tundra.** This is a vast treeless region extending across northern Canada and northern Siberia. The **Alpine zone,** above the tree line on mountains, corresponds to the tundra. The characteristic plants are sphagnum and other mosses, reindeer moss and other lichens, grasses, and a few small shrubby bushes. The flowering herbaceous plants of the tundra are noted for their showy blossoms. Musk oxen, reindeer, snow hares, caribou, Arctic foxes, and wolves are characteristic mammals. In the brief summer, insects (especially flies and mosquitos) and nesting migratory birds abound. In New Hampshire, the tundra or Alpine zone begins on Mount Washington at less than 2,000 meters, while in the Rockies the tree line is up at the 3,500 to 4,000-meter level. On Mount Popocatepetl in Mexico the tree line is at some 5,000 meters or more. Not all the factors that determine the position of the tree line are understood. Temperature is paramount, but its action is subject to considerable modification. Sometimes isolated areas of trees grow far north of the tree line. The line may be sharp, or there may be a transition zone of stunted trees.

Taiga

South of the tundra is a belt of conifers, spruce, firs, and pines called the **taiga**. Birch and aspens are scattered through the taiga and are usually the first trees to grow after a fire. The moose is the characteristic large mammal; rodents and mink are abundant small ones. Intermediate in size and also abundant are lynxes, black bears, wolves, and martens.

Deciduous Forest

South of the taiga is the familiar **deciduous forest** of hardwoods. In the northern part of the zone, beeches and maples tend to predominate; in the southern part, oaks and hickory. The Virginia deer, the black bear, and the opossum are characteristic, although the yellow-spotted black salamander, *Ambystoma maculatum*, of the woodlands is used as an "index animal" for this zone in the eastern

Fig. 36-3. Alpine terrain in the Wenatchee National Forest, Washington. The area shown is between 2,000 and 2,800 meters (6,000-8,500 ft.) elevation and most is at the upper limits of tree growth. (Photo By A. Mills. Courtesy U. S. Forest Service.)

United States. In some places, when soil and moisture conditions are right the deciduous zone is replaced by a southern coniferous region.

Grassland

In many parts of the world **grasslands**, variously termed prairies, steppes, veldts, or pampas, commonly occur south of the deciduous zone. It is in this kind of environment that the evolution of horses, antelopes, and kangaroos took place. In some regions the prairie extends northward to the coniferous taiga, without any intervening deciduous zone. Rainfall is the major environmental factor governing the extent of grassland vegetation. Grasslands replace forests as annual rainfall decreases. Such a transition is apparent to anyone driving westward through the central United States.

Fig. 36-4. Grazing land in Utah. Note coniferous taiga in background. (Courtesy U. S. Department of Agriculture.)

Tropical Rainforest

The **tropical rainforest** makes a sixth major life zone. Such regions are found in Central America, northern South America, equatorial Africa, and southern Asia and the East Indies. The pattern of trees is radically different from that in other types of forest. Few northern forests are composed of more than a dozen species, and great stands can be found composed of but one or two species. In tropical rainforests, however, several hundred kinds of trees are commonly found intermixed. Moreover, the vegetation is stratified. Canopies of great ironwood, banak, and other trees tower to a height of 40 meters (125 feet) or more. Below them, in partial shade, there is a second layer of trees, and beneath these, in turn, is a mass of small trees, bushes, and plants of many kinds. It is little wonder that in this environment many of the animals are tree dwellers. In British Guiana 31 out of 59 species of mammals are arboreal.

Desert

A seventh zone is **desert**, which may occur at almost any latitude. Thorny bushes with waxy leaves characterize deserts everywhere. In the Americas the typical plant is the cactus. Animal life is sparse and mostly nocturnal. A traveler going westward across the great plains of the United States can observe the transition from grassland to desert vegetation in regions of decreasing average rainfall.

Minor Biomes

Particular climatic and soil conditions have produced special biomes in various parts of the world. For example, the northwest coast of the United States has a nontropical rainforest. **Chaparral** biomes of dwarf evergreen oaks and other shrubby trees are found in temperate and subtropical regions with a rainy winter and a hot, dry summer. Some 6,000,000 acres of California are covered with chaparral. The shores of the Mediterranean are also covered with chaparral in many places.

The character of a biome is determined by climate, which includes temperature, rainfall,

Fig. 36-5. Sagebrush and Joshua trees are typical plants of the Mojave desert, California.

amount of sunlight, and the character of the soil. The relative importance of climate and soil in controlling the living community varies from one region to another. No tropical rainforest can arise in the Great Plains area of the central United States, no matter what the soil conditions. In other regions, as in some parts of southeastern United States, soil seems to be the determining factor, within, of course, the limits set by climate.

Microclimates

The climate that matters for any animal or plant is obviously the climate in which it lives. This may be very different from the general climate of the geographical region constituting the range of the species. An insect living in the tree tops lives in a veritable desert compared with other species living on the forest floor where evaporation is only about 7 per cent of that near the tops of the trees. Conditions under stones, on the north side of boulders, or a few inches under the sand of a beach are almost always quite different from those of the general surroundings. This means that temperature and moisture readings recorded on equipment within a standard Weather Bureau louvered box set on a post one meter above ground level cannot be directly related to the animals, plants or microorganisms in the vicinity. Those species may and often do enjoy a very different set of conditions.

ATMOSPHERE

All of the physical factors of the environment in which plants and animals live have been modified in greater or lesser degree by the presence of living organisms. We live at the bottom of an ocean of air surrounding our planet. Approximately 80 per cent of the air is nitrogen, which has persisted since the ages before life appeared on earth. The story of oxygen is very different.

It is now well established that atmospheric oxygen was not always present but instead is a product of photosynthesis accumulated since the first appearance of aquatic green plants. Few physical factors are of greater importance. Without oxygen active animal life as we know it would be impossible. Without oxygen in the atmosphere the amount of ultraviolet radiation reaching the earth's surface would be so great that life on land would be impossible.

The amount of moisture and CO_2 in the air are also influenced by plant and animal life.

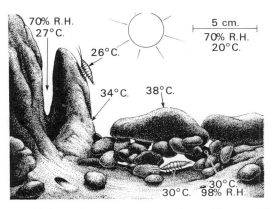

Fig. 36-6. Beach crustaceans encounter a wide range of microhabitats and microclimates. Note the differences in temperature and humidity above, at, and under the surface.

The composition of the ocean is continually being modified by the swarming myriads of organisms within it. Think of the millions of tons of lime that have been removed from the sea by the protozoans whose shells formed chalk deposits. Even the amount of light and the temperature are modified, though to a slight degree, by the presence of plants and their effects on the amount of moisture, carbon dioxide, and oxygen in the atmosphere.

SOILS

The soil is a very special case among the physical factors of the environment, partly because it has not merely been modified by living organisms but has actually been made by them. Without a covering of plants, the surface of the earth would be bare rock, washed by wind and rain. Because soil is of such crucial importance in agriculture and because it can be improved or easily damaged beyond repair, a special soil science, **pedology**, has developed. The Russians and Scandinavians have been leaders in this field and, consequently, a number of the terms are derived from their languages.

Soil Structure

As can be seen in any freshly cut bank, soil consists of a series of layers, called **horizons**. The uppermost horizons are commonly known as **topsoil**, technically the **A horizons**. Topsoil contains a surface layer (A_{00}) of undecomposed debris consisting of leaves, twigs, animal remains, and the like called **litter**. Beneath this is a layer (A_0) of more or less decomposed organic matter, called **leaf mold**. True soil begins with a dark horizon (A_1) of **humus**. This layer is composed of a high content of organic matter, thoroughly mixed with miscellaneous rock particles. Like the two layers above, it is rich in microscopic and semi-microscopic animals and plants. Below the humus is a lighter colored layer, the A_2 horizon, from which most of the soluble materials have been leached, *i.e.,* washed out, by sinking rainwater. In prairie soils the deep black chernozems, the leached A_2 layer, is said to be missing or, if listed as present, is relatively unleached.

The subsoil, or **B horizon**, lacks appreciable amounts of organic matter and contains few soil organisms, whether animals, plant roots, fungi, or bacteria. It is said to be mineralized. The upper layer of the subsoil, termed the B_1 horizon, is dark with the minerals leached from the topsoil. The deeper B_2 horizon is a tightly packed mineral soil, which may show a bottom layer of lime or other mineral deposits just above the C_1 **horizon** of weathered bedrock.

Soil Types

Soils are classified as **podzols** or **chernozems**, depending on the relative amount of leaching that has occurred. Podzols are commonly much richer in aluminum and iron and deficient in calcium, and hence are reddish or brown in color. For this reason they are often termed **pedalfers** (*pedon*, ground, + aluminum, + *ferum*, iron). Such soils are characteristic of the eastern half of the United States. Chernozems are characteristic of the drier great plains and western regions and are rich in calcium. Consequently they are termed **pedocals** (*pedon*, ground, + calcium).

In various regions having different temperatures and rainfalls the relative thickness of the soil horizons, as well as other soil characteristics, differ greatly. Under the coniferous forest of the taiga, the humus or A_1 horizon is very shallow and very sharply demarcated from the leached A_2 layer. Such a soil is called **mor**. The humus is very low in calcium, lacking in nitrates, and definitely acid (pH ranging from 3.0 to 6.5). Fungi play a prominent role in its decomposition. Earthworms, which are important soil formers, are rare and limited to acid-tolerant species, mainly unfamiliar surface dwellers such as *Bimastus eiseni* and *Dendrobaena octaedra*, which seldom burrow deeper than a few inches.

Under the temperate deciduous forest the A_1 horizon is much thicker and the A_2 less thoroughly leached and richer in organic matter. This type of soil is called **mull**. It is much richer in calcium than is mor, nitrates are present, and the pH ranges from acid to slightly alkaline (4.5 to 8.0). Bacteria play a prominent role in its decomposition. Earthworms are relatively abundant, and deep-burrowing species

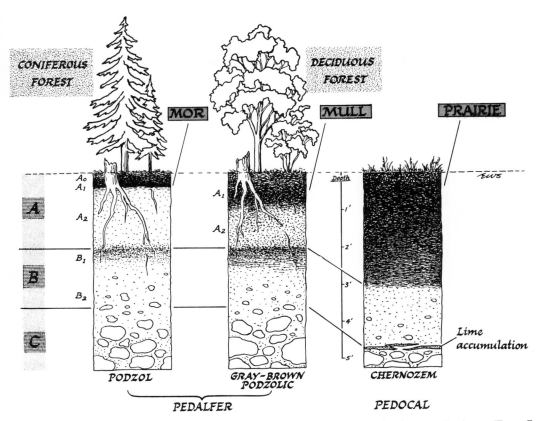

Fig. 36-7. Diagrammatic summary of the structure of major types of soil showing layers, or horizons. (From G. B. Moment: *General Zoology,* 2nd ed. Houghton Mifflin Co., Boston, 1967.)

such as the familiar night crawler, *Lumbricus terrestris*, and the "garden earthworm," *Helodrilus caliginosus*, are present.

Prairie soil, or **chernozem**, contrasts markedly with both mor and mull. Physically the A horizon of humus is extremely thick and fairly uniform. Because of limited rainfall, leaching is minimal. Most of the water is absorbed by the humus and the grass roots in the A horizon. Capillarity during surface drying and ascent of sap in roots returns dissolved minerals to the upper layers. As someone has said, the water in this type of soil is "hung from the top, like Monday's wash."

ECOLOGICAL SUCCESSION AND SOIL DEVELOPMENT

In most parts of the world today the community of plants and animals in a region remains constant decade after decade, presumably for many centuries. Such a stable fauna and flora constitute an ecological **climax** or a climax community. Each of the major biomes represents such a climax, which remains stable unless marked climatic changes take place. Whenever a climax community is upset or destroyed, as by cutting the trees and cultivating the soil, building a road, or flooding an area to make a freshwater pond, then the climax is restored through a regular series of stages known as a **sere**. Each stage of a sere is characterized by its own typical plants and animals.

In the deciduous forest zone of the temperate regions of the Northern Hemisphere, annual herbaceous plants appear first after a severe fire as they did in the rubble areas of bombed cities after World War II. These are followed in a year or two by various kinds of grasses, then shrubs,

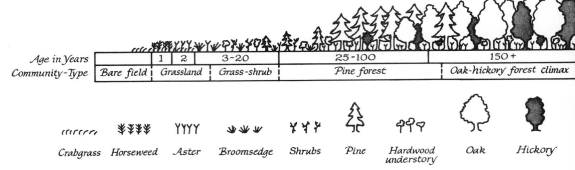

Age in Years						
		1	2	3-20	25-100	150+
Community-Type	Bare field	Grassland		Grass-shrub	Pine forest	Oak-hickory forest climax

Crabgrass   Horseweed   Aster   Broomsedge   Shrubs   Pine   Hardwood understory   Oak   Hickory

Fig. 36-8. Stages of plant succession in a sere leading from a bare field abandoned after "overfarming" to a deciduous forest climax. The entire process requires at least 150 years. (After E. P. Odum and H. T. Odum: *Fundamentals of Ecology,* 2nd ed. W. B. Saunders Co., Philadelphia, 1959.)

and later by pines, birches, or perhaps scrub oaks. These are at last superseded by the beech-oak-hickory climax. Various other seres have been studied, notably the one observed when vegetation begins to grow over a sand dune or to fill up a freshwater pond.

The **Great Succession** in the Northern Hemisphere is the reforestation of the land following the last glaciation of the Pleistocene era. This process is only now beginning to draw to a close. The retreating glaciers left tundra conditions. Even in the southern United States, remains of musk oxen, mastodons, and other animals, now restricted to the far north, have been found. In time the tundra became coniferous taiga and this in turn gave way to the deciduous forest which is today's climax community in most of eastern and central North America and Europe.

The causes of ecological successions are complex and are different in different cases. The situation following the destruction of a forest by a fire is not the same as after its destruction by a glacier. Climatic factors are important in all cases, but after a fire the climate commonly remains the same. The change from grassland to bushes to trees of various kinds depends largely on the effects of the plants and to some extent of the animals. One community may change things in such a way that some other group of plants and animals are able to replace them. The shade on the forest floor may finally become so deep that the seedlings of the trees casting the shade cannot develop, although seedlings of a different kind of tree can thrive.

Changes in the character of the soil play a key role in the movement of the stages of the sere in time. Some ecologists believe that deep burrowing earthworms are the chief agents in converting the mor soils of the taiga with their thin, sharply limited layer of humus into the mull soils of the deciduous forests with their thick and much less sharply delimited humus or topsoil. Darwin's extended account of the way earthworms cultivate the soil showed that all the humus of the soil of England passed many times through the bodies of earthworms. Recent measurements have substantiated his findings. As already noted, earthworms burrow into both the topsoil and the subsoil, pass it through their bodies, and leave it in castings on the surface of the ground. In light dry soil, such castings may weigh two to three tons per acre per year; in moist pastures they may be 10 tons; while in certain tropical localities amounts of up to 100 tons per acre per year have been measured. The soil also becomes finer and its colloidal properties modified in passing through earthworms. They continually deepen the topsoil, both by bringing subsoil to the surface and by dragging leaves into their burrows. Their burrows open the way for water and air to penetrate the soil. There can be no question that earthworms, once they have become abundant, play a very important role in soil dynamics.

Earthworms are by no means the only animals of the soil. The fauna of the soil is at least as abundant and often more abundant than the fauna of freshwater lakes or most parts of the ocean. The A_0 horizons of litter and leaf mold abound with invertebrates. In addition to a vast horde of insects, adults and larvae, the soil is the normal environment for teeming populations of nematodes, many very destructive to crops, plus enormous numbers of minute arachnids, mites, crustaceans, microannelids, rotifers, and even protozoans (especially amoebas and flagellates).

Also of great importance for soil building are the enormous numbers and kinds of microorganisms present. The fungi and bacteria play an essential role in the decay of dead plants and animals by releasing bound minerals and the end products of the breakdown of organic molecules, CO_2 and H_2O. Furthermore, a small number of bacterial species, by their ability to fix atmospheric nitrogen, provide a continuing supply of ammonia and nitrate, the only forms of nitrogen that can be absorbed through the roots of higher plants.

THE STUDY OF LAKES

Types of Lakes

The scientific study of lakes is known as **limnology**. Lakes are classified into three types on the basis of the amount of nutrition they provide for fish and other inhabitants. **Eutrophic** lakes are relatively shallow (10 meters or less) and contain an abundant supply of nutrients for algae and pond weeds, and therefore for crustaceans, and therefore for small fish, and therefore for larger fish. Lake Mendota in Wisconsin and many of the lakes in Scandinavia are well-studied examples of eutrophic lakes. The deeper parts of such lakes are poor in oxygen. **Oligotrophic** lakes are deep, relatively poor in nutrients and animals, and rich in oxygen. The Great Lakes of Middle North America, except Lake Erie, which is shallow, and the Finger Lakes of New York are typical oligotrophic bodies of water. **Dystrophic** lakes are usually parts of bogs. The nutrients are present but various organic acids and other substances inhibit the growth of plants and animals.

Eutrophication

In the course of thousands of years lakes evolve from oligotrophic to eutrophic. The rapid industrialization and urbanization of the United States in this century has led to massive and uncontrolled release of sewage, industrial wastes, and excess fertilizers into waterways and lakes. A consequence of this buildup of nutrients has been a pathological acceleration of eutrophication. Parts of Lake Erie have been severly damaged in this way. Population explosions of algae, so-called blooms, occur and can cause disastrous oxygen depletion in a lake. The algal cells in the uppermost layers of the water screen out the sunlight so that those deeper down no longer produce a great excess of oxygen over and above their own respiration. Furthermore, such dense populations of algae contain progressively more and more dead cells which are decomposed by oxygen-using bacteria and fungi. Such oxygen depletion often results in massive fish kills which in turn decay, releasing even more potential nutrients for further overgrowth of plants. The resulting imbalance has literally "killed" many lakes and streams in the United States in recent decades.

Farming a Lake

Up to a point eutrophication can be highly beneficial. In recent years the United States government has developed a farm pond program through the Soil Conservation Service. Over 350,000 ponds have been made by damming small streams. They are routinely stocked with bluegill sunfish and with bass at the rate of 1,500 bluegills per acre of pond surface and 100 bass if the pond is to be fertilized with either manure or chemicals, and about half as many if not. At the end of a year, the sunfish on which the bass feed average about 4 ounces each and the bass about 1 pound.

Unfertilized ponds yield from 40 to 150 pounds of fish an acre, while fertilized ponds yield from 200 to 400 pounds an acre. Both represent eutrophic conditions. The fish feed on each other and on plankton. In these ponds the fertilizer feeds the algae which constitute the phytoplankton. These tiny aquatic plants serve as food for *Daphnia, Cyclops*, and other crustaceans which are the most important members of the zooplankton. The crustaceans in turn are eaten by the smaller fish which are then eaten by the larger fish.

The amount of fertilizer added must be enough to produce a luxuriant growth, a so-called "bloom," of algae early in the summer. However, there must not be such a thick soup of algae that when they die off in the winter the processes of decay will use up all the oxygen and kill the fish. The number of fish initially introduced into the pond is also a very important factor. When bluegills are added at the rate of 1,500 per acre, they attain a weight of about 4 ounces in 12 months, but if 180,000 per acre are added, they average only 0.02 ounces. At the rate of 1,500 per acre without added fertilizer, the average weight is 1.1 ounce.

Annual Turnover

Most lakes in the temperate zone undergo a very important complete turnover of the water every spring and fall. Because this thoroughly mixes the dissolved oxygen and mineral nutrients in the lake, it stimulates plant and animal growth. This peculiar behavior of the water can be understood by remembering that water is most dense, *i.e.*, heaviest, at 4° C (about 39° F.). As water becomes still colder, it expands and becomes lighter. In the fall, as the surface water cools, it becomes heavier and sinks. Warmer water rises, becomes cold, and in turn sinks. This process continues until the entire lake is at 4° C.

In the winter, as the water cools to below 4° C., it no longer sinks but remains on the surface and ultimately forms a layer of ice. Except for a very thin layer immediately under the ice, lake water never falls below 4° C. This is a doubly fortunate circumstance for living things. It means that in the water they are protected from freezing temperatures. It also means that the lakes are available as aqueous environments, for if water continued to contract during cooling, ice would begin to form on the bottom of lakes and gradually fill them with ice. Under these circumstances ice would never leave the bottoms of lakes except in the southern edge of the temperate regions.

In spring, after the ice melts, the thin layer of water at the surface (which is below 4° C.) warms, becomes heavier, and sinks until all the lake again becomes 4° C. As the surface water, heated by the sun, comes warmer than 4° C., it does not sink but forms a warm surface layer. This is pushed by the prevailing wind against one shore of the lake. Water at that shore sinks (rather than piling up) while, at the same time, at the opposite shore, cooler water rises to replace that blown away. As a result, a countercurrent is set up in slightly deeper water. Since water at the top is warmer, *i.e.*, lighter, than deeper water in the lake, it forms a superficial layer called the **epilimnion**. Immediately below the epilimnion is a thin layer of water where the temperature falls very sud-

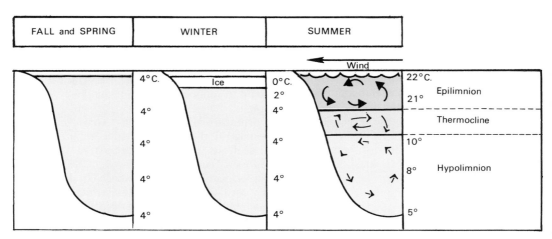

Fig. 36-9. Seasonal turnover of water in lakes. The density of water is greatest at 4°C. and decreases as it cools below 4°C. Thus ice forms at the top of a body of water. Wind and gravity are the major factors in producing the reverse temperature gradients characteristic of summer.

denly, the **thermocline**. The depth at which the thermocline occurs depends on season, winds, and other factors. In some lakes the thermocline begins within six feet of the surface and can be very noticeable to swimmers. In others, such as Lake Cayuga, one of the deep, narrow Finger Lakes in New York, it may be 30 feet below the surface. Below the thermocline is the **hypolimnion**. This is a zone of relatively uniform and cold temperature extending to the bottom. Just as the current along the undersurface of the epilimnion produces a current in the thermocline, so the current in the thermocline produces a slight current in the hypolimnion. The result is a very gradual and slight warming of the hypolimnion.

In deep oligotrophic lakes, the hypolimnion has a supply of oxygen and is hence inhabited by cold water species such as *Mysis*, a kind of shrimp, and by lake trout. In eutrophic lakes there is virtually no oxygen in the hypolimnion, partly because the algae in the epilimnion prevent the sunlight necessary for oxygen-producing photosynthesis to occur and partly because the continual "rain" of decaying dead organisms from above uses up the available oxygen. The fauna on the bottom of such lakes consists of insect larvae, some of which are red with hemoglobin, and other organisms adapted for low oxygen concentrations. Fish are scarce.

STREAMS, RIVERS, AND ESTUARIES

Flowing Water

Streams and rivers present special problems and opportunities for plants and animals. In the rapids, animals are adapted for holding firm to rocks and often possess flattened, streamlined bodies. Filamentous algae often abound attached to rocks on the bottom by means of holdfasts. Unless the waters are polluted with sewage, oxygen is abundant. Stream salamanders commonly lack lungs and breathe merely through skin capillaries.

Estuaries

Estuaries are tidal zones where fresh water enters the sea. In a sense, such regions are ecotones where tension exists between the freshwater and marine biomes. For a variety of reasons such environments are extremely favorable for life. Crabs, oysters, clams, fish, jellyfish, ctenophores, and many planktonic species abound. Plants are plentiful, a mixture of phytoplankton, larger algae, and aquatic higher plants, that provide an abundant input of organic materials for the food chain. Estuaries such as the Chesapeake Bay are rich sources of food for human consumption.

OCEANS

Marine ecology has already yielded information of major importance for the fisheries industry. Enthusiasts believe that in the future we may be able to farm the oceans as we now farm the land. Because nearly 75 per cent of the earth's surface is ocean, this is indeed an exciting idea. Careful studies have been made of the yield of diatoms and other minute green plants which constitute the basic food in the sea. The annual crop in terms of pounds of dry weight per acre varies, according to the part of the ocean, from an amount equal to that of a forest to that of grass on a semi-arid plain. The large-scale scientific study of the oceans began with the world survey cruise of the *H.M.S. Challenger* from 1872 to 1876 with an international staff of scientists. The 50 enormous volumes which resulted from that voyage form the basis of our knowledge of marine ecology.

Ocean Currents

Key factors in the life of the sea are the wind-driven **ocean currents**. North of the equator the trade winds blow continually from the northeast toward the southwest. South of the equator the trade winds blow continually from the southeast toward the northwest. As a result great equatorial currents flow westward on either side of the equator in both the Atlantic and the Pacific oceans. When these westward currents strike the continents they are deflected north and south. This is the source of the Gulf Stream, flowing north along the east coast of North America, and of the Kuroshio (Japan) current, flowing northward from the Philippines along the coasts of China and Japan. The westerly winds in the temperate zones of both the northern and southern hemispheres help propel the Gulf Stream and the Kuroshio current, respectively, to the west coasts of Europe and of North America. The final result is a circular motion. In the Southern Hemisphere the Peru or Humboldt current, flowing northward along the west coast of South America, and the Benguela current, flowing northward along the west coast of Africa, are due to similar forces. It is the cold Peru current that enables penguins to live close to the equator off the coast of South America.

There is another very different kind of current known as the **Arctic creep**. In the seas around both North and South poles the frigid waters sink and flow slowly along the bottom toward the equator. Here the water gradually rises to replace that which is lost by evaporation under the tropical sun.

Ocean Regions

The oceans are divided into several general regions depending on depth. Most continents are surrounded by a **continental shelf**, a more or less flat plain under about 200 meters (roughly 600 feet) of water but in some places much less. Off the east coast of Florida directly opposite the Bahamas the continental shelf is only a few miles wide and relatively non-existent when compared to the west coast of Florida and New England where the shelf extends approximately 300 kilometers. The shelf itself and the waters over the shelf constitute the **neritic zone**. The part of the neritic zone near shore with water to about 50 meters deep, usually strong wave action, and enough light for plant growth is termed **littoral**.

Beyond the neritic zone is the **oceanic zone**. This is the part of the ocean which the navies of the world call "blue water." The bottom slopes rapidly down from the continental shelf to depths of 3,000 meters and more. The regions of these great depths are the **abyssal regions** where light never penetrates. The temperature is virtually constant at 3° C. (about 37.4° F.). Part of the ocean bottom in this abyssal region is a flat plain but there are great mountain ranges, mile-deep trenches exceeding the Grand Canyon of the Colorado River in dimensions, and extensive regions of hills and what resemble river valleys.

The animals and plants in each region of the oceans are characteristically different. The animals and plants that live near shore are called **littoral**. Those which live at sea, especially far out in the open ocean, are called **pelagic**. Those myriads of floating or feebly swimming organisms that live near the surface of the ocean, either in the neritic or oceanic zone, are called **plankton—zooplankton** if animals, **phytoplankton** if plants. Strong swimming animals are called **nekton**, but those which live on the bottom, especially at great depths, are called **benthos**. Plants (mostly algae) are restricted to the upper layers through which sunlight can penetrate.

The most important members of the plankton are the diatoms, which constitute the basic plant food of the sea, and the copepods, which feed on the diatoms. Every animal phylum of marine plankton is represented in the form of floating eggs, larvae, adults, or as all three. Very often members of the zooplankton are transparent and have special adaptations for floating.

Animals living at great depths in eternal cold and darkness are bizarre by any standards.

An interesting and economically very important difference between the marine populations near the poles and those near the equator is that in the polar seas there are enormous numbers of individuals but relatively few species. In ecological terms the **biomass** is very great. In tropical waters the situation is reversed; there are relatively few individuals but a great variety of different species. Hence, the biomass is small. The explanation of this contrast is obscure, but it will be recalled that a similar contrast, insofar as number of species is concerned, exists between northern and tropical forests.

Although oceans cover nearly 75 per cent of the earth's surface, almost all the fishing done by man takes place over the continental shelves. Thus, only a tiny fraction of the vast ocean is used for the production of human food. The only significant exception is the whaling industry but many of the whales are captured near shore. It is worth noting that the baleen group of whales live exclusively on plankton, especially swarms of small oceanic crustaceans, while the toothed whales feed on giant squid and other members of the nekton.

Instrumentation of Oceanography

The classic tools of the oceanographer are a set of nets and trawls (which can be towed through the water at fixed depths or along the bottom), a number of Nansen bottles, and a good pair of sea legs.

In preparation for the coming modern age of ocean exploration and use, a wide variety of new instruments are being developed. Buoys adjusted to float at particular levels and

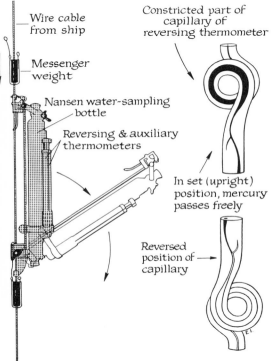

Wire cable from ship

Messenger weight

Nansen water-sampling bottle

Reversing & auxiliary thermometers

Constricted part of capillary of reversing thermometer

In set (upright) position, mercury passes freely

Reversed position of capillary

Fig. 36-10. Nansen water sampling bottle. When the bottle is lowered, it is open at both ends and water passes through. Alongside the bottle is a reversing mercury thermometer. When bottle and thermometer have reached the desired depth, a metal weight is sent down the wire. When this "messenger" hits the catch holding the upper end of the bottle to the wire, the catch opens, the bottle falls over, and valves at each end of the bottle close, trapping a water sample. This action also causes the mercury column in the thermometer to break thus recording the temperature of the water at the depth of the sample. (After A. Vine, in *International Science and Technology*, December, 1965.)

equipped with powerful radio signaling devices make it possible to detect currents and record their temperatures. Transistor radios attached to whales and fish enable scientists to track these animals over long periods of time. With sonar and radar devices, bottom depths and contours can be mapped and schools of fish, swarms of crustaceans, and other animals can be located. Stable floating work "platforms" have been constructed which resemble elongated tubular boats, one end of which can be filled with water to up-end it to a vertical position. The upper end with the workrooms is then supported by a deep probe extending many feet down into the water so as to be free of wave action. Most important of all are the newly equipped ocean-going research ships and the large laboratories for analyzing their findings which have recently been built in the United States and abroad.

MATTER AND ENERGY PATHWAYS

The major matter and energy pathways for animals begin with photosynthesis in the green plants, with the fixation of nitrogen by microorganisms, and with the absorption of various minerals from the soil and the sea by plants. In photosynthesis the energy of the sun is trapped and used to form carbohydrates. Carbon dioxide from the atmosphere and water are the raw materials for photosynthesis and free molecular oxygen is released as a waste product (see Chapter 15). The familiar over-all equation:

$$6CO_2 + 6H_2O \xrightarrow[\text{chlorophyll}]{\text{sunlight}} C_6H_{12}O_6 + 6O_2$$

gives the facts of primary importance to the ecologist. He is concerned with the sources of raw materials and energy, their amounts, and what happens to them after they have been processed by plants rather than with the biochemical machinery which enables chlorophyll to convert light into chemical energy. He is interested in the effects that the withdrawal of all this CO_2 may have on the entire living community and what effects the release of oxygen may have. He needs to know how much

Fig. 36-11. A group of female aquanauts 15 meters beneath the Caribbean Sea near St. John in the Virgin Islands. The TEKTITE II program was designed by the National Aeronautics and Space Administration to test the behavior of small groups living and working in a stressful environment resembling future space missions. Aquanauts carried on studies of marine life while sumberged. (Courtesy National Aeronautics and Space Administration.)

energy and matter a given community requires, how productive a given geographical region is, and how productive it might become. He wants to known where all the matter and energy come from, where they go, and how they get there.

A basic difference is at once apparent between the pathways for matter and for energy. The elements of which protoplasm is built, both the "big five" (carbon, oxygen, hydrogen, nitrogen, and phosphorus) and the various mineral elements (like iron, sulfur, sodium, calcium, potassium, iodine, and others required in small or even trace amounts), are used by different living things over and over again. Their pathways are cyclical.

In marked contrast, energy moves on a one-way street. It comes from the sun, is trapped by chlorophyll, and is utilized by living things in the three well-recognized ways, *i.e.,* to construct complex molecules, to do the work of moving substances across membranes, and to bring about muscular and other types of protoplasmic motion. Ultimately the complex

large molecules are broken down, by one agent or another, and the energy is dissipated as heat. In terms of thermodynamics, living things therefore require a continual source of energy if they are to maintain their improbable structures and the activities that are part of them. Life on this planet is thus dependent on an outside source, the sun, for its energy, while it can reuse its supply of matter over and over again.

The Carbon Cycle

The carbon cycle begins with the formation of carbohydrates in photosynthesis. Carbohydrate may be built into the body of a plant directly or may be utilized to form lipids and proteins and other more complex substances, such as nucleic acids or hormones. If the plant is eaten by an animal, its carbon becomes incorporated (in the most literal possible sense of that word) into the substance of the animal. Carbon is returned to the atmosphere as CO_2 in the respiration of plants and animals. Any carbon caught in the corpse of a plant or animal is utilized by the microorganisms of decay and they in their respiration return it to the air as CO_2. The cycles of oxygen and of hydrogen are obviously closely linked to the carbon cycle.

There has been fear that we, including our industries, are using up our supply of oxygen so fast that we will all suffocate. However, recent careful and extensive independent studies at Columbia University and at the National Bureau of Standards show that during the period from 1967 to 1970, there has been no detectable decline in the amount of atmospheric O_2. Since 1910 there has been no change in concentration of O_2 from 20.996 per cent by volume of dry air, at least it is so very small that it cannot be detected by available methods. Yet the amount of industrialization and the number of automobiles (which account for about 50 per cent of the oxygen consumed in urban areas) have undergone an enormous increase. Actually, simple arithmetic will show that if we were to burn all known fossil fuel reserves (coal, oil, natural gas), we would use up less than 3 per cent of the atmospheric O_2 supply. The CO_2 released would stimulate rates of photosynthesis and thereby increase oxygen production. Pollution may cut down oxygen supplies to disastrously low limits. In freshwater rivers and lakes the use of O_2 by bacteria metabolizing wastes may reduce the O_2 concentration far below the survival limits for many fish and other aerobic organisms.

The Nitrogen Cycle

The primary source of nitrogen is the atmosphere. Although the air is roughly 80 per cent nitrogen, most organisms cannot utilize the gaseous form of this essential element because it is extremely inert. Lightning fixes nitrogen, that is, forces it into chemical combination with other elements. This fixed nitrogen is washed out of the air by rain and reaches the earth as ammonia, NH_3, or nitrate, NO_3^-. However, most nitrogen fixation is carried on by special nitrogen-fixing bacteria in the sea or the soil. The nodules on the roots of peas, beans, and other legumes are specialized structures inhabited by nitrogen-fixing bacteria and are the most effective agents known for this purpose. Between 1 and 6 pounds of atmospheric nitrogen are fixed per acre per year, depending on the type of soil and the number of legumes. Certain species of soil bacteria and blue-green algae also are nitrogen fixers.

The nitrates, from bacterial or other sources, are absorbed by plants and converted into amino acids and proteins. Plant proteins are the ultimate source of all animal proteins. Nitrogen is returned to the soil or sea water by the bacteria of decay, which convert animal and plant proteins and urea to ammonia, or by animals which excrete ammonia instead of urea or uric acid. The ammonia is utilized by nitrite bacteria both in the soil and in the sea, and they convert it to nitrites (NO_2^-). These, in turn, are converted into nitrates by nitrate bacteria, after which nitrogen is again available for plants. These are the main features of the nitrogen cycle. Certain denitrifying bacteria utilize fixed nitrogen as a source of energy and release free nitrogen as waste. Many plants are able to absorb and utilize nitrites even better than nitrates. Animals become part of the nitrogen cycle only when they consume the nitrogen-containing compounds of plants or other animals.

The Phosphorus Cycle

Phosphorus is a rare element compared with nitrogen, but because it is involved in the energy transfer mechanisms within cells it is equally important. The phosphates dissolved in soil and in the sea are derived from erosion of phosphate-containing rocks. Phosphates are absorbed by plants and built into plant compounds that are in turn consumed by animals. The phosphorus is returned to the soil and the ocean by decay bacteria. It seems unlikely that the phosphate content of agricultural soil is being maintained. Soils in the Mississippi Valley that have been under cultivation for 50 years show a 36 per cent reduction of their phosphates, compared with virgin soil of the same type. Soil fertility can be maintained by the addition of phosphate fertilizers. Up to a point such application can promote crop yields. Too much zeal in the application of fertilizers can have undesirable effects as we have already indicated in our discussion of eutrophication of lakes. Generally, crops respond to added phosphate and other fertilizers in a predictable way that was summarized over a century ago by the German chemist Liebig in his **law of the minimum.** According to this widely applicable rule of biology, it is the essential element present in the least amount relative to an organism's needs that limits its growth. Thus, if an adequate amount of phosphorus (or any other element essential for plants) is present, adding more will not increase yields.

Food and Energy Chains

When food chains or webs are actually studied, they turn out to be pyramids leading from the basic plant source to herbivores and then up through a series of carnivores of increasing size but decreasing numbers. Such pyramids are often referred to as Eltonian pyramids, after the first scientist to point out this relationship. The principle can be illustrated by a more or less hypothetical Eskimo who eats nothing but polar bears, which eat seals, which eat large fish, which eat small fish, which eat crustaceans and other plankton, which eat diatoms. The total mass of protoplasm, the so-called biomass, becomes less and less in each ascending level of the pyramid. Because animals live within the laws of con-

servation of matter and energy, regardless of how efficient any animal is, catching food and all the other activities of living require expenditures of energy. Actually the amount of energy represented at any one trophic level of such a

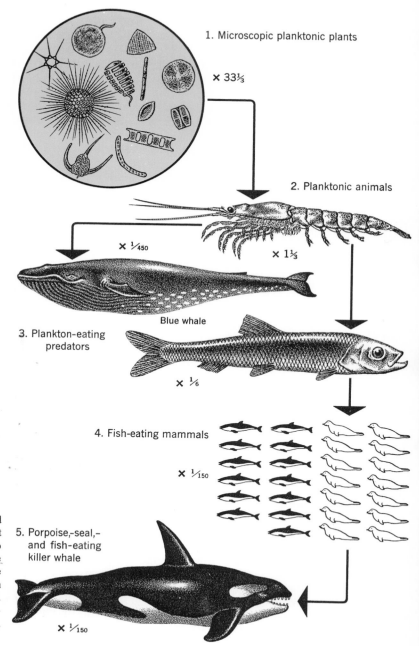

1. Microscopic planktonic plants

$\times 33\frac{1}{3}$

2. Planktonic animals

$\times \frac{1}{450}$

$\times 1\frac{1}{3}$

Blue whale

3. Plankton-eating predators

$\times \frac{1}{6}$

4. Fish-eating mammals

$\times \frac{1}{150}$

5. Porpoise,-seal,- and fish-eating killer whale

$\times \frac{1}{150}$

Fig. 36-12. Two related food chains. Observe that the largest animals ever to appear on this planet are only one step from the primary producers. (From R. H. MacArthur and J. H. Connell: *Biology of Populations.* John Wiley & Sons, Inc., New York, 1967.)

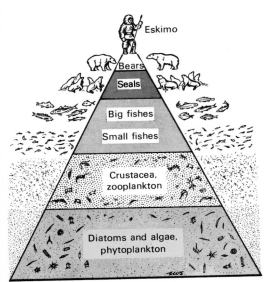

Fig. 36-13. The Eltonian pyramid of numbers. The closer a species of animals stands to the ultimate green plant food source, the larger its biomass. (From G. B. Moment: *General Zoology,* 2nd ed. Houghton Mifflin Co., Boston, 1967.)

pyramid is only about 1/10 of that represented at the adjacent lower level. Not only is there a pyramid of biomass and of energy but there is also a pyramid of individuals. All three factors narrow sharply as one trophic level is placed on top of another.

A good example of this **pyramid of numbers** of individuals is seen in an acre of Kentucky bluegrass. Here we find about 6,000,000 plants, 700,000 herbivorous invertebrates, 350,000 carnivorous invertebrates (such as spiders, ants, and predatory beetles), and 3 carnivorous vertebrates (such as birds and moles). A square mile of Arizona rangeland which was studied supported 1 coyote, 2 hawks and 2 owls, 45 jackrabbits, 8,000 wood and kangaroo rats, and 18,000 mice. This case illustrates another aspect of nutritional relationship between animals. When two species occupy the same nutritional level in a food-population pyramid, the species with the smaller body size will generally have the larger number of individuals, but will not necessarily have the larger biomass.

A certain Wisconsin lake can serve as a more or less typical example of a **pyramid of biomass.** There were, one summer, about 1,000 kilograms of plants (the primary producers) per

hectare (1 hectare = 2.47 acres), 114 kilograms of herbivorous animals on the second trophic level, and 38 kilograms of carnivorous animals on the third major trophic level.

A bog lake in Minnesota can illustrate a **pyramid of energy** represented in the three major trophic levels. The level of primary producers, actually the photosynthetic primary receivers of energy, yielded a total of 70 calories from the plants of a certain unit of volume of bog; the second major trophic level, that of the herbivores, yielded 7.0 calories for the same volume; and the third major level, that of the carnivores, yielded only 1.3 calories per unit of volume.

Measurement of Energy Content

The amount of energy stored in a sample of coal or plants or animals can be measured in a **bomb calorimeter.** The plant or animal is homogenized and then completely dried; the resulting powder is weighed and placed within a metal container that can be immersed in water. The "bomb" is filled with oxygen, then immersed in a carefully measured amount of water at an accurately known temperature. The oxygen is ignited by an electric spark. The heat liberated by the resulting explosive total combustion of the dried material is absorbed by the water surrounding the metal container and as a result the temperature of the water is raised. Because it takes 1 calorie to raise the temperature of water 1°C., the number of calories present in a given weight of dried animal or plant can be calculated, once the weight of the water is known and the number of degrees through which its temperature is raised is determined.

Productivity

An ecological question of paramount importance for mankind is how to determine the potential productivity of any particular type of environment (terrestrial, freshwater, or marine) and then to discover how the maximum productive potential can be attained. It is obvious from a consideration of the facts about ecological pyramids of numbers, biomass, and energy that the closer any species lives to the primary producers at the bottom of the pyramid, the greater its supply of matter and energy will be in any environment.

The actual productivity of an area of land or water in terms of the biomass of plants or animals it can or does produce turns out to be difficult and rather tricky to measure, except under rather special conditions. Four methods are in common use.

Methods for Determining Productivity

The **harvest method** is the most obvious and works well enough in certain simple situations. In a field where only one crop of the plant you happen to be interested in can be grown per year, 100 per cent of the crop can be harvested and its amount measured. The situation in a lake or forest or swamp, not to mention the ocean, is far different. To obtain a 100 per cent harvest is often extremely difficult. Fish in a lake may be killed with a chemical such as rotenone and those which float to the surface counted and weighed. Skin divers can attempt to make a census of the far larger numbers, perhaps the vast majority, which sink to the bottom. The activities involved in making such a census may profoundly disturb the water of the lake, the bodies of the dead fish will add nutrients, the oxygen and carbon dioxide content will be changed, so that a second census a year later may be taken under such different circumstances that it can hardly confirm the first.

An even greater difficulty lies in the fact that the mass of animals or plants existing at any one time is only a very rough indication of actual productivity and even less an indication of potential productivity. Two lakes or two parts of the ocean may contain the same biomass, but in one the standing population may have been reached only after a relatively long period of time and, if much were removed, would require a long time to regain its original size. In the other lake the turnover of population may be very rapid, or could and would become so if any significant number of animals or plants were removed. Hence, if the same number were harvested from each lake, it would require a far longer period of time for the population in the less productive lake to regain its original size than in the more

productive one. How is one to know how often to harvest to get a true measure of productivity? How often to harvest raises the question, to be discussed in the following section on population dynamics, of the drastically different results of "harvesting" animals such as fish or deer at different points on their population curves. In taking a census of plants in a given land area, one would expect the number of trees to remain constant but expect new tree seedlings and annual plants to germinate during each growing season.

It has even been thought by some ecologists that differences in rate of turnover rather than differences in actual productivity can explain the ecological paradox that the biomass present in polar seas is so much greater than in the sunny tropical oceans where you might expect the greater productivity.

A second method of measuring productivity is to determine **oxygen production**, which is a measure of the amount of photosynthesis. This is more difficult than it sounds. Plants themselves as well as microorganisms and the animals in a lake use oxygen. Some ingenious workers have measured the rate at which oxygen is depleted from the hypolimnion during the summer. It will be recalled that during the summer there is no appreciable exchange between the water above the thermocline of a lake (where oxygen from the air and that produced by green plants in the water is available) and the deep water below the thermocline where there are no green plants and no access to atmospheric oxygen. The more productive the lake, the more organic debris will fall down into the hypolimnion, be metabolized by organisms there, and lead to exhaustion of the oxygen supply. This undoubtedly is true but only in certain special lakes.

The **disappearance of minerals** from a body of water has also been used as a measure of the rate of productivity. It has to be remembered that, in any closed body of water, the fauna and flora will reuse the minerals already present and that new supplies are being washed in by rivers and streams at various rates. The chief place where this method has been useful, and it is an important place, is in certain parts of the ocean. There are places where nitrogen and phosphorus accumulate in the water during the winter and are utilized by phytoplankton in the spring. Their rate of disappearance from the water can be determined.

Radioactive tracers, specifically phosphorus-32 which has a half-life of about two weeks and hence is fairly safe to use in the sense that it introduces no persistent radioactive contamination, have been introduced into ponds and their rates of incorporation into phytoplankton determined. This method seems to hold real promise for lake studies and possibly marine studies as well.

In summary, productivity depends on many very different kinds of factors. The amount of sunlight is, of course, basic. So also are temperature, moisture, sources of mineral nutrients, and the position of the organisms on the population curve when harvesting is done. Easily the most important of all is the **genetic constitution** of the members of the population. This has been dramatically demonstrated in the case of such crops as hybrid corn and new varieties of wheat, but it is most certainly true of animals as well. Those who would cultivate the oceans must look at the genetic make-up of the plants and animals as well as at currents and temperatures.

THE DYNAMICS OF POPULATIONS

The study of populations has been a major concern of biologists ever since Charles Darwin enunciated his theory of evolution based on natural selection. The tremendous reproductive potential of plants and animals is one of the most powerful of evolutionary forces. The study of populations is central to all the problems of ecology, and no part of ecology has more important practical aspects than a knowledge of the causes and characteristics of the rise and fall of populations. From a practical point of view, the problem is how to minimize undesirable populations and maximize desirable ones.

The major factors controlling population size can be more or less arbitrarily divided into three categories—habitat, predation, and competition within species. Some ecologists emphasize the role of population density on population growth and speak of density-dependent and density-independent factors. Others deny the existence of truly density-independent factors.

Effect of Habitat

The habitat includes both the non-living features, such as climate, periodicity and intensity of sunlight, soil, and topographical features, as well as the resident plants and animals. Tree squirrels do not live on treeless plains. Often a whole ecological community of animals depends on a plant community. When the plant biome disappears, the animal community must also disappear. This happened along the Atlantic coast of the United States when the fields of eelgrass were killed by disease. The populations of scallops, certain polychaetes, jellyfish, and many other animals shrank almost to zero when a strange protozoan parasite killed the eelgrass meadows in the shallow water where these animals lived or bred.

Effect of Predation

The second factor influencing population size is predation by carnivores and by bacteria or other parasitic plants and animal organisms. It is difficult to determine accurately how important predation by carnivores or even by disease organisms really is. It is entirely possible that there are cases in nature, perhaps many cases, where the number of the individuals eaten is so small in proportion to the total population of the prey that predation makes no measurable difference. This seems highly probable when the predator population is limited by some factor other than the supply of prey. For example, ospreys live by catching fish out of the ocean. The supply of possible prey is therefore enormous, if not absolutely unlimited, but ospreys are held in check by the limited number of suitable nesting sites which must be isolated from the birds' own enemies and fairly close to the water. In some populations, such as those of fish in lakes, it has been shown that removing small fish will permit those left to grow faster to a larger size than would have been possible under the initial crowded conditions. Recent studies of several lakes in Oklahoma show that it is practically impossible to deplete a healthy population of fish by predation with a hook and line. Of course if the lake is fished with a seine, the entire population could be caught.

There are other well-established cases where predation has drastically reduced and virtually exterminated the prey. This can be rather easily demonstrated in the laboratory. If a ciliate called *Didinium,* which attacks and eats parameciums, is placed in a culture of its prey, the population of didiniums will increase and, after a time, the population of parameciums will decrease until final extinction. At this point, of course, the predators are faced with an irremediable famine and become extinct also. If the fingerbowl world of parameciums and didiniums is complicated by the introduction of obstacles which will partially hide prey from predator, then irregular fluctuations of population abundance and scarcity will be observed. The more complex the environment, the longer the cycles continue. Clearly, the more complex situation more closely resembles the condition of nature.

The likelihood, however, is that in the vast majority of prey-predator relationships, a relatively stable steady state has been attained in the course of evolution. Were this not so, either the prey, the predator, or both would have become extinct. This is why the introduction of new predators produces such spectacular results. It is also why the elimination of the restraints imposed by predators may produce disastrous results for the prey.

The case of the mule deer on the Kaibab plateau, an isolated area of about 700,000 acres in Arizona, is often cited as an instance of a population explosion when wolves and other predators are removed only to be followed by a population crash. Something of the sort apparently did happen but accurate data are either lacking or contradictory. The causes are more complicated than was once believed. Lack of predators can hardly have been the only factor involved. It is now remembered that when the carnivores were removed, 150,000 to 200,000 sheep and cattle were also removed. Since there are supposed to have been only about 4,000 deer at that time, removal of competition probably had a large impact on the deer. The extent of the population explosion is also in doubt, for estimates of the peak vary all the way from 100,000 to fewer than 30,000. Furthermore, at the time nothing was known of the endocrine negative feedbacks (to be discussed below) which operate to limit populations under crowded conditions.

Public attention is often focused on populations which fluctuate at more or less regular intervals. The classic case is that of the snowshoe hare and the Canadian lynx. Records kept from about 1800 by the Hudson's Bay Company reveal oscillations with about a 10-year periodicity. In general, when hares are abundant so are the lynxes and, as would be expected, the population peaks for the lynxes tend to come slightly later than those for the hares. A number of such prey-predator cycles are known among lions and zebras, snowy owls and lemmings, and other animals.

It was long supposed that there was a causal relationship between the numbers of the prey and their food. Volterra and Lotka described these **prey-predator** cycles in mathematical terms; so they are commonly referred to as **Volterra-Lotka** cycles. The oscillations certainly exist, but there is now good reason to question whether in all cases the number of the predators merely rides up and down with the fluctuations in the numbers of the prey. For example, on Anticosti, a large island about 100 miles long situated in the Gulf of St. Lawrence, the population of snowshoe hares rises and falls as it does on the mainland of Canada, despite the fact that there are no lynxes at all on Anticosti. The fluctuations in the hares are perhaps due to several cooperating factors such as competition for food plants, increased incidence of disease with crowding, or the effects of fighting among themselves.

Competition Within Species

In addition to the general habitat and predation, a third important factor in controlling population size is competition within a species. Students of evolution commonly believe that this type of competition is the most severe of all. Sparrows do not compete with ducks, much less with turtles or sunfish. The more different any two organisms, the less they compete. Nor do all seed-eating birds, for example, compete with all other seed-eaters. As anyone who has fed wild birds or has raised finches or canaries knows, birds have very strong food preferences, often based on bill size. Only two individuals of the same species will choose exactly the same seeds.

Owing to recent studies, intra-specific competition in house mice is better understood than in any other species. The method is to place two or three breeding pairs in rooms with nesting boxes and food. As long as there is plenty of food and enough nesting sites, the population increases with virtually no emigration from the room. When the population reaches a level where all the food is eaten up every day, then enough mice emigrate to keep the population stationary at this high level. The birth rate remains as before. It is a curious fact that the mice which emigrate represent a typical cross section of the colony, not just the young, the old, or the most vigorous, but some of both sexes and all ages except nurslings.

If there is no escape from the room when the population catches up with the food supply, the picture is very different. First, the nurslings die. Then the birth of new litters stops. The older mice gradually die off, presumably of old age. As this happens, the total number of mice in the colony falls, but the average weight of the individuals gradually increases so that the total weight of the colony remains constant. The amount of food given per day is adequate for a certain biomass of mice, irrespective of whether it is divided among a larger number of thin mice or a smaller number of fatter ones.

What mechanism brings reproduction to a standstill in such a mouse colony? Apparently a very important factor is the emotional effect of the continual fighting caused by the food crisis. The adrenal glands become noticeably enlarged, and the reproductive endocrines are in some way thrown off balance. Thus there is in these mice a kind of **negative feedback** mechanism which sets limits to the population when it begins to press against the food supply. Signs of adrenal damage have also been found in crowded deer populations as well as in woodchucks and other species. Negative feedback is the kind usually found among living organisms. Many actions produce some counteraction which checks the results of the first and restores the organism to equilibrium.

Positive, or "runaway," **feedbacks** have also been reported in animals. It is as though a thermostat, once the temperature had risen to a given point, instead of turning the heat down, turned on more heat, and then when the temperature climbed still higher, turned on the heat still more. Positive feedbacks soon lead to disaster. For example, in several sea birds such as gulls and guillemots, the smaller the nesting colony, the fewer eggs hatch and the fewer young are raised. This leads to a vicious downward spiral. It is as though every time the temperature fell, the thermostat turned the heat down even more.

Population Growth Curves

The free growth of a population in an open environment can be regarded as a positive feedback situation. The more individuals, the faster the population grows, and the faster the population grows, the more individuals, and so on. As Malthus pointed out in 1798, populations tend to grow exponentially, like money at compound interest. At first, in a new environment, a population increases faster and faster, and then in a finite environment (and all environments, whether test tubes or continents, are finite) growth begins to slow down more and more. The result is an **S-shaped** or **logistic curve.**

This curve of population growth has the widest possible application, for the very simple but basic reason that all animals reproduce in a geometrical rather than an arithmetical series. In other words, in a unit of time a population does not add a fixed number of new individuals regardless of its own size, but the new individuals become part of the reproductive capital and add their own share of new individuals. The curves illustrated here are taken from studies of yeast cells and fruit flies, but they hold similarly for mice and deer and men. More precisely, the increase in population is proportional to N, the total number of individuals in the population. The actual rate of increase depends on the reproductive ability, r, of the individuals. This factor is very large for oysters, moderate for mice, and smaller for elephants. The rate of increase is not only determined by $r \times N$, but is correlated with how close N is to the upper possible limit in any finite environment—test tube, milk bottle, or Kaibab plateau. This factor is $(K - N)/K$,

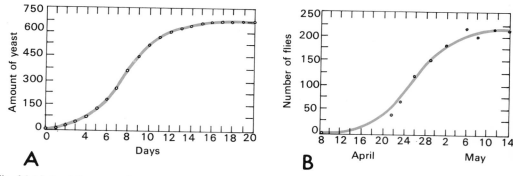

Fig. 36-14. Logistic curve of population growth in a finite environment. A, growth of a population of yeast cells (relative number of cells measured as optical density of suspension); B, population of fruit flies.

where K is the maximum possible population. In the beginning, when N is very small, this factor is approximately equal to one. As N increases, this factor becomes smaller and smaller until $N = K$, when $(K - N)/K = 0$, and increase in size ceases. In terms of calculus, the growth rate at any point $dN/dt = rN (K - N)/K$.

Practical Applications of Growth Curves

One important practical application of the facts expressed by this formula concerns the best time to harvest from a population of fish or deer or other organisms. Removing a certain number of individuals when a population is close to the lower end of its curve will produce a marked depressing effect. Removing the same number at the upper end of the curve will either have no appreciable effect or will allow those left to attain a somewhat greater size. Because of this principle emigration from overpopulated countries furnishes no permanent cure for the overcrowding.

Not only can much larger harvests be reaped when the harvesting is done close to the top of the population curve, but a harvest of any given size can be taken with far less effort. This has been demonstrated in the case of the codfish. If one population is about three times as large as another, it will require only about one-third the fishing effort (time towing nets, etc.) to harvest the same amount of fish from the large as from the small population.

From the human point of view the objective is often harvests in reverse, the diminution or elimination of an undesirable population. Here the strategy must be to press hard with the control measures when the population curve is at its lowest, if permanent results are to be achieved. The older chemical pesticides such as DDT and various neurological and enzymatic poisons can be very effective in controlling insects, rodents, and sometimes even nematodes. But they have very serious and undesirable side effects. Most of these poisons are very long-lasting, if not indestructible. They get into food chains and are transmitted from plants to herbivores and then up the chain of carnivores. The bald eagle has picked up enough of these poisons so that its eggs often fail to hatch even though the adults are not killed. Moreover, these poisons, when used over a forest or a swamp, indiscriminately kill many if not all, wildlife species. DDT in the ocean could eventually depress the oxygen-forming capacof oceanic plants with disastrous results to the atmosphere. Very promising results have been achieved in developing compounds that are toxic to insects but will disintegrate spontaneously on contact with water and air.

Nevertheless, many unsolved problems remain with such insecticides. Consequently, there has been a determined hunt for biological means of pest control. These methods are generally far safer and some, such as the use of sterilized males, are entirely safe. Methods of this kind were advocated by Rachel Carson in her book, *Silent Spring*. However, some biological methods carry their own potential dangers, *e.g.*, when a parasite against an unwanted animal is imported from another country.

Methods of Population Control

Three types of methods of population control deserve special mention. Perhaps the most ingenious is the use of **sterilized males.** Unfortunately, although this method has been a spectacular success, it is expensive and is effective only against species in which the females mate only once. The U. S. Department of Agriculture used this method, for example, against the highly destructive screwworm fly which produces maggots that attack the flesh of cattle. In one campaign they raised and sterilized by X-rays 100 million male flies per week. These were set loose from planes flying over the infested country. Within two years the screwworm fly was completely eliminated over wide areas. Over $200 million worth of damage of cattle is reliably estimated to have been averted within a three-year period. The recent resurgence of the screwworm appears to be due

to a change in cattle breeding by which the highly vulnerable calves are born throughout the summer instead of only in the colder months.

A second method is a very neat combination of chemical and biological techniques. **Sex attractants** have been extracted from female insects, analyzed chemically, and their molecular structures determined. Such compounds have been manufactured commercially and used to lure male insects into traps or to poisoned bait, a far safer method than spraying poisons on whole plants or whole forests. The sex attractant of the female gypsy moth, a species highly destructive to both coniferous and deciduous forests and to fruit trees, has been identified and is sold under the name of Gyplure. It will be of great interest to see how successful its use will be. The gypsy moth was introduced into Massachusetts in about 1868 by an entomologist said to be an amateur. In

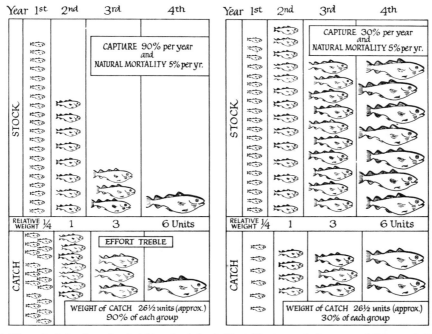

Fig. 36-15. The principle of maximal yield and minimal effort. Upper symbols represent the populations of one-, two-, three-, and four-year-old fish in stock populations. Lower symbols represent the fish caught. The same number of pounds of fish are harvested in both cases, both three times as many hours are required to catch this weight of fish from the smaller population and more of the fish are small in size. (After T. Park in W. C. Allee, A. E. Emerson, O. Park, T. Park, and K. P. Schmidt: *Principles of Animal Ecology,* W. B. Saunders Co., Philadelphia, 1949.)

any case, he "didn't known it was loaded." Millions of dollars of destruction has been the result and more millions have been spent spraying orchards with lead arsenate, introducing a small beetle which eats the gypsy moth larvae, and spraying forests by plane with DDT. Perhaps sex attractants will be an answer. A similar chemical attractant from cockroaches has been identified by drawing air over thousands of females and washing out their odors.

A third method of pest control is the introduction of **bacterial** or **viral diseases** in order to devastate an unwanted population. This carries obvious dangers. One must be certain that the disease will not attack other, desirable species. This method has been very successful against the great rabbit plagues of Australia where all other known methods had failed. The pathogen used was the myxomatosis virus which is transmitted from one rabbit to another by mosquitos. Two bacterial diseases have been very successful in controlling and all but eradicating the Japanese beetle in the United States after the various insects which parasitize this beetle in Japan had been introduced to no avail.

Competition Between Species

Many different species of animals and plants inhabit the same habitat—a spruce forest, a grassy meadow, or a pond. Do any two species occupy the same ecological niche within their habitat? Darwin suggested long ago that this was highly unlikely because either one or the other species would, in all probability, be at least slightly better adapted and hence, over a long period of time, able to outbreed the less well adapted and finally eliminate it. In more recent years J. Grinnell, on the basis of a long study of California birds, came to the conclusion that no two species did in fact occupy the same ecological niche. Within the following decade the same idea was supported by V. Volterra in Italy and G. F. Gause in Russia and came to be called the **principle of competitive exclusion**. The application of this principle makes it possible to see some rationale in the pattern of sympatric species.

It is commonplace to find two or many species living together and apparently competing, *i.e.,* occupying the same niche. Careful scrutiny in every case so far has shown that the different species actually occupy different although sometimes slightly overlapping niches. A well-known case investigated by R. H. MacArthur is that of the four species of warblers which live during the nesting season in the spruce forests of the north. He found that each species had a different zone of the tree where it specialized in hunting for food, although there was a certain amount of overlap. One species hunted for insects almost exclusively among the fresh growth at the tips of branches near the top of the tree. Another species specialized in the foliage over a year old beneath the new spring growth, another specialized in hunting on the leafless branches close to the trunk and in the upper part of the tree, and another chose a similar zone but close to the ground. Not only is the location of food hunting different, but there certainly would also be a difference, though not an absolute one, between the insects and spiders available in these various regions of the trees.

One of the most spectacular examples of species occupying separate niches within the same area of the earth's surface is found in the intertidal zones at the edges of the oceans. Here there is a distinct zonation of the algae with the greens, browns, and reds located in that order from average high to low tide levels. Such zonation is a consequence of the adaptations of each species to specific conditions of light, exposure to drying, temperature, and osmotic fluctuations.

Mathematical Models

At present ecologists are attempting to apply mathematical models to biological situations since a mathematical approach should help answer some old problems and open up new ones. There seems to be a relation, for example, between total population size, the number of related species, and the various sizes of the populations of the different species. But exactly what this relationship is remains unclear. In many cases pairs of sympatric species are closely similar save that one is roughly twice the weight of the other. Why not roughly three or four times the weight of the smaller species? Such pairs are seen in the downy and hairy woodpeckers, weasels, flycatchers, terns, frogs, and cats. When there is a group of similar species, such as the wild mice, which are all approximately the same size, then they have been found to live in different habitats.

GROWTH OF HUMAN POPULATIONS— A POTENTIAL ECOLOGICAL CRISIS

Modern human populations, whether agrarian or industrial, have wrought profound changes in the face of the earth. Cutting down a forest to clear the land for cultivation of crops has just as devastating an effect on the existing balance of plants and animals as does the later construction of a highway through the cleared fields. The myriad ways that man and all other living things are interdependent is one of the basic themes of biology.

In recent years there has been increasing public concern about the deleterious effects of the ever more numerous human occupants of this planet. The effluents of modern industrial societies are all too obvious: smogs caused by automobiles and industrial fumes; sewage and fertilizers flowing into bodies of water causing excessive eutrophication; accumulation of poisons, whether industrial wastes or pesticides, in the biome; radioactive, thermal, and even noise pollution. At the same time resources, from minerals to fossil fuels to living space, are rapidly shrinking. Many vocal environmentalists are viewing the situation with alarm. Does the earth already have too many people? Are there lessons to be learned from our knowledge of the population dynamics of animals and microorganisms? There is no arguing with the fact that the earth provides a finite environment for man and that there is an upper limit to the number of human beings that this environment can support. To a large extent, the limit depends on how people wish to exist. Present world populations would undoubtedly have to be reduced considerably if all people insisted on consuming diets comparable to those of the present inhabitants of the United States. Several times our present numbers could be fed if we all chose to exist as vegetarians. Thus, the upper limit of the earth's capacity to support human populations very much depends on what people expect in food and life styles.

No matter what estimate or conditions we are willing to accept, past and present population figures exhibit no trends except a geometric increase in the number of humans. Obviously, a leveling off to zero population growth must occur sometime. It might have occurred already had it not been for the overwhelming success of medicine during the past century in reducing infant mortality, conquering infectious diseases, and generally extending human life expectancies. During this time, when death rates have been lowered, there has been no balancing decrease in birth rates among most segments of humanity and consequently the total world population has soared upward during the 20th century. If there is no change in birth rates, the number of people will double between now and the year 2000.

Besides this trend of increasing numbers, there has been a decided shift in the patterns of distribution of people. Throughout the world,

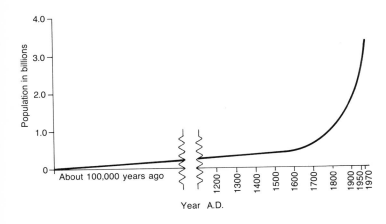

Fig. 36-16. Curve of human population growth. (From A. Turk, J. Turk, and J. T. Wittes: *Ecology, Pollution, Environment.* W. B. Saunders Co., Philadelphia, 1972.)

Fig. 36-17. Smog over Manhattan Island, January, 1966. A warning that will be heeded or a vision of the future? (Courtesy U.S. Department of Agriculture. Reproduced by permission of the New York Times.)

this century has been a time of urbanization with an accelerating migration of people from farm to city. Over 70 per cent of the population of the United States now live in less than 2 per cent of its area. This has accentuated urban problems that have existed for centuries: atmospheric pollution, waste disposal, congested transportation systems, shortages of housing, crime, and other forms of social disorganization. The decay of the central city and the flight of the middle class to the suburbs is only a minor perturbation in a massive movement of humans into urban areas. In a sense the real problem is that cities have become too large to be governable according to existing political systems. Whether these problems can best be solved by newly developed regional systems of government, by limiting the size of cities and building new ones, or other means, remains to be seen.

The problems created by the accelerating growth of human populations and the increasing disruption of our planetary ecosystem are based in biological facts but extend far beyond them into the areas of economics, political theories, and social philosophies. Who is to say what the optimal population is and when zero growth is desirable? Is it the same in all parts of the world and in all cultures? What life styles should be encouraged or even enforced? Obviously biologists alone cannot provide the answers but it is their urgent task to proclaim loud and clear what the dangers and the opportunities are. Biologists should discover and make available to all mankind the knowledge necessary to make possible humane and wise decisions. Only then can we continue to be a successful form of life.

USEFUL REFERENCES

Carson, R. L. *The Edge of the Sea*. Houghton Mifflin Co., Boston, 1955.

Cox, G. W. (ed.) *Readings in Conservation Ecology*. Appleton-Century-Crofts, New York, 1969.

Ehrlich, P. R., and Ehrlich, A. H. *Population Resources Environment*. W. H. Freeman & Co., San Francisco, 1970.

Ehrlich, P. R., Holdren, J. P., & Holm, R. W. *Man and the Ecosphere: Readings from the Scientific American.* W. H. Freeman & Co., San Francisco, 1971.

Hardin, G. (ed.) *Population, Evolution, and Birth Control, 2nd ed.* W. H. Freeman & Co., San Francisco, 1969.

Hardy, A. *The Open Sea.* Houghton Mifflin Co., Boston, 1969.

Kinne, O. *Marine Ecology,* Vol. 1. Wiley-Interscience Press, New York, 1970.

Kormondy, E. J. *Concepts of Ecology.* Prentice-Hall, Inc., Englewood Cliffs, N.J., 1969.

MacArthur, R. H., and Connell, J. H. *The Biology of Populations.* John Wiley & Sons, Inc., New York, 1967.

Odum, E. P. *Fundamentals of Ecology, 3rd ed.* W. B. Saunders Co., Philadelphia, 1971.

Reid, G. *Ecology of Inland Waters and Estuaries.* Reinhold Publishing Co., N.Y., 1961.

Revelle, R., *et al. Rapid Population Growth: Consequences and Policy.* Vol. 1: Summary and Recommendations. Office of the Foreign Secretary, National Academy of Sciences, Johns Hopkins Press, Baltimore, 1971.

Singer, S. F. (ed.) *Is there an Optimum Level of Population?* McGraw-Hill Book Co., New York, 1971.

Sladen, B. K., and Bang, F. B. (eds.) *Biology of Populations.* American Elsevier Publishing Co., New York, 1969.

Turk, A., Turk, J., and Wittes, J. T. *Ecology, Pollution, Environment.* W. B. Saunders Co., Philadelphia, 1972.

Ward, B., and Dubos, R. *Only One Earth: The Care and Maintenance of a Small Planet.* W. W. Norton & Co., Inc., New York, 1972.

Classification of the Plant Kingdom

According to the current Rules of International Botanical Nomenclature the largest taxonomic units in the Plant Kingdom are known as divisions, which are comparable to the phyla of the Animal Kingdom.

During the past century, botanists have accumulated evidence that the large groupings of plants used in earlier systems of classification (for example, the division Thallophyta, which included all the algae and fungi) implied a relatedness that is not real and ignored the diversity of forms and distinctly separate evolutionary development of these groups. The following table of plant classification retains a few of the earlier larger groupings and draws on a number of sources. Terminology for the lower groups is largely that used by Bold (*Morphology of Plants,* revised edition, Harper & Row, New York, 1967). The classification of the flowering plants is that proposed by Cronquist (*The Evolution and Classification of Flowering Plants,* Houghton Mifflin, Boston, 1968).

The prokaryotic blue-green algae (Cyanophyceae) and fungi (Schizomycetes) plus the Microtatobiotes (or smallest living things, including the rickettsias and viruses) have been placed in a separate division, the Prophyta, to point out the uniqueness of cellular organization or the acellular nature of these groups. This has been done in spite of the absence of valid taxonomic reasons for such classifications.

Division **Prophyta** (prokaryotic algae and fungi, rickettsias and viruses)
    Class **Cyanophyceae** (blue-green algae)
        Order Chroococcales
        Order Chamaesiphonales
        Order Oscillatoriales
           Suborder Oscillatorineae
           Suborder Nostochineae
    Class **Schizomycetes** (bacteria)
        Order Pseudomonadales
        Order Chlamydobacteriales
        Order Hyphomicrobiales
        Order Eubacteriales
        Order Actinomycetales (mold-like bacteria)
        Order Caryophanales
        Order Beggiatoales

Order Myxobacteriales⎫ (protozoa-like bacteria)
Order Spirochaetales ⎭
Order Mycoplasmatales
   Class **Microtatobiotes** (smallest living things)
Order Rickettsiales (intracellular parasites)
Order Virales (viruses)
Division *Algae*
  Subdivision *Chlorophyta*
   Class **Chlorophyceae** (green algae)
Order Volvocales
Order Chlorococcales
Order Ulotrichales
Order Ulvales
Order Cladophorales
Order Siphonales
Order Oedogoniales
Order Zygnematales
  Subdivision *Euglenophyta*
   Class **Euglenophyceae** (euglenoids)
Order Euglenales
Order Colaciales (Euglenocapoales)
  Subdivision *Charophyta*
   Class **Charophyceae** (stoneworts)
  Subdivision *Phaeophyta*
   Class **Phaeophyceae** (brown algae)
Subclass Isogeneratae
Order Ectocarpales
Sublcass Heterogeneratae
Order Laminariales
Subclass Cyclosporeae
Order Fucales
  Subdivision *Rhodophyta*
   Class Rhodophyceae (red algae)
Order Bangiales
Order Nemalioniales
Order Ceramiales
  Subdivision *Chrysophyta*
   Class Xanthophyceae (yellow-green algae)
Order Heterococcales
Order Heterotrichales
Order Heterosiphonales
   Class **Chrysophyceae** (golden-brown algae)
Order Chrysomonadales
   Class **Bacillariophyceae** (diatoms)
Order Centrales
Order Pennales
  Subdivision *Pyrrophyta*
   Class **Cryptophyceae** (cryptomonads)
   Class **Dinophyceae** (dinoflagellates)
Division *Fungi*
  Subdivision *Myxomycophyta* (slime molds)
   Class **Myxomycetes**
Order Physarales

        Order Stemonitales
        Order Liceales
        Order Trichales
     Class **Acrasiomycetes**
        Order Acrasiales
     Class **Plasmodiophoramycetes**
        Order Plasmodiophorales
   Subdivision *Phycomycophyta*
     Class **Phycomycetes** (water molds and black bread molds)
        Order Chytridiales
        Order Blastocladiales
        Order Saprolegniales
        Order Peronosporales
        Order Mucorales
   Subdivision *Ascomycophyta*
     Class **Ascomycetes**
        Order Endomycetales (yeasts)
        Order Aspergillales (brown, green molds)
        Order Sphaeriales (pink molds) *Neurospora*
        Order Erysiphales (powdery mildews)
        Order Pezizales (cup fungi)
   Subdivision *Basidiomycophyta*
     Class **Hemibasidiomycetes**
        Order Uredinales (rusts)
        Order Ustilaginales (smuts)
        Order Tremellales (jelly fungi)
        Order Auriculariales (ear fungus)
     Class **Holobasidiomycetes**
      Subclass Hymenomycetes
        Order Agaricales (gilled fungi)
        Order Polyporales (pore fungi)
      Subclass Gasteromycetes
        Order Lycoperdales (puffballs)
Division *Hepatophyta*
   Class **Hepadopsida** (liverworts)
        Order Marchantiales
        Order Jungermanniales
   Class **Anthoceropsida** (horned liverworts)
        Order Anthocerales
Division *Bryophyta*
   Class **Sphagnopsida** (peat mosses)
        Order Sphagnales
   Class **Andreaeopsida** (granite mosses)
        Order Andreaeales
   Class **Mnionopsida** (true mosses)
        Order Funariales
        Order Eubryales
        Order Polytrichales
Division *Psilophyta*
   Class **Psilopsida** (only class with living members)
Division *Microphyllophyta* (*Lycopsida*)

Class **Aglossopsida**
    Order Lycopodiales (ground pines and club mosses)
Class **Glossopsida**
    Order Selaginellales (spike mosses)
    Order Isoetales (quillworts)
Division *Arthrophyta (Sphenopsida)*
    Class **Arthropsida** (horsetails)
        Order Equisetales
Division *Pteridophyta (Pteropsida)* (ferns)
    Class **Eusporangiopsida**
        Order Ophioglossales
        Order Marattiales
    Class **Leptosporangiopsida**
        Order Filicales
        Order Marsileales
        Order Salviniales
Division *Spermatophyta* (seed plants)
  Subdivision *Gymnospermae* (gymnosperms)
    Class **Cycadopsida** (cycads)
        Order Cycadales
    Class **Ginkgopsida**
        Order Ginkgoales (maidenhair tree only living species)
    Class **Coniferopsida** (conifers)
        Order Coniferales
            Family Abietaceae (Pinaceae) (pine, cedar, larch, spruce, hemlock, fir)
            Family Taxodiaceae (redwoods, bald cypress)
            Family Cupressaceae (junipers, cypress)
            Family Araucariaceae (monkey puzzle)
            Family Podocarpaceae (*Podocarpus*)
            Family Taxaceae (yew and Torreya)
    Class **Gnetopsida**
        Order Ephedrales (Mexican tea)
        Order Gnetales (gnetum)
        Order Welwitschiales (Welwitschia)
  Subdivision *Angiospermae* (flowering plants)
    Class **Dicotyledoneae** (dicots, Magnoliatae)
      Sublcass **Magnoliidae**
        Order Magnoliales (magnolias)
        Order Piperales
        Order Aristolochiales
        Order Nymphaeales
        Order Ranales (buttercups)
        Order Papaverales
      Subclass **Hamamelidae**
        Order Trochodendrales
        Order Hamamelidales
        Order Eucommiales
        Order Urticales
        Order Leitneriales
        Order Juglandales
        Order Myricales

      Order Fagales
      Order Casuarinales
      Order Plumbaginales
    Subclass **Caryophyllidae**
      Order Caryophyllales
      Order Batales
      Order Polygonales
      Order Plumbaginales
    Subclass **Dilleniidae**
      Order Dilleniales
      Order Theales
      Order Malvales
      Order Lecythidales
      Order Sarraceniales
      Order Violales (violets)
      Order Salicales
      Order Capparales
      Order Ericales
      Order Diapensiales
      Order Ebenales
      Order Primulales (primroses)
    Subclass **Rosidae**
      Order Rosales (roses)
      Order Podostemales
      Order Haloragales
      Order Myrtales
      Order Proteales
      Order Cornales
      Order Santalales
      Order Rafflesiales
      Order Celastrales
      Order Euphorbiales
      Order Rhamnales
      Order Sapindales
      Order Ceraniales (geraniums)
      Order Linales
      Order Polygaiales
      Order Umbellales (carrot)
    Subclass **Asteridae**
      Order Gentianales
      Order Polemoniales
      Order Lamiales
      Order Plantaginales
      Order Scrophulariales (snapdragon)
      Order Campanulales
      Order Rubiales
      Order Dipsaceles
      Order Asterales (asters)
Class **Monocotyledoneae** (monocots, Liliatae)
    Subclass **Allsmatidae**
      Order Alismatales
      Order Hydrocharitales

Order Naiadales
Order Triuridales
Subclass **Commelinidae**
    Order Commelinales
    Order Eriocaulales
    Order Restionales
    Order Juncales (rushes)
    Order Cyperales (sedges and grasses)
    Order Typhales
    Order Bromeliales (pineapple, Spanish moss)
    Order Zingiberales
Subclass **Arecidae**
    Order Arecales (palms)
    Order Cycianthales
    Order Pandanales
    Order Arales (jack-in-the-pulpit, skunk cabbage, duckweed)
Subclass **Liliidae**
    Order Lilliales (lily, onion, iris)
    Order Orchidales (orchids)

Classification of the Animal Kingdom

The following table is based on that found in R. W. Pennak's *Collegiate Dictionary of Zoology*. The Protozoa are classified according to the recommendations of an international committee (Honigberg Committee) of the American Society of Protozoologists. Standardized endings for classes, orders, and families are used as recommended by a committee on taxonomy of the American Association for the Advancement of Science. This recommendation includes the use of *-iformes* (which can be readily anglicized as *-iform*) as the ending for all orders. Fossil groups and a few others are omitted.

Phylum *Protozoa*
  Subphylum *Sarcomastigophora*
    Superclass **Mastigophora (Flagellata)**
      Class **Phytomastigorphora**
          Order Chrysomonadiformes (*Synura*)
          Order Silicoflagelliformes
          Order Dinoflagellatiformes (*Noctiluca*)
          Order Eugleniformes (*Euglena*)
          Order Chloromonadiformes
          Order Volvociformes (*Volvox, Chlamydomonas*)
      Class **Zoomastigophorea**
          Order Choanoflagellatiformes (*Codosiga*)
          Order Rhizomastigiformes
          Order Kinetoplastiformes
             Suborder Trypanosomatina (*Trypanosoma*)
          Order Trichomonadiformes (*Trichomonas*)
          Order Hypermastigiformes (*Trichonympha*)
    Superclass **Opilinata**
          Order Opaliniformes (*Opalina ranarum*)
    Superclass **Sarcodina**
      Class **Rhizopodea**
          Order Amoebiformes (*Amoeba*)
          Order Arcelliniformes (*Arcella*)
          Order Foraminiferiformes (*Globigerina*)
          Order Acrasiformes (*Dictyostelium*)

Order Eumycetozoiformes (*Physarum*)
Order Labyrinthuliformes (*Labyrinthula*)
Class **Piroplasmea**
Class **Actinopodea**
  Subclass **Radiolaria** (*Thalassicola*)
  Subclass **Heliozoia** (*Actinophrys*)
  Subclass **Proteomyxidia**
Subphylum *Sporozoa*
  Class **Telosporea**
    Subclass **Gregarinia**
    Subclass **Coccidia** (*Plasmodium malariae, Eimeria*)
Subphylum *Cnidospora*
Subphylum *Ciliophora*
  Class **Ciliatea**
    Subclass **Holotrichia** (*Paramecium, Tetrahymena*)
    Subclass **Peritrichia** (*Vorticella*)
    Subclass **Suctoria**
    Subclass **Spirotrichia**
      Order Heterotrichiformes (*Stentor*)
      Order Oligotrichiformes (*Halteria*)
      Order Hypotrichiformes (*Euplotes*)
      Order Entodiniomorphiformes (*Diplodinium*)

Phylum *Porifera*
  Class **Calcarea**
    Order Homocoeliformes
    Order Heterocoeliformes
  Class **Hexactinellida**
    Order Hexasterophoriformes
    Order Amphidiscophoriformes
  Class **Demospongiae**
    Subclass **Tetractinellida**
    Subclass **Monaxonida**
    Subclass **Keratosa**
Phylum *Coelenterata* (**Cnidaria**)
  Class **Hydrozoa**
    Order Hydroidiformes
      Suborder Anthomedusae (*Hydra, Tubularia*)
      Suborder Leptomedusae (*Obelia*)
      Suborder Limnomedusae (*Gonionemus*)
    Order Hydrocoralliniformes
      Suborder Milleporina
      Suborder Stylasterina
    Order Trachyliniformes
      Suborder Trachymedusae (*Liriope*)
      Suborder Narcomedusae
      Suborder Pteromedusae
    Order Siphonophoriformes (*Physalia*)
  Class **Scyphozoa**
    Order Stauromedusiformes
    Order Cubomedusiformes

Order Coronatiformes
Order Discomedusiformes (*Aurelia, Cyanea*)
Class **Anthozoa**
Subclass **Alcyonaria**
Order Alcyonaceiformes (sea pens)
Order Coenothecaliformes
Order Gorgonaceiformes (sea fans, red coral)
Order Pennatulaceiformes
Order Stoloniferiformes
Order Telestaceiformes
Sublcass **Zoantharia**
Order Actiniariformes (sea anemones)
Order Madreporariformes (reef corals)
Order Zonanthideiformes
Order Antipathariformes
Order Cerianthariformes
Phylum *Ctenophora*
Class **Tentaculata**
Order Cestidiformes
Order Cydippidiformes
Order Lobatiformes
Order Platycteneiformes
Class **Nuda**

PROTOSTOMA

Phylum *Platylhelminthes*
Class **Turbellaria** (turbellarians)
Order Acoeliformes (*Polychoerus*)
Order Rhabdocoeliformes
Order Alloeocoeliformes
Order Tricladidiformes (planarians)
Order Polycladidiformes
Class **Trematoda** (flukes)
Order Monogeneiformes
Order Digeneiformes (liver and blood flukes)
Order Aspidocotyleiformes
Class **Cestoidea** (tapeworms)
Subclass **Cestodaria**
Subclass **Cestoda**
Order Aporideiformes
Order Cyclophyllideiformes
Order Diphyllideiformes
Order Lecanicephaloideiformes
Order Nippotaenideiformes
Order Proteocephaloideiformes
Order Tetraphyllideiformes
Order Trypanorhynchiformes
Order Pseudophyllideiformes
Phylum *Mesozoa*
Class **Dicyema**
Class **Orthonectida**

Phylum *Nemertinea*
    Class **Anopla**
        Order Palaeonemertineiformes
        Order Heteronemertineiformes
    Class Enopla
        Order Hoplonemertineiformes
        Order Bdellonemertineiformes
Phylum *Aschelminthes*
  Subphylum *Nematoda* (nematodes)
    Class **Aphasmidea**
        Order Chromadoridiformes
        Order Enoplidiformes
    Class **Phasmidea**
        Order Tylenchidiformes
        Order Rhabditidiformes
        Order Spiruridiformes
  Subphylum *Nematomorpha* (horsehair worms, etc.)
    Class **Gordioidea**
    Class **Nectonematoidea**
  Subphylum *Trochelminthes* (rotifers)
    Class **Digononta**
        Order Seisonideiformes
        Order Bdelloideiformes
    Class **Monogononta**
        Order Flosculariaceiformes
        Order Collothecaceiformes
        Order Ploimiformes
  Subphylum *Gastrotricha*
  Subphylum *Kinorhyncha*
  Subphylum *Priapulida*
Phylum *Entoprocta*
Phylum *Acanthocephala*
Phylum *Bryozoa*
    Class **Phylactolaemata**
    Class **Gymnolaemata**
    Class **Stenolaemata**
Phylum *Brachiopoda*
    Class **Inarticulata**
    Class **Articulata** (lampshells)
Phylum *Phoronidea*
Phylum *Tardigrada*
Phylum *Annelida*
    Class **Archiannelida**
    Class **Polychaeta** (polychaetes)
        Order Errantiformes (clamworms)
        Order Sedentariformes (fanworms)
    Class **Oligochaeta** (oligochaetes)
        Order Plesioporiformes (naids, *Tubifex*)
        Order Opisthoporiformes (earthworms)
        Order Prosoporiformes (branchiobdellids)
    Class **Hirudinea** (leeches)
        Order Acanthobdelliformes

          Order Rhynchobdelliformes
          Order Arhynchobdelliformes
     Class **Myzostoma**
Phylum *Onychophora*
Phylum *Sipunculoidea*
Phylum *Echiuroidea*
Phylum *Mollusca*
     Class **Amphineura** (chitons)
     Class **Monoplacophora** (shield shells)
     Class **Scaphopoda** (elephant tusk shells)
     Class **Gastropoda** (snails)
          Order Prosobranchiformes
          Order Opisthobranchiformes
          Order Pulmonatiformes
     Class **Pelecypoda** (clams, oysters)
          Order Protobranchiformes
          Order Filibranchiformes
          Order Eulamellibranchiformes
          Order Septibranchiformes
     Class **Cephalopoda** (cephalopods)
          Order Tetrabranchiformes (nautilus)
          Order Dibranchiformes (squids, octopuses)
Phylum *Anthropoda*
     Class **Merostomata**
          Order Xiphosuriformes (horsehoe crabs)
     Class **Pycnogonida** (sea spiders)
     Class **Arachnida**
          Order Scorpioidiformes (scorpions)
          Order Amblypygiformes (tarantulas)
          Order Schizomidiformes (whip-scorpions, pedipalps)
          Order Palpigradiformes (palpigrads)
          Order Araneiformes (spiders)
          Order Solpugidiformes (solpugids)
          Order Pseudoscorpioidiformes (pseudoscorpions)
          Order Phalngidiformes
          Order Ricinuleiformes (ricinules)
          Order Acariniformes (ticks, mites)
              Suborder Trombidoidea
              Suborder Hydrachnoidea
              Suborder Ixodoidea
              Suborder Parasitoidea
              Suborder Oribatoidea
              Suborder Acaroidea
              Suborder Demodicoidea
          Order Opilioniformes (harvestmen, daddy longlegs)
          Order Solifugiformes (wind scorpions)
     Class **Crustacea**
    Subclass **Branchiopoda** (branchiopods)
     Division Eubranchiopoda
          Order Anostraciformes (Artemia, brine shrimps)
          Order Notostraciformes (tadpole shrimps)
          Order Conchostraciformes (clam shrimps)

Division Oligobranchiopoda
    Order Cladoceriformes (water fleas)
Subclass **Copepoda** (copepods)
    Order Eucopepodiformes
        Suborder Calanoida
        Suborder Harpacticoida
        Suborder Cyclopoida
        Suborder Notodelphyoida
        Suborder Monstrilloida
        Suborder Caligoida
        Suborder Lernaeopodoida
    Order Branchiuriformes
Sublcass **Mystacocarida**
Subclass **Cephalocarida**
Subclass **Ostracoda** (ostracods)
    Order Podocopiformes
    Order Myodocopiformes
    Order Cladocopiformes
    Order Platycopiformes
Subclass **Cirripedia** (barnacles)
    Order Thoraciciformes
    Order Acrothoraciciformes
    Order Apodiformes
    Order Rhizocephaliformes
    Order Ascothoraciciformes
Subclass **Malacostraca**
  Division Leptostraca
    Order Nebaliaceiformes
  Division Eumalacostraca
   Subdivision Syncarida
    Order Anaspidaceiformes
    Order Bathynellaceiformes
   Subdivision Peracarida
    Order Mysidaceiformes (mysid shrimps)
    Order Thermosbaenaceiformes
    Order Spelaeogriphaceiformes
    Order Cumaceiformes
    Order Tanaidaceiformes
    Order Isopodiformes (pill-bugs, beach lice)
    Order Amphipodiformes (beach fleas)
   Subdivision Hoplocarida
    Order Stomatopodiformes
   Subdivision Eucarida
    Order Euphausiaceiformes (krill)
    Order Decapodiformes
      Section Natantia
        Suborder Penaeidea (shrimps, prawns)
        Suborder Caridea (shrimps, prawns)
        Suborder Stenopodidea (banded shrimps)
      Section Reptantia
        Suborder Macrura (lobsters, crayfish)
        Suborder Anomura (hermit, sand crabs)
        Suborder Brachyura (crabs)

Class **Diplopoda** (millipedes)
- Order Pselapognathiformes
- Order Limacomorphiformes
- Order Oniscomorphiformes
- Order Polydesmoideiformes
- Order Namatomorphoideiformes
- Order Juliformiformes
- Order Colobognathiformes

Class **Chilopoda** (centipedes)
- Order Scutigeromorphiformes
- Order Lithobiomorphiformes
- Order Scolopendromorphiformes
- Order Geophilomorphiformes

Class **Pauropoda**

Class **Symphyla**

Class **Insecta**

  Subclass **Apterygota**
- Order Proturiformes (proturans)
- Order Thysanuriformes (silverfish, bristletails)
- Order Collemboliformes (springtails)
- Order Apteriformes (apterans)

  Subclass **Pterygota**
- Order Orthopteriformes (grasshoppers, crickets, cockroaches)
- Order Grylloblattodeiformes (grylloblattids)
- Order Phasmidiformes (walking sticks)
- Order Mantodeiformes (praying mantis)
- Order Dermapteriformes (earwigs)
- Order Plecopteriformes (stone flies)
- Order Isopteriformes (termites)
- Order Zorapteriformes (zorapterans)
- Order Embiopteriformes (embiids)
- Order Corrodentiformes (book lice)
- Order Mallophagiformes (chewing and bird lice)
- Order Anopluriformes (sucking lice)
- Order Ephemeriformes (mayflies)
- Order Odonatiformes (dragonflies, damselflies)
- Order Thysanopteriformes (thrips)
- Order Hemipteriformes (bugs)
    - Suborder Cryptocerata (mostly aquatic)
    - Suborder Gymnocerata (mostly terrestrial)
- Order Homopteriformes (aphids, cicadas)
- Order Mecopteriformes (scorpion, flies)
- Order Megalopteriformes (dobson flies)
- Order Neuropteriformes (lacewings, ant lions)
- Order Trichopteriformes (caddis flies)
- Order Raphidiodeiformes (snake flies)
- Order Lepidopteriformes
    - Suborder Frenatae (moths)
    - Suborder Rhopalocera (butterflies)
- Order Dipteriformes (flies, mosquitos, gnats)
    - Suborder Orthorrhapha
    - Suborder Cyclorrhapha

Order Siphonapteriformes (fleas)
Order Coleopteriformes (beetles)
   Suborder Adephaga
   Suborder Polyphaga
Order Strepsipteriformes (*Stylops*)
Order Hymenopteriformes (ants, bees)
Phylum *Linguatulida*

DEUTEROSTOMA

Phylum *Echinodermata* (living groups only)
  Subphylum *Pelmatozoa*
    Class **Crinoidea** (sea lilies)
  Subphylum *Eleutherozoa*
    Class **Holothuroidea** (sea cucumbers, trepang)
    Class **Echinoidea** (sea urchins, sand dollars)
    Class **Asteroidea** (starfish)
    Class **Ophiuroidea** (brittle stars)
Phylum *Chaetognatha*
Phylum *Pogonophora*
Phylum *Chordata*
  Subphylum *Hemichordata* (*Enteropneusta*) (acorn worms)
  Subphylum *Pterobranchiata*
    Class **Rhabdopleuridea**
    Class **Cephalodiscidea**
  Subphylum *Tunicata* (*Urochorda*)
    Class **Ascidiacea**
    Class **Thaliacea**
    Class **Larvacea**
  Subphylum *Cephalochordata* (*Amphioxus*)
  Subphylum *Vertebrata*
    Class **Agnatha** (jawless fishes)
      Subclass **Ostracodermi**
      Subclass **Cyclostomata** (lampreys)
    Class **Ichthyognatha** (jawed fishes)
      Subclass **Placodermi**
      Subclass **Chondrichthyes** (elasmobranchs)
        Order Selachiformes
          Suborder Squali (sharks)
          Suborder Batoidea (skates, sawfish)
        Order Holocephaliformes (ratfish)
    Subclass **Actinopterygii** (ray-finned fishes)
      Superorder Chondrostei
        Order Cladistiformes (*Polypterus*)
        Order Chondrosteiformes (sturgeons, paddlefish)
      Superorder Holostei
        Order Protospondyliformes (*Amia*, fresh-water dogfish)
        Order Ginglymodiformes (gar pikes)
      Superorder Teleostei
        Order Isospondyliformes (herring, salmon)
        Order Haplomiformes (pikes)
        Order Bathyclupeiformes (deep-sea herring)

Order Iniomiformes (lantern fish)
Order Ateleopiformes
Order Giganturoideiformes (giganturans)
Order Lyomeriformes (gulpers)
Order Mormyriformes (African Electric eels)
Order Ostariophysiformes (catfish, electric eels, carp)
Order Apodiformes (eels)
Order Colocephaliformes (moray eels)
Order Heteromiformes (spiny eels)
Order Synentognathiformes
Order Cyprinodontiformes (killifish)
Order Salmoperciformes (troutperches)
Order Solenichthyiformes (sea horses, pipe fish)
Order Anacanthiniformes (cods, hakes, pollacks)
Order Allotriognathiformes (ribbon fish, oar fish)
Order Berycomorphiformes (squirrel fish)
Order Zeomorphiformes (John dorries)
Order Percomorphiformes (perch, bass, tuna)
Order Scleropareiformes (sticklebacks)
Order Hypostomidiformes (dragan fish)
Order Heterosomatiformes (halibuts, flounders)
Order Discocephaliformes (ramoras)
Order Plectognathiformes (puffer fish)
Order Malacichthyiformes (ragfish)
Order Chaudhureiformes (Burmese eels)
Order Xenopterygiformes (cling fish)
Order Haplodociformes (toadfish, midshipman)
Order Pediculatiformes (angler fish, sea devils)
Order Opisthomiformes (snout eels)
Order Symbranchiformes (tropical swamp eels)
Subclass **Sarcopterygii** (muscle-finned fishes)
Superorder Crossopterygii
Order Coelacanthiniformes (*Latimeria chalumnae*)
Superorder Dipnoi (lungfishes)
Order Sirenoideiformes
Class **Amphibia**
Order Stegocephaliformes
Order Gymnophioniformes (Apodiformes) (caecilians)
Order Caudatiformes (Urodeliformes) (Salamanders)
Order Salientiformes (Anuriformes) (frogs, toads)
Class **Reptilia**
Subclass **Anapsida**
Order Cheloniformes (turtles, tortoises)
Subclass **Lepidosauria**
Order Rhynchocephaliformes (*Sphenodon*, tuatara)
Order Squamatiformes
Suborder Lacertilia (Sauria) (lizards)
Suborder Serpentes (Ophidia) (snakes)
Order Crocodiliformes (Loricatiformes) (alligators)
Order Saurischiformes (large dinosaurs)
Order Caudatiformes (Urodeliformes) (salamanders)
Order Pterosauriformes (extinct flying reptiles)

Subclass **Synapsida**
>Order Therapsidiformes (mammal-like dinosaurs)

Class **Aves**
>Subclass **Archeornithes** (Archaeopteryx)
>Subclass **Neornithes**
>>Superorder Palaeognathae
>>>Order Apterygiformes (kiwis)
>>>Order Casuariformes (cassowaries, emus)
>>>Order Rheiformes (rheas)
>>>Order Struthioniformes (ostriches)
>>>Order Tinamiformes (tinamous)
>>Superorder Neognathae
>>>Order Sphenisciformes (penguins)
>>>Order Gaviformes (loons)
>>>Order Podicipitiformes (grebes)
>>>Order Procellariformes (albatrosses, petrels)
>>>Order Pelecaniformes (pelicans, gannets)
>>>Order Ciconiformes (storks, flamingos)
>>>Order Anseriformes (ducks, geese, swans)
>>>Order Falconiformes (vultures, eagles)
>>>Order Galliformes (grouse, chickens)
>>>Order Gruiformes (cranes, coots, rails)
>>>Order Charadriformes (gulls, killdeers)
>>>Order Columbiformes (pigeons, doves)
>>>Order Cuculiformes (cuckoos, roadrunners)
>>>Order Psittaciformes (parrots)
>>>Order Strigiformes (owls)
>>>Order Caprimuligiformes (whip-poor-wills)
>>>Order Micropodiformes (swifts)
>>>Order Coliformes (mouse-birds)
>>>Order Trogoniformes (trogons)
>>>Order Coraciformes (kingfishers)
>>>Order Piciformes (woodpeckers, toucans)
>>>Order Passeriformes (perching birds)

Class **Mammalia**
>Subclass **Prototheria** (monotremes)
>>Order Monotrematiformes (duckbilled platypus)
>Subclass **Metatheria** (marsupials)
>>Order Polyprotodontiformes (opossums)
>>Order Coenolestoiformes
>>Order Diprotodontiformes (kangaroos, kaolas)
>Subclass **Eutheria** (placentals)
>>Order Insectivoriformes (moles, hedgehogs)
>>Order Dermopteriformes (flying lemurs)
>>Order Chiropteriformes (bats)
>>Order primatiformes
>>>Suborder Tupaioidea (tree shrews)
>>>Suborder Lemuroidea (lemurs)
>>>>Family Indridae
>>>>Family Lemuridae
>>>Suborder Daubentonioidea (aye-aye)
>>>Suborder Lorisoidea (loris, pottos)

Family Galagidae
Family Lorisidae
Suborder Tarsoidea (*Tarsius*)
Suborder Anthropoidea
    Superfamily Ceboidea (platyrrhini) (New World Monkeys)
        Family Callithricidae (marmosets)
        Family Cebidae (spider, howler, capuchin monkeys)
    Superfamily Cercopithecoidea (Catarhinni) (Old World monkeys)
        Family Cercopithecidae (baboons, macaques, langurs)
        Family Colobidae (sakis, titis)
    Superfamily Hominoidea
        Family Hominidae (man)
        Family Simiidae (anthropoid apes)
Order Edentatiformes (armadillos, sloths)
Order Pholidotiformes (pangolins)
Order Lagomorphiformes (rabbits, pikas)
Order Rodentiformes (rats, beavers, squirrels)
Order Cetaceiformes
    Suborder Odontoceti (toothed whales, porpoises)
    Suborder Mysticeti (whalebone whales)
Order Carnivoriformes
    Suborder Fissipedia (dogs, cats, bears)
    Suborder Pinnipedia (seals, walruses)
Order Tubulidentatiformes (aardvarks)
Order Proboscidiformes (elephants)
Order Hyracoidiformes (conies)
Order Sireniformes (sea-cows, manatees)
Order Perissodactyliformes (horses, rhinos)
Order Artiodactyliformes (cattle, sheep, pigs, giraffes, hippos)

Index

Page numbers in **boldface** in the index indicate the location of definitions or major discussions of terms and concepts. An *f* after a page number indicates a figure or a table.

The secret of success in science
is the combination of
a bold imagination and
a willingness to grub for the facts.

ALFRED NORTH WHITEHEAD